Innovative Bio-Based Technologies for Environmental Remediation

Innovative Bio-Based Technologies for Environmental Remediation

Edited by
Pardeep Singh
Chaudhery Mustansar Hussain
and Mika Sillanpää

CRC Press
Taylor & Francis Group
Boca Raton London New York

CRC Press is an imprint of the
Taylor & Francis Group, an **informa** business

First edition published 2022
by CRC Press
6000 Broken Sound Parkway NW, Suite 300, Boca Raton, FL 33487-2742

and by CRC Press
2 Park Square, Milton Park, Abingdon, Oxon, OX14 4RN

© 2022 Taylor & Francis Group, LLC

CRC Press is an imprint of Taylor & Francis Group, LLC

Reasonable efforts have been made to publish reliable data and information, but the author and publisher cannot assume responsibility for the validity of all materials or the consequences of their use. The authors and publishers have attempted to trace the copyright holders of all material reproduced in this publication and apologize to copyright holders if permission to publish in this form has not been obtained. If any copyright material has not been acknowledged please write and let us know so we may rectify in any future reprint.

Except as permitted under U.S. Copyright Law, no part of this book may be reprinted, reproduced, transmitted, or utilized in any form by any electronic, mechanical, or other means, now known or hereafter invented, including photocopying, microfilming, and recording, or in any information storage or retrieval system, without written permission from the publishers.

For permission to photocopy or use material electronically from this work, access www.copyright.com or contact the Copyright Clearance Center, Inc. (CCC), 222 Rosewood Drive, Danvers, MA 01923, 978-750-8400. For works that are not available on CCC please contact mpkbookspermissions@tandf.co.uk

Trademark notice: Product or corporate names may be trademarks or registered trademarks and are used only for identification and explanation without intent to infringe.

Library of Congress Cataloguing-in-Publication Data
Names: Singh, Pardeep, editor. | Mustansar Hussain, Chaudhery, editor. | Sillanpää, Mika E. T., editor.
Title: Innovative bio-based technologies for environmental remediation / editied by Pardeep Singh, PGDAV College, University of Delhi, New Delhi, India, Chaudhery Mustansar Hussain, New Jersey Institute of Technology (NJIT), Newark, New Jersey, USA, Mika Sillanpää, Department of Biological and Chemical Engineering, Aarhus University, Nørrebrogade 44, 8000 Aarhus C, Denmark.
Description: First edition. | Boca Raton : CRC Press, 2022. | Includes bibliographical references and index.
Identifiers: LCCN 2021041391 | ISBN 9780367436032 (hardback) | ISBN 9781032187570 (paperback) | ISBN 9781003004684 (ebook)
Subjects: LCSH: Nanostructured materials. | Nanostructured materials--Environmental aspects. | Pollutants.
Classification: LCC TA418.9.N35 I348 2022 | DDC 620.1/15--dc23/eng/20211028
LC record available at https://lccn.loc.gov/2021041391

ISBN: 978-0-367-43603-2 (hbk)
ISBN: 978-1-032-18757-0 (pbk)
ISBN: 978-1-003-00468-4 (ebk)

DOI: 10.1201/9781003004684

Typeset in Times
by MPS Limited, Dehradun

 Printed in the United Kingdom
by Henry Ling Limited

Contents

Editors vii
Contributors ix

1 Bio-Based Technology for Environmental Management 1
  Cevat Yaman and Ayse Burcu Yaman

2 Advances in Biotechnology for the Bioremediation of Contaminated Ecosystem 25
  Alvina Farooqui, Gyanendra Tripathi, Nishi Aara, Suhail Ahmad, Arbab Husain,
  Adeeba Shamim, and Sadaf Mahfooz

3 Bioremediation: Tools and Techniques for Wastewater Reclamation 51
  P.P. Mirshad, Nair G. Sarath, A.M. Shackira, and Jos T. Puthur

4 Aquatic Plants Biosorbents for Remediation in the Case of Water Pollution as Future
  Prospectives 71
  Jyoti Mehta, Moharana Choudhury, Anu Sharma, and Arghya Chakravorty

5 Biobased Technologies for Remediation: Green Technology for
  Environmental CleanUp 89
  Anubhuti Singh, Gurudatta Singh, Priyanka Singh, and Virendra Kumar Mishra

6 Process Intensification in Bio-Based Approaches for Environmental Remediation 109
  Kailas L. Wasewar

7 Bio-Based Technologies and Combination of Other Technologies 125
  Yong Chen and Steplinpaulselvin Selvinsimpson

8 Low-cost Bioremediation Technologies for Transforming Waste to Wealth 133
  Rupika Sinha, Shipra Dwivedi, Sukhendra Singh, Singh Divakar, and Pradeep Srivastava

9 Phytoremediation: An Eco-Friendly, Sustainable Solution for Indoor and Outdoor Air
  Pollution 155
  Manisha Sarkar, Sujit Das, Randeep Rakwal, Ganesh Kumar Agrawal, and Abhijit Sarkar

10 Phytoremediation: Importance and General Mechanisms 197
  Surbhi Singla, Aastha Baliyan, and Baljinder Singh

11 Phytoremediation Mechanisms of Heavy Metal Removal: A Step Towards a Green and
  Sustainable Environment 207
  Nathaniel A. Nwogwu, Oluwaseyi A. Ajala, Fidelis O. Ajibade, Temitope F. Ajibade,

Bashir Adelodun, Kayode H. Lasisi, Adamu Y. Ugya, Pankaj Kumar, Ifeoluwa F. Omotade,
Toju E. Babalola, James R. Adewumi, and Christopher O. Akinbile

12 **Phytoremediation: A Sustainable Technology for Pollution Control and Environmental Cleanup** 237
Poonam Yadav, Anwesha Chakraborty, Sudhakar Srivastava, Shalini Sahani, and Pardeep Singh

13 **Nanophytoremediation: A Promising Strategy for the Management of Environmental Contaminants** 251
Nair G. Sarath, P. Pravisya, A.M. Shackira, and Jos T. Puthur

14 **Approaches of Overproduction and Purification of *Pleurotus* Laccase for the Treatment of Sugarcane Vinasse** 265
Joberson Alves Junior, Débora da Silva Vilar, Carlos Eduardo Maynard Santana, Ram Naresh Bharagava, Muhammad Bilal, Sikandar I. Mulla, Pankaj Kumar Arora, Álvaro Silva Lima, Ranyere Lucena de Souza, and Luiz Fernando Romanholo Ferreira

15 **Enhanced CO_2 Assimilation by Engineered *Escherichia coli* (*E. coli*)** 281
Navamallika Gogoi, Moharana Choudhury, Anwesha Gohain, and Anu Sharma

16 **Role of Biopolymers in Development of Sustainable Remediation Technologies** 295
Monika Yadav, Manita Das, Sonal Thakore, and R.N. Jadeja

17 **Nanocatalyst Synthesis by the Green Route: Mechanism and Application** 321
Naresh K Sethy, Zeenat Arif, PK Mishra, P. Kumar, and Rajesh Saha

18 **Bio-based Polymeric Material for Environmental Remediation** 343
Zeenat Arif, Naresh K Sethy, Pradeep Kumar Mishra, Pradeep Kumar, and Pratiksha Pandey

19 **Cyanide: Sources, Health Issues, and Remediation Methods** 357
Priyanka Yadav, Manisha Verma, and Vishal Mishra

20 **Biosurfactant: An Alternative Towards Sustainability** 377
Sanchayita Rajkhowa and Jyotirmoy Sarma

Index 403

Editors

Dr Pardeep Singh is presently working as an assistant professor in the Department of Environmental Studies, PGDAV College, University of Delhi New Delhi India. He obtained his master's degree from the Department of Environmental Science Banaras Hindu University, Varanasi India. He received his doctorate from the Indian Institute of Technology (Banaras Hindu University) IIT(BHU) Varanasi. He is working in the areas of waste management and wastewater treatment through the biological as well as photocatalytic route. He has published more than 65 papers in the international journals of waste management and edited more than 30 books with Elsevier, Wiley, CRC, and Springer.

Chaudhery Mustansar Hussain, PhD is an adjunct professor and director of labs in the Department of Chemistry and Environmental Sciences at the New Jersey Institute of Technology (NJIT), Newark, New Jersey, USA. His research is focused on the applications of nanotechnology and advanced technologies and materials, analytical chemistry, environmental management, and various industries. Dr. Hussain is the author of numerous papers in peer-reviewed journals as well as prolific author and editor of several (more than 50 books) scientific monographs and handbooks in his research areas published with Elsevier, Royal Society of Chemistry, John Wiley & sons, CRC, Springer, etc.

Mika Sillanpää received his MSc (Eng.) and DSc (Eng.) from Helsinki University of Technology, Finland (now known as Aalto University), where he also completed an MBA in 2013. He completed his master's studies in the Department of Chemical Engineering at Helsinki University of Technology. Since 2000, he has been a full professor at the University of Oulu, University of Kuopio (currently University of Eastern Finland), LUT University, and Aarhus University. He has peer-reviewed articles for over 250 academic journals, many of which are ranked top in their fields as well as serving on a number of editorial boards for several publications. Having an h-index of 92, his publications have been cited over 43,000 times (Google Scholar).

Mika Sillanpää has received numerous awards for research and innovation. Among these, he is the first Laureate of Scientific Committee on the Problems of the Environment (SCOPE)'s Young Investigator Award, which was delivered at the UNESCO Conference in Shanghai 2010, for his "significant contributions, outstanding achievements and research leadership in Environmental Technological Innovations to address present water pollution problems worldwide, especially with regard to wastewater treatment and reuse". In 2011, he was invited to act as a Principal Scientific Reviewer in the GEO-5 report of the United Nations Environmental Programme (UNEP).

Contributors

Nishi Aara
Department of Bioengineering
Integral University
Lucknow-226026
Uttar Pradesh, India

Bashir Adelodun
Department of Agricultural and Biosystems
Engineering
University of Ilorin, PMB 1515
Ilorin, Nigeria

James R. Adewumi
Department of Civil and Environmental
Engineering
Federal University of Technology
Akure, PMB 704, Nigeria

Ganesh Kumar Agrawal
Faculty of Health and Sport Sciences
University of Tsukuba, 1-1-1 Tennodai,
Tsukuba, Ibaraki 305-8574, Japan and
Research Laboratory for Biotechnology and
Biochemistry (RLABB), PO Box 13265
Kathmandu 44600, Nepal

Suhail Ahmad
Department of Bioengineering
Integral University
Lucknow-226026
Uttar Pradesh, India

Oluwaseyi A. Ajala
Department of Zoology and
Environmental Sciences
Punjabi University
Patiala-147002, Punjab, India and
Department of Chemistry, Faculty of Science
University of Ibadan
Ibadan, Nigeria

Fidelis O. Ajibade
Research Centre for Eco-Environmental Sciences
Chinese Academy of Sciences
Beijing 100085, PR China and
University of Chinese Academy of Sciences
Beijing 100049, PR China

Temitope F. Ajibade
University of Chinese Academy of Sciences
Beijing 100049, PR China

Christopher O. Akinbile
Department of Agricultural and
Environmental Engineering
Federal University of Technology
Akure, PMB 704, Nigeria

Joberson Alves Junior
Graduated Program in Process Engineering
Tiradentes University (UNIT)
Av. Murilo Dantas, 300, Farolândia
49032-490
Aracaju-Sergipe, Brazil

Shackira AM
Department of Botany
Sir Syed College
Taliparamba, Kannur
Kerala-670142, India

Zeenat Arif
Indian Institute of Technology (BHU)
Varanasi, India

Pankaj Kumar Arora
Department of Environmental Microbiology
Babasaheb Bhimrao Ambedkar University
Lucknow, India

Toju E. Babalola
Department of Water Resources
Management & Agro-Meteorology
Federal University mOye-Ekiti
Oye, Ekiti State, Nigeria

Aastha Baliyan
Department of Biotechnology
Panjab University
Chandigarh-160014, India

Ram Naresh Bharagava
Laboratory for Bioremediation and
Metagenomics Research (LBMR)
Department of Environmental
Microbiology (EM)

Babasaheb Bhimrao
Ambedkar University
  (A Central University)
Vidya Vihar
Raebareli Road
Lucknow-226025
Uttar Pradesh, India

Muhammad Bilal
School of Life Science and
  Food Engineering
Huaiyin Institute of Technology
Huaian 223003, China

Arghya Chakravorty
School of Biosciences and Technology
  Vellore Institute of Technology
Vellore-632014, India

Yong Chen
School of Environmental Science and
  Engineering
Huazhong University of Science and
  Technology
Wuhan 430074, China

Moharana Choudhury
Voice of Environment (VoE)
Guwahati-781034
Assam, India

Manita Das
Department of Chemistry
  Faculty of Science
The Maharaja Sayajirao University of Baroda
Vadodara-390002, India

Sujit Das
Laboratory of Applied Stress Biology
  Department of Botany
University of Gour Banga
Malda-732 103
West Bengal, India

Débora da Silva Vilar
Graduated Program in Process Engineering
Tiradentes University (UNIT)
  Av. Murilo Dantas,
  300, Farolândia, 49032-490
Aracaju-Sergipe, Brazil

Ranyere Lucena de Souza
Institute of Technology and Research (ITP)
  Tiradentes University (UNIT)
  Av. Murilo Dantas
  300, Farolândia
Aracaju, Sergipe 49032-490, Brazil

Singh Diwakar
School of Biochemical Engineering
  Indian Institute of Technology
(Banaras Hindu University)
Varanasi, U.P., India

Shipra Dwivedi
Department of Biochemical Engineering
  School of Chemical Technology
Harcourt Butler Technical University
Kanpur, U.P., India

Alvina Farooqui
Department of Bioengineering
Integral University
Lucknow-226026
Uttar Pradesh, India

Luiz Fernando Romanholo Ferreira
Institute of Technology and Research (ITP)
Tiradentes University (UNIT)
  Av. Murilo Dantas
  300, Farolândia Aracaju,
Sergipe 49032-490, Brazil

Navamallika Gogoi
Department of Chemistry
Arunachal University of Studies
Namsai-792103
Arunachal Pradesh, India

Anwesha Gohain
Department of Botany
Arunachal University of Studies
Namsai-792103
Arunachal Pradesh, India

Arbab Husain
Department of Bioscience
Integral University
Lucknow
Uttar Pradesh, 226026, India

Pankaj Kumar
Department of Agricultural Civil
  Engineering
Kyungpook National University
Daegu, Korea

Pradeep Kumar
Indian Institute of Technology (BHU)
Varanasi, India

Contributors

Kayode H. Lasisi
University of Chinese Academy of Sciences
Beijing 100049, PR China

Sadaf Mahfooz
Department of Bioscience
Integral University
Lucknow-226026, Uttar Pradesh, India

Jyoti Mehta
Department of Environmental Sciences
Central University of Jharkhand
Brambe-835205, Ranchi
Jharkhand, India

Pradeep Kumar Mishra
Indian Institute of Technology (BHU),
Varanasi, India

Virendra Kumar Mishra
Institute of Environment and
Sustainable Development Banaras Hindu
University
Varanasi, 221005, India

Vishal Mishra
School of Biochemical Engineering
IIT (BHU)
Varanasi, India

Sikandar I. Mulla
Department of Biochemistry
School of Applied Sciences
REVA University
Bangalore-560064, India

Nathaniel A. Nwogwu
Department of Agricultural and
Bioresources Engineering
Federal University of Technology
Owerri, Nigeria
Research Centre for Eco-Environmental Sciences
Chinese Academy of Sciences
Beijing 100085, PR China and
University of Chinese Academy of Sciences
Beijing 100049, PR China

Ifeoluwa F. Omotade
Department of Agricultural and
Environmental Engineering
Federal University of Technology
Akure, PMB 704, Nigeria

Pravisya P
Department of Botany
Malabar Christian College
Calicut, Kerala-673001, India

Pratiksha Pandey
Indian Institute of Technology (BHU)
Varanasi, India

Mirshad PP
Department of Botany
Sir Syed College
Taliparamba
Kannur, Kerala-670142, India

Jos T Puthur
Plant Physiology and Biochemistry Division
Department of Botany
University of Calicut
C.U. Campus P.O.
Kerala-673635, India

Sanchayita Rajkhowa
Department of Chemistry
Jorhat Institute of Science & Technology
Jorhat
Assam, India

Randeep Rakwal
Research Laboratory for Biotechnology
and Biochemistry (RLABB)
PO Box 13265, Kathmandu 44600
Nepal and Global Research Arch
for Developing Education (GRADE)
Academy Pvt. Ltd.
Birgunj 44300, Nepal

R.N. Jadeja
Department of Environmental Studies
Faculty of Science
The Maharaja Sayajirao University of Baroda
Vadodara-390002, India and
Department of Chemistry
Faculty of Science
The Maharaja Sayajirao University of Baroda
Vadodara-390002, India

Rajesh Saha
Department of Chemical Engineering
IIT (BHU), Varanasi, 221005
UP, India

Carlos Eduardo Maynard Santana
Graduated Program in Process
Engineering
Tiradentes University (UNIT)
Av. Murilo, Dantas
300, Farolândia
49032-490
Aracaju-Sergipe, Brazil.

Nair G Sarath
Plant Physiology and Biochemistry Division
Department of Botany
University of Calicut
C.U. Campus P.O.
Kerala-673635, India

Abhijit Sarkar
Laboratory of Applied Stress Biology
Department of Botany
University of Gour Banga
Malda-732 103
West Bengal, India

Manisha Sarkar
Laboratory of Applied Stress Biology
Department of Botany
University of Gour, Banga
Malda-732 103
West Bengal, India

Jyotirmoy Sarma
Department of Chemistry
Kaziranga University
Jorhat, Assam, India

Steplinpaulselvin Selvinsimpson
School of Environmental Science and
 Engineering
Huazhong University of
 Science and Technology
Wuhan 430074, China

Naresh K Sethy
Indian Institute of Technology (BHU)
 Varanasi, India

Adeeba Shamim
Department of Bioscience
Integral University
Lucknow-226026
Uttar Pradesh, India

Anu Sharma
Govt. Degree College, Bhaderwah,
 Doda
University of Jammu Union Territory of
 J & K, India

Álvaro Silva
Lima Institute of Technology and
 Research (ITP)
Tiradentes University (UNIT)
Av. Murilo Dantas
300, Farolândia
Aracaju
Sergipe 49032-490, Brazil.

Anubhuti Singh
Institute of Environment and Sustainable
 Development
Banaras Hindu University
Varanasi 221005, India

Baljinder Singh
Department of Biotechnology
Punjab University
Chandigarh-160014, India

Gurudatta Singh
Institute of Environment and Sustainable
 Development
Banaras Hindu University
Varanasi 221005, India

Priyanka Singh
Institute of Environment and Sustainable
 Development
Banaras Hindu University
Varanasi 221005, India

Sukhendra Singh
Department of Alcohol Technology and Biofuels
 Vasantdada Sugar Institute Pune,
Maharashtra, India

Surbhi Singla
Department of Biotechnology
Punjab University
Chandigarh-160014, India

Rupika Sinha
Department of Biotechnology
 Motilal Nehru National Institute of
 Technology Allahabad, Prayagraj
U.P. India

Pradeep Srivastava
School of Biochemical Engineering
 Indian Institute of Technology
(Banaras Hindu University), Varanasi
U.P., India

Sonal Thakore
Department of Chemistry
 Faculty of Science
The Maharaja Sayajirao
 University of Baroda
Vadodara-390002, India

Gyanendra Tripathi
Department of Bioengineering
Integral University
Lucknow-226026
Uttar Pradesh, India

Adamu Y. Ugya
Department of Environmental Management
Kaduna State University
Kaduna State, Nigeria and
Key Lab of Groundwater Resources and
 Environment of Ministry of Education
Key Lab of Water Resources and
 Aquatic Environment of Jilin Province
College of New Energy and Environment
Jilin University
Changchun, People's Republic of China

Manisha Verma
School of Biochemical Engineering
 IIT (BHU)
Varanasi, India

Kailas L. Wasewar
Advance Separation and Analytical Laboratory
 (ASAL), Department of Chemical
 Engineering
Visvesvaraya National
 Institute of Technology (VNIT)
Nagpur-440010 (MH), India

Monika Yadav
Department of Environmental Studies
 Faculty of Science
The Maharaja Sayajirao
 University of Baroda
Vadodara-390002, India

Priyanka Yadav
School of Biochemical Engineering IIT (BHU)
Varanasi, India

Ayse Burcu Yaman
Environmental Engineering Department
Imam Abdulrahman Bin Faisal University
Dammam, Saudi Arabia

Cevat Yaman
Environmental Engineering Department
Imam Abdulrahman Bin Faisal University
Dammam, Saudi Arabia

1
Bio-Based Technology for Environmental Management

Cevat Yaman and Ayse Burcu Yaman
Environmental Engineering Department, Imam Abdulrahman Bin Faisal University, Dammam, Saudi Arabia

CONTENTS

1.1 Phytoremediation 1
    1.1.1 Introduction 1
    1.1.2 Phytoremediation of Acid Mine Drainage (AMD) Water 2
    1.1.3 Phytoremediation for Brownfield Redevelopment 2
    1.1.4 Phytoremediation for Heavy Metals 5
    1.1.5 Phytoremediation for Oil Contamination 6
1.2 Bioaugmentation and Biostimulation 7
    1.2.1 Bioaugmentation and Biostimulation for Hydrocarbon Remediation 7
    1.2.2 Bioaugmentation for the Treatment of Industrial Wastewater 8
    1.2.3 Bioaugmentation for Composting 9
1.3 Bioventing 10
    1.3.1 Designing Bioventing Systems 13
        2.3.1.1 Airflow Rate 13
        1.3.1.2 Soil Moisture 13
        1.3.1.3 Nutrients 13
1.4 Biosparging 13
    1.4.1 Application, Advantages, and Disadvantages of Biosparging 14
1.5 Composting 15
    1.5.1 Aerobic Composting 15
        1.5.1.1 Important parameters for composting 15
    1.5.2 Anaerobic Composting 17
        1.5.2.1 Wet Fermentation 18
        1.5.2.2 Dry Fermentation 18
    1.5.3 Microbiology of Composting 18
References 18

1.1 Phytoremediation

1.1.1 Introduction

Phytoremediation is defined as the use of plants for remediation of contaminated soil and groundwater. The phytoremediation method is usually used to remediate inorganic and organic pollutants. Usually, the bioremediation of these pollutants takes place in the rhizosphere, which is also called the root zone. Inorganic contaminants are remediated by two mechanisms: phytostabilisation and phytoextraction. The phytostabilisation method, which is an eco-friendly alternative phytotechnology, includes the use of plants to stabilise the contaminated soil. In this method, contaminants, such as metals, are immobilised into a less soluble form, which makes the soil more productive. Phytoextraction, on the other hand, is the removal of inorganic pollutants by using the aboveground parts of the plants. If the upper portion of the plants is harvested, the pollutants are removed in concentrated form in the plant body. Municipal and industrial wastewater can be treated in constructed wetlands by using free-floating aquatic plant species and their associated microorganisms. Recently, some special fast-growing plants have been used in wastewater treatment. Specific plants can be used to extract and assimilate or decompose target organic contaminants. For instance, plant tissues are known to bioaccumulate heavy metals. Some inorganic contaminants, actually, are life-sustaining nutrients, which can be taken by the plant roots for growth. Furthermore, organic contaminants like PAHs and pesticides can also be removed by plants.

Phytoremediation has some advantages, which can be listed as convenient, low cost, suitability, aesthetic benefits, and minimum contaminant leaching. Operation of phytoremediation is also not costly as it involves only fertilisation and watering for growth. Additional operational costs will be from harvesting and disposing of plants polluted with heavy metals. One of the important disadvantages of phytoremediation is that it cannot extract and metabolise the pollutants found below rooting depth. The other disadvantage worth mentioning is that the removal of contaminants can take several years for certain contaminants to attain the desired concentrations. Some of the pollutants that can be remediated by phytoremediation are BTEX compounds, chlorinated contaminants, PAHs, nitrate, ammonium, phosphate, and heavy metals.

1.1.2 Phytoremediation of Acid Mine Drainage (AMD) Water

The activities of mining are known to produce acidic water that has the potential to pollute nearby water bodies. This acid-containing water is called AMD. AMD is formed after contact between water and pyrite, which will result in producing acids (Blowes et al., 2003). AMD water with low pH contains sulfates, iron, and other metals. The negative impacts of AMD will vary depending on local conditions, geomorphology, and the extent of the AMD-generating deposits (Mang and Ntushelo, 2019). The use of lime, caustic soda, calcium carbonate, hydrated lime, and soda ash for AMD neutralisation will produce voluminous sludge, which will cause more problems (Fang et al., 2003).

Compared to other treatment methods, phytoextraction and phytostabilisation methods were found to be more economic for AMD treatment (Pilon-Smits, 2005). One of the major setbacks of phytoremediation is that they require specific plants and metals (Hasan et al., 2007; Rahman et al., 2007). Some of the species, such as *Eichhornia crassipes, Spirodela polyrhiza,* and *Pistia stratiotes,* can metabolise heavy metals and toxic metals from AMD. It was shown that *Spirodela polyrhiza* species were sensitive at low pH values in water environments that are similar to AMD (Cruz et al., 2019). Moreover, it was reported that *E. crassipes* is easily adapted to a low pH environment, as shown in Figure 1.1. Table 1.1 summarises the bioremediation potentials of plants for AMD.

Remediation of AMD with phytoremediation is a promising strategy. However, due to the cost of the land and the time required for remediation, the phytoextraction approach can be limited in the scope of application. On the other hand, phytostabilisation seems to be a more suitable option for AMD remediation because of its lower cost and easy implementation. As an example, *Eichhornia crassipes, Chrysopogon zizanioides,* and *Brassica juncea* have been shown to be successful in the removal of AMD (Mang and Ntushelo, 2019).

1.1.3 Phytoremediation for Brownfield Redevelopment

The United States alone has more than 450,000 brownfield sites, which pose great threat to human health and the environment (O'Connor et al., 2019). Trillions of dollars have been spent to mitigate and clean up these sites. However, cleaning up these brownfield sites causes significant environmental and socio-economic impacts (Hou et al., 2017; Hou and Al-Tabbaa, 2014). Traditional remediation technologies for brownfield site remediation are usually based on the use of chemicals, fossil fuels, and

FIGURE 1.1 Toxicity of AMD in aquatic floating macrophytes. (A–B, *E. crassipes*; C–D, *P. stratiotes*; E–F, *S. polyrhiza*; Cruz et al., 2019).

electricity, which are subsequently associated with environmental footprints (Hou et al., 2018; Wang et al., 2019). Therefore, using nature-based solutions (NBSs) seems to be promising technologies to render energy-efficient remediation (Song et al., 2019; Zhu et al., 2017). One of these nature-based solutions is phytoremediation, which can offer many environmental and economic benefits compared to traditional remediation approaches (Figure 1.2).

Phytoremediation can be selected as an NBS because xenobiotic compounds can be adsorbed via evapotranspiration from stems and leaves of plants (Doucette et al., 2013; Huang et al., 2014). The mechanism to remove contaminants such as PCE from the subsurface by plants is that the roots release compounds such as carbohydrates and enzymes, which then accelerate microbial growth (RTDF, 2005). Selection of the appropriate trees in a brownfield cleanup is an important issue. Local species such as *eucalyptus* trees are generally preferred due to their better adaptation to the conditions. It should be taken into consideration that *eucalyptus* trees can be affected negatively under saline conditions (O'Connor et al., 2019).

TABLE 1.1

List of plant species for phytoextraction and phytostabilisation of metals in AMD (Mang and Ntushelo, 2019)

Plant Species	Family	Remediation Method	References
Chrysopogon zizanioides	Poaceae	Phytoextraction	(Melato et al., 2016)
Brassica juncea	Brassicaceae	Phytoextraction	(Blaylock et al., 1997; Begonia et al., 1998; Liang Zhu et al., 1999)
Eichhornia crassipes	Pontederiaceae	Phytoextraction	(Cruz et al., 2019; H., M., M. N., and A.F.F. F., 2011)
Waltheria indica	Malvaceae	Phytoextraction	(Rajakaruna and Bohm, 2002)
Cassia kleinii	Fabaceae	Phytoextraction	(Rajakaruna and Bohm, 2002; M., I.M.C. et al., 2006)
Thlaspi caerulescens	Brassicaceae	Phyotextraction	(Baker et al., 1994; Zhao et al., 2003)
Seberita acuminata	Sapotaceae	Phytoextraction	(Jaffré et al., 1979)
Rinorea benghalensis	Violaceae	Phytoextraction	(Wither and Brooks, 1977)
Cichorium intybus L.	Asteraceae	Phytoextraction	(Rio-Celestino et al., 2006)
Cynodon dactylon (L.) Pers.	Poaceae	Phytoextraction	(Rio-Celestino et al., 2006)
Armeria arenaria (Pers.) Schult., Festuca arvernensis Auquier, and Koeleria vallesiana (Honck.)	Plumbaginaceae, Poaceae, and Poaceae respectively	Phytoextraction	(Frérot et al., 2006)
Crotalaria biflora, Evolvulus alsinoides, and Hybanthus enneaspermus	Fabaceae, Convolvulaceae, and Violaceae respectively	Phytoextraction	(Rajakaruna and Bohm, 2002)
Clerodendrum infortunatum, Croton bonplandianus, Geniosporum tenuiflorum, Tephrosia villosa, and Waltheria indica	Verbenaceae, Euphorbiaceae, Lamiaceae, Fabaceae, and Sterculiaceae respectively	Phytoextraction	(Rajakaruna and Bohm, 2002)
Atriplex halimus L.	Amaranthaceae	Phytoextraction	(Lutts et al., 2004)
Lygeum spartum L.	Poaceae	Phytoextraction	(Conesa et al., 2007)
Festuca rubra and Agrostis tenuis Sibthorp	Both belong to Poaceae	Phytostabilisation	(Smith and Bradshaw, 1979)
Alnus viridis subsp. sinuate (Regel) and Anaphalis margaritacea (L.)	Betulaceae and Asteraceae respectively	Phytostabilisation	(Kramer et al., 2000)
Atriplex lentiformis	Amaranthaceae	Phytostabilisation	(Mendez et al., 2007)
Atriplex canescens	Chenopodiaceae	Phytostabilisation	(Sabey et al., 1990)

The contaminant removal from the brownfield is achieved by natural attenuation and transpiration. First, the roots of trees uptake chlorinated solvents from groundwater. Then, the absorbed contaminants are lifted up to surface and discharged into the atmosphere via leaves and stems (Ma and Burken, 2003). Also, the growth of tree can accelerate the volatilisation from the soil because of low moisture and the formation of pathways in the roots (Marr et al., 2006). One common contaminant encountered in land remediation is chlorinated solvents. With little energy and material use, a great environmental benefit can be obtained by

FIGURE 1.2 Phytoremediation for brownfield redevelopment (O'Connor et al., 2019).

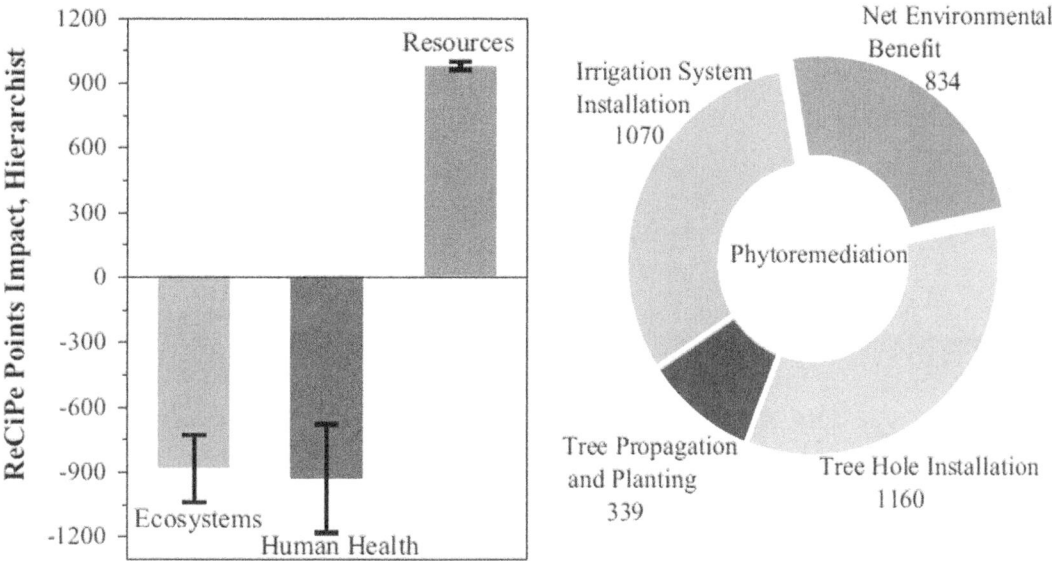

FIGURE 1.3 Life cycle assessment results and process contribution (O'Connor et al., 2019).

applying phytoremediation as an NBS. As illustrated in Figure 1.3, phytoremediation can provide a benefit in land remediation (O'Connor et al., 2019). Furthermore, the phytoremediation process is economical compared to traditional remediation methods. For instance, phytoremediation would cost ~$0.3 MM, whereas the cost of in-situ remediation would be ~$1.3 MM (O'Connor et al., 2019).

1.1.4 Phytoremediation for Heavy Metals

A variety of heavy metals can cause soil contamination. Some of the most commonly encountered heavy metals in soil remediation are Cd, As, Zn, Cu, and Pb. Most remedial activities deal with only one heavy metal, and only a limited research studies investigated the phytoremediation of multiple heavy metals. Therefore, the accumulation, enrichment, and phytoextraction of multiple heavy metals in plants are critical (Zhang et al., 2019). The mechanism of heavy metal contamination and remediation is very complex and includes several steps (Blamey et al., 1986; Rojjanateeranaj et al., 2017; Zhang et al., 2019). For example, Ryegrass (*Lolium perenne L.*) can be used for heavy metal remediation and grows

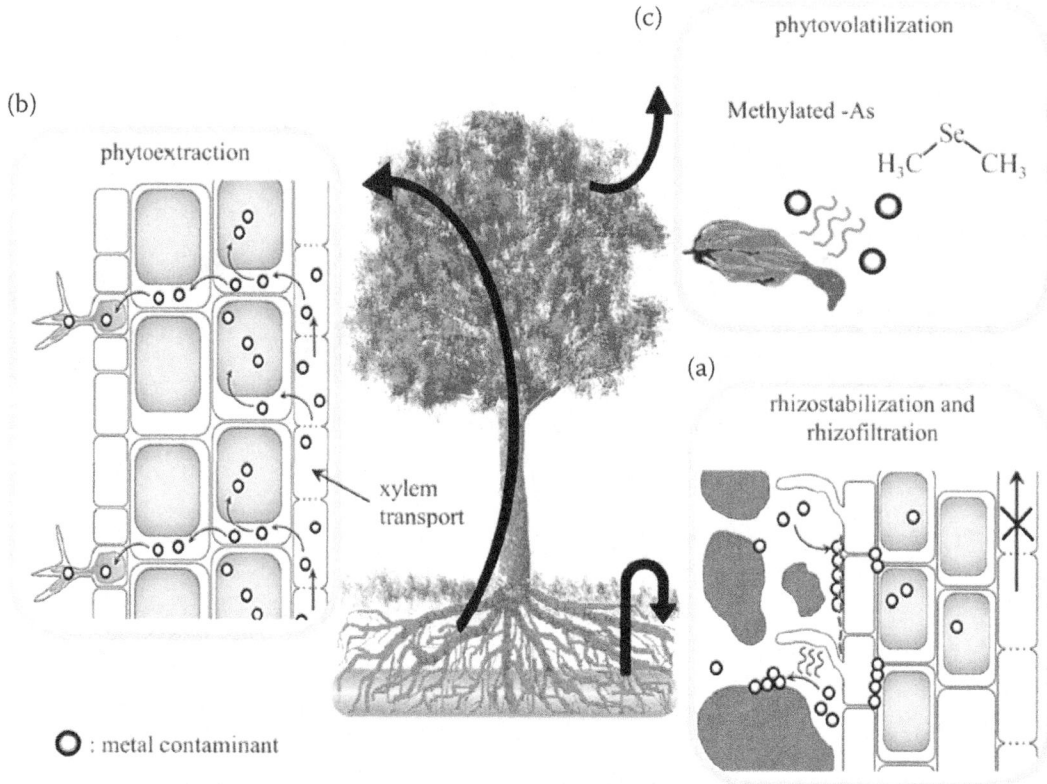

FIGURE 1.4 The main aspects of phytoremediation: (a) metal adsorption on soil particles, (b) metal accumulation in aerial organs, (c) leaf metabolism allows volatilisation of the toxic compound (DalCorso et al., 2019).

fast (Vigliotta et al., 2016; Xie et al., 2018). As stated in the literature studies, ryegrass can absorb Cd, Pb, and Zn from soil (Khalid et al., 2016). Plants and the rhizosphere employ mechanisms responsible for heavy metal removal. Some of these mechanisms are phytodegradation, phytovolatilisation, phytoextraction, rhizo stabilisation, and rhizo filtration, as shown in Figure 1.4. Zhang and Yang (Zhang et al., 2019) reported that the accumulation of various heavy metals in ryegrass was significantly high. The level of heavy metal accumulation in the ryegrass root was Cd > Pb > Zn, in the leaves it was Zn > Pb > Cd, and total extraction efficiency was Cd > Zn > Pb (Zhang et al., 2019).

Carrión and Mendoza (2019) investigated the phytoremediation capacity of *Amaranthus hybridus, Brassica rapa,* and *Amaranthus spinosus* for Ni, Cd, Pb, Zn, and Cr. The results showed that Zn, Pb, Cr, and Ni were accumulated in the roots of *Brassica rapa* at high concentrations. High concentrations of these heavy metals were observed on stems. Finally, they were observed at high concentrations in leaves. *Amaranthus hybridus*'s tissue had a better phytoremediation capacity of Zn.

1.1.5 Phytoremediation for Oil Contamination

For many years, subsurface contamination has been a major problem in the world due to its negative impacts on human and the environment. Traditional remediation technologies are sometimes not sufficient, very expensive, complicated, and not welcomed by the public. This directed the researchers to develop new alternative technologies, especially those that use plants and microorganisms. For instance, rhizoremediation is one of the alternative technologies currently available for in-situ soil and water bioremediation. Rhizoremediation is a phytoremediation method which is responsible for the degradation of organic pollutants in the plants' roots. Selection criteria of a plant for rhizoremediation includes rapid growth, long root system, tolerance to contaminants (Pilon-Smits, 2005), tolerance to varying environmental conditions, and suitability for different types of soils (Cook and Hesterberg, 2013).

Ubogu and Odokuma (Ubogu et al., 2019) investigated the rhizoremediation of a swamp contaminated with crude oil by using two different plant species, *Phragmites australis* and *Eichhornia crassipes*. The results of the study indicated that the remaining soil TPH concentrations in the rhizospheres of these two species indicated a decrease from 7,190 mg/kg to a final concentration of 4,000 mg/kg. The aerobic hydrocarbon-degrading microorganisms increase in uncontaminated rhizospheres of these two species can be related with the rise in root transudation as the plant aged (Ubogu et al., 2019).

1.2 Bioaugmentation and Biostimulation

Hydrocarbon contamination is a universal environmental and health concern. Crude oil, gasoline, diesel fuel, and underground storage tanks (USTs) are some of the main sources of hydrocarbon contamination. These hazardous oily contaminants are on the EPA's priority pollutants list. In addition, some chemicals in petroleum-based fuels stay in the environment for many years. Soil properties are affected physically, chemically, and ecologically due to the presence of hydrocarbons (Petrov et al., 2016; Ramadass et al., 2015). One of the most common remediation technologies is bioremediation, which uses microorganisms to biodegrade hydrocarbon. Bioremediation is commonly applied to cleaning up soil-contaminated sites and is a cost-effective and environmentally friendly remediation technology.

1.2.1 Bioaugmentation and Biostimulation for Hydrocarbon Remediation

If a contaminant that contains hydrocarbon spills on soil, biodegradation by indigenous bacteria advances belatedly because of low microbial population and activity (Abed et al., 2014). Bioremediation is widely known as a cost-effective cleanup technology for the remediation of soils contaminated with oil (Cerqueira et al., 2014). As an example, the two main bioremediation technologies known as bioaugmentation and biostimulation are customarily applied to clean up contaminated soils (Jiang et al., 2016; Polyak et al., 2018; Ramadass et al., 2018; Safdari et al., 2018). Bioaugmentation technology works by adding exogenous microorganisms to the contaminated soil or water (Maria et al., 2011; Taccari et al., 2012; Wu et al., 2013). Microorganisms may be introduced together with particularly cultivated bacteria that have the capability to degrade a selected contaminant. Biostimulation on the other hand is defined as the introduction of nutrients, like nitrogen and phosphorus, to stimulate the existing microorganisms (Kauppi et al., 2011; Sayara et al., 2011; Yu et al., 2005). Conversely, natural attenuation technology involves biodegradation, volatilisation, and adsorption.

Microorganisms that can degrade petroleum-based contaminants are specified as either eukaryotic or prokaryotic microorganisms (Balba et al., 1988). Various research works have been carried out about the biodegradation of petroleum-derived contaminants by bacteria (Jiang et al., 2016; Nwankwegu and Onwosi, 2017; Polyak et al., 2018; Ramadass et al., 2018; Safdari et al., 2018; Wu et al., 2016, 2017). Some examples of these bacteria are from genera Acinetobacter, *Pseudomonas, Mucor, Flavobacterium, Achromobacter, Bacillus, Alcaligenes, Rhodococcus, Candida, Penicillium,* and *Aspergillus* (Nwankwegu and Onwosi, 2017; Polyak et al., 2018; Wu et al., 2017). Alexander (1999) recommended that a C:N:P ratio of 100:10:2 is ideal for petroleum hydrocarbon remediation. Zhang and Cheng (Zhang et al., 2006) and Alavi and Mesdaghinia (Alavi et al., 2014) showed that a C:N:P ratio of 100:5:1 is ideal. For soils with low N rates, the optimum C:N:P ratio is achieved by introducing proper amounts of NH_4Cl and KH_2PO_4.

In bioremediation technology, microorganisms are used to convert organic pollutants such as diesel fuel into the non-toxic end products CO_2 and H_2O. All bacteria species can occur naturally and need specialised nutrients. Generally, these bacteria don't have adequate concentrations to provide full biodegradation. Therefore, there is an opportunity to advance and speed up the natural biodegradation by adding new bacteria and nutrients. Bacteria will break down the molecular structure of the contaminants when introduced as a slurry to contaminated soil. Due to the high level of interface between bacteria and the contaminant, the degradation rate seems to be rather fast at first, but slowly decreases as the more biodegradable contaminant is digested.

Bioaugmentation alone may not be effective to improve the remediation of hydrocarbon-contaminated soils. Some research studies showed that bioaugmentation improves biodegradation effectiveness temporarily and biostimulation seems to be a highly preferred bioremediation technology to reach sufficient cleanup (Polyak et al., 2018; Wu et al., 2016; Yang et al., 2015). Microbial populations are important in the remediation of petroleum hydrocarbon. Molecular methodologies provide help to understand the microbial community structure in remediation systems (Zhang et al., 2019). Improvement of hydrocarbon biodegradation is possible by applying different bioremediation technologies such as bioaugmentation and biostimulation alone or in combination. Biostimulation seems to be a highly efficient bioremediation methodology for hydrocarbon removal compared to bioaugmentation alone (Wu et al., 2019).

1.2.2 Bioaugmentation for the Treatment of Industrial Wastewater

Industrial wastewaters are responsible for a great proportion of pollution in freshwater bodies. For this reason, it is required that every country must monitor and regulate their industrial wastewaters. For instance, industrial wastewaters are regulated under the Clean Water Act in the United States, whereas they are regulated under the Industrial Emissions Directive (IED) in Europe (EC, 2015; EPA, U., 2015). Contaminants under these regulations for industrial wastewaters are different for each industrial process. For instance, an activated sludge treatment system is accepted as one of the best treatment technologies for different industrial wastewaters (Table 1.2) (Raper et al., 2018).

The microbial biomass in an activated sludge process carries out the treatment of organic materials through the biological reactions of the responsible microorganism (Zhang et al., 2014). Most industrial wastewaters contain compounds that are very difficult to remove; these wastewaters therefore persist in effluents of ASP. Bioaugmentation, the addition of exogenous microorganisms, improves the performance of ASPs (Semrany et al., 2012). Industrial wastewaters are the most challenging waters and bioaugmentation can be selected as a process to improve the treatment efficiency. Bioaugmentation was reported as unpredictable, but some operational factors such as strain selection and knowledge of the molecular biology can improve bioaugmentation (Herrero and Stuckey, 2015; Thompson et al., 2005). Table 1.3 summarises some bioaugmentation examples applied to contaminants found in industrial wastewaters.

TABLE 1.2

Emission limits for an activated sludge process as the best available technique (Raper et al., 2018)

Wastewater Source	BAT Emission Limit (mg/L)	Reference
Produced Water (Oil and gas wastewater)	Hydrocarbon oil index: 0.1–2.5 COD: 30–125 TN: 1–25	(Commission, 2014)
Food and Milk	Oil and grease: <10	(Commission, 2006)
	COD: <125	
	BOD5: <25	
	TN: <10	
Glass Manufacturing	COD: <5–130	(Commission, 2012)
	Total hydrocarbons: <15 Ammonia (as NH_4): <10 Phenol: <1	
Coke Making Wastewater	COD: <220	(Commission, 2013a)
	BOD5: <20	
	SCN: <4	
	PAHs[a]: 0.05	
	Phenols: 0.5	
	TN: <15–50	

Note
[a] Sum of fluoranthene, benzo[b]fluoranthene, benzo[k]fluoranthene, benzo[a]pyrene, indeno [1,2,3-cd] pyrene, and benzo[g,h,i]perylene.

TABLE 1.3

Examples of bioaugmentation applied to contaminants found in industrial wastewaters (Raper et al., 2018)

Contaminant	Examples		
Nitrogen	Scale	Application	Results
(Onyia et al., 2001)	Laboratory	Palm oil effluent	100% increase in nitrification. 20% reduction in land requirement.
(Shan and Obbard, 2001)	Laboratory	Prawn aquaculture wastewaters	Immobilised bacteria provided total ammoniacal nitrogen reduction of 83% maintaining ponds to remain at optimal conditions.
(Ma et al., 2009)	Full-scale	Petrochemical wastewaters	Discharge limits were maintained in 20 days. Effluent ammonia concentrations fell 67%.
Aromatic compounds	Scale	Application	Results
(Qu et al., 2011)	Laboratory	Synthetic alkaline wastewaters	*Pseudomonas sp.* provided 90% removal.
(Fang et al., 2013)	Laboratory	Coal gasification wastewater	Bioaugmentation increased removal rates from 66% to 80%.
(Duque et al., 2011)	Laboratory	2-fluorophenol	2-fluorophenol degrading species provided treatment of wastewaters subjected to shock loads.
(Martín-Hernández et al., 2012)	Laboratory	p-nitrophenol (PNP) wastewaters	Bioaugmentation provided rapid removal of shock loads of contaminant.
(Straube et al., 2003)	Laboratory and pilot-scale	PAH contaminated soil	Pseudomonas aeruginosa increased PAH degradation from 23% to 34%.
(Sun et al., 2014)	Pilot	Former coke works contaminated soil	Total PAH removal rates increased by 24% in the control, 35.9% with bioaugmentation, and 59% with biostimulation.

1.2.3 Bioaugmentation for Composting

Organic wastes such as manures, food waste, and yard waste can be treated effectively by aerobic composting. Overproduction of livestock causes air, water, and soil pollution (Petersen et al., 2007). An excess amount of manure will cause nutrient surpluses in the agricultural fields, particularly when nutrient demand exceeds the fertilisation needs. A proper nutrient management should involve volume reduction, proper design and operation, and a clean transportation system. Manure composting should follow the relevant regulations. The raw manure volume is reduced via aerobic biological and chemical processes (Bernal et al., 2009; Pardo et al., 2015). When the product is applied as a fertiliser, the threat of pathogens and weeds is greatly reduced, and the soil quality can be improved (Onwosi et al., 2017). The composting process is generally optimised by the C:N ratio of the composting materials. The average C:N ratio recommended is between 25 and 35, of which carbon is used as the energy source and nitrogen is used as the nutrient source for microorganisms. Pardo and Moral (Pardo et al., 2015) and Bernal and Alburquerque (Bernal et al., 2009) showed that gaseous emissions from a composting process account for 60–70% of the initial nitrogen and carbon content, respectively. The most commonly encountered reason for nitrogen loss is because of the volatilisation of ammonia (Martins and Dewes, 1992; Parkinson et al., 2004; Sommer, 2001).

During composting, several bacteria are responsible for the transformation of N and few fungi and bacteria species enhance the solubilisation of P and K (Sánchez-Monedero et al., 2017). Some composting bacteria are *Burkholderia, Pseudomonas, Zymomonas,* and *Xanthomonas. Nitrosomonas* and *Nitrospira* play a role as well in composting processes similar to that of nitrification bacteria (Anastasi et al., 2005; Insam and de Bertoldi, 2007; Ryckeboer et al., 2003). Fungi also play a role in compost processes. Fungi is usually present in compost piles through the first and final phases of the composting process. Some of the most common fungi species are *Aspergillus, Fusarium, Mortierella, Penicillium, Chrysosporium,* and *Trichoderma* (Anastasi et al., 2005).

Microbiological activities during composting produce heat, which causes an increase in the temperature of piles. There are four steps in the composting process based on the changes in temperature. The mesophilic stage is the first step, which takes place when the temperature in piles changes from ambient to 45°C. During the mesophilic stage, microorganisms grow and biodegrade the organic compounds, generating heat energy. Thermophilic stage starts when the temperature of the pile increases above 45°C and reaches up to 60°C. Then the cooling phase starts as microbial activity decreases because of nutrient depletion. The final stage is called maturation, in which various chemical reactions break down the phytotoxins.

Sahu and Manna (Sahu et al., 2019) used thermophilic fungi on different crop residues such as soybean, sugarcane, maize, pigeon pea, and cotton. The quality of the product was measured by using different indices of maturity and stability. The phases, in order, were mesophilic, thermophilic, second mesophilic, cooling, and humification. At the end of the study, the C:N ratio, biodegradability index, and nitrification index were greatly reduced. On the other hand, ash content, organic matter loss, and the ratios of cellulose biodegradation and lignin/cellulose increased. It was thus found that fungal species of *Rhizomucor pusillus, Tricoderma viride, Aspergillus awamori,* and *Aspergillus flavus* could reduce the decomposition period from approximately 365 days to 120 days.

1.3 Bioventing

Total petroleum hydrocarbons consist of three fractions, which are aliphatic, aromatic, and polar hydrocarbon. Diesel products primarily include C10 through C19 hydrocarbons, which have 64% of aliphatic compounds, 1 to 2% of olefinic compounds, and 35% of aromatic compounds. All of the abovementioned fuels include polycyclic aromatic compounds less than 5% (PAHs). Biodegradation of petroleum hydrocarbons by microorganisms have been studied by scientists for many years. Microorganisms degrading crude and refined petroleum hydrocarbons are described as either eukaryotic or prokaryotic. Some examples to the species responsible for biodegradation of petroleum-based contaminants are *Pseudomonas, Nocardia, Flavobacterium, Acinetobacter, Arthrobacter, Alcaligenes, Mycobacterium, Bacillus, Aspergillus, Fusarium,* and *Penicillium.* Bioremediation method can be selected depending on the properties of the polluting hydrocarbons, soil characteristics, and the groundwater where the pollutant is spilled. It is preferred that the hydrocarbons contain a carbon chain length of between C10 and C20, because these carbon chains are easier to break. However, hydrocarbons with C20-C40 carbon chains are known as hydrophobic, less soluble in water and resistant to biodegradation. Bioremediation is a technology that aims at cleaning a contaminated region by microorganisms and their enzymes. Specific contaminants such as chlorine-containing pesticides can be also biodegraded by microorganisms. Microorganisms that degrade these hydrocarbons can be bacteria, protozoa, nematodes, and fungi. Bioremediation of a contaminated site by adding specifically selected microorganisms is called bioaugmentation. Bioaugmentation technology is used effectively in soils and groundwater where there is not sufficient microbial population available for biodegradation. Bioremediation of a contaminated site by adding nutrients is called biostimulation. Some of the benefits of bioremediation can be summarised as effective, economic, environmentally friendly, less toxic by-products, applicable directly on the contaminated area and friendly with the existing flora.

Subsurface contamination by hydrocarbons and organic solvents is a major issue in developed countries. Spills, leaks, and releases from pits, lagoons, and USTs have contributed to subsurface contamination problem. Soil vapor extraction (SVE) technology seems to be very effective technology for removing volatile organic compounds from the vadose zone. A typical SVE system includes extraction wells installed in the unsaturated zone (vadose zone), blowers, air injection wells, air and water separator, and an off-gas removal system (Figures 1.5 and 1.6). Airflow is created in the vadose zone by creating a pressure gradient via the injection of air in the subsurface. The extracted air, which contains volatile organic compounds (VOCs), is usually treated by adsorption on granular activated carbon (GAC) or catalytic oxidation.

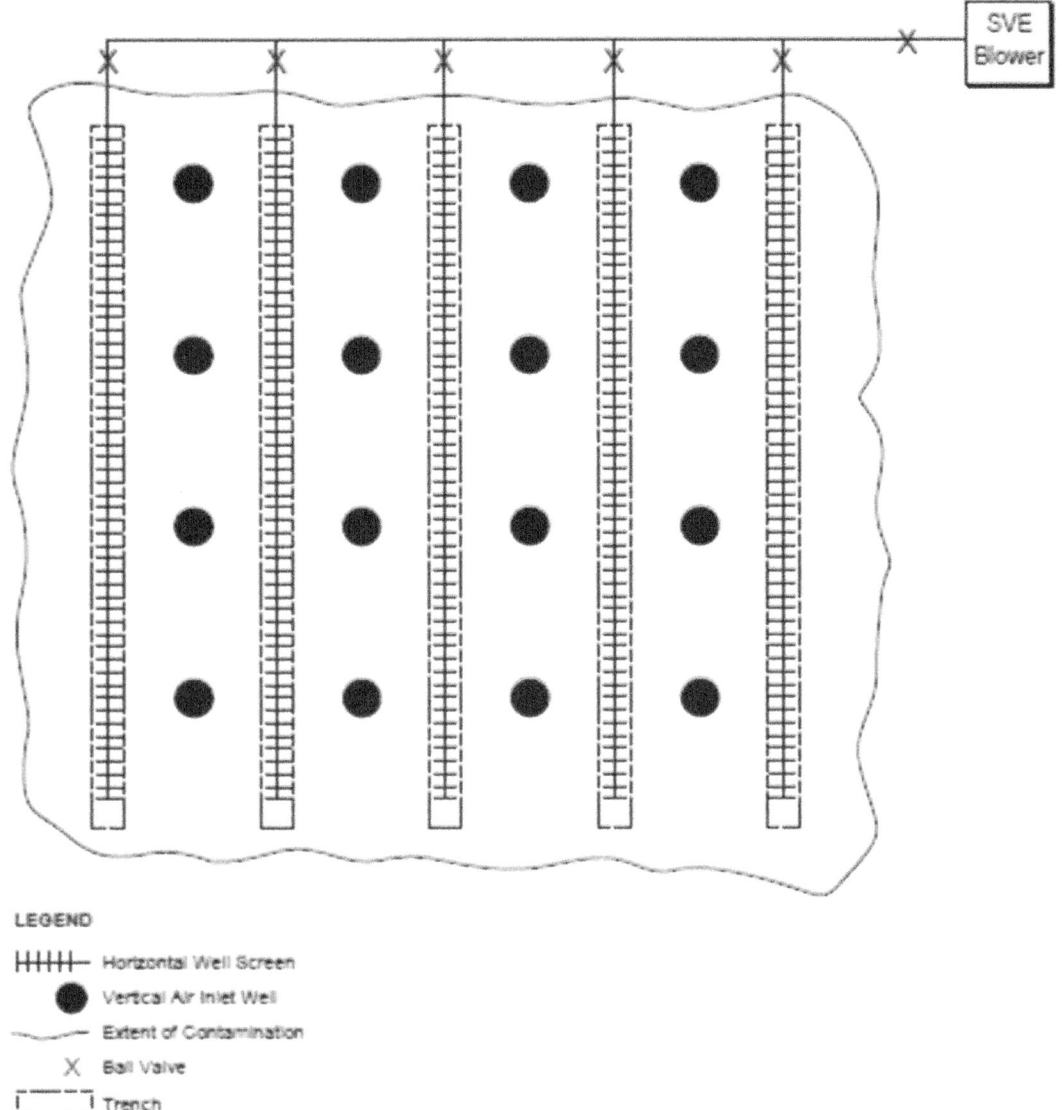

FIGURE 1.5 Cross section of an SVE system Army (Army, 2002).

Bioventing (BV) is similar to SVE in that the air is introduced to the subsurface for the removal of contaminants in-situ rather than aboveground. Indigenous microorganisms in the vadose zone require oxygen to biodegrade the organic contaminants. Therefore, sufficient flowrates with optimum residence times should be provided to the microorganisms in the subsurface. Since BV does not depend on volatilisation, it is more appropriate for biodegradation of semi-volatile organic compounds (SVOCs) rather than VOCs. A BV system includes extraction or injection wells placed in the vadose zone, blowers or vacuum pumps, and air injection wells (Figure 1.7; Army, 2002).

Generally, SVE is used for VOC removal from the vadose zone using volatilisation. However, air circulation in the unsaturated zone will improve the VOC biodegradation. BV also helps to bring down the cost of off-gas treatment and provides removal of SVOCs. In BV, the aerobic biodegradation processes dominate. In BV systems, the same blowers used in SVE systems can be used to supply air to the unsaturated zone to stimulate the microorganisms. For instance, benzene degradation occurs according to Equation 1.1.

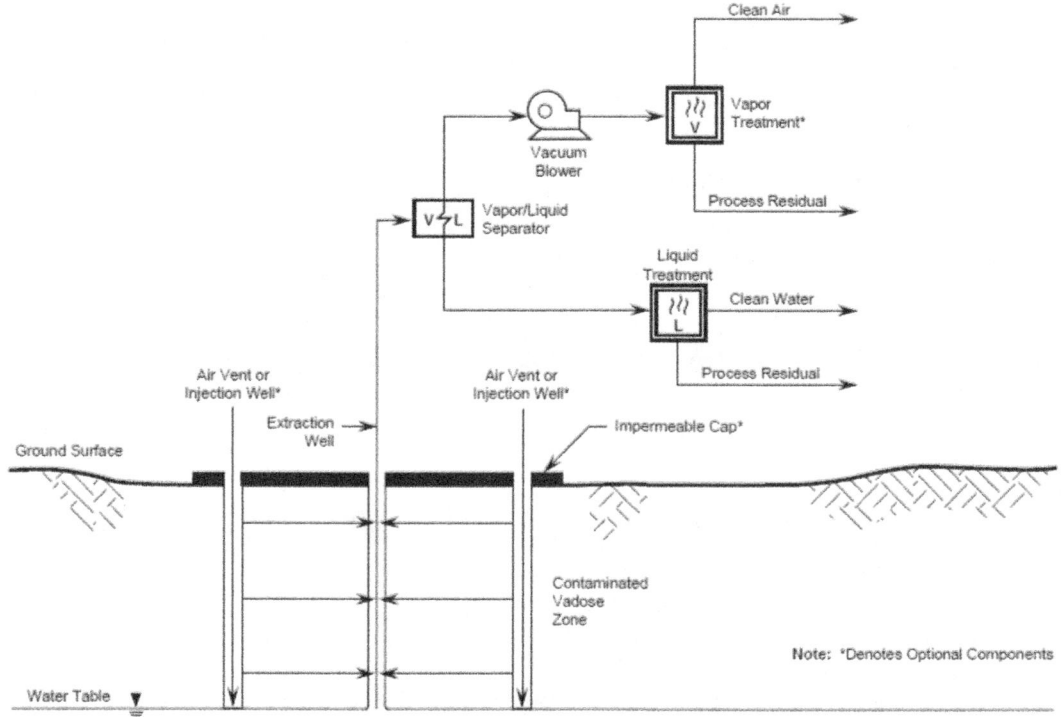

FIGURE 1.6 Generic soil vapor extraction system Army (Army, 2002).

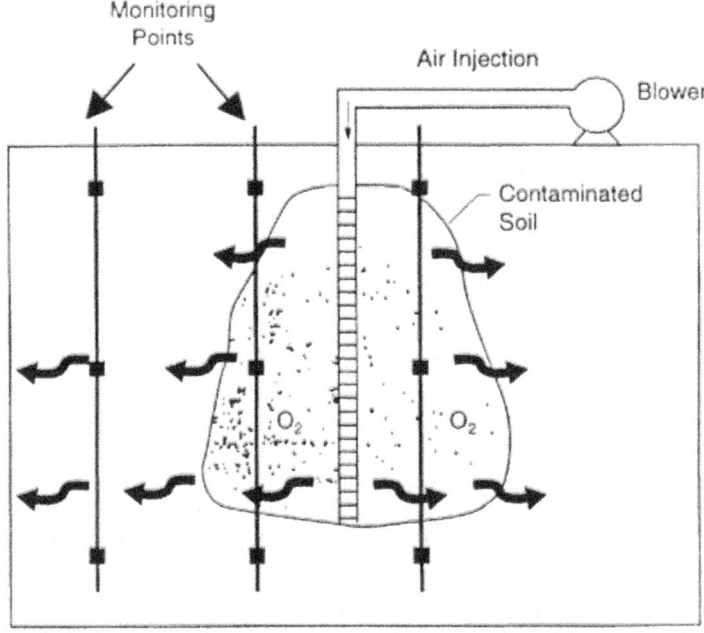

FIGURE 1.7 Typical bioventing system.
Source: **(Army, 2002).**

$$C_6H_6 + 7.5O_2 \rightarrow 6CO_2 + 6H_2O + biomass \qquad (1.1)$$

According to the benzene biodegradation equation, around 3.5 g of O_2 is needed for each gram of benzene biodegraded. In the implementation of BV system, gases such as O_2, CO_2, and CH_4 in the vadose zone are usually monitored to make sure that the aerobic conditions are present. Increased levels of CO_2 and decreased levels of O_2 indicate a higher level of microbial activity. Also, soil pH and alkalinity can increase or decrease the O_2 levels. For instance, soils with high pH and high alkalinity can include little CO_2 in the soil gas due to the presence of carbonates.

1.3.1 Designing Bioventing Systems

Designing bioventing systems includes three basic conditions.

1. Optimal oxygen concentrations in the contaminated soils for aerobic biodegradation.
2. Optimal soil moisture for microbial activity.
3. Optimal nutrients needed for microbial populations.

To maintain these requirements, various design and operating parameters should be identified. These are summarised below.

2.3.1.1 Airflow Rate

A BV system should provide optimum airflow to maintain maximum oxygen consumption by the subsurface microorganisms. Therefore, the rate of airflow required is usually much lower than the SVE airflow. As a common rule, the air in the contaminated vadose zone pore volume should be exchanged at least once every other day. Aerobic biological activity is specifically observed in the contaminated vadose zone where the oxygen concentrations are greater than 2%. When the oxygen level of the air is below 2%, biological activity will be terminated.

1.3.1.2 Soil Moisture

In the design of a BV system, it is critical to provide an optimal soil moisture, which varies between 40 and 60% to maintain biological activity in the contaminated zone. When the moisture content of soil increases above 60%, then the remaining pore sizes will shrink, reducing the air permeability. Therefore, it is essential to involve in the design moisture addition to the contaminated layer.

1.3.1.3 Nutrients

Nitrogen, phosphorous, and the micronutrients are usually present in the vadose zone. Hence, the need for these nutrients is critically important in the early phases of microbial growth.

1.4 Biosparging

Accidental releases of petroleum-based hydrocarbons from transmission lines and USTs are some of the frequently encountered causes of groundwater contamination. Petroleum-derived compounds include benzene, toluene, ethylbenzene, and xylene isomers, which are carefully monitored and regulated by many countries. At several spill sites, BTEX is present in a liquid phase, usually known as non-aqueous-phase liquids (NAPLs).

Introducing air into the contaminated groundwater zone is crucial due to the importance of oxygen supply to the microorganisms for in-situ bioremediation of a polluted saturated zone. Non-strippable and readily biodegradable compounds will be appropriate for the application of biosparging to remediate a dissolved plume. Air injection at low flow rates (0.5–3 cubic feet per minute) into contaminated groundwater to accelerate biodegradation is called biosparging. However, it should be taken into consideration that the time required to raise the oxygen concentrations in the saturated zone depends on the time needed for the diffusion of oxygen. Only about 0.5% of the oxygen in the injected air will be dissolved in water during air sparging. Therefore, evaluating and monitoring the changes in dissolved oxygen (DO) concentrations is critical after the initiation of biosparging. Biosparging is a proven technology for removing VOCs and BTEX (Adams and Ready, 2003; Wu et al., 2005; Brar et al., 2006).

Unlike vadose zone soil, a saturated zone does not include an air phase and, therefore, oxygen availability as an electron acceptor is significantly reduced. This low level of DO concentrations will result in anaerobic conditions in a saturated zone. However, some anaerobic microorganisms are able to degrade contaminants by using sulfate, nitrate, or ferric iron as electron acceptors (Lovley, 1997; Wiedemeier et al., 1999).

As an example, chlorobenzene is an important groundwater pollutant. The compound is known by its persistence in oxygen limited environments for years but can also be biodegraded if oxygen becomes available (Dermietzel and Vieth, 2002). Some studies showed that biodegradation of chlorobenzene is reduced under anoxic conditions (Kaschl et al., 2005; Nijenhuis et al., 2007). Therefore, natural attenuation of chlorobenzene cannot degrade chlorobenzene due to the anoxic environment within the saturated zone.

In the application of biosparging technology, contaminant-free air is introduced into the saturated zone. This will form a conic area of fine, branched channels. It is preferred that these small channels and the conic area are formed. The injection pressure must be greater than the pressure needed to overcome the water standing above the injection point. It is also important to note that an intermittent injection can be advantageous on the formation of branching of the channels. Biosparging method can be selected only when the air injection point is placed below the contamination layer. If a sufficient groundwater flow is desired, areas outside the conic area must also be oxygenated (ICSS, 2006).

Biosparging allows for the in-situ volatilisation of VOCs, contaminants desorption, and the biodegradation by increasing the DO concentrations in the contaminated aquifer. For contaminants with high volatility, stripping plays the critical role during the removal of pollutants. Nevertheless, it is complicated to quantify the part taken by the biological processes. In aquifers contaminated with chlorinated chloro-aliphatic that are not easily biodegradable under aerobic conditions, in-situ stripping or air stripping can be considered for the treatment. For non-volatile but biodegradable compounds such as mineral oil hydrocarbons, air stripping will play a major role by supplying the oxygen required for biodegradation (biosparging).

1.4.1 Application, Advantages, and Disadvantages of Biosparging

Applicability ICSS (ICSS, 2006):

- Possible to remove all volatile organic compounds.
- Aquifer must be homogenous.
- Permeability of the aquifer should be greater than 10^{-4} m/s.

Advantages ICSS (ICSS, 2006):

- Cost-effective technology.
- Introducing oxygen to the unsaturated zone will promote aerobic biodegradation.

Disadvantages ICSS (ICSS, 2006):

- Inhomogeneous aquifer will reduce the efficiency of the process.
- If the contamination layer is very deep, then a very high pressure will be needed to overcome the water column.

- If the dissolved iron is present at high concentrations, the aquifer will be blocked due to the formation of iron oxides.

1.5 Composting

Composting is a technology where microorganisms stabilise organic matters under aerobic or anaerobic conditions. In aerobic composting, microorganisms consume oxygen to biodegrade the organics, producing CO_2, water vapor, and heat (Chardoul et al., 2015). About half of the initial organic matter is lost to CO_2 and water vapor during aerobic composting. Therefore, composting processes can produce a high-quality soil conditioner from organic wastes such as food waste, while decreasing the volume and the mass of the input material. When the optimum environmental conditions for the growth of microorganisms are attained, composting processes occur fast. Some of the most important conditions for composting are listed as follows:

– Appropriate C:N ratio of the input material.
– Sufficient oxygen for the microorganisms.
– Adequate moisture content to improve the biological activity.
– Thermophilic temperature range for a strong microbial activity.

1.5.1 Aerobic Composting

In the aerobic composting process, first the input materials are mixed and then adequate air is introduced to the system to initiate the process. As the amount of oxygen in the compost pile is consumed by the microorganisms, decomposition slows down and, if the oxygen is not provided any longer, the process stops. Two types of aeration exist in aerobic composting: passive aeration and pressurised aeration. In passive aeration, windrowing the compost pile is required because the oxygen in the pile will be consumed rapidly. Because the generation of heat depends on biological reactions, temperature should be monitored as an indicator for the process effectiveness. The temperature of the compost pile usually reaches 48–60°C and stays at that level at least for two weeks (Chardoul et al., 2015). Then the temperature first drops to 38°C and then to the atmospheric temperature as the active composting process slows down.

If the oxygen concentration decreases during the active composting process, microbial activities and the temperature will decrease as well. Temperature levels can be increased again by mixing, turning, or by pressurised aeration. If there is sufficient oxygen and microbiological reactions available in the environment, the temperature in the compost mass can be increased above 60°C easily. At this temperature range, most of the pathogens are killed. The compost curing period starts after the active composting process is complete. During the curing period, composting of the organic materials continues, but at a slower rate. Decomposition continues until all the organic material is consumed by the microorganisms to produce water and carbon dioxide. Once the composting and curing processes are complete, C:N ratio, oxygen requirement, temperature, and odor parameters need to be determined to decide if the compost quality is acceptable.

1.5.1.1 Important parameters for composting

The main parameters that affect the process are aeration, nutrients, water content, porosity, consistency, pH, particle size, composition, and time. Table 1.4 shows recommended values of parameters for fast composting.

1.5.1.1.1 Oxygen and Aeration

Excess oxygen is utilised during aerobic composting. In the first week of the composting process, readily biodegradable organic compounds are metabolized faster. Thus, the need for oxygen and heat generation are highest at the beginning of the process and decreases as the composting continues. When

TABLE 1.4

Parameters for fast composting

Parameter	Range	Recommended Range
C:N	20:1–40:1	25:1–30:1
Moisture content	40–65%	50–60%
Oxygen concentration	>5%	>>5%
Particle diameter (cm)	0.32–1.27	Variable
pH	5.5–9.0	6.5–8.0
Temperature (°C)	43–65	54–60

the source of oxygen or air is reduced, composting will slow down. Oxygen concentrations in the pores of the compost pile should be at least 5%, assuming the oxygen content in the ambient air is 21%.

1.5.1.1.2 Nutrients

The primary nutrients (C, N, P, and K) are needed in sufficient amounts for microorganisms. Carbon is used for energy and growth and nitrogen is used for protein and reproduction. A balanced C:N ratio indicates that there are sufficient nutrients for microorganisms. Input materials are mixed at a C:N ratio varying between 15:1 and 30:1; however, a 20:1–40:1 ratio is better for a high-quality compost.

1.5.1.1.3 Moisture Content

Moisture is needed for microorganisms to carry out their metabolic processes. Water carries nutrients and microorganisms, and it provides optimum environmental conditions for chemical reactions during composting. If the moisture content in the organic matter drops to less than 15%, then the biological activity stops completely. The ideal moisture content required for composting is around 45–60%. Once the moisture content is below 40%, biological activity will slow down. However, if the moisture content rises above 60%, then the water will fill the pores of the compost pile, and thus replace the air in the pores. This will prevent air circulation, causing anaerobic conditions in the compost pile.

1.5.1.1.4 Porosity, Composition, and Particle Size

Porosity, composition, and particle size are physical parameters for the compost input material. These physical parameters affect the aeration of the compost pile. These properties can be optimised by raw material selection, mixing, and shredding. Porosity can be defined as a pore/air space in the compost material. Larger uniform particles increase the porosity. Composition refers to the material types utilised during composting. Microorganisms use oxygen on the surface of the particles. Therefore, aerobic degradation is higher for the smaller particles due to their higher surface area. Thus, particle sizes should be between 3.0–13 mm.

1.5.1.1.5 pH

The optimum pH in composting processes should be between 6.5 and 8. pH is important for materials that have high nitrogen content because at pH levels greater than 8.5, nitrogen compounds are converted to NH_3. Therefore, the pH should be less than 8 to prevent NH_3 generation. The levels of pH will change during composting due to different biodegradation reactions. For instance, during the first stages of composting, high concentrations of acids will reduce the pH. However, concentration of NH_3 produced from nitrogen-containing compounds will increase the pH.

1.5.1.1.6 Temperature

Composting occurs under mesophilic (10–40°C) and thermophilic (>40°C) temperatures. Although composting can occur under mesophilic temperatures, it is usually preferred that the composting temperature should be kept between 43–65°C. Temperatures in the thermophilic range are desired because thermophilic temperatures provide rapid composting and eliminate pathogens, weed seeds, and fly larvae. The threshold temperature value is 55°C to destroy pathogens that are harmful to humans. Most of the pathogens are killed at this temperature. For weeds, the critical temperature is 63°C.

The composting process itself causes a temperature increase. At some point, the temperature in the compost material reaches 71°C, killing most of the microorganisms. At this level of temperature rise, the composting process cannot proceed and will stop. The composting process will not start again until the microorganism population has reached a sufficient level. Therefore, the temperature should be monitored continuously. If the temperature exceeds 60°C, pressurised aeration or turning should be employed to accelerate the heat loss. Compost material moisture content must be greater than 40% because most of the heat loss during composting is due to water vaporisation.

1.5.1.1.7 Time

The time required for composting is a function of the material used, temperature, moisture content, frequency of aeration, and customer requests. Optimum moisture content, appropriate carbon to nitrogen ratio, and frequent aeration accelerate the composting time. Low moisture content, high carbon to nitrogen ratio, reduced temperature, limited aeration, and high wood-containing input materials slow down the composting process. Under optimum conditions, biodegradation and stabilisation can occur in 2 or 3 weeks, but for the best results at least a 2-month time period is needed. Although it is possible to make compost in less than 1 week with vessel or reactor-type compost systems, it is usually preferred that at least 4–6 weeks of curing time is needed before using the product. Different application time intervals are given in Table 1.5.

1.5.2 Anaerobic Composting

Anaerobic composting is usually preferred with wet materials containing high levels of nitrogen. However, usually most organic materials can be used in an anaerobic reactor such as cardboard, grass cuttings, paper waste, food waste, and manure. It is critical to avoid including an excess amount of materials rich in carbon such as sawdust, leaves, and yard waste. The composting process will slow down when the dry carbon-containing materials are present at high concentrations in the compost material.

Unlike aerobic composting, anaerobic composting requires low temperatures; thus, the temperature of the pile will not increase. Therefore, any pathogens or weeds found in the input material cannot be destroyed in the process. However, the pathogens will be slowly destroyed due to the harsh environment.

Anaerobic composting should occur in a sealed and air-tight reactor, which provides the ideal environment for the microorganisms to grow and degrade the organic compounds in the anaerobic

TABLE 1.5

Composting times for different applications

Method	Input Material	Range	Typical	Curing Time
Passive composting	Manure	6 months–2 years	1 year	
Windrow composting (infrequent aeration)[a]	Manure	4–8 months	6 months	
Windrow composting (frequent aeration)[b]	Manure	1–4 months	2 months	4 months
Windrow (passive aeration)	Manure	10–12 weeks	11 weeks	1–2 months
Aerated static piles	Sewage + wood pieces	3–5 weeks	4 weeks	1–2 months
Rectangular mixed bed	Sewage + yard waste or manure + sawdust	2–4 weeks	3 weeks	1–2 months
Rotating drum	Sewage sludge or MSW	3–8 days		2 months[c]
Vertical silo	Sewage sludge or MSW	1 week		2 months[c]

Notes
[a] Using front loader.
[b] Using special windrow turner.
[c] Requires secondary composting such as windrow or aerated piles.

environment. The anaerobic compost reactor must be air-tight and oxygen-free. Anaerobic digesters are classified as dry and wet.

1.5.2.1 Wet Fermentation

Wet anaerobic digesters utilise manures, slurries, and green materials. The moisture content of the material used in composting is usually higher than 80%. Wet fermentation transforms liquid waste into a renewable energy source, called biogas. This biogas is composed of CH_4 and CO_2, which are used for energy generation and the digestate. This digestate is then separated into a liquid fraction, which can be used as a liquid fertiliser, and a nutrient-rich solid portion as a soil fertiliser.

1.5.2.2 Dry Fermentation

Dry anaerobic digesters don't utilise any liquid to enhance the biological process. The input materials containing optimal moisture such as grass, plants, and food waste are shredded and placed in air-tight reactors for biodegradation. The biodegradation time of the input material depends upon the material type and the process temperature. During the selection of solid waste material for dry fermentation, the C:N ratio and the biodegradability of the input materials should be carefully evaluated. Once finished, the compost product should be odor free, with no excess liquid generated.

1.5.3 Microbiology of Composting

The major groups of microorganisms responsible for the composting process include bacteria, fungi, and actinomycetes. There exist other microorganisms that also play a role in the composting process, including nematodes, protozoa, and micro-arthropods. Some macroorganisms that take part in composting include beetles, larger nematodes, earthworms, and some insects. Each of these organisms is different, and they adapt under different environmental conditions (Chardoul et al., 2015).

The groups of bacteria are identified depending on the temperature values in which they can be active. These bacteria groups are classified as psychrophilic bacteria, which are active at temperatures between −15°C and 10°C, mesophilic bacteria that are active at temperatures from 10°C to 50°C, and thermophilic bacteria that are active at temperatures between 45°C and 70°C.

REFERENCES

Abed, R.M.M., et al. (2014). Characterization of hydrocarbon-degrading bacteria isolated from oil-contaminated sediments in the Sultanate of Oman and evaluation of bioaugmentation and biostimulation approaches in microcosm experiments. *International Biodeterioration & Biodegradation*, 89, 58–66.

Adams, J.A., & Ready, K.R. (2003). Extent of Benzene Biodegradation in Saturated Soil Column During Air Sparging. *Groundwater Monitoring & Remediation*, 23(3), 85–94.

Alavi, N., et al. (2014). Biodegradation of petroleum hydrocarbons in a soil polluted sample by oil-based drilling cuttings. *Soil and Sediment Contamination: An International Journal*, 23(5), 586–597.

Alexander, M. (1999). *Biodegradation and Bioremediation*. San Diego, Calif.; London: Academic Press.

Anastasi, A., Varese, G., & Marchisio, V. (2005). Isolation and identification of fungal communities in compost and vermicompost. *Mycologia*, 97, 33–44.

Army, U. (2002). *Engineering and Design: Soil Vapor Extraction and Bioventing*. Engineer Manual EM 1110-1-4001. US Army Corps of Engineers.

Baker, A.J.M., Reeves, R.D., & Hajar, A.S.M. (1994). Heavy metal accumulation and tolerance in British populations of the metallophyte *Thlaspi caerulescens* J. & C. Presl (Brassicaceae). *New Phytologist*, 127(1), 61–68.

Balba, M.T., Al-Awadhi, N., & Al-Daher, R. (1988). Bioremediation of oil-contaminated soil: microbiological methods for feasibility assessment and field evaluation. *Journal of Microbiological Methods*, 32, 155–164.

Begonia, G.B., et al. (1998). Growth responses of Indian mustard [*Brassica juncea* (L.) Czern.] and its phytoextraction of lead from a contaminated soil. *Bulletin of Environmental Contamination and Toxicology*, *61*(1), 38–43.

Bernal, M.P., Alburquerque, J.A., & Moral, R. (2009). Composting of animal manures and chemical criteria for compost maturity assessment. A review. *Bioresource Technology*, *100*(22), 5444–5453.

Blamey, F.P.C., et al. (1986). Role of trichomes in sunflower tolerance to manganese toxicity. *Plant and Soil*, *91*(2), 171–180.

Blaylock, M.J., et al. (1997). Enhanced accumulation of Pb in Indian mustard by soil-applied chelating agents. *Environmental Science & Technology*, *31*(3), 860–865.

Blowes, D.W., et al. (2003). 9.05 – the geochemistry of acid mine drainage. In Holland, H.D. & Turekian, K.K. (Eds.), Treatise on Geochemistry. Pergamon: Oxford. pp. 149–204.

Brar, S.K., et al. (2006). Bioremediation of hazardous wastes – a review. *Practice Periodical of Hazardous, Toxic, and Radioactive Waste Management*, *10*(2), 59–72.

Carrión, C., & Mendoza, W. (2019). Potential phytoremediator of native species in soils contaminated by heavy metals in the garbage dump Quitasol-Imponeda Abancay. *Journal of Sustainable Development of Energy Water and Environment Systems*, *7*(4).

Cerqueira, V.S., et al. (2014). Comparison of bioremediation strategies for soil impacted with petrochemical oily sludge. *International Biodeterioration & Biodegradation*, *95*, 338–345.

Chardoul, N., et al. (2015). *Best Management Practices for Commercial Scale Composting Operations Operator Guidebook*. MI, USA: Michigan Recycling Coalition.

Commission, E. (2006). Integrated pollution prevention and control reference document on best available techniques in the food, drink and milk industries [29 November 2019]; Available from: https://eippcb.jrc.ec.europa.eu/reference/BREF/fdm_bref_0806.pdf.

Commission, E. (2012). (BAT) conclusions under Directive 2010/75/EU of the European Parliament and of the Council on industrial emissions for the manufacture of glass [29 November 2019]; Available from: https://eur-lex.europa.eu/legal-content/EN/TXT/?uri=CELEX%3A32012D0134.

Commission, E. (2013a). Best available techniques (BAT) reference document for iron and steel production, industrial emissions directive 2010/75/EU. [29 November 2019]; Available from: https://op.europa.eu/en/publication-detail/-/publication/eaa047e8-644c-4149-bdcb-9dde79c64a12/language-en.

Commission, E. (2014). Best available technique conclusions, under directive 2010/75/EU of the European Parliament and of the Council on Industrial Emissions, for the Refining of Mineral Oil and Gas. [29 November 2019]; Available from: https://eur-lex.europa.eu/legal-content/EN/TXT/?uri=OJ:JOL_2014_307_R_0009.

Conesa, H.M., et al. (2007). Growth of *Lygeum spartum* in acid mine tailings: response of plants developed from seedlings, rhizomes and at field conditions. *Environmental Pollution*, *145*(3), 700–707.

Cook, R.L., & Hesterberg, D. (2013). Comparison of trees and grasses for rhizoremediation of petroleum hydrocarbons. *International Journal of Phytoremediation*, *15*(9), 844–860.

Cruz, M., Aguiar, R., & Mello, J. (2010). Phytoremediation of acid mine drainage by aquatic floating macrophytes. NCT-ACQUA – Annual Report – 2010. Institute of Science and Technology for Mineral Resource, Water and Biodiversity.

DalCorso, G., et al. (2019). Heavy metal pollutions: state of the art and innovation in phytoremediation. *International Journal of Molecular Sciences*, *20*(14), 3412.

Dermietzel, J., & Vieth, A. (2002). Chloroaromatics in groundwater: chances of bioremediation. *Environmental Geology*, *41*(6), 683–689.

Doucette, W., et al. (2013). Volatilization of trichloroethylene from trees and soil: measurement and scaling approaches. *Environmental Science & Technology*, *47*(11), 5813–5820.

Duque, A.F., et al. (2011). Bioaugmentation of a rotating biological contactor for degradation of 2-fluorophenol. *Bioresource Technology*, *102*(19), 9300–9303.

EC (2015). European Commission, The Industrial Emissions Directive [29 November 2019]; Available from: http://ec.europa.eu/environment/industry/stationary/ied/legislation.htm(Accessed.

EPA, U. (2015). Summary of the Clean Water Act [29 November 2019]; Available from: https://www.epa.gov/laws-regulations/summary-clean-water-act.

Fiset, J.F., Zinck, J.M., & Nkinamubanzi, P.C. (2003). Chemical stabilization of metal hydroxide sludge, *Proceeding of the X International Conference of Tailings and Mine Waste*, Vail, CO, USA, AA Balkema, 329–333.

Fang, F., et al. (2013). Bioaugmentation of biological contact oxidation reactor (BCOR) with phenol-degrading bacteria for coal gasification wastewater (CGW) treatment. *Bioresource Technology, 150*, 314–320.

Frérot, H., et al. (2006). Specific interactions between local metallicolous plants improve the phytostabilization of mine soils. *Plant and Soil, 282*(1), 53–65.

Hasan, S.H., Talat, M., & Rai, S. (2007). Sorption of cadmium and zinc from aqueous solutions by water hyacinth (*Eichchornia crassipes*). *Bioresource Technology, 98*(4), 918–928.

Herrero, M., & Stuckey, D.C. (2015). Bioaugmentation and its application in wastewater treatment: a review. *Chemosphere, 140*, 119–128.

Hou, D., et al. (2017). Incorporating life cycle assessment with health risk assessment to select the 'greenest' cleanup level for Pb contaminated soil. *Journal of Cleaner Production, 162*, 1157–1168.

Hou, D., et al. (2018). Climate change mitigation potential of contaminated land redevelopment: a city-level assessment method. *Journal of Cleaner Production, 171*, 1396–1406.

Hou, D., & Al-Tabbaa, A. (2014). Sustainability: a new imperative in contaminated land remediation. *Environmental Science & Policy, 39*, 25–34.

Huang, B., et al. (2014). Chlorinated volatile organic compounds (Cl-VOCs) in environment — sources, potential human health impacts, and current remediation technologies. *Environment International, 71*, 118–138.

ICSS (2006). *Manual for Biological Remediation Techniques*. Dessau Germany: International Centre for Soil and Contaminated Sites.

Insam, H., de Bertoldi, M. (2007). Microbiology of the composting process. In L.F. Diaz, M. de Bertoldi, W. Bidlingmaier, & E. Stentiford (Eds.), *Waste Management Series*. Elsevier, Volume 8, pp. 25–48.

Jaffré, T., et al. (1979). Nickel uptake by Flacountiaceae of New Caledonia. *Proceedings of the Royal Society of London. Series B. Biological Sciences, 205*(1160), 385–394.

Jiang, Y., et al. (2016). Insights into the biodegradation of weathered hydrocarbons in contaminated soils by bioaugmentation and nutrient stimulation. *Chemosphere, 161*, 300–307.

Kaschl, A., et al. (2005). Isotopic fractionation indicates anaerobic monochlorobenzene biodegradation. *Environmental Toxicology and Chemistry, 24*(6), 1315–1324.

Kauppi, S., Sinkkonen, A., & Romantschuk, M. (2011). Enhancing bioremediation of diesel-fuel-contaminated soil in a boreal climate: Comparison of biostimulation and bioaugmentation. *International Biodeterioration & Biodegradation, 65*(2), 359–368.

Khalid, S., et al. (2016). A comparison of technologies for remediation of heavy metal contaminated soils. *Journal of Geochemical Exploration, 182*, 247–268.

Kramer, P.A., et al. (2000). Native plant restoration of copper mine tailings: I. Substrate effect on growth and nutritional status in a Greenhouse study. *Journal of Environmental Quality, 29*(6), 1762–1769.

Liang Zhu, Y., et al. (1999). Overexpression of glutathione synthetase in Indian mustard enhances cadmium accumulation and tolerance. *Plant Physiology, 119*(1), 73–80.

Lovley, D.R. (1997). Potential for anaerobic bioremediation of BTEX in petroleum-contaminated aquifers. *Journal of Industrial Microbiology and Biotechnology, 18*(2), 75–81.

Lutts, S., et al. (2004). Heavy metal accumulation by the halophyte species Mediterranean saltbush. *Journal of Environmental Quality, 33*(4), 1271–1279.

Ma, F., et al. (2009). Application of bioaugmentation to improve the activated sludge system into the contact oxidation system treating petrochemical wastewater. *Bioresource Technology, 100*(2), 597–602.

Ma, X., & Burken, J.G. (2003). TCE diffusion to the atmosphere in phytoremediation applications. *Environmental Science & Technology, 37*(11), 2534–2539.

Maria, L., et al. (2011). Bioremediation of soil polluted with fuels by sequential multiple injection of native microorganisms: fieldscale processes in Poland. *Ecological Engineering, 37*(11), 1895–1900.

Marr, L.C., et al. (2006). Direct volatilization of naphthalene to the atmosphere at a phytoremediation site. *Environmental Science & Technology, 40*(17), 5560–5566.

Mang, K., & Ntushelo, K. (2019). Phytoextraction and phytostabilisation approaches of heavy metal remediation in acid mine drainage with case studies: a review. *Applied Ecology and Environmental Research, 17*(3), 6129–6149.

Martín-Hernández, M., Suárez-Ojeda, M.E., & Carrera, J. (2012). Bioaugmentation for treating transient or continuous p-nitrophenol shock loads in an aerobic sequencing batch reactor. *Bioresource Technology, 123*, 150–156.

Melato, F.A., Mokgalaka, N.S., & McCrindle, R.I. (2016). Adaptation and detoxification mechanisms of Vetiver grass (Chrysopogon zizanioides) growing on gold mine tailings. *International Journal of Phytoremediation, 18*(5), 509–520.

Martins, O., & Dewes, T. (1992). Loss of nitrogenous compounds during composting of animal wastes. *Bioresource Technology, 42*(2), 103–111.

Mendez, M.O., Glenn, E.P., & Maier, R.M. (2007). Phytostabilization potential of quailbush for mine tailings. *Journal of Environmental Quality, 36*(1), 245–253.

Mokhtar, H., Morad, N., Fizani, F., & Fizri, A. (2011). Hyperaccumulation of copper by two species of aquatic plants. In *International Conference on Environment Science and Engineering IPCBEE*. Singapore: IACSIT Press.

Nijenhuis, I., et al. (2007). Sensitive detection of anaerobic monochlorobenzene degradation using stable isotope tracers. *Environmental Science & Technology, 41*(11), 3836–3842.

Nwankwegu, A.S., & Onwosi, C.O. (2017). Bioremediation of gasoline contaminated agricultural soil by bioaugmentation. *Environmental Technology & Innovation, 7*, 1–11.

O'Connor, D., et al. (2019). Phytoremediation: climate change resilience and sustainability assessment at a coastal brownfield redevelopment. *Environment International, 130*, 104945.

Onwosi, C.O., et al. (2017). Composting technology in waste stabilization: on the methods, challenges and future prospects. *Journal of Environmental Management, 190*, 140–157.

Onyia, C.O., et al. (2001). Increasing the fertilizer value of palm oil mill sludge: bioaugmentation in nitrification. *Water Science and Technology, 44*(10), 157–162.

Pardo, G., et al. (2015). Gaseous emissions of solid waste: a systematic review. *Global Change Biology, 21*(3), 1313–1327.

Parkinson, R., et al. (2004). Effect of turning regime and seasonal weather conditions on nitrogen and phosphorus losses during aerobic composting of cattle manure. *Bioresource Technology, 91*(2), 171–178.

Petersen, S.O., et al. (2007). Recycling of livestock manure in a whole-farm perspective. *Livestock Science, 112*(3), 180–191.

Petrov, A.M., et al. (2016). Dynamics of ecological and biological characteristics of soddy-podzolic soils under long-term oil pollution. *Eurasian Soil Science, 49*(7), 784–791.

Pilon-Smits, E. (2005). Phytoremediation. *Annual Review of Plant Biology, 56*(1), 15–39.

Polyak, Y.M., et al. (2018). Effect of remediation strategies on biological activity of oil-contaminated soil – a field study. *International Biodeterioration & Biodegradation, 126*, 57–68.

Qu, Y., et al. (2011). Bioaugmentation with a novel alkali-tolerant Pseudomonas strain for alkaline phenol wastewater treatment in sequencing batch reactor. *World Journal of Microbiology and Biotechnology, 27*(8), 1919–1926.

Rahman, M.A., et al. (2007). Arsenic accumulation in duckweed (*Spirodela polyrhiza* L.): a good option for phytoremediation. *Chemosphere, 69*(3), 493–499.

Rajakaruna, N., & Bohm, B. (2002). Serpentine and its vegetation: a preliminary study from Sri Lanka. *Journal of Applied Botany, 76*, 20–28.

Ramadass, K., et al. (2015). Ecological implications of motor oil pollution: earthworm survival and soil health. *Soil Biology and Biochemistry, 85*, 72–81.

Ramadass, K., et al. (2018). Bioavailability of weathered hydrocarbons in engine oil-contaminated soil: impact of bioaugmentation mediated by *Pseudomonas* spp. on bioremediation. *Science of the Total Environment, 636*, 968–974.

Raper, E., et al. (2018). Industrial wastewater treatment through bioaugmentation. *Process Safety and Environmental Protection, 118*, 178–187.

Rio-Celestino, M.D., et al. (2006). Uptake of lead and zinc by wild plants growing on contaminated soils. *Industrial Crops and Products, 24*(3), 230–237.

Rojjanateeranaj, P., Sangthong, C., & Prapagdee, B. (2017). Enhanced cadmium phytoremediation of *Glycine max* L. through bioaugmentation of cadmium-resistant bacteria assisted by biostimulation. *Chemosphere, 185*, 764–771.

RTDF (2005). *Evaluation of Phytoremediation for Management of Chlorinated Solvents in Soil and Groundwater*. Cincinnati, OH, USA: United States Environmental Protection Agency.

Ryckeboer, J., et al. (2003). Microbiological aspects of biowaste during composting in a monitored compost bin. *Journal of Applied Microbiology*J Appl Microbiol, *94*(1), 127–137.

Sabey, B.R., Pendleton, R.L., & Webb, B.L. (1990). Effect of municipal sewage sludge application on growth of two reclamation shrub species in copper mine spoils. *Journal of Environmental Quality, 19*(3), 580–586.

Safdari, M.S., et al. (2018). Development of bioreactors for comparative study of natural attenuation, biostimulation, and bioaugmentation of petroleum-hydrocarbon contaminated soil. *Journal of Hazardous Materials, 342*, 270–278.

Sahu, A., et al. (2019). Thermophilic ligno-cellulolytic fungi: the future of efficient and rapid bio-waste management. *Journal of Environmental Management, 244*, 144–153.

Sánchez-Monedero, M., et al. (2017). Role of biochar as an additive in organic waste composting. *Bioresource Technology, 247*, 1155–1164.

Sayara, T., et al. (2011). Bioremediation of PAHs-contaminated soil through composting: Influence of bioaugmentation and biostimulation on contaminant biodegradation. *International Biodeterioration & Biodegradation, 65*(6), 859–865.

Semrany, S., et al. (2012). Bioaugmentation: possible solution in the treatment of Bio-Refractory Organic Compounds (Bio-ROCs). *Biochemical Engineering Journal, 69*, 75–86.

Shan, H., & Obbard, J. (2001). Ammonia removal from prawn aquaculture water using immobilized nitrifying bacteria. *Applied Microbiology and Biotechnology, 57*(5), 791–798.

Smith, R.A.H., & Bradshaw, A.D. (1979). The use of metal tolerant plant populations for the reclamation of metalliferous wastes. *Journal of Applied Ecology, 16*(2), 595–612.

Sommer, S.G. (2001). Effect of composting on nutrient loss and nitrogen availability of cattle deep litter. *European Journal of Agronomy, 14*(2), 123–133.

Song, Y., et al. (2019). Nature based solutions for contaminated land remediation and brownfield redevelopment in cities: a review. *Science of The Total Environment, 663*, 568–579.

Straube, W., et al. (2003). Remediation of polyaromatic hydrocarbons (PAHs) through landfarming with biostimulation and bioaugmentation. *Acta Biotechnologica, 23*, 179–196.

Sun, G.-D., et al. (2014). Isolation of a high molecular weight polycyclic aromatic hydrocarbon-degrading strain and its enhancing the removal of HMW-PAHs from heavily contaminated soil. *International Biodeterioration & Biodegradation, 90*, 23–28.

Taccari, M., et al. (2012). Effects of biostimulation and bioaugmentation on diesel removal and bacterial community. *International Biodeterioration & Biodegradation, 66*(1), 39–46.

Thrall, A.J., et al. (2006). Plant–soil relations of a serpentine site in the southern coast of Sri Lanka. In *Fifth International Conference on Serpentine Ecology*, Siena, Italy.

Thompson, I.P., et al. (2005). Bioaugmentation for bioremediation: the challenge of strain selection. *Environmental Microbiology, 7*(7), 909–915.

Ubogu, M., Odokuma, L.O., & Akponah, E. (2019). Enhanced rhizoremediation of crude oil-contaminated mangrove swamp soil using two wetland plants (*Phragmites australis* and *Eichhornia crassipes*). *Brazilian Journal of Microbiology: [Publication of the Brazilian Society for Microbiology], 50*(3), 715–728.

Vigliotta, G., et al. (2016). Effects of heavy metals and chelants on phytoremediation capacity and on rhizobacterial communities of maize. *Journal of Environmental Management, 179*, 93–102.

Wang, Y., et al. (2019). Green synthesis of nanoparticles for the remediation of contaminated waters and soils: constituents, synthesizing methods, and influencing factors. *Journal of Cleaner Production, 226*, 540–549.

Wiedemeier, T.H., Rifai, H.S., Newell, C.J., & Wilson, J.T. (1999). *Natural Attenuation of Fuels and Chlorinated Solvents in the Subsurface*. New York, NY: John Wiley & Sons, Inc.

Wither, E.D., & Brooks, R.R. (1977). Hyperaccumulation of nickel by some plants of Southeast Asia. *Journal of Geochemical Exploration, 8*(3), 579–583.

Wu, Y.W., et al. (2005). Separation of petroleum hydrocarbons from soil and groundwater through enhanced bioremediation. *Energy Sources, 27*(1–2), 221–232.

Wu, M., et al. (2013). Degradation of polycyclic aromatic hydrocarbons by microbial consortia enriched from three soils using two different culture media. *Environmental Pollution, 178*, 152–158.

Wu, M., et al. (2016). Bioaugmentation and biostimulation of hydrocarbon degradation and the microbial community in a petroleum-contaminated soil. *International Biodeterioration & Biodegradation, 107*, 158–164.

Wu, M., et al. (2017). Bacterial community shift and hydrocarbon transformation during bioremediation of short-term petroleum-contaminated soil. *Environmental Pollution, 223*, 657–664.

Wu, M., et al. (2019). Effect of bioaugmentation and biostimulation on hydrocarbon degradation and microbial community composition in petroleum-contaminated loessal soil. *Chemosphere, 237*, 124456.

Xie, H., Zhu, L., & Wang, J. (2018). Combined treatment of contaminated soil with a bacterial Stenotrophomonas strain DXZ9 and ryegrass (*Lolium perenne*) enhances DDT and DDE remediation. *Environmental Science and Pollution Research, 25*(32), 31895–31905.

Yang, Q., et al. (2015). Effects and biological response on bioremediation of petroleum contaminated soil. *Huan jing ke xue= Huanjing kexue, 36*(5), 1856–1863.

Yu, K.S.H., et al. (2005). Natural attenuation, biostimulation and bioaugmentation on biodegradation of polycyclic aromatic hydrocarbons (PAHs) in mangrove sediments. *Marine Pollution Bulletin, 51*(8), 1071–1077.

Zhang, X.-X., et al. (2006). Microbial PAH-degradation in soil: degradation pathways and contributing factors. *Pedosphere, 16*, 555–565.

Zhang, X., et al. (2014). New insight into the biological treatment by activated sludge: the role of adsorption process. *Bioresource Technology, 153*, 160–164.

Zhang, J., et al. (2019). Effects of the combined pollution of cadmium, lead and zinc on the phytoextraction efficiency of ryegrass (*Lolium perenne* L.). *RSC Advances, 9*(36), 20603–20611.

Zhao, F.-J., Lombi, E., & McGrath, S. (2003). Assessing the potential for zinc and cadmium phytoremediation with the hyperaccumulator *Thlaspi caerulescens*. *Plant and Soil, 249*, 37–43.

Zhu, Q., Feng, Y., & Choi, S.-B. (2017). The role of customer relational governance in environmental and economic performance improvement through green supply chain management. *Journal of Cleaner Production, 155*, 46–53.

2

Advances in Biotechnology for the Bioremediation of Contaminated Ecosystem

Alvina Farooqui[1], Gyanendra Tripathi[1], Nishi Aara[1], Suhail Ahmad[1], Arbab Husain[2], Adeeba Shamim[2], and Sadaf Mahfooz[2]
[1]Department of Bioengineering, Integral University, Lucknow, U.P., India
[2]Department of Bioscience, Integral University, Lucknow, U.P., India

CONTENTS

2.1 Introduction 26
2.2 Contaminants in the Ecosystem 27
    2.2.1 Types of Contaminants 27
        2.2.1.1 Natural Contaminants 28
        2.2.1.2 Anthropogenic Contaminants 28
    2.2.2 Contaminants Impact on the Ecosystem 29
    2.2.3 Ecological Risk Assessment (ERA) of Contaminated Soil 29
        2.2.3.1 ERA Approach 30
        2.2.3.2 Ecology Survey 30
        2.2.3.3 Ecotoxicity Testing 30
        2.2.3.4 Chemical Analysis 30
        2.2.3.5 Biological Testing 31
        2.2.3.6 Decision Making 31
2.3 Bioremediation 31
    2.3.1 Bioremediation Strategies 31
        2.3.1.1 Ex-Situ Bioremediation 32
        2.3.1.2 In-situ Bioremediation 34
    2.3.2 Factors of Bioremediation 35
        2.3.2.1 Microbial Population 35
        2.3.2.2 Environmental Factors 35
2.4 Advances in Biotechnology for Bioremediation 35
    2.4.1 Biosorption 36
    2.4.2 Biomineralisation 37
    2.4.3 Dendroremediation 37
    2.4.4 Rhizoremediation 37
    2.4.5 Biostimulation 38
    2.4.6 Mycoremediation 38
    2.4.7 Genoremediation 39
    2.4.8 Cyanoremediation 39
2.5 Applications of Bioremediation 40
    2.5.1 Applications 40

DOI: 10.1201/9781003004684-2

2.6 Future Aspects 41
2.7 Conclusion.. 42
References.. 43

2.1 Introduction

In early times, numerous resources were found on earth and we had a better quality of the environment to sustain, which was inextricably linked to the good health of the living beings. But as human beings evolved with time, they exploited resources for their benefits and ease of living, which knocked down the quality of our ecosystem. Anthropogenic activities for economic development has led to widespread contamination of the ecosystem (Bilal and Iqbal, 2019). The bioaccumulation of contaminants like pesticides, greenhouse gases, hydrocarbons, nuclear waste, and heavy metals have had a deleterious impact on human health and the environment (López-Pacheco et al., 2019; Bilal and Iqbal, 2019). These contaminants may be carcinogenic, mutagenic, or may even interfere with the endocrine functions, leading to health issues in humans and animals (Ahmed et al., 2017; McCallum et al., 2017).

Earlier, the health effects and environmental effects related to bioaccumulation from manufacturing and disposal after usage were not acknowledged as they are today. As of now, it is seen that the contaminated sites are a potential threat to living beings, which alerted the attention of researchers, leading to collaborative efforts to cope with the adverse effects.

Waste treatment techniques like solvent extraction, composting, landfilling, incineration, pyrolysis, precipitation, gas-phase chemical reduction, electro-dialysis, and alkali metal reduction biosorption were used for the deterioration of persistent pollutants generated as a result of indiscriminate industrial activities (Shivajirao, 2012). Conventional treatment methods suffer from limitations not only in terms of cost, space, and energy requirements but also in terms of efficiency and reliability as they are unable to degrade the contaminant completely. Use of toxic chemicals in the procedure and the potentially toxic by-product iformation are some of the other drawbacks of the traditional methods. Keeping in view the potential drawbacks of the conventional methods of waste treatment, the development of methods based on green technology that is more eco-friendly, cost effective, and efficient are the needs of the hour.

Presently, green technologies such as bioremediation and phytoremediation are being recognised by researchers for remediation of industrial waste because of their cost effectiveness, environmental friendliness, and efficiency in the dumping of hazardous waste. However, bioremediation had its golden era in the late 1980s and early 1990s. Unlike conventional techniques, bioremediation is an environmentally sound technique and can probably disrupt various contaminants through biological processes. Bioremediation is the process in which the contamination is being cleaned up with the help of the metabolic activities of the microorganisms under controlled conditions. In other words, it can be explained as the approach to detoxify organic waste by using microbes under controlled conditions under the limits of regulatory authorities. The important factor of bioremediation is microbes, as they potentially detoxify the environment and obtain energy from the process (Tang et al., 2007). The prime microorganisms included in bioremediation are archaea, bacteria, and fungi (Strong and Burgess, 2008). Microbes are crucial in the contaminant's degradation and are present in soil, water, and sediments to help reinstate the original natural ecosystem and avert further contaminant release into the surroundings.

Although every microorganism is proficient in eliminating contaminants from the site of contamination to an extent, a rare amount of particular or engineered microorganisms is selected broadly for efficient contamination removal. The microorganisms used in bioremediation include *Bacillus, Corynebacterium, Staphylococcus, Alcaligenes, Streptococcus, Shigella, Klebsiella, Enterobacter Acinetobacter,* and *Escherichia,* amongst which *Bacillus* species are extensively used in organic pollutant remediation (Haritash and Kaushik, 2009).

There are several advantages of bioremediation over conventional techniques. The major advantage is that it is worthwhile in terms of cost. In the United States, the conventional technique used for remediation of all contaminated sites like incineration would roughly cost around $1.7 trillion (Kuiper et al., 2004). Further counting on advantage, it is done in a non-sterile, open environment and with low technical techniques, i.e., the expertise is not always required to carry out the bioremediation process. Moreover, it can be carried out on-site, unlike conventional techniques. The bioremediation technique does not involve hazardous chemicals. The methods involved in bioremediation are not technically complex and are relatively easy to implement. Bioremediation is now being opted for worldwide, especially in the United States, where it is gaining limelight. However, the efficiency of bioremediation is based on different aspects, such as the nature of chemicals and the quantity of the contaminants per square meter, the physicochemical characteristics of the surroundings, and their availability to microbes (El Fantroussi and Agathos, 2005).

Biotechnology comes forth by using the chemistry of living beings. It is used as a tool for cell manipulation to create an alternative and innovative approach that aims at effective ways to develop traditional products, considering the maintenance of the natural ecosystem. To date, much advancements have been made and many more are under study regarding future perspectives for the advancement in biotechnology for bioremediation. Several scientists have recognised the application of metagenomics, GM microbes (genetically modified microbes), microbial fuel cells, nanomaterials, biofilms, and participated in constructing wetlands for the breakdown of persistent organic pollutants from various sources. Microbial degradation, along with meta-approaches and genetic engineering, allows the deletion of persistent organic pollutants in an environmentally friendly manner.

For the bioremediation approach, biotechnological processes avail in in-situ treatment and ex-situ treatments for eradicating contaminants that are generally based on the metabolic activities of the microbes. The process of biotechnology has numerous advantages over traditional methods. In many cases when the biotechnological processes are carried out, they destroy hazardous waste. But in some cases, bioremediation has a limiting factor as it has a slow rate of degradation of the compounds (Chakrabarty, 1982). The slow degradation at the contamination site restricts the use of microorganisms in bioremediation. To overcome this challenge, genetic engineering comes into the limelight. Molecular techniques led to the enhanced rate of reaction by the addition of a protein or enzyme in a microorganism (Chakrabarty, 1986). For a desired genetically engineered strain, it is mandatory to select a microorganism that has mastery over the needed degradative enzymatic machinery (Dua et al., 2002).

Thus, with all the advancements being pursued in biotechnology, bioremediation is the superior option to choose as it transforms the contaminant to its smaller state, i.e., up to molecular constituent, rather than a remediation approach that does convert waste from one form to other (Kuiper et al., 2004).

2.2 Contaminants in the Ecosystem

In tandem with rising industrialisation and urbanisation, the consumption of chemicals has increased by manifolds as it is hard to pursue living without these chemicals. Different industries like paper and pulp industries, leather industries, and pharmaceutical industries fulfil contemporary living. The step-up in the utilisation of these chemicals has contributed to the rise of contamination, i.e., the adverse health effects are caused by using these chemicals (Harrison, 2015). Pharmaceuticals and stimulants are grouped under 'emerging contaminants' as their concentration is found to be higher (in the range from ng/L to μg/L) in many contaminated zones (Tripathi et al., 2020).

2.2.1 Types of Contaminants

Generally, the contaminants are classified into two categories: natural and anthropogenic volcanic eruptions and emissions of volatile organic compounds from plants. Volcanic eruptions, vegetation fire, windblown dust, and sea-salt spray organic compounds from plants are the natural sources contributing 29–55% of annual particulate matter in urban locations; whereas anthropogenic sources include a wide range of types, as man has aggravated the problem of pollution by indiscriminate activities like industrialisation, use of rail and land automobiles, agricultural practices, nuclear explosions, quarrying, and mining (Gaur et al., 2014). The accelerated use of metals nanoparticles in catalysis, environmental remediation, sensors, and their utilisation for commercial products is raising serious problems over their toxic impact on the environment (Mahfooz et al., 2018) and these contaminants are usually non-biodegradable and, as a result, they persist in the environment. The problem usually worsens because of their bioaccumulation propensity. When contaminants get involved in the food chain, they rack up in humans' fatty tissues, further resulting in adverse health effects along with environmental degradation (Rasheed et al., 2019; Venkatesan and Halden, 2014; Kallenborn et al., 2012).

2.2.1.1 Natural Contaminants

Natural contaminants may not be the ones that come to mind while considering contaminants and damages associated with them. Yet these contaminants are real and are liberated at their maximum during natural disasters such as blizzards, volcanic eruptions, etc. Natural disasters signify meteorologic or geologic phenomena that seriously strike the community and require extraordinary attempts to cope with them (Gunn, 1989; Timberlake et al., 1984; World Health Organization (WHO), 1980). The contamination was analogous to a particular compound that relies upon the toxicity level along with the released concentration. Natural disasters turn out to be a robust and eminent mechanism of contaminant release either directly or indirectly.

Concern about contaminant release from the natural disaster has been augmented due to the increase of natural events as well as the population explosion in disaster-prone areas (Noji, 1997). Uncontrolled release of contaminants may pose major soil, air, and water contamination, resulting in adverse health outcomes (Sanderson, 1992). Some of the common pollutants that are released from natural sources are dioxin, particulate matter, hydrocarbons, acetylene, nitrous oxide, cadmium, PAHs, etc.

2.2.1.2 Anthropogenic Contaminants

Anthropogenic contaminants are the type of contaminants that are discharged continuously into the environment because of various human activities. Sources of anthropogenic contaminants include industrial discharges, effluent disposals, and wastewater (Rhind, 2009). The sources are denoted as point sources and non-point sources. The point sources are the ones that release contaminants from a specific location and often these source loadings are concentrated. On the other hand, non-point sources refer to the contaminants from indistinct and disperse sources that take place over a big area. Non-point sources are otherwise known as diffuse sources. Far-reaching anthropogenic contaminants are found in groundwater, surfacewater, soil, etc. (López-Pacheco et al., 2019).

2.2.1.2.1 Pesticides

These are the chemical compounds that contain sulfur, oxygen, nitrogen, chlorine, bromine, and phosphorus and some heavy metals like arsenic, copper, sulfates, lead, and mercury (Stuart et al., 2012). Pesticides are used in the agricultural field for the elimination and control of pests that harm crops (Kolpin et al., 1998). Despite the advantage of pesticides in crop productivity and yield, when it is used indiscriminately, it causes negative effects in the form of environmental pollution, especially water pollution. When water is contaminated by pesticides, water becomes harmful for the living organisms that either live in it or consume it (Lapworth et al., 2012). The pesticides considered harmful are categorised as contaminant sources. Examples include acetochlor oxanilic acid (OA), acephate, 3-Hydroxycarbofuran, acetochlor ethane sulfonic acid (ESA), clethodim, etc. (EPA, U., 2016).

2.2.1.2.2 Pharmaceuticals

According to a study conducted by the U.S. Geological Survey (USGS) from 2004 to 2009, it was suggested that pharmaceutical production industries are major sources of environmental pollution. Ten to 1,000 times higher concentration of pharmaceutical components were found in the effluent from a wastewater treatment plant that was receiving a pharmaceutical discharge in comparison to the other 24 wastewater treatment plants across the nation that do not have pharmaceutical waste discharge (Loper et al., 2007; Phillips et al., 2010). The source of pharmaceutical waste in the water is not just the manufacturing plants. The drugs and antibiotics used by the livestock industries, and the water course receiving water runoff from animal feeding operations, have also contributed to the water contamination. Drugs like caffeine, acetaminophen, cotinine, carbamazepine, and diphenhydramine have been found in water sources by USGS studies (Loper et al., 2007).

2.2.1.2.3 Personal Care Compound

These products are compounds that are used in daily life to improve the quality of living (Boxall et al., 2012). Products like diclofenac, musks, carbamazepine and iopamidol, clofibric acid, triclosan, bisphenol, and phthalates are the emerging contaminants present in water. Caffeine and nicotine are among the 'lifestyle compounds' that are identified in groundwater through sewage effluent (Stuart et al., 2012). Triclosan, which is an active ingredient in soap, toothpaste, and anti-microbial products, is a personal care compound found in the wastewater treatment plant (Dhillon et al., 2015). These chemical compounds adversely affect aquatic life (WHO, 2015).

2.2.1.2.4 By-Products of Water Treatment

The by-product of water treatment refers to the chemical, inorganic, and organic substances that are formed during the process of water treatment because of the reaction between the naturally present organic substances (like humid acid and fulvic acid) and disinfectant (like, chlorine). N-nitrosodimethylamine (NDMA) is a compound present during wastewater chlorination and it is carcinogenic (Mitch et al., 2003). Similarly, other by-products contaminate the drinking water. The capability of wastewater treatment plants for the elimination of emerging pollutants is different, which is generally low and negative (Chen et al., 2017; Ben et al., 2018).

2.2.2 Contaminants Impact on the Ecosystem

Industrialisation has concentrated people in urban areas. Utilisation or overuse of natural resources and development of synthetic products for the benefit of living beings has raised the concentration of contaminants in the environment by burning fossil fuels and disposing organic pollutants into sewage. This causes disease, illness in human beings, and kills fish and wildlife (Halliday, 1999; López-Pacheco et al., 2019; Rodríguez-Delgado et al., 2016).

The technologies and the use of resources are developing day by day, which unfortunately develops more chemicals and compounds, resulting in the addition of compounds that potentially cause environmental threats to living beings. Industrial waste, pharmaceutical chemicals, personal care products, and various other chemicals that are earmarked as endocrine disrupters, are not metabolized and sent out directly into sewers or wastewater treatment plants. These chemical compounds have become a challenge while designing a treatment plant for their eradication (Monteiro et al., 2016; Tripathi et al., 2021).

Soil is among the essential natural resources that is irreplaceable. It acts as a mediator among different components such as air, water, bedrock, and biota which are responsible for making up our environment, and the interaction of these components fulfills our need for food, fuel, and fiber for living. The soil becomes contaminated either from the point source or the diffuse source. Point sources like industrial households and diffuse sources like mobilisation and transportation by floodwaters degrade the quality and functioning of soil, which has a long-term impact on living organisms (Defra, 2009). Several strategies have been considered to resolve soil contaminants; generally conventional remediation techniques (ex-situ or in-situ) or bioremediation (Gomes et al., 2013). Air is a chief factor as it is responsible for carrying dust particles or harmful gases that affect the regional environment and human health as we breathe in such an environment. Moreover, it poses a significant impact on the global biogeochemical cycle (Han et al., 2007; Guoshun et al., 2002).

2.2.3 Ecological Risk Assessment (ERA) of Contaminated Soil

It is a process of collection, organisation, and analysation of environmental parameters data for the estimation of associated risk and undesired effect on living beings caused by human activities. During

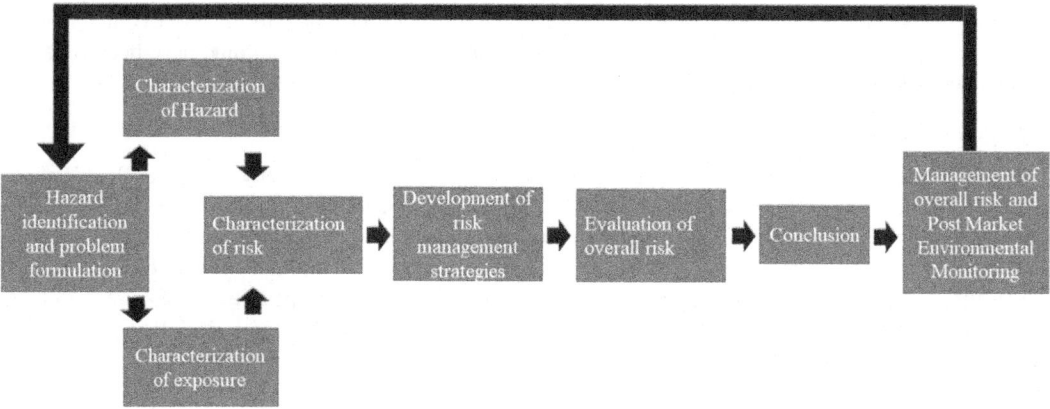

FIGURE 2.1 Steps involved in the ERA.

the management of the contaminated land, risk assessment appears to be an important part in the process of decision making. These assessments, therefore, can be used to study environmental problems that occur due to historical or ongoing activities, and sometimes, the activities are associated with the future, although usually future activities are overlooked (Caeiro et al., 2005; Long and MacDonald, 1998). The ERA involves several steps that are interconnected. These steps have their importance and none of it can be ignored while making a decision (Figure 2.1).

2.2.3.1 ERA Approach

The ERA definition is given by the U.S. EPA (Environmental Protection Agency, 1998) as "a process that evaluates the likelihood that adverse ecological effects may occur or are occurring as a result of exposure to one or more stressors. ERA approach governs to look forwards the adverse effects on organisms, population, and communities from chemicals present in the environment" (Environment Agency, 2003).

2.2.3.2 Ecology Survey

Ecological survey or assessment can turn out to be important in judging the biological status of the site at the biological organisation. However, ecological surveys depict the ecological diversity at the site and the quantity of the species. Ecological surveys can help to detect different sorts of taxa such as taxa which are sensitive to change (Environment Agency, 2003).

2.2.3.3 Ecotoxicity Testing

The ecotoxicity test is designed to test the toxic effect of any chemical compound added to the soil and to allow regulatory risk assessment. However, the ecotoxicity tests are used as well to get access to the toxicity data for the deviation of the chemical threshold. Different bioassays (tests of a single species' toxicity) are present to measure the contaminants and their effect on the soil. The ecological tests are to be performed either ex-situ, i.e., inside the laboratory or in-situ, i.e., inside its original environment; further, its results can be taken to study the antagonistic effects of the contaminants on the component and ecological receptor of the soil (Environmental Protection Agency, 1998).

2.2.3.4 Chemical Analysis

The chemical analysis comes into the role when a certain substance is highlighted with a good amount in the first step but in the biological assessment or toxicity survey substance does not appear (Figure 2.1).

So, by further sequestration, it could be explained with the help of chemicals that within a certain level these substances do not possess adverse effects and are not bioavailable in the environment (Boopathy, 2000). For metals, in most cases it is the toxicity of the free ion that is the greatest concern; however, the proportion present in this form is generally very small compared with total concentration.

2.2.3.5 Biological Testing

The biological test helps to understand contaminants and their biological significance that pertains to the first step and the results of ecological surveys (Figure 2.1). Initially, the existing ecological data is reviewed. Further, based on this data, it is concluded whether this data alone proves to be sufficient evidence to the risk to species diversity, etc., or not. But usually such data are unavailable or do not have convincing evidence of the impact, and the soil ecotoxicity test at this point is employed. The biological tests can therefore only be considered a sub-element of calculations that could potentially be used at the site of contamination (Boopathy, 2000; Environmental Protection Agency, 1998).

2.2.3.6 Decision Making

Decision making is based on aspects of risk assessment and testing. Considering the proposed ERA framework, a decision of acceptability or not is based on the risk noted at the initial step. In this assessment of risk, the assumptions made for the contamination level or measured with the help of testing are compared. While both exposure and effects are integrated into one appraisal effectively when biological testing is used, unlike the chemical-specific approach (Boopathy, 2000). Therefore, biological tests become a question of whether they can reliably detect adverse effects and whether they are sufficient for further investigation. In reality, a site is supposed to be harmless unless evidence is available for contaminants. The determination of risk, its magnitude, and associated uncertainties can be carried out by subsequent risk characterisation (Environmental Protection Agency, 1998).

2.3 Bioremediation

As we have discussed earlier, the bioremediation approach can be defined as the process of degrading harmful substances or compounds into an innocuous state with the help of the living microorganisms under controlled conditions (Sylvia et al., 2005). The microbes used in the bioremediation process can be indigenous to the site of contamination or brought from the outside to the site of contamination. Bioremediation is part of the metabolic activities of the microbes. The effectiveness of bioremediation depends upon the microorganism; the microorganisms should be enzymatically active to attack the contaminants and detoxify them. Further, for making bioremediation effective, environmental manipulation is done to support microbial growth and enhance the degradation process. The aerobic condition is favourable for the bioremediation process (Le Borgne et al., 2008).

Keeping in mind bioremediation is technology, it also has limitations like other technologies. High aromatic hydrocarbons, chlorinated organic compounds, etc., are some examples that are unaffected by the microbial attack. So, either bioremediation is unable to clean up such contaminants or cleans up at a really slow rate that cannot be even predicted. However, bioremediation still proves to be a better option when compared to traditional methods, like incineration, and other contaminants can be cleaned up on-site, which reduces further exposure risk of contaminants to the environment. On-site, bioremediation reduces the risk of transportation accidents and is accepted more by the public (Colberg and Young, 1995).

2.3.1 Bioremediation Strategies

The restoration of a polluted environment in an environmentally friendly manner at low costs can be achieved by the process of bioremediation over the years. The pollutant removal process depends mostly on the type of the contaminant that includes chlorinated compounds, agrochemicals, greenhouse gases,

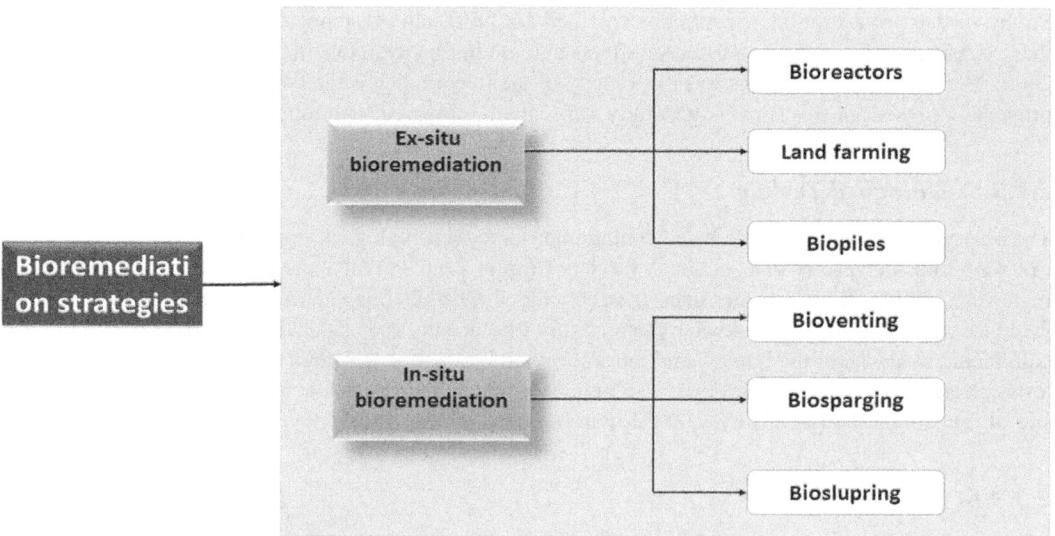

FIGURE 2.2 Strategies of bioremediation.

dyes, heavy metals, hydrocarbons, nuclear waste, and sewage and plastics. Bioremediation techniques are either in-situ or ex-situ, taking into account the site of application (Figure 2.2). Environmental policies, pollutant type, depth and concentration of pollution, location, environment condition, and rate are major selection criteria mainly considered for bioremediation techniques (Smith et al., 2015). In addition to the basis of selection, the basis of performance (nutrient concentrations, oxygen, temperature, and pH) that decides to carry out the bioremediation processes are considered before the bioremediation project. The major contaminants of groundwater and soil are due to hydrocarbons, so most of the bioremediation techniques focus on hydrocarbon contamination. Also, other corrective techniques (Pavel and Gavrilescu, 2008) that are more efficient and economical to implement during demonetisation are thought to be used at contaminate sites with contaminants separated by repairing. In addition, pollution from hydrocarbons due to crude oil pollution can be easily prevented and controlled by bioremediation techniques. Major sources of energy such as petroleum-based products are main sources of pollution (Khudur et al., 2015).

2.3.1.1 Ex-Situ Bioremediation

Ex-situ bioremediation is a technique that relies on transferring the contaminant from the contaminated site to another site for treatment. This technique is based on the depth of contamination, cost of treatment, types of contaminants, the geographical location of the contaminated location, and geology. Performance criteria of the technique also govern the need for ex-situ techniques for bioremediation (Philp and Atlas, 2005).

2.3.1.1.1 Bioreactors

In ex-situ bioremediation, reactors (aqueous reactor or slurry reactor) are useful for the treatment of soil and water contamination. Bioremediation carried out inside the reactor involves processing of the contaminated sludge, soil, water, or sediment, with the help of an engineered contaminant system. Bioreactors have different operating modes, including fed-batch, batch, sequencing batch, multistage, and continuous. The conditions of a bioreactor assist the process of cells, simulating them and providing an unnatural environment to provide adaptability for unnatural growth. Samples that are contaminated are fed into a bioreactor as a solution or dry matter. The parameters controlling the bioprocess step, such as temperature, pH, agitation, substrate, inoculum concentrations and radiation rate, are the major benefits of bioremediation based on a bioreactor. In a bioreactor, the control and manipulation ability of process parameters means that biological responses can be optimised that will ultimately decrease the bioremediation time. However, controlled bioremediation, nutrient addition, mass transfer (interaction between pollutants and microbes), and contaminant bioavailability are different limiting factors of the bioremediation process. These factors can be effectively maintained in bioreactors and make it more efficient (Godheja et al., 2019). Also, it is used for the treatment of contaminated soil and water with organic compounds of volatile nature, including benzene, ethylbenzene, xylene, and toluene. The bioreactor system proves to be of great use for a good rate of biodegradation. In a bioreactor, the soil needs to be pre-treated or, on the other hand, the contaminant can be stripped from the soil by vacuum extraction, etc. before taking into a bioreactor (Abuabdou et al., 2020).

2.3.1.1.2 Land Farming

Land farming is considered among the simplest techniques for bioremediation. In land farming, a well-maintained bed that is prepared beforehand, is used to spread contaminated soil excavated over it and continued with periodical tilling until the contaminants are degraded. The motive of this procedure is to augment the growth of indigenous microbes and facilitate the biodegradation of contaminants in aerobic conditions. Land farming is considered for both techniques of bioremediation, i.e., in-situ and ex-situ. This is due to the treatment tricks. Depth of the contaminant plays asignificant role, as to whether land cultivation can be carried out in-situ or ex-situ. In the cultivation of land, the excavation of contaminated soil from the treatment site determines the bioremediation type. When excavated contaminated soil is treated on the same site, it considered in-situ. It is stated that when the pollutant is lying less than 1 m below ground level, bioremediation is done without quarrying, and when the pollutant is lying at more than 1.7 m, then it needs to be moved on the ground surface for enhancing the effectiveness of bioremediation (Nikolopoulou et al., 2013). Recently, land farming has witnessed much attention because it is profitable, requires low maintenance, and reduces monitoring along with cleanup liabilities. However, it has a limit to treat superficial, i.e., 10–35 cm of soil. Landfills have some limitations that include large operational space, reduced microbial activity due to adverse environmental conditions, extra costs due to digging, and elimination of inorganic pollutants (Khan et al., 2004). Furthermore, it isn't suitable for the contaminated soil treatment with toxins with volatile nature because of its mechanism and design for pollution removal (evaporation), mainly for areas with hot (tropical) climates.

2.3.1.1.3 Biopiles

Generally, biopiles are known as the mixed form of composting and land farming. For aerated composite piles, the engineered cells are crucial. And these engineered cells are generally available for surface-contamination treatment with petroleum hydrocarbons. Biopiles led to control of the damage occurring from contaminants by volatilisation and leaching. Biopiles are certainly a refined version of land farming and provide favourable conditions for indigenous aerobic and anaerobic microbes (von Fahnestock and Wickramanayake, 1998). The constituents of this procedure are irrigation, aeration, nutrient treatment bed, and leachate collection systems. The procedure of this specific ex-situ technique is attributed to its physiological characteristics, which enables effective biodegradation in this situation with the temperature, aeration, and nutrients effectively controlled (Whelan et al., 2015). Use of biopiles at contaminated sites can prevent the destabilisation of contaminants with low molecular weight. It is used effectively to remove pollutants from very cold areas (Gomez and Sartaj, 2013). Besides, it has been stated that biopiles are used for treatment of large plants with contaminated soil in confined spaces. The setup of biopiles can be easily extended to pilot systems to achieve the uniformity during laboratory studies. The biopiles' efficiency depend on the compaction of contaminated soil before processing. Bulging agents such as organic materials, sawdust, straw, wood chips, and bark are added to the biopile concentrate to enhance the therapeutic process (Kuppusamy et al., 2016).

However, compared to other field-based ex-situ bioremediation techniques, biopile systems conserve space, land cultivation, maintenance, robust engineering, and operation costs, especially power supplies to remote sites in soil contaminated with air pumps. This process will enable uniform distribution of air. Biopiles do have some limitations. Furthermore, excessive heating of the air can cause soil drying through bioremediation, which results in the inhibition of microbial activities and promotes evaporation (Sanscartier et al., 2009).

2.3.1.2 In-situ Bioremediation

In-situ bioremediation is defined as the process applied to soil and groundwater at the site with the least disturbance (Vidali, 2001). This is the most desirable bioremediation because it is profitable, and the treatment is done at the site, removing the problem of transportation of contaminants. The depth of the soil is limited in in-situ bioremediation in order to be effective; however, up to 60 cm depth is treated effectively by in-situ bioremediation. In some cases, it is effective only up to 30 cm into the soil due to the desired rate of oxygen diffusion. An example of in-situ bioremediation is oil displaced by a biosurfactant produced from *C. tropicalis* MTCC230, an adaptive strain observed to be further used for the bioremediation of sea oil spilled (Ashish and Debnath, 2018). Following are some important land treatments.

2.3.1.2.1 Bioventing

Bioventing is the most commonly used in-situ bioremediation technique. It stimulates the indigenous microbes by providing nutrients as well as air through wells to the site of contamination. Bioventing provides a smaller rate of airflow and concentration of oxygen required at the site for biodegradation. So, it can be used where the contamination has a good depth and it works for hydrocarbons (Mrozik and Piotrowska-Seget, 2010). In bioventing, microbial transformation of pollutants is achieved in a harmless form (Philp Atlas, 2005) with the ultimate goal to increase bioremediation by adding nutrients and moisture by modification. This technique gained attention mainly in restoring contaminated sites with lightly dispersed petroleum products (Ho onsinhener and PICS2014). Even though the design of bioventing is intended to provide aeration in the unsaturated zone, this makes it available for the anaerobic bioremediation process, especially in the treatment of polluted vadose zones containing compounds with chlorine, recalcitrant to aerobic conditions. The next step, a mixture of nitrogen with carbon dioxide in low concentrations and hydrogen in exchange for air or pure oxygen, can be injected to produce a reduction in chlorinated vapor with hydrogen acting as an electron donor (Shah et al., 2001). In low permeable soil, pure oxygen is injected that can cause a higher oxygen concentration than air injection. In addition, oxidation is useful for partial oxidation of recursive compounds to enhance the biodegradation rate (Philp and Atlas, 2005).

2.3.1.2.2 Biosparging

Unlike bioventing, biosparging is defined as the air is injected with pressure under the water table so that the groundwater-oxygen concentration is enhanced. This increases the biodegradation process of contaminants with the help of naturally occurring microbes. Biosparging establishs good contact between soil and groundwater as it enhances the mixing. It is cost effective as small-diameter air injectors are installed, which shows flexibility in designing and constructing the system (Vidali, 2001). In this process, at the saturated zone the injection of air is done to promote biodegradation by stimulating upward movement of the organic compounds with a volatile nature. The efficacy of biosparging relies on two chief issues, specifically permeability of soil, which regulates pollutant bioavailability to microorganisms and the biodegradability of the petroleum constituents that will tells about the rate of constituents degradation by microorganisms (Philp and Atlas, 2005). The technique of biosparging is analogous in operation with the in-situ air-sparging technique that is a closely related technique to biosparging. The in-situ technique of airsparging depends on a high airflow rate to get pollutant volatilisation; however, biosparging endorses biodegeradation. Correspondingly, the mechanism for the removal of pollutants is not equally exclusive for both techniques. This technique has been extensively used in treating aquifers polluted with petroleum products, specifically kerosene and diesel (Vidali, 2001).

2.3.1.2.3 Bioslurping

The technique of bioslupring combines soil extraction, vacuum-enhanced pumping, and bio-water combination to attain groundwater and soil abatement by the indirect provision of oxygen and contaminated biodegradation (Godheja et al., 2016). This technique is mainly used for the recovery of free products like mild, non-aqueous phase liquids. The remodelling of capillaries in saturated and unsaturated regions is done. It is also used to remove organic contaminants that are semi-volatile and volatile in nature. The method uses a "slurp" that spreads into the product layer, which draws fluid (earthen gases and free products) from this layer. The pumping process brings the upper part of the light non-aqueous phase liquid (LNAPL) to the surface, where it is detached from air and water. After removing completely free products, the system can be easily built to be controlled as a conventional bioventing system to accomplish the therapeutic process. Installation of a vacuum at a water table and a highly permeable site that can saturate soil lenses that are particularly difficult to be airtight are among the major concerns of in-situ technology (Shackelford and Jefferis, 2000).

2.3.2 Factors of Bioremediation

There are different factors that are involved in the control and optimisation of bioremediation (Lu et al., 2012; Ishii et al., 2017). These factors comprise

- Population availability of microbes that are capable of degrading contaminants.
- The availability of the contaminants to the microbial population (Ahmad et al., 2017).
- Other environmental parameters are pH, temperature, type of soil, the presence of nutrients, and oxygen (Liang et al., 2013).

2.3.2.1 Microbial Population

Microorganisms are present in almost every environmental condition, which means they can be isolated easily. Microorganisms are capable of growing at sub-zero temperatures, in excess heat, in drought conditions, in water, in the absence of oxygen, in the presence of oxygen, in a contaminated environment, or any waste stream. Carbon and energy sources are the main requirements of microorganisms. And, because of their adapting nature, microorganisms can be used for degrading contaminants, which results in detoxifying the environment (Ahmad et al., 2017; Biffinger et al., 2007). For the bioremediation process, the microorganism must come in contact with the contaminant, which is not an easy task, as both the microbes and contaminants are unevenly distributed in the environment. Although some microorganisms are mobile and can move towards contaminants, some fungi move by growing in filamentous form. Therefore, to enhance the mobilisation, sodium dodecyl sulphate (SDS), a surfactant, is used. *Rhodococcus, Alcaligenes, and Phanaerochaetechrysosporium* are among the microbes that are generally used for bioremediation (Cody, 2000).

2.3.2.2 Environmental Factors

Microbes are isolated in stress conditions and they optimally grow in a thin range, so it becomes crucial to obtain optimal conditions (Lu et al., 2018). The rate of biochemical reaction is affected by the temperature and many reaction rates double up for every 10° rise in temperature and cells die above a particular temperature. Water is essential for maintaining optimal moisture conditions. A high pH can be controlled by adding lime (Liang et al., 2013).

Availability of oxygen helps to evaluate whether the system is anaerobic or aerobic. The soil structure has control over the delivery of water, air, and nutrients. Organic matter or gypsum is applied to improve the soil structure; although the microbes present at the contaminated site might not be sufficient enough for bioremediation. So, biostimulation comes into the role here, i.e., nutrients and oxygen are added for supporting the growth of microbes as nutrients act as the basic building block to create the necessary enzymes for the breakdown of the contaminants (Vidali, 2001).

2.4 Advances in Biotechnology for Bioremediation

The ongoing research and high scope of biotechnology for the advancement of bioremediation in the treatment of contaminants have evolved in many novel bioremediation techniques. It has also advanced

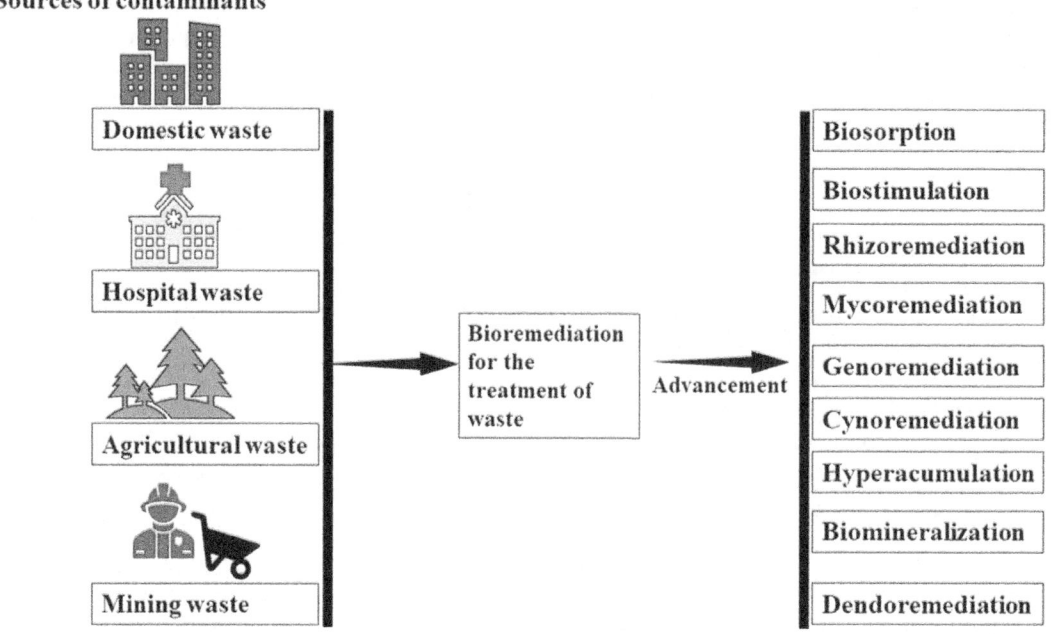

FIGURE 2.3 Advancement in bioremediation for the treatment of contamination.

many traditional techniques by modifying them for better efficiency of treatment. Figure 2.3 presents different advanced technologies that are used for the treatment of contaminants from different sources.

2.4.1 Biosorption

The features of organic material to collect heavy metals from contaminated water or wastewater through physico-chemical pathways or metabolically mediated uptake create a process known as biosorption. In this process, the metal ions that are known as biosorbates bind with biosorbents, originated from a biological process (Mrvčić et al., 2012). Natural evacuation incorporates the utilisation of microorganisms, plant inferred materials, farming or modern squanders, biopolymers, etc. Biosorption is reversible phenomenon wherein particles are attached lossely to the pores on surface of the biosorbent. This technique is in contrast with the oxidation of waste water through aerobic or anaerobic digestion (Davis et al., 2003). The benefits of this process incorporate a simple activity; there is no extra supplement necessary, low amount of sludge generation, cost-effective procedure with high effectiveness, recovery of biosorbent, and no expansion of chemical oxygen demand (COD) of water. It is generally a very conventional and effective procedure (Chojnacka, 2010). Biosorption can eliminate contaminants even in low concentrations and has unique importance regarding heavy metal evacuation attributable to harmfulness at parts per billion levels. In the process of biosorption, farming waste, industrial waste, and other microbes (dead and live) can be utilised as biosorbents. This process is reversible, is a single-stage process, has no risk of toxic impacts and cell development, doesn't need additional supplements, it is not controlled by metabolism, and permits moderate harmony concentration of metal particles to make it a very effective and beneficial process (Chojnacka, 2009). Biosorbents contain functional and chemical groups such as phosphodiester, thioether, carbonyl, sulfhydryl, amine, sulfonate, phenolic amide, imidazole, carboxyl, phosphate, and imine; they do help in the attraction and separation of metal ions. There are many natural products that are used as biosorbents involved in the metal ion binding through physical or chemical (electrostatic interaction or van der Waals forces) binding, complex chelation, precipitation, and reduction. For example, an intraparticle diffusion model for biosorption of Zn (II) was studied by employing VMSDCM, a Zn sequestering bacterium (Mishra et al., 2014).

2.4.2 Biomineralisation

The major source of contamination of heavy metals in the atmosphere is mining activities. Several studies have reported metals are present at a higher level around industrial areas and mining areas. The biggest concern in this scenario is the contaminants of metals that are present around agricultural land or soil due to metal mining and industrialisation. Metallic soils provide an unfavourable domain for plants. Many plants are genetically modified due to absorption or contamination of metals like Ni, Ar, Zn, Cu, Pb, and Cd. Development of metal resistance happens in each presented region (Ernst, 2006). Soil shows its metal resistance ability due to the presence of biological compounds, metalloids, and minerals. Complicated structured inorganic materials play an important function produced by the natural pathway in biomineralisation. Nature gives fantastic instances of organically shaped minerals; biomineralisation, for example, MSPs, which have developed as engaging transporters for controlled anticancer medication conveyance attributable to their minimum poisonousness, high surface zone, and enormous open-pore volumes (Kim et al., 2011). Now, research continues to find new uses of biomineralisation to produce inorganic-biological hybrids for more poisonous components (Achal et al., 2012; Chen et al., 2013). Separation or isolation of heavy metals from the stabilised solid phase is offered an efficient way through this process.

2.4.3 Dendroremediation

The forest has provided food and shelter to our progenitors for many ages. In the past, humans have grown trees for many uses, including building materials, production of energy, paper, furniture, rubber, and fruits. Trees are used in dendroremediation for the removal or treatment of contaminated soil. The term dendroremediation *dendron* means "tree" and *remediate* means "reuse" (Dickmann et al., 1996; Schoenmuth and Pestemer, 2004a, 2004b). Dendroremediation is a process of phytoremediation (Komives and Gullner, 2006; Pilon-Smits, 2005; Dietz and Schnoor, 2001). It is a method to degrade, treat, and isolate the inorganic or organic pollutants from the environment by using living plants (Dickmann et al., 1996; Schoenmuth and Pestemer, 2004a). The dendroremediation process is an efficient and effective way to treat different types of contaminants and is used to treat contaminated soil with landfill leachates, pesticides, polycyclic aromatic hydrocarbons, heavy metals, solvents, and crude oil (Abdel-Shafy and Mansour, 2018).

Wood has many features that play an essential role and may be important for efficient dendroremediation; for example, transportation of nutrients and water and storage of gases and organic compounds (Mani and Kumar, 2014). Tree roots release organic chemicals that are useful in the rhizosphere and decontaminate the contaminated rhizosphere. Enhancement of microbial growth by trees in their rhizosphere plays an important role that is helpful in the achievement of dendroremediation by inciting the presence of the contaminant for use by the root system of the plant, as well as decomposing some biological contaminants (Jordahl et al., 1997; Tesar et al., 2002). Recently, *Populus* (aspens, poplars, and cottonwoods) and *Salix* (willows and osiers) have become very effective approaches for dendroremediation (Roberts and Marston, 2011).

2.4.4 Rhizoremediation

Rhizoremediation is a process used for the treatment of contaminants, where microorganisms are removed or treat the contamination in the rhizosphere; the rhizosphere is a part of the plant's root. Remediation indicates the process of degrading organic contaminants; rhizome denotes the root. Some organic chemicals are released by the plant's root-like sterols, monosaccharides, and also organic acids (Jeevanantham et al., 2019). These chemicals are used by the microorganism for the removal of contaminants. Some microorganisms help in nitrogen fixation whereas other microorganisms help to defend the attack of pathogenic microbes (Stephenson and Black, 2014). It is a pollution-free and economical process that is used for the degradation of heavy metals and is an economical and environmentally friendly process with no addition of chemicals.

There are a few preferences for rhizoremediation because it is environmentally friendly, economical, and uses no addition of any synthetics for its evacuation, so the result is negligible or nothing. Yet in the event of a chemical treatment procedure like a chemical oxidation process, the volatilisation process influences the work or operation of our atmosphere or environment. In this process, there is no need for any type of input as only efficient microbial stains are provided.

The rhizoremediation process using various trees and microorganisms to degrade the contaminants from the atmosphere and environment, as reported by many researchers. This process is completed by three steps, including the production of organic acid, the formation of biofilm, and the production of biosurfactants. There are various types of microbes used for the rhizoremediation process, like *Paenibaciluus* species, *Mycobacterium* species, *Pseudomonas* species, *Mesorhizobium* species, *Rhodococcus* species, *Bacillus* species, *Staphylococcus* species, *Aspergillus* species, *Flavobacterium* species, and species. There are various types of microorganisms used for the rhizoremediation process, like *Paenibaciluus* species, *Mycobacterium* species, *Pseudomonas* species, *Mesorhizobium* species, *Rhodococcus* species, *Staphylococcus* species, *Bacillus* species, *Aspergillus* species, *Flavobacterium* species, and species. Researchers reported that many factors, including characteristics of soil, types of contaminants, variety of plants, microorganisms and their populations, and availability of supplements' concentration of contaminants for rhizosphere microorganisms could influence the rustication and degradation of contaminants from the rhizosphere zone (Idris et al., 2004; Chopra et al., 2007; Abou-Shanab et al., 2008; Aboudrar et al., 2007; Sriprang et al., 2003; Vivas et al., 2003; Abhilash et al., 2011; Kumar and Philip, 2006).

2.4.5 Biostimulation

The biostimulation process is defined as stimulating or enhancing the population of microorganisms by adding electron donors and acceptors and nutrients (Scow and Hicks, 2005). Those microorganisms, which are especially known as a recyclers of nature, are fungi and bacteria. They are capable of converting synthetic and natural chemicals in the form of energy (Tang et al., 2007) and raw substances for their consumption and growth suggests that the replacement of costly physical or chemical remediation methods should be replaced by biological remediation methods that are economical or cost effective and environmentally friendly. Infusion of nutrients and other supplementary compounds to the native microbial population to enhance and spread at the highest rate (biostimulation) is one of the most common approaches for in-situ bioremediation of accidental spreading and chronically contaminated sites worldwide (Tyagi et al., 2011; Cheng et al., 2009). The main purpose of biostimulation is to achieve the complete reduction of contaminants from the contaminated zone and to reduce the time of bioremediation. In this process enhancement, the microorganisms' degrading capacity by the electron exchangers and supplements are infused into the polluted or contaminated site. Concerning lab-scale tests, supplements are added as non-living salts and as explained chemical species, while at the field scale, supplements are often added in the form of inorganic fertilisers, organic wastes, and agro-wastes. In this method, inorganic fertilisers, agro-wastes, and organic wastes are frequently added in the mode of nutrients. The key benefits of this method are the utilisation of homegrown microbes, lower cost, and without the urge of modifications needed by allochthonous species. Inorganic fertilisers, agro-wastes, and organic wastes are continuously added in this process in the mode of nutrients.

2.4.6 Mycoremediation

The term mycoremediation is given by Stamets, a kind of bioremediation; this method is used for the degradation or isolation and also immunisation of the week immune system of the atmosphere. Mycofiltration is an identical method that uses fungal mycelia to remove or separate the microbes and hazardous waste from contaminated water in the soil via the incitement of microbial and enzymatic action. The endophytic, mycorrhizal, and saprophytic growths are equipped for the restoration of the soil's water environments and adjusting the organic community by breaking down cellulose and lignin through the acids and extracellular enzymes, which are secreted by mycelium. The approach of mycoremediation is picking the correct fungal community to recognise a particular contaminant (Dudhane et al., 2012). The best approach to mycoremediation is choosing the right parasitic breed which generates an exclusive toxin. An example of this approach is the bioremediation of Aluminium metal via emission of glomalin around the Gmelina plants. Some fungal species for example; Cladosporium resinae, Aureobasidium pullulans, Aspergillus niger, Penicillium species, Funaliatrogii, Ganoderma lucidum (Say et al., 2003; Loukidou et al., 2003) are capable to recover metals from the contaminated environment. *Aspergillus fumigates* is used for the degradation of Pb(II) ions from the aqueous mixture of e-waste (Kumar Ramasamy et al., 2011).

2.4.7 Genoremediation

Genoremediation is a process and also a type of bioremediation that is applied for the absorption of metals such as mercury from the environment, volatilisation of contaminants, and degradation of organic contaminants through genetically modified plants. The resistance capacity of different plant species like cottonwood, yellow poplar, rice, and tobacco is ten times greater against the concentration of mercury compared to those that kill non-transgenic control. In this process, genetically modified plants play a crucial role and take part in the remediation of contaminants from the contaminated field. Expression of both merB and merA genes by cottonwood and yellow poplar demonstrated the ability of phytovolatilisation of mercury in wetlands (Lyyra et al., 2007). Uses of genetically modified and metal-resistant plants are enhancing the procedure of transporting this technique from the experimental site to the fields and agricultural land. The genoremediation process is becoming popular and this process is being adopted in many countries. Accumulation of metals and the ability of resistance can be enhanced by up-regulation of modified or natural gene-encoding antioxidant enzymes or those that take part in the formation of phytochelatins and glutathione.

In the comparison of the wild-type plant, the accumulation potential of Cd and Se and greater biomass properties is shown by the transgenic *Brassica juncea* plant (Bañuelos et al., 2005). Metal transporter genes are introduced in genoremediation (gene-targeted bioremediation) and also genes that facilitate chelate production are included. The catabolic possibilities for remediation of wastes have been found in transgenic bacteria. Additionally, on account of components that may be volatile, genes that encourage conversion to volatile forms have been over-communicated (Dhankher et al., 2012). Enhancement in decomposition or degradation of organic contaminants by genetically cloned plants that contain bacterial genes also facilitates the production of chelator, volatilisation of contaminants, and metal absorption.

2.4.8 Cyanoremediation

Cyanoremediation is a process of bioremediation in an environmentally friendly, extremely useful, and most efficient procedure for the decontamination of chemically synthesised pesticides (pest killer) from aquatic and agroecosystems; bioremediation is microbes that mediate the removal, degradation, and transformation of contaminants into less hazardous or nontoxic components (Singh et al., 2011). The use of different organisms, such as cyanobacteria, actinomycetes, bacteria, and algae, and their effects in the remediation of contaminants and pesticides have been testified (Singh and Gupta, 2016). Cyanobacterial strains are very useful and effective in the degradation of pesticides and they are also known as the downfall of pesticides (Megharaj et al. 1994). Cyanobacteria are formerly known as blue-green algae, which are among the most primitive photosynthetic prokaryotes on this planet. Cyanobacteria have diverse structures and morphology including filamentous colonies and they are unicellular (Potts and Whitton, 2000; Burja et al., 2001).

Cyanoremediation is a process where cyanobacterial strains are used for the treatment or degradation and transformation of contaminants like pesticides, dyes, and heavy metals from polluted wastewater. There are many examples of cyanobacterial strains that are strongly applied for the bioremediation of synthetic chemicals or pesticides, e.g., *Aulosira fertilissima, Anabaena fertilissima*, and *Westiellopsis prolific* used for the treatment of 2,4-d dichlorophenoxyacetic acid (Kumar et al., 2013). According to Hatzios (1991), there are three steps for the degradation of pesticides:

I. Phase I involves hydrolysis, reduction, and oxidation. In this stage, oxygenation is the essential or important step to the removal of pesticides and oxidative enzymes that contribute to many oxidative reactions such as peroxidases, polyphenol oxidase, and cytochrome P450s.

II. Phase II includes the formation of a pesticide or the metabolites of pesticides to a sugar, glutathione, or amino acid, which improves the solubilisation of water and decreases the harmfulness contrasted with parent pesticide constituents.

III. The metabolites are converted into secondary conjugates in this phase and they are also non-poisonous. In the degradation cycle, other dynamic oxygen species and single oxygen are produced by pesticides in different regions of the photosynthetic electron transport chain. The cellular systems work as scavengers of active oxygen species with the help of raising antioxidative machineries like peroxidase, catalase, and superoxide dismutase (Palanisami et al., 2009).

2.5 Applications of Bioremediation

There is a plethora of work that has been done to remove contaminants present in the environment in the biochemical cycle, though we need to search for new ways for the removal of pollutants completely. Currently, the newest application of biotechnology is in bioremediation, as it became the most needed arena for the renewal of pollutants from our environment. "Bioremediation is a biological process, which is chiefly used to remove the pollutants from the environment in a sustainable manner" (Bhatia et al., 2020). Nowadays, many bioremediation approaches are used to treat wastes and contaminated localities. Many researchers are taking advantage of bioremediation (Ortiz-Hernández et al., 2018).

It is not only a method for pollutant removal from the environment, but also an effective and environmentally friendly approach. In this process, the microorganism is used to detoxify or eliminate contaminants from the soil and water (Singh and Tripathi, 2007; Wasi et al., 2008). Bioremediation's sole motive is to use environmentally friendly microbes to create a pollution-free environment. This approach can be divided into two broad groups, i.e., in-situ and ex-situ bioremediation. Treatments are not done at the same location in the case of ex-situ bioremediation. In this approach, the polluted soil is first dug up and then treated at a different location. The in-situ treatment is performed on the polluted site without digging up and transporting contaminated water or soil. In this approach, an aqueous solution containing bacteria is used to degrade the organic matter for groundwater and soil remediation (Singh and Tiwari, 2014).

2.5.1 Applications

Composting, which is a form of in-situ bioremediation, is used for cellulose, hemicellulose, and lignin degradation by using cellulase, manganese peroxidase, lignin peroxidase, and white-rot fungi (Rao et al., 2010). Besides this, composting is also used to treat organic materials from municipal solid waste by maintaining low pressure and 75% moisture (Maken et al., 2013).

Bioventing, which is a type of in-situ bioremediation, is used for the hydrocarbons, petroleum, and non-fuel hydrocarbons like acetone removal from the polluted area by using a compressor blower to inject the air for 15 months at a low-flow rate (Niu et al., 2009; Conesa et al., 2004).

Bioaugmentation, which is also a type of in-situ bioremediation, is used for the management of soil, wastewater, and the elimination of chlorinated organics. This is mainly used for the removal of soluble toxic chemicals by using nitrogen as an essential component, indigenous microbes with regular supplementation and batch, and continuous-flow activated sludge reactors (Bouchez et al., 2000; Boon et al., 2002).

Land farming is an inexpensive, simple, and self-heating process, which is also a type of in-situ bioremediation. Land farming with an aerobic process is used for organic materials, whereas an anaerobic process is used to convert organic solids to humus (Antizar-Ladislao et al., 2008).

Biosparging is the most efficient and non-invasive method of in-situ bioremediation, which is used for the removal of hydrocarbons. In biosparging, indigenous microbes are expedient in the presence of metals. It is mainly used for the treatment of aquifers contaminated with oil spinoffs such as diesel and kerosene (da Silva et al., 2020).

Bioremediation is also used for the degradation of solid waste like rubber, which constitutes approximately 12% of total solid waste. Rubber cannot be degraded or recycled effortlessly because of its physical configuration (Conesa et al., 2004). The toxic part of rubber is removed by using fungi such as *Recinicium bicolour* before the degradation. Thereafter, oxidising or sulfur-reducing bacteria such as *Thiobacillus ferroxidans* and *Pyrococcus furiosus* are used to devulcanise the rubber so that it can be reused effortlessly (Keri et al., 2008). Subsequently, the best waste management for rubber is controlled combustion, and released heat can be used for energy production (Conesa et al., 2004).

Bioremediation is chiefly used for the elimination of heavy metals. These toxic heavy metals pollute both soil and water. The elimination of these metals is crucial because of their ability to be accumulated into the food chain, which further leads to accumulation in the human body with adverse effects. The toxic metals can be eliminated by using different living microorganisms such as algae, fungi, bacteria, and cyanobacteria. These biological agents work as biosorbents for toxic metal sequestering (Igiri et al., 2018). The reaction among the cell membrane of microorganisms, which are negatively charged, and heavy metals with positive charges, transport heavy metals to the cytoplasm of cells to become bioaccumulated, known as biosorption. Nearly all toxic metals can be eliminated using *Aspergillus*.

Bioremediation is also used for the deprivation of some compounds, which are organic in nature, known as xenobiotics. These xenobiotics tend to accumulate in the environment as they degrade gradually and sometimes they accumulate permanently due to very slow deprivation. Employing white-rot fungus in the presence of certain enzymes can be used for the degradation of xenobiotics (Azaizeh et al., 2011).

Bioremediation can also be used to transform the organic and agricultural waste that are high in nutritional value and can be converted into valuable yields. Every year, tons of organic waste are produced worldwide; therefore, the management of this waste has become of utmost importance globally. For this purpose, vermicomposting can be used. In this process, microorganisms and earthworms work together. Microbes degrade the organic materials and then the earthworms initiate the acclimatising of the substrate and modify the biological action (Dominguez, 2004; Suthar, 2007).

By using vermicomposting, bulk amounts of organic pollutants can be transformed in the biofertiliser. This is a rapid method to get a huge amount of humus and phytotoxicity to a reduced extent (Lorimor et al., 2001). Vermicomposting is also used to convert cattle dung and crop remains to get value-added end products for sustainable crop production (Suthar, 2009).

Thus, using several bioremediation approaches simultaneously could lead to the enhanced remediation ability and reduce expenses concurrently (Cassidy et al., 2015; Banitz et al., 2016). Genetically modified microorganisms (GMOs) for targeted contaminant degradation have also been used as a promising approach to improve bioremediation proficiency (Paul et al., 2005).

2.6 Future Aspects

The rapid increase in anthropogenic activities has challenged ecosystems with the abundant release of toxic compounds into the environment. Awareness of this fact amalgamated the international efforts in searching for potent environmental cleanup assessments. Environmental purification is a tremendously challenging job, as it is hard to remove certain contaminants or transform them into non-toxic products. Conventional methods like physical removal are principally the first response option; however, they hardly achieve total cleanup since the physical methods reclaim only up to 15% of the environmental pollutant, which means more radical perspectives are needed.

Much of the attention has been relocated from conventional to biological methods, for instance, bioremediation, which is efficacious, socio-economically palatable, and environmentally sustainable. Bioremediation has been utilised by humans for a long time; farmers have depended on composting to decay solid waste caused by plants and animals. In terms of novel biotechnology, bioremediation has been explored since the 1940s as a route to speed up natural biodegradation processes. Bioremediation is the biotechnology of exploiting metabolic activities of microorganisms', to deteriorate the environmental pollutants into non-toxic forms. It employs biological agents like plants, fungi, or bacteria to counteract or degrade toxic pollutants.

Conclusively, the practice of bioremediation, which is eco-friendly, potent, and cost effective, will lessen health risks, restore biodiversity heritage, and refurbish the damaged ecosystem naturally. Besides, indirect or direct occupation opportunities could be generated through well-planned bioremediation schemes; it will also enable sustainable management of polluted soils. All of this is encouraging and will assist the incorporation of new sustainable socio-economic green undertakings.

Bioremediation has the efficiency to reinstate polluted environments economically, yet effectually. A cause of uncertainty is created due to a lack of adequate knowledge about the effect of several environmental factors on the degree and rate of biodegradation. It is imperative to mention that numerous field tests have not been appropriately designed, well controlled, or properly evaluated, leading to ambiguity when choosing response options. Hence, future field investigations should endow serious efforts in implementing scientifically appropriate methodologies and obtaining the best possible quality data.

Furthermore, the exploration of an inclusive diversity of microbes with detoxification capabilities is still lacking. The scant knowledge about microorganisms and their natural role in the environment could influence the adequacy of their uses. Comprehensive knowledge of microbial diversity in a polluted environment is vital to get a deeper insight into potent degraders and to understand their biochemistry and genetics that will result in developing proper bioremediation approaches; thereby preserving the enduring sustainability of marine and terrestrial ecosystems. However, metabolomics, proteomics, and genomics of microbes associated with concerned bioremediation will be advantageous in exploring possible solutions pointing to particular contaminants. Subsequently, the following areas associated with bioremediation shall be explored:

1. Identification and comparison of gene and protein sequences in the proficient elimination of pollutants.
2. Diversity and phylogenetic studies of essential gene and protein sequences related to bioremediation processes.
3. Application of vital genes in plant biotechnology.

2.7 Conclusion

For the successful completion of bioremediation, the characterisation of the site is the most important step. It helps in the development of the most appropriate and practicable technique for bioremediation either in-situ or ex-situ. Ex-situ techniques of bioremediation are expensive because of the transportation and excavation costs, whereas in-situ techniques have no additional cost. This chapter gives brief information about the recent advancements in biotechnological techniques and tools for plants and microbe-assisted bioremediation of the pollutants present in the environment. It reveals the various mechanisms and enzymatic reactions used by microbes and plants in the efficient remediation of contaminants from the environment. It also describes the application of bioremediation as a substitute for the contaminants' removal, compared to chemical and physical methods that are less efficient and non-economical. The microbes and plants use different processes like biosorption, precipitation, enzymatic transformation, phytoremediation, and complexation. These processes are environment-dependent because the microbes used are temperature and pH-sensitive. In the same way, to carry out remediation using hyperaccumulator plants, the type of soil and concentration of contaminants play major roles. Identification of plants with high growth rates and phytoextraction potential should be carried out for the remediation of contaminants from water and soil. Development of microbial consortia

Advances in Biotechnology

by using synthetic biology technology with the ability to remediate contaminants using their metabolic properties is trending. Cooperation between environmentalists and researchers is required to explore efficient remediation of contaminants using bioremediation.

REFERENCES

Abdel-Shafy, H.I., & Mansour, M.S. (2018). Phytoremediation for the elimination of metals, pesticides, PAHs, and other Pollutants from wastewater and soil. In *Phytobiont and Ecosystem Restitution*, 101–136. Singapore: Springer.

Abhilash, P.C., Srivastava, S., Srivastava, P., Singh, B., Jafri, A., & Singh, N. (2011). Influence of rhizospheric microbial inoculation and tolerant plant species on the rhizoremediation of lindane. *Environmental and Experimental Botany*, 74, 127–130.

Aboudrar, W., Schwartz, C., Benizri, E., Morel, J.L., & Boularbah, A. (2007). Soil microbial diversity as affected by the rhizosphere of the hyperaccumulator Thlaspicaerulescens under natural conditions. *International journal of phytoremediation*, 9(1), 41–52.

Abou-Shanab, R.A., Ghanem, K., Ghanem, N., & Al-Kolaibe, A. (2008). The role of bacteria on heavy-metal extraction and uptake by plants growing on multi-metal-contaminated soils. *World Journal of Microbiology and Biotechnology*, 24(2), 253–262.

Abuabdou, S.M., Ahmad, W., Aun, N.C., & Bashir, M.J. (2020). A review of anaerobic membrane bioreactors (AnMBR) for the treatment of highly contaminated landfill leachate and biogas production: effectiveness, limitations and future perspectives. *Journal of Cleaner Production*, 255, 120215.

Achal, V., Pan, X., & Zhang, D. (2012). Bioremediation of strontium (Sr) contaminated aquifer quartz sand based on carbonate precipitation induced by Sr resistant Halomonas sp. *Chemosphere*, 89(6), 764–768.

Ahmed, M.B., Zhou, J.L., Ngo, H.H., Guo, W., Thomaidis, N.S., & Xu, J. (2017) Progress in the biological and chemical treatment technologies for emerging contaminant removal from wastewater: a critical review. *Journal of Hazardous Materials*, 323, pp. 274–298.

Ahmed, M.B., Zhou, J.L., Ngo, H.H., Guo, W., Thomaidis, N.S., & Xu, J. (2017). Progress in the biological and chemical treatment technologies for emerging contaminant removal from wastewater: a critical review. *Journal of Hazardous Materials*, 323, 274–298.

Antizar-Ladislao, B., Spanova, K., Beck, A.J., & Russell, N.J. (2008). Microbial community structure changes during bioremediation of PAHs in an aged coal-tar contaminated soil by in-vessel composting. *International Biodeterioration & Biodegradation*, 61(4), 357–364.

Ashish, & Debnath, M. (2018). Application of biosurfactant produced by an adaptive strain of *C. tropicalis* MTCC230 in microbial enhanced oil recovery (MEOR) and removal of motor oil from contaminated sand and water. *Journal of Petroleum Science and Engineering*, 170, 40–48.

Azaizeh, H., Castro, P.M., & Kidd, P. (2011). Biodegradation of organic xenobiotic pollutants in the rhizosphere. In *Organic Xenobiotics and Plants*, 191–215. Dordrecht: Springer.

Banitz, T., Frank, K., Wick, L.Y., Harms, H., & Johst, K. (2016). Spatial metrics as indicators of biodegradation benefits from bacterial dispersal networks. *Ecological Indicators*, 60, 54–63.

Bañuelos, G., Terry, N., LeDuc, D.L., Pilon-Smits, E.A., & Mackey, B. (2005). Field trial of transgenic Indian mustard plants shows enhanced phytoremediation of selenium-contaminated sediment. *Environmental Science & Technology*, 39(6), 1771–17

Ben, W., Zhu, B., Yuan, X., Zhang, Y., Yang, M., & Qiang, Z. (2018) Occurrence, removal and risk of organic micropollutants in wastewater treatment plants across China: comparison of wastewater treatment processes. *Water Research*, 130, 38–46.

Bhatia, R.K., Sakhuja, D., Mundhe, S., & Walia, A. (2020). Renewable energy products through bioremediation of wastewater. *Sustainability*, 12(18), 7501.

Bilal, M., Iqbal, H.M. (2019). An insight into toxicity and human-health-related adverse consequences of cosmeceuticals – a review. *Science of the Total Environment*, 670, 555–568. 10.1016/j.scitotenv.2019.03.261.77.

Boon, N., De Gelder, L., Lievens, H., Siciliano, S.D., Top, E.M., & Verstraete, W. (2002). Bioaugmenting bioreactors for the continuous removal of 3-chloroaniline by a slow release approach. *Environmental Science & Technology*, 36(21), 4698–4704.

Boopathy, R. (2000). Factors limiting bioremediation technologies. *Bioresource Technology*, *74*(1), 63–67.

Bouchez, T., Patureau, D., Dabert, P., Juretschko, S., Dore, J., Delgenes, P., ... & Wagner, M. (2000). Ecological study of a bioaugmentation failure. *Environmental Microbiology*, *2*(2), 179–190.

Boxall, A.B., Rudd, M.A., Brooks, B.W., Caldwell, D.J., Choi, K., Hickmann, S., Innes, E., Ostapyk, K., Staveley, J.P., Verslycke, T., & Ankley, G.T. (2012). Pharmaceuticals and personal care products in the environment: what are the big questions?. *Environmental Health Perspectives*, *120*(9), 1221–1229.

Burja, A.M., Banaigs, B., Abou-Mansour, E., Burgess, J.G., & Wright, P.C. (2001). Marine cyanobacteria – a prolific source of natural products. *Tetrahedron*, *57*(46), 9347–9377.

Caeiro, S., Costa, M.H., Ramos, T.B. (2005). Assessing heavy metal contamination in sado Estuary sediment: An index analysis approach. *Ecological Indicators*, *5*, 151–169.

Caplan, J.A. (1993). The worldwide bioremediation industry: prospects for profit. *Trends in Biotechnology*, *11*(8), 320–323.

Cassidy, D.P., Srivastava, V.J., Dombrowski, F.J., & Lingle, J.W. (2015). Combining in situ chemical oxidation, stabilization, and anaerobic bioremediation in a single application to reduce contaminant mass and leachability in soil. *Journal of Hazardous Materials*, *297*, 347–355.

Chakrabarty, A.M. (1982). Genetic mechanisms in the dissimilation of chlorinated compounds. In *Biodegradation and Detoxification of Environmental Pollutants*, 127. FL: CRC Boca Raton.

Chakrabarty, A.M. (1986). Genetic engineering and problems of environmental pollution. *Bio/Technology*, *8*, 515–530.

Chen, H., Peng, H., Yang, M., Hu, J., & Zhang, Y. (2017) Detection, occurrence, and fate of fluorotelomer alcohols in municipal wastewater treatment plants. *Environmental Science and Technology*, *51*, 8953–8961

Chen, Z., Li, Z., Lin, Y., Yin, M., Ren, J., & Qu, X. (2013). Biomineralization inspired surface engineering of nanocarriers for pH-responsive, targeted drug delivery. *Biomaterials*, *34*(4), 1364–1371.

Cheng, S.S., Hsieh, T.L., Pan, P.T., Gaop, C.H., Chang, L.H., Whang, L.M., & Chang, T.C. (2009). Study on biomonitoring of aged TPH-contaminated soil with bioaugmentation and biostimulation. In *10th International In Situ and On-Site Bioremediation Symposium, In Situ and On-Site Bioremediation-2009*.

Chojnacka, K. (2009). *Biosorption and Bioaccumulation-New Tools for Separation Technologies of 21 Century*. Nova Science Publishers.

Chojnacka, K. (2010). Biosorption and bioaccumulation – the prospects for practical applications. *Environment International*, *36*(3), 299–307.

Chopra, B.K., Bhat, S., Mikheenko, I.P., Xu, Z., Yang, Y., Luo, X., ... & Zhang, R. (2007). The characteristics of rhizosphere microbes associated with plants in arsenic-contaminated soils from cattle dip sites. *Science of the Total Environment*, *378*(3), 331–342.

Cody, R.J. (2000). Soil Vapor Sampling and Analysis: Sources of Random and Systematic Error in the Characterization of Low Level Organohalide Sources. In Proceedings of National RCRA Corrective Action Meeting, August 2000, Washington, D.C. http://www.clu-in.org/eiforum2000.

Colberg, P.J., & Young, L.Y. (1995). Anaerobic degradation of nonhalogenated homocyclic aromatic compounds coupled with nitrate, iron, or sulfate reduction. In *Microbial Transformation and Degradation of Toxic Organic Chemicals*, 307330.

Conesa, J.A., Martin-Gullon, I., Font, R., & Jauhiainen, J. (2004). Complete study of the pyrolysis and gasification of scrap tires in a pilot plant reactor. *Environmental Science & Technology*, *38*(11), 3189–3194.

da Silva, S., Gonçalves, I., Gomes de Almeida, F.C., Padilha da Rocha e Silva, N.M., Casazza, A.A., Converti, A., & AsforaSarubbo, L. (2020). Soil bioremediation: overview of technologies and trends. *Energies*, *13*(18), 4664.

Davis, T.A., Volesky, B., & Mucci, A. (2003). A review of the biochemistry of heavy metal biosorption by brown algae. *Water Research*, *37*(18), 4311–4330.

Defra, U.K. (2009). *Soil Strategy for England; Supporting Evidence Paper*. London: DEFRA.

Dhankher, O.P., Pilon-Smits, E.A., Meagher, R.B., & Doty, S. (2012). Biotechnological approaches for phytoremediation. In *Plant Biotechnology and Agriculture*, 309–328. Academic Press.

Dhillon, G.S., Kaur, S., Pulicharla, R., Brar, S.K., Cledón, M., Verma, M., & Surampalli, R.Y. (2015). Triclosan: current status, occurrence, environmental risks and bioaccumulation potential. *International Journal of Environmental Research and Public Health*, *12*(5), 5657–5684.

Dickmann, D.I., Nguyen, P.V., & Pregitzer, K.S. (1996). Effects of irrigation and coppicing on above-ground growth, physiology, and fine-root dynamics of two field-grown hybrid poplar clones. *Forest Ecology and Management, 80*(1–3), 163–174.

Dietz, A.C., & Schnoor, J.L. (2001). Advances in phytoremediation. *Environmental health perspectives, 109*(Suppl. 1), 163–168.

Dominguez, J. (2004). 20 state-of-the-art and new perspectives on vermicomposting research. In C.A. Edwards (Ed.), *Earthworm Ecology*, 2nd Edition, pp. 401–424. CRC Press. doi:10.1201/9781420039719.ch20

Dua, M., Singh, A., Sethunathan, N., & Johri, A. (2002). Biotechnology and bioremediation: successes and limitations. *Applied Microbiology and Biotechnology, 59*(2–3), 143–152.

Dudhane, M., Borde, M., & Jite, P.K. (2012). Effect of aluminium toxicity on growth responses and antioxidant activities in Gmelina arboreaRoxb. inoculated with AM fungi. *International Journal of Phytoremediation, 14*(7), 643–655.

El Fantroussi, S., & Agathos, S.N. (2005). Is bioaugmentation a feasible strategy for pollutant removal and site remediation?. *Current Opinion in Microbiology, 8*(3), 268–275.

[EA] Environment Agency (2003). Ecological risk assessment. A public consultation on a framework and methods for assessing harm to ecosystems from contaminants in soil.

EPA, U. (2016). Contaminant Candidate List (CCL) and Regulatory Determination-Final CCL 4 Chemical Contaminants.

Ernst, W.H. (2006). Evolution of metal tolerance in higher plants. *For Snow Landsc Res, 80*(3), 251–274.

García-Delgado, C., Alfaro-Barta, I., & Eymar, E. (2015). Combination of biochar amendment and mycoremediation for polycyclic aromatic hydrocarbons immobilization and biodegradation in creosote-contaminated soil. *Journal of Hazardous Materials, 285*, 259–266.

Gaur, N., Flora, G., Yadav, M., & Tiwari, A. (2014). A review with recent advancements on bioremediation-based abolition of heavy metals. *Environmental Science: Processes & Impacts, 16*(2), 180–193.

Godheja, J., Shekhar, S.K., Siddiqui, S.A., & Modi, D.R. (2016). Xenobiotic compounds present in soil and water: a review on remediation strategies. *Journal of Environmental & Analytical Toxicology, 6*(392), 2161-0525.

Godheja, J., Modi, D.R., Kolla, V., Pereira, A.M., Bajpai, R., Mishra, M.,... & Shekhar, S.K. (2019). Environmental remediation: microbial and nonmicrobial prospects. In *Microbial Interventions in Agriculture and Environment*, 379–409. Singapore: Springer.

Gomes, H.I., Dias-Ferreira, C., & Ribeiro, A.B. (2013). Overview of in situ and ex situ remediation technologies for PCB-contaminated soils and sediments and obstacles for full-scale application. *Science of the Total Environment, 445*, 237–260.

Gomez, F., & Sartaj, M. (2013). Field scale ex-situ bioremediation of petroleum contaminated soil under cold climate conditions. *International Biodeterioration & Biodegradation, 85*, 375–382.

Gunn, S.W.A. (1989). *Multilingual Dictionary Of Disaster Medicine And International Relief: English, Français, Español*. Springer Science & Business Media.

Guoshun, Z., Ronghul, H., & Mingxing, W. (2002). Great progress in study on aerosol and its impact on the global environment. *Progress in Natural Science, 12*, 407–413.

Halliday, S. (1999). The Great Stink of London: Sir Joseph Bazalgette and the Cleansing of the Victorian Metropolis. 2013 ed.

Han, L., Zhuang, G., Cheng, S., Wang, Y., & Li, J. (2007). Characteristics of re-suspended road dust and its impact on the atmospheric environment in Beijing. *Atmospheric Environment, 41*(35), 7485–7499.

Haritash, A.K., & Kaushik, C.P. (2009). Biodegradation aspects of polycyclic aromatic hydrocarbons (PAHs): a review. *Journal of hazardous materials, 169*(1–3), 1–15.

Hatzios, K.K. (1991) "Biotransformation of herbicides in higher plants." *Environmental chemistry of herbicides 2* : 141–185.

Hong, S., & Xingang, L.I. (2011). Modeling for volatilization and bioremediation of toluene-contaminated soil by bioventing. *Chinese Journal of Chemical Engineering, 19*(2), 340–348.

Harrison, R.M. (Ed.). (2015). *Pollution: Causes, Effects and Control*. Royal Society of Chemistry.

Hedges, S.B., Chen, H., Kumar, S., Wang, D.Y., Thompson, A.S., & Watanabe, H. (2001). A genomic timescale for the origin of eukaryotes. *BMC Evolutionary Biology, 1*(1), 1–10.

Idris, R., Trifonova, R., Puschenreiter, M., Wenzel, W.W., & Sessitsch, A. (2004). Bacterial communities associated with flowering plants of the Ni hyperaccumulator Thlaspi goesingense. *Applied and Environmental Microbiology*, *70*(5), 2667–2677.

Igiri, B.E., Okoduwa, S.I., Idoko, G.O., Akabuogu, E.P., Adeyi, A.O., & Ejiogu, I.K. (2018). Toxicity and bioremediation of heavy metals contaminated ecosystem from tannery wastewater: a review. *Journal of Toxicology*, *2018*.

International Federation of Red Cross, Red Crescent Societies, & Centre for Research on the Epidemiology of Disasters. (2002). *World Disasters Report*. International Federation of Red Cross and Red Crescent Societies.

Ishii, S.I., Suzuki, S., Yamanaka, Y., Wu, A., Nealson, K.H., & Bretschger, O. (2017). Population dynamics of electrogenic microbial communities in microbial fuel cells started with three different inoculum sources. *Bioelectrochemistry*, *117*, 74–82.

Jeevanantham, S., Saravanan, A., Hemavathy, R.V., Kumar, P.S., Yaashikaa, P.R., & Yuvaraj, D. (2019). Removal of toxic pollutants from water environment by phytoremediation: A survey on application and future prospects. *Environmental Technology & Innovation*, *13*, 264–276.

Jordahl, J.L., Foster, L., Schnoor, J.L., & Alvarez, P.J. (1997). Effect of hybrid poplar trees on microbial populations important to hazardous waste bioremediation. *Environmental Toxicology and Chemistry: An International Journal*, *16*(6), 1318–1321.

Kallenborn, R., Halsall, C., Dellong, M., & Carlsson, P. (2012). The influence of climate change on the global distribution and fate processes of anthropogenic persistent organic pollutants. *Journal of Environmental Monitoring*, *14*(11), 2854–2869.

Keri, S., Bethan, S., & Adam, G.H. (2008). Tire rubber recycling and bioremediation. *Bioremediation Journal*, *12*, 1–11.

Khan, F.I., Husain, T., & Hejazi, R. (2004). An overview and analysis of site remediation technologies. *Journal of environmental management*, *71*(2), 95–122.

Khudur, L.S., Shahsavari, E., Miranda, A.F., Morrison, P.D., Nugegoda, D., & Ball, A.S. (2015). Evaluating the efficacy of bioremediating a diesel-contaminated soil using ecotoxicological and bacterial community indices. *Environmental Science and Pollution Research*, *22*(19), 14809–14819.

Kim, T.W., Slowing, I.I., Chung, P.W., & Lin, V.S.Y. (2011). Ordered Mesoporous polymer– silica hybrid nanoparticles as vehicles for the intracellular controlled release of macromolecules. *ACS nano*, *5*(1), 360–366.

Kolpin, D.W., Thurman, E.M., & Linhart, S.M. (1998). The environmental occurrence of herbicides: the importance of degradates in ground water. *Archives of Environmental Contamination and Toxicology*, *35*(3), 385–390.

Komives, T., & Gullner, G. (2006). Dendroremediation: the use of trees in cleaning up polluted soils. In *Phytoremediation Rhizoremediation*, 23–31. Dordrecht: Springer.

Kuiper, I., Lagendijk, E.L., Bloemberg, G.V., & Lugtenberg, B.J. (2004). Rhizoremediation: a beneficial plant–microbe interaction. *Molecular Plant–Microbe Interactions*, *17*(1), 6–15.

Kumar, M., & Philip, L. (2006). Enrichment and isolation of a mixed bacterial culture for complete mineralization of endosulfan. *Journal of Environmental Science and Health Part B*, *41*(1), 81–96.

Kumar, V., Shahi, S.K., & Singh, S. (2018). Bioremediation: an eco-sustainable approach for restoration of contaminated sites. In *Microbial Bioprospecting for Sustainable Development*, 115–136. Singapore: Springer.

Kumar, J.N., Amb, M.K., Kumar, R.N., Bora, A., & Khan, S.R. (2013). Studies on biodegradation and molecular characterization of 2, 4-D Ethyl Ester and Pencycuron induced Cyanobacteria by using GC-MS and 16S rDNA sequencing. *Proceedings of the International Academy of Ecology and Environmental Sciences*, *3*(1), 1.

Kumar Ramasamy, R., Congeevaram, S., & Thamaraiselvi, K. (2011). Evaluation of isolated fungal strain from e-waste recycling facility for effective sorption of toxic heavy metal Pb (II) ions and fungal protein molecular characterization-a mycoremediation approach. *Asian Journal of Experimental Biological Sciences*, *2*, 342–347.

Kuppusamy, S., Palanisami, T., Megharaj, M., Venkateswarlu, K., & Naidu, R. (2016). Ex-situ remediation technologies for environmental pollutants: a critical perspective. In *Reviews of Environmental Contamination and Toxicology, Vol. 236*, 117–192. Cham: Springer.

Lapworth, D.J., Baran, N., Stuart, M.E., & Ward, R.S. (2012). Emerging organic contaminants in groundwater: a review of sources, fate and occurrence. *Environmental Pollution, 163*, 287–303.

Le Borgne, S., Paniagua, D., & Vazquez-Duhalt, R. (2008). Biodegradation of organic pollutants by halophilic bacteria and archaea. *Journal of Molecular Microbiology and Biotechnology, 15*(2–3), 74–92. doi: 10.1159/000121323

Liang, F., Xiao, Y., & Zhao, F. (2013). Effect of pH on sulfate removal from wastewater using a bioelectrochemical system. *Chemical Engineering Journal, 218*, 147–153.

Long, E.R., & MacDonald, D.D. (1998). Recommended uses of empirically derived, sediment quality guidelines for marine and estuarine ecosystems. *Human and Ecological Risk Assessment, 4*(5), 1019–1039.

Loper, C.A., Crawford, J.K., Otto, K.L., Manning, R.L., Meyer, M.T., & Furlong, E.T. (2007). *Concentrations of selected pharmaceuticals and antibiotics in south-central Pennsylvania waters, March through September, 2006*(No. 300). US Geological Survey.

López-Pacheco, I.Y., Silva-Núñez, A., Salinas-Salazar, C., Arévalo-Gallegos, A., Lizarazo-Holguin, L.A., Barceló, D., Iqbal, H.M.N., Parra-Saldívar, R. (2019) Anthropogenic contaminants of high concern: Existence in water resources and their adverse effects. *Science of The Total Environment, 690*, 1068–1088. ISSN 0048-9697. 10.1016/j.scitotenv.2019.07.052.

Lorimor, J., Fulhage, C., Zhang, R., Funk, T., Sheffield, R., Sheppard, C., & Newton, G. L. (2001). Manure management strategies/technologies, White paper on animal agriculture and the environment for national center for manure and animal waste management. *MWPS, Ames, IA, 52.*

Loukidou, M.X., Matis, K.A., Zouboulis, A.I., & Liakopoulou-Kyriakidou, M. (2003). Removal of As (V) from wastewaters by chemically modified fungal biomass. *Water Research, 37*(18), 4544–4552.

Lu, L., Xing, D., & Ren, N. (2012). Bioreactor performance and quantitative analysis of methanogenic and bacterial community dynamics in microbial electrolysis cells during large temperature fluctuations. *Environmental Science & Technology, 46*(12), 6874–6881.

Lu, S., Lepo, J.E., Song, H.-X., Guan, C.-Y., & Zhang, Z.-H. (2018). Increased rice yield in long-term crop rotation regimes through improved soil structure, rhizosphere microbial communities, and nutrient bioavailability in paddy soil. *Biology and Fertility of Soils, 54*, 909–923.

Lyyra, S., Meagher, R.B., Kim, T., Heaton, A., Montello, P., Balish, R.S., & Merkle, S.A. (2007). Coupling two mercury resistance genes in Eastern cottonwood enhances the processing of organomercury. *Plant Biotechnology Journal, 5*(2), 254–262.

Macek, T., Macková, M., & Káš, J. (2000). Exploitation of plants for the removal of organics in environmental remediation. *Biotechnology Advances, 18*(1), 23–34.

Mahfooz, S., Jahan, S., Shamim, A., Husain, A., & Farooqui, A. (2018). Oxidative stress and response of antioxidant system in Nostoc muscorum exposed to different forms of Zinc. *Turkish Journal of Biochemistry, 43*(4), 352–361.

Maken, A., Assobhei, O., & Mountadar, M. (2013). Effect of initial moisture content on the in vessel composting under air pressure oforganic fraction of municipal solid waste in Morocco. *Journal of Environmental Health Science & Engineering, 10*(1), 10–15.

Mani, D., & Kumar, C. (2014). Biotechnological advances in bioremediation of heavy metals contaminated ecosystems: an overview with special reference to phytoremediation. *International Journal of Environmental Science and Technology, 11*(3), 843–872.

McCallum, E.S., Du, S.N., Vaseghi, M., Shanjani, Choi, J.A., Warriner, T.R., Sultana, T., Scott, G.R., & Balshine, S. (2017) In situ exposure to wastewater effluent reduces survival but has little effect on the behaviour or physiology of an invasive Great Lakes fish. *Aquatic Toxicology, 184*, 37–48.

Megharaj, M., Madhavi, D.R., Sreenivasulu, C., Umamaheswari, A., & Venkateswarlu, K. (1994). Biodegradation of methyl parathion by soil isolates of microalgae and cyanobacteria. *Bulletin of Environmental Contamination and Toxicology, 53*, 292–297. doi:10.1007/BF00192047.

Mishra, V., Balomajumder, C., & Agarwal, V.K. (2014). Biological removal of heavy metal zinc from industrial effluent by Zinc sequestering bacterium Vmsdcm. *Clean Technologies and Environmental Policy, 16*(3), 555–568. 10.1007/s10098-013-0655-x.

Mitch, W.A., Sharp, J.O., Trussell, R.R., Valentine, R.L., Alvarez-Cohen, L., & Sedlak, D.L. (2003). N-nitrosodimethylamine (NDMA) as a drinking water contaminant: a review. *Environmental Engineering Science, 20*(5), 389–404.

Monteiro, S.S., Pereira, A.T., Costa, E., Torres, J., Oliveira, I., Bastos-Santos, J., et al. (2016). Bioaccumulation of trace element concentrations in common dolphins (*Delphinus delphis*) from Portugal. *Marine Pollution Bulletin 113*, 400–407. doi: 10.1016/j.marpolbul.2016.10.033.

Mrozik, A., & Piotrowska-Seget, Z. (2010). Bioaugmentation as a strategy for cleaning up of soils contaminated with aromatic compounds. *Microbiological Research, 165*(5), 363–375. Retrieved March 13, 2016, from http://www.sciencedirect.com/science/article/pii/S0944501309000585.

Mrvčić, J., Stanzer, D., Šolić, E., & Stehlik-Tomas, V. (2012). Interaction of lactic acid bacteria with metal ions: opportunities for improving food safety and quality. *World Journal of Microbiology and Biotechnology, 28*(9), 2771–2782.

Nikolopoulou, M., Pasadakis, N., Norf, H., & Kalogerakis, N. (2013). Enhanced ex situ bioremediation of crude oil contaminated beach sand by supplementation with nutrients and rhamnolipids. *Marine Pollution Bulletin, 77*(1-2), 37–44.

Niu, G.L., Zhang, J.J., Zhao, S., Liu, H., Boon, N., & Zhou, N.Y. (2009). Bioaugmentation of a 4-chloronitrobenzene contaminated soil with Pseudomonas putida ZWL73. *Environmental Pollution, 157*(3), 763–771.

Noji, E. K. (1997). *The Nature of Disaster: General Characteristics and Public Health Effects*, 3–20. Oxford, United Kingdom: Oxford University Press.

Ortiz-Hernández, M.L., Castrejón-Godínez, M.L., Popoca-Ursino, E.C., Cervantes-Dacasac, F.R., & Fernández-Lópezd, M. (2018). Strategies for biodegradation and bioremediation of pesticides in the environment. *Strategies for Bioremediation of Organic and Inorganic Pollutants*, 95–115.

Palanisami, S., Prabaharan, D., & Uma, L. (2009). Fate of few pesticide-metabolizing enzymes in the marine cyanobacterium Phormidiumvalderianum BDU 20041 in perspective with chlorpyrifos exposure. *Pesticide Biochemistry and Physiology, 94*(2-3), 68–72.

Paul, D., Pandey, G., Pandey, J., & Jain, R.K. (2005). Accessing microbial diversity for bioremediation and environmental restoration. *TRENDS in Biotechnology, 23*(3), 135–142.

Pavel, L.V., & Gavrilescu, M. (2008). Overview of ex situ decontamination techniques for soil cleanup. *Environmental Engineering & Management Journal (EEMJ), 7*(6).

Phillips, P.J., Smith, S.G., Kolpin, D.W., Zaugg, S.D., Buxton, H.T., Furlong, E.T., Esposito, K., & Stinson, B. (2010). Pharmaceutical formulation facilities as sources of opioids and other pharmaceuticals to wastewater treatment plant effluents. *Environmental Science & Technology, 44*(13), 4910–4916.

Philp, J.C., & Atlas, R.M. (2005). Bioremediation of contaminated soils and aquifers. In *Bioremediation: Applied Microbial Solutions for Real-World Environmental Cleanup*, 139–236. American Society of Microbiology.

Pilon-Smits, E. (2005). Phytoremediation. *Annual Review of Plant Biology, 56*, 15–39.

Potts, M., & Whitton, B.A. (2000). The biology and ecology of cyanobacteria.

Prpich, G.P., Rehmann, L., & Daugulis, A.J. (2008). On the use, and reuse, of polymers for the treatment of hydrocarbon contaminated water via a solid–liquid partitioning bioreactor. *Biotechnology Progress, 24*(4), 839–844.

Rao, M.A., Scelza, R., Scotti, R., & Gianfreda, L. (2010). Role of enzymes in the remediation of polluted environments. *Journal of Soil Science and Plant Nutrition, 10*(3), 333–353.

Rasheed, T., Li, C., Nabeel, F., Huang, W., & Zhou, Y. (2019) Self-assembly of alternating copolymer vesicles for the highly selective, sensitive and visual detection and quantification of aqueous Hg2+. *Chemical Engineering Journal, 358*, 101–109.

Rhind, S.M. (2009). Anthropogenic pollutants: a threat to ecosystem sustainability? *Philosophical Transactions of the Royal Society B, 364*, 3391–3401. 10.1098/rstb.2009.0122.

Richardson, M.L., & Bowron, J.M. (1985). The fate of pharmaceutical chemicals in the aquatic environment. *Journal of Pharmacy and Pharmacology, 37*(1), 1–12.

Roberts, J., & Marston, F. (2011). *Water regime for wetland and floodplain plants: a source book for the Murray-Darling Basin*. Canberra: National Water Commission.

Rodríguez-Delgado, M., Orona-Navar, C., García-Morales, R., Hernandez-Luna, C., Parra, R., Mahlknecht, J., Ornelas-Soto, N. (2016). Biotransformation kinetics of pharmaceutical and industrial micropollutants in groundwaters by a laccase cocktail from Pycnoporus sanguineus CS43 fungi. *International Biodeterioration and Biodegradation, 108*, 34–41. 10.1016/j.ibiod.2015.12.003.

Sanderson, L.M. (1992). Toxicologic disasters: natural and technologic. *Hazardous Materials Toxicology: Clinical Principles of Enviromental Health*, 326–331. Baltimore, MD: Williams & Wilkins.

Sanscartier, D., Zeeb, B., Koch, I., & Reimer, K. (2009). Bioremediation of diesel-contaminated soil by heated and humidified biopile system in cold climates. *Cold Regions Science and Technology, 55*(1), 167–173.

Say, R., Yilmaz, N., & Denizli, A. (2003). Removal of heavy metal ions using the fungus Penicillium canescens. *Adsorption Science & Technology, 21*(7), 643–650.

Schoenmuth, B.W., & Pestemer, W. (2004b). Dendroremediation of trinitrotoluene (TNT) Part 2: Fate of radio-labelled TNT in trees. *Environmental Science and Pollution Research, 11*(5), 331–339.

Schoenmuth, B.W., & Pestemer, W. (2004a). Dendroremediation of trinitrotoluene (TNT) Part 1: Literature overview and research concept. *Environmental Science and Pollution Research, 11*(4), 273–278.

Scow, K.M., & Hicks, K.A. (2005). Natural attenuation and enhanced bioremediation of organic contaminants in groundwater. *Current Opinion in Biotechnology, 16*(3), 246–253.

Shackelford, C.D., & Jefferis, S.A. (2000, November). Geoenvironmental engineering for in situ remediation. In *ISRM International Symposium*. International Society for Rock Mechanics and Rock Engineering.

Shah, J.K., Sayles, G.D., Suidan, M.T., Mihopoulos, P., & Kaskassian, S. (2001). Anaerobic bioventing of unsaturated zone contaminated with DDT and DNT. *Water Science and Technology, 43*(2), 35–42.

Shivajirao, P.A. (2012). Treatment of distillery wastewater using membrane technologies. *International Journal of Advanced Engineering Research and Studies, 1*(3), 275–283.

Singh, S.P., & Tiwari, G. (2014). Application of bioremediation on solid waste management: a review. *Journal of Bioremediation & Biodegradation, 5*(6).

Singh, J.S., & Gupta, V.K. (2016). Degraded land restoration in reinstating CH4 sink. *Frontiers in Microbiology, 7*, 923.

Singh, J.S., Singh, D.P., & Dixit, S. (2011). Cyanobacteria: an agent of heavy metal removal. *Bioremediation of Pollutants*. New Delhi: IK International Publisher, 223–243.

Singh, S.N., & Tripathi, R.D. (Eds.). (2007). *Environmental Bioremediation Technologies*. Springer Science & Business Media.

Singh, J.S., Abhilash, P.C., Singh, H.B., Singh, R.P., & Singh, D.P. (2011). Genetically engineered bacteria: an emerging tool for environmental remediation and future research perspectives. *Gene, 480*(1–2), 1–9.

Singh, D.P., Khattar, J.I.S., Nadda, J., Singh, Y., Garg, A., Kaur, N., & Gulati, A. (2011). Chlorpyrifos degradation by the cyanobacterium Synechocystis sp. strain PUPCCC 64. *Environmental Science and Pollution Research, 18*(8), 1351–1359.

Singh, V., Tripathi, G., & Mishra, V. (2021). Biological treatment advancements for the remediation of selenium from wastewater. *Selenium Contamination in Water*, 228–251.

Smith, E., Thavamani, P., Ramadass, K., Naidu, R., Srivastava, P., & Megharaj, M. (2015). Remediation trials for hydrocarbon-contaminated soils in arid environments: evaluation of bioslurry and biopiling techniques. *International Biodeterioration & Biodegradation, 101*, 56–65.

Sriprang, R., Hayashi, M., Ono, H., Takagi, M., Hirata, K., & Murooka, Y. (2003). Enhanced accumulation of Cd2+ by a Mesorhizobium sp. transformed with a gene from Arabidopsis thaliana coding for phytochelatin synthase. *Applied and Environmental Microbiology, 69*(3), 1791–1796.

Stephenson, C., & Black, C.R. (2014). One step forward, two steps back: the evolution of phytoremediation into commercial technologies. *Bioscience Horizons: The International Journal of Student Research, 7*.

Strong, P.J., & Burgess, J.E. (2008). Treatment methods for wine-related and distillery wastewaters: a review. *Bioremediation Journal, 12*(2), 70–87.

Stuart, M., Lapworth, D., Crane, E., & Hart, A. (2012). Review of risk from potential emerging contaminants in UK groundwater. *Science of the Total Environment, 416*, 1–21.

Suthar, S. (2007). Nutrient changes and biodynamics of epigeic earthworm Perionyx excavatus (Perrier) during recycling of some agriculture wastes. *Bioresource Technology, 98*(8), 1608–1614.

Suthar, S. (2009). Bioremediation of agricultural wastes through vermicomposting. *Bioremediation Journal, 13*(1), 21–28.

Sylvia, D.M., Fuhrmann, J.F., Hartel, P.G., & Zuberer, D.A. (2005). *Principles and Applications of Soil Microbiology*. New Jersey: Pearson Education Inc.

Tang, C.Y., Fu, Q.S., Criddle, C.S., & Leckie, J.O. (2007). Effect of flux (transmembrane pressure) and membrane properties on fouling and rejection of reverse osmosis and nanofiltration membranes treating perfluorooctane sulfonate containing wastewater. *Environmental Science & Technology, 41*(6), 2008–2014.

Tang, C.Y., Fu, Q.S., Criddle, C.S., & Leckie, J.O. (2007). Effect of flux (transmembrane pressure) and membrane properties on fouling and rejection of reverse osmosis and nanofiltration membranes treating perfluorooctane sulfonate containing wastewater. *Environmental Science & Technology*, *41*(6), 2008–2014.

Tesar, M., Reichenauer, T.G., & Sessitsch, A. (2002). Bacterial rhizosphere populations of black poplar and herbal plants to be used for phytoremediation of diesel fuel. *Soil Biology and Biochemistry*, *34*(12), 1883–1892.

Timberlake, L., O'Keefe, P., Tinker, J., Cheney, B., & McCormack, C. (1984). *Natural Disasters: Acts of God-Or Acts of Man?* Earthscan.

Tripathi G., Yadav V.K., Singh J., Mishra V. (2020) Analytical methods of water pollutants detection. In Pooja D., Kumar P., Singh P., Patil S. (Eds.), *Sensors in Water Pollutants Monitoring: Role of Material. Advanced Functional Materials and Sensors*. Singapore: Springer. 10.1007/978-981-15-0671-0_5

Tripathi, G., Husain, A., Ahmad, S., Hasan, Z., & Farooqui, A. (2021). Contamination of water resources in industrial zones. In *Contamination of Water*, pp. 85–98. Academic Press.

Tyagi, M., da Fonseca, M.M.R., & de Carvalho, C.C. (2011). Bioaugmentation and biostimulation strategies to improve the effectiveness of bioremediation processes. *Biodegradation*, *22*(2), 231–241.

United States. Environmental Protection Agency. Office of Ground Water, & Drinking Water. (1998). *Empirically based models for predicting chlorination and ozonation by-products: Trihalomethanes, haloacetic acids, chloral hydrate, and bromate*. US Environmental Protection Agency.

Venkatesan, A.K., & Halden, R.U. (2014). Wastewater treatment plants as chemical observatories to forecast ecological and human health risks of manmade chemicals. *Scientific Reports*, *4*(1), 1–7.

Vidali, M. (2001). Bioremediation. an overview. *Pure and Applied Chemistry*, *73*(7), 1163–1172.

Vivas, A., Vörös, I., Biró, B., Campos, E., Barea, J.M., &Azcón, R. (2003). Symbiotic efficiency of autochthonous arbuscular mycorrhizal fungus (G. mosseae) and Brevibacillus sp. isolated from cadmium polluted soil under increasing cadmium levels. *Environmental Pollution*, *126*(2), 179–189.

von Fahnestock, F.M., & Wickramanayake, G.B. (1998). *Biopile design, operation, and maintenance handbook for treating hydrocarbon-contaminated soils*.

Wang, Q., Chen, M., Shan, G., Chen, P., Cui, S., Yi, S., Zhu, L. (2017). Bioaccumulation and biomagnification of emerging bisphenol analogues in aquatic organisms from Taihu Lake, China. *Science of the Total Environment*, *598*, 814–820. 10.1016/j.scitotenv.2017.04.167.

Wasi, S., Jeelani, G., & Ahmad, M. (2008). Biochemical characterization of a multiple heavy metal, pesticides and phenol resistant Pseudomonas fluorescens strain. *Chemosphere*, *71*(7), 1348–1355.

Whelan, M.J., Coulon, F., Hince, G., Rayner, J., McWatters, R., Spedding, T., & Snape, I. (2015). Fate and transport of petroleum hydrocarbons in engineered biopiles in polar regions. *Chemosphere*, *131*, 232–240.

Whitton, B.A., & Potts, M. (Eds.). (2007). *The Ecology of Cyanobacteria: Their Diversity in Time and space*. Springer Science & Business Media.

World Health Organization. (1980). Emergency care in natural disasters. Views of an international seminar. *WHO Chron*, *34*, 96–100.

World Health Organization. (2015). Antibiotic resistance: Multi-country public awareness survey.

3

Bioremediation: Tools and Techniques for Wastewater Reclamation

P.P. Mirshad[1], Nair G. Sarath[2], A.M. Shackira[1], and Jos T. Puthur[2]
[1]*Department of Botany, Sir Syed College, Taliparamba, Kannur, Kerala, India*
[2]*Plant Physiology and Biochemistry Division, Department of Botany, University of Calicut, Kerala, India*

CONTENTS

3.1 Introduction 51
3.2 Microbial Remediation 52
      3.2.1 Mechanism of Microbial Remediation 54
3.3 Mycoremediation 54
      3.1 Mechanism of Mycoremediation 54
3.4 Phycoremediation 55
      3.4.1 Remediation Using Microalgae 55
      3.4.2 Remediation Using Macroalgae 57
      3.4.3 Mechanism of Phycoremediation 58
3.5 Phytoremediation of Wastewater 58
      3.5.1 Mechanism of Phytoremediation 60
3.6 Future Prospectus and Conclusion 61
References 61

3.1 Introduction

Exploitation of natural resources and uncontrolled anthropogenic activities cause serious threats to water bodies and result in the scarcity of fresh water for domestic and other purposes. This ever-increasing environmental problem is faced in many countries around the world and has attracted more researchers for finding ways and means for sustainable reclamation methods of wastewater (Zeng et al., 2020). The untreated wastewater released from industries such as textiles, oil refineries, and chemical industries including nuclear plants rapidly contaminate water bodies and also cause high health hazards to ecosystems and affect the groundwater quality (Singh et al., 2017). Besides, the sewage waste discharge into the water stream contains enormous amounts of organic nutrients like nitrogen (N), phosphorus (P), potassium (K), etc. that cause eutrophication and result in the reduction of oxygen content in the water bodies, thereby disturbing the living diversity of aquatic organisms (Emparan et al., 2019). The excess increase in nutrient content of water bodies also causes foaming, discoloration, and enhanced growth of toxic algal species (algal blooms), as reported from several parts of the world, and it will diminish the survival of entire aquatic ecosystems (Driscoll et al., 2003).

    Remediation strategies, which were conventionally practiced, have several disadvantages like high cost, harmful effect to the freshwater biota, complex implementation strategies, less efficiency, etc.

Hence, planning and implementation of a low-cost technique for the proper management of wastewater and its recuperation by natural means have a better importance nowadays. This promising, ecofriendly, cost effective, highly efficient strategy of remediation in which biological agents are exploited for the successful degradation/transformation of pollutants is known as bioremediation. The major objective of this chapter is to highlight the role of major bioremediating agents for the remediation of contaminant from wastewater and there by safeguarding drinking water.

3.2 Microbial Remediation

Microbes, algae, fungi, and higher plants are regarded as effective bioremediating agents for the remediation of pollutants from wastewater (Figure 3.1) and are being used for the restoration of polluted water (Table 3.1). The microorganisms like bacteria enhance the degradation of xenobiotics deposited in wastewaters through various metabolic pathways (Hesnawi et al., 2014). Bacteria are being exploited for the synthesis of various valuable products such as enzymes, probiotics, and biofuels and these microorganisms' main role is to decontaminate toxic wastewater. Bacteria and various hydrolytic enzymes produced within bacterial cells are mainly involved in the biodegradation of pollutants in the wastewater (Rani et al., 2019). Application of photosynthetic bacteria in the decontamination of wastewater is considered one of the most potential methods and is increasingly utilised for the removal of H_2S from anaerobic waste treatment (Kobayashi et al., 1983). Heavy metals disposed from the industrial effluents have deleterious effects on the surface and groundwater resources. The bacterial strains like *E. coli* and *B. subtilis* effectively remove toxic metal ions like cadmium (Cd), lead (Pb), and chromium (Cr) from aqueous solutions (Huang et al., 2001). Likewise, the biomass of *Bacillus* sp. was effectively utilised for the remediation of Cu, nickel (Ni), and Cr from wastewater (Al-Daghistani, 2012). Cu and Cd from contaminated domestic-industrial as well as effluents were effectively removed by *Enterobacter* sp. and *Stenotrophomonas* sp. (Bestawy et al., 2013).

The bacterial consortium, such as *Bacillus pumilus, Brevibacterium* sp., and *Pseudomonas aeruginosa*, acts in a synergistic way for degrading the organic compounds present in sewage wastewater. *Acinetobacter* sp. and *Rhodococcus* sp. are used for the decontamination of industrial wastewater and are capable of removing pollutants such as chlorophenol, 2,4-dichlorophenol, and pentachlorophenol (Paisio et al., 2014). Phenol-contaminated wastewater was effectively treated with bacterial species such as *Acinetobacter* sp., *Bacillus* sp., and *Pseudomonas* sp. These species reduce the wastewater toxicity by

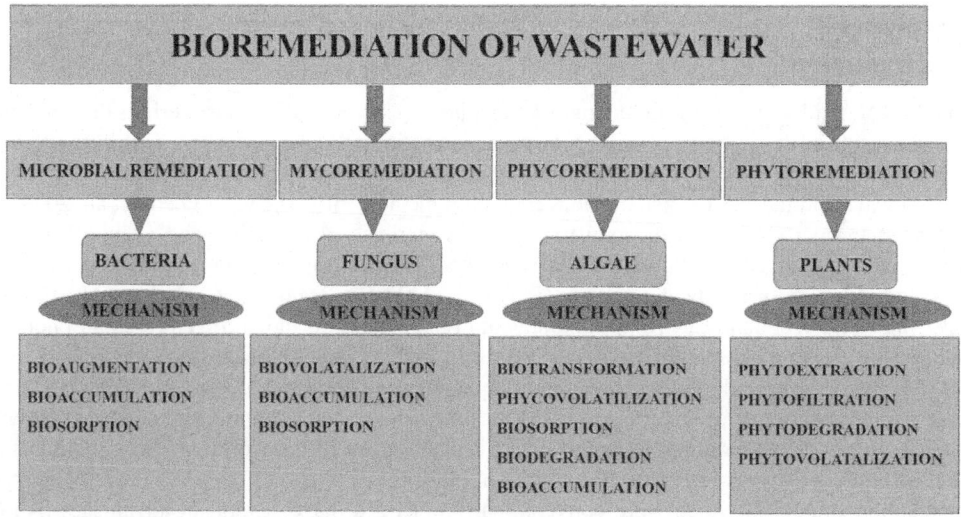

FIGURE 3.1 Bioremediation of wastewater.

TABLE 3.1

List of bacteria and fungi used for bioremediation of various pollutants

	Bacteria	
Species Name	Type of Pollutant	Reference
Bacillus cereus	Hg	Gupta et al. (1999)
Microbacterium sp.	Cr	Pattanapipitpaisal et al. (2001)
Bacillus subtilis	Pb, Cd and Cr	Huang et al. (2001)
Bacillus circulans and *Bacillus megaterium*	Cr	Srinath et al. (2002)
Pseudomonas putida	Zn	Green-Ruiz et al. (2008)
Bacillus coagulans	Cr	Vijayaraghavan and Yun (2008)
Bacillus cereus	Zn	Joo et al. (2010)
Bacillus pumilus	Pb	Colak et al. (2011)
Trametes versicolor	Cu	Subbaiah et al. (2011)
Bacillus sp.	Cu, Ni, and Cr	Al-Daghistani (2012)
Bacillus thioparans	Cu and Pb	Rodríguez-Tirado et al. (2012)
Bacillus licheniformis	Cr, Fe and Cu	Samarth et al. (2012)
Enterobacter sp. and *Stenotrophomonas* sp.	Cu and Cd	Bestawy et al. (2013)
Acinetobacter sp. and *Rhodococcus* sp.	Chlorophenol	Paisio et al. (2014)
Pseudomonas stutzeri and *Pseudomonas putida*	Hg and Pb	Maulin (2017)
Acinetobacter sp., *Bacillus* sp., and *Pseudomonas* sp.	Phenol	Poi et al. (2017)
Pseudomonas diminuta, P. pseudoalcaligenes, and *Escherichia* sp.	Grease and oil	Rani et al. (2019)
Pseudomonas putida	Catechol, phenol, and cresols	Michalska et al. (2020)
	Fungi	
Species Name	Type of Pollutant	Reference
Phanerochaete chrysosporium	Ni and Pb	Çeribasi and Yetis (2001)
Aspergillus terreus	Cu	Gulati et al. (2002)
Aspergillus niger	Cu	Dursun et al. (2003)
Saccharomyces cerevisiae	Pb and Cr	Özer and Özer (2003)
Penicillium canescens	As, Hg, Cd, and Pb	Say et al. (2003)
Lentinus sajor	Cr	Arica and Bayramoğlu (2005)
Saccharomyces cerevisiae	Cr	Ksheminska et al. (2005)
Penicillium chrysogenum	Ni	Haijia et al. (2006)
Pencillium simpliccium	Cd and Zn	Fan et al. (2008)
Aspergillus fumigatus	Pb	Kumar Ramasamy et al. (2011)
Pleurotus ostreatus	Cu and Ni	Javaid et al. (2011)
Pleurotus ostreatus	Cr	Arbanah et al. (2013)
Aspergillus terreus and *Aspergillus niger*	Nitrate and phosphate	Kshirsagar (2013)
Trametes versicolor	Ni	Subbaiah and Yun (2013)
Aspergillus niger	Pb	Iram et al. (2015)
Aspergillus flavus	Cu	Iram et al. (2015)
Phlebia acerina	Congo Red	Kumar et al. (2018)

degrading phenolic compounds and lowering chemical oxygen demand of contaminated water (Poi et al., 2017). Bioaugmentation of the activated sludge is regarded as a promising decontamination strategy for removing contaminants from wastewater. *Pseudomonas putida* is being effectively utilised for degrading aromatic compounds such as primarily catechol, phenol, and cresols from contaminated wastewater (Michalska et al., 2020). *Pseudomonas stutzeri* and *Pseudomonas putida* have a role in the bioremediation of toxic metals from wastewater (Maulin, 2017). *Pseudomonas diminuta, P. pseudoalcaligenes,* and *Escherichia* sp. were used for the breakdown of grease and oil from contaminated water (Rani et al., 2019).

3.2.1 Mechanism of Microbial Remediation

Bacterial treatment is carried out based on the properties of the biodegradability of organic pollutants present in wastewater. Bioaccumulation is a process in which the metal ions concentrate in the living cells. Uptake and transportation of the metal ions were the two important steps in the bioaccumulation process. *Pichia stipitis* was able to bioaccumulate Cu and Cr ions (Yilmazer and Saracoglu, 2009). Bioaugmentation is considered a cheap and ecofriendly strategy for remediating recalcitrant contaminants from the contaminated water. Bioaugmentation involves the addition of some selected bacterial strains to wastewater and enhances the catabolism of specific compounds and thus enhances removal efficiency of pollutants from wastewater (Herrero and Stuckey, 2015). The strategies undertaken by microbes to flourish in the polluted water include efflux mechanism, precipitation, extracellular sequestration, biosorption, changes in cell morphology, and increased production of siderophores (Naik and Dubey, 2013). Contaminated water samples can be cleaned up with the utilisation of siderophores, which can be used as indicators for the elevated iron content in the environment (Saha et al., 2016). Biosorption is the binding of metal ions in a reversible manner to the functional groups present on the different biosorbent (Kanamarlapudi et al., 2018).

3.3 Mycoremediation

Mycoremediation is a form of remediation in which fungi are used to clean up the polluted area, whether it is land or water; the use of fungi for removing contaminants from wastewater was developed as a sustainable method over conventional techniques (Table 3.1). According to Sankaran et al. (2010), the fungal biomass formed as a part of reclamation of contaminated water has a high demand compared with the bacterial-activated sludge process. *Aspergillus fumigatus* is considered a suitable candidate for biosorption of heavy metal and can absorb Pb ions, even up to a concentration of 100 g/L (Kumar Ramasamy et al., 2011). *Aspergillus terreus* and *Aspergillus niger* also show a remarkable capability for purifying the polluted water and they effectively removed nitrates and phosphates from wastewater (Ezeronye and Okerentugba, 1999; Kshirsagar, 2013). White rot fungi are ubiquitous in distribution and their adaptability to extreme conditions makes them ideal candidates for wastewater reclamation. For example, *Phlebia acerina*, a white wood rot fungus, has been used to disintegrate the toxic wastewater pollutants such as Congo Red and Eriochrome Black T (Kumar et al., 2018). *Penicillium, Aspergillus,* and *Rhizopus* have been proven efficient candidates for the removal of heavy metals from polluted water (Shahid et al., 2020). *Aspergillus fumigatus* has been effectively utilised for decolourisation and distillation of polluted water (Mohammad et al., 2006).

3.1 Mechanism of Mycoremediation

Fungi is considered a promising and effective microorganism for the bioremediation of wastewater. Biosorption, bioaccumulation, and biovolatilisation are the major strategies of fungi to clear contaminated areas by the biotransformation of xenobiotics. Biovolatilisation involves the alteration of the pollutants into their volatile compounds by enzymes (Akhtar and Mannan, 2020). The breakdown of pollutants involves the action of intracellular enzymes like cytochrome P450 and extracellular enzymes such as manganese peroxidase, laccase, and versatile peroxidase (Silva et al., 2019). Fungi have the capability to sequester metal ions from aqueous solutions. They also transform complex organic compounds by the catalytic action of special degrading enzymes. Fungi remediate significant amounts of organic pollutants from contaminated water through the process of adsorption. It is also considered a most effective biosorbent for the removal of toxic metals, such as cadmium (Cd), chromium (Cr), copper (Cu), mercury (Hg), nickel (Ni), and lead (Pb) from wastewaters (Sankaran et al., 2010). Biosorption is a passive adsorption mechanism by which heavy metals can accumulate in from wastewater. This complex mechanism is affected by different factors like physiological state of cell, change in pH, temperature, duration of contact time, ionic strength, concentration of metal, and cell wall composition (Coelho et al., 2015). Manganese was removed by *Pleurotus ostreatus* from polluted water with surfactants. The activity of different surfactants helped the efficient removal of manganese by increasing the surface area for metal binding on the hypha (Wu et al., 2016). The lignin-degrading enzymes such as laccase, manganese, and peroxidase were able to degrade the dyes by the action of these enzymes. Moreover, *Aspergillus flavus* was able to absorb and degrade congo red dye (Bhattacharya and Das, 2011).

3.4 Phycoremediation

The method of remediation of toxic components from water using algae (macro and micro) is generally termed phycoremediation (Table 3.2). According to Rao et al. (2011), phycoremediation is an algal-based remediation technology to eliminate or transform environmental contaminants, including nutrients and toxic substances from wastewater; CO_2 sequestration coupled with biomass production was also activated. Phycoremediation was employed widely for the remediation of various contaminants like organic and inorganic mineral elements, coliform bacteria, heavy metal ions, radionuclides, petroleum compounds, herbicides, pesticides, etc. from the wastewater (Raouf et al., 2012). Algae are aquatic, eukaryotic, or prokaryotic (blue green algae) photosynthetic organisms that can endure changes in temperature, salinity, pH, level of oxygen, and light intensities (Barsanti et al., 2008). The eco-friendliness, cost effectiveness, and easy methods of culturing signifies its prominence from other detoxification mechanisms and this inexpensive rehabilitation strategy now has great attention from researchers (Glick, 2010). Besides the remediation potential, algae from the remediation cultures are exploited for the biosynthesis of biofuels (biodiesel, bioethanol, and biomethanol). The microalgae with almost 50% lipid content and dry mass of 50 $g/m^2/day$ has the potentiality to produce about 10,000 gallons oil in an acre/year (Singh et al., 2016), indicating the dual role of microbes as the best candidates of bioremediation and as contributors of renewable energy sources (Brar et al., 2020).

3.4.1 Remediation Using Microalgae

Microalgae are fast-growing aquatic, photosynthetic, unicellular, or simple multicellular and microscopic organisms that rapidly increase their biomass by using solar energy, CO_2, and nutrients. The prokaryotic microalgae include cyanobacteria (blue green algae) and diatoms and some members of green algae are also found under the category of eukaryotic microalgae. The algal biomass consists of sugars, oils, lipids, and other carbon compounds. Because of the fast-growing biomass production and metabolites synthesised, they are widely used for the purposes of human and animal feed production, biofuel production, cosmetics, and pharmaceuticals (Mostafa, 2012). Besides this, microalgae are the best candidates for the production of bioactive compounds, antioxidants, and secondary metabolites. *Phaeocystis* sp., *Dunaliella* sp., *Spirulina*, *Anabaena*, *Nostoc*, and *Oscillatoria* produce a wide variety of bioactive compounds. Compounds like cyanovirin (inhibitor of replication) and cryptophycin (antiproliferative) are isolated from the cellular extract of *Nostoc* strains and glycoproteins from *Chlorella vulgaris* (anti cancerous) are industrially important (Li et al., 2007; Tyagi et al., 2010). The CO_2 fixation capability of the microalgae is much higher than terrestrial plants. So, the remediation using microalgae has ecological as well as biorefinery importance. Microalgae-based remediation is nowadays widely used for treating household sewage, manufacturing effluents, agro-chemical wastes, livestock wastes, and in the swine industry (Raouf et al., 2012). A list of algal candidates involved were represented in Table 3.2.

Microalgae-mediated remediation was exploited to clear polluted wastewater to reduce the toxic contaminants and it also mediated the absorption of minerals and other organic components. Microalgae is fast growing in nature and, because of this, they need a significant amount of nutrients like nitrogen, phosphorus, etc. for their growth, protein, and nucleic acid biosynthesis (Rao et al., 2011). Sydney et al.

TABLE 3.2

List of algal candidates involved in the bioremediation of contaminated water

Species Name	Type of Pollutant	Reference
Chlorella miniata	Ni	Wong et al. (2000)
Chlorella vulgaris BCC 15	Hg	Inthorn et al. (2002)
Laminaria japonica	Zn	Davis et al. (2003)
Chlorella fisca	Bisphenol A	Hirooka et al. (2003)
Laminaria japonica	Al	Lee et al. (2004)
Anabaena cylindrica	Cu	Tien et al. (2005)
Chlamydomonas reinhardtii	Hg	Tüzün et al. (2005)
Oscillatoria angustissima	Co	Mehta and Gaur (2005)
Enterobacter sp.	Pb and Cu	Lu et al. (2006)
Chlorella vulgaris	Fe	Romera et al. (2006)
Lessonia nigrescens	As	Hansen et al. (2006)
Prototheca zopfii	Aliphatic and polycyclic aromatic hydrocarbons	Ueno et al. (2006)
Spirulina sp.	Cr	Doshi et al. (2007)
Kappaphycus alvarezii	Cd and Co	Kumar et al. (2007)
Spirogyra sp.	Pb	Gupta and Rastogi (2008)
Skeletonema costatum	Phenanthrene	Hong et al. (2008)
Palmaria palmate	Cr	Murphy et al. (2008)
Fucusvesiculosus	Cr	Murphy et al. (2008)
Nitella pseudoflabellata	Cr	Gomes and Asaeda (2009)
Spirulina platensis	Cu	Celekli et al. (2010)
Oedogonium hatei	Ni	Gupta et al. (2010)
Maugeotia genuflexa	Ar	Luo and Xiao (2010)
Scenedesmus obliquus	Crude-oil degradation	Tang et al. (2010)
Ascophyllum nodosum	Cd	Vogel et al. (2010)
Chlorella sp.	PO_4^{3-}	Wang et al. (2010)
Chlamydomonas mexicana	Acephate	Kumar et al. (2011)
Cladophora sp.	Pb and Cu	Lee and Chang (2011)
Chlamydomonas reinhardtii	Fluroxypyr	Zhang et al. (2011)
Chlamydomonas reinhardtii	Isoproturon	Bi et al. (2012)
Monoraphidium braunii	Bisphenol	Gattullo et al. (2012)
Chlamydomonas reinhardtii	Prometryne	Jin et al. (2012)
Pediastrum tetras	Mesotrione	Moro et al. (2012)
Scenedesmus rubescens	NO_3^-, NO_2^-, and PO_4^{3-}	Su et al. (2012)
Spirogyra hyalina	Co	Kumar and Oommen (2012)
Ulva lactuca sp.	Cd	Lupea et al. (2012)
Cystoseira barbata	Cd, Ni, and Pb	Yalçın et al. (2012)
Sargassum bebanom	Cr	Javadian et al. (2013)
Chlorella sp.	NH_4^+	Udom et al. (2012)
Stoechospermum marginatum	Cr	Koutahzadeh et al. (2013)
Scenedesmus obliquus	Nonylphenol and octylphenol	Zhou et al. (2013)
Spirulina platensis	Cu	Al-Homaidan et al. (2014)
Chlamydomonas Mexicana and Chlorella vulgaris	Biophenol A	Ji et al. (2014)
Chlamydomonas mexicana	Atrazine	Kabra et al. (2014)

TABLE 3.2 (Continued)
List of algal candidates involved in the bioremediation of contaminated water

Species Name	Type of Pollutant	Reference
Microcystis sp. and *Scenedesmus quadricauda*	Mesotrione	Ni et al. (2014)
Sargassum muticum	Sb	Ungureanu et al. (2015)
Chlamydomonas sp.	NH_4^+	Ding et al. (2016)
Cladophora sp.	Cd	Sargın et al. (2016)
Pelvetia canaliculata, Fucus spiralis, Ascophyllum nodosum, and *Laminaria hyperborea*	Petrochemical wastewater	Cechinel et al. (2016)
Saccharina japonica and *Sargassum horneri*	Sr	Wang et al. (2018)
Sargassum vulgare	Fe	Benaisa et al. (2019)

(2011) have reported the potential application of microalgae for the removal of heavy metals, phosphorus, nitrogen, and ammonium from the wastewater. The effective CO_2 sequestration and ability to tolerate the harsh water environment help them to flourish in the treatment water and thereby reduce the toxic contamination (Zeng et al., 2012). Munoz and Guieysse (2006) reported that the photosynthetic algae release oxygen into the treatment system (wastewater) and help in the aeration and thereby reduce the BOD (biological oxygen demand) and COD (chemical oxygen demand).

Polyvalent metal ions are present in excess levels in the industrial effluents and the microalgae growing in it significantly reduces the polyvalent metal ions by binding with them (De-Bashan and Bashan, 2010). During the growth phase, the microalgae synthesise peptides and these peptides bind with the toxic heavy metals and form organic metal complexes and compartmentalise within the vacuoles and thereby decrease the negative impact of the heavy metals. According to studies of Lee and Chang (2011) on *Spirogyra* sp. and *Cladophora* sp., the candidates that have potent remediation potential for Pb and Cu are *Spirogyra* sp., as they showed higher adsorption capacity even when both species had the same types of functional groups on the cell surface.

3.4.2 Remediation Using Macroalgae

Macroalgae are fast-growing, photosynthetic algae, which are commonly referred to as seaweeds and are widely distributed and constitute numerous varieties and forms. The larger ones have complex structures and they all have a variety of economic importance. Most of them are used as food for human as well as fodder, and like microalgae they are also important in biofuel production, pharmaceuticals, mining, and bioremediation (Porse and Rudolph, 2017; Moutinho et al., 2018). Macroalgae are the major sources of many macro- and micronutrients important for the human diet and are also rich in bioactive compounds with antioxidant activities (Liu et al., 2017). The major three classes of macroalgae are green, brown, and red, named on the basis of the major pigment present in them. Recently, many scientists have been interested in applications of macroalgae for the decontamination of toxicants from water bodies, especially in sea water. Seaweed acts as chief biomonitors of heavy metals and are used in the remediation of Cd, Cu, Zn, and Pb (Chakraborty et al., 2014).

Luo et al. (2020) reported the remediation potential of seaweed such as *Gracilaria lemaneiformis, Ulva lactuca,* and *Sargassum horneri* against various heavy metals. Macroalgae species *Gracilaria edulis* and *Gracilaria changii* are used as biofilters with potential application in the remediation of contaminants like nitrate and ammonia from the shrimp wastewater. Mawi et al. (2020) concluded that this macroalgal biofilter is an ecologically sustainable solution for the remediation of wastewater from the shrimp industry. The toxic ions of metals, including Cu^{2+}, Cd^{2+}, Pb^{2+}, and Zn^{2+}, are bioremediated by using *Caulerpa lentillifera*. They synthesise and utilise some functional groups in response to toxic metal stress and absorb heavy metals through biosorption (Pavasant et al., 2006). The enhanced synthesis of phytochelatins are found in marine algae *Thalassiosira pseudonana* and *Thalassiosira weissflogii* in response to metal toxicity, and phytochelatins have the potential to bind the toxic metal ions and thereby reduce the toxicity (Ahner et al., 2002). The algal cell wall components are negatively charged and have an affinity to heavy metal cations and thereby adsorb the metals. The plant body of the algae produce the ligands such malate, citrate, phylate, histidine, etc. to tolerate and detoxify the pollutant present in their habitats (Sekabira et al., 2011). The brown macroalgae *Sargassum* sp. and *Padina* sp. have the ability to endure cations like Cd^{2+} and Cr^{3+} and anion $Cr_2O_7^{2-}$ (Sheng et al., 2007). Wang et al. (2018) concluded that the algae *Saccharina japonica* and *Sargassum horneri* have the potential to remediate the strontium (Sr) from nuclear polluted water. Because of the potent biosorption potential, it can be recommended as a biosorbent for Sr remediation. *Sargassum vulgare* was used for the removal of Fe^{3+} through biosorption. Its natural abundance, high metal binding capacity, and varied functional groups make it a perfect candidate of Fe^{3+} removal (Benaisa et al., 2019). The in vitro experiments with the economically important red algae elkhorn sea moss (*Kappaphycus alvarezii*) showed its tolerance toward the heavy metals Cd and Co. They efficiently remove about 3.064 mg/100 g FW and 3.365 mg/100 g FW Cd and Co from the growth medium (Kumar et al., 2007). The brown algae such as *Fucus spiralis, Ascophyllum nodosum,* and *Laminaria hyperborea* thrive and live in the petroleum contaminated water, and among these *L. hyperborea* showed a higher level of tolerance, so it can serve as a suitable candidate and as a natural cation exchanger in water contaminated with petrochemical wastes (Cechinel et al., 2016).

3.4.3 Mechanism of Phycoremediation

Mechanism of phycoremediation is a topic of serious research nowadays as the exact mechanism is not clear yet and it may vary even out from species to species. However, it has been proved that the cell wall receptor molecules bind to the toxic compound and thereby reduce the toxicity. Modulation of biochemical pathways and active uptake of the toxicant to their plant body, converting it into a nontoxic form are the major strategies operating in response to environmental polllutants (Gautam et al., 2014; Singh et al., 2017; Balzano et al., 2020). Algae utilise the mechanisms such as biosorption, biotransformation, biodegradation, phycovolatilisation, complexation, or bioaccumulation as modes of remediation strategies.

In the case of toxic metals, the algal cells (both living and non-living) make an initial non-metabolic approaches for handling the toxicant. These fast and reversible metal removal processes include metal complexation, adsorption, ion exchange, chemisorption, chelation with various agents, and diffusion. After the initial level of tolerance, metabolic processes are further modified in accordance with the pollutant and reduce its toxicity (Chabukdhara et al., 2017). Binding of pollutants with cell surface ligands, binding with algal metabolites, excretion of toxic compounds, volatilisation of metals, and enzymatic conversion of toxic pollutants to less toxic ones are the strategies adopted by the algae to withstand and survive in harsh living conditions (Monteiro et al., 2012). Algae, like plants, synthesise compounds that have the capacity to bind with toxic metal ions present in water. These compounds are synthesised in algal cells as a defensive mechanism. It includes metallothioneins (MTs), phytochelatins (PCs), and polyphosphate compounds. These finite mechanisms in algae help them to easily thrive in the toxic environmental contaminants in their habitats. So, development and research need to focus more on the bioremediation potential aspects of algae for the sustainable and economically feasible management of pollutants in water bodies.

3.5 Phytoremediation of Wastewater

Phytoremediation is considered a plant-associated strategy for the decontamination of xenobiotics from polluted areas (Table 3.3). Decontamination of wastewater using higher plants has many advantages, such as being less expensive, having no secondary consequences, and also is a sustainable approach. The plants used for phytoremediation must obviously be fast growing, have a high biomass production rate, are stress resistant, and easily harvestable. In phytoremediation, plants absorb pollutants through their roots and further translocate this contaminant to the metabolically active shoot system.

TABLE 3.3

List of plants utilised for the removal of pollutants from wastewater

Species Name	Type of Pollutant	Reference
Brassica juncea and *Helianthus annuus*	Cu, Cd, Cr, Ni, Pb, and Zn	Dushenkov et al. (1997)
Hydrilla verticillata	Chlordane	Chaudhry et al. (2002)
Azolla pinnata	Ni	Arora et al. (2004)
Eichhornia crassipes	Cu, Pb, Zn, and Cd	Liao and Chang (2004)
Schoenoplectus californicus	DDT and chlordane	Miglioranza et al. (2004)
Pista stratiotes, Spirodela intermedia, and *Lemna minor*	Fe, Cu, Zn, Mn, Cr, and Pb	Miretzky et al. (2004)
Lemna gibba	U and As	Mkandawire et al. (2004)
Ipomoea aquatica	Cd	Wang et al. (2008)
Eichhornia crassipes	Cr, Cd, and Pb	Agunbiade et al. (2009)
Salvinia natans	Cr	Dhir (2009)
Eichhornia crassipes	Cr and Zn	Mishra and Tripathi (2009)
Lemna minor	Flazasulfuron and dimethomorph	Olette et al. (2009)
Wolffia globosa	As	Zhang et al. (2009)
Hydrilla verticillata	As	Srivastava et al. (2010)
Pistia stratiotes	Chlorpyrifos	Prasertsup and Ariyakanon (2011)
Azolla filiculoides	As	Vesely et al. (2011)
Pistia stratiotes	Pb	Vesely et al. (2011)
Carex pendula	Pb	Yadav et al. (2011)
Potamogeton pusillus	Cu and Cr	Monferrán et al. (2012)
Iris pseudacorus	Chlorpyrifos	Wang et al. (2013)
Pistia stratiotes	Cr, Ni, and Pb	Abubakar et al. (2014)
Arundo donax	Cd and Zn	Dürešová et al. (2014)
Eichhornia crassipes	Cd, Cr, Cu, Zn, Fe, and B	Elias et al. (2014)
Pteris cretica and *Pteris vittata*	As	Farraji (2014)
Ceratophyllum submersum	Aldrin and endosulfan	Guo et al. (2014)
Phragmites communis	α-HCH and β-HCH	Guo et al. (2014)
Cyperus rotundus	Triazophos	Li et al. (2014)
Nymphaea amazonum	Cyhalothrin	Mahabali and Spanoghe (2014)
Phragmites australis and *Typha latifolia*	Al, Cu, Pb, Fe, Co, B, and Zn	Morari et al. (2015)
Phragmites australis	Cu, Pd, and Cd	Salman et al. (2015)
Potamogeton pectinatus	Cd and Cu	Salman et al. (2015)
Salix sp. and *Populus* sp.	C_2HCl_3 and C_2Cl_4	Limmer and Burken (2016)
Pistia stratiotes, Azolla pinnata, and *Salvinia molesta*	Fe, Cu and Mn	Manjunath and Kousar (2016)
Juncus effusus	Tebuconazole	Lv et al. (2016)
Acorus calamus	Chlorpyrifos	Wang et al. (2016)
Eichhornia crassipes, Ludwigia stolonifera, Echinochloa stagnina, and *Phragmites australis*	Cd, Ni, and Pb	Eid et al. (2020)
Ipomoea aquatica and *Centella asiatica*	Al and Fe	Hanafiah et al. (2020)
Typha latifolia and *Thelypteris palustris*	Zn and Cu	Hejna et al. (2020)
Helianthus sp.	Cr	Mallhi et al. (2020)
Cynodon dactylon, Chloris virgata, and *Desmostachya bipinnata*	Cr, Pb, and Cd	Mishra et al. (2020)

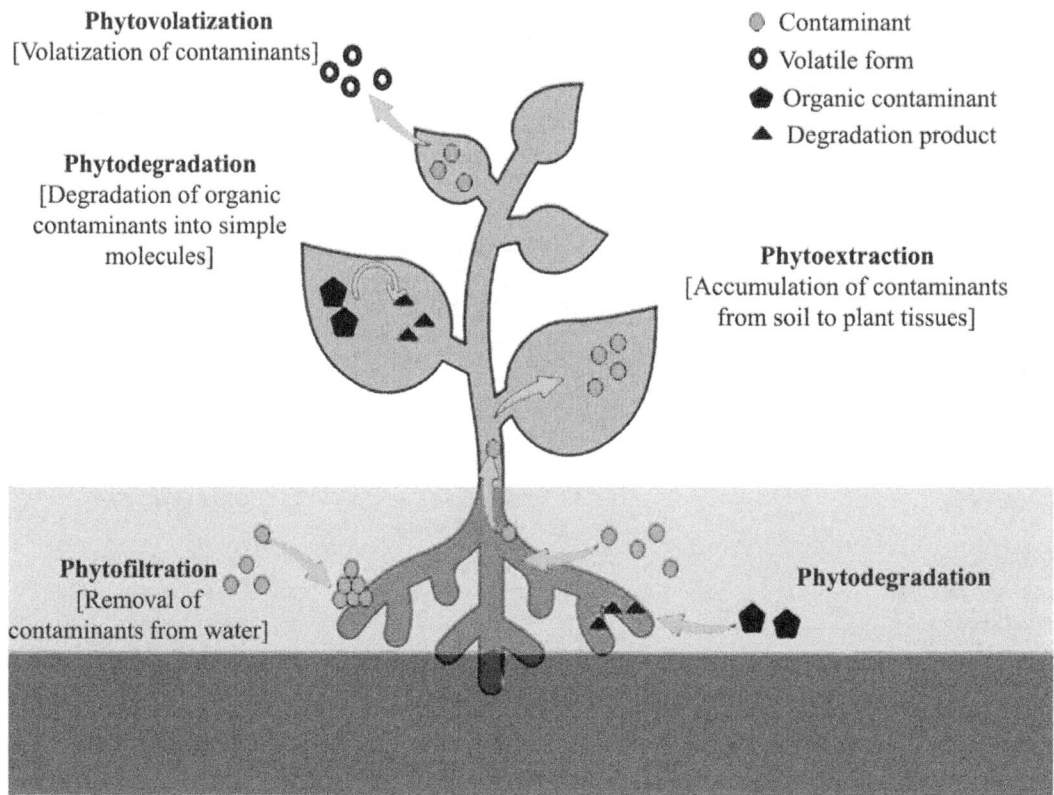

FIGURE 3.2 Mechanism of phytoremediation.

3.5.1 Mechanism of Phytoremediation

Phytoremediation of wastewater by plants involves phytoextraction, rhizofiltration, phytodegradation, and phytovolatilisation (Figure 3.2). Phytoextraction involves the absorption, accumulation, and translocation of xenobiotics through plant roots into the aboveground portions (shoots). A number of aquatic plants have been characterised as ideal candidates for phytoextraction of different contaminants. Aquatic plant species such as *Elodea densa*, *Sagittaria montevidensis*, *Salvinia auriculata*, *Pistia stratiotes*, and *Eichhornia crassipes* are used for the remediation of Hg from contaminated water reservoirs and *Salvinia herzogii* and *Pistia stratiotes* efficiently removed Cr from water bodies (Maine et al., 2004; Molisani et al., 2006). *Pistia stratiotes*, *Spirodela intermedia*, and *Lemna minor* were also used for the effective remediation of different toxic metal ions such as Fe, Cu, Zn, Mn, Cr, and Pb from contaminated water bodies (Miretzky et al., 2004). *Eichornia crassipes*, an important phytoremediation plant, significantly cleans up Cr, Cd, and Pb in shoots and roots (Agunbiade et al., 2009). *Eichhornia crassipes* is considered an ideal plant for reclamation of polluted water contaminated with Cu, Pb, Zn, and Cd (Liao and Chang, 2004; Mishra and Tripathi, 2009). *Lemna gibba*, a free-floating weed, is considered an excellent choice for the phytoaccumulation and phytoextraction of Zn (up to 25 mg/g dry matter) from contaminated water. They are fast-growing plants and show tolerance to cold, heat, and various other stresses (Khellaf and Zerdaoui, 2009). Likewise, *Ipomoea aquatica* is a promising plant to remediate heavy-metal-contaminated wastewater, and they bioaccumulate a significant amount of Cd, i.e., 375–2,227/kg for roots and 45–144/kg for shoots, respectively (Wang et al., 2008). *Potamogeton pusillus* accumulates a significant amount of Cu^{2+} and Cr^{6+} from polluted water bodies (Monferrán et al., 2012). Submerged aquatic plants such as *Ceratophylum demersum*, *Cabomba piauhyensis*, *Egeria densa*, *Myriophylum spicatum*, and *Hydrilla verticillata* were proposed to be used for the decontamination of contaminated sites (Abu Bakar et al., 2013; Pat-Espadas et al., 2018). *Phragmites australis* and

Typha latifolia were effectively used for the phytoremediation of urban wastewater and this can be effectively reutilised for the irrigation purposes. They showed a higher accumulation rate for heavy metals such as Al (96%), Cu (91%), Pb (88%), Fe (44%), Co (31%), B (40%), and Zn (85%) (Morari et al., 2015). *Pistia stratiotes, Azolla pinnata,* and *Salvinia molesta* were very much capable for the accumulation of Fe, Cu, and Mn from the textile effluent (Manjunath and Kousar, 2016). A list of plant candidates potentially involved in phytoremediation is presented in the Table 3.3.

Phytofiltration, or rhizofiltration, is the removal of pollutants by exploiting the potential of plant roots to uptake the xenobiotics from wastewater or surface water (Ahmadpour, 2011). Reclamation of wastewater through rhizofiltration or phytofiltartion has the capacity to entrap and filter toxicants such as metal ions and organic pollutants from polluted water resources. Rhizofiltration is used for the remediation of Pb, Cd, Cu, Ni, Zn, and Cr from contaminated water by aquatic or land plants. *Carex pendula* is a common wetland plant effectively utilised for in-situ remediation of Pb from polluted water resources (Yadav et al., 2011). Water hyacinth is effectively utilised for the removal of 70% of heavy metals such as Cd, Cr, Cu, Zn, Fe, and B in ceramic wastewater (Elias et al., 2014). *Pistia stratiotes* is regarded as an excellent rhizofiltrating agent and is able to cope up the high amount of contaminants, such as Cr, Ni, and Pb, without showing any visible symptoms (Abubakar et al., 2014). *Arundo donax* is appropriate for the decontamination of Cd and Zn from wastewater, even under conditions of continuous flow systems (Dürešová et al., 2014). Aquatic pteridophytes, such as *Pteris cretica* and *Pteris vittata,* are considered hyperaccumulators of arsenic (As) through rhizofiltration (Farraji, 2014). According to Dushenkov et al. (1997), *Brassica juncea* and *Helianthus annuus* efficiently removed different kinds of metals, such as Cu, Cd, Cr, Ni, Pb, and Zn, from aqueous solutions.

The phytodegradation technique is used for the treatment of river sediments, groundwater, and surface water. In this technique, the organic pollutants are degraded into simpler molecules by the enzymatic activity of plants and their associated microorganisms. Unlike microorganisms, which metabolize organic contaminants to carbon dioxide and water, phytodegradation depends on plant enzymes required to metabolize the contaminants completely into carbon dioxide and water. The aquatic plant *Azolla filiculoides* has been reported to have the potential for the removal of bisphenol A from aqueous solutions (Zazouli et al., 2014).

Phytovolatilisation is the conversion of contaminants into a volatile compound and its release into the atmosphere through plants' stomata. This mechanism is chiefly applied for the remediation of water pollutants, such as Se, Hg, and As, and organic compounds, such as benzene, nitrobenzene, phenol, and antrazine (Materac et al., 2015). Trichloroethylene and tetrachloroethylene, two major groundwater contaminants, were effectively phytovolatilised by plants such as willow (*Salix* sp.) and hybrid poplar (*Populus* sp.) (Limmer and Burken, 2016).

3.6 Future Prospectus and Conclusion

Bioremediation is regarded as a low cost and ecofriendly strategy that involves the use of various organisms to clean the polluted environments. Bioremediation of wastewater is a relatively new technology that has undergone more research in recent times. Anthropogenic activities release numerous toxic pollutants into the natural water bodies, polluting both freshwater resources as well as marine water. Different methods have been proposed for efficient wastewater treatment by using different bioagents, including microbes and higher plants. Bioremediation is a cost-effective method that does not entail superior infrastructures, and can be used to treat bulk amounts of polluted water. In developed countries, bioremediation has an important role in removing, transforming, or degrading various kinds of pollutants found in wastewater.

REFERENCES

Abu Bakar, A.F., Yusoff, I., Fatt, N.T., Othman, F., & Ashraf, M.A. (2013). Arsenic, zinc, and aluminium removal from gold mine wastewater effluents and accumulation by submerged aquatic plants (*Cabomba piauhyensis, Egeria densa,* and *Hydrilla verticillata*). *BioMed Research International, 2013,* 1–7.

Abubakar, M.M., Ahmad, M.M., & Getso, B.U. (2014). Rhizofiltration of heavy metals from eutrophic water using Pistia stratiotes in a controlled environment. *Journal of Environmental Science, Toxicology and Food Technology, 8*(6), 27–34.

Agunbiade, F.O., Olu-Owolabi, B.I., & Adebowale, K.O. (2009). Phytoremediation potential of Eichornia crassipes in metal-contaminated coastal water. *Bioresource Technology, 100*(19), 4521–4526.

Ahmadpour, P. (2011). *Evaluation of Four Plant Species for Phytoremediation of Cadmium-and Copper-contaminated Soil* (Doctoral dissertation, Universiti Putra Malaysia).

Ahner, B.A., Wei, L., Oleson, J.R., & Ogura, N. (2002). Glutathione and other low molecular weight thiols in marine phytoplankton under metal stress. *Marine Ecology Progress Series, 232*, 93–103.

Akhtar, N., & Mannan, M.A.U. (2020). Mycoremediation: expunging environmental pollutants. *Biotechnology Reports*, e00452.

Al-Daghistani, H. (2012). Bio-remediation of Cu, Ni and Cr from rotogravure wastewater using immobilized, dead, and live biomass of indigenous thermophilic Bacillus species. *The Internet Journal of Microbiology, 10*(1), 1–10.

Al-Homaidan, A.A., Al-Houri, H.J., Al-Hazzani, A.A., Elgaaly, G., & Moubayed, N.M. (2014). Biosorption of copper ions from aqueous solutions by Spirulina platensis biomass. *Arabian Journal of Chemistry, 7*(1), 57–62.

Arbanah, M., Miradatul Najwa, M.R., & Ku Halim, K.H. (2013). Utilization of Pleurotus ostreatus in the removal of Cr (VI) from chemical laboratory waste. *International Refereed Journal of Engineering and Science, 2*(4), 29–39.

Arıca, M.Y., & Bayramoğlu, G. (2005). Cr (VI) biosorption from aqueous solutions using free and immobilized biomass of Lentinus sajor-caju: preparation and kinetic characterization. *Colloids and Surfaces A: Physicochemical and Engineering Aspects, 253*(1-3), 203–211.

Arora, A.N.J.U., Sood, A.N.J.U.L.I., & Singh, P.K. (2004). Hyperaccumulation of cadmium and nickel by Azolla species. *Indian journal of plant physiology, 3*, 302–304.

Balzano, S., Sardo, A., Blasio, M., Chahine, T.B., Dell'Anno, F., Sansone, C., & Brunet, C. (2020). Microalgae metallothioneins and phytochelatins and their potential use in bioremediation. *Frontiers in Microbiology, 11*, 517.

Barsanti, L., Frassanito, A.M., Passarelli, V. and Gualtieri, P. (Eds.). (2008). *Algal Toxins: Nature, Occurrence, Effect and Detection*. Dordrecht, The Netherlands: Springer Science & Business Media.

Benaisa, S., Arhoun, B., Villen-Guzman, M., El Mail, R., & Rodriguez-Maroto, J.M. (2019). Immobilization of brown seaweeds Sargassum vulgare for Fe 3+ removal in batch and fixed-bed column. *Water, Air, and Soil Pollution, 230*(1), 19.

Bestawy, E.E., Helmy, S., Hussien, H., Fahmy, M., & Amer, R. (2013). Bioremediation of heavy metal-contaminated effluent using optimized activated sludge bacteria. *Applied Water Science, 3*(1), 181–192.

Bhattacharya, S., & Das, A. (2011). Mycoremediation of Congo red dye by filamentous fungi. *Brazilian Journal of Microbiology, 42*(4), 1526–1536.

Bi, Y.F., Miao, S.S., Lu, Y.C., Qiu, C.B., Zhou, Y., & Yang, H. (2012). Phytotoxicity, bioaccumulation and degradation of isoproturon in green algae. *Journal of Hazardous Materials, 243*, 242–249.

Brar, A., Kumar, M., Singh, R.P., Vivekanand, V., & Pareek, N. (2020). Phycoremediation coupled biomethane production employing sewage wastewater: energy balance and feasibility analysis. *Bioresource Technology*, 123292.

Cechinel, M.A., Mayer, D.A., Pozdniakova, T.A., Mazur, L.P., Boaventura, R.A., de Souza, A.A.U., de Souza, S.M.G.U., & Vilar, V.J. (2016). Removal of metal ions from a petrochemical wastewater using brown macro-algae as natural cation-exchangers. *Chemical Engineering Journal, 286*, 1–15.

Çelekli, A., Yavuzatmaca, M., & Bozkurt, H. (2010). An eco-friendly process: predictive modelling of copper adsorption from aqueous solution on Spirulina platensis. *Journal of Hazardous Materials, 173*(1–3), 123–129.

Çeribasi, I.H., & Yetis, U. (2001). Biosorption of Ni (II) and Pb (II) by Phanerochaete chrysosporium from a binary metal system–kinetics. *Water SA, 27*(1), 15–20.

Chabukdhara, M., Gupta, S.K., & Gogoi, M. (2017). Phycoremediation of heavy metals coupled with generation of bioenergy. In *Algal Biofuels*, 163–188. Cham: Springer.

Chakraborty, S., Bhattacharya, T., Singh, G., & Maity, J.P. (2014). Benthic macroalgae as biological indicators of heavy metal pollution in the marine environments: a biomonitoring approach for pollution assessment. *Ecotoxicology and Environmental Safety, 100*, 61–68.

Chaudhry, Q., Schröder, P., Werck-Reichhart, D., Grajek, W., & Marecik, R. (2002). Prospects and limitations of phytoremediation for the removal of persistent pesticides in the environment. *Environmental Science and Pollution Research*, 9(1), 4.

Coelho, L.M., Rezende, H.C., Coelho, L.M., de Sousa, P.A., Melo, D.F., & Coelho, N.M. (2015). Bioremediation of polluted waters using microorganisms. *Advances in Bioremediation of Wastewater and Polluted Soil*, 10, 60770.

Çolak, F., Atar, N., Yazıcıoğlu, D., & Olgun, A. (2011). Biosorption of lead from aqueous solutions by Bacillus strains possessing heavy-metal resistance. *Chemical Engineering Journal*, 173(2), 422–428.

Davis, T.A., Volesky, B., & Mucci, A. (2003). A review of the biochemistry of heavy metal biosorption by brown algae. *Water Research*, 37(18), 4311–4330.

De-Bashan, L.E., & Bashan, Y. (2010). Immobilized microalgae for removing pollutants: review of practical aspects. *Bioresource technology*, 101(6), 1611–1627.

Dhir, B. (2009). Salvinia: an aquatic fern with potential use in phytoremediation. *Environment and We An International Journal of Science and Technology*, 4, 23–27.

Ding, G.T., Yaakob, Z., Takriff, M.S., Salihon, J., & Abd Rahaman, M.S. (2016). Biomass production and nutrients removal by a newly-isolated microalgal strain Chlamydomonas sp in palm oil mill effluent (POME). *International Journal of Hydrogen Energy*, 41(8), 4888–4895.

Doshi, H., Ray, A., & Kothari, I.L. (2007). Bioremediation potential of live and dead Spirulina: spectroscopic, kinetics and SEM studies. *Biotechnology and Bioengineering*, 96(6), 1051–1063.

Driscoll, C.T., Whitall, D., Aber, J., Boyer, E., Castro, M., Cronan, C., Goodale, C.L., Groffman, P., Hopkinson, C., Lambert, K., & Lawrence, G. (2003). Nitrogen pollution in the northeastern United States: sources, effects, and management options. *BioScience*, 53(4), 357–374.

Dürešová, Z., Šuňovská, A., Horník, M., Pipíška, M., Gubišová, M., Gubiš, J., & Hostin, S. (2014). Rhizofiltration potential of Arundo donax for cadmium and zinc removal from contaminated wastewater. *Chemical Papers*, 68(11), 1452–1462.

Dursun, A.Y., Uslu, G., Tepe, O., Cuci, Y., & Ekiz, H.I. (2003). A comparative investigation on the bioaccumulation of heavy metal ions by growing Rhizopus arrhizus and Aspergillus niger. *Biochemical Engineering Journal*, 15(2), 87–92.

Dushenkov, S., Vasudev, D., Kapulnik, Y., Gleba, D., Fleisher, D., Ting, K.C., & Ensley, B. (1997). Removal of uranium from water using terrestrial plants. *Environmental Science & Technology*, 31(12), 3468–3474.

Eid, E.M., Galal, T.M., Sewelam, N.A., Talha, N.I., & Abdallah, S.M. (2020). Phytoremediation of heavy metals by four aquatic macrophytes and their potential use as contamination indicators: a comparative assessment. *Environmental Science and Pollution Research*, 1–14.

Elias, S.H., Mohamed, M.A.K.E.T.A.B., Ankur, A.N., Muda, K., Hassan, M.A.H.M., Othman, M.N., & Chelliapan, S. (2014). Water hyacinth bioremediation for ceramic industry wastewater treatment-application of rhizofiltration system. *Sains Malaysiana*, 43(9), 1397–1403.

Emparan, Q., Harun, R., & Danquah, M.K. (2019). Role of phycoremediation for nutrient removal from wastewaters: a review. *Applied Ecology and Environmental Research*, 17(1), 889–915.

Ezeronye, O.U., & Okerentugba, P.O. (1999). Performance and efficiency of a yeast biofilter for the treatment of a Nigerian fertilizer plant effluent. *World Journal of Microbiology and Biotechnology*, 15(4), 515–516.

Fan, T., Liu, Y., Feng, B., Zeng, G., Yang, C., Zhou, M., Zhou, H., Tan, Z., & Wang, X. (2008). Biosorption of cadmium (II), zinc (II) and lead (II) by Penicillium simplicissimum: isotherms, kinetics and thermodynamics. *Journal of Hazardous Materials*, 160(2-3), 655–661.

Farraji, H. (2014). Wastewater treatment by phytoremediation methods. *Wastewater Engineering: Advanced Wastewater Treatment Systems*, 194.

Gattullo, C.E., Bährs, H., Steinberg, C.E., & Loffredo, E. (2012). Removal of bisphenol A by the freshwater green alga Monoraphidium braunii and the role of natural organic matter. *Science of the Total Environment*, 416, 501–506.

Gautam, R.K., Sharma, S.K., Mahiya, S., & Chattopadhyaya, M.C. (2014). Contamination of heavy metals in aquatic media: transport, toxicity and technologies for remediation. In Sanjay Sharma (Ed.), *Water: Presence, Removal and Safety*, pp. 1–24. The Royal Society of Chemistry.

Glick, B.R. (2010). Using soil bacteria to facilitate phytoremediation. *Biotechnology Advances*, 28(3), 367–374.

Gomes, P.I., & Asaeda, T. (2009). Phycoremediation of Chromium (VI) by Nitella and impact of calcium encrustation. *Journal of Hazardous Materials, 166*(2-3), 1332–1338.

Green-Ruiz, C., Rodriguez-Tirado, V., & Gomez-Gil, B. (2008). Cadmium and zinc removal from aqueous solutions by Bacillus jeotgali: pH, salinity and temperature effects. *Bioresource Technology, 99*(9), 3864–3870.

Gulati, R., Saxena, R.K., & Gupta, R. (2002). Fermentation waste of Aspergillus terreus: a potential copper biosorbent. *World Journal of Microbiology and Biotechnology, 18*(5), 397–401.

Guo W, Zhang H, Huo S (2014) Organochlorine pesticides in aquatic hydrophyte tissues and surrounding sediments in Baiyangdian wetland, China. *Ecological Engineering, 67*, 150–155

Gupta, A., Phung, L.T., Chakravarty, L., & Silver, S. (1999). Mercury resistance in *Bacillus cereus* RC607: transcriptional organization and two new open reading frames. *Journal of Bacteriology, 181*(22), 7080–7086.

Gupta, V.K., & Rastogi, A. (2008). Biosorption of lead from aqueous solutions by green algae Spirogyra species: kinetics and equilibrium studies. *Journal of Hazardous Materials, 152*(1), 407–414.

Gupta, V.K., Rastogi, A., & Nayak, A. (2010). Biosorption of nickel onto treated alga (Oedogonium hatei): application of isotherm and kinetic models. *Journal of Colloid and Interface Science, 342*(2), 533–539.

Haijia, S., Ying, Z., Jia, L., & Tianwei, T. (2006). Biosorption of Ni^{2+} by the surface molecular imprinting adsorbent. *Process Biochemistry, 41*(6), 1422–1426.

Hanafiah, M.M., Zainuddin, M.F., Mohd Nizam, N.U., Halim, A.A., & Rasool, A. (2020). Phytoremediation of Aluminum and Iron from Industrial Wastewater Using Ipomoea aquatica and Centella asiatica. *Applied Sciences, 10*(9), 3064.

Hansen, H.K., Ribeiro, A., & Mateus, E. (2006). Biosorption of arsenic (V) with Lessonia nigrescens. *Minerals Engineering, 19*(5), 486–490.

Hejna, M., Moscatelli, A., Stroppa, N., Onelli, E., Pilu, S., Baldi, A., & Rossi, L. (2020). Bioaccumulation of heavy metals from wastewater through a Typha latifolia and Thelypteris palustris phytoremediation system. *Chemosphere, 241*, 125018.

Herrero, M., & Stuckey, D.C. (2015). Bioaugmentation and its application in wastewater treatment: a review. *Chemosphere, 140*, 119–128.

Hesnawi, R., Dahmani, K., Al-Swayah, A., Mohamed, S., & Mohammed, S.A. (2014). Biodegradation of municipal wastewater with local and commercial bacteria. *Procedia Engineering, 70*, 810–814.

Hirooka, T., Akiyama, Y., Tsuji, N., Nakamura, T., Nagase, H., Hirata, K., & Miyamoto, K. (2003). Removal of hazardous phenols by microalgae under photoautotrophic conditions. *Journal of Bioscience and Bioengineering, 95*(2), 200–203.

Hong, Y.W., Yuan, D.X., Lin, Q.M., & Yang, T.L. (2008). Accumulation and biodegradation of phenanthrene and fluoranthene by the algae enriched from a mangrove aquatic ecosystem. *Marine Pollution Bulletin, 56*(8), 1400–1405.

Huang, M.S., Pan, J., & Zheng, L.P. (2001). Removal of heavy metals from aqueous solutions using bacteria. *Journal of Shanghai University (English Edition), 5*(3), 253–259.

Inthorn, D., Sidtitoon, N., Silapanuntakul, S., & Incharoensakdi, A. (2002). Sorption of mercury, cadmium and lead by microalgae. *ScienceAsia, 28*(3), 253–261.

Iram, S., Shabbir, R., Zafar, H., & Javaid, M. (2015). Biosorption and bioaccumulation of copper and lead by heavy metal-resistant fungal isolates. *Arabian Journal for Science and Engineering, 40*(7), 1867–1873.

Javadian, H., Ahmadi, M., Ghiasvand, M., Kahrizi, S., & Katal, R. (2013). Removal of Cr (VI) by modified brown algae Sargassum bevanom from aqueous solution and industrial wastewater. *Journal of the Taiwan Institute of Chemical Engineers, 44*(6), 977–989.

Javaid, A., Bajwa, R., Shafique, U., & Anwar, J. (2011). Removal of heavy metals by adsorption on Pleurotus ostreatus. *Biomass and Bioenergy, 35*(5), 1675–1682.

Ji, M.K., Kabra, A.N., Choi, J., Hwang, J.H., Kim, J.R., Abou-Shanab, R.A., Oh, Y.K., & Jeon, B.H. (2014). Biodegradation of bisphenol A by the freshwater microalgae *Chlamydomonas mexicana* and *Chlorella vulgaris*. *Ecological Engineering, 73*, 260–269.

Jin, Z.P., Luo, K., Zhang, S., Zheng, Q., & Yang, H. (2012). Bioaccumulation and catabolism of prometryne in green algae. *Chemosphere, 87*(3), 278–284.

Joo, J.H., Hassan, S.H., & Oh, S.E. (2010). Comparative study of biosorption of Zn2+ by Pseudomonas aeruginosa and Bacillus cereus. *International Biodeterioration and Biodegradation, 64*(8), 734–741.

Kabra, A.N., Ji, M.K., Choi, J., Kim, J.R., Govindwar, S.P., & Jeon, B.H. (2014). Toxicity of atrazine and its bioaccumulation and biodegradation in a green microalga, Chlamydomonas mexicana. *Environmental Science and Pollution Research, 21*(21), 12270–12278.

Kanamarlapudi, S.L.R.K., Chintalpudi, V.K., & Muddada, S. (2018). Application of biosorption for removal of heavy metals from wastewater. *Biosorption, 18*, 69.

Khellaf, N., & Zerdaoui, M. (2009). Phytoaccumulation of zinc by the aquatic plant, Lemna gibba L. *Bioresource Technology, 100*(23), 6137–6140.

Kobayashi, H.A., Stenstrom, M., & Mah, R.A. (1983). Use of photosynthetic bacteria for hydrogen sulfide removal from anaerobic waste treatment effluent. *Water Research, 17*(5), 579–587.

Koutahzadeh, N., Daneshvar, E., Kousha, M., Sohrabi, M.S., & Bhatnagar, A. (2013). Biosorption of hexavalent chromium from aqueous solution by six brown macroalgae. *Desalination and Water Treatment, 51*(31-33), 6021–6030.

Ksheminska, H., Fedorovych, D., Babyak, L., Yanovych, D., Kaszycki, P., & Koloczek, H. (2005). Chromium (III) and (VI) tolerance and bioaccumulation in yeast: a survey of cellular chromium content in selected strains of representative genera. *Process Biochemistry, 40*(5), 1565–1572.

Kshirsagar, A.D. (2013). Application of bioremediation process for wastewater treatment using aquatic fungi. *International Journal of Current Research, 5*(7), 1737–1729.

Kumar Ramasamy, R., Congeevaram, S., & Thamaraiselvi, K. (2011). Evaluation of isolated fungal strain from e-waste recycling facility for effective sorption of toxic heavy metal Pb (II) ions and fungal protein molecular characterization—A mycoremediation approach. *Asian Journal of Experimental Biological Sciences, 2*, 342–347.

Kumar, J.N., & Oommen, C. (2012). Removal of heavy metals by biosorption using freshwater alga Spirogyra hyalina. *Journal of Environmental Biology, 33*(1), 27.

Kumar, K., Dasgupta, C.N., Nayak, B., Lindblad, P., & Das, D. (2011). Development of suitable photo-bioreactors for CO2 sequestration addressing global warming using green algae and cyanobacteria. *Bioresource Technology, 102*(8), 4945–4953.

Kumar, K.S., Ganesan, K., & Rao, P.S. (2007). Phycoremediation of heavy metals by the three-color forms of Kappaphycus alvarezii. *Journal of Hazardous Materials, 143*(1-2), 590–592.

Kumar, R., Negi, S., Sharma, P., Prasher, I.B., Chaudhary, S., Dhau, J.S., & Umar, A. (2018). Wastewater cleanup using Phlebia acerina fungi: an insight into mycoremediation. *Journal of Environmental Management, 228*, 130–139.

Lee, H.S., Suh, J.H., Kim, I.B., & Yoon, T. (2004). Effect of aluminum in two-metal biosorption by an algal biosorbent. *Minerals Engineering, 17*(4), 487–493.

Lee, Y.C., & Chang, S.P. (2011). The biosorption of heavy metals from aqueous solution by Spirogyra and Cladophora filamentous macroalgae. *Bioresource Technology, 102*(9), 5297–5304.

Li, H.B., Cheng, K.W., Wong, C.C., Fan, K.W., Chen, F., & Jiang, Y. (2007). Evaluation of antioxidant capacity and total phenolic content of different fractions of selected microalgae. *Food Chemistry, 102*(3), 771–776.

Li, Z., Xiao, H., Cheng, S., Zhang, L., Xie, X., & Wu, Z. (2014). A comparison on the phytoremediation ability of triazophos by different macrophytes. *Journal of Environmental Sciences, 26*(2), 315–322.

Liao, S., & Chang, W.L. (2004). Heavy metal phytoremediation by water hyacinth at constructed wetlands in Taiwan. *Photogrammetric Engineering and Remote Sensing, 54*, 177–185.

Limmer, M., & Burken, J. (2016). Phytovolatilization of organic contaminants. *Environmental Science and Technology, 50*(13), 6632–6643.

Liu, X., Yuan, W., & Meng, X. (2017). Extraction and quantification of phlorotannins from edible brown algae. *Transactions of the ASABE, 60*(1), 265–271.

Lu, W.B., Shi, J.J., Wang, C.H., & Chang, J.S. (2006). Biosorption of lead, copper and cadmium by an indigenous isolate Enterobacter sp. J1 possessing high heavy-metal resistance. *Journal of Hazardous Materials, 134*(1-3), 80–86.

Luo, H., Wang, Q., Liu, Z., Wang, S., Long, A., & Yang, Y. (2020). Potential bioremediation effects of seaweed Gracilaria lemaneiformis on heavy metals in coastal sediment from a typical mariculture zone. *Chemosphere, 245*, 125636.

Luo, J.M., & Xiao, X.I.A.O. (2010). Biosorption of cadmium (II) from aqueous solutions by industrial fungus Rhizopus cohnii. *Transactions of Nonferrous Metals Society of China, 20*(6), 1104–1111.

Lupea, M., Bulgariu, L., & Macoveanu, M. (2012). Biosorption of Cd (II) from aqueous solution on marine green algae biomass. *Environmental Engineering and Management Journal (EEMJ)*, *11*(3).

Lv, T., Zhang, Y., Casas, M.E., Carvalho, P.N., Arias, C.A., Bester, K., & Brix, H. (2016). Phytoremediation of imazalil and tebuconazole by four emergent wetland plant species in hydroponic medium. *Chemosphere*, *148*, 459–466.

Mahabali, S., & Spanoghe, P. (2014). Mitigation of two insecticides by wetland plants: feasibility study for the treatment of agricultural runoff in Suriname (South America). *Water, Air, and Soil Pollution*, *225*(1), 1771.

Maine, M.A., Suñé, N.L., & Lagger, S.C. (2004). Chromium bioaccumulation: comparison of the capacity of two floating aquatic macrophytes. *Water Research*, *38*(6), 1494–1501.

Mallhi, A.I., Chatha, S.A.S., Hussain, A.I., Rizwan, M., Bukhar, S.A.H., Hussain, A., Mallhi, Z.I., Ali, S., Hashem, A., Abd_Allah, E.F., & Alyemeni, M.N. (2020). Citric acid assisted phytoremediation of chromium through sunflower plants irrigated with tannery wastewater. *Plants*, *9*(3), 380.

Manjunath, S., & Kousar, H. (2016). Phytoremediation of Textile Industry Effluent using free floating macrophyte Azolla pinnata. *International Journal of Environmental Science*, *5*, 68–71.

Materac, M., Wyrwicka, A., & Sobiecka, E. (2015). Phytoremediation techniques of wastewater treatment. *Environmental Biotechnology*, *11*, 10–19.

Maulin, P.S. (2017). Environmental bioremediation of industrial effluent. *Journal of Molecular Biology and Biotechnology*, *2*(1), 1–3.

Mawi, S., Krishnan, S., Din, M.F.M., Arumugam, N., & Chelliapan, S. (2020). Bioremediation potential of macroalgae *Gracilaria edulis* and *Gracilaria changii* co-cultured with shrimp wastewater in an outdoor water recirculation system. *Environmental Technology and Innovation*, *17*, 100571.

Mehta, S.K., & Gaur, J.P. (2005). Use of algae for removing heavy metal ions from wastewater: progress and prospects. *Critical reviews in Biotechnology*, *25*(3), 113–152.

Michalska, J., Piński, A., Żur, J., & Mrozik, A. (2020). Selecting bacteria candidates for the bioaugmentation of activated sludge to improve the aerobic treatment of landfill leachate. *Water*, *12*(1), 140.

Miglioranza, K.S., de Moreno, J.E., & Moreno, V.J. (2004). Organochlorine pesticides sequestered in the aquatic macrophyte Schoenoplectus californicus (CA Meyer) Sojak from a shallow lake in Argentina. *Water Research*, *38*(7), 1765–1772.

Miretzky, P., Saralegui, A., & Cirelli, A.F. (2004). Aquatic macrophytes potential for the simultaneous removal of heavy metals (Buenos Aires, Argentina). *Chemosphere*, *57*(8), 997–1005.

Mishra, T., Pandey, V.C., Praveen, A., Singh, N.B., Singh, N., & Singh, D.P. (2020). Phytoremediation ability of naturally growing plant species on the electroplating wastewater-contaminated site. *Environmental Geochemistry and Health*, 1–11.

Mishra, V.K., & Tripathi, B.D. (2009). Accumulation of chromium and zinc from aqueous solutions using water hyacinth (*Eichhornia crassipes*). *Journal of Hazardous Materials*, *164*(2-3), 1059–1063.

Mkandawire, M., Taubert, B., & Dudel, E.G. (2004). Capacity of Lemna gibba L.(Duckweed) for uranium and arsenic phytoremediation in mine tailing waters. *International Journal of Phytoremediation*, *6*(4), 347–362.

Mohammad, P., Azarmidokht, H., Fatollah, M., & Mahboubeh, B. (2006). Application of response surface methodology for optimization of important parameters in decolorizing treated distillery wastewater using Aspergillus fumigatus UB2 60. *International Biodeterioration and Biodegradation*, *57*(4), 195–199.

Molisani, M.M., Rocha, R., Machado, W., Barreto, R.C., & Lacerda, L.D. (2006). Concentração de mercúrio em macrófitas aquáticas em duas represas no sistema Paraíba do Sul: Rio Guandu, SE do Brasil. *Brazilian Journal of Biology*, *66*, 101–107.

Monferrán, M.V., Pignata, M.L., & Wunderlin, D.A. (2012). Enhanced phytoextraction of chromium by the aquatic macrophyte Potamogeton pusillus in presence of copper. *Environmental Pollution*, *161*, 15–22.

Monteiro, C.M., Castro, P.M., & Malcata, F.X. (2012). Metal uptake by microalgae: underlying mechanisms and practical applications. *Biotechnology Progress*, *28*(2), 299–311.

Morari, F., Dal Ferro, N., & Cocco, E. (2015). Municipal wastewater treatment with Phragmites australis L. and Typha latifolia L. for irrigation reuse. Boron and heavy metals. *Water, Air, and Soil Pollution*, *226*(3), 56.

Moro, C., Bricheux, G., Portelli, C., & Bohatier, J. (2012). Comparative effects of the herbicides chlortoluron and mesotrione on freshwater microalgae. *Environmental Toxicology and Chemistry*, *31*(4), 778–786.

Mostafa, S.S. (2012). Microalgal biotechnology: prospects and applications. *Plant Science*, *12*, 276–314.

Moutinho, S., Linares, F., Rodríguez, J.L., Sousa, V., & Valente, L.M.P. (2018). Inclusion of 10% seaweed meal in diets for juvenile and on-growing life stages of Senegalese sole (Solea senegalensis). *Journal of Applied Phycology*, *30*(6), 3589–3601.

Munoz, R., & Guieysse, B. (2006). Algal–bacterial processes for the treatment of hazardous contaminants: a review. *Water Research*, *40*(15), 2799–2815.

Murphy, V., Hughes, H., & McLoughlin, P. (2008). Comparative study of chromium biosorption by red, green and brown seaweed biomass. *Chemosphere*, *70*(6), 1128–1134.

Naik, M.M., & Dubey, S.K. (2013). Lead resistant bacteria: lead resistance mechanisms, their applications in lead bioremediation and biomonitoring. *Ecotoxicology and Environmental Safety*, *98*, 1–7.

Ni, Y., Lai, J., Wan, J., & Chen, L. (2014). Photosynthetic responses and accumulation of mesotrione in two freshwater algae. *Environmental Science: Processes and Impacts*, *16*(10), 2288–2294.

Olette, R., Couderchet, M., & Eullaffroy, P. (2009). Phytoremediation of fungicides by aquatic macrophytes: toxicity and removal rate. *Ecotoxicology and Environmental Safety*, *72*(8), 2096–2101.

Özer, A., & Özer, D. (2003). Comparative study of the biosorption of Pb (II), Ni (II) and Cr (VI) ions onto S. cerevisiae: determination of biosorption heats. *Journal of Hazardous Materials*, *100*(1–3), 219–229.

Paisio, C.E., Quevedo, M.R., Talano, M.A., González, P.S., & Agostini, E. (2014). Application of two bacterial strains for wastewater bioremediation and assessment of phenolics biodegradation. *Environmental Technology*, *35*(14), 1802–1810.

Pat-Espadas, A.M., Loredo Portales, R., Amabilis-Sosa, L.E., Gómez, G., & Vidal, G. (2018). Review of constructed wetlands for acid mine drainage treatment. *Water*, *10*(11), 1685.

Pattanapipitpaisal, P., Brown, N., & Macaskie, L. (2001). Chromate reduction and 16S rRNA identification of bacteria isolated from a Cr (VI)-contaminated site. *Applied Microbiology and Biotechnology*, *57*(1-2), 257–261.

Pavasant, P., Apiratikul, R., Sungkhum, V., Suthiparinyanont, P., Wattanachira, S., & Marhaba, T.F. (2006). Biosorption of Cu^{2+}, Cd^{2+}, Pb^{2+}, and Zn^{2+} using dried marine green macroalga Caulerpa lentillifera. *Bioresource Technology*, *97*(18), 2321–2329.

Poi, G., Aburto-Medina, A., Mok, P.C., Ball, A.S., & Shahsavari, E. (2017). Bioremediation of phenol-contaminated industrial wastewater using a bacterial consortium—from laboratory to field. *Water, Air, and Soil Pollution*, *228*(3), p.89.

Porse, H., & Rudolph, B. (2017). The seaweed hydrocolloid industry: 2016 updates, requirements, and outlook. *Journal of Applied Phycology*, *29*(5), 2187–2200.

Prasertsup, P., & Ariyakanon, N. (2011). Removal of chlorpyrifos by water lettuce (*Pistia stratiotes* L.) and duckweed (*Lemna minor* L.). *International Journal of Phytoremediation*, *13*(4), 383–395.

Rani, N., Sangwan, P., Joshi, M., Sagar, A., & Bala, K. (2019). Microbes: a key player in industrial wastewater treatment. In *Microbial Wastewater Treatment*, 83–102. Elsevier.

Rao, P., Kumar, R.R., Raghavan, B.G., Subramanian, V.V., & Sivasubramanian, V. (2011). Application of phycoremediation technology in the treatment of wastewater from a leather-processing chemical manufacturing facility. *Water Sa*, *37*(1).

Raouf, N., Al-Homaidan, A.A., & Ibraheem, I.B.M. (2012). Microalgae and wastewater treatment. *Saudi Journal of Biological Sciences*, *19*(3), 257–275.

Rodríguez-Tirado, V., Green-Ruiz, C., & Gómez-Gil, B. (2012). Cu and Pb biosorption on Bacillus thioparans strain U3 in aqueous solution: kinetic and equilibrium studies. *Chemical Engineering Journal*, *181*, 352–359.

Romera, E., Gonzalez, F., Ballester, A., Blazquez, M.L., & Munoz, J.A. (2006). Biosorption with algae: a statistical review. *Critical Reviews in Biotechnology*, *26*(4), 223–235.

Saha, M., Sarkar, S., Sarkar, B., Sharma, B. K., Bhattacharjee, S., & Tribedi, P. (2016). Microbial siderophores and their potential applications: a review. *Environmental Science and Pollution Research*, *23*(5), 3984–3999.

Salman, J.M., Amrin, A.R., Hassan, F.M., & Jouda, S.A. (2015). Removal of congo red dye from aqueous solution by using natural materials. *Mesopotamia Environmental Journal*, *1*(3), 82–89.

Samarth, D.P., Chandekar, C.J., & Bhadekar, R.K. (2012). Biosorption of heavy metals from aqueous solution using Bacillus licheniformis. *International Journal of Pure and Applied Sciences and Technology*, *10*(2), 12.

Sankaran, S., Khanal, S.K., Jasti, N., Jin, B., Pometto III, A.L., & Van Leeuwen, J.H. (2010). Use of filamentous fungi for wastewater treatment and production of high value fungal byproducts: a review. *Critical Reviews in Environmental Science and Technology, 40*(5), 400–449.

Sargın, I., Arslan, G., & Kaya, M. (2016). Efficiency of chitosan–algal biomass composite microbeads at heavy metal removal. *Reactive and Functional Polymers, 98*, 38–47.

Say, R., Yilmaz, N., & Denizli, A. (2003). Removal of heavy metal ions using the fungus Penicillium canescens. *Adsorption Science and Technology, 21*(7), 643–650.

Sekabira, K., Origa, H.O., Basamba, T.A., Mutumba, G., & Kakudidi, E. (2011). Application of algae in biomonitoring and phytoextraction of heavy metals contamination in urban stream water. *International Journal of Environmental Science and Technology, 8*(1), 115–128.

Shahid, M.J., AL-surhanee, A.A., Kouadri, F., Ali, S., Nawaz, N., Afzal, M., Rizwan, M., Ali, B., & Soliman, M.H. (2020). Role of microorganisms in the remediation of wastewater in floating treatment wetlands: a review. *Sustainability, 12*(14), 5559.

Sheng, P.X., Ting, Y.P., & Chen, J.P. (2007). Biosorption of heavy metal ions (Pb, Cu, and Cd) from aqueous solutions by the marine alga Sargassum sp. in single-and multiple-metal systems. *Industrial and Engineering Chemistry Research, 46*(8), 2438–2444.

Silva, A., Delerue-Matos, C., Figueiredo, S.A., & Freitas, O.M. (2019). The use of algae and fungi for removal of pharmaceuticals by bioremediation and biosorption processes: a review. *Water, 11*(8), 1555.

Singh, A.K., Sharma, N., Farooqi, H., Abdin, M.Z., Mock, T., & Kumar, S. (2017). Phycoremediation of municipal wastewater by microalgae to produce biofuel. *International Journal of Phytoremediation, 19*(9), 805–812.

Singh, V., Tiwari, A., & Das, M. (2016). Phyco-remediation of industrial waste-water and flue gases with algal-diesel engenderment from micro-algae: a review. *Fuel, 173*, 90–97.

Srinath, T., Verma, T., Ramteke, P.W., & Garg, S.K. (2002). Chromium (VI) biosorption and bioaccumulation by chromate resistant bacteria. *Chemosphere, 48*(4), 427–435.

Srivastava, S., Mishra, S., Dwivedi, S., & Tripathi, R.D. (2010). Role of thiol metabolism in arsenic detoxification in Hydrilla verticillata (Lf) Royle. *Water, Air, and Soil Pollution, 212*(1-4), 155–165.

Su, Y., Mennerich, A., & Urban, B. (2012). Coupled nutrient removal and biomass production with mixed algal culture: impact of biotic and abiotic factors. *Bioresource Technology, 118*, 469–476.

Subbaiah, M.V., & Yun, Y.S. (2013). Biosorption of Nickel (II) from aqueous solution by the fungal mat of Trametes versicolor (rainbow) biomass: equilibrium, kinetics, and thermodynamic studies. *Biotechnology and Bioprocess Engineering, 18*(2), 280–288.

Subbaiah, M.V., Vijaya, Y., Reddy, A.S., Yuvaraja, G., & Krishnaiah, A. (2011). Equilibrium, kinetic and thermodynamic studies on the biosorption of Cu (II) onto Trametes versicolor biomass. *Desalination, 276*(1-3), 310–316.

Sydney, E.D., Da Silva, T.E., Tokarski, A., Novak, A.D., De Carvalho, J.C., Woiciecohwski, A.L., Larroche, C., & Soccol, C.R. (2011). Screening of microalgae with potential for biodiesel production and nutrient removal from treated domestic sewage. *Applied Energy, 88*(10), 3291–3294.

Tang X, He LY, Tao XQ, Dang Z, Guo CL, Lu GN, Yi XY (2010). Construction of an artificial microalgal-bacterial consortium that efficiently degrades crude oil. *Journal of Hazardous Materials, 181*, 158–162.

Tien, C.J., Sigee, D.C., & White, K.N. (2005). Copper adsorption kinetics of cultured algal cells and freshwater phytoplankton with emphasis on cell surface characteristics. *Journal of Applied Phycology, 17*(5), 379–389.

Tüzün, I., Bayramoğlu, G., Yalçın, E., Başaran, G., Celik, G., & Arıca, M.Y. (2005). Equilibrium and kinetic studies on biosorption of Hg (II), Cd (II) and Pb (II) ions onto microalgae Chlamydomonas reinhardtii. *Journal of Environmental Management, 77*(2), 85–92.

Tyagi, S., Singh, G., Sharma, A., & Aggarwal, G. (2010). Phytochemicals as candidate therapeutics: an overview. *International Journal of Pharmaceutical Sciences Review and Research, 3*, 53–55.

Udom, I., Halfhide, T., Gilliea, B., Dalrymple, O., Zaribaf, B.H., Zhang, Q., & Ergas, S.J. (2012). Harvesting Microalgae Grown on Wastewater. *Proceedings of the Water Environment Federation, 2012*(13), 3646-3658.

Ueno, R., Wada, S., & Urano, N. (2006). Synergetic effects of cell immobilization in polyurethane foam and use of thermotolerant strain on degradation of mixed hydrocarbon substrate by Prototheca zopfii. *Fisheries Science, 72*(5), 1027–1033.

Ungureanu, G., Santos, S., Boaventura, R., & Botelho, C. (2015). Biosorption of antimony by brown algae S. muticum AND A. nodosum. *Environmental Engineering and Management Journal (EEMJ)*, *14*(2).

Upadhyay, A.K., Singh, N.K., Bankoti, N.S., & Rai, U.N. (2017). Designing and construction of simulated constructed wetland for treatment of sewage containing metals. *Environmental Technology*, *38*(21), 2691–2699.

Vesely, M.D., Kershaw, M.H., Schreiber, R.D., & Smyth, MJ (2011). Natural innate and adaptive immunity to cancer. *The Annual Review of Immunology*, *29*, 235–271.

Veselý, T., Tlustoš, P., & Száková, J. (2011). The use of water lettuce (Pistia stratiotes L.) for rhizofiltration of a highly polluted solution by cadmium and lead. *International Journal of Phytoremediation*, *13*(9), 859–872.

Vijayaraghavan, K., & Yun, Y.S. (2008). Bacterial biosorbents and biosorption. *Biotechnology Advances*, *26*(3), 266–291.

Vogel, M., Günther, A., Rossberg, A., Li, B., Bernhard, G., & Raff, J. (2010). Biosorption of U (VI) by the green algae Chlorella vulgaris in dependence of pH value and cell activity. *Science of the Total Environment*, *409*(2), 384–395.

Wang, K.S., Huang, L.C., Lee, H.S., Chen, P.Y., & Chang, S.H. (2008). Phytoextraction of cadmium by Ipomoea aquatica (water spinach) in hydroponic solution: effects of cadmium speciation. *Chemosphere*, *72*(4), 666–672.

Wang, L., Min, M., Li, Y., Chen, P., Chen, Y., Liu, Y., Wang, Y., & Ruan, R. (2010). Cultivation of green algae Chlorella sp. in different wastewaters from municipal wastewater treatment plant. *Applied Biochemistry and Biotechnology*, *162*(4), 1174–1186.

Wang, Q., Yang, J., Li, C., Xiao, B., & Que, X. (2013). Influence of initial pesticide concentrations in water on chlorpyrifos toxicity and removal by Iris pseudacorus. *Water Science and Technology*, *67*(9), 1908–1915.

Wang, Q., Li, C., Zheng, R., & Que, X. (2016). Phytoremediation of chlorpyrifos in aqueous system by riverine macrophyte, Acorus calamus: toxicity and removal rate. *Environmental Science and Pollution Research*, *23*, 16241–16248.

Wang, X., Shan, T., & Pang, S. (2018). Phytoremediation potential of Saccharina japonica and Sargassum horneri (Phaeophyceae): biosorption study of strontium. *Bulletin of Environmental Contamination and Toxicology*, *101*(4), 501–505.

Wong, J.P.K., Wong, Y.S., & Tam, N.F.Y. (2000). Nickel biosorption by two chlorella species, C. Vulgaris (a commercial species) and C. Miniata (a local isolate). *Bioresource Technology*, *73*(2), 133–137.

Wu, M., Xu, Y., Ding, W., Li, Y., & Xu, H. (2016). Mycoremediation of manganese and phenanthrene by Pleurotus eryngii mycelium enhanced by Tween 80 and saponin. *Applied Microbiology and Biotechnology*, *100*(16), 7249–7261.

Yadav, B.K., Siebel, M.A., & van Bruggen, J.J. (2011). Rhizofiltration of a heavy metal (lead) containing wastewater using the wetland plant *Carex pendula*. *CLEAN–Soil, Air, Water*, *39*(5), 467–474.

Yadav, S., & Chandra, R. (2011). Heavy metals accumulation and ecophysiological effect on Typha angustifolia L. and Cyperus esculentus L. growing in distillery and tannery effluent polluted natural wetland site, Unnao, India. *Environmental Earth Sciences*, *62*(6), 1235–1243.

Yalçın, S., Sezer, S., & Apak, R. (2012). Characterization and lead (II), cadmium (II), nickel (II) biosorption of dried marine brown macro algae Cystoseirabarbata. *Environmental Science and Pollution Research*, *19*(8), 3118–3125.

Yilmazer, P., & Saracoglu, N. (2009). Bioaccumulation and biosorption of copper (II) and chromium (III) from aqueous solutions by Pichia stipitis yeast. *Journal of Chemical Technology and Biotechnology: International Research in Process, Environmental & Clean Technology*, *84*(4), 604–610.

Zazouli, M.A., Mahdavi, Y., Bazrafshan, E., & Balarak, D. (2014). Phytodegradatiopotential of bisphenolA from aqueous solution by Azolla Filiculoides. *Journal of Environmental Health Science and Engineering*, *12*(1), 66.

Zeng, X., Danquah, M.K., Zheng, C., Potumarthi, R., Chen, X.D., & Lu, Y. (2012). NaCS–PDMDAAC immobilized autotrophic cultivation of Chlorella sp. for wastewater nitrogen and phosphate removal. *Chemical Engineering Journal*, *187*, 185–192.

Zeng, Y., Cai, Y., Tan, Q., & Dai, C. (2020). An integrated modeling approach for identifying cost-effective strategies in controlling water pollution of urban watersheds. *Journal of Hydrology*, *581*, 124373.

Zhang, S., Qiu, C.B., Zhou, Y., Jin, Z.P., & Yang, H. (2011). Bioaccumulation and degradation of pesticide fluroxypyr are associated with toxic tolerance in green alga Chlamydomonas reinhardtii. *Ecotoxicology, 20*(2), 337–347.

Zhang, X., Zhao, F.J., Huang, Q., Williams, P.N., Sun, G.X., & Zhu, Y.G. (2009). Arsenic uptake and speciation in the rootless duckweed Wolffia globosa. *New Phytologist, 182*(2), 421–428.

Zhou, G.J., Peng, F.Q., Yang, B., & Ying, G.G. (2013). Cellular responses and bioremoval of nonylphenol and octylphenol in the freshwater green microalga Scenedesmus obliquus. *Ecotoxicology and Environmental Safety, 87*, 10–16.

4

Aquatic Plants Biosorbents for Remediation in the Case of Water Pollution as Future Prospectives

Jyoti Mehta[1], Moharana Choudhury[2], Anu Sharma[3], and Arghya Chakravorty[4]
[1]Department of Environmental Sciences, Central University of Jharkhand, Brambe, Ranchi, Jharkhand, India
[2]Voice of Environment (VoE), Guwahati, Assam, India
[3]Govt. Degree College, Bhaderwah, Doda, University of Jammu, Union Territory of J&K, India
[4]School of Biosciences and Technology, Vellore Institute of Technology, Vellore, India

CONTENTS

4.1 Introduction 71
    4.1.1 Indications of Substantial Metals in Plants and Animals 72
    4.1.2 Aquatic Plants Used as Biosorbents 73
4.2 Ordering of Aquatic Macrophytes 74
    4.2.1 Macrophytes 74
    4.2.2 Requirement for Phytoremediation Technology 75
        4.2.2.1 Phytoremediation 75
        4.2.2.2 System of Phytoremediation 75
    4.2.3 Ongoing Practices Applied for Heavy Metal Removal Treatment 75
    4.2.4 Essential Role of Phytoremediation and Aquatic Plants 76
4.3 Types of Aquatic Plants Generally Used in Phytoremediation 76
    4.3.1 Pistia stratiotes L. 76
    4.3.2 Salvinia auriculatais 77
    4.3.3 Eichhornia crassipes 78
    4.3.4 Duckweed 78
4.4 Application of Aquatic Plants in Phytoremediation for Inorganic Pollutants from Water 79
4.5 Application of Aquatic Plants for Organic Water Contaminants in Phytoremediation 79
4.6 Favourable Circumstances in Use of Aquatic Macrophytes 79
4.7 Possible Systems with Respect to the Phytoremediation of Ecological Contaminants 80
4.8 Advantages of Phytoremediation 81
4.9 Downsides or Restrictions and Phytoremediation Challenges 81
4.10 Conclusion and Future Projections 81
References 82

4.1 Introduction

The problem of environmental degradation has been an issue in the last few decades. The changing lifestyle, technology, and use of new and hazardous chemicals have been a precursor to the growing problem of pollution these days. In this scenario of degradation and uncertainty, we are searching for the solution to the pollution. The solution should be sustainable and in consonance with the design of nature and should be able to follow the principles of green chemistry. We have no room for the more toxic and hazardous products and by-products. So in these testing times bioremediation has emerged as one of the fields where we see a hope of solving the pollution issues in a sustainable manner. We have seen damages by inorganic and natural toxic substances in water bodies. The water quality is debased because of general population development, industrial development, sub-urbanisation, and excess use of common water assets (WHO and UNICEF, 2006; CDC, 2016; Enyoh et al., 2018). Polluted ocean conditions upset the whole aquatic environment, which changes the lifestyle of living creatures including plants, animals, and microbes. The slowdown effects consistently happen at the level of species. Water pollution happens mostly because of farming manures, everyday wastewaters, corrosive heavy wash, heavy metals, insecticides, oils, and numerous added non-living and natural synthetic mixes (Verla et al., 2018).

The "biosorption" function, a producer in the field, has the property of some biomolecules (or biomass types) to bind and concentrate fluid arrangements of selected particles or different atoms. The biosorption measure involves a solid biosorbent or sorbent stage and a fluid step (dissolvable, normal H_2O) containing a sorbent (sorbate, metal particle). Biosorbent materials have a metal separated character and can be used to decrease the grouping of large metal particles by the parts per million (ppm) to parts per billion (ppb) stage in arrangement. The point of interest in innovation is that it exists above ordinary ones and integrates its ease, advance competence, minimisation of compound or potentially organic ooze, no additional supplement requirement, biosorbent recovery, and possibility of adsorption-following metal improvement (Gadd, 1993). Water lettuce, water hyacinth, *Lemna minor*, and water spinach have a biosorbent capacity.

4.1.1 Indications of Substantial Metals in Plants and Animals

A collection of weighty metals is watched as often as possible in living animals because of their exceptionally responsive and penetrable nature that in the end causes irreversible harm to the strength of biota. Indeed, even a little fixation may cause harsh inconveniences to people just as the earth utilised X-beam fluorescence to decide the heavy metal fixation in water and soil tests gathered from the encompassing territories of some iron and steel workshops. A high metal focus was seen in the dirt gathered from the neighbouring regions of this workshop when contrasted with the removed area from the workshop. In this way, it tends to be gotten that expanding good ways from the contamination locales may have less adverse consequences for the natural frameworks or biota. Essentially, detailed substantial metal defilement was found in the dirt samples and these metals were moved from the debased soil to the wheat crop, which was developed close to the streets in Turkey. According to their discoveries, the degree of lead was recognised to be more prominent than normally endorsed for soils. Strangely, the degrees of Cu, Fe, Mn, and Ni were discovered to be higher in the wheat plants before washing when contrasted with those of the purged wheat plants. The more elevated level of metal pollution was ascribed to the weighty traffic load just as the impact of wind course. The gathering of metals by a few harvest plants included grain, clover, grapes, spinach, wheat, and so on. The examples were extricated from the consumable parts of these yields as these parts are devoured by people or other living creatures. Roots and leaves are the principle parts of any plant through which hefty metals get consumed from the sullied soils. Yet in another study, Byers et al. revealed that heavy metal substances in algal plants utilise compact vitality dispersive X-beam fluorescence (EDXRF) because this instrument can be utilised for on-location assurance of metallic substances in the examples. Also, nuclear ingestion spectrometry (AAS), inductively coupled plasma-nuclear emanation spectrometry (ICP-AES), inductively coupled plasma-nuclear mass spectrometry (ICP-MS), and frequency dispersive XRF (WD-XRF) are likewise a portion of the significant strategies for metal assurance in different tests. Fish also contain weighty metal pollutants. Scientists took note that types of fish found alongside Algerian coasts like sardine (*Sardina pilchardus*) and swordfish likewise had more significant levels of Cd, Pb, and Hg than recommended by specialists of Algeria and Europe (Figure 4.1).

Aquatic Plants as Biosorbents

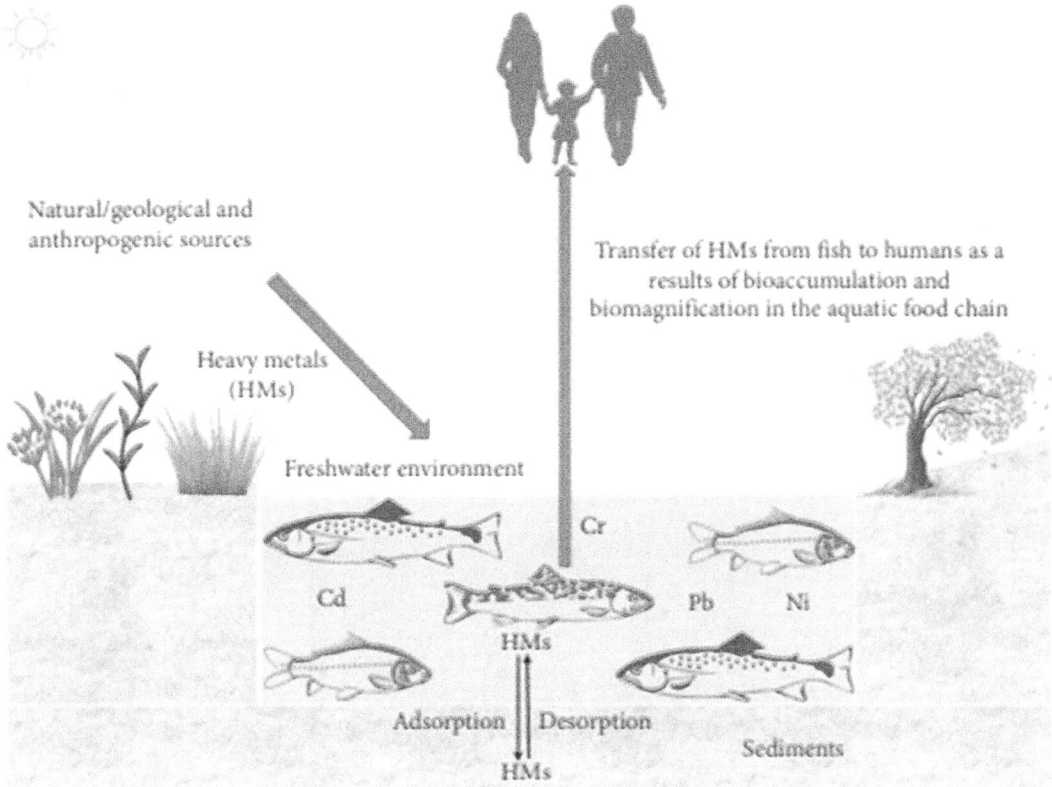

FIGURE 4.1 Trophic transfer of heavy metals from freshwater fish to humans in the human food chain used with permission of CC. BY. 4.0 License, Hindawi (Ali et al., 2019).

4.1.2 Aquatic Plants Used as Biosorbents

Aquatic floatables are characterised as plants swimming over the surface of water, generally through lowered roots. These water plants have been ideal for the treatment of water because they have a huge limit on the retention of supplements and other substances from the water, thus reducing the pollutant burden. It was further noted that the primary course of immense metal take-up in aquatic plants was through the roots. This is attributed to the growth of surface floater plants, while roots are involved in removing heavy metals and supplements in reduced plants in a similar manner as leaves (Dhote, 2007).

Wetland plants are also known to play a unique character in cleaning up the aquatic pollution through the absorption of pollutants contained in the water bodies. Each species plays a special part in the process of bioremediation. Wetland plant species – particularly free-gliding, lowered (established), semi-water floating/growing (established) species – recognise global significance as they demonstrate extreme capabilities in removing a range of contaminants, including heavy metals, radionuclides, explosives, as well as natural wastewater pollution (Wolf et al., 1991; Salt et al., 1995; Cardwell et al., 2002). A few examples of wetland plants assuming a crucial role in metal absorption from polluted water include *Wolffia globosa, Azolla pinnata, Elodea densa, Vallisneria spiralis,* and *Salvinia minima* (Figure 4.2).

FIGURE 4.2 Three free-floating plants (hyper-accumulator) in wastewater tanks. (a) Water hyacinth, (b) water lettuce, and (c) duckweed used with permission of Elsevier (Rezania et al., 2016).

4.2 Ordering of Aquatic Macrophytes

4.2.1 Macrophytes

The water plants (macrophytes) are ordinarily classified among seven divisions: *Rhodophyta, Cyanobacteria, Pteridophyta, Spermatophyta, Chlorophyta, Bryophyta,* and *Xanthophyta.*

On the basis of development design, oceanic plants are sorted as four significant gatherings. These include:

- Set I: These are called developing water plants. They establish in soil and the plant development transcends in the water. Instances of such plants include *Typha latifolia* and *Phragmites australis*.
- Set II: These are ordinarily called floating aquatic plants. These plant gatherings are generally situated on wet grounds and also originate at water profundities around 0.5–3.0 m. This set incorporates angiosperm plants; an example is *Potamogeton pectinatus*.
- Set III: These are comprised of lower aquatic plants, generally found beneath river surfaces. This set incorporates angiosperms, charophytes, pteridophytes, and mosses.
- Set IV: This set incorporates free-floating aquatic plants that are non-rooted. This gathering is exceptionally expanded in its living spaces and qualities. The water species in correlation with terrestrial plant species are comparatively much more appropriate for curing wastewaters. These macrophytes have numerous unmistakable physical features like development rate, and enormous creations of plant form, contamination take-up capacity, and better refining impacts because of a straight connection with contaminated water. These aquatic plants perform significant capacities

Aquatic Plants as Biosorbents 75

at basic and practical degrees of water biological systems. A portion of basic-level capacities are variations in water development and shelter to fish and different spineless creatures of aquatic water living spaces; furthermore they are a decent food source. The macrophytic plant species have the tendency to modify the nature of water by adjusting oxygen, supplement cycle, and weighty metal growth that is very useful.

These water plants can gather heavy metals. This trademark renders these plants more interesting for study, especially for the management of mechanical and domestic discarded water. The capability of water plants in phytoremediation is predominantly subject to acknowledgment of the plant types and distinction of metal take-up or capacity prospective for a similar weighty metal. In phytoremediation, a few ecological components ought to be kept up like substance species, beginning convergence of the metal, interface of various weighty metals, temperature, pH, redox potential, and saltiness. Phytotechnology utilises the floating macrophytes for treating water where various sorts of duckweeds and water hyacinths are being utilised. Plants along the root zone could likewise be utilised for treating small bulks of sewage water.

4.2.2 Requirement for Phytoremediation Technology

4.2.2.1 Phytoremediation

The enormous cost problem unlocked the door for creative innovation. As of late, the elimination of significant metals with the help of biological agents attracted a great deal of civic responsiveness and innovative work funding. Eccles (1999) has shown that, in comparison to conventional developments, organic steps for the removal of large metals are less costly by means of a money-saving advantage study. The extension of this discovery work has made phytoremediation environmentally friendly and financially sustainable engineering. The use of flora in metal remediation presents an appealing substitute as they are powered by the sun and can be achieved locally, minimising the costs and human deployment (Salt et al., 1995, 1998). The usage of plants in ecological sanitation is encouraged by phytoremediation. Phytoreconciliation is generally a modern way in which natural xenobiotics, heavy metals, and radionuclides pollute the waste water, groundwater, and soil. Although the term is defined in various ways, the use of plants to extract impurities from the earth can be summarised with a very simple definition (GardeaTorresdey, 2003).

4.2.2.2 System of Phytoremediation

Phytoremediation innovation includes phytoextraction, in which metal plant aggregations are used for the transfer of bulk metals through root media in the plants from the soil (Kumar et al., 1995). Rhizofiltration, where roots of the plants assimilate, absorb pollutants and lethal metals (Dushenkov et al., 1995). Another factor used to minimise pollutant motions in soil is phytostabilisation. This cycle uses the ability of plant roots to change their soil conditions; for example, pH and soil humidity. Several root exudates trigger metals, so bioavailability is reduced (Susarla et al., 2002).

4.2.3 Ongoing Practices Applied for Heavy Metal Removal Treatment

In reducing significant metal contamination by mechanical effluents, particulate exchange saps or electrodialysis are used. The particulate trading, chromatography component, and the capabilities of engineered particulate exchangers have been given due importance by writing on diagnostic science in the light of cross-connected styrene divinyl benzene copolymer to support significant metal particles (Pohl, 2006). Pascal et al. (2007) conceived yet another idea for the treatment of water wastes that was determined by neighbourhood and electrochemically regulated variations of pH in the polymer film by immediate treatment of toxins by dynamic surfaces. The isolation of the cached effluent in the pitches and the regeneration of the gum for further use still yield significant amounts of auxiliary effluents from washing the gums with compound reagents. In the electro deionising (EDI) cell, the layer stack was strengthened to prevent precipitation of bivalent metal hydroxide by the expulsion from profluent electric plating (Feng et al., 2007). A regular period in industries is compound precipitation. The conventional metal control strategies are based on a precipital concoction combined with pre- or post-oxidation and filtration to concentrate on the intrigue forms. The key drawback of these techniques is the production of solid deposits (oozes) with harmful aggravations, the last removal of which is by large sites. In addition (Mauchauffee and Meux, 2007), sodium was used, as opposed to calcium hydroxide, to minimise ooze calculations. Regardless of the growing number of hazardous water destinations poisoned in metals, the most commonly used methods to contaminate heavy metals remain an unbelievably exorbitant evacuation period. This financial approach is beneficial to the need for elective knowledge-based innovation. With evaluations obtained from various foundations, cleaning up hazardous squanders is expanded to cost $400 billion in the United States alone (Salt et al., 1995). Regular developments (e.g., concoction precipitation, sedimentation, particle trafficking, or mini-filtration) are not ecological, because the results of various reactions and related contamination of amphibian species pose genuine danger to oceanic existence.

4.2.4 Essential Role of Phytoremediation and Aquatic Plants

Use of aquatic flora in plant remediation covers a large, cost-efficient and environmentally friendly polluted area. As the characteristic safeguard for toxins and substantive metals, aquatic plants are used (Pratas et al., 2014). The most powerful and beneficial technique is to expel various sensitive metals alongside various pollutants using submerged plants (Ali et al., 2013; Guittonny-Philippe et al., 2015). Wetlands formed alongside oceanic plants are used worldwide for the treatment of wastewater (Fritioff and Greger, 2003; Gorito et al., 2017). The aquatic species of plants are known as an essential factor in the improvement of phytoremediation of heavy metal collection (Fritioff and Greger, 2003). Water plants have become increasingly popular over the long term because of their ability to clean up contaminated destinations worldwide (Gopal 2003). The water-submerged plants are continuously constructing large roots, making the plants the perfect choice in their underlying foundations and shoots for the selection of pollutants (Mays and Edwards, 2001; Stoltz and Greger, 2002). The production of aquatic plants and their growth are slow, which can limit phytoremediation's development interest (Said et al., 2015). In any experience, the amount of interest that creativity has on wastewater treatment (Syukor et al., 2014; Koźmińska et al., 2018) underlies this shortage (Figure 4.3).

4.3 Types of Aquatic Plants Generally Used in Phytoremediation

Water plants may be associated with phosphorus and nitrogen fixation in water and are therefore favoured for their advancement by high temperatures. When floating plants with these qualities, e.g., high critical quality, are routinely found in waters with moderate river speed and are limited by changes in water levels, the plant population of such species might soon disappear. The coastal macrophytes are organised in several categories such as *Vallisneria* sp., *Pistia stratiotes, Eichhornia crassipes, Lemna minor,* and *Salvinia auriculata.*

4.3.1 Pistia stratiotes L.

The name *Pistia stratiotes* L. is also known as water lettuce; the araceae/arum family belongs to it. The other basic names of these plants include water cabbage, water lettuce, jalkhumbhi, and shellflower and are found in reservoirs, streams, and wetlands (Quattrocchi, 2012). The *Pistia stratiotes* are light green with 20 cm in width and 10–20 gm in weight. The lower surface of the plant is whitish fur submerged under the leaves (Dipu et al., 2011) and it has a hanging structure. Water salt has outstanding resistance to large pH and temperature ranges (Lima et al., 2013). With the development of little girls' plants the expansion and multiplication of water salad takes place. *P. stratiotes* also offers water-based seeds, which germinate in the wet seasons (Acevedo and Nicolson, 2005). A magnificent rival for the plant

Aquatic Plants as Biosorbents

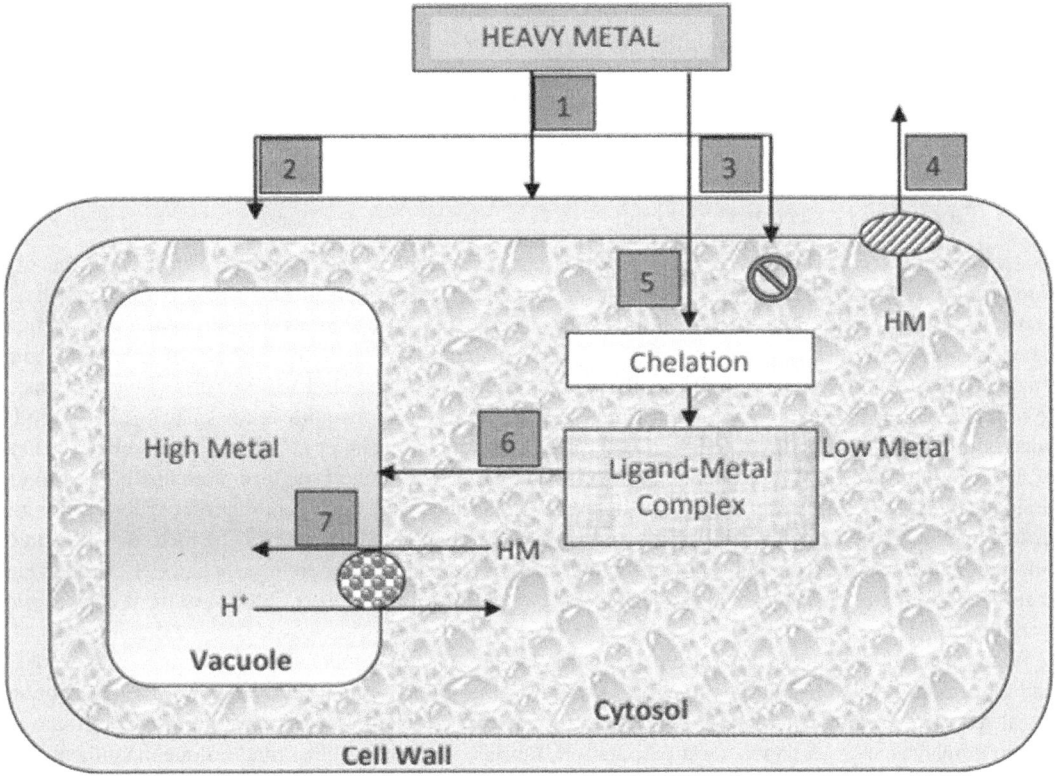

FIGURE 4.3 A cellular mechanism for detoxification and tolerance of heavy metals: (1) Restriction of metal movement to roots by mycorrhiza; (2) binding to cell wall and root exudates; (3) reduced influx by plasma membrane; (4) active efflux into apoplast; (5) chelation of metals by various ligands such as phytochelatins (PCs), metallothioneins (MTs), and acids in cytosol; (6) formation of ligand-metal complex; (7) transport of ligand-metal complex to the vacuole; and (8) transport and accumulation of metals in the vacuole used with permission of Elsevier (Yadav et al., 2018).

remediation of pollutants is water lettuce (*P. stratiotes*). It is more inclined than other waters (Forni et al., 2006; Yasar et al., 2018) For instance, natural oxygen requests (BOD), a requirement for substance oxygen (COD), oxygen broken down (DO), pH, all out of Kjeldahl Nitrogen (TKn), alkaline (NH3), nitrite ($NO2^{-00}$), nitrate ($NO3^-$) in addition, phosphate (PO_4^{3-}), drinking water, storm water, sewage water, and mechanical wastes. *P. stratiotes* appear different than hyacinth water and show that the limit for zinc and mercury from modern wastewater streams is best removed (Skinner et al., 2007). The Cu, Zn, Fe, Cr, and Cd take-up has no uncertain effect on the plant that enables *P. stratiotes* to be used as a plant for the calculation of natural pollutants by large-scale wastewater metals (Mishras and Tripathi, 2008). *P. stratiotes* biomass reduces the sullied arrangement of the investigation by over 70% of zinc and cadmium (Rodrigues et al., 2017).

4.3.2 Salvinia auriculatais

The tiny free-glide macrophyte, known as a water fern, is commonly dissipated in watery environments. It can easily reproduce and immediately resolve limitless settlements in zones. In a fair setting, *Salvinia* could duplicate its population within 3–5 days (Henry-Silva and Camargo, 2006). The use of *Salvinia* as a bio-pointer to the contamination file, as well as to phytoremedy, is upheld at a substantial rate of growth, easy management, and broad appropriateness and affectability to various toxic substances (Gardner and Al-Hamdani, 1997). Water distance species *(Salvinia)*, particularly *S. natans*, are thoughtfully used in pest control because they have an enormous threshold with respect to their evacuation due to the rapid rate of growth and resistance to toxins (Dhir et al., 2009, 2011). Furthermore, scaffolding created inside the built wetland has the capability to be applied in the management of various types of wastewater with success (Abd-Elnaby and Egorov, 2012). *Salvinia* practicalities have a higher rate of selection of metals. Phosphate convergence increases with sulphur expansion as weighty metal intake increases (Hoffmann et al., 2004).

4.3.3 Eichhornia crassipes

Water hyacinth (*Eichhornia crassipes*) is a free-coast water plant that is linked firmly with the lily family inthe Pontedericeae group. The most inescapable intrusive herb in the world is the water hyacinth. It has a big, dim, blue root structure with straight leaves. The roots include a stolen plant that develops new plants (Rezania et al., 2015). Water hyacinth is especially capable of producing and gradually eliminating toxins in highly polluted environments (Maine et al., 2001). It is gravely inclined to remedy various pollutants, like natural materials, heavy metals, absolute solids suspended; all solids and complements broken down (Kumar et al., 2010; Muthunarayanan et al., 2011; Yadav et al., 2011). Water hyacinth *(Eichhornia crassipes)* is recommended for treatment of modern drainage, home-grown drainage, wastewater effluents, and water-borne lakes as it has (1) a high assimilation capacity amongst different natural and inorganic pollutants and (2) can be maintained by a lot of interaction with water. Stronger capacities for remediation of pollutants such as arsenic, zinc, mercury, nickel, coffee, and lead that result from contemporary and domestic wastewater (Dixit and Tiwari, 2007) are from *Eichhornia crassipes*.

Water hyacinth has shown a high selection limit for various heavy metals such as copper, nickel, zinc, and chromium. It also has an advantage in providing the tiniest organic biosorbent spills that support metal recovery (Mahmood et al., 2010). The provision of colour pollutants in situations contribute to large companies such as paper, food preparation, fabrics, veal, maquillage, and colour manufacturing. The colour is usually clear and strongly opposed to oxidation experts, who essentially boost water pollution. Water hyacinths help to accumulate receptive colours sufficiently by the far-reaching resistance towards these colours (Shah et al., 2010; Priya and Selvan, 2017). Water hyacinths display core-clearing skills in various ongoing tests, including Cd, Pb, Cu, Zn, Fe, As, Mn, Cr, As, Al, and Hg (Shuvaeva et al., 2013; Aurangzeb et al., 2014; Fazal et al., 2015; Mahalakshmi et al., 2019). Cr and Cu were removed by 99.98%; 99.96% when introduced to tannery effluents by the water shooting powder hyacinth (Sarkar et al., 2017).

4.3.4 Duckweed

Duckweed is an open-skimming water plant on the coast of tranquil and quiet water. The plant is in the Araceae family and is nevertheless usually located in the Lemnoideae subfamily. This group consists of five genera: (1) *Wolffia*, (2) *Wolffiella*, (3) *Spirodela*, (4) *Lemna*, and (5) *Landoltia*, which have not been perceived by under 40 species (Les et al., 2002). Otherwise, duckweed is referred to as a water centre. It is found abundantly in trenchs, wetlands, and lakes; it is a smaller plant on earth and grows rapidly. They can be duly generated in a high pH (3.5–10.5) and 7–35°C (Radic et al., 2010). The capability of duckweed to generate colossal pH, temperature, and supplement levels in polluted sites makes them strong for use in plant regeneration. A broad variety of significant metals, inorganic and natural chemicals, pesticides, supplements, mechanical, and home-grown wastewater (Mkandawire & Dudel, 2007) can be wiped out by duckweed. Duckweed will certainly repress green growth and species in different lakes, as its large production rate allows it to cover lakes. Furthermore, it eliminates nitrogen from these lakes by scent of salt and denitrification (Alaerts et al., 1996). The reduction of the additional content by using duckweed biomass helps to update the essence of water and water corruption. Duckweed has a higher composite oxygen request for evacuation, organic oxygen request (BOD), all out of suspended wastewater fat (TSS). Various heavy metals have a higher end like As, Cr, Cu, Zn, Ag, Hg, Pb, and Cd, like *L. minor*, which has been accomplished by various forms of duckweed. Wastewater *Gibba* (Rahman and Hasegawa, 2011) (S. Obek and Sasmaz, 2011). Publicy, *L. minor* and *L. gibba* have

Aquatic Plants as Biosorbents

been tested for their efficiency of boron, arsenic, and uranium remediation. A good plant remediator for arsenic was explored and revealed to be *Spirodela polyrhiza* (Rahman et al., 2007). *L. gibba* has been observed to be sufficient in the treatment of boron with a minimum centralised concentration of 2 mg L^{-1} and no harm to biomass (Marín and Oron, 2007). Also it can incorporate uranium (120%), boron (40%), and arsenic (133%) (Sasmaz and Obek, 2009). *L. minor* has been found to be a brilliant arsenic cure competitor (Alvarado et al., 2008). Duckweed is the most appropriate plant for remediation when tested with different macrophytes. Past studies have included the use of duckweed for the remediation of mechanical and horticultural residue pollution and heavy metals (Mohedano et al., 2012; Zhang et al., 2014); and this stands documented. Several scientists have announced that duckweed, *L. minor*, for example nickel, manganese, zinc, plutonium, arsenic, and copper, will take up a massive convergence of hefty metals (Böcük et al., 2013). *Lemna minor* demonstrates an expansion of 6.1% chromium absorption, 26.5%, 20.5%, and 20.2% at the alternating chromium stress conversion.

4.4 Application of Aquatic Plants in Phytoremediation for Inorganic Pollutants from Water

The waters of hexavalent chromium affected (CrVI) are also used to fix plants (Saha et al., 2017). The *Lemna minor* may remove solvent Pb, Ni (Axtell et al., 2003), and Cr (VI) (Uysal, 2013). Furthermore, the aggregate potential of *L. minor* has been announced. A high level of metal convergences is obtained at marina roots and leaves of *Phragmites communis* and *Najas* (Baldantoni et al., 2004). The tiny aquatic plant, *Azolla caroliniana*, revealed that plants could potentially decontaminate water from mercury to chromium (Bennicelli et al., 2004). *Eichhornia crassipes* are used to boost water quality borders (Hussain et al., 2010; Shirinpur-Valadi et al., 2019). Outcomes of several explorations showed that hyacinth is a modest collector of cadmium and zinc.

4.5 Application of Aquatic Plants for Organic Water Contaminants in Phytoremediation

The aquatic plants have floating and movement activity in waters. They are having the capacity to remove pollutants from organic contaminants like these plants (*Lemna, Salvinia, Pisa,* and *Eichhornia*). *Myriopyllum picatum, Chara vulgaris,* and *Lemna minor* (Wu et al., 2010, Liu & Wu 2018, Zhou et al., 2018, Liu et al., 2019) were proposed as the means of suppressing alkyl benzene sulfonate. *Lemna gibba* is used in different metabolic processes; for example, in plants with BOD, suspended compounds, ammonium nitrate and phosphate, odourous salts, and total nitrogen (Ekperusi et al., 2019). Furthermore, *Lemna minor* (Yaseen and Scholz, 2017; Neag et al., 2018; Ekperusi et al., 2019) will increase quality of water polluted by blue and material colours.

4.6 Favourable Circumstances in Use of Aquatic Macrophytes

Though substantial metal absorption (Jackson, 1998; Keskinkan et al., 2003) and expulsion are of great benefit to aquatic macrophytes, the ability to remove nutrients from water bodies and to acclimate them into their organisms (Reddy and DeBusk, 1985; Tripathi and Shukla, 1991; Tripathi et al., 1991) is the most prominent advantage of plant remediation. Seagoing macrophytes with acclimatised supplements and significant metals from water bodies must periodically be obtained for legal administration of phytoremediation. In the future, enhancing progress towards turning overabundance or squandering biomass into substantive materials is a critical factor in encouraging the use of the treatment system. Innovative feed products such as fertilisers and methane gas have been transformed into major feed products and innovative feeds, although the innovations therein are currently not being realised. Amphibian macrophytes, like most plants on earth, build up sugar such as starch polymers like

TABLE 4.1

Aquatic plants' role in contaminant removal from water

S. No.	Name of Plants	Contaminant's Response	Reference
1.	*Ipomonea aquatica*	Rise in size of root fall in root length	
2.	*Eichhornia crassipes*	Compact growth and wilting and chlorosis observed reduced plant height and root length, death of the plant	
3.	*Echinodorus amazonicus*	Decline of growth in plant height and root length reduced, chlorosis	
4.	*Pistia stratiotes L.*	Decrease in the root volume, chlorosis, cell membrane damage, decline in growth rate, photosynthesis, rise in enzyme activity especially catalase (CAT), as superoxide dismutase (SOD), peroxidase (POX), and ascorbate peroxidase (APX)	
5.	*Lemna minor L.*	Compact photosynthetic efficiency, enzymatic activity, reduced chlorophyll, roots, and shoots growth	Radic et al. (2010)
6.	*Isoetes taiwanensis*	Reserve of root and shoot growth	Li et al. (2005)

cellulose, hemicellulose, and starch. Fermented sugars can quickly become ethanol, corrosion-corrosive lactic, and other significant ageing products in the event of starch hydrolysis, and thus provide a promising new alternative to the assets of biomass of seahorses. Using water hyacinth and water lettuce, sugars have been performed as hydrolysis enzymes for two aquatic plants ideal for water decontamination. (*Eichhornia crassipes*) Post-treatments were the best means by which sodium and hydrogen peroxide were used to enhance enzyme hydrolysis of water hyacinths and salad leaves with soluble/oxidant pretreatment. The decrease in sugar from the enzymatic hydrolysate in water leaves was around 1.8 times higher than water leaves in hyacinths; water leaves are therefore more attractive than water hyacinths in any respect. While some polysaccharides, for example, cellulose and hemicellulose, were present at these plant basements, monosaccharides were less than leaves (Mishima et al., 2006).

Inorganic feed additives suggest poor bioavailability and high toxicity. Natural feed supplements are now being used, including amino acid chelates. This dramatically increases bioavailability and decreases toxicity. These products are therefore also costly (cost is higher than inorganic salts). The multibasic ICP-OES Analysis analyses for minerals and feeds additional substances based on *Lemna minor*'s biomass, the formation period, and the adequacy of the biosorption and bioaccumulation periods of microelements by *L. minor* that proved capable of being animal feed were considered.

4.7 Possible Systems with Respect to the Phytoremediation of Ecological Contaminants

The small maintenance and higher biomass could thrive well in unfavourable circumstances (Obinna and Ebere, 2019). In comparison to non-accumulators, the hyper-accumulators accumulate heavy metals such as Cd, Zn, Co, Ni, Pb, and Mn, which may reach 100–1,000 times. The bioavailable portion of a metal element is extended to include preparatory parts, which often play a vital part in dispensing natural pollutants, like microorganisms such as parasites and microscopic organisms in near association to roots, for example, rhizosphere. Any hypothesis to explain the potential water treatment frameworks is now being proposed. Water macrophytes change the physical state of the body in water bodies. Other amphibians will decrease the amount of disintegrated CO_2 in water in high photosynthetic motion in the presence of other autotrophs. Aquatic plants preserve pollutants and store their biomass (Obinna and Ebere 2019). This increases the disintegration of oxygen into the water that activates expanded water pH.

4.8 Advantages of Phytoremediation

As a characteristic cycle, phytoremediation offers numerous advantages:

a. The cycle does not upset the nearby condition and keeps up the scene.
b. Most helpful at undipped and lower-level polluted locales.
c. A wide-ranging assortment of natural pollutants are be able to be dealt with.
d. The thought is tastefully acceptable and has community adequacy. It is reasonable for those regions where different strategies are not substantial. It is financially more practical than the other remediation methods. Phytoremediation has less upkeep and establishment costs in contrast with different methods.
e. The estate on polluted soils can forestall metal draining and disintegration. Quick development and enormous biomass-creating plants can likewise be utilised for vitality creation.
f. Reuse and recuperation of important metals can be done using the process of phytoremediation.
g. These cause the least impact to nature and encompassing individuals.

4.9 Downsides or Restrictions and Phytoremediation Challenges

Despite the fact that phytoremediation is a situation agreeable cycle, it has negative angles also:

i. It is nearly a drawn-out remediation measure.
ii. It can have a poisonous impact to nature by moving toxins from the water or ground to searching creatures.
iii. High metal pollution is likely to be unsafe to the flora; however, a few animal groups are exceptionally effectual in evacuation of harmfulness.
iv. Due to buildup of poisonous metals and foreign substances, plants are found to be destructive to animals and the overall population so there ought to be confined admittance to the site.
v. It is not appropriate for exceptionally despoiled lands since plant life can aggregate low to direct degrees of pollutants from water and soil region.
vi. Phytoremediation is an environmentally friendly technique. It includes numerous challenges later on, like improvement of nearby limits and to build up compelling administrative approaches. There is an absence of involvement utilising phytoremediation, less accessible information, execution principles, and cost-benefit examination.

4.10 Conclusion and Future Projections

With much explained, we can say that, we need to explore all the possibilities to ensure the water bodies are cleaned with natural agents rather than using chemicals or other mechanical means that either leave the harmful by-products or prove physically or genetically harmful for the aquatic flora and fauna. A significant function for the survival of life has been played by water. The use of amphibious macrophytes may be a financially informed bioremediation strategy for the treatment of defiled water. The feasible remedial structures enable plants to grow rapidly and collect more biomass. Some plants are more resilient and can be seen as strong resistance to the transfer of contaminants to the food networks, particularly the specific types of wild aquatic weeds. The audit concerns various parts of the application of aquatic plants to the phytoremediation of various inorganic and natural seagoing pollutants. In plant remediation, these aquatic plants played important roles in eliminating toxins from water. The aquatic plants for the treatment and management of defiled water bodies are highly recommended. In the aquaculture industry, or in field experiments, for example, the most popular applications for amphibian macrophytes (rising, decreased, and free streaming) were created. The phytoremediation frameworks include various aquatic plants like water hyacinth, water lettuce etc., as they are universal in nature and possess extraordinary bioaccumulation power, an invasive mechanism, and an irregular regenerative limit (Ekperusi et al., 2019). The managers like water plants used in the remediation are another limitation in terms of danger to the aquatic frameworks, as an aggressive species. The oceanic macrophytes should also be regarded as a colossal problem as obtrusive organisms, close to their value in the phytoremediation industry. Managerial approaches to monitor the interference of oceanic macrophytes during phytoremedation must refer to the mechanical, physical, organic, and concoction processes. Ability and amassing of heavy metals by seaside macrophytes are handled by the complete mix, transport, and chelator exercises. Hereditary nature improves the aggregation and durability of plants, which demonstrates that it works exceptionally well to enhance phytoremediation adequacy. Diverse progress has been evaluated in plants on the atomic level, which favours transgenic techniques in order to pursue the conversion of metals into plants. Hereditary plants display a high degree of resistance and metal consumption and, consequently, in earthly plants quality control was affected. The hereditary building of amphibian plants is under consideration to upgrade their significant metal consumption cap. Later, biomass removal can act as a source for the production of biogas and also can be used as a feed for livestock. In addition, after filtration, the use of water plants is not compulsory, as in other conventional methods, and can be easily recycled for the use of large quantities of contaminated water and soil. The benefits of aquatic plants in contaminant treatment in current research are enormous. Thus, we find that the process of phytoremediation as a particular case of bioremediation has a great potential to act as the solution to cleaning the contaminated water bodies. The plant species present in the water bodies act as the pollution degraders in the water bodies. The plant species used can be either the indigenous ones or can be introduced depending upon the nature of the pollutants to be degraded. Therefore, aquatic plants as the agents of phytoremediation have tremendous potential for cleaning the contaminated sites and act against various toxic substances.

REFERENCES

Abd-Elnaby, A.M., & Egorov, M.A. (2012). Efficiency of different particle sizes of dried *Salvinia natans* in the removing of Cu (II) and oil pollutions from water. *Journal of Water Chemistry and Technology*, *34*(3), 143–146.

Acevedo, R.P., & Nicolson, D.H. (2005). *Araceae. Contributions from the US National Herbarium*, *52*, 44.

Akinbile, C.O., & Yusoff, M.S. (2012). Assessing water hyacinth (*Eichhornia crassipes*) and lettuce (*Pistia stratiotes*) effectiveness in aquaculture wastewater treatment. *International Journal of phytoremediation*, *14*(3), 201–211.

Alaerts, G.J., Mahbubar, R., & Kelderman, P. (1996). Performance analysis of a full-scale duckweed-covered sewage lagoon. *Water Research*, *30*(4), 843–852.

Ali, H., Khan, E., & Ilahi, I. (2019). Environmental chemistry and ecotoxicology of hazardous heavy metals: environmental persistence, toxicity, and bioaccumulation. *Journal of Chemistry*, *2019*.

Alvarado, S., Guedez, M., Lue-Meru, M.P., Nelson, G., Alvaro, A., Jesus, A.C., & Gyula, Z. (2008). Arsenic removal from waters by bioremediation with the aquatic plants Water Hyacinth (*Eichhornia crassipes*) and Lesser Duckweed (*Lemna minor*). *Bioresource Technology*, *99*(17), 8436–8440.

Aurangzeb, N., Nisa, S., Bibi, Y., Javed, F., & Hussain, F. (2014). Phytoremediation potential of aquatic herbs from steel foundry effluent. *Brazilian Journal of Chemical Engineering*, *31*(4), 881–886.

Axtell, N.R., Sternberg, S.P., & Claussen, K. (2003). Lead and nickel removal using Microspora and *Lemna minor*. *Bioresource Technology*, *89*(1), 41–48.

Baldantoni, D., Alfani, A., Di Tommasi, P., Bartoli, G., & De Santo, A.V. (2004). Assessment of macro and microelement accumulation capability of two aquatic plants. *Environmental Pollution*, *130*(2), 149–156.

Bennicelli, R., Stępniewska, Z., Banach, A., Szajnocha, K., & Ostrowski, J. (2004). The ability of *Azolla caroliniana* to remove heavy metals (Hg (II), Cr (III), Cr (VI)) from municipal waste water. *Chemosphere*, *55*(1), 141–146.

Bhaskaran, K., Nadaraja, A.V., Tumbath, S., Shah, L.B., & Veetil, P.G.P. (2013). Phytoremediation of perchlorate by free floating macrophytes. *Journal of Hazardous Materials*, *260*, 901–906.

Böcük, H., Yakar, A., & Türker, O.C. (2013). Assessment of *Lemna gibba* L. (duckweed) as a potential ecological indicator for contaminated aquatic ecosystem by boron mine effluent. *Ecological Indicators*, 29, 538–548.
Cardwell, A.J., Hawker, D.W., & Greenway, M. (2002). Metal accumulation in aquatic macrophytes from southeast Queensland, Australia. *Chemosphere*, 48(7), 653–663.
CDC. (2016). Global WASH Fast Facts. Global Water, Sanitation, & Hygiene (WASH). https://www.cdc.gov/healthywater/global/wash_statistics.html Accessed 30/7/2019.
Chen, G., Huang, J., Fang, Y., Zhao, Y., Tian, X., Jin, Y., & Zhao, H. (2019). Microbial community succession and pollutants removal of a novel carriers enhanced duckweed treatment system for rural wastewater in Dianchi Lake basin. *Bioresource Technology*, 276, 8–17.
Chen, J.C., Wang, K.S., Chen, H., Lu, C.Y., Huang, L.C., Li, H.C.,... & Chang, S.H. (2010). Phytoremediation of Cr (III) by *Ipomonea aquatica* (water spinach) from water in the presence of EDTA and chloride: effects of Cr speciation. *Bioresource Technology*, 101(9), 3033–3039.
Daud, M.K., Ali, S., Abbas, Z., Zaheer, I.E., Riaz, M.A., Malik, A. & Zhu, S.J. (2018). Potential of duckweed (*Lemna minor*) for the phytoremediation of landfill leachate. *Journal of Chemistry*, 2018.
Dhir, B. (2009). *Salvinia*: an aquatic fern with potential use in phytoremediation. *Environment & We an International Journal of Science & Technology*, 4, 23–27.
Dhir, B., Sharmila, P., & Saradhi, P.P. (2009). Potential of aquatic macrophytes for removing contaminants from the environment. *Critical Reviews in Environmental Science and Technology*, 39(9), 754–781.
Dhir, B., Sharmila, P., Saradhi, P.P., Sharma, S., Kumar, R., & Mehta, D. (2011). Heavy metal induced physiological alterations in *Salvinia natans*. *Ecotoxicology and Environmental Safety*, 74(6), 1678–1684.
Dhote, S. (2007). Role of Macrophytes in improving water quality of an aquatic eco-system. *Journal of Applied Sciences and Environmental Management*, 11(4), 133–135.
Dipu, S., Kumar, A.A., and Thanga, V.S.G. (2011). Phytoremediation of dairy effluent by constructed wetland technology. *The Environmentalist*, 31(3), 263–278.
Dixit, S., & Tiwari, S. (2007). Effective utilization of an aquatic weed in an eco-friendly treatment of polluted water bodies. *Journal of Applied Sciences and Environmental Management*, 11(3).
Dushenkov, V., Kumar, P.B.A.N., Motto, H., & Raskin I. (1995). Rhizofiltration – the use of plants to remove heavy metals from aqueous streams. *Environmental Science and Technology*, 29(5), 1239–1245.
Eccles, H. (1999). Treatment of metal-contaminated wastes: why select a biological process? *Trends in Biotechnology*, 17, 462–465.
Ekperusi, A.O., Sikoki, F.D., & Nwachukwu, E.O. (2019). Application of common duckweed (*Lemna minor*) in phytoremediation of chemicals in the environment: state and future perspective. *Chemosphere*, 223, 285–309.
Enyoh, C.E., Verla, A.W., & Egejuru, N.J. (2018). pH variations and chemometric assessment of borehole water in Orji, Owerri Imo State, Nigeria. *Journal of Environmental Analytical Chemistry*, 5(2), 1–9.
Erdei, L. (2005). Phytoremediation as a program for decontamination of heavy-metal polluted environment. *Acta Biologica Szegediensis*, 49(1-2), 75–76.
Fazal, S., Zhang, B., & Mehmood, Q. (2015). Biological treatment of combined industrial wastewater. *Ecological Engineering*, 84, 551–558.
Feng, X., Wu, Z., & Chen, X. (2007). Removal of metal ions from electroplating effluent by EDI process and recycle of purified water. *Separation and Purification Technology*, 57(2), 257–263.
Forni, C., Patrizi, C., & Migliore, C. (2006). Floating aquatic macrophytes as a decontamination tool for antimicrobial drugs. In *Soil and Water Pollution Monitoring, Protection and Remediation*, 467–477. Dordrecht: Springer.
Fritioff, A., & Greger, M. (2003). Aquatic and terrestrial plant species with potential to remove heavy metals from storm water. *International Journal of Phytoremediation*, 5(3), 211–224.
Gadd, G.M. (1993). Interactions of fungi with toxic metals. In *The Genus Aspergillus*, 361–374. Boston, MA: Springer.
Galal, T.M., Eid, E.M., Dakhil, M.A., & Hassan, L.M. (2018). Bioaccumulation and rhizofiltration potential of *Pistia stratiotes* L. for mitigating water pollution in the Egyptian wetlands. *International Journal of Phytoremediation*, 20(5), 440–447.
GardeaTorresdey, J.L. (2003). Phytoremediation: where does it stand and where will it go? *Environmental Progress*, 22(1), A2–A3.

Gardner, J.L., & Al-Hamdani, S.H. (1997). Interactive effects of aluminum and humic substances on *Salvinia*. *Journal of Aquatic Plant Management*, *35*, 30–34.

Gonzales-Gustavson, E., Rusinol, M., Medema, G., Calvo, M., & Girones, R. (2019). Quantitative risk assessment of norovirus and adenovirus for the use of reclaimed water to irrigate lettuce in Catalonia. *Water Research*, *153*, 91–99.

Gopal, B. (2003). Perspectives on wetland science, application and policy. *Hydrobiologia*, *490*(1-3), 1–10.

Gorito, A.M., Ribeiro, A.R., Almeida, C.M.R., & Silva, A.M. (2017). A review on the application of constructed wetlands for the removal of priority substances and contaminants of emerging concern listed in recently launched EU legislation. *Environmental Pollution*, *227*, 428–443.

Guittonny-Philippe, A., Petit, M.E., Masotti, V., Monnier, Y., Malleret, L., Coulomb, B. & Laffont-Schwob, I. (2015). Selection of wild macrophytes for use in constructed wetlands for phytoremediation of contaminant mixtures. *Journal of Environmental Management*, *147*, 108–123.

Hadi, F., Ali, N., & Ahmad, A. (2014). Enhanced phytoremediation of cadmium-contaminated soil by *Parthenium hysterophorus* plant: effect of gibberellic acid (GA3) and synthetic chelator, alone and in combinations. *Bioremediation Journal*, *18*(1), 46–55.

Henry-Silva, G.G., & Camargo, A.F.M. (2006). Efficiency of aquatic macrophytes to treat Nile tilapia pond effluents. *Scientia Agricola*, *63*(5), 433–438.

Hoffmann, T., Kutter, C., & Santamaria, J. (2004). Capacity of Salvinia minima Baker to tolerate and accumulate As and Pb. *Engineering in Life Sciences*, *4*(1), 61–65.

Hussain, S.T., Mahmood, T., & Malik, S.A. (2010). Phytoremediation technologies for Ni++ by water hyacinth. *African Journal of Biotechnology*, *9*(50), 8648–8660.

Jafari, N. (2010). Ecological and socio-economic utilization of water hyacinth (*Eichhornia crassipes* Mart Solms). *Journal of Applied Sciences and Environmental Management*, *14*(2).

Koźmińska, A., Wiszniewska, A., Hanus-Fajerska, E., & Muszyńska, E. (2018). Recent strategies of increasing metal tolerance and phytoremediation potential using genetic transformation of plants. *Plant Biotechnology Reports*, *12*(1), 1–14.

Krishna, K.B., & Polprasert, C. (2008). An integrated kinetic model for organic and nutrient removal by duckweed-based wastewater treatment (DUBWAT) system. *Ecological Engineering*, *34*(3), 243–250.

Kumar, H.S., Bose, A., Raut, A., Sahu, S.K., & Raju, M.B.V. (2010). Evaluation of Anthelmintic Activity of *Pistia stratiotes* Linn. *Journal of Basic and Clinical Pharmacy*, *1*(2), 103.

Kumar, H.S., Bose, A., Raut, A., Sahu, S.K., & Raju, M.B.V. (2010). Evaluation of Anthelmintic Activity of *Pistia stratiotes* Linn. *Journal of Basic and Clinical Pharmacy*, *1*(2), 103.

Kumar, P.B.A.N., Dushenkov, V., Motto, H., & Raskin, I. (1995). Phytoextraction – the use of plants to remove heavy metals from soils. *Environmental Science and Technology*, *29*, 1232–1238.

Leão, G.A., de Oliveira, J.A., Felipe, R.T.A., Farnese, F.S., & Gusman, G.S. (2014). Anthocyanins, thiols, and antioxidant scavenging enzymes are involved in Lemna gibba tolerance to arsenic. *Journal of Plant Interactions*, *9*(1), 143–151.

Les, D.H., Crawford, D.J., Landolt, E., Gabel, J.D., & Kimball, R.T. (2002). Phylogeny and systematics of Lemnaceae, the duckweed family. *Systematic Botany*, *27*(2), 221–240.

Li, H., Cheng, F., Wang, A., & Wu, T. (2005, September). Cadmium removal from water by hydrophytes and its toxic effects. In *Proc. of the International Symposium of Phytoremediation and Ecosystem Health*.

Lima, L.K., Pelosi, B.T., da Silva, M.G.C., & Vieira, M.G. (2013). Lead and chromium biosorption by *Pistia stratiotes* biomass. *Chemical Engineering Transactions*, *32*, 1045–1050.

Liu, N., & Wu, Z. (2018). Toxic effects of linear alkylbenzene sulfonate on *Chara vulgaris* L. *Environmental Science and Pollution Research*, *25*(5), 4934–4941.

Liu, Y., Liu, N., Zhou, Y., Wang, F., Zhang, Y., & Wu, Z. (2019). Growth and Physiological Responses in *Myriophyllum spicatum* L. Exposed to Linear Alkylbenzene Sulfonate. *Environmental Toxicology and Chemistry*, *38*(9), 2073–2081.

Mahalakshmi, R., Sivapragasam, C., & Vanitha, S. (2019). Comparison of BOD 5 Removal in Water Hyacinth and Duckweed by Genetic Programming. In *Information and Communication Technology for Intelligent Systems*, 401–408. Singapore: Springer.

Mahamadi, C. (2011). Water hyacinth as a biosorbent: a review. *African Journal of Environmental Science and Technology*, *5*(13), 1137–1145.

Mahmood, T., Malik, S.A., & Hussain, S.T. (2010). Biosorption and recovery of heavy metals from aqueous solutions by *Eichhornia crassipes* (water hyacinth) ash. *BioResources*, *5*(2), 1244–1256.

Maine, M.A., Duarte, M.V., & Sune, N.L. (2001). Cadmium uptake by floating macrophytes. *Water Research*, *35*(11), 2629–2634.

Marín, C.M.D.C., & Oron, G. (2007). Boron removal by the duckweed *Lemna gibba*: a potential method for the remediation of boron-polluted waters. *Water Research*, *41*(20), 4579–4584.

Mauchauffee, S., & Meux, E. (2007). Use of sodium decanoate for selective precipitation of metals contained in industrial wastewater. *Chemosphere*, 69, 763–768.

Mays, P.A., & Edwards, G.S. (2001). Comparison of heavy metal accumulation in a natural wetland and constructed wetlands receiving acid mine drainage. *Ecological Engineering*, *16*(4), 487–500.

Mesa, J., Mateos-Naranjo, E., Caviedes, M.A., Redondo-Gómez, S., Pajuelo, E., & Rodríguez-Llorente, I.D. (2015). Scouting contaminated estuaries: heavy metal resistant and plant growth promoting rhizobacteria in the native metal rhizoaccumulator Spartina maritima. *Marine Pollution Bulletin*, *90*(1–2), 150–159.

Mishra, V.K., & Tripathi, B.D. (2008). Concurrent removal and accumulation of heavy metals by the three aquatic macrophytes. *Bioresource Technology*, *99*(15), 7091–7097.

Mkandawire, M., & Dudel, E.G. (2007). Are *Lemna* spp. effective phytoremediation agents. *Bioremediation, Biodiversity and Bioavailability*, *1*(1), 56–71.

Mohedano, R.A., Costa, R.H., Tavares, F.A., & Belli Filho, P. (2012). High nutrient removal rate from swine wastes and protein biomass production by full-scale duckweed ponds. *Bioresource Technology*, *112*, 98–104.

Muthunarayanan, V., Santhiya, M., Swabna, V., & Geetha, A. (2011). Phytodegradation of textile dyes by water hyacinth (*Eichhornia crassipes*) from aqueous dye solutions. *International Journal of Environmental Sciences*, *1*(7), 1702–1717.

Neag, E., Malschi, D., & Maicaneanu, A. (2018). Isotherm and kinetic modelling of toluidine blue (TB) removal from aqueous solution using *Lemna minor*. *International Journal of Phytoremediation*, *20*(10), 1049–1054.

Obek, E., & Sasmaz, A. (2011). Bioaccumulation of aluminum by *Lemna gibba* L. from secondary treated municipal wastewater effluents. *Bulletin of Environmental Contamination and Toxicology*, *86*(2), 217–220.

Obinna, I.B., Ebere, E.C. (2019). Phytoremediation of polluted water bodies with aquatic plants: recent progress on heavy metal and organic pollutants. *Analytical Methods in Environmental Chemistry Journal*. 10.20944/preprints201909.0020.v1.

Pascal, V., Laetitia, D., Joel, L., Marc, S., & Serge, P. (2007). New concept to remove heavy metals from liquid waste based on electrochemical pH-switchable immobilized ligands. *Applied Surface Science*, *253*(6), 3263–3269.

Pohl, P. (2006). Application of ion-exchange resins to the fractionation of metals in water. *TrAC Trends in Analytical Chemistry*, *25*(1), 31–43.

Polomski, R.F., Taylor, M.D., Bielenberg, D.G., Bridges, W.C., Klaine, S.J., & Whitwell, T. (2009). Nitrogen and phosphorus remediation by three floating aquatic macrophytes in greenhouse-based laboratory-scale subsurface constructed wetlands. *Water, Air, and Soil Pollution*, *197*(1-4), 223–232.

Pratas, J., Paulo, C., Favas, P.J., & Venkatachalam, P. (2014). Potential of aquatic plants for phytofiltration of uranium-contaminated waters in laboratory conditions. *Ecological Engineering*, *69*, 170–176.

Priya, E.S., & Selvan, P.S. (2017). Water hyacinth (*Eichhornia crassipes*)–An efficient and economic adsorbent for textile effluent treatment–A review. *Arabian Journal of Chemistry*, *10*, S3548–S3558.

Quattrocchi, U. (2012). *CRC World Dictionary of Medicinal and Poisonous Plants: Common Names, Scientific Names, Eponyms, Synonyms, and Etymology (5 Volume Set)*. United States: CRC Press.

Radic, S., Babic, M., Skobic, D., Roje, V., & Pevalek-Kozlina, B. (2010). Ecotoxicological effects of aluminum and zinc on growth and antioxidants in *Lemna minor* L. *Ecotoxicology and Environmental Safety*, *73*(3), 336–342.

Radic, S., Stipanicev, D., Cvjetko, P., Mikelic, I.L., Rajcic, M.M., Sirac, S. & Pavlica, M. (2010). Ecotoxicological assessment of industrial effluent using duckweed (*Lemna minor* L.) as a test organism. *Ecotoxicology*, *19*(1), 216.

Rahman, M.A., & Hasegawa, H. (2011). Aquatic arsenic: phytoremediation using floating macrophytes. *Chemosphere*, *83*(5), 633–646.

Rahman, M.A., Hasegawa, H., Ueda, K., Maki, T., Okumura, C., & Rahman, M.M. (2007). Arsenic accumulation in duckweed (*Spirodela polyrhiza* L.): a good option for phytoremediation. *Chemosphere*, *69*(3), 493–499.

Reddy, K.R., & DeBusk, W.F. (1985). Nutrient removal potential of selected aquatic macrophytes. *Journal of Environmental Quality*, 14, 459–462.

Rezania, S., Din, M.F.M., Taib, S.M., Dahalan, F.A., Songip, A.R., Singh, L., & Kamyab, H. (2016). The efficient role of aquatic plant (water hyacinth) in treating domestic wastewater in continuous system. *International Journal of Phytoremediation*, *18*(7), 679–685.

Rezania, S., Ponraj, M., Din, M.F.M., Songip, A.R., Sairan, F.M., & Chelliapan, S. (2015). The diverse applications of water hyacinth with main focus on sustainable energy and production for new era: an overview. *Renewable and Sustainable Energy Reviews*, *41*, 943–954.

Rezania, S., Taib, S.M., Din, M.F.M., Dahalan, F.A., & Kamyab, H. (2016). Comprehensive review on phytotechnology: heavy metals removal by diverse aquatic plants species from wastewater. *Journal of Hazardous Materials*, *318*, 587–599.

Rodrigues, A.C.D., do Amaral Sobrinho, N.M.B., dos Santos, F.S., dos Santos, A.M., Pereira, A.C.C., & Lima, E.S.A. (2017). Biosorption of toxic metals by water lettuce (*Pistia stratiotes*) biomass. *Water, Air, & Soil Pollution*, *228*(4), 156.

Saha, P., Shinde, O., & Sarkar, S. (2017). Phytoremediation of industrial mines wastewater using water hyacinth. *International Journal of Phytoremediation*, *19*(1), 87–96.

Said, M., Cassayre, L., Dirion, J.L., Nzihou, A., & Joulia, X. (2015). Behavior of heavy metals during gasification of phytoextraction plants: thermochemical modelling. In *Computer Aided Chemical Engineering* (Vol. 37), 341–346). Netherlands: Elsevier.

Salt, D.E., Blaylock, M., Kumar, P.B.A.N., Dushenkov, V., Ensley, D., Chet, I., & Raskin, I. (1995). Phytoremediation: a novel strategy for the removal of toxic elements from the environment using plants. *Bio/Technology*, 13, 468–474.

Salt, D.E., Smith, R.D., & Raskin, I. (1998). Phytoremediation. *Annual Review of Plant Physiology and Plant Molecular Biology*, 49, 643–648.

Sarkar, M., Rahman, A.K.M.L., & Bhoumik, N.C. (2017). Remediation of chromium and copper on water hyacinth (*E. crassipes*) shoot powder. *Water Resources and Industry*, *17*, 1–6.

Sarwar, T., Shahid, M., Khalid, S., Shah, A.H., Ahmad, N., Naeem, M.A. & Bakhat, H.F. (2019). Quantification and risk assessment of heavy metal build-up in soil–plant system after irrigation with untreated city wastewater in Vehari, Pakistan. *Environmental Geochemistry and Health*, 1–17.

Sasmaz, A., & Obek, E. (2009). The accumulation of arsenic, uranium, and boron in *Lemna gibba* L. exposed to secondary effluents. *Ecological Engineering*, *35*(10), 1564–1567.

Shah, R.A., Kumawat, D.M., Singh, N., & Wani, K.A. (2010). Water hyacinth (*Eichhornia crassipes*) as a remediation tool for dye effluent pollution. *International Journal of Science and Nature*, *1*(2), 172–178.

Shirinpur-Valadi, A., Hatamzadeh, A., & Sedaghathoor, S. (2019). Study of the accumulation of contaminants by *Cyperus alternifolius*, *Lemna minor*, *Eichhornia crassipes*, and *Canna generalis* in some contaminated aquatic environments. *Environmental Science and Pollution Research*, *26*(21), 21340–21350.

Shuvaeva, O.V., Belchenko, L.A., & Romanova, T.E. (2013). Studies on cadmium accumulation by some selected floating macrophytes. *International Journal of Phytoremediation*, *15*(10), 979–990.

Singh, D., Tiwari, A., & Gupta, R. (2012). Phytoremediation of lead from wastewater using aquatic plants. *Journal of Agricultural Technology*, *8*(1), 1–11.

Skinner, K., Wright, N., & Porter-Goff, E. (2007). Mercury uptake and accumulation by four species of aquatic plants. *Environmental Pollution*, *145*(1), 234–237.

Sricoth, T., Meeinkuirt, W., Saengwilai, P., Pichtel, J., & Taeprayoon, P. (2018). Aquatic plants for phytostabilization of cadmium and zinc in hydroponic experiments. *Environmental Science and Pollution Research*, *25*(15), 14964–14976.

Stoltz, E., & Greger, M. (2002). Accumulation properties of As, Cd, Cu, Pb and Zn by four wetland plant species growing on submerged mine tailings. *Environmental and Experimental Botany*, *47*(3), 271–280.

Susarla, S., Medina, V.F., & McCutcheon, S.C. (2002). Phytoremediation: an ecological solution to organic chemical contamination. *Ecological Engineering*, 18, 647–658.

Syukor, A.A., Zularisam, A.W., Ideris, Z., Ismid, M.M., Nakmal, H.M., Sulaiman, S., & Nasrullah, M. (2014). Performance of Phytogreen Zone for BOD5 and SS Removal for Refurbishment Conventional

Oxidation Pond in an Integrated Phytogreen System. *World Academy of Science, Engineering and Technology, 8,* 159.

Tripathi, B.D., & Shukla, S.C. (1991). Biological treatment of wastewater by selected aquatic plants. *Environmental Pollution,* 69, 69–78.

Tripathi, B.D., Srivastava, J., & Misra, K. (1991). Nitrogen and phosphorus removal capacity of four chosen aquatic macrophytes in tropical freshwater ponds. *Environmental Conservation, 18,* 143–147.

UNICEF (2006). Meeting the MDG drinking water and sanitation target: the urban and rural challenge of the decade. In *Meeting the MDG Drinking Water and Sanitation Target: The Urban and Rural Challenge of the Decade,* 41.

Uysal, Y. (2013). Removal of chromium ions from wastewater by duckweed, *Lemna minor* L. by using a pilot system with continuous flow. *Journal of Hazardous Materials, 263,* 486–492.

Vajpayee, P., Rai, U.N., Sinha, S., Tripathi, R.D., & Chandra, P. (1995). Bioremediation of tannery effluent by aquatic macrophytes. *Bulletin of Environmental Contamination and Toxicology, 55*(4), 546–553.

Verla, A.W., Verla, E.N., Amaobi, C.E., & Enyoh, C.E. (2018). Water pollution scenario at river Uramurukwa flowing through Owerri metropolis, Imo state, Nigeria. *International Journal of Applied Scientific Research, 3*(3), 40–46.

Wang, Z., Zhang, J., Song, L., Li, E., Wang, X., & Xiao, B. (2015). Effects of linear alkylbenzene sulfonate on the growth and toxin production of *Microcystis aeruginosa* isolated from Lake Dianchi. *Environmental Science and Pollution Research, 22*(7), 5491–5499.

Wolf, S.D., Lassiter, R.R., & Wooten, S.E. (1991). Predicting chemical accumulation in shoots of aquatic plants. *Environmental Toxicology and Chemistry, 10,* 655.

Wu, Z., Yu, D., Li, J., Wu, G., & Niu, X. (2010). Growth and antioxidant response in Hydrocharis dubis (Bl.) Backer exposed to linear alkylbenzene sulfonate. *Ecotoxicology, 19*(4), 761–769.

Yadav, K.K., Gupta, N., Kumar, A., Reece, L.M., Singh, N., Rezania, S., & Khan, S.A. (2018). Mechanistic understanding and holistic approach of phytoremediation: a review on application and future prospects. *Ecological Engineering, 120,* 274–298.

Yadav, S.B., Jadhav, A.S., Chonde, S.G., & Raut, P.D. (2011). Performance Evaluation of Surface Flow Constructed Wetland System by Using *Eichhornia crassipes* for Wastewater Treatment in an Institutional Complex. *Universal Journal of Environmental Research & Technology, 1*(4).

Yasar, A., Zaheer, A., Tabinda, A.B., Khan, M., Mahfooz, Y., Rani, S., & Siddiqua, A. (2018). Comparison of Reed and Water Lettuce in Constructed Wetlands for Wastewater Treatment: Yasar et al. *Water Environment Research, 90*(2), 129–135.

Yaseen, D.A., & Scholz, M. (2017). Comparison of experimental ponds for the treatment of dye wastewater under controlled and semi-natural conditions. *Environmental Science and Pollution Research, 24*(19), 16031–16040.

Zhan, F., Li, B., Jiang, M., Li, T., He, Y., Li, Y., & Wang, Y. (2019). Effects of arbuscular mycorrhizal fungi on the growth and heavy metal accumulation of bermuda grass [*Cynodon dactylon* (L.) Pers.] grown in a lead–zinc mine wasteland. *International Journal of Phytoremediation, 21*(9), 849–856.

Zhang, K., Chen, Y.P., Zhang, T.T., Zhao, Y., Shen, Y., Huang, L., & Guo, J.S. (2014). The logistic growth of duckweed (*Lemna minor*) and kinetics of ammonium uptake. *Environmental Technology, 35*(5), 562–567.

Zhou, J., Wu, Z., Yu, D., Pang, Y., Cai, H., & Liu, Y. (2018). Toxicity of linear alkylbenzene sulfonate to aquatic plant *Potamogeton perfoliatus* L. *Environmental Science and Pollution Research, 25*(32), 32303–32311.

5 Biobased Technologies for Remediation: Green Technology for Environmental Cleanup

Anubhuti Singh, Gurudatta Singh, Priyanka Singh, and Virendra Kumar Mishra
Institute of Environment and Sustainable Development, Banaras Hindu University, Varanasi, U.P., India

CONTENTS

5.1 Introduction	89
5.2 Exploiting Existing Life Forms for Environmental Remediation	91
5.3 Plant-Assisted Remediation (Phytoremediation)	93
5.3.1 Concept and Definition	93
5.3.2 Phytoremediation Strategies	93
5.3.2.1 Phytoextraction	93
5.3.2.2 Phytostabilisation	94
5.3.2.3 Phytodegradation	94
5.3.2.4 Rhizofiltration	94
5.3.2.5 Phytovolatilisation	95
5.3.3 Contaminant Uptake and Breakdown Mechanism by Plants	95
5.3.3.1 Adsorption and Diffusion	95
5.3.3.2 Breakdown	96
5.3.3.3 Conjugation	98
5.3.3.4 Sequestration	98
5.3.3.5 Advantages and Limitations of Phytoremediation	98
5.4 Microbe-Assisted Remediation (Bioremediation)	98
5.4.1 Concept and Definition	98
5.4.2 In-Situ Bioremediation Strategies	99
5.4.2.1 Bioaugmentation	99
5.4.2.2 Bioventing	100
5.4.2.3 Biostimulation	101
5.4.2.4 Biosparging	101
5.4.3 Ex-Situ Bioremediation Strategies	101
5.5 Factors Affecting Bioremediation	102
5.6 Conclusion and Future Projections	102
References	103

5.1 Introduction

As human civilisation originates there were the innumerable declining impact of anthropogenic activities arises, which are tremendously assaulting the excellence of the whole of the inhabitable environments of the earth like soil, oceans, atmosphere, and lives present in the era. Contaminants releasing in the soil, shallow water, groundwater, as well as in the air with hazardous and also poisonous chemicals are the main issues faced due to the industrial world today (Pilon-Smits 2005). Rising anthropogenic actions, to its highest level such as industrialisation, agriculture, and urbanisation leads to the release of various inorganic and organic pollutants, both of which are highly accountable for deterioration in the quality of soil as well as water (Mishra and Tripathi, 2008). Extensive use of the capitals and in-attention in the utilisation of resources leading to numerous ecological risks likewise rising temperature globally, rise in contaminating metals such as heavy metals in topsoil as well as water, bio-diversity losses, and rising health issues in human beings. Pollutants that are generally responsible for deteriorating the excellence of environments are either inorganic in nature and comprise of heavy metals such as Se, Cu, Cr, Pb, Ni, Cd, and Co; metalloids like As, Se, and Hg (Pandey et al., 2015); and radionuclides U, Sr, and Cs (Dushenkov 2003; Cerne et al., 2011), or organic in nature such as hydrocarbons, herbicides, trichloroethylene (TCE), and trinitrotoluene (Hong et al., 2001; Davis et al., 2002; Table 5.1).

Remediating these contaminants at any site by means of the traditional engineering strategies such as physicochemical approaches like filtration, incineration, oxidation, evaporation, and decreasing, electrochemical treatments, reverse osmosis, as well as ion exchange approaches have been practiced (Erdem and Özverdi, 2011; Vasudevan and Oturan, 2014). However, these approaches are facing challenges due to technical and economic perspectives. Because of their high need for labor, environmentally unfriendly, non-cost effective, high chemical requirements, bulk accumulation, release of secondary environmental contaminants, and many other limitations, these technologies are totally unacceptable to reinstate contaminated soil (Meuser 2012; Yao et al., 2012). The biobased technologies, i.e., use of biological agents towards the treatment of major environmental pollutants, have shown greater applicability in comparison to traditional technologies. A comparison between conventional and biobased technologies is shown in Table 5.2.

Developing our country in a more sustained manner with the sustained environment and agriculture growth, remediation of contaminated sites and abatement of entering hazardous chemicals in the food-chain is essential to be reduced. Regarding this, remediating practice based on plants called phytoremediation has grown much over the last two eras. It is a modest, active, profitable, less labor intensive, extensively suitable, well suited, environmentally friendly, maintainable, and consistent, which can be validated in huge areas, chiefly when natural, environmentally, and socio-economically valued vegetations are utilised for the remediation of the polluted places (Pandey et al., 2015, 2016). Including phytobial technology, there are also various microbiological bioremediation practices, which can be applied by adding molecular microbiological approaches that are approved as a revolution in the microbiological and biotechnology realm, leading to quick and high-output approaches for self-determining evaluation and exploiting microbes exist in contaminated settings.

TABLE 5.1

Different sources of environmental pollutants and their impact on the environment

Sources	Contaminants	Impact on the Environment
Industries	Dyes, petrochemicals, gases, pigments, chemicals, oils	• Water sources polluted by chemicals such as heavy metals, leads, pesticides, and hydrocarbons.
Agriculture	Organic and inorganic pesticides	• The explosive growth of algae depletes oxygen in the water.
Transportation	Petrol, diesel, and noxious gases	• Pesticides also killed the non-targeted ones and polluted river bodies and soil.
Trading activities	Packaging materials, and plastics	
Residences	Metals, excreta, sewage, paper, paints, and solvents	• Gases released into the air are carcinogenic, biologically active, or radioactive adversely impact well-being.
		• Persistent toxic compounds and chemicals salts adversely affect the fertility of soil.

TABLE 5.2

Comparison between conventional and biobased techniques of remediation

Parameters	Conventional Technique	Biobased Techniques
Cost	Require high cost for remediation	Cost-effective remediation technique
Skills	Demand efficient skills and techniques	No technical tools required
Transport	Transportation of pollutant to the site of removal	No soil excavation, transport, and storage required
Energy	Depends upon thermal energy	Applied naturally by utilising the sun's energy
Environmental Impact	Release of secondary pollutants	Qualitative excellence of the site recovered
Toxicity	Incineration techniques release noxious gases	Reduction in wind and water erosion via plants
Feasibility	Requires specific location to operate	Feasible at any place and location
Labour	Need high labor work	All work performed by a biological agent itself
Chemicals	Require high demands of chemicals	All process occurs naturally (no chemicals required)

This chapter represents an impression of the diverse applications of biobased methods, including phytobial and microbial technology in bioremediations of the inorganic as well as organic pollutants in an ecological monarchy and outlined the current advancement in the applications of such techniques.

5.2 Exploiting Existing Life Forms for Environmental Remediation

There are numerous living biological forms that had been utilised as a tool for remediating various contaminants from different matrices. Some of the most pronounced ones are plants, bacteria, fungi, algae, and their various engineered forms (Hadacek, 2002), some of such studies are described in Table 5.3. In the plantae kingdom, numerous families are characterised by comparatively greater

TABLE 5.3

Various life forms utilised for the removal of contaminants

Life Forms	Groups	Examples	References
Plants	Bryophytes	*Azolla caroliniana, Riccia fluitans, Fontinalis antipyretica, Callitriche stagnalis*	Favas et al., 2014
	Pteridophytes	*Davallia griffithiana, Pteris vittata, Actiniopteris radiat, Adiantum philippense*	Drăghiceanu et al., 2014
	Angiosperms	*Broussonetia papyrifera, Viburnum awabuki,*	Kang et al., 2018
	Gymnosperms	*Pseudotsuga menziesie, Eriophorum angustifolium*	Bergqvist and Greger, 2012
Bacteria		*Pannonibacter phragmitetus, Thiobacillus ferrooxidans, Deinococcus radiodurans*	Valls and De Lorenzo, 2002
Algae	Chlorophyta	*Ulva ohnoi*	Bastos et al., 2019
		Spirogyra hyaline	Kumar and Oommen, 2012
		Platymonas subcordiformis	Mei et al., 2006
	Cyanophyta (Blue green algae)	*Nostoc carneum, Oscillatoria geminate, Spirulina laxissima,* and *Anabaena ambigua*	Vanhoudt et al., 2018
Fungi	AMF (Arbuscular Mycorrhizal Fungi)	*Glomus Zac-19*	Hernández-Ortega et al., 2012
GEMs	Bacteria	*Pseudomonas fluroscence* HK44, *Sphingomonas* sp. GY2B	Ripp et al., 2000
			Lu et al., 2013

numbers of species known as good candidates for phytoremediation. Plants produce numerous secondary plant metabolites (SPMEs) (Hadacek 2002), proving they are a resource for use as phytoremediating agents. In their study, Jha et al. (2015) showed that SPMEs like cymene, carvone, limonene, and pinene improve degrading PCBs (polychlorinated biphenyls). In another study, Narasimhan et al. (2003) applied the metabolomics of the rhizosphere-driven approaches. Diverse classes of microbes belonging to the bacteria, archaea, or eukarya, occur naturally and can also be utilised in various bioremediation processes; for example, *Pseudomonas putida* PCL1444, from the rhizosphere of *Lolium multiflorum* cv. Barmultra growing into soil contaminated with PAH (polyaromatic hydrocarbon) defends plants from the contaminants by utilising root exudates for their development (Kuiper et al., 2002). Experimentations conducted on oil-spilled beaches result in the presence of phylotypes associated with the α subclass of proteobacteria (α-Proteobacteria) appear in DGGE fingerprints found only on the site contaminated by oil but not in those that are uncontaminated (MacNaughton et al., 1999). Pollutants such as alkylbenzenes and n-alkanes are degraded by a denitrifying microbial community analyzed by fluorescence in-situ hybridisation (Fish) with group-specific rRNA (Rabus et al., 1999). Proteobacteria were reported in groundwater contaminated with petroleum below a crude oil storing cavity under the surface (Watanabe et al., 2000). Recently, genetic engineering of heavy-metal-tolerant *Ralstonia eutropha* was performed that demonstrates that the inoculating Cd^{2+}-contaminated soil with the *Ralstonia* that is subjected to be engineered, expressively reduced the noxious property of the heavy metal on the growing rate of tobacco plants (Valls et al., 2000). Plants can be modified genetically for improving their capability of utilising a phytoremediation agent acting as a host (Tripathi et al., 2020) to various microbial groups (Basu et al., 2018; Kuffner et al., 2008; Hussain et al., 2018), closely related to it. The process of engineering genetically engineered plants with microbes to stimulate the potential of phytomicrobial remediation is nurtured into the host plant, performed in the following steps:

- Firstly, isolate bacteria and amalgamation of genes, leading to production of particular enzymes accountable for pollutant degradation, following the transfer of this modified bacteria to the host (Valls and De Lorenzo, 2002).
- Secondly, augmentation in the quantity of this remediation microbe is inside the host (Abou-Shanab et al., 2007; Li et al., 2007).

The single genetically engineered microbe accepted until now for its field application in the USA for utilising as an agent of bioremediation was *Pseudomonas fluorescens* HK44, owning naphthalene catabolic plasmid (pUTK21), mutated by transposon insertion of lux genes (Ripp et al., 2000) (Figure 5.1).

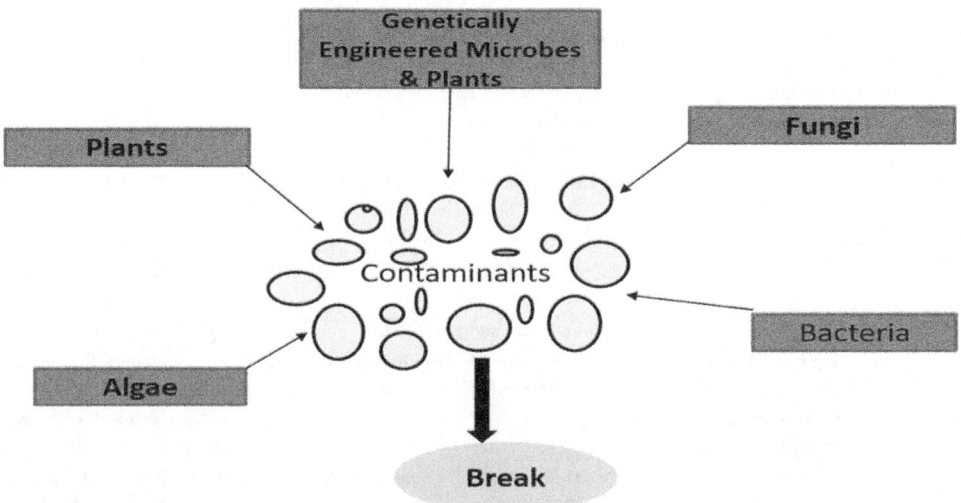

FIGURE 5.1 Living entities utilised in the breakdown of contaminants.

5.3 Plant-Assisted Remediation (Phytoremediation)

5.3.1 Concept and Definition

Phytoremediation is the one and only cheapest and environmentally friendly approach for removing pollutants from soil and water by using plants as remediating agents (EPA 2000). The method used for phytoremediation seems relatively simpler, but it is a complicated procedure in which the pollutants override over any of the degradation pathways of oxidation or reduction to their simplest form and accumulated either within vacuoles or volatilised in the atmosphere. Hyperaccumulation is the accumulation of contaminants inside plants. Accumulator plants gather pollutants inside the plant to an extent that are not enough noxious for them, while hyperaccumulators can accumulate to 1,000 mg/kg dry weight of the pollutants (Reeves, 2003). The most popular illustration of hyperaccumulation by *Pteris vittata* is described with a hyperaccumulation capacity of arsenic not more than 7,500 mg/kg without little toxic property on the plant (Ma et al., 2001).

5.3.2 Phytoremediation Strategies

Phytoremediation utilises remediating contaminants, pollutants, or heavy metals as well as metalloids from soil or water that is economic in terms of cost, and is feasible as well as a competent method when linked to various engineered technology like excavation, soil incineration, soil washing, flushing, solidification, etc. (Ali et al., 2013; Wang et al., 2015). The ideas relate to remediating contaminants by the ease of plants proposed by Chaney (1983). This thought appears quite pleasing and receives high societal acceptance, which is able to be exploited on a broad range and locates where other techniques of the remediations were not too operative and economic. Phytoremediation technologies are divided into five types, depending upon the nature of the contaminant to be removed:

- Phytoextraction/phytoaccumulation
- Phytostabilisation
- Phytodegradation
- Rhizofiltration
- Phytovolatilisation

Apart from these techniques, the establishment of the plant on impure topsoil is cost effective in many other forms: (i) Phytostabilisation; (ii) phytoextraction of expensive metals like Hg, Ag, and Ni; and (iii) sustained land managements (Ali et al., 2013). This idea is utilised for green plants for contaminant removal achieves a high status as green remediations of the harmful metals as well as metalloids as a useful alternative to various traditional physicochemical remediation procedures (Ali et al., 2013). Succeeding paragraphs discuss different techniques of remediation broadly.

5.3.2.1 Phytoextraction

In this method of removal of contaminants, there is an absorption of contaminants from polluted sites, either soil, water, or sediment, by the roots, and then it is translocated and accumulated in aerial plant parts. It is the furthermost significant remediation method for the removal of pollutants, metals, and metalloids from polluted soil, water, biosolids, and sediments (Ali et al., 2013; He et al., 2005; Seth 2012). Phytoextraction is the most appropriate method exploited at the commercial scale when relating various phytoremediation methods (Sun et al., 2015). There were various features that affect the competence of phytoextraction that includes property of soil, bioavailability of metal, metal speciation, and characters of plants to be used. Plants chosen for phytoextraction have numerous features like (Ali et al., 2013; Tong et al., 2004; Shabani and Sayadi, 2012) quick rate of growth, high biomass, hyperaccumulator to heavy metals, extensive dispersion, translocation of metals from its roots to shoots is easy, highly tolerable to hazardous pollutants, highly resist to phytopathogens and pests, highly adapted to changing climate scenarios, ease in cultivation and harvesting, and unattractive for herbivores avoiding the plant to be entered in the food chain. It has been reported by Baker and Brooks (1989), plants from the Brassicaceae, Lamiaceae, Asteraceae, and Euphorbiaceae families are more appropriate to be used for the phytoextraction.

5.3.2.2 Phytostabilisation

Phytostabilisation or phytoimmobilisation denotes using plants in the reduction of movement or/and availability of contaminants in the soil water and sediments and to avoid leakage in groundwater and blocking access in the food chain, which comprises the process of adsorption by roots, complexation, and precipitation in the root zone (Rew, 2007). The chief measure taken in this process is to amend the contaminated site to alter it by chemically and biologically by enhancing organic matter content of the site and rising cation exchange capacity. Along with organic contaminants and heavy metals, phytostabilisation can also be used to remediate sites with contamination of radioactive compounds presence at a low level at any site (Lasat et al., 1998; Watt et al., 2002).

5.3.2.3 Phytodegradation

Phytodegradation is the process of breaking down the pollutants by various means of enzymes from more toxic forms to fewer toxic forms either in the rhizosphere region before uptake of roots or after taking up of pollutants into the root zone followed by further synthesis. It is the most significant remediation procedure and is also called phytotransformation. In this mechanism of remediation of xenobiotics in the plant cells, there are three stages performed. The first phase is activation, described as the phase when reactive polar functional groups are associated with the lipophilic organics, which leads to reducing the lipophilic activity of pollutants and this leads to enhanced solubilisation in the liquified environment of the cell. Enzymes like cytochrome P450 and carboxylesterases are acting as catalysts for this reaction. The second phase is conjugation, which includes a reaction catalyzed by glutathione S-transferases (GST) and glucosyltransferases (GT) enzymes, the reaction of joining the altered complexes with endogenous compounds like amino acids, sugars, or glutathione to reduce their phytotoxicity. The last stage, sequestration/compartmentalisation, is the removal of inactivated pollutants from the cytosolic sections to apoplastic compartments occurs (Truu et al., 2015).

Effectiveness of plant degradation to contaminants is lowered only in the case when plants are more sensitive to the high concentrations of a contaminant that hampered its growth (Doty, 2008). The availability of pollutants to plant also affects their degradation rate (Pilon-Smits, 2005). Another drawback of the procedure is that most report transformation of toxic compounds that lead to forming metabolites more toxic than the previous ones (Marecik et al., 2006).

5.3.2.4 Rhizofiltration

This technique of removal of pollutants, rhizofiltration, is remediation in the region of roots, i.e., rhizosphere zone. In this procedure, decontamination of the aqueous ecosystem is done either using aquatic plants or terrestrial plants. In this procedure, remediation is done either on the site, called in-situ, or away from the site of contamination, called ex-situ. The plant absorbs pollutants by the aid of their roots. Pollutants are captivated by the aided plant roots until the point of saturation and, ultimately, the plant is picked off with its attached roots from the region. In this process, there is no translocation of contaminants from the root to shoot that occurs so there is very little chance of environmental pollution after remediation (Dushenkov, 2003). Generally, terrestrial plants have characteristic features such as high growth rate and development and have the bulk of fibrous roots, making them suitable agents for rhizome filtration. Indian mustard, rye, sunflower, tobacco, and spinach are some of the highly cultivated plants to be utilised for rhizofilteration (Henry, 2000).

5.3.2.5 Phytovolatilisation

Phytovolatilisation is the technique that involves the extraction of a pollutant from the region of contamination performed by the help of the plant's root and the pollutant is then converted from high toxicity into a reduced toxic form within the plant and there is a release of a less toxic pollutant into the atmosphere through the plant's leaves. This technique involves the evapotranspiration mechanisms used for Hg, Se, and organic solvent contamination (Bizily et al., 1999; Karami and Shamsuddin, 2010). By utilisation, wooded plants tritium (3H), an isotope of hydrogen that is radioactive in nature, is reported to be remediated by utilising this technique (Murphy, 2001). Phytovolatilisation efficacy is influenced by the climatic condition of the region in which remediation has taken place and even on the genetic constituents of the plants to be utilised for remediation purposes. In relation to this nature of the contaminant, its capability to translocate and its absorbing frequency is also affected by the concentration at the site of remediation (Meagher, 2000). Reports from the Environmental Protection Agency, United States, reported alfalfa, black locust, canola, and Indian mustard are good phytovolatilising plants that are highly (EPA, 2000) recommended for this process. The process of phytoremediation is represented in Figure 5.2 with different steps of phytoremediation that occurs in different parts of plants.

Types of phytoremediation and its goal for selecting the plant's species for removing the contaminants are discussed in Table 5.4.

5.3.3 Contaminant Uptake and Breakdown Mechanism by Plants

Plants possess a different mechanism for organic and inorganic contaminant breakdown that is described in Figure 5.3. Pollutants have the diverse mechanisms of degradation (1) with microbial community resides in the rhizosphere zone (rhizo-degradation), (2) exudations by the plant (phytodegradation), and (3) in plants (Pilon-Smits 2005; Vangronsveld et al., 2009; Stephenson and Black, 2014). The frequently stated mechanism of phytoremediation in plants generally includes steps of adsorption, conjugation, and sequestration, which is elaborated on.

5.3.3.1 Adsorption and Diffusion

Plants absorbed pollutants with the uptake of water during their metabolism, reducing the chance of leaching at the level of groundwater (Vangronsveld et al., 2009). Managing polluted soil by applying the phytoremediation strategy has an essential step that requires the diffusion and mobility of water along with air through the breakdown of soil particles and producing passageways physically achieved mainly by growing root of remediating plants (Gerhardt et al., 2009). Growing roots by penetrating soil causes cracking in the arrangement of soil particles and not only alters the physical property but the entire biology of microbes that reside in the root region of the plant (Batty and Dolan, 2013).

Diverse biochemicals like organic acids (citric acid), bicarbonate anions, protons, and additional cations are root exudates of the rhizosphere zone of plants that affect the property of the soil chemically, which ultimately effects the movement and microbial availability for degradation of various soil pollutants (Pilon-Smits 2005; Padmavathiamma and Li, 2012; Stephenson and Black, 2014). As inorganic pollutants are not degraded by means of chemicals, certain plants are able to arrest contaminants in their rhizosphere region or accumulations of the contaminants in the aerial part of the plants and the accumulated plant is extracted with contaminants. Soil contaminated with phosphorus and carbonates responsible for its high pH also need to be removed (Tak et al., 2013). For increasing the availability of these pollutants to biological entity, i.e., plants, firstly it is necessary to grow vegetation to polluted land as protons and acids are secreted as exudates from plants as the metabolism of microbes takes place, converting the insoluble complexes to ions that are in the form of free ions (Padmavathiamma and Li, 2012; Stephenson and Black, 2014).

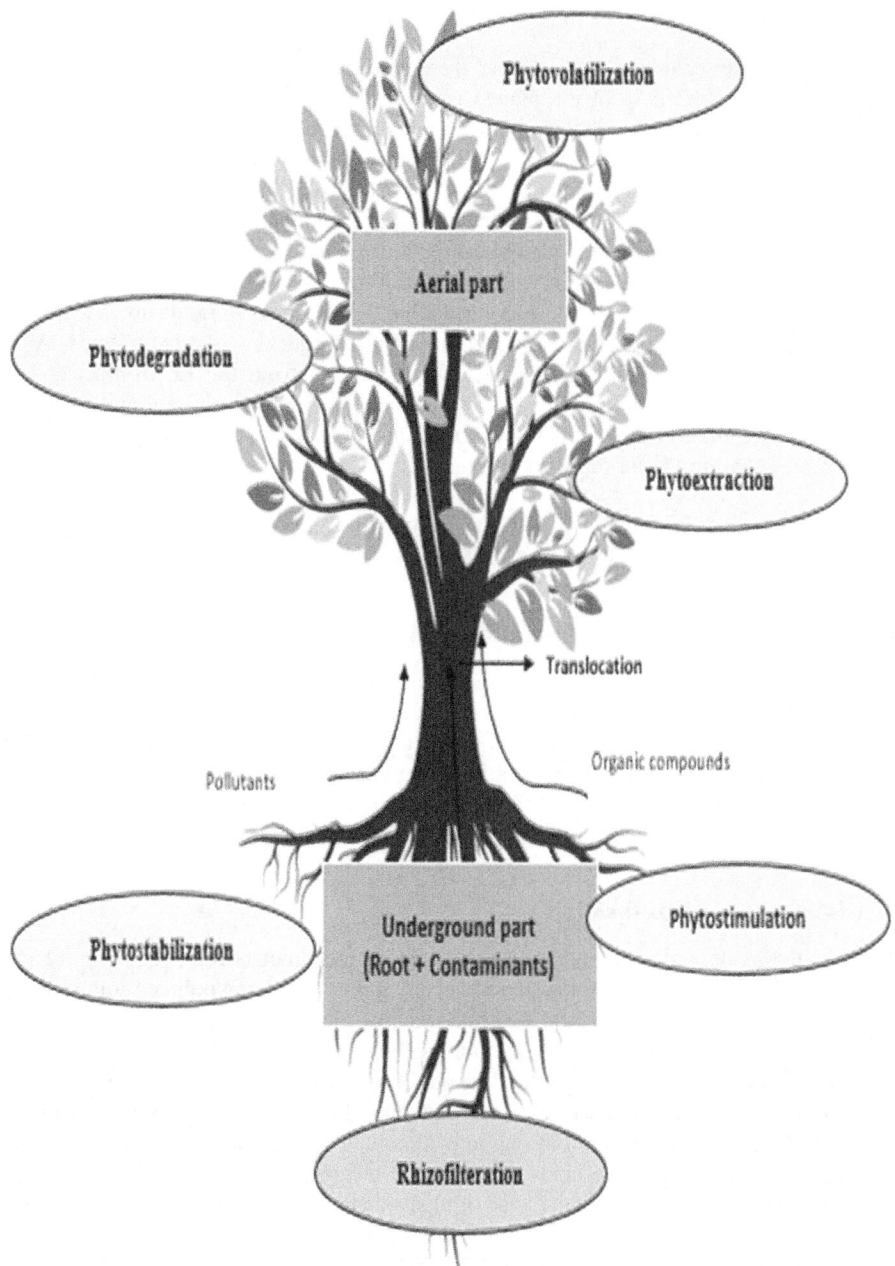

FIGURE 5.2 Process of phytoremediation in the aerial and underground plant parts.

5.3.3.2 Breakdown

Minimisation in hazardous effects of diverse contaminant residues in the form of organics done by the process of reduction, conversion, and catabolising these harmful pollutants inside the cells (Sandermann 1994) approved as a "green liver model." The very first step toward the degradation of the organic contaminants involves the breakdown of the aromatic chain, mainly if the pollutant is a derivative of benzene mediated by the activity of oxygenase enzyme cytochrome P450 monooxygenases (Sandermann 1994). After degradation, functional groups are added to the pollutants to enhance their

TABLE 5.4
Phytoremediation process and its characteristics with examples

Techniques	Goal	Selection Criteria for Plant Species	Examples of Plant (With Family)
Phytoextraction	Extract contaminant to root and then translocate to shoot and then harvest it	Able to tolerate extremities, high accumulation rate, high growth rate, easily harvestable, maximum translocation	*Rhizobium meliloti, Glomus caledonium* (Arbuscular mycorrhizal fungi) (Teng et al., 2010)
Phyto stabilisation	Immobilisation of pollutant in the root zone	Plentiful root arrangements, lower translocation factor	*Medicago sativa* L. (Leguminosae), *Noccaea caerulescens, Thlaspi caerulescens* (Brassicaceae) (Ouvrard et al., 2011)
Phyto-volatilisation	Converting pollutant for volatilisation and releasing it to surroundings	Tolerance, high translocation factor, rapid growth rate	*Populus nigra* var. italica Munchh. (Salicaceae) (Macci et al., 2013)
Rhizofiltration	Absorption accumulation and precipitation of pollutant from roots	Tolerance, high adsorption surface, low translocation factor	*Pistia stratiotes, Eichhornia crassipes, Lemna minor,* and *Hydrocotyle umbellate* (Dushenkov, 2003; Henry 2000)
Phytodegradation	Degrading pollutants by the help of hydrolytic enzymes and metabolites	Tolerance, high accumulation capability	*Medicago sativa* L. (Leguminosae), *Festuca arundinacea* Schreb. (Poaceae) (Sun et al., 2015)

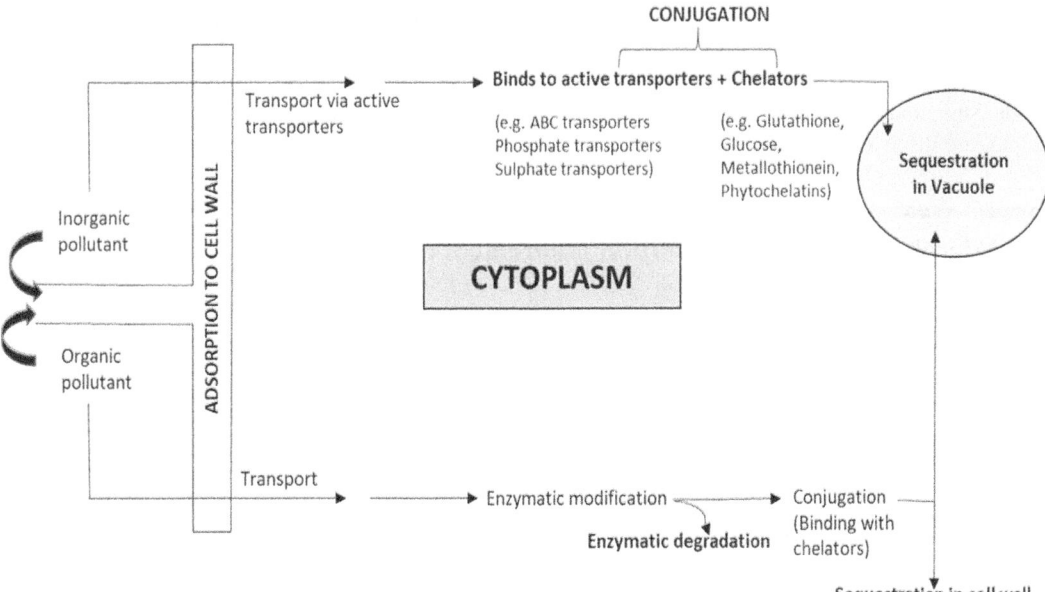

FIGURE 5.3 Mechanism of phytoremediation for organic and inorganic pollutants.

activity and solubility if their polarity is good (Komives and Gullner, 2005); other groups of the enzymes can also assess these reactions like N- or O-dealkylation, ester hydrolysis, and alkyl hydroxylation (Sun et al., 2015; Schwitzguébel, 2017).

5.3.3.3 Conjugation

The active complexes of organic pollutants are then conjugated with the plant's polar molecules like malonic acid, glutathione, carbohydrate, and sulfate. This conjugation includes peptide conjugation and ether, ester, or thioether linkages and is quickened by the transferase enzymes (glycosyltransferases and GST) (Aken et al., 2009). Formed products of these conjugates are either soluble or bonded. In the case of inorganic pollutants, the free ionic species enhances their bioavailability for biosurfactants and chelators (Maestri et al., 2010; Tak et al., 2013). After binding with chelators, these species are able to be passed easily from cell membranes (Nowack et al., 2006; Johnson and Singhal, 2015). Transportation pathways for this kind of chemical are similar to the nutrient transport pathway: (I) apoplastic and (II) symplastic or transmembrane pathways (Robinson et al., 1997; Maestri et al., 2010; Singh et al., 2011; Padmavathiamma and Li, 2012).

5.3.3.4 Sequestration

The solubilisations of conjugates broken completely into CO_2 and H_2O molecules can be in use for later reactions, while the bonded conjugates are excluded outside the cell to the apoplast, called exocytosis, and later assimilated into the cell wall (Komossa et al., 1995). Salts like NaCl are mostly seized within vacuoles (Hasegawa 2013; Manousaki and Kalogerakis, 2010), whereas metals are seized inside cell walls, organelles, vacuoles, or the Golgi complex (Singh et al., 2011).

5.3.3.5 Advantages and Limitations of Phytoremediation

Among various traditional remediation techniques, phytoremediation is verifiably advantageous in many perspectives (Pilon-Smits and Freeman, 2006; Susarla et al., 2002; Chaudhry et al., 2005). Traditional ex-situ approaches of phytoremediation include excavating the polluted topsoil, storing it away from the sites, washing the contaminated soils, and then covering it for further stabilisation, which is non-economic and requires high cost for remediation compared to the in-situ method of remediation (Pilon-Smits and Freeman, 2006). Comparative limitations and advantages of phytoremediation technology are discussed in Table 5.5.

5.4 Microbe-Assisted Remediation (Bioremediation)

5.4.1 Concept and Definition

Bioremediation is the technique that exploits the metabolism of microbes to remediate contamination that exists in the environment, making it unpleasant. Showing characters to be utilised as bioremediating agents across some essential ones, a significant characteristic feature of bioremediation is that it can be performed in an unsterile, exposed environment that consists of numerous microorganisms (Chandra and Dubey, 2020). There are various kinds of microorganisms that include bacteria, fungi, and algae that have a high counteracting ability to neutralise pollutants reported (Shukla et al., 2014). Out of different microbes, bacteria perform a dominant role in the process of bioremediation, while other microbes like fungi and grazing protozoa also have some influence in the process of remediation (Chandra and Dubey, 2020). Microorganisms have driven this transformation to fulfill their energetic needs, while purifying the contaminants, or maybe coincidental in nature called co-metabolism (Eevers et al., 2017). Microbes are present in all aspects of the environment universally due to this reason, and their ubiquitous presence and huge biomass when compared to other living creatures on the earth, broader range, diverse in their catalytic machineries, and even enormous capacity to tolerate the non-availability of oxygen and other life-threatening environmental states (Mishra and Tripathi, 2008; Watanabe et al., 2000), Sometimes

TABLE 5.5
Advantages and limitations of phytoremediation

Advantages	Limitations
• Easy to implement. • No technical tool required. • Utilises the sun's energy for remediation (Schwitzguébel 2017). • Organic compounds and nutrients are added to the soil; in addition, enhance soil quality (Schwitzguébel 2017; Wiszniewska et al., 2016). • Roots aid in stabilisation of soil and reduce erosion. • Feasible at any place. • More public acceptance.	• Plant intolerance in nature if there is a high concentration of pollutant. • Require adequate agronomic practices. • Insufficiency of the site in terms of the size of plants (Schwitzguébel 2017). • Sometimes it takes a longer duration for remediation if the plant's growth rate is low. • There is a need for careful disposal of contaminant accumulated plants.

this bioremediation approach of utilising microorganisms is not always successful. In some cases, there is the failure of these tactics; the reason behind this is usually little adaptation, more time consumption, microbes not compete, and little connectivity between target pollutants and microbes (Singh et al., 2017). There is ongoing research to develop degradation-efficient microbes by considering their inheritances and biochemistry, and developing approaches for utilising it in the field is demanding more significant human efforts. There is a need for well-made, efficient microbial bioreactors to be recognised as well-organised techniques for ensuring that the microbe's growth and its metabolic process occur in a precise situation and definite environmental conditions that provide the essential optimal environment settings for effective decontamination (Singh et al., 2020).

5.4.2 In-Situ Bioremediation Strategies

Variations in bioremediation that mainly occur in nature are well defined by the site where this process happens. "In situ" is utilised in various contexts and always means "on-site" when referring to the region or location where the contaminants are generated. In the framework of the bioremediation technique, in situ designates the site of the remediation that occurs at the same site of contamination where it arises with no translocation from one place to another, while ex situ includes removing the contaminant for remediation elsewhere (Aggarwal et al., 1990). There are three kinds of in-situ bioremediation techniques that are more pronounced in nature:

1. **Bioattenuation** is subjected to the natural process of degradation of the pollutant.
2. **Biostimulation** is the process of stimulating the degradation of pollutants by adding water, nutrients, and electron donors or acceptors.
3. **Bioaugmentation** is the process where the microbes that are proven to be efficient in degradation are added to the site of contamination (Madsen 1991).

A specific bacterial remediation method is determined by numerous features, such as site environment, native microbiological communities, and the kind of pollutant, its numbers, and its toxic nature. In-situ bioremediation is also an attractive selection for groundwater with low pollutant concentration because the treatment occurs in the subsurface aquifer directly (Figure 5.4).

5.4.2.1 Bioaugmentation

When microorganisms native to the site of contamination are not showing much effectiveness for the purpose of bioremediation, bioaugmentation is applied. In this process, genetically engineered microbes or healthy microbes are introduced at the location of contamination for enhanced remediation of pollutants. Introducing microbes from the external environment into the soil has been utilised as an attempt

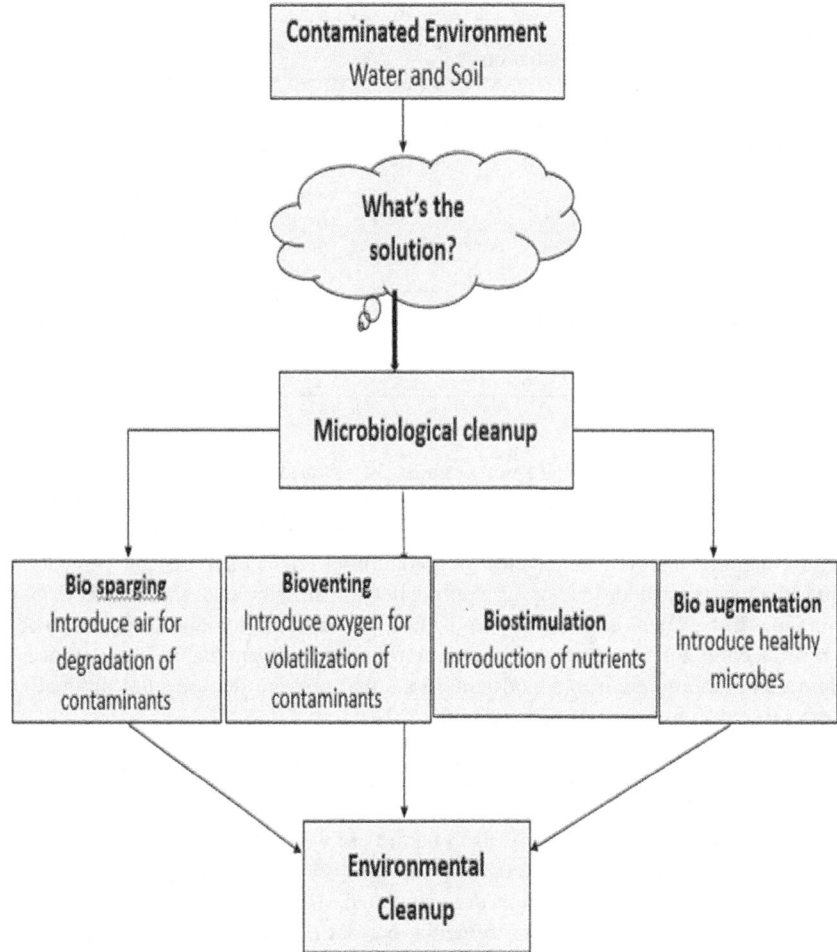

FIGURE 5.4 Process of bioremediation for removing contaminants.

to hasten the bioremediation process. It is required to examine the activity of microbes introduced in any environment for removal of the contaminants at regular intervals. When a genetically modified organism is used as a bioaugmenting agent, the rate of bioaugmentation is enhanced because of the establishment of trans-conjugants that are able to degrade the contaminants relative to the direct influence of injecting microbes (Dejonghe et al., 2000).

5.4.2.2 Bioventing

Bioventing is another approach in remediating contaminated sites by the enrichment of oxygen for enhancing the activity of aerobic degraders of the site. Researchers report that venting boosts the in-situ biodegradation of POLs (petroleum-oil-lubricants) by introducing oxygen to microbes present in the soil. In this approach, there is a direct supply of oxygen provided to the contaminants. When compared to soil vapour vacuum extraction (SVE), bioventing utilises little air flow rate, which provides adequate oxygen supply to retain the activity of microbes in action. The optimum flow of oxygen at a steady rate takes full advantage of biodegradation as vapours pass gradually over biological active soils while decreasing volatilisation of pollutants for effective degradation. A basic venting arrangement comprises a well and a blower, which pumps air through the well and into the soil (Arshad et al., 2007). Bioventing comprises coupling of soil venting along with bioremediation, which is frequently regarded as the most active resource to supply oxygen to the vadose zone soil.

5.4.2.3 Biostimulation

In various circumstances, the advancement in bioactivity for increasing removal of contaminants, there is a need to stimulate the whole environment of the site, which is achieved by adding nutrients and water intake along with increasing oxygen flow, which optimises the rate of removal, rather than venting alone. (Das and Dash, 2014). Stimulating microbial degradation by providing their nutrition to the soil increases in the form of fertilisers; water-solubles like KNO_3, $NaNO_3$, NH_3NO_3, K_2HPO_4, and $MgNH_4PO_4$ and slow-release compounds like customblen, IBDU, max-bac, and oleophilic like Inipol EAP22, F1, MM80, S200 are added to stimulate the rate of bioremediation (Nikolopoulou and Kalogerakis, 2008).

5.4.2.4 Biosparging

Biosparging involves the biodegradation approach rather than volatilisation stimulated by the introduction of air. Bacteria occur in nature that are responsible for degrading pollutants can enhance their rate by availability of oxygen into the vadose zone by injecting air under adequate pressure below the ground. Biosparging upsurges the intermixing of soil and water in the saturated zone by increasing contact between both. This technique can be handled with ease and is cost effective because it requires a small diameter of injection points that allow ease and less installation cost; even the requirement of air injection at smaller points is significantly flexible relative to the design and creation of this structure. In a simple arrangement, in-situ air sparging (IAS) involve the inoculation of air inside the aquifer with intending to remediate any trapped undissolved pollutants (source zones) or dissolved pollutant clouds. As the basic mechanisms of an IAS system are not complex like blowers/compressors, piping, and air injection wells, and because it is simply combined with additional remediation techniques like soil vapour extraction (SVE), this knowledge has expanded worldwide for its use and response to the environment. Creating a casual survey of underground storage tank (UST) regulators, it is now possible to be the most skillfully engineered in-situ remediation option when targeting the treatment of hydrocarbon-contaminated aquifers.

5.4.3 Ex-Situ Bioremediation Strategies

Ex-situ bioremediation is also performed by microbes, depending upon their metabolic biological activity in which there is a need to dig the soil by excavation and transport it to an aboveground treatment area followed by aeration, which is the process to improve the rate of degradation of carbon-based pollutants through native microorganism populations. Under the aerobic settings, microorganisms can use organic pollutants like petroleum hydrocarbon mixtures, polycyclic aromatic hydrocarbons (PAH), phenols, cresols, and some pesticides as the sources of carbon and energy and reduce it eventually to CO_2 as well as water. There is an array of hydrocarbons presents in the environment that is able to be degraded by utilising ex-situ bioremediation: general hydrocarbons, kerosene, phenols, cresols, PAH, semi-volatile organic compounds, diesel range hydrocarbons, lubricating oils, straight-chain aliphatic.

Composting is the most commonly practiced technique of ex-situ remediation that is generally performed away from the contamination site. It is a procedure in which organic contaminants are degraded by microbes, typically at a higher temperature. Most representative compost temperatures range between 55°C and 65°C. This amplified temperature results from the heat exacerbated by the microbes during degrading of the carbon-based compounds present in the waste. Conventionally, the exercise of implementing compost is proposed to decrease the bulk and liquid content of organic waste, to abolish pathogens, and to remove complexes that produce odor. Bioremediating contaminated soil by adding compost to the polluted site presents a sustained technique because the recyclable organics in the compost material are being exploited as fertilisers for zwitterionic agriculture. In addition, composting also recovers the soil structure, nutrient status, and microbial activities of the soil (Table 5.6).

TABLE 5.6
Methods representing the types of bioremediation

Method	Techniques	Process	Limitations
In Situ			
	Biosparging	Introduction of air for rising degradation of pollutants by soil and water intermixing in the saturated zone.	Possibility of migration of pollutants; not suitable in every environment.
	Biostimulation	Introduction of nutrients to improve the environmental condition for efficient microbial activity in the unsaturated zone.	There is a problem in algal blooms developing in the water.
	Bioventing	Introduction of oxygen to the vadose zone to enhance volatilisation of pollutants.	Diffusion of oxygen may be limited; extremely low soil moisture content may limit the effectiveness of bioventing.
	Bioaugmentation	Introduction of efficient microbial communities to the polluted soil.	Competition between invasive and indigenous ones.
Ex Situ			
	Composting	Anaerobic, converts solid organic wastes into humus-like material via high-temperature treatment.	The additional cost of excavation.

5.5 Factors Affecting Bioremediation

The required remediation technique to be utilised for remediation either is in situ or ex situ and it is necessary to determine the optimum environmental factors that are most important to develop bioremediation technologies (Calvo et al., 2009; Das and Chandran, 2011). Out of them, some are discussed as follows:

1. **Bioavailability**

    In topsoil masses, which are the small "composite units" in the mixture of soil components, the availability of pollutants to degrade is limited by their transportation to the microbial cell, i.e., distribution of contaminant out of the composite unit to the microbial cell bound to the exterior of the aggregate (Tang et al., 1998). Reports suggest that the biological availability of pollutants also varies species to species and their capability to unbind the pollutant. Features like organic matter content, cation exchange capacity, micropore volume, soil texture, and surface area affect the capture of contaminants (Chung and Alexander, 2002).

2. **Environmental factors**

    Environment settings such as temperature, pH, oxygen availability, electron, and salinity affect the growth; metabolism of microbes and, to some level, the behaviour of the contaminant such as solubilisation and volatilisation, is also affected (Naik and Duraphe, 2012). It is necessary to optimise the factors related to environmental conditions required for efficient biodegradation (Tekere et al., 2001). By establishing the optimum growth surroundings like pH, aeration, and nutrients for optimal degradation activity and growth of microbes, it is found that microbes show high-degradation ability and hasten enzymatic activity at optimum environmental conditions (Table 5.7).

5.6 Conclusion and Future Projections

The worldwide necessity for a solution in environmental difficulties includes numerous remediation procedures, but an advance measure to remediate the environment due to its numerous advantages over other remediation procedures is the process of bioremediation. With its numerous fruitful ecofriendly, cost effective, and productive advantages, bioremediation is restricted in use due to some of its

TABLE 5.7

Environmental aspects and optimal conditions for the microbial activity for soil or water bioremediation.

Environmental Aspects	The Condition Required for Microbial Activity	Effects
Soil moisture	25–85% water holding capacity	Transportation of contaminants and degradation products; reduction of pollutants.
Contaminants	Not too toxic	Influence bioavailability and increase/hinder biodegradation.
Oxygen	>0.2 mg/L, aerobic	Optimum activity of aerobes.
Nutrients	N and P for microbial growth	Growth and reproduction of microorganisms.
Temperature	15–45°C	Interacting population living together on the contaminated site and rapidity in the process of degradation; contaminants have high survivability at a lower temperature.
Redox potential		Concentrations and proportions of electron donors or acceptors to determine degradation pathway and efficiency of degradation.

unfriendly impacts because it is applicable only to those compounds that are naturally biodegradable. Every contaminant that exists in the environment does not have a quick and comprehensive degradation and many times there is a need for careful measures to be taken for proper disposal of remediates. Emerging technologies focus on diverse interdisciplinary research and the possibility of overcoming numerous hindrances in present techniques, as well as respecting the moral, lawful, and societal issues, will stand ahead in cleaning the environment in the forthcoming era.

Further research is needed to develop rhizosphere microbes living in symbiosis with host plants, and their examination is necessary in order to check their compatibility of degrading the polluted substrates; it is vital to explore the worth of utilising the technique as well as assistance to define the suitability of the usage. The capability of the phytobial organisations to abolish composite carbon-based complexes desires to be the conventional, gives the contamination originates on lands usually due to industries comprise an excessive array of pollutants. Many different phytoremediation methods, such as chemical-aided phytoextractions as well as the bacterial-aided methods of phytoremediation, can also be utilised to purify contaminated top soil in huge measures. Supplementary exploration is desirable in the fields of gene manufacturing to expand the phytoremediation capacities of transgenic plants, and also to comprehend the mechanism and efficiency of the phytoremediation technique in the direction to make these techniques more operative and time saving as well as economically viable.

REFERENCES

Abou-Shanab, R.A.I., Angle, J.S., & Van Berkum, P. (2007). Chromate-tolerant bacteria for enhanced metal uptake by Eichhornia crassipes (Mart.). *International Journal of Phytoremediation, 9*(2), 91–105.

Aggarwal, P.K., Means, J.L., Hinchee, R.E., Headington, G.L., & Gavaskar, A.R. (1990). *Method to select chemicals for in-situ biodegradation of fuel hydrocarbons*. Final report, November 1988-January 1990 (No. AD-A-227541/0/XAB). OH, USA: Battelle Columbus Labs.

Aken, B.V., Correa, P.A., & Schnoor, J.L. (2009). Phytoremediation of polychlorinated biphenyls: new trends and promises. *Environmental Science & Technology, 44*(8), 2767–2776. Pandey, V.C., Singh, J.S., Singh, R.P., Singh, N., & Yunus, M., 2011. Arsenic hazards in coal fly ash and its fate in Indian scenario. *Resources, Conservation and Recycling, 55*(9–10), 819–835.

Ali, H., Khan, E., & Sajad, M.A. (2013). Phytoremediation of heavy metals – concepts and applications. *Chemosphere, 91*(7), 869–881.

Antizar-Ladislao, B. (2010). Bioremediation: working with bacteria. *Elements, 6*(6), 389–394.

Arshad, M., Saleem, M., & Hussain, S. (2007). Perspectives of bacterial ACC deaminase in phytoremediation. *TRENDS in Biotechnology, 25*(8), 356–362.

Baker, A.J.M., & Brooks, R. (1989). Terrestrial higher plants which hyperaccumulate metallic elements. A review of their distribution, ecology and phytochemistry. *Biorecovery*, *1*(2), 81–126.

Bastos, E., Schneider, M., de Quadros, D.P.C., Welz, B., Batista, M.B., Horta, P.A., Rörig, L.R., & Barufi, J.B. (2019). Phytoremediation potential of Ulva ohnoi (Chlorophyta): Influence of temperature and salinity on the uptake efficiency and toxicity of cadmium. *Ecotoxicology and Environmental Safety*, *174*, 334–343.

Basu, S., Rabara, R.C., Negi, S., & Shukla, P. (2018). Engineering PGPMOs through gene editing and systems biology: a solution for phytoremediation?. *Trends in Biotechnology*, *36*(5), 499–510.

Batty, L.C., & Dolan, C. (2013). The potential use of phytoremediation for sites with mixed organic and inorganic contamination. *Critical Reviews in Environmental Science and Technology*, *43*(3), 217–259.

BBSRC (1999). *A Joint Research Council Review of Bioremediation Research in the United Kingdom*. Swindon: BBRRC, EPSRC and NERC

Bergqvist, C., & Greger, M. (2012). Arsenic accumulation and speciation in plants from different habitats. *Applied Geochemistry*, *27*(3), 615–622.

Bizily, S.P., Rugh, C.L., Summers, A.O., & Meagher, R.B. (1999). Phytoremediation of methylmercury pollution: merB expression in Arabidopsis thaliana confers resistance to organomercurials. *Proceedings of the National Academy of Sciences*, *96*(12), 6808–6813.

Calvo, C., Manzanera, M., Silva-Castro, G.A., Uad, I., & González-López, J. (2009). Application of bioemulsifiers in soil oil bioremediation processes. Future prospects. *Science of the Total Environment*, *407*(12), 3634–3640.

Cerne, M., Smodis, B., & Strok, M. (2011). Uptake of radionuclides by a common reed (Phragmites australis (Cav.) Trin. ex Steud.) grown in the vicinity of the former uranium mine at Zirovski vrh. *Nuclear Engineering and Design*, *241*(4), 1282–1286.

Chandra, H., & Dubey, R.C. (2020). 10 bioremediation – with. *Bioremediation Technology: Hazardous Waste Management*, 215.

Chaney, R.L. (1983). Plant uptake of inorganic waste. In J.F. Parr, P.B. Marsh, & J.M. Kla (Eds.), *Land treatment of hazardous wastes*. Park Ridge: Noyes Data Corporation, p. 50–76.

Chaudhry, M.S., Batool, Z., & Khan, A.G. (2005). Preliminary assessment of plant community structure and arbuscular mycorrhizas in rangeland habitats of Cholistan desert, Pakistan. *Mycorrhiza*, *15*(8), 606–611.

Chung, N., & Alexander, M. (2002). Effect of soil properties on bioavailability and extractability of phenanthrene and atrazine sequestered in soil. *Chemosphere*, *48*(1), 109–115.

Das, N., & Chandran, P. (2011). Microbial degradation of petroleum hydrocarbon contaminants: an overview. *Biotechnology Research International*, 2011.

Das, S., & Dash, H.R. (2014). Microbial bioremediation: a potential tool for restoration of contaminated areas. In *Microbial Biodegradation and Bioremediation*, 1–21. USA: Elsevier.

Davis, L.C., Castro-Diaz, S., Zhang, Q., & Erickson, L.E. (2002). Benefits of vegetation for soils with organic contaminants. *Critical Reviews in Plant Sciences*, *21*(5), 457–491.

Dejonghe, W., Goris, J., El Fantroussi, S., Höfte, M., De Vos, P., Verstraete, W., & Top, E.M. (2000). Effect of dissemination of 2, 4-dichlorophenoxyacetic acid (2, 4-D) degradation plasmids on 2, 4-D degradation and on bacterial community structure in two different soil horizons. *Applied and Environmental Microbiology*, *66*(8), 3297–3304.

Doty, S.L. (2008). Enhancing phytoremediation through the use of transgenics and endophytes. *New Phytologist*, *179*(2), 318–333.

Drăghiceanu, O.A., Dobrescu, C.M., & Soare, L.C. (2014). Applications of pteridophytes in phytoremediation. *Current Trends in Natural Sciences*, *3*(6), 68–73.

Dushenkov, S. (2003). Trends in phytoremediation of radionuclides. *Plant and Soil*, *249*(1), 167–175.

Eevers, N., White, J.C., Vangronsveld, J., & Weyens, N. (2017). Bio-and phytoremediation of pesticide-contaminated environments: a review. In *Advances in Botanical Research* (*Vol. 83*), 277–318. Academic Press.

EPA (2000). *Introduction to Phytoremediation. National Risk Management Research Laboratory*. US Environmental Protection Agency. Available from: http://www.clu-in.org/download/remed/introphyto.pdf.

EPA. (2000). National Primary Drinking Water Regulations: arsenic and clarifications to compliance and new source contaminants monitoring; Proposed Rule (40 CFR Parts 141 and 142). *Federal Register*, *65*, 38888–38983.

Erdem, M., & Özverdi, A. (2011). Environmental risk assessment and stabilization/solidification of zinc extraction residue: II. Stabilization/solidification. *Hydrometallurgy*, *105*(3-4), 270–276.

Favas, P.J., Pratas, J., Varun, M., D'Souza, R., & Paul, M.S. (2014). Accumulation of uranium by aquatic plants in field conditions: prospects for phytoremediation. *Science of the Total Environment*, *470*, 993–1002.

Gargouri, B., Karray, F., Mhiri, N., Aloui, F., & Sayadi, S. (2011). Application of a continuously stirred tank bioreactor (CSTR) for bioremediation of hydrocarbon-rich industrial wastewater effluents. *Journal of Hazardous Materials*, *189*(1–2), 427–434.

Gerhardt, K.E., Huang, X.D., Glick, B.R., & Greenberg, B.M. (2009). Phytoremediation and rhizoremediation of organic soil contaminants: potential and challenges. *Plant Science*, *176*(1), 20–30.

Hadacek, F. (2002). Secondary metabolites as plant traits: current assessment and future perspectives. *Critical Reviews in Plant Sciences*, *21*(4), 273–322.

Hasegawa, P.M. (2013). Sodium (Na1) homeostasis and salt tolerance of plants. *Environmental and Experimental Botany*, *92*, 1931

He, Z.L., Yang, X.E., & Stoffella, P.J. (2005). Trace elements in agroecosystems and impacts on the environment. *Journal of Trace Elements in Medicine and Biology*, *19*(2–3), 125–140.

Henry, J.R. (2000). *An Overview of the Phytoremediation of Lead and Mercury*, 1–31. Washington, DC: National Network of Environmental Management Studies (NNEMS). Environmental Protection Agency.

Hernández-Ortega, H.A., Alarcón, A., Ferrera-Cerrato, R., Zavaleta-Mancera, H.A., López-Delgado, H.A., & Mendoza-López, M.R. (2012). Arbuscular mycorrhizal fungi on growth, nutrient status, and total antioxidant activity of Melilotus albus during phytoremediation of a diesel-contaminated substrate. *Journal of Environmental Management*, *95*, S319–S324.

Hong, M.S., Farmayan, W.F., Dortch, I.J., Chiang, C.Y., McMillan, S.K., & Schnoor, J.L. (2001). Phytoremediation of MTBE from a groundwater plume. *Environmental Science & Technology*, *35*(6), 1231–1239.

Hussain, I., Aleti, G., Naidu, R., Puschenreiter, M., Mahmood, Q., Rahman, M.M., Wang, F., Shaheen, S., Syed, J.H., & Reichenauer, T.G. (2018). Microbe and plant assisted-remediation of organic xenobiotics and its enhancement by genetically modified organisms and recombinant technology: a review. *Science of the Total Environment*, *628*, 1582–1599.

Jha, P., Panwar, J., & Jha, P.N. (2015). Secondary plant metabolites and root exudates: guiding tools for polychlorinated biphenyl biodegradation. *International Journal of Environmental Science and Technology*, *12*(2), 789–802.

Johnson, A., & Singhal, N. (2015). Increased uptake of chelated copper ions by Lolium perenne attributed to amplified membrane and endodermal damage. *International Journal of Molecular Sciences*, *16*(10), 25264–25284.

Kang, W., Bao, J., Zheng, J., Xu, F., & Wang, L. (2018). Phytoremediation of heavy metal contaminated soil potential by woody plants on Tonglushan ancient copper spoil heap in China. *International Journal of Phytoremediation*, *20*(1), 1–7.

Karami, A., & Shamsuddin, Z.H. (2010). Phytoremediation of heavy metals with several efficiency enhancer methods. *African Journal of Biotechnology*, *9*(25), 3689–3698.

Komives, T., & Gullner, G. (2005). Phase I xenobiotic metabolic systems in plants. *Zeitschrift für Naturforschung C. A Journal of Biosciences*. *60*, 179–185.

Komossa, D., Langebartels, C., & Sandermann, J. (1995). In S. Trapp & J. McFarlane (Eds.), *H. Metabolic Processes for Organic Chemicals in Plants. Plant Contamination: Modeling and Simulation of Organic Chemical Processes*. USA.

Kuffner, M., Puschenreiter, M., Wieshammer, G., Gorfer, M., & Sessitsch, A. (2008). Rhizosphere bacteria affect growth and metal uptake of heavy metal accumulating willows. *Plant and Soil*, *304*(1–2), 35–44.

Kuiper, I., Kravchenko, L.V., Bloemberg, G.V., & Lugtenberg, B.J. (2002). Pseudomonas putida strain PCL1444, selected for efficient root colonization and naphthalene degradation, effectively utilizes root exudate components. *Molecular Plant-Microbe Interactions*, *15*(7), 734–741.

Kumar, J.N., & Oommen, C. (2012). Removal of heavy metals by biosorption using freshwater alga Spirogyra hyalina. *Journal of Environmental Biology*, *33*(1), 27.

Laha, S., Tansel, B., & Ussawarujikulchai, A. (2009). Surfactant–soil interactions during surfactant-amended remediation of contaminated soils by hydrophobic organic compounds: a review. *Journal of Environmental Management*, *90*(1), 95–100.

Lasat, M.M., Fuhrmann, M., Ebbs, S.D., Cornish, J.E., & Kochian, L.V. (1998). Phytoremediation of a radiocesium-contaminated soil: evaluation of cesium-137 bioaccumulation in the shoots of three plant species. *Journal of Environmental Quality, 27*(1), 165–169.

Li, W.C., Ye, Z.H., & Wong, M.H. (2007). Effects of bacteria on enhanced metal uptake of the Cd/Zn-hyperaccumulating plant, Sedum alfredii. *Journal of Experimental Botany, 58*(15e16), 4173e4182

Lu, J., Guo, C., Li, J., Zhang, H., Lu, G., Dang, Z., & Wu, R. (2013). A fusant of Sphingomonas sp. GY2B and Pseudomonas sp. GP3A with high capacity of degrading phenanthrene. *World Journal of Microbiology and Biotechnology, 29*(9), 1685–1694.

Ma, L.Q., Komar, K.M., & Kennelley, E.D., University of Florida (2001). Methods for removing pollutants from contaminated soil materials with a fern plant. U.S. Patent 6,302,942.

Macci, C., Doni, S., Peruzzi, E., Bardella, S., Filippis, G., Ceccanti, B., & Masciandaro, G. (2013). A real-scale soil phytoremediation. *Biodegradation, 24*(4), 521–538.

MacNaughton, S.J., Stephen, J.R., Venosa, A.D., Davis, G.A., Chang, Y.J., & White, D.C. (1999). Microbial population changes during bioremediation of an experimental oil spill. *Applied and Environmental Microbiology, 65*(8), 3566–3574.

Madsen, E.L. (1991). Determining in situ biodegradation. *Environmental Science & Technology, 25*(10), 1662–1673.

Maestri, E., Marmiroli, M., Visioli, G., & Marmiroli, N. (2010). Metal tolerance and hyperaccumulation: costs and trade-offs between traits and environment. *Environmental and Experimental Botany, 68*(1), 1–13.

Manousaki, E., & Kalogerakis, N. (2010). Halophytes present new opportunities in phytoremediation of heavy metals and saline soils. *Industrial & Engineering Chemistry Research, 50*(2), 656–660.

Marecik, R., Króliczak, P., & Cyplik, P. (2006). Fitoremediacja–alternatywa dla tradycyjnych metod oczyszczania œrodowiska. *Biotechnologia, 3*(74), 88–97.

Meagher, R.B. (2000). Phytoremediation of toxic elemental and organic pollutants. *Current Opinion in Plant Biology, 3*(2), 153–162.

Mei, L.I., Xitao, X., Renhao, X.U.E., & Zhili, L.I.U. (2006). Effects of strontium-induced stress on marine microalgaePlatymonas subcordiformis (Chlorophyta: Volvocales). *Chinese Journal of Oceanology and Limnology, 24*(2), 154–160.

Meuser, H. (2012). *Soil Remediation and Rehabilitation: Treatment of Contaminated and Disturbed Land (Vol. 23)*. USA: Springer Science & Business Media.

Mishra, V.K., & Tripathi, B.D. (2008). Concurrent removal and accumulation of heavy metals by the three aquatic macrophytes. *Bioresource Technology, 99*(15), 7091–7097.

Murphy, C.E.J. (2001). *An Estimate of the History of Tritium Inventory in Wood Following Irrigation With Tritiated Water.* Aiken, SC: Westinghouse Savannah River Company. Available from: https://www.srs.gov/general/srs-home.

Naik, M.G., & Duraphe, M.D. (2012). Review paper on–parameters affecting bioremediation. *International Journal of Life Science and Pharma Research, 2*(3), L77–L80.

Narasimhan, K., Basheer, C., Bajic, V.B., & Swarup, S. (2003). Enhancement of plant-microbe interactions using a rhizosphere metabolomics-driven approach and its application in the removal of polychlorinated biphenyls. *Plant Physiology, 132*(1), 146–153.

Nikolaivits, E., Dimarogona, M., Fokialakis, N., & Topakas, E. (2017). Marine-derived biocatalysts: importance, accessing, and application in aromatic pollutant bioremediation. *Frontiers in Microbiology, 8*, p.265.

Nikolopoulou, M., & Kalogerakis, N. (2008). Enhanced bioremediation of crude oil utilizing lipophilic fertilizers combined with biosurfactants and molasses. *Marine Pollution Bulletin, 56*(11), 1855–1861.

Nowack, B., Schulin, R., & Robinson, B.H. (2006). Critical assessment of chelant-enhanced metal phytoextraction. *Environmental Science & Technology, 40*(17), 5225–5232.

Ouvrard, S., Barnier, C., Bauda, P., Beguiristain, T., Biache, C., Bonnard, M., Caupert, C., Cébron, A., Cortet, J., Cotelle, S., & Dazy, M. (2011). In situ assessment of phytotechnologies for multicontaminated soil management. *International Journal of Phytoremediation, 13*(sup. 1), 245–263.

Padmavathiamma, P.K., & Li, L.Y. (2012). Rhizosphere influence and seasonal impact on phytostabilisation of metals – a field study. *Water, Air, & Soil Pollution, 223*(1), 107–124.

Pandey, V.C., Bajpai, O., & Singh, N. (2016). Energy crops in sustainable phytoremediation. *Renewable and Sustainable Energy Reviews, 54*, 58–73.

Pandey, V.C., Pandey, D.N., & Singh, N. (2015). Sustainable phytoremediation based on naturally colonizing and economically valuable plants. *Journal of Cleaner Production, 86*, 37–39.

Pilon-Smits, E. (2005). Phytoremediation. *Annual Review of Plant Biology, 56*, 15–39.

Pilon-Smits, E.A., & Freeman, J.L. (2006). Environmental cleanup using plants: biotechnological advances and ecological considerations. *Frontiers in Ecology and the Environment, 4*(4), 203–210.

Rabus, R., Wilkes, H., Schramm, A., Harms, G., Behrends, A., Amann, R., & Widdel, F. (1999). Anaerobic utilization of alkylbenzenes and n-alkanes from crude oil in an enrichment culture of denitrifying bacteria affiliating with the b-subclass of Proteobacteria. *Environmental Microbiology, 1*(2), 145–158.

Reeves, R.D. (2003). Tropical hyperaccumulators of metals and their potential for phytoextraction. *Plant and Soil, 249*(1), 57–65.

Rew, A. (2007). Phytoremediation: an environmentally sound technology for pollution prevention, control and remediation in developing countries. *Educational Research and Reviews, 2*(7), 151–156.

Ripp, S., Nivens, D.E., Ahn, Y., Werner, C., Jarrell, J., Easter, J.P., Cox, C.D., Burlage, R.S., & Sayler, G.S. (2000). Controlled field release of a bioluminescent genetically engineered microorganism for bioremediation process monitoring and control. *Environmental Science & Technology, 34*(5), 846–853.

Robinson, B.H., Chiarucci, A., Brooks, R.R., Petit, D., Kirkman, J.H., Gregg, P.E.H., & De Dominicis, V. (1997). The nickel hyperaccumulator plant Alyssum bertolonii as a potential agent for phytoremediation and phytomining of nickel. *Journal of Geochemical Exploration, 59*(2), 75–86.

Rosen, M.J. (1989). *Surfactant and Interfacial Phenomena*. New York, NY: John Willey and Sons.

Sandermann, J.H. (1994). Higher plant metabolism of xenobiotics: the'green liver'concept. *Pharmacogenetics, 4*(5), 225–241.

Schwitzguébel, J.P. (2017). Phytoremediation of soils contaminated by organic compounds: hype, hope and facts. *Journal of Soils and Sediments, 17*(5), 1492–1502.

Seth, C.S. (2012). A review on mechanisms of plant tolerance and role of transgenic plants in environmental clean-up. *The Botanical Review, 78*(1), 32–62.

Shabani, N., & Sayadi, M.H. (2012). Evaluation of heavy metals accumulation by two emergent macrophytes from the polluted soil: an experimental study. *The Environmentalist, 32*(1), 91–98.

Sharma, S. (2012). Bioremediation: features, strategies and applications. *Asian Journal of Pharmacy and Life Science*, ISSN, 2231, 4423.

Shukla, S.K., Mangwani, N., Rao, T.S., & Das, S. (2014). Biofilm-mediated bioremediation of polycyclic aromatic hydrocarbons. In *Microbial Biodegradation and Bioremediation*, 203–232. USA: Elsevier.

Singh, B.R., Gupta, S.K., Azaizeh, H., Shilev, S., Sudre, D., Song, W.Y., Martinoia, E., & Mench, M. (2011). Safety of food crops on land contaminated with trace elements. *Journal of the Science of Food and Agriculture, 91*(8), 1349–1366.

Singh, P., Jain, R., Srivastava, N., Borthakur, A., Pal, D.B., Singh, R., Madhav, S., Srivastava, P., Tiwary, D., & Mishra, P.K. (2017). Current and emerging trends in bioremediation of petrochemical waste: A review. *Critical Reviews in Environmental Science and Technology, 47*(3), 155–201.

Singh, P., Singh, V.K., Singh, R., Borthakur, A., Madhav, S., Ahamad, A., Kumar, A., Pal, D.B., Tiwary, D., & Mishra, P.K. (2020). Bioremediation: a sustainable approach for management of environmental contaminants. In *Abatement of Environmental Pollutants*, 1–23. USA: Elsevier.

Sivaperumal, P., Kamala, K., & Rajaram, R. (2017). Bioremediation of industrial waste through enzyme producing marine microorganisms. In *Advances in Food and Nutrition Research (Vol. 80)*, 165–179. Academic Press.

Stephenson, C., & Black, C.R. (2014). One step forward, two steps back: the evolution of phytoremediation into commercial technologies. *Bioscience Horizons: The International Journal of Student Research*, 7.1–15

Sun, J., Wu, X., & Gan, J. (2015). Uptake and metabolism of phthalate esters by edible plants. *Environmental Science & Technology, 49*(14), 8471–8478.

Susarla, S., Medina, V.F., & McCutcheon, S.C. (2002). Phytoremediation: an ecological solution to organic chemical contamination. *Ecological Engineering, 18*(5), 647–658.

Tak, H.I., Ahmad, F., & Babalola, O.O. (2013). Advances in the application of plant growth-promoting rhizobacteria in phytoremediation of heavy metals. In *Reviews of Environmental Contamination and Toxicology (Vol. 223)*, 33–52. New York, NY: Springer.

Tang, W.C., White, J.C., & Alexander, M. (1998). Utilization of sorbed compounds by microorganisms specifically isolated for that purpose. *Applied Microbiology and Biotechnology*, *49*(1), 117–121.

Tekere, M., Mswaka, A.Y., Zvauya, R., & Read, J.S. (2001). Growth, dye degradation and ligninolytic activity studies on Zimbabwean white rot fungi. *Enzyme and Microbial Technology*, *28*(4–5), 420–426.

Teng, Y., Luo, Y., Sun, X., Tu, C., Xu, L., Liu, W., et al. (2010). Influence of arbuscular mycorrhiza and Rhizobium on phytoremediation by alfalfa of an agricultural soil contaminated with weathered PCBs: a field study. *International Journal of Phytoremediation*, *12*, 516533.

Tong, Y.P., Kneer, R., & Zhu, Y.G. (2004). Vacuolar compartmentalization: a second-generation approach to engineering plants for phytoremediation. *Trends in Plant Science*, *9*(1), 7–9.

Tripathi, S., Singh, V.K., Srivastava, P., Singh, R., Devi, R.S., Kumar, A., & Bhadouria, R. (2020). Phytoremediation of organic pollutants: current status and future directions. In *Abatement of Environmental Pollutants*, 81–105. USA: Elsevier.

Truu, J., Truu, M., Espenberg, M., Nõlvak, H., & Juhanson, J. (2015). Phytoremediation and plant-assisted bioremediation in soil and treatment wetlands: a review. *The Open Biotechnology Journal*, *9*(1), 85–92.

Valls, M., & De Lorenzo, V. (2002). Exploiting the genetic and biochemical capacities of bacteria for the remediation of heavy metal pollution. *FEMS Microbiology Reviews*, *26*(4), 327–338.

Valls, M., Atrian, S., de Lorenzo, V., & Fernández, L.A. (2000). Engineering a mouse metallothionein on the cell surface of Ralstonia eutropha CH34 for immobilization of heavy metals in soil. *Nature Biotechnology*, *18*(6), 661.

Vangronsveld, J., Herzig, R., Weyens, N., Boulet, J., Adriaensen, K., Ruttens, A., Thewys, T., Vassilev, A., Meers, E., Nehnevajova, E., & van der Lelie, D. (2009). Phytoremediation of contaminated soils and groundwater: lessons from the field. *Environmental Science and Pollution Research*, *16*(7), 765–794.

Vanhoudt, N., Vandenhove, H., Leys, N., & Janssen, P. (2018). Potential of higher plants, algae, and cyanobacteria for remediation of radioactively contaminated waters. *Chemosphere*, *207*, 239–254.

Vasudevan, S., & Oturan, M.A. (2014). Electrochemistry: as cause and cure in water pollution – an overview. *Environmental Chemistry Letters*, *12*(1), 97–108.

Wang, H.Q., Zhao, Q., Zeng, D.H., Hu, Y.L., & Yu, Z.Y. (2015). Remediation of a magnesium-contaminated soil by chemical amendments and leaching. *Land Degradation & Development*, *26*(6), 613–619.

Watanabe, K., Watanabe, K., Kodama, Y., Syutsubo, K., & Harayama, S. (2000). Molecular characterization of bacterial populations in petroleum-contaminated groundwater discharged from underground crude oil storage cavities. *Applied and Environmental Microbiology*, *66*(11), 4803–4809.

Watt, N.R., Willey, N.J., Hall, S.C., & Cobb, A. (2002). Phytoextraction of 137Cs: the effect of soil 137Cs concentration on 137Cs uptake by Beta vulgaris. *Acta Biotechnologica*, *22*(1–2), 183–188.

Wiszniewska, A., Hanus-Fajerska, E., Muszyńska, E., & Ciarkowska, K. (2016). Natural organic amendments for improved phytoremediation of polluted soils: a review of recent progress. *Pedosphere*, *26*(1), 1–12.

Yao, Z., Li, J., Xie, H., & Yu, C. (2012). Review on remediation technologies of soil contaminated by heavy metals. *Procedia Environmental Sciences*, *16*, 722–729.

6 Process Intensification in Bio-Based Approaches for Environmental Remediation

Kailas L. Wasewar
Advance Separation and Analytical Laboratory (ASAL), Department of Chemical Engineering, Visvesvaraya National Institute of Technology (VNIT), Nagpur, MH, India

CONTENTS

- 6.1 Introduction 110
- 6.2 Process Intensification 110
- 6.3 Bio-Based Approaches 111
- 6.4 Ex-Situ Bioremediation 111
- 6.5 In-Situ Bioremediation 112
- 6.6 Natural/Intrinsic Attenuation 112
- 6.7 Composting 112
- 6.8 Land Farming 112
- 6.9 Culture-Based Approach 113
- 6.10 Omics-Based Approach 113
- 6.11 Bioaugmentation 113
- 6.12 Biostimulation 114
- 6.13 Phytoremediation 114
- 6.14 Biopile 114
- 6.15 Windrows 115
- 6.16 Bioreactors 115
- 6.17 Bioventing 115
- 6.18 Bioslurping 116
- 6.19 Biosparging 116
- 6.20 Biosorption 116
- 6.21 Microbial Electrochemical Technologies 116
- 6.22 Hybrid Approaches 117
- 6.23 Permeable Reactive Barrier 117
- 6.24 Monoliths 117
- 6.25 Membrane Technology 118
- 6.26 Membrane Bioreactors 118
- 6.27 Hybrid Membrane Bioreactors 118
- 6.28 Nanobioremediation 119
- 6.29 Conclusion 119
- References 120

6.1 Introduction

As the world is heading towards rapid urbanisation and technological advancements, more undesirable and unwanted activities by humans is raising major environmental issues like global warming and imbalance in soil ecological processes, leading to lower agricultural yield, drastic climate change, etc. (Goswami et al., 2018). Rapid industrialisation and population explosion have resulted in the generation and dumping of various contaminants into the environment. These harmful compounds deteriorate human health as well as the surrounding environments. The harmful contaminants spread in the environment in any form are responsible for many adverse effects on living things as well as non-living things. Bioremediation may be the promising approach for the removal or treatment of these contaminants. Sustainable development revolves around three major factors: social, economic, and environmental.

Remarkable efforts have been taken in the last few decades for the development and implementation of sustainable and environmentally friendly technologies and processes in industrial and other sectors. Soil contamination with petroleum hydrocarbons, persistent organic pollutants, halogenated organic chemicals, and toxic metal(loid)s is a serious global problem affecting human and ecological health (Megharaj and Naidu, 2017). There has been an increasing burden on the industry to have safe emissions and disposals, which forced them to look for greener and more sustainable approaches in the industry. The minimisation of waste at the source is the best way for more sustainable and environmentally friendly approaches in industry. Process intensification with innovative approaches has been successfully employed to overcome such challenges (Boodhoo and Harvey, 2013).

The various bio-based approaches used for environmental remediation are decarbonation of energy, biofilters, biosensors, bioadsorption, horizontal gene transfer, etc. Bioremediation is defined as a process that relies on biological mechanisms to reduce (degrade, detoxify, mineralise, or transform) the concentration of pollutants to an innocuous state (Azubuike et al., 2016). The conventional bio-based approaches for environmental remediation have certain limitations; hence, it is now a requirement to develop new approaches (Mishra et al., 2020). The developments in various new equipment and methods may expect enhancement in efficiency and eco-friendliness. Process intensification plays a significant role for developing such approaches. In the present chapter, various aspects of process intensification, conventional bio-based approaches, and intensifying bio-based approaches for environmental remediations are discussed. Also, a few applications of bio-based intensifying approaches are presented.

6.2 Process Intensification

In general, process intensification targets a significant reduction in plant volume; hence, a decrease in cost and environmental impact and an increase in safety. There are numerous definitions stated in literature that may be summarised as drastic improvements in technologies and processes with a substantial decrease in overall volume, less requirement of energy, higher efficiency, almost zero waste or well-treated waste, cheaper and safer innovative, and sustainable approaches for the benefit of society (Boodhoo and Harvey, 2013; Cross and Ramshaw, 1986; Stankiewicz and Moulijn, 2000; Gerven and Stankiewicz; 2009). Process intensification is classified as mainly two categories of equipment and methods. Equipment includes with reactions and without reactions as spinning disk reactor, static mixer reactor, monolithic reactors, microreactors, static mixers, rotating equipment, etc. The methods comprise multifunctional reactors, hybrid separations, alternative energy sources, and other methods (Stankiewicz and Moulijn, 2000).

Process intensification has benefits for the business, process, and environment (Boodhoo and Harvey, 2013). Business benefits can be achieved by the responsive processing, such as miniaturised plants, reduced capital cost, reduced operating cost, distributed manufacturing, and faster introduction of new products to market. The process benefits include higher selectivity, higher product purity, higher reaction rates, improved product properties, improved process safety, and wider processing conditions.

These can result in faster, safer, and greener processing. Environmental benefits of process intensification are the sustainable processing with reduced energy use, reduced waste, reduced solvent use, smaller plants, and less obstructions in landscapes (Boodhoo and Harvey, 2013).

Process intensification approaches can be implemented successfully in industries subjected to overcome few challenges. Still, there are numerous reasons for the lack of implementation of process intensification. Boodhoo and Harvey (2013) have stated these challenges as perception of risk, lack of champions within industry, control/monitoring, and limitations of process intensifications.

6.3 Bio-Based Approaches

The soil on earth is an important resource and also supports functioning of the ecosystem (Rafael et al., 2020). The health of soil has been damaged by many domestic, industrial, agricultural, and other activities. To understand the soil health, various indicators must be assessed: physical, chemical, and biological. Physical indicators are soil properties (pH, organic matter, redox potential, texture, and nutrient content), chemical indicators are contaminant concentrations, and biological indicators are biological communities (biomass, activity, biodiversity) and ecotoxicity bioassays (plant, animal, microbial bioassays) (Rafael et al., 2020).

Most of the traditional methods for soil treatments are suffered by economical infeasibility and are not eco-friendly. These methods are mainly classified as phytoremediation, bioremediation, and vermiremediation.

Anthropogenic activities such as industrial, mining, and military processes are the major sources that contribute to widespread contamination of the environment throughout the world, with numerous chemicals including petroleum hydrocarbons, polyaromatic hydrocarbons (PAHs), polychlorinated biphenyls (PCBs), halogenated dibenzodioxins/furans, chlorinated solvents, pesticides, and toxics (Megharaj and Naidu, 2017).

In biological approaches, microorganisms (bacteria, fungi) are used for the contaminated or polluted soil remediation (Rafael et al., 2020). These approaches can be used for treatment of inorganic and organic contaminants (Megharaj et al., 2011; Fingerman and Nagabhushanam, 2016; Park et al., 2011). There are mainly three bio-approaches for remediation of soil: natural attenuation (native microbial used), bioaugmentation (based on inoculation of microbial strains), and biostimulation (change of conditions) (Bento et al., 2005; Rafael et al., 2020). The addition of nutrients and other additives (inorganic, organic, sewage sludge, manure, compost, etc.) enhance the degradation rate (Park et al., 2011; Ramadass et al., 2018; Ros et al., 2010; Lee et al., 2008).

6.4 Ex-Situ Bioremediation

Environmental pollution has been on the rise in the past few decades, due to increased human activities on energy reservoirs, unsafe agricultural practices, and rapid industrialisation. Amongst the pollutants that are of environmental and public health concerns due to their toxicities are heavy metals, nuclear wastes, pesticides, greenhouse gases, and hydrocarbons. Remediation of polluted sites using microbial processes (bioremediation) has proven effective and reliable due to its eco-friendly features (Azubuike et al., 2016). Bioremediation can either be carried out ex situ or in situ, depending on several factors that include, but are not limited to, cost, site characteristics, and type and concentration of pollutants.

In the case of ex-situ bioremediation, contaminants are excavated from contaminated sites and transported to another location for treatment. These methods are usually employed based on different factors such as cost of treatment, type and concentration of pollutants, geographical location, and geology of the contaminated site (Azubuike et al., 2016; Philp and Atlas 2005). Biopile, windrow, bioreactors, and land farming are the methods for ex-situ bioremediation. Generally, ex-situ techniques are more expensive compared to in-situ techniques as a result of additional cost attributed to excavation. However, the cost of on-site installation of equipment and inability to effectively visualise and control the subsurface of polluted sites are major concerns when carrying out in-situ bioremediation. Therefore, choosing the appropriate bioremediation technique, which will effectively reduce pollutant concentrations to an innocuous state, is crucial for a successful bioremediation project (Azubuike et al., 2016).

6.5 In-Situ Bioremediation

In the case of in-situ bioremediation, the contaminated site is treated on-site with a suitable bio-based method. In this method, excavation is not needed and hence there is no disturbance to the soil structure. This method is less expensive compared to ex-situ remediation as the extra cost for excavation is not needed in this case. There are two types of in-situ methods: natural or intrinsic attenuation and enhanced remediation (Azubuike et al., 2016). The microbial activities can be enhanced by various methods such as bioventing, biosparging, and phytoremediation. In-situ bioremediation has been successfully employed for the treatment of chlorinated solvents, dyes, heavy metals, and hydrocarbon polluted sites (Folch et al., 2013; Kim et al., 2014; Frascari et al., 2015; Roy et al., 2015).

6.6 Natural/Intrinsic Attenuation

In this process, the contaminants or pollutants are degraded/decomposed to the harmless products by the natural processes of microbial degradation, volatilisation, sorption, and immobilisation (Megharaj and Naidu, 2017; Azubuike et al, 2016). Both microbial aerobic and anaerobic processes are used to biodegrade polluting substances, including those that are recalcitrant. The absence of an external force implies that the technique is less expensive compared to other in-situ techniques (Azubuike et al, 2016). This is generally considered for petroleum hydrocarbon polluted sites. Natural healing with time is the best way for remediation of the environment, but it is very slow. It was successfully employed for a few hydrocarbons such as benzene, toluene, ethylbenzene, and xylene, but is not suitable for persistent organic pollutants (Megharaj and Naidu, 2017).

6.7 Composting

The contaminated soil can be remediated by using compost or composting as it is a very economical treatment method that can efficiently increase the fertility of the soli with enhanced bioremediation (Megharaj and Naidu, 2017). There have been numerous investigations in this area for the treatment of soil from various contaminants such as toxic metals, PAHs, and pesticides (Semple et al.,2001; Tandy et al., 2009; Zeng et al., 2011). Degradation of organic contaminants in soil is often difficult due to their low bioavailability. The addition of surfactants to soil can increase the bioavailability of some organic pollutants (Chan 2008). A short summary of this method has been presented in literature (Megharaj and Naidu, 2017).

6.8 Land Farming

Land farming is one of the cheapest and simplest bioremediation techniques. It requires less equipment for operation and also can be considered as in-situ or ex-situ bioremediation, which can be decided based on the treatment site where depth of contaminants plays a significant role. Excavation and/or tilted of contaminated soils are considered in land farming bioremediation (Azubuike et al., 2016). Overall, the land farming bioremediation technique is very simple to design and implement, requires low capital inputs, and can be used to treat large volumes of polluted soil with minimal environmental impact and energy requirements (Maila and Colete 2004). But, this bioremediation has certain limitations including large operating space, reduction in microbial activities due to unfavourable environmental conditions, additional cost due to excavation, reduced efficacy in inorganic pollutant removal, not suitable for treating soil polluted with toxic volatiles, time consuming, and less efficient (Azubuike et al., 2016; Khan et al., 2004; Maila and Colete 2004).

6.9 Culture-Based Approach

Environmental contamination with persistent organic pollutants has emerged as a serious threat of pollution. Bioremediation is a key to eliminate these harmful pollutants from the environment and has gained the interest of researchers during the past few decades (Mishra et al., 2020) including culture-based and omics-based approaches. In the culture-based approach, the microorganisms present in the ecosystem have been used for treatment (Varjani and Upasani, 2017). The microbiological techniques are employed for the isolation and cultivation of microorganisms from the pollutant side. These techniques include enumeration, liquid media, and biolog plates (Varjani and Upasani, 2013; Wilkes et al., 2016). The intensification in the development of suitable microbial strains for the specific application of degradation or removal of various organic pollutants has been performed. The various applications of developed intensified strains for the degradation and removal of various organic pollutants are crude oil; diesel oil; petroleum oil; petroleum hydrocarbons; marine oil spills; 4-bromophenol; 2,4-dichlorophenoxy acetic acid; chlorothalonil; propiconazole; thiabendazole; atrazine; chloroanilines; 2,4,5-trichlorophenoxyacetic acid; pendimethalin; 2,4-dichlorophenoxyacetic acid; beta-cypermethrin; glyphosate; acetamiprid; organophosphorus; chlorpropham; 2-chloro-4-nitrophenol; di(2-ethylhexyl) phthalate; methyl parathion; and malachite green (Mishra et al., 2020).

6.10 Omics-Based Approach

The omics-based approach for microbial treatment of organic pollutants in a polluted site has been discussed by Mishra et al., (2020). It is a kind of next-generation approach called a cultivation-independent technique, which comprises various steps of the selection and characterisation of microbial communities in reservoirs. These approaches have various methods, such as metagenomics, meta-transcriptomics, metaproteomics, and metabolomics. The information for monitoring the microbial communities is provided by metagenomics to the functional gene in polluted sites (Thomas and Gilbert, 2012). Metatranscriptomics is used to characterise different microbial communities for ecosystem applications including carbon cycling, bioremediation, and agriculture (Starke et al., 2017; Bastida et al., 2016; Schneider et al., 2012). The functioning of the ecosystem for phylogenetic and functional information is obtained by the use of metaproteins in polluted sites. The various application of the omics-based approach for microbial treatment of organic pollutants in polluted sites are presented in literature (Mishra et al., 2020; Malla et al., 2018).

6.11 Bioaugmentation

The biological approach for environmental remediation can be performed in two different ways by the introduction of efficient microbial strains (bioaugmentation) and by the addition of rate-limiting nutrients to the soil (biostimulation) (Goswami et al., 2018). In bioaugmentation, a microbial strain of autochthonous or allochthonous wild type or genetically modified microorganisms is employed for the treatment of hazardous waste sites and removal of pollutants (Megharaj and Naidu, 2017; Goswami et al., 2018, Mrozik and Piotrowska-Seget, 2010). It can be mostly considered for the treatment of oil-contaminated environments. Simply, bioaugmentation enhances the degradation rate of the pollutants by the addition of microorganisms (Leahy and Colwell, 1990; Adams et al., 2015). The bioaugmentation depends on the adaptation of the microbial consortia to the site and various abiotic factors (Godleads et al., 2015). It can be employed for the soils with less microorganisms and need multistage remediation (Mrozik and Piotrowska-Seget, 2010, Forsyth et al., 1995). A number of microorganisms have been used for bioaugmentation for various applications, such as four chlorobenzoic acid; chlorobenzoates; 2,4-dinitrotoluene; napthalene; etc. using *Pseudomonas putida, Cupriavidus necator, Pseudomonas fluorescens, Pseudomonas putida,* etc., respectively (Massa et al., 2009; Monti et al., 2005; Filonov et al., 2005; Goswami et al., 2018).

6.12 Biostimulation

Biostimulation is a remediation technique that is highly efficient, cost effective, and eco-friendly in nature. Biostimulation refers to the addition of rate-limiting nutrients like phosphorus, nitrogen, oxygen, and electron donors to severely polluted sites to stimulate the existing bacteria to degrade the hazardous and toxic contaminants (Megharaj and Naidu, 2017; Tyagi et al., 2010; Elektorowicz, 1994; Goswami et al., 2018). In biostimulation, rate-limiting nutrients are added to improve the degradation capability of microorganisms and this method is the most efficient for hydrocarbon remediation (Adams et al., 2015; Nikolopoulou and Kalogerakis, 2009). The rate of remediation by biostimulation is affected by moisture content, pH, temperature, and environmental physiology. Biostimulation can be employed successfully for the degradation of sulphates, petroleum hydrocarbons, and polyester polyurethanes. Goswami et al., (2018) have summarised the application of biostimulation for the degradation of various compounds such as atrazine, alachlor, metolachlor, trifluralin, cyanazine, fluometuron, methabenzthiazuron, etc. using various nutrients including animal manure, sewage sludge, cellulose, straw, compost, ammonium nitrate, potassium nitrate, and ammonium phosphate, etc.

6.13 Phytoremediation

Plants can be used for the remediation of wastewater as well as soil by phytoremediation. It is one of the cost-effective alternatives to other tradition physiochemical methods. It can be performed as phytostabilisation and phytoaccumulation. Phytoremediation has emerged as a promising strategy for in-situ removal of a wide variety of contaminants (Gerhardt et al., 2009; Megharaj and Naidu, 2017; Azubuike et al., 2016). Depending on the pollutant type (elemental or organic), there are several mechanisms (accumulation or extraction, degradation, filtration, stabilisation, and volatilisation) involved in phytoremediation. Plants in association with microbes seem to be more effective for removal/degradation of organic contaminants from impacted soils. The disadvantages of phytoremediation are that it is a slow process requiring several years and more crop harvests and the challenge is that there are stressors (variation in temperature, nutrients, precipitation, herbivory, plant pathogens, and competition by weeds) that affect phytoremediation in the field but are not encountered in the greenhouse (Megharaj and Naidu, 2017). One of the major advantages of using plants to remediate polluted sites is that some precious metals can bioaccumulate in some plants and recover after remediation, a process known as phytomining. Azubuike et al., (2016) summarised various plants with the potential for removal of various contaminants from soil and wastewater. A few of the contaminants are gasoline, diesel, brominated diphenyl ethers, PCBs, anthracene, fluoranthene, PAHs, As, Cu, Pb, Zn, Cd, Fe, B, Cr, and Ni; silver nanoparticles have been treated using various plants as *Ludwigia octovalvis, Aegiceras corniculatum, Spartina maritime, Arundo donax, Eichhorina crassipes, Phragmites australis, Plectranthus amboinicus, Luffa acutangula, Dracaena reflexa, Sparganium* sp., etc. (Azubuike et al., 2016).

6.14 Biopile

In the biopile remediation technique, contaminated soil is excavated and piled above ground. The nutrients are added in the pile and also sometimes aerated to increase the bioremediation through enhanced microbial activities (Azubuike et al., 2016). This step in this remediation involves aeration, irrigation, nutrient and leachate collection systems, and a treatment bed. It is considered for most applications due to cost effectiveness and other constructive features where operating conditions such as nutrient, temperature, and aeration are effectively controlled (Azubuike et al., 2016; Whelan et al., 2015). Heating systems can be incorporated in biopiles to enhance the activities of microbes and hence biodegradability of contaminants (Aislabie et al., 2006). There are humidified biopiles that have optimal moisture content, reduced leaching, and minimal volatilisation of less degradable contaminants compared to heated and passive biopiles (Sanscartier et al., 2009). It can be efficiently employed for the treatment of large volumes of contaminated soil with limited space. Other additives, such as bulking agents that include straw, saw dust, bark, or wood chips and other organic materials, can be added to increase the rate of biodegradation in biopile remediation (Rodrı́guez-Rodrı́guez et al., 2010).

6.15 Windrows

In windrows, an ex-situ bioremediation process, periodic turning of contaminated soil piles is performed to enhance the biodegradation performance of hydrocarbonoclastic bacteria available in the soil (Azubuike et al., 2016). In this method, apart from turning piles, the addition of water is carried out to obtain higher aeration, uniform pollutant distribution, and nutrient and microbial performance, which further enhances the bioremediation rate through assimilation, biotransformation, and mineralisation (Barr, 2002). It has a high degradation rate compared to biopiling. This method is not suitable for soil contaminated with toxic volatiles due to periodic turning (Hobson et al., 2005).

6.16 Bioreactors

A bioreactor is used to process certain raw material to convert it into products via biological reactions. The bioreactors can be operated in batch, fed-batch, sequencing batch, continuous and multistage which depends on the operating and capital investment, quality and quantity of the material. Bioreactor has several advantages as compared to other ex situ bioremediation as excellent control of process parameters (temperature, pH, agitation and aeration rates, substrates, and inoculum concentrations). The factors for the design of ex-situ bioremediation considered effective are controlled bioaugmentation, nutrient addition, increased pollutant bioavailability, and mass transfer (contact between pollutant and microbes) (Azubuike et al., 2016).

Various types of bioreactors as stirred tank, granular sludge bed, packed bed, slurry reactor, and membrane bioreactor have been employed for the treatment of different pollutants such as petroleum, polyaromatic, saturated, aromatic hydrocarbons, linear alkylbenze sulfonate, benzene, toluene, ethylbeneze, xylenenaphthalene, dibromoeopentyl glycol, nano fullerenes, nanosilver, carbofuran, and tetryl from contaminated soil (Azubuike et al., 2016; Chikere et al., 2016; Bhattacharya et al., 2015; Delforno et al., 2015; Firmino et al., 2015; Jua´rez-Ramı́rez et al., 2015; Mustafa et al., 2015; Saravanan et al., 2015; Xu et al., 2015; Yang et al., 2015; Zangi-Kotler et al., 2015; Plangklang and Alissara Reungsang, 2010; Fuller et al., 2003).

6.17 Bioventing

Bioventing is one of the intensified in-situ bioremediation techniques used for the treatment of contaminated soil. The stimulation of airflow is controlled by supplying oxygen to the unsaturated (vadose) zone to intensify bioremediation through enhanced activities of microbes (Azubuike et al., 2016; Philp and Atlas, 2005). It is more suitable for the light-spilled petroleum products contaminated soil as compared to other in-situ techniques (Ho¨hener and Ponsin, 2014). The biodegradation of contaminants are intensified by the number of air injection points for providing uniform air distribution. The pure oxygen can be injected to get higher concentration of oxygen as compared to air injection which accelerate biodegradation (Philp and Atlas, 2005). It can be observed that bioventing is prolonged due to certain environmental factors in scaling up, which can be overcome by a combination of bioventing and biotrickling filters that further reduce the exit level of contaminants and reduction in time (Burgess et al., 2001; Magalhã~es et al., 2009; Azubuike et al., 2016).

6.18 Bioslurping

Bioslurping is one of the in-situ bioremediation techniques for the treatment of contaminants. In this technique, vacuum-enhanced pumping, soil vapour extraction, and bioventing are combined to perform groundwater and soil remediation (Azubuike et al., 2016; Gidarakos and Aivalioti, 2007). It is designed and employed for light non-aqueous phase liquids and volatile and semi-volatile organic compounds. The oxygen transfer rate is decreased in this method due to limits in air permeability by excessive soil moisture. This will reduce the biological activities of the microbes. It is not suitable for low permeability soil but less groundwater is needed, which saves costs and hence minimises storage, treatment, and disposal costs (Philp and Atlas, 2005).

6.19 Biosparging

Biosparging is similar to bioventing, where air is injected into soil subsurface to stimulate microbial activities to degrade the contaminants from the site. In this technique, air is injected at the saturated zone for upward movement of VOCs to the unsaturated zone to intensify the biodegradation of contaminants. Soil permeability and pollutant biodegradability are the key parameters to decide the effectiveness of biosparging (Azubuike et al., 2016; Philp and Atlas, 2005). Biosparging has been employed for the treatment of contaminants such as diesel, kerosene, benzene, toluene, ethylbenzene, and xylene (Azubuike et al., 2016; Kao et al., 2008).

6.20 Biosorption

Microorganisms have the ability to accumulate metals from aqueous solutions; hence, they can be used as biosorbents compared to conventional technologies. The ability of microorganisms to bind metals from an aqueous solution in some cases is selectively known as biosorption and microorganisms (bacteria, fungi, and algae) responsible for this process are called biosorbents (Sen and Chakrabarti, 2009). The modes by which the microorganisms remove metal ions from a solution are (i) extracellular accumulation/precipitation, (ii) cell surface sorption or complexation, and (iii) intracellular accumulation. Among these, process (ii) can occur whether the organism is living or dead; process (i) may be facilitated by microbial viability, while process (iii) requires microbial activity. Biosorbents that are used for biosorption of metals from solutions could be classified into three broad categories: (1) exopolysaccharides, (2) living cultures, and (3) non-living biomass and preparations (Sen and Chakrabarti, 2009).

6.21 Microbial Electrochemical Technologies

Microbial electrochemical technology is an emerging area applied in the field of interdisciplinary subjects with biological interventions that have been used in the last decade (Mishra et al., 20202). Broadly, it is applied in the fields of microbial fuel cell technology, bioelectrochemical treatment system, microbial electrolysis cell, and microbial electrosynthesis system. Microbial electrochemical technology has many emerging applications, such as for the degradation and removal of α-ethynylestradiol sulfate; 2,4,6-trichlorophenol; petroleum hydrocarbon; methyl orange; phenol; sulfide; benzene; methyl-tertbutyl ether; ammonium; copper; sulfadiazine; p-nitrophenol; PAH; cefazolin sodium; p-chloronitrobenzene; etc.

6.22 Hybrid Approaches

For remediation processes, one single method may not be suitable to achieve desired degradation of contaminants; hence, a combination of technologies as hybrid approaches may be considered. The Fenton oxidation and bioremediation hybrid approach can be a promising method for the enhanced removal of POP (Megharaj and Naidu, 2017). In the integrated hybrid method, fast and aggressive oxidation by Fenton and then degradation of contaminants by microbial action is achieved. More than 90% efficiency of this process was observed (Gan and Ng, 2012). The pretreatment by Fenton oxidation has the advantages to decrease pollutant concentrations to less toxic level, improve the bioavailability, prevent incomplete mineralisation of partially oxidised contaminants, and release oxygen.

6.23 Permeable Reactive Barrier

A permeable reactive barrier is a physical technique for remediation of groundwater, but many researchers have mentioned it as a biological process also (Thiruvenkatachari et al., 2008; Obiri-Nyarko et al., 2014). This process has the biological reaction with degradation, precipitation, and sorption of contaminants. A permanent or semi-permanent reactive barrier (medium) is submerged in the trajectory of polluted groundwater. As polluted water flows through the barrier under its natural gradient, pollutants become trapped and undergo a series of reactions resulting in clean water in the flow (Azubuike et al., 2016; Garcı´a et al., 2014; Zhou et al., 2014; Thiruvenkatachari et al., 2008; Obiri-Nyarko et al., 2014). Ideally, the barriers are usually reactive enough to trap pollutants, permeable to allow the flow of water but not pollutants, passive with little energy input, inexpensive, readily available, and accessible (De Pourcq et al., 2015). The effectiveness of this technique depends mostly on the type of media used, which is influenced by pollutant type, biogeochemical and hydrogeological conditions, environmental and health influence, mechanical stability, and cost (Obiri-Nyarko et al., 2014; Liu et al., 2015). In general, PRB is an in-situ technique used for remediating groundwater polluted with different types of pollutants including heavy metals and chlorinated compounds including PAHs, 1, 2-dichloroethane, landfill leachate, nitrate and nitrite, benzene, toluene, ethylbenze and xylene, chlorinated volatile organic compounds, orange G dye, Al, Zn, Cu, Cr, etc. using various permeable reactive barrier clay, oxygen reactive compound and clinoptilolite, natural pyrite, zero-valent iron coupled with polyhydroxybutyrate, mixture of zero-valent iron, zeolite and activated carbon, bio-barriers (arthrobacter viscosus, trametes versicolor, white-rot fungi, *Mycobaterium* sp., and *Pseudomonas* sp. immobilised bead), organic substrates and zero-valent irons, granular oxygen-capturing materials (ZVI powder, sodium citrate, and inorganic salts), and granular activated carbon, granular iron, etc. (Azubuike et al., 2016).

6.24 Monoliths

Monolith-based structures, reactors, and other equipment have a significant role in environment remediation where reaction, separation, and heat transfer can take place simultaneously (Wood, 2013). Apart from chemical and other processes, bio-process can also be performed using monoliths and this is the emerging area for research and its applications (Wood, 2013; Ebrahimi et al., 2006).

For better performance of bioreactors, gas–liquid mass transfer is needed with less energy dissipation. Such a kind of reduction in energy dissipation can be observed in monoliths, which can be an attractive alternative for conventional bioreactors (Ebrahimi et al., 2006; Wood, 2013). The clogging of the monolith channels/pores due to formation of biofilms may be the major limitation of monoliths that can be overcome by controlling the high hydraulic retention time and rinsing of tap water to remove the films (Ebrahimi et al., 2006; Wood, 2013).

The simultaneous removal of SO_2 and NO_x from flue gases can be obtained with the integration of biological and physiochemical processes using monolith reactor/support by reactive absorption and enhanced mass transfer (Maas et al., 2006). The low-concentration (<5 g/m^3) pollutants and VOCs from gases can be removed effectively and efficiently by using monolithic bioreactors and biofiltrations (Rene et al., 2010). Monolithic bioreactors have been employed for the removal of styrene from the gas phase by the yeast, like the thermally dimorphic fungus *Sporothrix variecibatus* (Rene et al., 2010). The trickling monolithic bioreactor was considered for the separation of methanol from air (Jin et al., 2008). The degradation of toluene from air was performed in a monolithic bioreactor (Jin et al., 2006).

6.25 Membrane Technology

Membrane technologies have been successfully employed for wastewater treatment (Mazzei et al., 2013a). The removal of heavy metals from wastewater can be performed by reverse osmosis (Zhang et al., 2009; Chan and Dudeney, 2008). The experimental investigation was performed for the removal of heavy metals in a pilot plant of a hybrid bioreactor with reverse osmosis that resulted in high removal efficiency (Dialynas and Diamadopoulos, 2009; Fu and Wang, 2011).

6.26 Membrane Bioreactors

Biological methods play a significant role in wastewater treatments. Membrane bioreactors are more popular due to several advantages. Membrane bioreactors have high retention of almost all suspended matter and soluble compounds, high-quality effluents, oxidation of high molecular compounds due to high residence time, less settling, etc. (Mazzei et al., 2013). Aerobic submerged membrane bioreactors are a more common configuration used for wastewater treatment (Lesjean and Huisjes, 2008). High-strength wastewater can be used in an anaerobic membrane bioreactor for treatment. Integrated aerobic–anaerobic membrane bioreactors are the more intensive configuration for wastewater treatment (Pozo and Diez, 2005; Zhang et al., 2005). The major drawbacks of the membrane bioreactors for wastewater treatments are fouling and high cost, which hinders them to employ for commercially in many cases. The oil mill wastewater containing phenolic compounds has been successfully treated in a membrane bioreactor to produce phytotherapic (Mazzei et al., 2013).

6.27 Hybrid Membrane Bioreactors

Hybrid membrane bioreactors are the combination of membrane bioreactors and wastewater treatment technologies (Mazzei et al., 2013). It may be a possible potential alternative to the conventional wastewater treatment technologies such as the activated sludge process. The biofilm membrane bioreactor is used for denitrification by biofilms growing on fluidised support (Lee et al., 2006; Downing and Nerenberg, 2008). The novel aerobic granular sludge membrane bioreactor is the combination of submerged hollow-fibre membrane filtration and a sequencing aerobic sludge blanket reactor having the advantage of better filtration characteristics with low compressibility (Tay et al., 2007). The elevated temperature is used to reject organics by evaporation in a hybrid membrane distillation in a submerged membrane bioreactor (Mazzei et al., 2013Phattaranawik et al., 2009). This configuration resulted in a low energy requirement with a very high permeate quality and a stable flux. The osmotic membrane bioreactor is another innovative configuration with forward osmosis and is technically and economically feasible (Mazzei et al., 2013).

6.28 Nanobioremediation

Nanotechnology is playing a significant role in environmental remediation. Nanobioremediation is a new and emerging technique for the remediation of pollutants using biosynthetic nanoparticles. It is still a new area but growing rapidly in the field of nanotechnology (Yadav et al., 2017). Biosynthesis of nanoparticles using microorganisms has emerged as a rapidly developing research area in green nanotechnology across the globe, with various biological entities employed in the synthesis of nanoparticles, constantly forming an impute alternative for conventional chemical and physical methods. The need for biosynthesis of nanoparticles arose due to the high cost of physical and chemical processes. Biosynthesis of nanoparticles is a bottom-up approach where the main reaction occurring is reduction/oxidation. The microbial enzymes or the plant phytochemicals with antioxidants or reducing properties are usually responsible for the reduction of metal compounds into their respective nanoparticles. Yadav et al. (2017) listed various plants (*Freshwater diatom Stauroneis* sp.,Persimmon*), Pelargonium graveolens, Hibiscus rosa sinensis, Coriandrum sativum, Emblica officinalis, Phyllanthium,* mushroom extract*, Elettaria cardamomom, Parthenium hysterophorus, Ocimum* sp.*, Euphorbia hirta, Nerium indicum, Azadirachta indica, Brassica juncea, Pongamia pinnata, Clerodendrum inerme, Gliricidia sepium, Desmodium triflorum, Opuntia ficus indica, Coriandrum sativum, Carica papaya (fruit), Pelargoneum graveolens,* aloe vera extract*, Capsicum annum, Avicennia marina, Rhizophora mucronata, Ceriops tagal, Rumex hymenosepalus, Pterocarpus santalinus, Sonchus asper, Terminalia catappa, Banana peel, Mucuna pruriens, Cinnamomum zeylanicum, Medicago sativa, Magnolia kobus, Dyopiros kaki, Allium cepa L., Azadirachta indica A. Juss, Camellia sinensis L., Chenopodium album L., Justicia gendarussa L., Macrotyloma uniflorum (Lam) Verde, Mentha piperita L., Mirabilis jalapa L., Syzygium aromaticum (L), Terminalia catappa L., Amaranthus spinosus, Brassica juncea, Medicago sativa, Helianthus annuus, Diopyros kaki, Ocimum sanctum L., Cinnamomum zeylanicum, Blume, Cinnamomum camphora L., Gardenia jasminoides, Ellis,* soybeanAloe vera), bacteria (*Bacillus cereus, Oscillatory willei, Escherichia coli, Pseudomonis stuzeri, Bacillus subtilis, Bacillus* sp.*, Bacillus cereus, Bacillus thuringiensis, Lactobacillus strains, Pseudomonas stutzeri, Corynebacterium, Staphylococcus aureus, Ureibacillus thermosphaericus, Magnetosirillium magneticum,* sulphate-reducing bacteria,sp.,strain*, Shewanella* sp.,sulphate-reducing bacteria of the family *Desulfobacteriaceae),* yeast and fungi (*Torilopsis* sp.*, Rhodospiridium dibovatum, Schizosacharomyces pombe, Candida glabrata, Schizosaccharomyce pombe,* silver tolerant yeast strains *MKY3, Cladosporium cladosporioides, Coriolus versicolor, Fusarium semitectum, Fusarium oxysporum, Phaenerochaete chrysosporium, Aspergillus flavus, Extremophillic yeast, Aspergillus niger, Aspergillus oryzae, Fusarium solani, Pleurotus sajor-caju, Trichoderma viride, Aspergillus flavus, A. furnigatus, A. terreus, A. nidulans, Verticillium* sp.,lichen fungi) for the production of different nanoparticles including silicon, germanium, gold, silver, platinum, nickel, cobalt, zinc, copper, lead, palladium, etc.

The removal of environmental contaminants (such as heavy metals and organic and inorganic pollutants) from contaminated sites using nanoparticles/nanomaterial formed by plants, fungi, and bacteria with the help of nanotechnology is called nanobioremediation (NBR). NBR is the emerging technique for the removal of pollutants for environmental cleanup. Current technologies for remediation of contaminated sites include chemical and physical remediation, incineration, and bioremediation (Yadav et al., 2017). *Noaea mucronata* and *Euphorbia macroclada* are a few plant species used for forming nanoparticles and employed for the removal and bioremediation of heavy metals, lead, zinc, copper, cadmium, and nickel (Mohsenzadeh and Rad, 2011).

6.29 Conclusion

Process intensification plays a significant role in environmental remediation. Various approaches presented are conventional and intensification and also new approaches can be significantly result in cost effectiveness including operational flexibility. The world population has been increasing to 8 billion;

hence, the load on environmental remediation will increase day by day due to industrialisation, agriculture, and domestic developments. The foremost step to a successful bioremediation is site characterisation, which helps establish the most suitable and feasible bioremediation technique. However, these methods can be used to treat a wide range of contaminants effectively. The cost of environmental remediation cannot be the major factor to decide suitable and effective bioremediation techniques to be employed. It may be decided based on soil type, pollutant depth and type, site location relative to human habitation, and performance characteristics of each bioremediation technique.

REFERENCES

Adams, G.O., Fufeyin, P.T., Okoro, S.E., et al., (2015). Bioremediation, biostimulation and bioaugmentation: a review. *International Journal of Environmental Bioremediation & Biodegradation*, *3*(1), 28–39.

Adams, G.O., Fufeyin, P.T., Okoro, S.E., et al., (2015). Bioremediation, biostimulation and bioaugmentation: a review. *International Journal of Environmental Bioremediation & Biodegradation*, *3*(1), 28–39.

Aislabie, J., Saul, D.J., & Foght, J.M. (2006). Bioremediation of hydrocarboncontaminated polar soils. *Extremophiles*, *10*, 171–179.

Azubuike, C.C., Chikere, C.B., & Okpokwasili, G.C. (2016). Bioremediation techniques – classification based on site of application: principles, advantages, limitations and prospects. *World Journal of Microbiology & Biotechnology*, *32*(180), 1–18.

Barr, D. (2002). *Biological Methods for Assessment and Remediation of Contaminated Land: Case Studies*. London: Construction Industry Research and Information Association

Bastida, F., Jehmlich, N., Lima, K., et al., (2016). The ecological and physiological responses of the microbial community from a semiarid soil to hydrocarbon contamination and its bioremediation using compost amendment. *Journal of Proteomics*, *135*, 162–169.

Bento, F.M., Camargo, F.A.O., Okeke, B.C., et al., (2005). Comparative bioremediation of soils contaminated with diesel oil by natural attenuation, biostimulation and bioaugmentation. *Bioresource Technology*, *96*, 1049–1055.

Bhattacharya, M., Guchhait, S., Biswas, D., & Datta, S. (2015). Waste lubricating oil removal in a batch reactor by mixed bacterial consortium: a kinetic study. *Bioprocess and Biosystems Engineering*, *38*, 2095–2106. doi:10.1007/s00449-015-1449-9.

Boodhoo, K., & Harvey, A. (2013). Process intensification: an overview of principles and practice. In Boodhoo, K., & Harvey, A. (Eds.), Process Intensification for Green Chemistry: Engineering Solutions for Sustainable Chemical Processing (1st Edition), 1–31, UK: John Wiley & Sons, Ltd.

Burgess, J.E., Parsons, S.A., & Stuetz, R.M. (2001). Developments in odour control and waste gas treatment biotechnology: a review. *Biotechnology Advances*, *19*, 35–63. doi:10.1016/S0734-9750(00)00058-6.

Chan, B.K.C., & Dudeney, A.W.L. (2008). Reverse osmosis removal of arsenic residues from bioleaching of refractory gold concentrates. *Minerals Engineering*, *21*, 272–278.

Chikere, C.B., Okoye, A.U., & Okpokwasili, G.C. (2016). Microbial community profiling of active oleophilic bacteria involved in bioreactorbased crude-oil polluted sediment treatment. *Journal of Applied & Environmental Microbiology*, *4*, 1–20. doi:10.12691/jaem-4-1-1.

Cross, W.T., & Ramshaw, C. (1986). Process intensification – laminar flow heat transfer. *Chemical Engineering Research and Design*, *64*(4), 293–301.

De Pourcq, K., Ayora, C., García-Gutiérrez, M., Missana, T., & Carrera, J. (2015). A clay permeable reactive barrier to remove Cs-137 from groundwater: column experiments. *Journal of Environmental Radioactivity*, *149*, 36–42. doi:10.1016/j.jenvrad.2015.06.029.

Delforno, T.P., Moura, A.G.L., Okada, D.Y., Sakamoto, I.K., & Varesche, M.B.A. (2015). Microbial diversity and the implications of sulfide levels in an anaerobic reactor used to remove an anionic surfactant from laundry wastewater. *Bioresource Technology*, *192*, 37–45. doi:10.1016/j.biortech.2015.05.050.

Dialynas, E., & Diamadopoulos, E. (2009). Integration of a membrane bioreactor coupled with reverse osmosis for advanced treatment of municipal wastewater. *Desalination*, *238*, 302–311.

Downing, L.S., & Nerenberg, R. (2008). Total nitrogen removal in a hybrid, membrane-aerated activated sludge process. *Water Research*, *42*, 3697–3708.

Ebrahimi, S., Kleerebezem, R., Kreutzer, M.T., Kapteijn, F., Moulijn, J.A., Heijnen, J.J., & van Loosdrecht, M.C.M. (2006). Potential application of monolith packed columns as bioreactors, control of biofilm formation. *Biotechnology & Bioengineering*, *93*, 238–245.

Elektorowicz, M. (1994). Bioremediation of petroleum-contaminated clayey soil with pretreatment. *Environmental Technology*, *15*(4), 373–380.

Filonov, A.E., Akhmetov, L.I., Puntus, I.F., et al., (2005). The construction and monitoring of genetically tagged, plasmid-containing, naphthalene-degrading strains in soil. *Microbiology*, *74*(4), 453–458.

Fingerman, M., & Nagabhushanam, R. (2016). *Bioremediation of Aquatic and Terrestrial Ecosystems*. Enfield: Science Publishers Inc.

Firmino, P.I.M., Farias, R.S., Barros, A.N., Buarque, P.M.C., Rodrı́guez, E., Lopes, A.C., & dos Santos, A.B. (2015). Understanding the anaerobic BTEX removal in continuous-flow bioreactors for ex situ bioremediation purposes. *Chemical Engineering Journal*, *281*, 272–280. doi:10.1016/j.cej.2015.06.106.

Folch, A., Vilaplana, M., Amado, L., Vicent, R., & Caminal, G. (2013). Fungal permeable reactive barrier to remediate groundwater in an artificial aquifer. *Journal of Hazardous Materials*, *262*, 554–560. doi:10.1 016/j.jhazmat.2013.09.004.

Forsyth, J.V., Tsao, Y.M., & Bleam, R.D. (1995). Bioremediation: when is augmentation needed? In Hinchee, R.E., Fredrickso, J., & Alleman, B.C. (Eds.), *Bioaugmentation for Site Remediation*. Columbus, OH: Battelle Press, 14 pp.

Frascari, D., Zanaroli, G., & Danko, A.S. (2015). In situ aerobic cometabolism of chlorinated solvents: a review. *Journal of Hazardous Materials*, *283*, 382–399. doi:10.1016/j.jhazmat.2014.09.041.

Fu, F., & Wang, Q. (2011). Removal of heavy metal ions from wastewater: a review. *Journal of Environmental Management*, *92*, 407–418.

Fuller, M.E., Kruczek, J., Schuster, R.L., Sheehan, P.L., & Arienti, P.M. (2003). Bioslurry treatment for soils contaminated with very high concentrations of 2,4,6-trinitrophenylmethylnitramine (tetryl). *Journal of Hazardous Materials*, *100*, 245–257. doi:10.1016/S0304-3894(03)00115-8.

Gan, V.S., & Ng, H.K. (2012). Current status and prospects of Fenton oxidation for the decontamination of persistent organic pollutants (POPs) in soils. *Chemical Engineering Journal*, *213*, 295–317.

García, Y., Ruiz, C., Mena, E., Villaseñor, J., Cañizares, P., & Rodrigo, M.A. (2014). Removal of nitrates from spiked clay soils by coupling electrokinetic and permeable reactive barrier technologies. *Journal of Chemical Technology & Biotechnology*, *90*, 1719–1726. doi:10.1002/jctb.4488.

Gerhardt, K.E., Huang, X.D., Glick, B.R., & Greenberg, B.M. (2009). Phytoremediation and rhizoremediation of organic soil contaminants: potential and challenges. *Plant Science*, *176*, 20–30.

Gerven, T. Van, & Stankiewicz, A. (2009). Structure, energy, synergy, time – the fundamentals of process intensification. *Industrial & Engineering Chemistry Research*, *48*(5), 2465–2474.

Gidarakos, E., & Aivalioti, M. (2007). Large scale and long term application of bioslurping: the case of a Greek petroleum refinery site. *Journal of Hazardous Materials*, *149*, 574–581. doi:10.1016/j.jhazmat.2 007.06.110.

Godleads, O.A., Prekeyi, T.F., Samson, E.O., et al., (2015) Bioremediation, biostimulation and bioaugmention: a review. *International Journal of Environmental Bioremediation & Biodegradation*, *3*(1), 28–39.

Goswami, M., Chakraborty, P., Mukherjee, K., Mitra, G., Bhattacharyya, P., Dey, S., & Tribedi, P. (2018). Bioaugmentation and biostimulation: a potential strategy for environmental remediation. *Journal of Microbiology & Experimentation*, *6*(5), 223–231.

Höhener, P., & Ponsin, V. (2014). In situ vadose zone bioremediation. *Current Opinion in Biotechnology*, *27*, 1–7. doi:10.1016/j.copbio.2013.08.018.

Hobson, A.M., Frederickson, J., & Dise, N.B. (2005). CH4 and N2O from mechanically turned windrow and vermincomposting systems following in-vessel pre-treatment. *Waste Management*, *25*, 345–352. doi: 10.1016/j.wasman.2005.02.015.

Jin, Y.M., Veiga, M.C., & Kennes, C. (2006). Development of a novel monolith-bioreactor for the treatment of VOC-polluted air. *Environmental Technology*, *27*, 1271–1277.

Jin, Y.M., Veiga, M.C., & Kennes, C. (2008). Removal of methanol from air in a low-pH trickling monolith bioreactor. *Process Biochemistry*, *43*, 925–931.

Juárez-Ramírez, C., Galíndez-Mayer, J., Ruiz-Ordaz, N., Ramos-Monroy, O., Santoyo-Tepole, F., & Poggi-Varaldo, H. (2015). Steady-state inhibition model for the biodegradation of sulfonated amines in a packed bed reactor. *New Biotechnology*, *32*, 379–386. doi:10.1016/j.nbt.2014.07.010.

Kao, C.M., Chen, C.Y., Chen, S.C., Chien, H.Y., & Chen, Y.L. (2008). Application of in situ biosparging to remediate a petroleumhydrocarbon spill site: field and microbial evaluation. *Chemosphere, 70*, 1492–1499. doi:10.1016/j.chemosphere.2007.08.029.

Khan, F.I., Husain, T., & Hejazi, R. (2004). An overview and analysis of site remediation technologies. *Journal of Environmental Management, 71*, 95–122. doi:10.1016/j.jenvman.2004.02.003.

Kim, S., Krajmalnik-Brown, R., Kim, J.-O., & Chung, J. (2014). Remediation of petroleum hydrocarbon-contaminated sites by DNA diagnosis-based bioslurping technology. *Science of the Total Environment, 497*, 250–259. doi:10.1016/j.scitotenv.2014.08.002.

Leahy, J.G., & Colwell, R.R. (1990). Microbial degradation of hydrocarbons in the environment. *Microbiology Reviews, 54*(3), 305–315.

Lee, S.-H., Oh, B.-I., & Kim, J. (2008). Effect of various amendments on heavy mineral oil bioremediation and soil microbial activity. *Bioresource Technology, 99*, 2578–2587.

Lee, W.-N., Kang, I.-J., & Lee, C.H. (2006). Factors affecting filtration characteristics in membranecoupled moving bed biofilm reactor. *Water Research, 40*, 1827–1835.

Lesjean, B., & Huisjes, E.H. (2008). Survey of the European MBR market: trends and perspectives. *Desalination, 231*, 71–81.

Liu, Y., Mou, H., Chen, L., Mirza, Z.A., & Liu, L. (2015). Cr(VI)-contaminated groundwater remediation with simulated permeable reactive barrier (PRB) filled with natural pyrite as reactive material: environmental factors and effectiveness. *Journal of Hazardous Materials, 298*, 83–90. doi:10.1016/j.jhazmat.2015.05.007.

Maas, P. van der, van den Brink, P., Utomo, S., Klapwijk, B., & Lens, P. (2006). NO removal in continuous BioDeNOx reactors: Fe(II)EDTA(2-) regeneration, biomass growth, and EDTA degradation. *Biotechnology & Bioengineering, 94*, 575–584.

Magalhães, S.M.C., Jorge, R.M.F., & Castro, P.M.L. (2009). Investigations into the application of a combination of bioventing and biotrickling filter technologies for soil decontamination processes—a transition regime between bioventing and soil vapour extraction. *Journal of Hazardous Materials, 170*, 711–715. doi:10.1016/j.jhazmat.2009.05.008.

Maila, M.P., & Colete, T.E. (2004). Bioremediation of petroleum hydrocarbons through land farming: are simplicity and cost-effectiveness the only advantages? *Reviews in Environmental Science and Bio/Technology, 3*, 349–360. doi:10.1007/s111157-004-6653-z.

Malla, M.A., Dubey, A., Yadav, S., Kumar, A., Hashem, A., & Abd_Allah, E.F. (2018). Understanding and designing the strategies for the microbe-mediated remediation of environmental contaminants using omics approaches. *Frontiers in Microbiology, 9*, 1132. doi: 10.3389/fmicb.2018.01132.

Massa, V., Infantino, A., Radice, F., et al., (2009). Efficiency of natural and engineered bacterial strains in the degradation of 4-chlorobenzoic acid in soil slurry. *International Biodeterioration & Biodegradation, 63*(1), 112–115.

Mazzei, R., Piacentini, E., Drioli, E., & Giorno, L. (2013). Membrane bioreactors for green processing in a sustainable production system. In Boodhoo, K., & Harvey A. (Eds.), *Process Intensification for Green Chemistry: Engineering Solutions for Sustainable Chemical Processing* (1st Edition), 227–250. John Wiley & Sons, Ltd.

Mazzei, R., Piacentini, E., Drioli, E., & Giorno, L. (2013a). Membrane separations for green chemistry. In Boodhoo K., &Harvey A. (Eds.), *Process Intensification for Green Chemistry: Engineering Solutions for Sustainable Chemical Processing* (1st Edition), 311–353. John Wiley & Sons, Ltd.

Megharaj, M., & Naidu, R. (2017). Soil and brownfield bioremediation. *Microbial Biotechnology, 10*(5), 1244–1249.

Megharaj, M., Ramakrishnan, B., Venkateswarlu, K., et al., (2011). Bioremediation approaches for organic pollutants: a critical perspective. *Environment International, 37*, 1362–1375.

Mishra, B., Varjani, S., Kumar, G., Awasthi, M.K., Awasthi, S.K., Sindhu, R., Binod, P., Rene, E.R., & Zhang, Z. (2020). Microbial approaches for remediation of pollutants: innovations, future outlook, and challenges. *Energy & Environment, 0*(0), 1–30, DOI: 10.1177/0958305X19896781.

Mohsenzadeh, F., & Rad, C.A. (2011). *Int. Confere. Nanotechnol. Biosensoors IACSIT*. Singapore: Press, 25, 16–20.

Mohsenzadeh, F., & Rad, C.A. (2012). Bioremediation of heavy metal pollution by nano-particles of Noaea Mucronata. *International Journal of Bioscience, Biochemistry and Bioinformatics 2*, 85–89.

Monti, M.R., Smania, A.M., Fabro, G., et al., (2005). Engineering *Pseudomonas fluorescens* for biodegradation of 2, 4-dinitrotoluene. *Applied and Environmental Microbiology*, *71*(12), 8864–8872.

Mrozik, A., & Piotrowska-Seget, Z. (2010). Bioaugmentation as a strategy for cleaning up of soils contaminated with aromatic compounds. *Microbiological Research*, *165*(5), 363–375.

Mustafa, Y.A., Abdul-Hameed, H.M., & Razak, Z.A. (2015). Biodegradation of 2,4-dichlorophenoxyacetic acid contaminated soil in a roller slurry bioreactor. *Clean-Soil Air Water*, *43*, 1115–1266. doi:10.1002/clen.201400623.

Nikolopoulou, M., & Kalogerakis, N. (2009). Biostimulation strategies for fresh and chronically polluted marine environments with petroleum hydrocarbons. *Journal of Chemical Technology & Biotechnology*, *84*(6), 802–807.

Obiri-Nyarko, F., Grajales-Mesa, S.J., & Malina, G. (2014). An overview of permeable reactive barriers for in situ sustainable groundwater remediation. *Chemosphere*, *111*, 243–259. doi:10.1016/j.chemosphere.2014.03.112.

Park, J.H., Lamb, D., Paneerselvam, P., et al., (2011). Role of organic amendments on enhanced bioremediation of heavy metal(loid) contaminated soils. *Journal of Hazardous Materials*, *185*, 549–574.

Phattaranawik, J., Fane, A.G., Pasquier, A.C.S., Bing, W., & Wong, F.S. (2009). Experimental study and design of a submerged membrane distillation bioreactor. *Chemical Engineering & Technology*, *32*, 38–44.

Philp, J.C., & Atlas, R.M. (2005). Bioremediation of contaminated soils and aquifers. In Atlas, R.M., & Philp, J.C. (Eds.), *Bioremediation: Applied Microbial Solutions for Real-world Environmental Cleanup*, 139–236. Washington: American Society for Microbiology (ASM) Press.

Plangklang, P., & Alissara Reungsang, A. (2010). Bioaugmentation of carbofuran by Burkholderia cepacia PCL3 in a bioslurry phase sequencing batch reactor. *Process Chemistry*, *45*, 230–238. doi:10.1016/j.procbio.2009.09.013.

Pozo, R. Del, & Diez, V. (2005). Integrated anaerobic–aerobic fixed-film reactor for slaughterhouse wastewater treatment. *Water Research*, *39*, 1114–1122.

Rafael, G.L., José, M.B., & Carlos, G. (2020). Biological methods of polluted soil remediation for an effective economically-optimal recovery of soil health and ecosystem services. *Journal of Environmental Science and Public Health*, *4*, 112–133.

Ramadass, K., Megharaj, M., Venkateswarlu, K., et al., (2018). Bioavailability of weathered hydrocarbons in engine oil-contaminated soil: impact of bioaugmentation mediated by *Pseudomonas* spp. on bioremediation. *Science of The Total Environment 636*, 968–974.

Rene, E.R., Lopez, M.E., Veiga, M.C., & Kennes, C. (2010). Performance of a fungal monolith bioreactor for the removal of styrene from polluted air. *Bioresource Technology*, *101*, 2608–2615.

Rodríguez-Rodríguez, C.E., Marco-Urrea, E., & Caminal, G. (2010). Degradation of naproxen and carbamazepine in spiked sludge by slurry and solid-phase Trametes versicolor systems. *Bioresource Technology*, *101*, 2259–2266. doi:10.1016/j.biortech.2009.11.089.

Ros, M., Rodríguez, I., García, C., et al., (2010). Microbial communities involved in the bioremediation of an aged recalcitrant hydrocarbon polluted soil by using organic amendments. *Bioresource Technology*, *101*, 6916–6923.

Roy, M., Giri, A.K., Dutta, S., & Mukherjee, P. (2015). Integrated phytobial remediation for sustainable management of arsenic in soil and water. *Environment International*, *75*, 180–198. doi:10.1016/j.envint.2014.11.010.

Sanscartier, D., Zeeb, B., Koch, I., & Reimer, K. (2009). Bioremediation of diesel-contaminated soil by heated and humidified biopile system in cold climates. *Cold Regions Science and Technology*, *55*, 167–173. doi:10.1016/j.coldregions.2008.07.004.

Saravanan, V., Rajasimman, M., & Rajamohan, N. (2015) Performance of packed bed biofilter during transient operating conditions on removal of xylene vapour. International Journal of *Environmental Science and Technology*, *12*, 1625–1634. doi:10.1007/s13762-014-0521-3.

Schneider, T., Keiblinger, K.M., Schmid, E., et al., (2012). Who is who in litter decomposition Metaproteomics reveals major microbial players and their biogeochemical functions? *ISME Journal*, *6*, 1749–1762.

Semple, K.T., Reid, B.J., & Fermor, T.R. (2001). Impact of composting strategies on the treatment of soils contaminated with organic pollutants. *Environmental Pollution*, *112*, 269–283.

Sen, R., & Chakrabarti, S. (2009). Biotechnology – applications to environmental remediation in resource exploitation. *Current Science, 97*(6), 768–775.

Stankiewicz, A.I., & Moulijn, J.A. (2000). Process intensification: transforming chemical engineering. *Chemical Engineering Progress, 96*(1), 22–34.

Starke, R., Bastida, F., Abad_1a, J., et al., (2017). Ecological and functional adaptations to water management in a semiarid agroecosystem: a soil metaproteomics approach. *Scientific Reports, 7*, 10221.

Tandy, S., Healey, J.R., Nason, M.A., Williamson, J.C., & Jones, D.L. (2009). Remediation of metal polluted mine soil with compost: co-composting verses incorporation. *Environmental Pollution, 157*, 690–697.

Tay, J.H., Yang, P., Zhuang, W.Q., Tay, S.T.L., & Pan, Z.H. (2007). Reactor performance and membrane filtration in aerobic granular sludge membrane bioreactor. *Journal of Membrane Science, 304*, 24–32.

Thiruvenkatachari, R., Vigneswaran, S., & Naidu, R. (2008). Permeable reactive barrier for groundwater remediation. *Journal of Industrial and Engineering Chemistry, 14*, 145–156. doi:10.1016/j.jiec.2007.10.001.

Thomas, T., & Gilbert, F.M. (2012). Metagenomics – a guide from sampling to dataanalysis. *Microbial Informatics and Experimentation, 2*, 3.

Tyagi, M., da Fronseca, M.M., & de Carvalho, C.C. (2010). Bioaugmentation and biostimulation strategies to improve the effectiveness of bioremediation processes. *Biodegradation, 22*(2), 231–241.

Varjani, S.J., & Upasani, V.N. (2017). A new look on factors affecting microbial degradation of petroleum hydrocarbon pollutants. *International Biodeterioration and Biodegradation, 120*, 71–83.

Varjani, S.J., & Upasani, V.N. (2013). Comparative studies on bacterial consortia for hydrocarbon degradation. *International Journal of Innovative Research in Science, Engineering and Technology, 2*, 5377–5383.

Whelan, M.J., Coulon, F., Hince, G., Rayner, J., McWatters, R., Spedding, T., & Snape, I. (2015). Fate and transport of petroleum hydrocarbons in engineered biopiles in polar regions. *Chemosphere, 131*, 232–240.

Wilkes, H., Buckel, W., Golding, B.T., et al., (2016). Metabolism of hydrocarbons in n-alkane utilizing anaerobic bacteria. *Journal of Molecular Microbiology and Biotechnology, 26*, 138–151.

Wood, J. (2013). Monolith reactors for intensified processing in green chemistry. In Boodhoo, K., & Harvey, A. (Eds.), *Process Intensification for Green Chemistry: Engineering Solutions for Sustainable Chemical Processing* (1st Edition), 175–197. John Wiley & Sons, Ltd.

Xu, P., Ma, W., Han, H., Jia, S., & Hou, B. (2015). Isolation of a naphthalene-degrading strain from activated sludge and bioaugmentation with it in a MBR treating coal gasification wastewater. *Bulletin of Environmental Contamination and Toxicology, 94*, 358–364. doi:10.1007/s00128-014-1366-7.

Yadav, K.K., Singh, J.K., Gupta, N., & Kumar, V. (2017). A review of nanobioremediation technologies for environmental cleanup: a novel biological approach. *Journal of Materials and Environmental Sciences, 8*(2), 740–757.

Yang, Y., Wang, Y., Hristovski, K., & Westerhoff, P. (2015). Simultaneous removal of nanosilver and fullerene in sequencing batch reactors for biological wastewater treatment. *Chemosphere, 125*, 115–121. doi:10.1016/j.chemosphere.2014.12.003.

Zangi-Kotler, M., Ben-Dov, E., Tiehm, A., & Kushmaro, A. (2015). Microbial community structure and dynamics in a membrane bioreactor supplemented with the flame retardant dibromoneopentyl glycol. *Environmental Science and Pollution Research (International), 22*, 17615–17624. doi:10.1007/s11356-015-4975-8.

Zeng, G.M., Yu, Z., Chen, Y.N., Zhang, J.C., Li, H., Yu, M., et al., (2011). Response of compost maturity and microbial community composition to pentachlorophenol (PCP)-contaminated soil during composting. *Bioresource Technology, 102*, 5905–5911.

Zhang, D., Lu, P., Long, T., & Verstraete, W. (2005). The integration of methanogensis with simultaneous nitrification and denitrification in a membrane bioreactor. *Process Biochemistry, 40*, 541–547.

Zhang, L.N., Wu, Y.J., Qu, X.Y., Li, Z.S., & Ni, J.R. (2009). Mechanism of combination membrane and electro-winning process on treatment and remediation of Cu2þ polluted water body. *Journal of Environmental Sciences, 21*, 764–769.

Zhou, D., Li, Y., Zhang, Y., Zhang, C., Li, X., Chen, Z., Huang, J., Li, X., Flores, G., & Kamon, M. (2014). Column test-based optimization of the permeable reactive barrier (PRB) technique for remediating groundwater contaminated by landfill leachates. *Journal of Contaminant Hydrology, 168*, 1–16. doi:10.1016/j.jconhyd.2014.09.003.

7 Bio-Based Technologies and Combination of Other Technologies

Yong Chen and Steplinpaulselvin Selvinsimpson
School of Environmental Science and Engineering, Huazhong University of Science and Technology, Wuhan, China

CONTENTS

7.1 Introduction 125
7.2 Bioremediation Technologies 126
7.3 Nanotechnology in Bioremediation 128
7.4 Future Research 129
7.5 Conclusion 129
References 129

7.1 Introduction

Large amounts of toxins are discharged into aquatic areas by industries, which may have an impact on both flora and wildlife. When compared to suspended strains, dyes, medications, and personal care products cause significant damage to the aquatic environment and provide greater tolerance to environmental fluctuations such as temperature, hazardous chemicals, and pH (Chen et al., 2017; Dursun and Tepe, 2005; Steplin Paul Selvin et al., 2018; Yan et al., 2013). As a result, researchers have proposed a bioremediation method to the degradation of environmental contaminants in recent years (Adikesavan and Nilanjana, 2016). Bioremediation is the process of pollution degradation in the environment using biological techniques and microorganisms' metabolic capability to degrade contaminants. The pollutant may be completely mineralised after treatment with bioremediation. Furthermore, because it is not an invasive approach, it does not harm the environment. Specificity, high selectivity, cost and energy economy, and minimal demand are some of the advantages of bioremediation over physicochemical approaches. Bioremediation, on the other hand, has a drawback in that it takes a long time to degrade a harmful substance, generally several months to over a year (Azubuike et al., 2016). Despite its many benefits, the processes may take longer and be less predictable than traditional approaches. Furthermore, its use is restricted in badly contaminated locations with high levels of toxic and hazardous pollutants (Perelo, 2010, Singh et al., 2020). Essentially, bioremediation employs microorganisms to remove pollutants from water samples and soil matrices (Saxena et al., 2019; Bharagava et al., 2017a, b; Gautam et al., 2017; Chandra et al., 2015; Saxena and Bharagava, 2017; Mosa et al., 2016).

Ex-situ and in-situ bioremediation procedures can both be effective depending on a variety of criteria such as site features, cost, and the kind and concentration of contaminants. Ex-situ procedures are typically more expensive than in-situ approaches; nevertheless, equipment installation costs on-site, visualisation, and control of the subsurface of polluted sites are all major considerations when using in-situ techniques. As a result, adopting the appropriate bioremediation strategy will aid in reducing pollutant concentrations to a safe level. Furthermore, bioaugmentation and biostimulation improved bioremediation, while environmental parameters that determine bioremediation effectiveness are maintained at optimal levels (Azubuike et al., 2016).

In order to increase the efficacy of bioremediation, numerous technologies, including bioremediation and nanotechnology, have been investigated in order to develop more efficient strategies for their removal from the environment. It is critical to have a comprehensive understanding of the interactions between the contaminant, microorganisms, and nanoparticles because both negative and positive outcomes are possible. Certain nanoparticles, for example, stimulate bacteria while others are harmful. As a result, making the right decision is crucial (Vázquez-Núñez et al., 2020).

Nanotechnology is also being used in a variety of industries, including textiles, medical, electronics, pharmaceuticals, optics, sports, and cosmetics. Nanotechnology is also used in environmental remediation for a variety of applications (Tratnyek and Johnson, 2006; Mueller and Nowack, 2010; Singh and Misra, 2014; Patil et al., 2016; Singh et al., 2020). Carbon nanotubes have recently attracted a lot of attention because of their unusual properties, such as a large edge plane/basal plane ratio, fast electrode kinetics, and high electronic conductivity. Furthermore, numerous research have found that including CNTs into the immobilisation of bovine serum albumin improves the immobilisation capacity and mechanical strength of alginate beads (Jiang et al., 2006). Furthermore, CNTs were impregnated into chitosan hydrogel beads for effective pollution removal (Chatterjee et al., 2010, Yan et al., 2013). In an another example, *S. oneidensis* MR-1, as a facultative gram-negative bacterium, is capable of reducing a variety of compounds, including Pd(II) (De Windt et al., 2006), Cr(VI) (Middleton et al., 2003), Mn(IV) (Myers and Nealson, 1988), Fe(III) (Myers and Nealson, 1990), U(VI) (Burgos et al., 2008), nitrate (Viamajala et al., 2002), nitrite (Krause and Nealson, 1997), Tc(VII) (Lloyd and Macaskie, 1996), nitrobenzene (Cai et al., 2012), thiosulfate, sulfite (Shirodkar et al., 2011), elemental sulfur (Moser and Nealson, 1996), etc. (Yan et al., 2013).

As a result, the main goal of this chapter was to investigate the behaviour of living creatures used to clean up contaminants in the presence of nanomaterials. It also focuses on bio-based nanotechnology for environmental pollution remediation, as well as providing a better grasp of bioremediation approaches, their principles, disadvantages, benefits, and perspectives.

7.2 Bioremediation Technologies

Bioremediation technology strategies often follow natural recovery, leaving pollutants in place and allowing natural processes such as biological and chemical transformation to break down the pollutant in situ, reducing its bioavailability (Magar, 2001). The most important step in bioremediation is site characterisation, which aids in determining the best ex-situ and in-situ bioremediation methods because these two bioremediation methods have their own advantages in specific ways. For instance, ex-situ bioremediation is more expensive due to its additional costs responsible for the excavation and transportation. On the other hand, they could be employed in a controlled manner to remediate a variety of contaminants. In-situ methods, on the other hand, have no additional excavation costs. However, the high cost of equipment installation on-site, as well as the inability to effectively visualise and manage the subsurface of a contaminated site, may render certain in-situ bioremediation procedures useless. The geological properties of contaminated sites, such as pollutant concentration and type, soil type, site position relative to human occupancy, and bioremediation technology effectiveness, are also important determining variables for effective contaminated site treatment (Azubuike et al., 2016).

Molecular approaches such as metabolomics, proteomics, transcriptomics, and genomes have also helped to better understand microbial identity, roles, and catabolic and metabolic pathways. The above-mentioned methods circumvent the limitations of microbial culture-based methodologies. Low population or absence of microbes, resource limitation with degradative capacities, and contaminant bioavailability are some of the primary obstacles that could stymie bioremediation success. Because bioremediation is based on microbial processes, there are two basic ways to speed up microbial activity in contaminated areas.

Biostimulation encourages the indigenous population through inducing the issues that affect the microbial growth. Further, bioaugmentation includes suitable kinds for the degradation of particular pollutants. Phytoremediation, which uses plants and algae to degrade toxins in the environment, is also employed (Perelo, 2010). The addition of nutrients/substrates to a polluted sample to stimulate the activities of autochthonous bacteria is known as biostimulation. Because microorganisms are abundant, it is clear that pollutant degraders are naturally present in contaminated areas, and their populations and metabolic activity may rise or decrease in response to contaminant concentration.

As a result, the use of agro-industrial wastes with sufficient nutrient composition, notably in terms of phosphorus, potassium, and nitrogen, will aid in resolving the problem of nutrient deficiency in the majority of contaminated areas. Nonetheless, it was shown that excessive stimulant additions resulted in microbial metabolic activity and diversity being reduced (Wang et al., 2012). Bioaugmentation, on the other hand, is a vital strategy for boosting microbial populations with degradative capacities. When compared to pure isolates, microbial consortiums have been shown to remove pollutants more effectively (Silva-Castro et al., 2012). This is due to metabolic differences across isolates, which may arise as a result of their isolation source, adaption process, or contaminant composition, and will result in synergistic effects, which may lead to complete and rapid destruction of pollutants when isolates are combined together (Bhattacharya et al., 2015).

Biological agents such as microalgae, bacteria, fungus, protists, plants, or their enzymes are utilised in bioremediation to reduce environmental contaminants into less hazardous forms. These bacteria are non-pathogenic and native to the contaminated environment. Aerobic bacteria are widely utilised because they are highly effective and easier to separate and manage during the biodegradation process (Mani and Kumar, 2014; Akcil et al., 2015; Dixit et al., 2015).

Bioremediation is usually accomplished in one of two ways: natural attenuation or induced bioremediation. Bioaugmentation, bioreactors, bioleaching, bioventing, biostimulation, composting, land farming, rhizofiltration, and phytoremediation are some of the most prevalent bioremediation processes (Li and Li, 2011; Thangadurai et al., 2020). For example, bacteria like *Aspergillus niger, Pseudomonas aeruginosa,* and *Rhodopseudomonas sphaeroides* have been used in wastewater treatment (Liu et al., 2015; Thangadurai et al., 2020). During their metabolisms, these bacteria may be able to break down contaminants in aquatic settings. The disadvantages of biological wastewater treatment include slow biodegradation and poor cell recovery due to substrate inhibition (Thangadurai et al., 2020).

Microorganisms frequently defend themselves against contaminants in two ways (Al-Rub et al., 2004; Abdel Hameed, 2006). One is the creation of degradative enzymes for the target pollutants, and the other is heavy metal resistance. Metal microbe interactions, biomineralisation, biosorption, bioaccumulation, bioleaching, and biotransformation are among the bioremediation methods. In general, bacteria use a variety of techniques to bind, immobilise, volatize, oxidise, and convert heavy metals. Microorganisms, such as microalgae or phytoplankton, for example, may bind to heavy metal ions used in the bioremocal of metallic ions existing in contaminated environments (Bitton, 2005).

Microorganisms are widely used in the remediation process, but in recent years the application of nanotechnology has become an advantage to resolve various issues because nanoparticles have more advantages compared to microorganisms (Ferroudj et al., 2013). Among the several emerging technologies, nanotechnology has proven to be a superb cover for wastewater treatment and a variety of other environmental challenges (Zare et al., 2013; Sadeghi et al., 2013; Thangadurai et al., 2020).

Because of their environmental benefits, biosurfactants are chosen over chemical counterparts. However, large-scale use of biosurfactants in contaminated areas is uneconomical due to high production costs and limited scalability. Biosurfactant yield could be increased by using agro-industrial wastes as nutrition sources for putative biosurfactant producers during fermentation. Simultaneous application of different bioremediation technologies during the remediation process will help boost remediation efficiency by reducing the weaknesses of individual techniques while also lowering costs (Cassidy et al., 2015; Martínez-Pascual et al., 2015).

As a result, nanoparticles could help reduce pollutant toxicity to microorganisms. Nanomaterials improve surface area and lower activation energy, allowing microorganisms to degrade pollutants more effectively, resulting in a reduction in remediation time and expense (Rizwan et al., 2014).

7.3 Nanotechnology in Bioremediation

Nanotechnologies utilised in bioremediation procedures are projected to drive technological advancements in developed and developing countries to improve environmental quality (Bartke et al., 2018; Medina-Pérez et al., 2019). Much research has been carried out to identify the mechanisms of degradation and remediation (Kumar and Gopinath, 2009; Vázquez-Núñez et al., 2020). Nanotechnologies are generally used in various applications such as engineering, cosmetics, agriculture, food, medical fields, space, and the environment. For wastewater treatment, numerous nanomaterials such as metal oxides, zeolites, fibres, carbon nanotubes, enzymes, noble metals, and nanoscale zero-valent iron are commonly used.

Nanomaterials have been integrated with biological processes in recent years to accelerate and promote the removal of dangerous pollutants from the environment (Kumari and Singh, 2016). A suitable interaction between nanomaterials and the living organisms should be essential since bioremediation uses living organisms. For example, it is generally understood that nanoparticle toxicity, size, and nanonutrition can all have an impact on live organisms, and hence on the bioremediation process as a whole. Further, Tan et al. (2018) show that the chemical and physical interaction among nanomaterials, biota, and the pollutants depend on various parameters including pollutants, media, types of organisms, temperature, and pH. Temperature and pH, for example, are important factors in the healthy development of living organisms. However, these variables may have an impact on the stability of nanomaterials as well as the pollutant. Wang et al. (2014), for example, claimed that Au NPs were stable in MilliQ water with a buffer. This stability, on the other hand, was lost at pH 4, 7, 8, and 10. As a result, more research and adequate experimental designs are required to fully comprehend the impact of the aforementioned characteristics on the nanobioremediation of contaminants (Vázquez-Núñez et al., 2020). The combination of nanomaterials and biotechnologies could result in a significant improvement in remediation capacities, avoiding process intermediates and increasing degradation speed (Kang, 2014; Fulekar and Pathak, 2017; Vázquez-Núñez et al., 2020).

In addition, in-situ nanoremediation approaches necessitate the use of reactive nanomaterials to change and eliminate pollutants in situ. To reduce pollutants, these nanomaterials have characteristics that allow for both catalysis and chemical reduction. No soil is moved to another location for disposal, and no groundwater is pumped out for aboveground treatment when these approaches are used. Nanosclae Fe particles, for example, are extremely successful in the cleanup and transformation of a wide range of environmental contaminants. Because pump and treat therapies are expensive and take a long time to operate, in-situ groundwater treatment solutions are becoming more popular. For example, the number of nZVI-based applications is rapidly growing. In the field, only a few projects have been reported. Despite the fact that the procedure is a beneficial replacement for current site remediation procedures, the possible dangers are poorly known. Furthermore, the variables and processes that contribute to ecotoxicity are complicated, and understanding of the potential impacts of manufactured nanoparticles in the environment on human health is scarce. The majority of public concerns revolve around the unknown dangers of employing nanoparticles for site remediation. However, nanoremediation has the potential to lower the overall costs of cleaning up large-scale polluted sites, reduce process time, reduce some pollutant concentrations to near-zero levels, eliminate the need for treatment and disposal of polluted dredged soil, and finally, it could be done in situ. Before this approach is employed on a large scale, proper evaluation and full-scale ecosystem research of the nanomaterials must be addressed to avoid any potential negative environmental repercussions (Karn et al., 2009).

Furthermore, the synergy between nanomaterials and microorganisms for the elimination of certain pollutants has been demonstrated in batch experiments; however, there is still a lack of information about the synergetic effect of nanomaterials and biotechnologies throughout a nano bioremediation process, as well as how these combined technologies react to pollutants of various types. It should be mentioned that no safety evidence on the long-term usage of nanomaterials with microorganisms has been provided to our knowledge. Bionanoparticles have a number of advantages over metallic nanoparticles, including biodegradability, which means they have a lower environmental impact. Nanotechnologies could be employed in remediation operations to decontaminate soil, air, or water at the moment, but new cost-effective production methods should emerge. The regulatory framework is an important consideration when using these types of materials.

Researchers may be able to contribute to a better understanding of the interactions between nanomaterials and bio-based technologies as remediation progresses in varying environmental conditions and, as a result, make recommendations for tighter control. Finally, nanobioremediation has the potential to significantly contribute to sustainability because it has environmental benefits and is less expensive than other technologies. Even a wide range of nanomaterial applications, in combination with biological therapies, have showed high efficacy in the breakdown of contaminants, opening up new avenues for dealing with environmental issues (Vázquez-Núñez et al., 2020).

7.4 Future Research

It is obvious that bioremediation strategies are varied and have proven to be quite efficient in treating contaminated sites containing a variety of pollutants. The interaction of bioremediation procedures with environmental elements that can stifle microbial activity is kept at an optimal level, giving a clear picture of contaminated locations. Because of their number, diversity, and community structure in contaminated areas, microorganisms play an important role in bioremediation. Simultaneous use of many bioremediation technologies during cleanup will improve remediation efficacy by decreasing single-technique flaws while also lowering costs. The efficacy of bioremediation could be improved by using a synthetic biology approach to design microorganisms with a target compound's degradative pathway. The use of nanoparticles may reduce the contaminant's toxicity to microorganisms. It increases surface area and lowers activation energy, enhancing microorganism efficacy in degrading harmful pollutants, resulting in a reduction in overall remediation time and expense. As a result, integrating bioremediation with nanotechnology will pave the road for advancement in wastewater treatment.

7.5 Conclusion

Contaminants' existence and persistence have significantly increased, posing a threat to human health and the environment. More systematic ways for their removal from environmental matrices have been investigated using bioremediation in conjunction with nanotechnology. Several technologies, including bioremediation in combination with nanotechnology, have been investigated to find more systematic techniques to removing them from environmental matrices. It's critical to understand the interaction between the pollutant, the microbe, and the nanomaterials because both good and negative consequences can occur. Some nanoparticles, for example, stimulate microbes while others are harmful. As a result, careful selection is essential. So, in the presence of bioremediation assisted by nanomaterials, the reaction of living organisms used to remediate toxins, as well as their interaction with environmental matrices, is critical for progress in the area.

REFERENCES

Adikesavan, S., & Nilanjana, D. (2016). Degradation of cefdinir by C andida Sp. SMN 04 and M g O nanoparticles – an integrated (Nano-Bio) approach. *Environmental Progress & Sustainable Energy*, *35*(3), 706–714.

Akcil, A., Erust, C., Ozdemiroglu, S., Fonti, V., & Beolchini, F. (2015). A review of approaches and techniques used in aquatic contaminated sediments: metal removal and stabilization by chemical and biotechnological processes. *Journal of Cleaner Production*, *86*, 24–36.

Al-Rub, F.A., El-Naas, M.H., Benyahia, F., & Ashour, I. (2004). Biosorption of nickel on blank alginate beads, free and immobilized algal cells. *Process Biochemistry*, *39*(11), 1767–1773.

Azubuike, C.C., Chikere, C.B., & Okpokwasili, G.C. (2016). Bioremediation techniques–classification based on site of application: principles, advantages, limitations and prospects. *World Journal of Microbiology and Biotechnology*, *32*(11), 180.

Bartke, S., Hagemann, N., Harries, N., Hauck, J., & Bardos, P. (2018). Market potential of nanoremediation in Europe – market drivers and interventions identified in a deliberative scenario approach. *Science of the Total Environment*, *619*, 1040–1048.

Bharagava, R.N., Chowdhary, P., Saxena, G. (2017b). Bioremediation: an ecosustainable green technology: its applications and limitations. In R.N. Bharagava (Ed.), *Environmental Pollutants and their Bioremediation Approaches* (1st Edition), 1–22. Boca Raton: CRC Press, Taylor & Francis Group.

Bharagava, R.N., Saxena, G., & Chowdhary, P. (2017a). Constructed wetlands: an emerging phytotechnology for degradation and detoxification of industrial wastewaters. In R.N. Bharagava (Ed.), *Environmental Pollutants and their Bioremediation Approaches*, 1st edn., 397–426. Boca Raton: CRC Press, Taylor & Francis Group.

Bhattacharya, M., Guchhait, S., Biswas, D., & Datta, S. (2015). Waste lubricating oil removal in a batch reactor by mixed bacterial consortium: a kinetic study. *Bioprocess and Biosystems Engineering*, *38*, 2095–2106.

Bitton, G. (2005). *Wastewater Microbiology*. New Jersey: John Wiley & Sons. https://www.wiley.com/en-us/Wastewater+Microbiology%2C+3rd+Edition-p-9780471717911

Burgos, W.D., McDonough, J.T., Senko, J.M., Zhang, G.X., Dohnalkova, A.C., Kelly, S.D., Gorby, Y., & Kemner, K.M. (2008). Characterization of uraninite nanoparticles produced by Shewanella oneidensis MR-1. *Geochimica et Cosmochimica Acta*, *72*, 4901–4915.

Cai, P.J., Xiao, X., He, Y.R., Li, W.W., Yu, L., Lam, M.H.W., & Yu, H.Q. (2012). Involvement of c-type cytochrome CymA in the electron transfer of anaerobic nitrobenzene reduction by Shewanella oneidensis MR-1. *Biochemical Engineering Journal*, *68*, 227–230.

Cassidy, D.P., Srivastava, V.J., Dombrowski, F.J., & Lingle, J.W. (2015). Combining in situ chemical oxidation, stabilization, and anaerobic bioremediation in a single application to reduce contaminant mass and leachability in soil. *Journal of Hazardous Materials*, *297*, 347–355.

Chandra, R., Saxena, G., & Kumar, V. (2015). Phytoremediation of environmental pollutants: an ecosustainable green technology to environmental management. In R. Chandra (Ed.), *Advances in Biodegradation and Bioremediation of Industrial Waste* (1st Edition), 1–30. Boca Raton: CRC Press, Taylor & Francis Group.

Chatterjee, S., Lee, M.W., & Woo, S.H. (2010). Adsorption of congo red by chitosan hydrogel beads impregnated with carbon nanotubes. *Bioresource Technology*, *101*, 1800–1806.

Chen, Y., Liang, J., Liu, L., Lu, X., Deng, J., Pozdnyakov, I.P., & Zuo, Y. (2017). Photosensitized degradation of amitriptyline and its active metabolite nortriptyline in aqueous fulvic acid solution. *Journal of Environmental Quality*, *46*(5), 1081–1087.

De Windt, W., Boon, N., Van den Bulcke, J., Rubberecht, L., Prata, F., Mast, J., Hennebel, T., & Verstraete, W. (2006). Biological control of the size and reactivity of catalytic Pd(0) produced by Shewanella oneidensis. *Antonie van Leeuwenhoek, International Journal of General and Molecular Microbiology*, *90*, 377–389.

Dixit, R., Malaviya, D., Pandiyan, K., Singh, U.B., Sahu, A., Shukla, R., Singh, B.P., Rai, J.P., Sharma, P.K., Lade, H., & Paul, D. (2015). Bioremediation of heavy metals from soil and aquatic environment: an overview of principles and criteria of fundamental processes. *Sustainability*, *7*(2), 2189–2212.

Dursun, A.Y., & Tepe, O. (2005). Internal mass transfer effect on biodegradation of phenol by Ca-alginate immobilized Ralstonia eutropha. *Journal of Hazardous Materials*, *126*(1-3), 105–111.

Ferroudj, N., Nzimoto, J., Davidson, A., Talbot, D., Briot, E., Dupuis, V., Bée, A., Medjram, M.S., & Abramson, S. (2013). Maghemite nanoparticles and maghemite/silica nanocomposite microspheres as magnetic Fenton catalysts for the removal of water pollutants. *Applied Catalysis B: Environmental*, *136*, 9–18.

Fulekar, M.H., & Pathak, B. (2017). *Environmental Nanotechnology*. Boca Raton, FL: CRC Book Press, Taylor & Francis Group. doi:10.1201/9781315157214

Gautam S., Kaithwas G., Bharagava R.N., & Saxena G. (2017). Pollutants in tannery wastewater, pharmacological effects and bioremediation approaches for human health protection and environmental safety. In R.N. Bharagava (Ed.), *Environmental Pollutants and Their Bioremediation Approaches* (1st Edition), 369–396. Boca Raton: CRC Press, Taylor & Francis Group.

Hameed, M.A. (2006). Continuous removal and recovery of lead by alginate beads, free and alginate-immobilized Chlorella vulgaris. *African Journal of Biotechnology*, *5*(19), 1819–1823.

Jiang, Z.Y., Xu, S.W., Lu, Y., Yuan, W.K., Wu, H., & Lv, C.Q. (2006). Nanotube-doped alginate gel as a novel carrier for BSA immobilization. *Journal of Biomaterials Science, Polymer Edition, 17*, 21–35.

Jiang, S., Lee, J.H., Kim, M.G., Myung, N.V., Fredrickson, J.K., Sadowsky, M.J., & Hur, H.G. (2009). Biogenic formation of As-S nanotubes by diverse Shewanella strains. *Applied and Environmental Microbiology, 75*, 6896–6899.

Kang, J.W. (2014). Removing environmental organic pollutants with bioremediation and phytoremediation. *Biotechnology Letters, 36*(6), 1129–1139.

Karn, B., Kuiken, T., & Otto, M. (2009). Nanotechnology and in situ remediation: a review of the benefits and potential risks. *Environmental Health Perspectives, 117*(12), 1813–1831.

Krause, B. & Nealson, K.H. (1997). Physiology and enzymology involved in denitrification by Shewanella putrefaciens. *Applied and Environmental Microbiology, 63*, 2613–2618.

Kumar, S.R., & Gopinath, P. (2009). Nano-bioremediation: Applications of nanotechnology for bioremediation. In K.L. Wang, S.M.-H. Wang, Y.-T. Hung, N.K. Shammas, & J.P. Chen (Eds.), *Handbook of Advanced Industrial and Hazardous Wastes Management* (1st Edition, Volume 1), 27–48. Boca Raton, FL, USA: CRC Press.

Kumari, B., & Singh, D.P. (2016). A review on multifaceted application of nanoparticles in the field of bioremediation of petroleum hydrocarbons. *Ecological Engineering, 97*, 98–105.

Li, Y.Y., & Li, B. (2011). Study on fungi-bacteria consortium bioremediation of petroleum contaminated mangrove sediments amended with mixed biosurfactants. In *Advanced Materials Research* (Vol. 183, pp. 1163–1167). Trans Tech Publications Ltd. doi: 10.4028/www.scientific.net/AMR

Liu, X., Xu, W., Pan, Y., & Du, E. (2015). Liu et al. suspect that Zhu et al.(2015) may have underestimated dissolved organic nitrogen (N) but overestimated total particulate N in wet deposition in China. *Science of the Total Environment, 520*, 300–301.

Lloyd, J.R., & Macaskie, L.E. (1996). A novel phosphorlmager-based technique for monitoring the microbial reduction of technetium. *Applied and Environmental Microbiology, 62*, 578–582.

Magar, V.S. (2001). Natural recovery of contaminated sediments. *Journal of Environmental Engineering, 127*(6), 473–474.

Mani, D., & Kumar, C. (2014). Biotechnological advances in bioremediation of heavy metals contaminated ecosystems: an overview with special reference to phytoremediation. *International Journal of Environmental Science and Technology, 11*(3), 843–872.

Martínez-Pascual, E., Grotenhuis, T., Solanas, A.M., & Viñas, M. (2015). Coupling chemical oxidation and biostimulation: effects on the natural attenuation capacity and resilience of the native microbial community in alkylbenzene-polluted soil. *Journal of Hazardous Materials, 300*, 135–143.

Medina-Pérez, G., Fernández-Luqueño, F., Vazquez-Nuñez, E., López-Valdez, F., Prieto-Mendez, J., Madariaga-Navarrete, A., & Miranda-Arámbula, M. (2019). Remediating polluted soils using nanotechnologies: environmental benefits and risks. *Polish Journal of Environmental Studies, 28*, 1–18.

Middleton, S.S., Bencheikh-Latmani, R., Mackey, M.R., Ellisman, M.H., Tebo, B.M., & Criddle, C.S. (2003). Cometabolism of Cr(VI) by Shewanella oneidensis MR-1 produces cell-associated reduced chromium and inhibits growth. *Biotechnology & Bioengineering, 83*, 627–637.

Mosa, K.A., Saadoun, I., Kumar, K., Helmy, M., & Dhankher, O.P. (2016). Potential biotechnological strategies for the cleanup of heavy metals and metalloids. *Frontiers in Plant Science, 7*, 303

Moser, D.P., & Nealson, K.H. (1996). Growth of the facultative anaerobe Shewanella putrefaciens by elemental sulfur reduction. *Applied and Environmental Microbiology, 62*, 2100–2105.

Mueller, N.C., & Nowack, B. (2010). Nanoparticles for remediation: solving big problems with little particles. *Elements, 6*(6), 395–400.

Myers, C.R., & Nealson, K.H. (1988). Bacterial manganese reduction and growth with manganese oxide as the sole electron-acceptor. *Science, 240*, 1319–1321.

Myers, C.R., & Nealson, K.H. (1990). Respiration-linked proton translocation coupled to anaerobic reduction of manganese(IV) and iron(III) in Shewanella putrefaciens MR-1. *Journal of Bacteriology, 172*, 6232–6238.

Patil, S.S., Shedbalkar, U.U., Truskewycz, A., Chopade, B.A., & Ball, A.S. (2016). Nanoparticles for environmental clean-up: a review of potential risks and emerging solutions. *Environmental Technology & Innovation, 5*, 10–21.

Perelo, L.W. (2010). In situ and bioremediation of organic pollutants in aquatic sediments. *Journal of Hazardous Materials, 177*(1-3), 81–89.

Rizwan, M., Singh, M., Mitra, C.K., & Morve, R.K. (2014). Ecofriendly application of nanomaterials: nanobioremediation. *Journal of Nanoparticles*. doi:10.1155/2014/431787

Sadeghi, R., Karimi-Maleh, H., Khalilzadeh, M.A., Beitollahi, H., Ranjbarha, Z., & Zanousi, M.B.P. (2013). A new strategy for determination of hydroxylamine and phenol in water and waste water samples using modified nanosensor. *Environmental Science and Pollution Research*, 20, 6584–6593.

Saxena, G., & Bharagava, R.N. (2017). Organic and inorganic pollutants in industrial wastes, their ecotoxicological effects, health hazards and bioremediation approaches. In R.N. Bharagava (Ed.), *Environmental Pollutants and their Bioremediation Approaches* (1st Edition), 23–56. Boca Raton: CRC Press, Taylor & Francis Group.

Saxena, G., Purchase, D., Mulla, S.I., Saratale, G.D., & Bharagava, R.N. (2019). Phytoremediation of heavy metal-contaminated sites: eco-environmental concerns, field studies, sustainability issues, and future prospects. *Reviews of Environmental Contamination and Toxicology*, 249, 71–131.

Shirodkar, S., Reed, S., Romine, M., & Saffarini, D. (2011). The octahaem SirA catalyses dissimilatory sulfite reduction in Shewanella oneidensis MR-1. *Environmental Microbiology*, 13, 108–115.

Silva-Castro, G.A., Uad, I., Gónzalez-López, J., Fandiño, C.G., Toledo, F.L., & Calvo, C. (2012). Application of selected microbial consortia combined with inorganic and oleophilic fertilizers to recuperate oil-polluted soil using land farming technology. *Clean Technologies and Environmental Policy*, 14, 719–726.

Singh, R., & Misra, V. (2014). Application of zero-valent iron nanoparticles for environmental cleanup. *Advanced Materials for Agriculture, Food, and Environmental Safety*, 385–420.

Singh, R., Behera, M., & Kumar, S. (2020). Nano-bioremediation: An innovative remediation technology for treatment and management of contaminated sites. In *Bioremediation of Industrial Waste for Environmental Safety*, 165–182. Singapore: Springer.

Steplin Paul Selvin, S., Ganesh Kumar, A., Sarala, L., Rajaram, R., Sathiyan, A., Princy Merlin, J., & Sharmila Lydia, I. (2018). Photocatalytic degradation of rhodamine B using zinc oxide activated charcoal polyaniline nanocomposite and its survival assessment using aquatic animal model. *ACS Sustainable Chemistry & Engineering*, 6(1), 258–267.

Tan, W., Peralta-Videa, J.R., & Gardea-Torresdey, J.L. (2018). Interaction of titanium dioxide nanoparticles with soil components and plants: current knowledge and future research needs – a critical review. *Environmental Science: Nano*, 5(2), 257–278.

Thangadurai, D., Sangeetha, J., & Prasad, R. (2020). *Nanotechnology for Food, Agriculture, and Environment*. Switzerland AG: Springer Nature.

Tratnyek, P.G., & Johnson, R.L. (2006). Nanotechnologies for environmental cleanup. *Nano Today*, 1(2), 44–48.

Vázquez-Núñez, E., Molina-Guerrero, C.E., Peña-Castro, J.M., Fernández-Luqueño, F., & de la Rosa-Álvarez, M. (2020). Use of nanotechnology for the bioremediation of contaminants: a review. *Processes*, 8(7), 826.

Viamajala, S., Peyton, B.M., Apel, W.A., Petersen, J.N. (2002). Chromate reduction in Shewanella oneidensis MR-1 is an inducible process associated with anaerobic growth. *Biotechnology Progress*, 18, 290–295.

Wang, X., Wang, Q., Wang, S., Li, F., & Guo, G. (2012). Effect of biostimulation on community level physiological profiles of microorganisms in field-scale biopiles composed of aged oil sludge. *Bioresource Technology*, 111, 308–315.

Wang, A., Ng, H.P., Xu, Y., Li, Y., Zheng, Y., Yu, J., Han, F., Peng, F., & Fu, L. (2014). Gold nanoparticles: synthesis, stability test, and application for the rice growth. *Journal of Nanomaterials*, 1, 1–6.

Yan, F.F., Wu, C., Cheng, Y.Y., He, Y.R., Li, W.W., & Yu, H.Q. (2013). Carbon nanotubes promote Cr (VI) reduction by alginate-immobilized Shewanella oneidensis MR-1. *Biochemical Engineering Journal*, 77, 183–189.

Zare, K., Najafi, F., & Sadegh, H. (2013). Studies of ab initio and Monte Carlo simulation on interaction of fluorouracil anticancer drug with carbon nanotube. *Journal of Nanostructure in Chemistry*, 3(1), 71.

8
Low-Cost Bioremediation Technologies for Transforming Waste to Wealth

Rupika Sinha[1], Shipra Dwivedi[2], Sukhendra Singh[3], Singh Divakar[4], and Pradeep Srivastava[4]
[1]Department of Biotechnology, Motilal Nehru National Institute of Technology Allahabad, Prayagraj, U.P., India
[2]Department of Biochemical Engineering, School of Chemical Technology, Harcourt Butler Technical University, Kanpur, U.P., India
[3]Department of Alcohol Technology and Biofuels, Vasantdada Sugar Institute, Pune, Maharashtra, India
[4]School of Biochemical Engineering, Indian Institute of Technology (Banaras Hindu University) Varanasi, U.P., India

CONTENTS

8.1	Introduction	134
8.2	Sources of Wastewater	134
	8.2.1 Household Wastewater	135
	8.2.2 Agriculture Wastewater	135
	8.2.3 Industrial Wastewater	135
8.3	Wastewater Conversion Technologies	135
8.4	Algae-Based Wastewater Treatment	135
	8.4.1 Growth Requirement of Algae	135
	8.4.2 Algal Wastewater Treatment and Biofuel	136
	8.4.3 Advantages and Disadvantages of Microalgae as a Substrate for Biofuel Production	137
8.5	Biofuel Production Using Wastewater	137
	8.5.1 Biogas Production	137
	8.5.1.1 Biogas Production Process	138
	8.5.1.2 Anaerobic Digestion Process	138
	8.5.1.3 Biogas Calorific Value	139
	8.5.2 Production of Biodiesel from Wastewater	139
	8.5.3 Production of Bioethanol from Wastewater	140
	8.5.3.1 The Process of Bioethanol Preparation	141
	8.5.3.2 Bioethanol Production Using Algae Grown in Wastewater	141
	8.5.3.3 Bioethanol Production Using Wastewater from the Food Industry	142
	8.5.3.4 Bioethanol Production Using Wastewater of the Soft Drink Industry	142
	8.5.4 Production of Biohydrogen Using Wastewater	143
	8.5.4.1 Biohydrogen Production Process	143
	8.5.4.2 Biohydrogen Production from Microalgae	144
	8.5.5 Production of Bioelectricity Using Wastewater	144
	8.5.6 Production of Biopolymers	144
	8.5.6.1 Production of L-Lactic Acid Using Wastewater Sludge	145
	8.5.6.2 Production of PHAs Using Wastewater Sludge	145

DOI: 10.1201/9781003004684-8

8.6	Analysis of Conversion Process Efficiency	145
	8.6.1 Process Efficiency for Biogas Production	145
	8.6.2 Process Efficiency for Bioethanol Production	146
	8.6.3 Process Efficiency for Biodiesel Production	146
	8.6.4 Process Efficiency for Biohydrogen Production	146
8.7	Large-Scale Technologies	147
	8.7.1 Integrated Biogas–Biodiesel Production Approach	147
	8.7.2 Biohydrogen Production Using Multi-stage Bioreactors	148
8.8	Conclusion	148
References		149

8.1 Introduction

Wastewater generated from various sources such as sewage, industries, and various runoffs needs proper management techniques to extract their maximum potential. The history of sewage management could be traced back to the period of Minoan kings around 3,000 BC (Gethin, 2021). Also, the study of the Indus valley civilisation shows the existence of flush tanks. The wastewater generated was carried to the nearest water body. Later, the treatment of wastewater in proper equipment using chemicals was proposed due to the serious concern of the pollutants. With the increase in global population, the use of water for households and industries as well as agricultural purposes has increased tremendously, which has caused the shortage of potable drinking and clean water, along with an increase in the concentration of pollutants in water bodies. Becoming one of the major global concerns, this has led to the enhancement of research for wastewater management. Some techniques have been suggested, which include treatment and recirculation of wastewater for various uses.

The World Health Organization has clearly defined various grades of water such as greywater, wastewater, reclaimed water, green water, and drinking water. It has also provided the guidelines for proper reuse of each of the water grades. These uses include sanitation, fire systems, irrigation, green spaces, industrial cooling, fire systems, and industrial processing. Efficient wastewater management also leads to the reduction of effluent disposal form sewage or industries.

Another approach could be the conversion of wastewater for the generation of useful products, thereby improving employment. One of the economical strategies would be the utilisation of waste for energy generation. The potential of algae has been evaluated for the bioremediation of wastewater; the algal biomass can then be utilised for the production of biofuels. This chapter summarises the production of biofuels by algae grown on wastewater. The products promised by algae can vary from methane, biodiesel, hydrogen, and bioethanol to biomass itself (Pittman et al., 2011). Algal biofuels have the added advantage over biofuel crops as the algae exhibit exponential growth without competing with food crops. It is an environmentally friendly approach for the removal of contaminants from wastewater with simultaneous energy generation. Various optimisations carried out by different researchers have been reviewed that can be employed for scaling up the process. This may enhance the employability of low-income countries such as India, along with dealing with the effluent hazards created due to large population density.

8.2 Sources of Wastewater

To generate energy using algal biomass that is grown on wastewater for the production of biogas and biodiesel, the identification of available wastewater sources and suitable algal strains are crucial steps. The composition of the wastewater affects the growth of algae. It may also influence the accumulation of lipids and starches for biofuel production.

8.2.1 Household Wastewater

The water utilised by urban and rural populations for household and other activities is known as household or municipal wastewater. It contains a lower amount of phosphorus and nitrogen compared to animal wastewater (Chen et al., 2015). On the other hand, it may contain a significant proportion of heavy metals, including zinc, copper, and lead (Cai et al., 2013). Such unit operations as carbon adsorption and membrane separation can be partially or completely replaced by microalgal cultivation.

8.2.2 Agriculture Wastewater

Agricultural runoffs contain different types of manures, insecticides, pesticides, herbicides, and weedicides. Moreover, it may also contain several inorganic compounds of nitrates and phosphates. The effluent from fields may contain pollutants in the form of ions, particles, molecules, compounds, or living organisms. High turbidity of agricultural wastewater reduces the penetration of light, which may affect the growth of microalgae (Chen et al., 2015).

8.2.3 Industrial Wastewater

Effluents from industries such as textiles, paint, electronics, iron steel plants, and tanneries may contain heavy metals such as lead, mercury, arsenic, cadmium, nickel, zinc, chromium, and copper. Shanab et al. studied the mercury, lead, and cadmium removal efficiency of three different microalgae. In this study, a cyanobacterium (*Pseudochlorococcum typicum*) was found to remove the highest percentage (Shanab et al., 2012). Previous reports suggest the use of wastewater from palm oil, carpet, and olive mill for microalgal growth, which can be employed for the production of biofuels (Chen et al., 2015).

8.3 Wastewater Conversion Technologies

Green technologies are being emphasised due to the rapid increase in global pollution. Thus, conventional treatment of wastewater has been coupled with the different process so that waste remediation is associated with economical advantages. This includes mass cultivation of algae and production of different biofuels from wastewater. Commercial products such as poly-L-lactic acid can also be synthesised in a cost-efficient manner using this strategy. Another polymer, known as polyhydroxyalkanoate (PHA), is also being studied for the technical and commercial feasibility of the production process using wastewater.

8.4 Algae-Based Wastewater Treatment

8.4.1 Growth Requirement of Algae

The algae are simple microorganisms with photosynthetic activity and can be grown in aquatic environments. Autotrophic microalgae utilise sunlight and nutrient sources like carbon, nitrogen, and phosphorous and convert them into adenosine triphosphate (ATP) required for their growth and oxygen. They produce biomass, which can be further used in the production of biodiesel, bioethanol, biohydrogen, and methane. As the heterotrophic microalgae do not perform photosynthesis, they utilise the organic compounds. On the other hand, mixotrophic algae acquire external organic nutrients as they perform photosynthesis.

The autotrophic algae utilise inorganic carbon sources including salts, CO_2, and light energy. Parameters such as temperature, pH, light intensity, carbon source, and nutrient source play an important role in algae growth. The optimum temperature required for algal growth is 25 °C to 30 °C and an increase in temperature above this range affects algal growth. Similarly, optimal pH required for algal growth is 7 but it has been reported that some algae can tolerate a pH as high as up to 10 and below up to 4 (Moheimani, 2005). Some researchers reported that the optimum nitrogen:phosphorous ratio should be in between 5:1 and 10:1 for algal growth. The accumulation of lipid decreases with the reduction in nitrogen and phosphorus concentration (Liang et al., 2013; Sharma et al., 2012).

8.4.2 Algal Wastewater Treatment and Biofuel

Nowadays, algae, especially microalgae, are used on a large scale for production of biofuels. Algae utilises greenhouse gases like carbon dioxide present in the environment and various pollutants of wastewater like nitrogen and phosphorous for their growth and produce a large amount of biomass (Schenk et al., 2008). The process of utilising algae for the treatment of wastewater is a more attractive method compared to the other treatment methods because they absorb nutrients like nitrogen and phosphorus and other harmful toxic metals that are major pollutants. These pollutants can originate from agriculture, domestic use, and industrial wastewater. Microalgae can be sustained in a nutrient-rich environment and can accumulate carbohydrates and lipids in their biomass for biofuel production. In addition to this, they accumulate various toxic metals from wastewater and render the wastewater treatment process more attractive. This property attributes to their bioremediation ability, as shown in Table 8.1.

The remediation property of algae can be attributed to the biosorption or bioaccumulation mechanisms. Biosorption is a passive process of adsorption on the surface of dead or living cells. On the other hand, the bioaccumulation requires energy for the transport of metal ions across the cell membrane.

Researchers discovered that polymeric material found on the external cell layer of exopolysaccharide producing microorganisms play a crucial role in binding of the metal ion (De Philippis et al., 2011). Others have studied the role of carboxyl groups of the polysaccharides in the biosorption process, specifically for heavy metals. Extracellular precipitation can also lead to bioremediation of metal pollutants (Olguín and Sánchez-Galván, 2012). This precipitation may or may not be related to the intracellular mechanism, can also occur along with biosorption. The metal uptake capacity (q) is determined in terms of amount of metal adsorbed per amount of dry cell and is given by the following equation (Olguín and Sánchez-Galván, 2012):

$$q = \frac{v(c_{mi} - c_{mf})}{x_m}$$

where c_{mi} (mg L^{-1}) and c_{mf} (mg L^{-1}) are the initial and final metal concentrations, respectively.

TABLE 8.1

List of some algae with metal remediation properties

Algae Used	Pollutants	Mechanism of Remediation	References
Anabaena cylindrica	Lead	Adsorption (Langmuir isotherm)	(Swift and Forciniti, 1997)
Aphanothece halophytica	Zinc	Adsorption (Langmuir isotherm)	(Incharoensakdi and Kitjaharn, 2002)
Nostoc PCC7936	Chromium	Adsorption by biomass and reduction of Cr(IV) to Cr (III)	(Colica et al., 2010)
Desmodesmus communis	Nitrogen and phosphorus	Utilisation for growth	(Samorì et al., 2013)
Nostoc sphaeroides	Lead and chromium	Adsorption (Langmuir isotherm)	(Jiang et al., 2016)
Lyngbya wollei	Copper	Biotransformation	(Bishop et al., 2018)
Chlorella vulgaris (modified algal residuals)	Mercury	Adsorption	(Peng et al., 2018)
Scenedesmus obliquus	Iron	Adsorption	(Bouzit et al., 2018)
Desmodesmus sp.	Copper, nickel, and phosphorus	Cu, Ni-Adsorption, P-assimilation	(Rugnini et al., 2018)

v = volume of sample (*L*)

x_m = dry cell mass(*g*)

This metal-adsorbing capacity of the cells can be used for selection of suitable algae for bioremediation. The distribution of the metal and uptake mechanism aids in determination of amount of possible recovery. Metal distribution on the surface (adsorption) and within the cells (accumulation) should be evaluated for recovery of the metal ions. This recovery is beneficial as it is the source of metals as well as regenerates the system for further bioadsorption.

The recovery of the adsorbed metal is carried out by either the ion-exchange process or use of chelating substances such as ethylenediamine tetraacetic acid (EDTA). Researchers have studied the variation in adsorption and accumulation of cadmium, copper, and lead by the algae *Chlorella kesslerii*. The amount of adsorbed metals was quantified using EDTA, whereas the internalised quantity was estimated using concentrated nitric acid. They also found that the addition of humic acid improved the proportion of adsorbed lead due to the formation of complex of lead, humic acid, and internalisation sites (Lamelas et al., 2009).

8.4.3 Advantages and Disadvantages of Microalgae as a Substrate for Biofuel Production

Microalgae utilisation for biofuel production and wastewater treatment poses several advantages as it provides a renewable source of energy like biohydrogen, bioethanol, biodiesel, and methane. Biofuel production from algae is both economically and environmentally sustainable. Microalgae have a high growth rate so even in low biomass concentration, they can proliferate rapidly and can be harvested around the year. Microalgae can effectively use the available inorganic carbon sources such as carbon dioxide for growth. They are tolerant to the high concentration of CO_2 that is responsible for the high rate of CO_2 mitigation. It was reported that 1 kg of dry algal biomass can use approximately 1.83 kg of CO_2 (Wang and Yin, 2018).

Algae reduce greenhouse gases such as carbon dioxide in the environment by using them for photosynthetic activity and assimilate in the form of carbohydrates and lipids. Microalgae cultivation does not require herbicides and pesticide applications. Microalgae lack hemicellulose and lignin, as found in other biomasses; thus, they require milder pretreatment methods. The cultivation process for microalgae is convenient in comparison to other biomasses. These benefits promote the utilisation of algae cultivation for biofuel production and wastewater treatment. The disadvantage associated with microalgae utilisation is that they are produced in low concentrations. Also, due to their small size, the harvesting of biomass is a difficult process on a large scale (Misra et al., 2016; Zhan et al., 2016). However, many researchers are working to overcome this issue (Zhao et al., 2019, Show et al., 2019).

8.5 Biofuel Production Using Wastewater

Production strategies of biofuels including biogas, biodiesel, bioethanol, and biohydrogen have been studied by many workers. This strategy has not only opened avenues for the "waste to energy" approach but also has significantly contributed in the reduction of carbon footprints generated by conventional fuels.

8.5.1 Biogas Production

Biogas is one of the economical and renewable energy resources that is widely used for electricity generation. In Germany, approximately 98% of the biogas is utilised for power supply in thermal and power plants (Weiland, 2003, 2010). A pure form of biogas primarily containing methane and free from impurities such as CO_2 can be used as vehicle fuel by substituting petrol and diesel. China has the highest number of domestic biogas digesters, which are used for cooking and lighting (Chen et al., 2014). Also, the microbially digested end product can be used as organic fertiliser for agricultural purposes. China has established many biogas plants for anaerobic digestion and gas supply for domestic purposes. In recent years, an increase in medium and large-scale plants with fermentation volume greater than 300 m³ have been observed due to urbanisation, modern agricultural techniques, and large-scale poultry farming, whereas a decline in growth rate of domestic biogas plants has been witnessed (Wang et al., 2016). Thus, some biogas plants have been installed near large-scale farms, which not only treat the waste but also generate clean energy.

Numerous research has been carried out in different types of reactors using a variety of wastewaters. Rao et al. reported in 1997 about biogas produced in an Upflow Anaerobic Sludge Blanket (UASB) reactor. The volume of the reactor was 29 L, and it utilised a high concentration of synthetic waste and produced biogas containing 72% methane. With this reactor, the organic loading rate in terms of chemical oxygen demand (COD) achieved was 47 Kg COD m^{-3} day^{-1} with a low hydraulic retention time (HRT). For every kg of COD removed, 0.29 m³ of methane was obtained (Rao et al., 1997). In another study by Ergüder in 2000, it was found that olive mill wastewater was anaerobically treated in batch reactors with about 90% efficiency, which led to the production of about 57 litres of methane per litre of olive mill wastewater. This means that 413 mL of methane was produced with 1 g of COD (Ergüder 2000).

8.5.1.1 Biogas Production Process

Biogas is produced by an anaerobic digestion process in which the organic material is decomposed by microorganisms under oxygen free environment. The solid waste requires more pretreatment for size reduction compared to the sludge of wastewater. Moreover, the cost of water required for the digestion process is also reduced when biogas is produced using wastewater.

8.5.1.2 Anaerobic Digestion Process

Biogas is produced using a three step process resulting in the production of methane and other by products such as hydrogen sulphide gas and carbon dioxide gas by digestion of sewage waste.

- Step 1: Hydrolysis of organic compounds such as proteins, lipids, and carbohydrates. In this step, the complex organic compounds are broken into simpler digestible molecules by the enzymes secreted by the bacteria present in the sludge.
- Step 2: These simple molecules are then converted into different types of fatty acids such as butyric acid, propionic acid, etc. This conversion is carried out by various anaerobic bacteria such as *Actinomycetes* sp., *Clostridium* sp., *Staphylococcus* sp., etc.
- Step 3: Methanogenesis (production of methane) by conversion of fatty acids into methane. This step is carried out by methanogenic bacteria in anaerobic conditions. These may include *Methanosarcina, Methanobacillus, Methanobacterium,* and *Methanococcus* sp. Methanogenesis is the rate-limiting step as it is the slowest step of biogas production. The rate of methanogenesis depends on environmental conditions that influence the growth of methane-producing bacteria. The optimum pH and temperature range of methanogenic bacteria vary 6.6–7.6 and 32–40°C, respectively. Thus, the growth of acid-producing bacteria within the reactor may decrease the pH of the environment, leading to a reduction in the growth of methanogenic bacteria. To achieve the mesophilic range of temperature for the growth of methanogens, pre-heating of waste is done before adding it to the digester (Eddy, 1986).

Generally, the percentage of reduction in volatile solids provides the estimate of gas production. Usually, it ranges from 0.75 to 1.1 m³ of biogas produced per kg of reduction of volatile solids. Methane yield calculated on a theoretical basis is 0.35 m³ of methane per kg of COD converted, though the experimental yield may vary depending on the environmental factors which control the growth of methanogenic bacteria. The constituents and biodegradation ability of the organic substrate affect the yield of biogas while the production rate will vary with factors such as temperature, the density of microbes, and their culture conditions. Production of biogas depends on factors such as the content of volatile solids of the sludge and enzymatic activity of microbes present in the digester. When the biogas production is calculated on a per capita basis, in primary plants the yield was found to range between 15 L per person per day and 22 L per person per day, whereas in secondary treatment plants, this range increased up to 28 L per person per day (Berktay and Nas, 2007).

The impurities present in the biogas, such as CO_2, H_2S, and excess water vapours, are required to be removed. The presence of H_2S in the biogas may cause corrosion to the engines and metal pipes. Thus, it is removed by scrubbing the gas with a gas scrubber or an iron oxide sponge. Another approach for the removal of this gas could be the addition of metals to the sludge before anaerobic digestion. This results in the formation of insoluble metal salts that are precipitated during the digestion process. Most of the CO_2 gas is removed, along with the removal of the H_2S gas; the remaining can be removed by absorption in an aqueous solution. This removal of CO_2 should be carried out mostly when the biogas is purified for commercial purposes as it is an expensive step. The purified biogas can be either utilised immediately or may be stored for later use. This biogas is primarily composed of methane. The biogas can be used for fire directly or in the internal combustion gas engine. Boilers, water pumps, electric generators, incinerators, and blowers can also use the calorific property of biogas. Another optimised on-site use of biogas can be to heat the influent of the anaerobic digester. The rmaining gas can be utilised for other purposes.

8.5.1.3 Biogas Calorific Value

The fraction of methane in the biogas depends on the anaerobic digestion process and may vary between 0.55 and 0.80. Other components may include hydrogen sulphide, carbon dioxide, and moisture. The energy content of pure methane is about 35,000 Btu m^{-3}(Berktay and Nas, 2007). Though the calorific value of methane is lower than natural gas, it is an excellent fuel (Berktay and Nas, 2007). Biogas can replace the natural gas by a few modifications in the equipment design. These design changes are incorporated due to the low calorific value of methane gas.

Hot water and steam can also be harvested when biogas is used for electricity generation (Fjørtoft et al., 2019). In such cases, cooling devices and engine exhaust are a source of hot water and steam. Thus, coupling steam and hot water collection with electricity generation can enhance the conversion efficiency to more than 80%. Both light and heavy motor vehicles can use compressed biogas in an efficient manner as a fossil fuel alternative. This requires scrubbing of biogas from its CO_2, H_2S, and moisture content. After removal of the impurities, biogas can be used as vehicle fuel like that of compressed natural gas (CNG) (Berktay and Nas, 2007). Co-digestion of sewage waste and algae grown in High Rate Algal Ponds (HARPs) has been found to improve the biogas yield by 25% as observed by some researchers (Vassalle et al., 2019).

8.5.2 Production of Biodiesel from Wastewater

Biodiesel has properties similar to that of diesel and produced from an oil source which should contain triglycerides or fatty acids. The conversion takes place by the process of transesterification. The first-generation biodiesel was produced by edible oils and lipids. The second-generation biodiesel is manufactured using waste cooking oil, non-edible oil, and animal fat. Third-generation biodiesel is produced from microalgae that accumulate lipids in their biomass (Leong et al., 2018). As compared to the petroleum diesel, the advantages associated with biodiesel are that it is less toxic, renewable, and biodegradable.

Moreover, the burning of biodiesel is clean compared to petroleum diesel. The burning of biodiesel is associated with reduced emissions of carbon dioxide, carbon monoxide, and particulate matters. Chemically, biodiesel is composed of alkyl esters of fatty acids that are catalyzed by either a base or an acid. The process involves trans-esterification of fatty acids with alcohols. Sewage sludge containing a significant proportion of lipids is a suitable substrate for biodiesel production. Sludge obtained from both domestic and industrial wastewater contains a significant proportion of long-chain fatty acids, oils, and grease. Sludge may also contain the phospholipid fraction of cell membranes. These properties of sludge can be employed for biodiesel production.

Sludge can be collected either as primary or activated sludge from the water treatment plant. Primary sludge is collected after the mechanical process of wastewater treatment that settles in the primary sedimentation basin. The primary sludge contains a large fraction of organic matter, such as kitchen waste, industrial waste, and faeces. In the activated sludge, the removal of the dissolved organic matter is carried out by microorganisms, which also consume the inorganic nutrients for growth in the presence of oxygen. This process results in the formation of activated sludge. It has been reported that the primary sludge is more suitable for biodiesel production compared to the activated sludge, due to the higher yield of fatty acid methyl esters (Capodaglio and Callegari, 2018).

The process of biodiesel production from sludge has a pre-treatment step. Municipal sewage contains a large quantity of water that poses a hindrance for the extraction of lipids. Thus, dewatering of sludge is carried out along with the removal of pathogens, which is also known as the pretreatment step. Dewatered sludge has a high viscosity, which inhibits the extraction process. Therefore, dried sludge has a better suitability. Purification processes such as centrifugation and filtration can be used to harvest the sludge.

Another step in biodiesel production is the extraction of lipids followed by their conversion process. Siddiquee and Rohani (2011) carried out the lipid extraction process using methanol and hexane as solvents for the production of biodiesel using wastewater sludge (Shin et al., 2018; Siddiquee and Rohani, 2011). On the other hand, Olkiewicz et al. carried out liquid-liquid extraction for the extraction of lipids from municipal sewage sludge. This extraction method avoids the expensive dewatering process. Using this method, about 91% of the lipids from the primary sludge were recovered (Olkiewicz et al., 2014). Hexane is commonly used for the extraction of lipids due to its superior lipid accumulating ability. Other solvents such as toluene, methanol, and diethyl ether can also be used for lipid extraction.

The conversion process can be carried out by using methods such as pyrolysis and transesterification. There are other methods also, such as micro-emulsification and direct use (Ma and Hanna, 1999). Pyrolysis is the conversion of lipids to fatty acid acyl esters at high temperatures in the absence of air or oxygen and the presence or absence of a catalyst. This process is effective and free of pollution compared to other cracking processes. A research team from South Korea has identified a novel process for the conversion of lipids found in sludge into biodiesel. This technology provides an economical approach for biodiesel production along with increased yield compared to the conventional methodology of production. They found that the yield obtained from the non-catalytic reaction was much higher in the case of sewage sludge compared to the biomass containing soy abean and microalgae (Kwon et al., 2012).

Transesterification is also one of the widely used conversion processes. In this process, usually methanol acts as an acyl acceptor. This process is catalysed by an acid or a base. Another process, known as in-situ transesterification, is the extraction of lipids coupled with the simultaneous conversion of the lipids into FAME. In this process, the acid or the base catalyst are added at the initiation of the reaction, which is followed by hexane extraction of biodiesel rather than lipids. The biodiesel is extracted and is then centrifuged. The upper phase of the supernatant is discarded, whereas the bottom layer contains the biodiesel. The process is time consuming, but still economical.

8.5.3 Production of Bioethanol from Wastewater

Bioethanol signifies ethanol that is produced by fermentation of starch found in different biological sources. These biological sources may vary from crops such as sugarcane, sweet sorghum, rice, corn, potato, and sweet potato (first-generation bioethanol) to lignocellulosic feedstock such as husk and straw (second-generation bioethanol). Third-generation bioethanol is produced by using the starch stored or accumulated in the cell mass of microalgae (Singh et al., 2018b). These microalgae do not compete with food crops or arable land for bioethanol production. *Saccharomyces cerevisiae* is the most common yeast used for the production of bioethanol. A bacterium known as *Zymomonas* sp. can also be utilised for bioethanol production. Bioethanol can be used by blending it with petrol. This reduces the emissions of toxic gases as well as reduces the use of fossil-derived petrol. Also, ethanol blending reduces the use of fossil fuel, which is more costly.

Moreover, ethanol has a higher octane rating that reduces the knocking characteristics of the engine. The blending of ethanol can be represented by EX, where E represents ethanol and X stands for the percentage (v/v) of ethanol in the fuel. Nowadays, E10 is the most common blend, but the design of flexible fuel vehicles has led to the use of a petroleum blend containing 85% of bioethanol (Nair et al., 2017). Thus, production of bioethanol needs consideration for both environmental and economic benefits.

8.5.3.1 The Process of Bioethanol Preparation

The bioethanol process can be subdivided into some steps:

- Pretreatment of substrate: The substrate is subjected to size reduction by the process of grinding and milling followed by treatment with acid, alkali, enzymes, or steam.
- Hydrolysis of substrate: The release of fermentable sugar from the biomass is facilitated by the use of enzymes produced by microorganisms such as fungi or bacteria.
- Fermentation: The hydrolyzed sugar is then consumed by the fermentation microorganisms such as *Saccharomyces cerevisiae* for bioethanol production.
- Distillation: The bioethanol produced is then purified using multiple distillation units, and the purity of the product is analyzed.

Traditionally, *Saccharomyces cerevisiae* has been used for bioethanol production by converting the starchy material into glucose that is then fermented in bioethanol (Singh et al., 2017). Studies have been carried out to adapt the microorganism so that it can also utilise the pentose sugars. This can be carried out by changes at the genetic level (Hong et al., 2014). The potential of other microorganisms for bioethanol fermentation has also been investigated and found that *Rhizopus, Mucor, Zymomonas mobilis, Fusarium,* and some recombinant strains of *Escherichia coli* are also efficient in its production (Nair et al., 2017). A few thermotolerant strains, such as *Candida acidithermophilium, Kluveromyces species, Cryptococcus tepidarius,* and have been developed for bioethanol production (Bharathiraja et al., 2014).

There are two strategies for bioethanol production. The first is to carry out both hydrolysis and fermentation steps in two distinct phases. This is known as Separate Hydrolysis and Fermentation (SHF). The second strategy is to carry out hydrolysis and fermentation simultaneously. This method is known as Simultaneous Saccharification and Fermentation (SSF).

Municipal sludge can be used for bioethanol production as it is an economical and feasible source of starch and cellulose. The wastewater collected from the paper industry contains a high cellulosic concentration, which makes it a suitable candidate for bioethanol production. Dubey et al. (2012) studied bioethanol production by hydrolyzing waste paper. The hydrolysate was found to contain about 70% carbohydrates, which was fermented using *Pichia stipitis*. This technique led to the production of 3.73 g/L of bioethanol, with a fermentation efficiency of about 77% (Dubey et al., 2012).

8.5.3.2 Bioethanol Production Using Algae Grown in Wastewater

It has been suggested that the conversion of algal starch to bioethanol is an economic process as it does not require arable land, but the growth of algae requires a supply of specific nutrients in the medium. Nitrogen and phosphorus are utilised by algae for the reproduction process. The supply of nutrients increases the cost of algae culture and production of bioethanol. The nitrogen and phosphorus can be supplemented by the use of fertilisers, which may again lead to a shortage of fertiliser supply for agricultural purposes. Thus, wastewaters from different domestic and industrial sources are capable of proving these macro- and micronutrient supplies for algal growth (Bibi et al., 2017). Studies have suggested that microalgae are very efficient in harvesting nitrogen and phosphorus metals from wastewater both in free as well as immobilised form (Pittman et al., 2011). *Scenedesmus* and *Chlorella* species are efficient in the removal of more than 80% of nitrates, ammonia, and phosphorus from treated wastewater (Martınez et al., 2000; Ruiz-Marin et al., 2010; Zhang et al., 2008). This nutrient removal property of algae indicates their dual advantage of wastewater treatment and bioethanol production. Thus, domestic and industrial effluents can be utilised for the cultivation of algae, which become a source of starch for bioethanol production.

Ellis et al. (2012) have demonstrated the butanol, acetone, and ethanol production efficiency of *Clostridium saccharoperbutylacetonicum* N1–4, which utilised algal biomass grown in wastewater. The feedstock for this study was 10% algae, which were pretreated with an acid or base. When this medium was supplemented with 1% glucose, the production increased up to 7.2 g L^{-1} of total acetone, butane, and ethanol production. The increase was further noticed when the pretreated media was supplemented with enzymes such as xylanase and cellulase. The addition of enzymes enhanced the production of the three organic compounds by 250%. Thus, engineering of the conventional methods by process optimisation techniques could lead to enhancement in the product yield (Ellis et al., 2012).

Other studies have shown that microalgae may also utilise nutrients in the form of ions such as ammonium, nitrate, and phosphate for growth (Craggs et al., 1997; Kim and Jeune, 2009). Moreover, the water required for the culture of algae need not be added. Thus, the use of wastewater can reduce the cost of algae culture. The research carried out to culture microalgae in wastewater for production of bioethanol is less than that for biodiesel. Lewis Oscar et al. evaluated microalgae known as *Stigeoclonium* sp., Kütz. BUM11007 for bioethanol production by cultivating it in domestic wastewater. The time required for consumption of nutrients by the algae was monitored. Removal of nutrients from the wastewater was accompanied by biomass accumulation. The sugar accumulated in the form of an algal biomass was consumed by *Saccharomyces cerevisiae* for bioethanol production. The efficiency of algae in depleting nutrients from wastewater and production of bioethanol was evaluated. It was found that an ethanol yield of 0.195 g g^{-1} was obtained. This study suggested that microalgae are a significant source of bioethanol production and the process becomes more economical when they are cultivated in wastewater. This coupling of two environmentally friendly processes is a sustainable approach towards energy generation. This study can be extended to the determination of the rate-limiting step and development of large-scale microalgal biorefinery (LewisOscar et al., 2015).

8.5.3.3 Bioethanol Production Using Wastewater from the Food Industry

Researchers have studied the effect of effluent from the food industry, such as the noodle-making industry, which contains a very high concentration of starch. Many such small-scale industries discharge the wastewater into water bodies without proper pretreatment. This causes a rise in oxygen demand in the water bodies. Thus, fermentation of the wastewater by using yeast cells by immobilising them in calcium alginate beads was investigated. The production of ethanol was carried out in a sequencing batch reactor with immobilised yeast cells. The entrapment of cells has several advantages, such as wastewater could be treated more efficiently with high cell loading and does not require the sedimentation of cells after the treatment process is over. *Siripattanakul* studied the effect of production of ethanol in a sequential batch reactor containing immobilised cells. This reactor can be applied as a small-scale on-site water treatment device. Experiments were carried out to determine an optimal cell loading and wastewater treating efficiency. It was also compared with a sequential batch reactor containing free cells (Siripattanakul-Ratpukdi, 2012).

8.5.3.4 Bioethanol Production Using Wastewater of the Soft Drink Industry

Another group of researchers evaluated the potential of wastewater of the soft drink industry for bioethanol production using yeast cells. *Saccharomyces cerevisiae* was used for fermenting lemon, cola, and orange-type soft drinks. The variation in cell mass, sugar consumption, and bioethanol concentration with respect to time was studied. It was also noticed that the addition of yeast extracts at a concentration of 15 g L^{-1} increased the production of bioethanol. The sugar concentration in the soft drinks varies from 10% to 12%, which started depleting within 12 hours of the beginning of the fermentation process. The inoculum size was kept at 2 g L^{-1}. It was found that the theoretical and experimental yields were similar. The process kinetics were evaluated using many models by determining their parameters. The model that considered the inhibition caused by ethanol was best found to match the

Low-Cost Bioremediation Technologies 143

experimental data. Researchers also investigated the effect of an additive such as sodium benzoate. Some products that were rejected during the quality check of the bottling process can also act as substrates of the yeast mediated bioethanol production process. Several parameters were optimised, including the addition of nutrients and composition of the soft drinks, which were intended to act as substrates. The variations in biomass generation, sugar consumption, and bioethanol production in wastewater and synthetic media fermentation were compared. The effect of nutrient supplementation and other constituents of soft drinks on the production process was examined. This led to an evaluation of the technical feasibility of the ethanol production process using wastewater from the soft drink industry (Isla et al., 2013).

8.5.4 Production of Biohydrogen Using Wastewater

Hydrogen gas is a promising alternative of fossil fuels with reduced emission of greenhouse gases (Lam et al., 2019; Show et al., 2019). It is preferred as fuel because it does not emit any toxic gases such as NOx and SOx. Moreover, it is a carbon-free fuel and considered to be clean. On burning, it only releases water vapours as by-products. It contains 122 kJ of energy per gram, which is much higher than the energy content of a hydrocarbon containing fuel. It is highly suitable for energy supply to vehicles, industries, and power stations. Hydrogen gas produced by biological means is known as biohydrogen.

8.5.4.1 Biohydrogen Production Process

The common methods of biohydrogen production are:

1. Photolysis of water by cyanobacteria or algae.
2. Decomposition of organic compounds in the presence of light by photosynthetic bacteria.
3. Hydrogen production in the dark by strictly or facultative anaerobic bacteria.
4. Microbial fuel cells.

The facultative anaerobic microorganisms that produce biohydrogen may include *Escherichia coli*, *Enterobacter aerogenes*, *Klebsiella pneumonia,* and *Methanobacterium formiccium*.

Biohydrogen production can also be carried out using wastewater sludge that contains mixed culture. Other sources of mixed culture could be aerobic or anaerobic sludge or compost. Many researchers have carried out biohydrogen production using mixed culture because of the low cost of the substrate, no requirement of sterile conditions and high yield of biohydrogen (Table 8.2). In order to adapt the microorganisms for a high yield of biohydrogen, various pretreatment methods and shocks have been studied including alkali, acid, heat, oxygen, and chemical treatment (Sivagurunathan et al., 2017). The

TABLE 8.2

Biohydrogen production from different wastewaters

Wastewater Type	Microorganisms Used	Hydrogen Production	References
Food waste and sewage sludge	*Clostridium* sp.	111.2 ml H_2/VSS/h	(Kim, Han et al., 2004)
Cheese whey	Mixed culture	10 mM/g COD	(Yang, Zhang et al., 2007)
Rice mill	*Enterobacter aerogenes*	1.74 mol H_2/mol reducing sugar	(Ramprakash and Muthukumar, 2014)
Distillery	*Enterobacter cloacae* IIT-BT 08	7.4 mol H_2/kg $COD_{reduced}$	(Mishra and Das, 2014)
Distillery	Mixed culture	10.95 mmol/g COD	(Gadhe, Sonawane et al., 2014)
Distillery	Mixed culture	15.30 mmol/g COD	(Gadhe, Sonawane et al., 2015)
Beverage	Enriched mixed culture	1.30 mol/mol hexose utilised	(Sivagurunathan and Lin, 2019)

effects of such shocks on the biohydrogen production ability of mixed consortia were analyzed. Some workers found that pretreatment with iron and nickel oxide nanoparticles enhanced the biohydrogen yield to about 17 mmol/g COD. It was proposed that this enhancement could be due to increased activity of ferrodoxin, ferrodoxin oxidoreductase, and hydrogenase (Gadhe et al. 2015).

8.5.4.2 Biohydrogen Production from Microalgae

Biohydrogen can be produced using microalgae as the feedstock. The commonly used species are *Scenedesmus* sp., *Chlorella* sp., and *Saccharina* sp. Algae were subjected to various pretreatment methods, and the effect on biohydrogen production was evaluated. It was found that heat and acid pretreatments were most effective in increasing the production of biohydrogen and the maximum yield was obtained when *Chlorella* sp. was used as the substrate (Wang and Yin, 2018).

8.5.5 Production of Bioelectricity Using Wastewater

Bioelectricity is the electricity produced by microbes in a Microbial Fuel Cell (MFC). An MFC can be described as "a biochemical-catalyzed system which generates electrical energy through the oxidation of biodegradable organic matter in the presence of either fermentative bacteria or enzyme under mild reaction conditions (ambient temperature and pressure)" (Mohan et al., 2008). Thus, the chemical energy of waste components is converted into electrical energy (Mohan et al., 2008; Rabaey et al., 2003; Singh et al., 2018a). The transfer of electrons is catalyzed by the microorganisms. Thus, the selection of a suitable strain is an important parameter in the efficient bioelectricity generation by the microbial fuel cell (Chaudhuri and Lovley, 2003; Logan et al., 2006). Many factors influence the generation of bioelectricity by the MFC (Mohan et al., 2008):

1. The design of MFCs should be effective.
2. Membrane electrodes should be assembled efficiently to minimise the resistance in proton transfer.
3. The turbulence in the MFC should be high, which increases the interaction of the microbes with the substrate.
4. Sufficient surface area should be present to support bacterial growth.
5. The cathode reaction should be efficient.
6. Selection of the microorganisms (mixed culture) should be monitored.

Bose et al. worked on a dual-chambered MFC in a fed-batch process. It contained sewage as the fuel or substrate, and an open-circuit analysis was carried out to determine the maximum value of electricity generated in terms of voltage. A calibrated multimeter was used for data recording. A similar experiment was carried out for bioelectricity generation, and the voltage was measured across a resistor after every 24 hours. The effect of bioelectricity generation on reduction in COD of the wastewater was also measured. It was found that power generation in the MFC supported a higher COD removal rate compared to the open circuit analysis (Bose et al., 2018). Thus, it can be proposed that MFCs are efficient in the generation of electricity along with treatment of wastewater.

8.5.6 Production of Biopolymers

Use of wastewater for the synthesis of biopolymers is aimed at reducing the cost of the polymer. The L-lactic acid-based polymer has found numerous applications in industries varying from paper coating, disposable materials and fibres, and packaging materials. They are also used in drug delivery, tissue implants, and surgical materials. Similarly, another polymer for bioplastics, known as polyhydroxy alkanoates (PHA), is being evaluated.

8.5.6.1 Production of L-Lactic Acid Using Wastewater Sludge

L-lactic acid is a precursor of biodegradable plastics. Thus, studies have been carried out to evaluate the conversion of cellulose found in sludge into L-lactic acids. This conversion is a process of sustainable development by reducing the use of non-biodegradable plastics along with treatment of wastewater. The process of conversion involves the breakdown of cellulose into a simple compound that is then used for L-lactic acid production. The high cost of its precursor hampers the sustainability of the utilisation of biodegradable plastics. This barrier can be removed by utilising substrates from wastewater as it has been estimated that about 40% of the cost of production of L-lactic acid is due to its raw materials (Nakasaki et al., 1999). Thus, it is proposed that the use of cellulose from wastewater may reduce the cost of production by a significant factor. Nakasaki et al. evaluated the conversion of cellulose obtained from the fish-processing industry and paper mills. Lactic acid bacteria utilise the cellulose of the sludge and were isolated from the sludge itself. An analysis of the glucose and polysaccharide content of both sources were performed. It was found that sludge obtained from the paper-processing industry had a higher concentration of cellulose; thus it was found to be a potent substrate for L-lactic acid production. It is also suggested that with this technology the cost of production of L-lactic acid may reduce to less than or equal to $1 per kg, which again may depend on the cost of collection of sludge and the L-lactic acid production process of hydrolysis (Nakasaki et al., 1999).

8.5.6.2 Production of PHAs Using Wastewater Sludge

PHAs (known as bioplastics) are produced by microorganisms under nutrient limiting conditions. Chemical modifications in these bioplastics can lead to the development of materials with a variety of properties. The conventional method for production of PHAs uses starch as the raw material. However, the use of starch-containing food material such as maize leads to an increased consumption of land for non-food purposes, resulting in a food shortage. Thus, attempts were made to utilise wastewater for PHA production. The production process can be carried out in wastewater treatment plants. The volatile fatty acids (VFA) produced in the anaerobic digester were converted into PHAs under aerobic conditions (Pittmann and Steinmetz, 2017).

Yan et al. (2006) studied the effect of different wastewaters in the production of PHAs. They collected activated sludge from different wastewater treatment plants of sources such as municipal, paper and pulp, cheese, and starch manufacturing industries. The activated sludge was used as a source of microorganisms for PHA production. Glucose, acetate, and wastewater from different effluents were used as carbon sources. It was found that the maximum concentration of PHAs was obtained when activated sludge from the paper and pulp industry was used and acetate was used as a carbon source. The presence of a high concentration of volatile fatty acids in the wastewaters was the cause of a high amount of accumulation of PHAs. It was found that about 40% conversion of COD into PHA was achieved. Thus, a load of pollution in the wastewater was also reduced (Yan et al., 2006).

8.6 Analysis of Conversion Process Efficiency

The process of energy generation using wastewater is an environmentally friendly approach. The process of biofuel production should be cost effective and sustainable. These techniques should be evaluated based on their technical and commercial feasibility.

8.6.1 Process Efficiency for Biogas Production

Morero et al. have examined the biogas production using sludge based on the Life Cycle Assessment (LCA). They determined the power and economic requirements of the biogas production process along with the effects on the environment (Morero et al., 2017). Similarly, others have described the reduction of emission of greenhouse gases in the two biogas plants and used a procedure mentioned in the Renewable Energy Directive, which again is based on the principle of LCA (Viskovic et al., 2017).

The shortcomings associated with these assessments were that the economics related to energy and material recovery were not analyzed. Moreover, comparative analysis of different types of plants should have been done. However, few studies have also incorporated both the aspects of efficiency and energy consumption of biogas digesters (Li et al., 2017). Some researchers have determined the economical efficiency of individual production processes in two different biogas plants by energy analysis (Viskovic et al., 2017). Others have demonstrated economic and technical analysis of a biogas producing plant that used palm-mill oil effluent as its raw material. They also studied the recovery of waste heat from a biogas-driven engine (Firdaus et al., 2017).

8.6.2 Process Efficiency for Bioethanol Production

As discussed earlier, the profit in algae production can be enhanced by the use of wastewater. This is ensured by the availability of water, CO_2 and other nutrients in the wastewater, which cost about 30% of the algal fuel cost. Moreover, the effective cost of waste treatment is reduced as it is utilised for energy generation, which results in lowering the power consumption from other sources by 20%. The selection of a strain of algae based on bioethanol yield also plays a crucial role in determining the cost of bioethanol production.

Moreover, a thermo-tolerant strain will involve less energy expenditure on the maintenance of temperature. Another important factor related to strain selection is the pretreatment cost required for recovery of the carbohydrates.

8.6.3 Process Efficiency for Biodiesel Production

The demand for biodiesel is rising with awareness of people towards the environment and limited availability of fossil fuels. On the contrary, the cost of vegetable and seed oil is dominant in increasing the cost of biodiesel production. Therefore, research is being carried out to minimise the cost of the process and enhance its efficiency. Use of co-solvents can enhance the efficiency of the lipid extraction process. The extraction process can be further accelerated by the use of a high shear mixing technique. In the transesterification step, an alkali-catalyzed reaction may enhance the rate of conversion compared to the acid catalyzed reaction.

It should be noted that feedstocks containing free fatty acids more than 1% should be the first acid catalyzed, followed by alkali catalysis. It prevents the formation of soap in such cases.

Similarly, the selection of sludge containing oil-producing microorganisms will enhance biodiesel production. Another problem associated with the use of wastewater for biodiesel production is the presence of pharmaceutical ingredients in the sludge. In such cases, the sludge should be treated for removal of therapeutic chemicals. These considerations are supportive for the optimisation of the production process to ensure a profitable future of biodiesel produced from wastewater (Kargbo, 2010).

8.6.4 Process Efficiency for Biohydrogen Production

The biohydrogen production process again has numerous constraints, some of which are still to be overcome. The problem is enhanced due to the mixed culture that contains a different variety of microorganisms. Thus, a complex and uncontrolled environment is found containing extracellular metabolites of the mixed culture. The activity of acidogenic hydrogen producing microorganisms should be examined for optimisation of the process. The inhibition posed by a mixed culture is attributed to multiple factors, including competition amongst the constituent microorganisms of the mixed culture for the substrate, presence of hydrogen-consuming microorganisms, and shift in metabolic flux towards metabolites other than nitrogen.

Moreover, the presence of some microorganisms can also result in the secretion of growth inhibitors, which may reduce the population size of the desired microorganism. Hydrogen is consumed by microorganisms such as lactic acid bacteria and sulfur-reducing bacteria. They also produce certain by-products that inhibit the growth of biohydrogen producers. Biohydrogen production is favoured by the supply of inorganic nutrients such as iron, nitrogen, phosphorus, and carbon. Therefore, wastewater can be more effective in increasing the production of biohydrogen as they contain these nutrients. HRT is an important process parameter that determines the product concentration. When HRT is long, the microbial population is dominated by archaea, which are hydrogen consumers, leading to reduction in yield, whereas when the HRT is short, the active hydrogen producers are also washed off. Thus, the optimisation of HRT is an important step to maximise production (Sivagurunathan et al., 2017).

8.7 Large-Scale Technologies

Production of biofuel from wastewater falls under the category of low-value, high-volume product. Therefore, the unit operations should be carried out at a large scale. The cultivation of algae on a large scale is carried out in either open-air systems or closed-air systems (Hwang et al., 2016).

As in the case of wastewater cultivation, there may not be a stringent selection of the strain. Therefore open-air systems can also be used. The disadvantages associated with such systems are that they are susceptible to contamination and the period of illumination is variable based on nature. Similarly, the temperature cannot be controlled. Such systems are location and time dependent. However, there are certain advantages associated with open-air algal cultivation, such as illumination cost is not incurred, the power consumption in mixing is very low, and small investments are required (Kumar et al., 2015). There are three main open-air systems used for cultivation of algae in wastewater, which include HRAP, big shallow ponds, and circular ponds. HRAPs are an economical method of cultivation of microalgae.

Moreover, the treatment of wastewater with microalgae has an economical advantage, as the need for mechanical aeration is diminished. This occurs due to the oxygen-releasing functionality of algae. They have not found a wide use for algal culture, but with the rise in awareness towards removal of nutrients from wastewater and commercial application of algae production, they may be used more in future (Park et al., 2011).

Closed-air systems are used when microalgae are required to be grown in a controlled environment, where environmental factors such as gas exchange, mixing, and light intensity and period can be controlled. It is possible to cultivate a pure culture using such photobioreactors (PRRs). The problems associated with PBR are difficulty in scale-up and increase in temperature of the reactor as it is a closed-air system. The increase in temperature may be lethal for the algae that are mostly heat sensitive. An additional provision of heat exchangers may solve this problem, but this increases the cost of the process and energy consumption (Pruvost et al., 2016). Another problem with large-scale PBR is the accumulation of dissolved oxygen, which has a negative effect on the growth of algae. Sparging of air can remove dissolved oxygen, which may again impact the operational cost. Some workers have suggested the use of perfluorocarbon nanoemulsions for this problem (Lee and Yeh, 2015). These are oxygen scavengers. Based on the requirement and land availability, many designs of PBRs have been proposed. Tubular and flattened plate-type PBR systems are the most common types (Pulz, 2001).

Production of algae as a biofilm is not as widely applicable as a suspension culture. This methodology of algal culture reduces the problem of harvesting of biomass. However, this strategy is not coupled with wastewater treatment for the removal of nutrients due to economical concerns (Kesaano and Sims, 2014).

8.7.1 Integrated Biogas–Biodiesel Production Approach

The wastewater from industries is sent to the primary settling tank. Sludge settles, whereas the greasy and oily contents float up to the surface. The sludge is separated and removed by the scrapers; on the other hand, the clarified water is sent for microbial degradation. The sludge is then carried to the biogas plant where it is hydrolyzed, and the biogas is collected, whereas the digested leftover is used as fertiliser. The grease and oils are then recollected for the process of saponification. In an algae-based biofuel production process (Figure 8.1), microalgae are grown in HRAPs supplied with the clarified and digested water. These microalgae are then transferred to the biodiesel plant (Krishna et al., 2012).

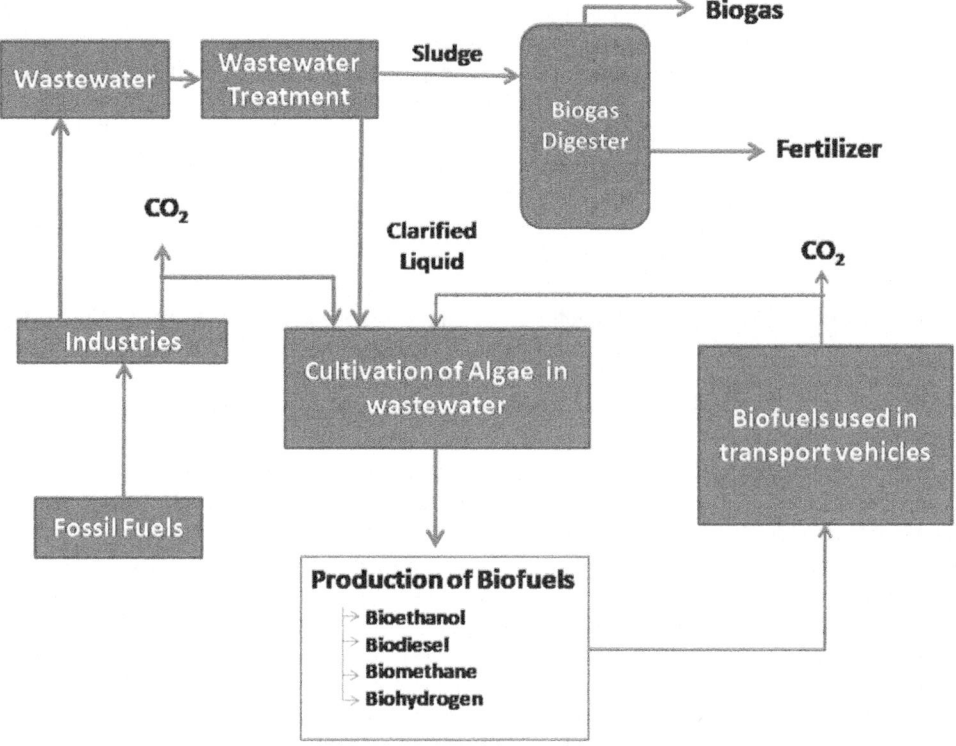

FIGURE 8.1 Integrated system for biogas and algal biofuel production.

8.7.2 Biohydrogen Production Using Multi-stage Bioreactors

The biohydrogen fermentation is a multi-step process, with different environmental specifications. Thus, a multi-stage bioreactor approach was adopted by a few researchers to enhance the production from feedstock (Lo et al., 2009; Luo et al., 2011). In this strategy, direct photolysis is carried out in the first reactor in which the sunlight is perlocated. This causes the growth of blue green algae. In the second stage, the photosynthetic microbes utilise the unfiltered infrared light. This reactor is known as a photo-fermentative reactor. In the third stage reactor, dark fermentation is carried out by the effluents of the second stage and the algae from the first stage. The substrate is converted by the bacteria into organic acids. The fourth stage reactor converts the organic acids into hydrogen in a light-independent reaction process. Thus, the four stages work maximally when separated from each other due to stringent light intensity requirements (Sharma and Arya, 2017).

8.8 Conclusion

The rise in global energy demand and increasing pollution are the two major challenges of modern society. As we know that due to increase in population, the demand for energy is also increasing and due to limited fossil fuel resources, we cannot rely on them to meet our future energy needs. The burning of fossil fuels causes a major threat to the environment in terms of increase in global temperature. Similarly, controlling the pollution caused by the effluents from different sources like industries, agriculture wastewater, and municipal wastewater are very challenging for society. The cost of wastewater treatment can be substantially reduced when it is coupled with energy generation in the form of renewable fuel. On the other hand, the energy generation process also becomes economical when wastewater becomes feedstock. Thus, the two processes symbiotically support each other. Utilisation of nutrient removal efficiency of microalgae is also beneficial for wastewater treatment as well as the growth of microalgae itself. Microbes mediated remediation offer several applications:

1. Waste treatment as the dried biomass is efficient in remediation of heavy metals from the environment.
2. Energy generation as the source of starch and lipids for production of bioethanol and biodiesel, respectively.
3. Some strains are a source of biochemicals such as fatty acids, hydrocarbons, pharmaceutical ingredients (with antitumour and antimicrobial properties), and dyes and colourants.
4. Source of feed for animals and fish.
5. CO_2 sequestration property is used in reducing air pollution.

Wastewater can also be utilised for production of chemicals like L-lactic acid and PHAs, which are commercially in demand. Thus, this integrative approach of coupling wastewater treatment with industrial processes is a "waste to wealth" strategy. Technical and economical feasibility of the process design needs consideration for scaling up the process from the lab to an industrial scale.

REFERENCES

Berktay, A., & Nas, B. (2007). Biogas production and utilization potential of wastewater treatment sludge. *Energy Sources, Part A: Recovery, Utilization, and Environmental Effects, 30*, 179–188.

Bharathiraja, B., Yogendran, D., Ranjith Kumar, R., Chakravarthy, M., & Palani, S. (2014). Biofuels from sewage sludge – a review. *International Journal of ChemTech Research, 6*, 4417–4427.

Bibi, R., Ahmad, Z., Imran, M., Hussain, S., Ditta, A., Mahmood, S., & Khalid, A. (2017). Algal bioethanol production technology: a trend towards sustainable development. *Renewable and Sustainable Energy Reviews, 71*, 976–985.

Bishop, W.M., Villalon, G.V., & Willis, B.E. (2018). Assessing copper adsorption, internalization, and desorption following algaecide application to control lyngbyawollei from Lake Gaston, NC/VA, USA. *Water, Air, & Soil Pollution, 229*, 152.

Bose, D., Dhawan, H., Kandpal, V., Vijay, P., & Gopinath, M. (2018). Bioelectricity generation from sewage and wastewater treatment using two-chambered microbial fuel cell. *International Journal of Energy Research, 42*, 4335–4344.

Bouzit, L., Jbari, N., El Yousfi, F., Alaoui, N.S., Chaik, A., & Stitou, M. (2018). Adsorption of Fe3+ by a living microalgae biomass of Scenedesmus obliquus. *Mediterranean Journal of Chemistry, 7*, 156–163.

Cai, T., Park, S.Y., & Li, Y. (2013). Nutrient recovery from wastewater streams by microalgae: status and prospects. *Renewable and Sustainable Energy Reviews, 19*, 360–369.

Capodaglio, A.G., & Callegari, A. (2018). Feedstock and process influence on biodiesel produced from waste sewage sludge. *Journal of environmental management, 216*, 176–182.

Chaudhuri, S.K., & Lovley, D.R. (2003). Electricity generation by direct oxidation of glucose in mediatorless microbial fuel cells. *Nature biotechnology, 21*, 1229.

Chen, G., Zhao, L., & Qi, Y. (2015). Enhancing the productivity of microalgae cultivated in wastewater toward biofuel production: a critical review. *Applied Energy, 137*, 282–291.

Chen, Y., Hu, W., Feng, Y., & Sweeney, S. (2014). Status and prospects of rural biogas development in China. *Renewable and Sustainable Energy Reviews, 39*, 679–685.

Colica, G., Mecarozzi, P.C., & De Philippis, R. (2010). Treatment of Cr (VI)-containing wastewaters with exopolysaccharide-producing cyanobacteria in pilot flow through and batch systems. *Applied Microbiology and Biotechnology, 87*, 1953–1961.

Craggs, R.J., McAuley, P.J., & Smith, V.J. (1997). Wastewater nutrient removal by marine microalgae grown on a corrugated raceway. *Water Research, 31*, 1701–1707.

De Philippis, R., Colica, G., & Micheletti, E. (2011). Exopolysaccharide-producing cyanobacteria in heavy metal removal from water: molecular basis and practical applicability of the biosorption process. *Applied Microbiology and Biotechnology, 92*, 697.

Dubey, A.K., Gupta, P., Garg, N., & Naithani, S. (2012). Bioethanol production from waste paper acid pretreated hydrolyzate with xylose fermenting Pichia stipitis. *Carbohydrate Polymers*, *88*, 825–829.

Eddy, I.S.M. (1986). *Wastewater Engineering: Treatment Disposal Reuse*. New York: McGraw-Hill Companies.

Ellis, J.T., Hengge, N.N., Sims, R.C., & Miller, C.D. (2012). Acetone, butanol, and ethanol production from wastewater algae. *Bioresource Technology*, *111*, 491–495.

Ergüder, T., Güven, E., & Demirer, G. (2000). Anaerobic treatment of olive mill wastes in batch reactors. *Process Biochemistry*, *36*, 243–248.

Firdaus, N., Prasetyo, B.T., Sofyan, Y., & Siregar, F. (2017). Part I of II: palm oil mill effluent (POME): biogas power plant. *Distributed Generation & Alternative Energy Journal*, *32*, 73–79.

Fjørtoft, K., Morken, J., Hanssen, J.F., & Briseid, T. (2019). Pre-treatment methods for straw for farm-scale biogas plants. *Biomass and Bioenergy*, *124*, 88–94.

Gadhe, A., Sonawane, S.S., & Varma, M.N. (2014). Evaluation of ultrasonication as a treatment strategy for enhancement of biohydrogen production from complex distillery wastewater and process optimization. *International Journal of Hydrogen Energy*, *39*, 10041–10050.

Gadhe, A., Sonawane, S.S., & Varma, M.N. (2015). Enhanced biohydrogen production from dark fermentation of complex dairy wastewater by sonolysis. *International Journal of Hydrogen Energy*, *40*, 9942–9951.

Gethin, M. (2021). *Wastewater treatment: problems and solutions [Online]*. Available: https://www.mwwatermark.com/wastewater-treatment-problems-and-solutions/ [Accessed 28/03/2021 2021].

Hong, J., Yang, H., Zhang, K., Liu, C., Zou, S., & Zhang, M. (2014). Development of a cellulolytic Saccharomyces cerevisiae strain with enhanced cellobiohydrolase activity. *World Journal of Microbiology and Biotechnology*, *30*, 2985–2993.

Hwang, J.-H., Church, J., Lee, S.-J., Park, J., & Lee, W.H. (2016). Use of microalgae for advanced wastewater treatment and sustainable bioenergy generation. *Environmental Engineering Science*, *33*, 882–897.

Incharoensakdi, A., & Kitjaharn, P. (2002). Zinc biosorption from aqueous solution by a halotolerant cyanobacterium Aphanothecehalophytica. *Current Microbiology*, *45*, 261–264.

Isla, Miguel A., Comelli, Raúl N., & Seluy, Lisandro G. (2013). Wastewater from the soft drinks industry as a source for bioethanol production. *Bioresource Technology*, *136*, 140–147. doi:10.1016/j.biortech.2013.02.089.

Jiang, J., Zhang, N., Yang, X., Song, L., & Yang, S. (2016). Toxic metal biosorption by macrocolonies of cyanobacterium NostocsphaeroidesKützing. *Journal of Applied Phycology*, *28*, 2265–2277.

Kargbo, D.M. (2010). Biodiesel production from municipal sewage sludges. *Energy & Fuels*, *24*, 2791–2794.

Kesaano, M., & Sims, R. (2014). Algal biofilm based technology for wastewater treatment. *Algal Research*, *5*, 231–240.

Kim, M.-K., & Jeune, K.-H. (2009). Use of FT-IR to identify enhanced biomass production and biochemical pool shifts in the marine microalgae, *Chlorella ovalis*, cultured in media composed of different ratios of deep seawater and fermented animal wastewater. *Journal of Microbiology and Biotechnology*, *19*, 1206–1212.

Kim, S.-H., Han, S.-K., & Shin, H.-S. (2004). Feasibility of biohydrogen production by anaerobic co-digestion of food waste and sewage sludge. *International Journal of Hydrogen Energy*, *29*, 1607–1616.

Krishna, A.R., Dev, L., & Thankamani, V. (2012). An integrated process for Industrial effluent treatment and Biodiesel production using Microalgae. *Research in Biotechnology*, *3*.

Kumar, K., Mishra, S.K., Shrivastav, A., Park, M.S., & Yang, J.-W. (2015). Recent trends in the mass cultivation of algae in raceway ponds. *Renewable and Sustainable Energy Reviews*, *51*, 875–885.

Kwon, E.E., Kim, S., Jeon, Y.J., & Yi, H. (2012). Biodiesel production from sewage sludge: new paradigm for mining energy from municipal hazardous material. *Environmental Science & Technology*, *46*, 10222–10228.

Lam, M.K., Loy, A.C.M., Yusup, S., & Lee, K.T. (2019). Biohydrogen production from algae. *Biohydrogen*, 219–245. Netherlands: Elsevier.

Lamelas, C., Pinheiro, J.P., & Slaveykova, V.I. (2009). Effect of humic acid on Cd (II), Cu (II), and Pb (II) uptake by freshwater algae: kinetic and cell wall speciation considerations. *Environmental Science & Technology*, *43*, 730–735.

Lee, Y.-H., & Yeh, Y.-L. (2015). Reduction of oxygen inhibition effect for microalgal growth using fluoroalkylated methoxy polyethylene glycol-stabilized perfluorocarbon nano-oxygen carriers. *Process Biochemistry, 50*, 1119–1127.

Leong, W.-H., Lim, J.-W., Lam, M.-K., Uemura, Y., & Ho, Y.-C. (2018). Third generation biofuels: a nutritional perspective in enhancing microbial lipid production. *Renewable and Sustainable Energy Reviews, 91*, 950–961.

LewisOscar, F., Praveenkumar, R., & Thajuddin, N. (2015). Bioethanol production using starch extracted from microalga Stigeoclonium sp. Kütz. BUM11007 cultivated in domestic wastewater. *Research Journal of Environmental Sciences, 9*, 216–224.

Li, J., Zhang, S., Kong, C., Duan, Q., Deng, L., Mei, Z., & Lei, Y. (2017). Power-generating capacity of manure-and wastewater-to-energy conversion systems for commercial viability in Yunnan, China. *Journal of Renewable and Sustainable Energy, 9*, 043103.

Liang, K., Zhang, Q., Gu, M., & Cong, W. (2013). Effect of phosphorus on lipid accumulation in freshwater microalga Chlorella sp. *Journal of Applied Phycology, 25*, 311–318.

Lo, Y.-C., Su, Y.-C., Chen, C.-Y., Chen, W.-M., Lee, K.-S., & Chang, J.-S. (2009). Biohydrogen production from cellulosic hydrolysate produced via temperature-shift-enhanced bacterial cellulose hydrolysis. *Bioresource Technology, 100*, 5802–5807.

Logan, B.E., Hamelers, B., Rozendal, R., Schröder, U., Keller, J., Freguia, S., Aelterman, P., Verstraete, W., & Rabaey, K. (2006). Microbial fuel cells: methodology and technology. *Environmental Science & Technology, 40*, 5181–5192.

Luo, G., Talebnia, F., Karakashev, D., Xie, L., Zhou, Q., & Angelidaki, I. (2011). Enhanced bioenergy recovery from rapeseed plant in a biorefinery concept. *Bioresource Technology, 102*, 1433–1439.

Ma, F., & Hanna, M.A. (1999). Biodiesel production: a review. *Bioresource Technology, 70*, 1–15.

Martınez, M., Sánchez, S., Jimenez, J., El Yousfi, F., & Munoz, L. (2000). Nitrogen and phosphorus removal from urban wastewater by the microalga Scenedesmus obliquus. *Bioresource Technology, 73*, 263–272.

Mishra, P., & Das, D. (2014). Biohydrogen production from Enterobacter cloacae IIT-BT 08 using distillery effluent. *International Journal of Hydrogen Energy, 39*, 7496–7507.

Misra, N., Panda, P., Parida, B., & Mishra, B. (2016). Way forward to achieve sustainable and cost-effective biofuel production from microalgae: a review. *International Journal of Environmental Science and Technology, 13*, 2735–2756.

Mohan, S.V., Mohanakrishna, G., Reddy, B.P., Saravanan, R., & Sarma, P. (2008). Bioelectricity generation from chemical wastewater treatment in mediatorless (anode) microbial fuel cell (MFC) using selectively enriched hydrogen producing mixed culture under acidophilic microenvironment. *Biochemical Engineering Journal, 39*, 121–130.

Moheimani, N.R. (2005). *The Culture of Coccolithophorid Algae for Carbon Dioxide Bioremediation*. Australia: Murdoch University.

Morero, B., Vicentin, R., & Campanella, E.A. (2017). Assessment of biogas production in Argentina from co-digestion of sludge and municipal solid waste. *Waste Management, 61*, 195–205.

Nair, R., Lennartsson, P.R., & Taherzadeh, M.J. (2017). Bioethanol production from agricultural and municipal wastes. *Current Developments in Biotechnology and Bioengineering*, 157–190. USA: Elsevier.

Nakasaki, K., Akakura, N., Adachi, T., & Akiyama, T. (1999). Use of wastewater sludge as a raw material for production of L-lactic acid. *Environmental Science & Technology, 33*, 198–200.

Olguín, E.J., & Sánchez-Galván, G. (2012). Heavy metal removal in phytofiltration and phycoremediation: the need to differentiate between bioadsorption and bioaccumulation. *New Biotechnology, 30*, 3–8.

Olkiewicz, M., Caporgno, M.P., Fortuny, A., Stüber, F., Fabregat, A., Font, J., & Bengoa, C. (2014). Direct liquid–liquid extraction of lipid from municipal sewage sludge for biodiesel production. *Fuel Processing Technology, 128*, 331–338.

Organization, W. H. & Council, W. P. (2006). *Health Aspects of Plumbing*. World Health Organization.

Park, J., Craggs, R., & Shilton, A. (2011). Wastewater treatment high rate algal ponds for biofuel production. *Bioresource Technology, 102*, 35–42.

Peng, Y., Liu, X., Gong, X., Li, X., Liu, Y., Leng, E., & Zhang, Y. (2018). Enhanced Hg (II) adsorption by monocarboxylic-acid-modified microalgae residuals in simulated and practical industrial wastewater. *Energy & Fuels, 32*, 4461–4468.

Pittman, J.K., Dean, A.P., & Osundeko, O. (2011). The potential of sustainable algal biofuel production using wastewater resources. *Bioresource Technology*, *102*, 17–25.

Pittmann, T., & Steinmetz, H. (2017). Polyhydroxyalkanoate production on waste water treatment plants: process scheme, operating conditions and potential analysis for German and European municipal waste water treatment plants. *Bioengineering*, *4*, 54.

Pruvost, J., Cornet, J.-F., & Pilon, L. (2016). Large-scale production of algal biomass: photobioreactors, *Algae Biotechnology*, 41–66. Switzerland: Springer.

Pulz, O. (2001). Photobioreactors: production systems for phototrophic microorganisms. *Applied Microbiology and Biotechnology*, *57*, 287–293.

Rabaey, K., Lissens, G., Siciliano, S.D., & Verstraete, W. (2003). A microbial fuel cell capable of converting glucose to electricity at high rate and efficiency. *Biotechnology Letters*, *25*, 1531–1535.

Ramprakash, B., & Muthukumar, K. (2014). Comparative study on the production of biohydrogen from rice mill wastewater. *International Journal of Hydrogen Energy*, *39*, 14613–14621.

Rao, A., Lata, K., Raman, P., Kishore, V., & Ramachandran, K. (1997). Studies of anaerobic treatment of synthetic waste in a UASB reactor. *Indian Journal of Environmental Protection*, *17*, 349–354.

Rugnini, L., Costa, G., Congestri, R., Antonaroli, S., di Toppi, L.S., & Bruno, L. (2018). Phosphorus and metal removal combined with lipid production by the green microalga Desmodesmus sp.: an integrated approach. *Plant Physiology and Biochemistry*, *125*, 45–51.

Ruiz-Marin, A., Mendoza-Espinosa, L.G., & Stephenson, T. (2010). Growth and nutrient removal in free and immobilized green algae in batch and semi-continuous cultures treating real wastewater. *Bioresource Technology*, *101*, 58–64.

Samorì, G., Samorì, C., Guerrini, F., & Pistocchi, R. (2013). Growth and nitrogen removal capacity of Desmodesmuscommunis and of a natural microalgae consortium in a batch culture system in view of urban wastewater treatment: part I. *Water Research*, *47*, 791–801.

Schenk, P.M., Thomas-Hall, S.R., Stephens, E., Marx, U.C., Mussgnug, J.H., Posten, C., Kruse, O., & Hankamer, B. (2008). Second generation biofuels: high-efficiency microalgae for biodiesel production. *Bioenergy Research*, *1*, 20–43.

Shanab, S., Essa, A., & Shalaby, E. (2012). Bioremoval capacity of three heavy metals by some microalgae species (Egyptian Isolates). *Plant Signaling & Behavior*, *7*, 392–399.

Sharma, A., & Arya, S.K. (2017). Hydrogen from algal biomass: a review of production process. *Biotechnology Reports*, *15*, 63–69.

Sharma, K.K., Schuhmann, H., & Schenk, P.M. (2012). High lipid induction in microalgae for biodiesel production. *Energies*, *5*, 1532–1553.

Shin, H.-Y., Shim, S.-H., Ryu, Y.-J., Yang, J.-H., Lim, S.-M., & Lee, C.-G. (2018). Lipid Extraction from Tetraselmis sp. Microalgae for Biodiesel Production Using Hexane-based Solvent Mixtures. *Biotechnology and Bioprocess Engineering*, *23*, 16–22.

Show, K.-Y., Yan, Y.-G., & Lee, D.-J. (2019). Biohydrogen production from algae: Perspectives, challenges, and prospects. *Biofuels from Algae*. Elsevier, 325–343.

Siddiquee, M.N., & Rohani, S. (2011). Experimental analysis of lipid extraction and biodiesel production from wastewater sludge. *Fuel Processing Technology*, *92*, 2241–2251.

Singh, H.M., Pathak, A.K., Chopra, K., Tyagi, V., Anand, S., & Kothari, R. (2018a). Microbial fuel cells: a sustainable solution for bioelectricity generation and wastewater treatment. *Biofuels*, 1–21.

Singh, S., Chakravarty, I., & Kundu, S. (2017). Mathematical modelling of bioethanol production from algal starch hydrolysate by Saccharomyces cerevisiae. *Cellular and Molecular Biology (Noisy-le-Grand, France)*, *63*, 83–87.

Singh, S., Chakravarty, I., Pandey, K.D., & Kundu, S. (2018b). Development of a process model for simultaneous saccharification and fermentation (SSF) of algal starch to third-generation bioethanol. *Biofuels*, 1–9.

Siripattanakul-Ratpukdi, S. (2012). Ethanol production potential from fermented rice noodle wastewater treatment using entrapped yeast cell sequencing batch reactor. *Applied Water Science*, *2*, 47–53.

Sivagurunathan, P., Kumar, G., Pugazhendhi, A., Zhen, G., Kobayashi, T., & Xu, K. (2017). Biohydrogen Production from Wastewaters. *Biological Wastewater Treatment and Resource Recovery*, 197.

Sivagurunathan, P., & Lin, C.-Y. (2019). Biohydrogen Production From Beverage Wastewater Using Selectively Enriched Mixed Culture. *Waste and Biomass Valorization*, 1–10.

Swift, D.T., & Forciniti, D. (1997). Accumulation of lead by Anabaena cylindrica: mathematical modeling and an energy dispersive X-ray study. *Biotechnology and Bioengineering, 55*, 408–418.

Vassalle, L., Díez-Montero, R., Machado, A. T. R., Moreira, C., Ferrer, I., Mota, C. R., & Passos, F. (2019). Upflow anaerobic sludge blanket in microalgae-based sewage treatment: co-digestion for improving biogas production. *Bioresource Technology, 300*, 122677.

Viskovic, M., Martinov, M., & Djatkov, D. (2017). Sustainability of biogas production and utilisation-case studies, Proceedings of the 45th International Symposium on Agricultural Engineering, Actual Tasks on Agricultural Engineering, 21–24 February 2017, Opatija, Croatia. University of Zagreb, Faculty of Agriculture, 407–415.

Wang, J., & Yin, Y. (2018). Fermentative hydrogen production using pretreated microalgal biomass as feedstock. *Microbial Cell Factories, 17*, 22.

Wang, X., Lu, X., Yang, G., Feng, Y., Ren, G., & Han, X. (2016). Development process and probable future transformations of rural biogas in China. *Renewable and Sustainable Energy Reviews, 55*, 703–712.

Weiland, P. (2003). Production and energetic use of biogas from energy crops and wastes in Germany. *Applied Biochemistry and Biotechnology, 109* 263–274.

Weiland, P. (2010). Biogas production: current state and perspectives. *Applied Microbiology and Biotechnology, 85*, 849–860.

Yan, S., Tyagi, R., & Surampalli, R. (2006). Polyhydroxyalkanoates (PHA) production using wastewater as carbon source and activated sludge as microorganisms. *Water Science and Technology, 53*, 175–180.

Yang, P., Zhang, R., McGarvey, J.A., & Benemann, J.R. (2007). Biohydrogen production from cheese processing wastewater by anaerobic fermentation using mixed microbial communities. *International Journal of Hydrogen Energy, 32*, 4761–4771.

Zhan, J., Zhang, Q., Qin, M., & Hong, Y. (2016). Selection and characterization of eight freshwater green algae strains for synchronous water purification and lipid production. *Frontiers of Environmental Science & Engineering, 10*, 548–558.

Zhang, E., Wang, B., Wang, Q., Zhang, S., & Zhao, B. (2008). Ammonia–nitrogen and orthophosphate removal by immobilized Scenedesmus sp. isolated from municipal wastewater for potential use in tertiary treatment. *Bioresource Technology, 99*, 3787–3793.

Zhao, F., Li, Z., Zhou, X., Chu, H., Jiang, S., Yu, Z., Zhou, X., & Zhang, Y. (2019). The comparison between vibration and aeration on the membrane performance in algae harvesting. *Journal of Membrane Science, 592*, 117390.

9

Phytoremediation: An Eco-friendly, Sustainable Solution for Indoor and Outdoor Air Pollution

Manisha Sarkar[1], Sujit Das[1], Randeep Rakwal[2], Ganesh Kumar Agrawal[3,4], and Abhijit Sarkar[1]
[1]*Laboratory of Applied Stress Biology, Department of Botany, University of Gour Banga, Malda, West Bengal, India*
[2]*Faculty of Health and Sport Sciences, University of Tsukuba, 1-1-1 Tennodai, Tsukuba, Ibaraki, Japan*
[3]*Research Laboratory for Biotechnology and Biochemistry (RLABB), Kathmandu, Nepal*
[4]*Global Research Arch for Developing Education (GRADE) Academy Pvt. Ltd., Birgunj, Nepal*

CONTENTS

9.1	Introduction	156
9.2	Air Pollution: Origin and Multi-fariousness	158
	9.2.1 Particulate Matters	158
	9.2.2 Trace Gases	159
9.3	Various Expedients for Reducing Air Pollution	159
9.4	Phytoremediation: Concept, Principle, Strategies	161
	9.4.1 Phytoextraction	161
	9.4.2 Phytodegradation	162
	9.4.3 Phytovolatilisation	162
	9.4.4 Phytostabilisation	162
	9.4.5 Rhizodegradation	163
	9.4.6 Rhizofiltration	163
9.5	Phytoremediation of Outdoor Air Pollution	163
	9.5.1 Phytoremediation of Various Outdoor Air Pollutants	163
	9.5.1.1 Phytoremediation of Airborne Particulate Matter	163
	9.5.1.2 Phytoremediation of Volatile Organic Compounds	164
	9.5.1.3 Phytoremediation of Inorganic Air Contaminants	164
9.6	Phytoremediation of Indoor Air Pollution	168
	9.6.1 Indoor Air Pollution: Grave Alarm?	168
	9.6.2 Common Indoor Pollutants: Matters of Concern	169
	9.6.3 Phytoremediation of Various Indoor Air Pollutants	170
	9.6.3.1 Phytoremediation of Indoor Airborne Particulates	170
	9.6.3.2 Phytoremediation of Indoor Volatile Organic Compounds	170
	9.6.3.3 Phytoremediation of Other Indoor Air Pollutants	173
	9.6.4 Limitations of Potted Plant/Static Chamber Experiments	173
	9.6.5 Active Botanical Biofiltration Technology with Functional Green Wall	174
	9.6.5.1 Potential of Active Green Wall for PM Removal	174
	9.6.5.2 Potential of Active Green Wall for VOC Removal	174

DOI: 10.1201/9781003004684-9

    9.6.5.3   Potential of Active Green Wall for CO_2 Removal 176
9.7   Concluding Remarks and Future Prospects 177
Acknowledgements 177
References 177

9.1 Introduction

An air pollutant is defined as any man-made or naturally occurring substance that has a significant adverse impact on the natural attributes of the earth's atmosphere (Das et al., 2017; Manisalidis et al., 2020). All the substances are not necessarily regarded as pollutants until they are subjected to imprudent usage, unplanned emissions, and/or technological asymmetry that renders them damaging to the environment. As a matter of fact, air pollution is inevitably being confronted as a necessary evil as a consequence of increasing unplanned industrialisation and rapid expansion of megacities, particularly in developing nations around the world, which may lead to multifarious environmental concerns such as heavy metal increment in soil, agricultural productivity loss, food crisis, freshwater shortage (Sarkar and Agrawal, 2010a,b; Sarkar et al., 2010; Sarkar et al., 2014; Srivastava et al., 2017; Sarkar et al., 2018), and also ruthless human health problems globally (Jamrozik and Musk, 2011; Faustini et al., 2013; Franklin et al., 2015; Kim et al., 2017). Apparently, air pollution remains a severe problem worldwide, having the effect that induces and aggravate diseases like cardiovascular and respiratory diseases, ischemia heart disease, chronic obstructive pulmonary diseases, acute respiratory infections, and even lung cancer (Brunekreef and Holgate, 2002; Vineis et al., 2006; Giles et al., 2011; Atkinson et al., 2013; WHO, 2014a; Kumar et al., 2015; Landrigan, 2017; Banerjee et al., 2017; Yu et al., 2019). In addition, air pollution imputes other adverse health issues in most of the organs and systems of the human body like disorders in intestines, kidney disorders, nervous system disorder, and myocardial infarction; thus, effectuating the increment of all causes of mortality, premature mortality mostly in children, and other acute diseases (Castaño-Vinyals et al., 2008; Genc et al., 2012; Kaplan et al., 2013; Kumar et al., 2015; Raziani and Raziani, 2021). As per the data elucidated by the WHO, around 92% of the global population belongs to an area with poor air quality (WHO, 2016). Air pollution exposure is reportedly accountable for 7 million fatalities globally in 2012 (one-eighth of the total global mortality), primarily due to heart ailments and strokes, which was considered to be more than twice prior estimates (WHO, 2014b). Recent studies of the WHO reported air pollution to be the world's single greatest environmental health concern and it also has detrimental impacts on animals, plants, and ecosystem productivity, including the ecological resourses such as water and soils (Smith et al., 2013; WHO, 2014b; Vallero, 2014; Duan et al., 2017). Also, air quality of the indoor environment as another source of pollution exposure has a significant impact on human well-being as most of the individuals spend 90% of their time primarily indoors either at home or at work (Richardson et al., 2005; Ho and Kuschner, 2012). It is important to mention that another major concern of morbity and mortality (3.8 million death annually) is driven by the cause of indoor air pollution due to the decreased indoor air quality (IAQ) and reckless emissions from household air pollutant sources such as smoking, burning of solid fuels, using typical energy-intensive stoves, cleaning, combustion, use of construction materials, and allergens from pets. (McCormack et al., 2008; Carrer et al., 2001; WHO, 2018; Pérez-Padilla et al., 2010; Liqun and Yanqun, 2011). Furthermore, prevalent indoor air pollutants such as PMs, which are outsourced from outdoors, the outgassing of VOCs from various synthetic polymers, and the respiratory output of carbon dioxide from the human body all have significant contributions toward IAQ degradation as well (Irga et al., 2018). Conventional emission reduction approaches that focus on specific technological methods have proven to be ineffective to maintain environmental or climatic sustainability.

    Over the previous three decades, the globe has seen unprecedented urban expansion, as well as an unmanageable increase in urban population at a rate of 2.3% per year on average, particularly in developing nations from 2000 to 2030 (United Nations, 2000, 2004; Brockherhoff, 2000; UNFPA, 2004; Mahendra and Seto, 2019). The rapid growth of urbanisation and economy simultaneously elevate energy consumption. As per the information is concerned, in the year 2012, China's energy consumption (primary fossil fuel) increased from 602.75 million in tones (in the year 1980) to 3,672 million tones due to the rapid urbanisation from 17.92% to 52.57% in the years 1978–2012 and the gross domestic products (GDPs) of China also increased by 51,894.2 billion Chinese yuan in 2012 from 454.6 million yuan in the year 1980 (Zhao and Wang, 2015). Pollutants in water, air, and soil have formed a stereo-network on this planet where nitrogen dioxide (NO_2), sulphur dioxide (SO_2), ozone (O_3), volatile organic compounds (VOCs), and particulate matters (PMs) have become prevalent as a result of increased combustion of fossil fuels having less effective emission control measures with low combustion efficiency. India is likewise combating the same scenario where urbanisation is also expected to be accompanied with increased economic development, greater energy expendature, and also severe air pollution in mega cities (Rizwan et al., 2013; Gurjar et al., 2016). One of the most hazardous predominant air pollutants of China is considered to be the PMs and, as an outcome, China had to experience severe haze pollution over the first quarter of 2013 (Huang et al., 2014). According to the International Agency for Research on Cancer (IARC), 223,000 people died in 2010 as a consequence of lung cancer throughout the world, with air pollution playing a crucial part in being a carcinogen (International Agency for Research on Cancer, 2013). Thereby, the negative impact of globalisation and its interventions are concurrently driving a shift in people's conventional behaviour as well as exhibiting an extreme decline in air quality, resulting in severe indoor and ambient air pollution with negative health consequences.

Various laws and policies have been framed in order to simply instruct the industrial farms and vehicle manufacturers to employ new technologies, which are beneficial in reducing the air pollution from both mobile as well as stationary resources. In order to attenuate the pollutant emission sources, several mechanical techniques have been incorporated as an implementation of air pollution abatement strategies (wet scrubbers, electromagnetic precipitators, mechanical collectors, fume incineration, fabric collectors); however, the issues regarding occasional failure of expensive devices, their high maintenance charges, and low output efficiency cannot be excluded in the industries (Singh and Verma, 2007). The medium to small-scale industries depend upon less efficient alternatives; as a result of that, an alarming proportion of pollutants are being exhaled and action is a massive concern for developing countries. In order to overcome all of these drawbacks, necessary actions should be taken and other adequate corrective measures for the immediate remedial recourse should be initiated (Singh and Verma, 2007). Keeping that fact in consideration, the expected solution would be to manipulate the emission sources and enhance the quantity of natural sinks. Phytoremediation is one of the plant-driven mechanisms for mitigating pollution from air, water, and soil either individually or in combination with the plant-associated microorganisms. This approach has been effectively subjected to remediate soil and water contaminants largely (Cunningham and Ow, 1996; Lasat, 2001; Schröder et al., 2002; Awa and Hadibarata, 2020). However, phytoremidiation has also been beneficial and marked as a potential mechanism to clean ambient air from the gaseous contaminants that get readily absorbed/adsorbed by means of intensive gas exchange mechanisms, mostly by the autotrophic plants (Gawronski et al., 2017). In phytoremediation technique, pollution from ambient air is absorbed by plants along with their allied microorganisms and subsequently get degraded and detoxified through a number of means that have already been established as cost-efficient, self-maintaining, environmentally friendly, and sustainable alternatives for abating air pollution in both outdoor and indoor environments (Weyens et al., 2015). In addition to its soil-stabilising, cost-effective, self-sustaining properties, this technology offers additional aesthetic value and greater energy conservation than any other sophisticated technologies for remediating a wide spectrum of (inorganic and organic) air pollutants.

Various reviews have reported plants' efficacy in absorbing certain pollutants under diverse environmental circumstances as well as the usage of higher plants with their own microbiome to abate the airborne contaminants in outdoor environments mostly. Because of the existing benefit of having abundant biologically functional surface areas, plants are adept in trapping air contaminants through direct (absorption/adsorption) processes or wet/dry deposition (Agrawal et al., 2019). Additionally, the plant-microbe interaction is of great significance in terms of a phytoremedial approach by promoting plant development while disintegrating, detoxing, or sequestering certain contaminants (Weyens et al., 2009a,b). Plants with different photosynthetic mechanisms (C3, C4, Crassulacean Acid Metabolism/CAM) influence the ambient air quality diversely in order to mitigate air contaminants by their own characteristic way. The matter of fact is that C4 plants exhibit the intensity of higher gaseous exchange; however, C3 plants are especially competent for exchanging higher amounts of CO_2 throughout the day only. To remediate indoor air pollution, CAM plants are considered to be more efficient for their capacity to exchange gases during the night, especially after being subjected to stress conditions (Winter and Holtum, 2014). The potential colonisation of microorganisms on leaves (phyllomicrobiomes) is considered to be a fruitful approach to degrade various harmful organic pollutants (De Kempeneer et al., 2004; Vorholt, 2012). The emerging air pollution abatement strategy with regard to the plant-soil-microbe system needs to be observed under intensive investigation. Accompanying the selection of plants for remediating proper groups of contaminants, the rate of efficient elimination of pollutants under varied environmental conditions requires sincere attention.

9.2 Air Pollution: Origin and Multi-fariousness

Any airborne foreign substance that has severe implications on human health and also on other key components of the environment is termed as air pollutant. Additionally, air contaminants can be generated through both natural and anthropogenic activities, but the transboundary flow of pollutant necessitates research into their origins and sinks in the atmosphere (DiGiovanni and Fellin, 2006; Mhawish et al., 2017; Agrawal et al., 2019). The natural sources of air pollutants could be particularly due to forest fires, volcanic eruptions, sandstorms, and chemical reactions in the presence of solar radiation and also the carbon cycle. However, burning of fossil fuels, generation of energy, transportation, industrial, residential, or agricultural interventions, on the other hand, contribute to the effective cause of anthropogenic pollution (Popescu and Ionel, 2010; Weyens et al., 2015; Kumar et al., 2016; Banerjee et al., 2015; Agrawal et al., 2019). Air pollutants are mainly categorised as primary and secondary, depending upon their principle of origin and whether they are released directly into the ambient or develope as a result of an interaction between primary pollutants (Sitaras and Siskos, 2008; Sarkar et al., 2012a; Das et al., 2017). One of the most toxic air contaminant aerosols, due to their chemical composition and microphysical properties, contributes extensive effects on regional climate change and human health (Ren-Jian et al., 2012; Mhawish et al., 2018).

9.2.1 Particulate Matters

PMs, a group of significant airborne contaminants, consist primarily of a heterogeneous combination of minute solid and aqueous particles floating in air that differ in size and chemical composition across space and time with aerodynamic diameters ranging from 0.001 to 10 μm (WHO, 2013; Das et al., 2021). The physical attributes (size, surface area, density), molecular composition, and distribution of size of atmospheric PMs are considered to be greatly influenced by broad varieties of emission sources (Perrino, 2010). As per the Environmental Protection Agency (EPA), the PMs can be broadly subdidived into two different subgroups depending on their ability to get into the lungs through penetration, i.e., one is the coarse particulate matter (PM_{10}) with an aerodynamic diameter of 10 μm and another one is the fine particulate matter ($PM_{2.5}$) with an aerodynamic diameter of 2.5 μm (Esworthy, 2013). The emission sources of PMs can be both natural and man-made (Atkinson et al., 2010; Das et al., 2021). Those naturally occurring particulate contaminants are produced by volcanoes, desert dust, growing vegetation, sea spray, and wood fires (Misra et al., 2001; Das et al., 2021), even including various biological sources like pollen, bacteria, fungal spores, etc. In contrast, anthropogenic particles that are derived from man-made activities are highly variable and generated from industrial and agricultural undertakings along with fossil fuel burning (i.e., coal, lignite, heavy oil, and plant biomass), erosion, road dust, river beds, sites of a construction, mining operations, and brake and tyre corrosion (Juda-Rezler et al., 2011; Srimuruganandam and Nagendra, 2012; Das et al., 2021). There is a dynamic range of constituents of PMs that are primarily subjected to contain inorganic ions (i.e., sulphates, nitrates, sodium, ammonium, etc.), organic and elemental carbon, metal, particle bound water, and polycyclic aromatic hydrocarbon (PAH) (Cheung et al., 2011; Das et al., 2021). According to Das et al. (2020),

Phytoremediation 159

86.18% of elemental carbon, 22.9% of sodium chloride, and 18.24% of secondary nitrates were observed as the constituets of $PM_{10.}$ Two fundamental methods, i.e., dry deposition and wet deposition, are responsible for the removal of PMs from the ambient. Mostly the tiny particles of PMs are subjected to suspension in the ambient because of their small size and, additionally for that very same reason, they are able to fluctuate their concentration greatly from one day to the next (Johansson et al., 2007). Apart from that, a huge range of our household activities are greatly responsible for the increased level of PM concentrations in indoor environments rather outdoor environments (Madureira et al., 2012).

9.2.2 Trace Gases

The earth's atmosphere constitutes a complex combination of several gases and thereby few observations are crucial for the understanding of sun-atmosphere-environment interactions (Das et al., 2017). Despite occupying 99% of the atmospheric volume by primary gases like oxygen, nitrogen, carbon dioxide, water, and inert gases, several trace amounts of gases (1% of atmospheric composition) that are responsible for climate change and several environmental problems have marked their position wisely. Despite the concentration of these gases varying from parts per trillion to parts per million (ppt to ppm), the role of trace gases in fluctuating atmospheric chemistry can be highlighted convincingly. Depending on their atmospheric lifetime, they are of two categories, i.e., short-lived trace gases such as carbon monoxide (CO), ozone (O_3), sulphur compounds (SO_x), nitrogen compounds (NO_x), aldehydes, non-methane hydrocarbons with a lifetime of hours, days, and several weeks; long-lived trace gases include chlorofluorocarbons (CFCs), hydrochlorofluorocarbon (HCFC), halons, tetrachloromethane (CCl_4), carbon tetrafluoride (CF_4). Nevertheless, VOCs, CO, and NO_x play a major contribution in altering global atmospheric chemistry (Kampa and Castanas., 2008). VOCs are highly reactive and act as a precursor molecule of O_3. Under favourable environmental conditions, VOCs react with NO_X and produce tropospheric O_3 (Sarkar and Agrawal, 2010a; Cho et al., 2011; Cho et al., 2013; Sarkar and Agrawal, 2012), which is recognised as the major greenhouse gas that plays an important role in changing global climate (Sarkar et al., 2010; Rai et al., 2012; Cho et al., 2013). Several VOCs (i.e., benzene, ethylbenzene, xylene, and toluene) apparently project adverse impacts on the environment, mostly in urban and industrial sectors (Sahu et al., 2020). The hydrocarbon compounds, including isoprene, along with mono terpenes like myrecene and ocemene, that have been observed to be emitted from forest areas, perform a crucial role in the production of photochemical oxidants (Agrawal et al., 2018).

9.3 Various Expedients for Reducing Air Pollution

Besides having the detailed insight into the underlyling causes of air pollution, it is fundamental to understand the movement and fate of those air contaminants in order to remediate air pollution. To control the reckless drive of air pollution, several actions can be taken, among which is to first eliminate the anthropogenic emissions worldwide and the next step is to scientifically remediate the rest of the pollutants present in the environment. However, crucial undertakings have been put forward as well as executed for subsiding air pollution (Macpherson et al., 2017). "Atmospheric Pollution Prevention and Control Action Plan" has been put in force by the government of China since the year 2013 in order to reduce PM 2.5 at the level 25% by 2017 relative to 2012, and also enforced strict restrictions on major pollution sources like transports, industrial sectors, and also power plants (Jind et al., 2016; Liu et al., 2016; Feng et al., 2019). The Clean Air Act has also been introduced in the USA similarly with a constant thought that these policies and legislations would be beneficial to avert the air quality from degrading (Kumar and Gupta, 2016). Several science-based technologies have also been projected like diesel particulate filters (Tsai et al., 2011) and photocatalysis for the removal of VOCs (Huang et al., 2016) (Table 9.1). Apart from that, for the removal of indoor formaldehyde, the catalytic oxidation and chemisorptions methods are of great importance (Pei and Zhang, 2011; Wang et al., 2013; Huang et al., 2014). However, apart from these technological aspects, the mitigation of air pollution through

TABLE 9.1

Technologies utilised in air pollution control

Technologies	Working strategy	Air contaminants	References
Fabric filters/Bag houses	Make use of the "physical barrier" mechanism and are aided by a degree of adsorption.	PMs	Wang et al., 2004a; Donovan, 2020; Darcovich et al., 1997
Melt-blown technique	The effective fabrication of micro/nano fibres in a single step may lead to the formation of unique filtration and separation materials to minimise air pollution.	PMs	Deng et al., 2019
Cyclone	It operates by forcing the gaseous suspension to flow spirally (thus it is named cyclone) within a confined radius causing the particles to be pushed against the vessel's walls by centrifugal force.	PMs	Coury et al., 2004; Darcovich et al., 1997; Bogodage and Leung, 2015
Electrostatic precipitation (ESP)	Charged solid particles are being eleminated from gas streams using electrostatic forces.	PMs, NO_2, SO_2, HCl, and NH_3	Guieysse et al., 2008; Zeng et al., 2020; Jafarinejad, 2017; Yuan and Shen, 2004
Wet scrubbing	Water or alkaline solution is used for the removal of particulates and/or gaseous pollutants.	SOx, PM, CO_2	Darcovich et al., 1997; Wang et al., 2004b; Qaroush et al., 2017; Jafarinejad, 2017
Dry/Sedimentary Scrubbing	Powdered alkali-absorbent material is injected in polluted acid gas streams, thereby producing salt.	SO_2 and HCL	Strömberg and Karlsson, 1988; Srivastava et al., 2001; Wang et al., 2004b; Jafarinejad, 2017
Condensation	Change of phase from gas to liquid leads to the extraction of VOCs from other vapour mixtures.	VOCs	Wang et al., 2004c; Gupta and Verma, 2002
Thermal oxidation	Complete combustion of combustible substances to carbon dioxide and water vapour.	VOCs	Wang et al., 2004d; Jafarinejad, 2017; Lewandowski, 1999
Catalytic oxidation/catalytic incineration	Converts the pollution by oxidation process with the aid of a catalyst to accelerate the process.	VOCs, SO_X, NO_X	Chu and Windawi, 1996; Kosusko and Nunez, 1990; Guo et al., 2001; Wang et al., 2004e; Jafarinejad, 2017;
Gas phase activated carbon adsorption	Contaminants are adsorbed on the surface of granular activated carbon beds in a selective manner.	VOCs	Wang et al., 2004g
Gas-phase biofiltration	Well-adapted microorganisms in biofilters are capable of biodegrading pollutants in the gas phase to innocuous end products.	VOCs, PMs, CO_2, NO_X, H_2S	Rene et al., 2018; Vigueras et al., 2008; Maia et al., 2012; Das et al., 2019; Guieysse et al., 2008; Huang et al., 2016
Ultraviolet (UV) photolysis	High-energy UV radiation helps to oxidise air contaminants and transforming it from harmful to nontoxic substance.	VOCs, O_3	Guieysse et al., 2008; Zeng et al., 2020
High efficiency particulate air filters	Particles are trapped (they stick to a fiber) through diffusion, interception, and impaction	PMs	Liu et al., 2015; Zhang et al., 2019; Wang et al., 2004f
Ozonation	Helps in oxidising air pollutants	VOCs,	Samarghandi et al., 2014; Guieysse et al., 2008
Adsorption	Removes contaminants from a gas stream by settling them on the surface of the absorbent	VOCs, NOx, CO_2, SO_2	Luengas et al., 2015; Fang et al., 2020; Gao et al., 2011; Guo et al., 2015

TABLE 9.1 (Continued)

Technologies utilised in air pollution control

Technologies	Working strategy	Air contaminants	References
Membrane separation	Filtering and separating particular contaminants using the membrane's micropores, and selective permeability along with concentration gradient as driving force	VOCs	Verstraete et al., 1994; Luengas et al., 2015; Zhang et al., 2002
Non-thermal plasma (NTP)	Produce ionised gas that binds to contaminants and decomposes it into non-toxic by-products.	PMs, VOCs, SO_2, H_2S, NOX, CO, CO_2, HC	Luengas et al., 2015; Francke et al., 2000; Roland et al., 2002; Adnan et al., 2017
Selective noncatalytic reduction (SNCR)	Reduce NO_X chemically with injection of nitrogen-type reducing reactant	NOX	Jafarinejad, 2017
Selective catalytic reduction (SCR)	It allows for nitrogen oxide (NOx) reduction to occur in an oxidising environment using ammonia as a reductant inside a catalyst system	NOX	Jafarinejad, 2017

biological means, i.e., biological remediation or bioremediation, is an efficient, eco-friendly alternative (Mueller et al., 1996). Plants have been considered a primary means of assimilation and transformation of harmful pollutants into non-toxic forms, a technique known as phytoremediation (Cunningham et al., 1995; Salt et al., 1995; Brilli et al., 2018; Sharma, 2018). Similarly, significant attention should be drawn towards another alternative way, microbial biodegradation, which is effective in transforming contaminants into components with reduced toxicity (Ma et al., 2016; Wei et al., 2017). Being a heterotrophic organism, the microbes mostly colonise into plant roots (rhizosphere) and shoots (phyllosphere) in order to be enacted actively (Weyens et al., 2015; Gawronski et al., 2017; Wei et al., 2017).

9.4 Phytoremediation: Concept, Principle, Strategies

Phytoremediation is known to be a highly eco-friendly plant-based biotechnology that efficiently degrades/detoxifies a potential amount of air contaminants either via scavenging or metabolizing them (Nowak et al., 2006; Brack, 2002). Though the entire functionality of plant-associated microbes are under investigation, their roles for supporting plants are to combat against biotic and abiotic stresses, to absorb significant amounts of water and nutrients, and also to induce particular plant hormones, inhibitory allelochemicals, and siderophores (Weyens et al., 2015). Plant leaves and stems that are much exposed to air are considered to be significant adsorbents of air contaminants. Thereafter, rainfall brings them back to the soil, and that is how the pollutants can come in the vicinity of plant rhizospheres and get subsided by plant-microbial interactions through phytoremediation. In view of air pollution abatement, the phyllospheric microorganisms that are colonised on plant surfaces can potentially detoxify the harmful contaminants by degradation, transformation, and sequestration methods (Weyens et al., 2015). The entire event of phytoremediation comprises six individual strategies: phytoextraction, phytovolatilisation, phytodegradation, phytostabilisation, rhizodegradation, and rhizofiltration.

9.4.1 Phytoextraction

In phytoextraction, both the aboveground part (phyllosphere) and below-ground part (rhizosphere) of plants are indulged, because after getting extracted from the contaminated site (i.e., soil, water) the pollutants translocated and accumulated in the phyllosphere (Pandey and Bajpai, 2019). The plants which that a genetic prospect to accumulate significant amounts of contaminants (hyperaccumulator plants) compared to normal plants are mainly utilized in this strategy to endure pollutants (Ramalho et al., 2006). There are a few alternative names that are also referred to for this remediating process i.e., phytoaccumulation, phytoabsorbtion, or phytosequestration (Favas et al., 2014) and phytomining or biomining (Pivetz, 2001). When it comes to pros, it is considered to be one of the most efficient practices where plants are capable of eliminating the pollutants from contaminated sites without imposing further adverse effects on the environment (EPA, 2000; Yanai et al., 2006). What makes it accepted is that this technique is super affordable compared to any other technologies. In view of extracting pollutants, plants that belong to the families Scrophulariaceae, Lamiaceae, Euphobiaceae, and Brassicaceae are reported to have significant importance (Baker and Brooks, 1989).

9.4.2 Phytodegradation

Phytodegradation, also termed phytotransformation, is the breakdown of organic pollutants sequestered by plants, either through internal metabolic functions or directly via the action of enzymes generated by the plant. In this decontamination technique, organic contaminants are transformed into a less toxic form followed by further synthesis either prior to their uptake in the rhizosphere or after that (Pandey and Bajpai, 2019). There are a wide range of contaminants like pesticides, munitions, solvents, and other inorganic substances that can be metabolized by some plants in order to detoxify them (Pivetz, 2001; Newman and Reynolds, 2004; Kvesitadze et al., 2006; Gerhardt et al., 2009). This technique causes plants to generate required enzymes independent of rhizospheric microbes to degrade/metabolize the contaminants. The enzymes that are used for metabolizing organic contaminants are dehalogenases, laccases, nitroreductases, nitrilases, phosphatises, and oxidoreductases (EPA, 2000; Favas et al., 2014).

9.4.3 Phytovolatilisation

Since plants can interact with various organic compounds, it becomes easier for plants to curve their transport and fate. Phytovolitisation is one of the most significant processes in which plants can uptake contaminants from contaminated sites and thereby transpire it into the air via stoma in the form of volatile compounds, sometimes as a result of phytotransformation. Plants are able to volatilise two ways, i.e., either direct phytovolatilisation through the leaves and stems or indirect phytovolatilisation from soil by the activity of plant roots (Sakakibara et al., 2010; Limmer and Burken, 2016). This evapotranspiration mechanism is considered to be efficient in several reports because of its remediating capacity for Hg, Se, and As contamination (Heaton et al., 1998; Bañuelos et al., 2000; Sakakibara et al., 2010; Karami and Shamsuddin, 2010; He et al., 2015). The regimes not only depend on the genetic capability of plant but also nature of contaminants as well as the climate of the remediation site is of significant importance.

9.4.4 Phytostabilisation

Phytostabilisation is another cost-effective, in-situ phytoremediating approach that develops a plant cover on the surface of polluted soils to decrease its susceptibility to wind, water, and direct human or animal contact with the aim to accumulate and restrict the contaminants within the vadose zone. Further, the process includes accumulation of contaminants through roots; immobilisation within the rhizosphere by certain strategies like controlling soil erosion, reducing leaching, persistant aerobic environment in the rhizosphere; and also includes organic substances to bind the contaminants (Bolan et al., 2011). Recently, the crucial role of this method involves the stabilisation of metals (i.e., As, Ce, Cu Cd, Hg, Zn) in terms of research and field practices. Additionally, the superficial application of soil amendments are known to become effective to sustain the immobility of contaminants only if it is applied on a regular basis (Bolan et al., 2011).

9.4.5 Rhizodegradation

Rhizodegradation, often known as phytostimulation, is another remediation technique that is predominant in the rhizospheric environment of plants that stimulates the microbial populations, followed by the degradation of pollutants in the soil, accompanied by a bioactive root-associated microbial community (Prasad, 2011). This process is stimulated by the microorganisms present in the rhizosphere. The growing roots of plants provide enough surface area for oxygen transfer to the microorganisms and promote their proliferation. Moreover, as a primary reservoir of carbon and energy, they utilise the plant exudates and metabolites to stimulate microorganisms' activity and degrade organic contaminants (Dzantor, 2007; Favas et al., 2014). Nevertheless, phytostimulation has been reported to have a significant impact in degrading petroleum hydrocarbons, PCBs, and PAHs (Pilon-Smits, 2005).

9.4.6 Rhizofiltration

Another promising phytoremediation technology that employs the approach of aiding either terrestrial plants and their root systems or aquatic plants as a biofilter in order to effectively sequester metals, metalloids, and radioactive elements from the water bodies is rhizofiltration. In this process, the plants are subjected to absorbing and concentrating particular amounts of contaminants through their root system/other submerged parts. To decontaminate the aquatic ecosystem, the selective plants are grown until the plant becomes saturated with the absorbed amount of contaminant and is finally harvested (Pandey and Bajpai, 2019). The chance of getting exposed to atmospheric contamination is minimum in this remediating method since the contaminants are not allowed to be transported from root zone to aerial part (Dushenkov et al., 1995). Rhizofiltration has been involved in the potential remediation of several harmful contaminants (EPA, 1998; Verma et al., 2006; Tomé et al., 2008; Cook et al., 2009; Juwarkar et al., 2010; Fulekar et al., 2010; Lee and Yang, 2010; Yadav et al., 2011).

9.5 Phytoremediation of Outdoor Air Pollution

Phytoremediation is an emerging plant-based idea to abate pollution. However, this plant-driven sequential methodology for abating air contaminants is not yet evident. Plant organs and the associative microbes are able to effectively degrade air pollutants by means of different phytoremediation strategies. The most prevalent source of outdoor air pollution is the emissions from several combustion activities (i.e., running motor vehicle, solid fuel burning, activities in industrial farms, etc.). Additionally, the other potential causes of outdoor air contamination incorporate fumes from forest fires, airborne dust particles, and biogenic release from plant sources such as pollen and spores. Thereafter, the contaminants come in the vicinity of the plant, along with its microbiomes, which are significantly efficient to curve their fate.

9.5.1 Phytoremediation of Various Outdoor Air Pollutants

9.5.1.1 Phytoremediation of Airborne Particulate Matter

In urban areas, the airborne particulate matter is reported to be removed from the atmosphere to some extent by particular plants through their foliage and also can suspend them on waxy layers (Beckett et al., 2000). The air filtration process of remediation was adopted by several countries that are mainly dependent on the plant-specific characteristic traits like size, texture, thickness and pubescence of leaves, amount of the epicuticular wax, and the type of trichome that helps to collect the dust. Plants have that mechanism to abate PMs to some extent, yet it can be altered depending upon the climatic impacts on the environment like rain, wind, volume, and chemical nature of PMs in the atmosphere (Weyens et al., 2015). The "i-Tree" model developed by Nowak is known to perfectly ascribe the urban green structure and its ecosystem services (Nowak et al., 2008). Several surveys have suggested the positive impact of urban forests in reducing a significant amount of PMs from the atmosphere from different parts of the world, i.e., China (Yang et al., 2005; Yin et al., 2014), UK (McDonald et al., 2007), USA (Nowak et al., 2006), and France (Selmi et al., 2016). Apparently the green belts are considered to be a potential way out in terms of reducing dispersion of PMs in industrial areas (Singh and Tripathi, 2007; Banerjee and Srivastava, 2010). Taking into account the enormous size and total surface area (leaf) of tree species is the most important parameter for phytoremediation as they provide large amounts of area to accumulate airborne particles (Fowler et al., 1989; Beckett et al., 2000; Freer-Smith et al., 2004; Dzierżanowski and Gawroński, 2011; Liu et al.,2018). Convincingly, trees are considered to be the best sink of PMs, followed by the herbaceous vegetations (Weber et al., 2014; Letter and Jäger, 2020). Vegetation is equally effective in reducing street-level concentrations of PMs by 60%, as deposition to vegetation can make a very modest improvement in urban air quality (Pugh et al., 2012). Plants that are involved in phytoremediation of PMs are shown in Table 9.2.

9.5.1.2 Phytoremediation of Volatile Organic Compounds

VOCs, a large group of organic compounds, are again considered to be another important component of that atmosphere that act as the precursor gas for the generation of the phytotoxic secondary air pollutant ozone (Sarkar et al., 2012a, b). The indoor and outdoor air qualities have been highly degraded by the presence of VOCs. Depending on the nature of sources, VOCs can be categorised into transportation and industry (anthropogenic) generated AVOCs; another one is trees and other plants that generate BVOCs (Ren et al., 2014). Apparently, for the remediation of VOCs from air, few plants have been recommended. Several plants act as potential sources of low molecular biogenic VOCs for their defense against any herbivores, pathogens, or environmental stress; therefore, to reduce the amount of VOCs from the atmosphere, plants with low-VOC emitting potential should be in consideration (Weyens et al., 2015; Agrawal et al., 2019). In view of phytoremediation, studies show that the openings (cuticle and stomata) of plant bodies are responsible for potential VOC uptake, predominantly via stomata during the daytime when the stomata are open, unless it is a CAM plant (Weyens et al., 2015). Additionally, the occupied amount of VOCs on the waxy layer determine the potential of cuticular absorption (Treesubsuntorn and Thiravetyan, 2012; Sriprapat and Thiravertyan 2013). Once inserted into the plant system, either the VOCs undergo sequential degradation through the plant itself thereby transforming them into harmless constituents or the excretion and storage method in case degradation fails to occur (Weyens et al., 2015). Taking this into account, a report has been published where plants absorb and eliminate benzene and toluene from the air (Porter, 1994). Plant leaves are also capable of eradicating lower concentrations of formaldehyde from the air, reported by Wolverton (1988). In that context, plant leaves are well-known VOC-detoxifying agents; however, recently there are a few reports of various VOC-degrading microorganisms present in the phyllosphere (Weyens et al., 2015). Giese et al. (1994) reported that *Chlorophytum comosum* L. (spider plant) and *Glycine max* L. (soybean plant) can also detoxify lower levels of formaldehyde to some extent by means of phytoremediation.

9.5.1.3 Phytoremediation of Inorganic Air Contaminants

Apart from the particulate matter and VOCs, there are several air contaminants like SO_2, CO_2, CO, NO_x, and O_3 that impart negative effects on the environment. However, it should be mentioned that some tolerant plants may take up and bioaccumulate these pollutants from the atmosphere as a potential sink in order to abate air pollution (Liu and Ding, 2008; Das et al., 2018; Molnar et al., 2020). The air quality of urban areas reportedly improved by the potency of urban shrubs and trees (Nowak et al., 2006). As per the research by Weber and collegues, herbaceous plants like *Achillea millefolium* (yarrow plant), *Polygonum aviculare* (common knotgrass), *Berteroa incana* (Hoary alyssum plant), *Flaveria trinervia* (clustered yellowtops), and *Aster gymnocephalus* (Texas tansyaster) are also impactful in order to remediate air contaminants (Weber et al., 2014). The plant epidermal system, i.e., stomata and cuticles, facilitate the entry of air pollutants into the plant system. To remediate automobile pollution, roadside plants are known to be highly efficient as they can undergo carbon sequestration. Carbon sequestration is another significant method of uptake and storing CO_2 in plants for longer periods of

TABLE 9.2

Name of plants with corresponding pollutants (outdoor)

Pollutant	Principle	Name of the Plants	References
Particulate matter	Accumulation on leaf surfaces and phytostabilisation in waxes	*Catalpa bignonioides* Walter *Corylus colurna* L. *Fraxinus pennsylvanica* Marsh. *Ginkgo biloba* L. *Platanus* × *hispanica* Mill. ex Muenchh. *Quercus rubra* L. *Tilia tomentosa* Moench 'Brabant' *Acer tataricum* subsp. *Ginnala* (Maxim.) Wesm. *Sambucus nigra* L. *Sorbaria sorbifolia* (L.) A.Br. *Spiraea japonica* L.f. *Syringa* C.K. Schneid. 'Palibin' *Viburnum lantana* L.	Popek et al., 2013
Particulate matter	Deposition and surface accumulation	*Amygdalus triloba* (Lindl.) Ricker *Euonymus japonicus* Thunb. *Lonicera maackii* (Rupr.)Maxim. *Prunus cerasifera* Ehrh. *Magnolia denudata* Desr. *Rhus typhina* Nutt. *Platanus occidentalis* L. *Fraxinus pennsylvanica* Marsh. *Populus tomentosa* Carr. *Ginkgo biloba* L. *Ulmus pumila* L. *Salix matsudana* Koidz. *Pinus tabulaeformis* Carr. *Platycladus orientalis* (L.) Franco *Pinus armandi* Franch. *Ailanthus altissima* (Mill.) Swingle *Sophora japonica* L.	Xu et al., 2018
Particulate matter	Green belt with efficient dust-capturing plants	*Albizia lebbeck* (L.) Benth. *Alstonia scholaris* L.R.Br. *Anthocephalus indicus* (Roxb.) Miq *Bougainvillea spectabilis* Wild. *Caesalpinea pulcherima* (L.) SW. *Cassia auriculata* L. *Cassia siamea* Lam. *Delonix regia* (Bojer ex Hook.) Raf. *Ficus religiosa* L. *Lagerstroemia speciosa* (L.) Pers. *Mimusops elengi* L. *Peltophorum inerme* (Roxb.) Navesex Fernandez Villar *Swietenia mahagoni* (L.) Lacq. *Tabebuia aurea* Benth and Hook.f.ex S.Moore *Thevetia nerifolia* Juss Ex. Steud	Das and Prasad, 2012
Particulate matter	Accumulation on leaf surfaces and phytostabilisation in waxes	*Acer campestre* L. *Fraxinus excelsior* L. *Platanus* × *hispanica* Mill. ex Muenchh. 'Acerifolia' *Tilia cordata* Mill. *Forsythia* × *intermedia* Zabel *Hedera helix* L. *Physocarpus opulifolius* (L.) Maxim. *Spiraea japonica* L.	Dzierzanowski et al., 2011

(*Continued*)

TABLE 9.2 (Continued)

Name of plants with corresponding pollutants (outdoor)

Pollutant	Principle	Name of the Plants	References
Particulate matter	Surface deposition	Sorbus aria Acer campestre Populus deltoides ×trichocarpa 'Beaupré' Pinus nigra var. Maritime Cupressocyparis leylandii	Beckett et al., 2000
Particulate matter	Surface deposition	Juniperus formosana Pinus bungeana Platycladus orientalis Pinus tabulaeformis Euonymus japonicus	Song et al., 2015
Particulate matter	Particle retention and entrapment in the leaves	Platanus × hispanica Tilia cordata Quercus ilex Quercus cerris Olea europaea	Blanusa et al., 2015
Particulate matter $PM_{2.5}$	Capture and retention of Particles	Eucommia ulmoides, Tilia tuan, Platanus occidentalis, Armeniaca sibirica, Malus micromalus, Ulmus pumila, Lonicera maackii, Parthenocissus thomsoni, Tilia tuan, Philadelphus pekinensis, Ginkgo biloba, Phyllostachys propinqua, Lonicera maackii, Armeniaca sibirica, Magnolia denudate, Eucommia ulmoides, Tilia tuan, Broussonetia papyrifera, Sophora japonica, Magnolia denudate, Ulmus pumila, Armeniaca sibirica, Philadelphus pekinensis, Ilex chinensis	Chen et al., 2017
Particulate matter	Surface diposition on leaves and shoots	Tilia cordata Mill.	Popek et al., 2017
Particulate matter	Accumulation on leaf surfaces	Acer campestre L. Acer platanoides L. Acer pseudoplatanus L. Aesculus hippocastanum L. Alnus × × spaethii Call. Aralia elata Seem. Berberis thunbergii DC. Betula pendula Roth. Carpinus betulus L. Catalpa speciosa Warder 'Pulverulenta' Cornus alba L. Cornus stolonifera Michx. Elaeagnus angustifolia L. Fagus silvatica L. Forsythia × × intermedia Zab. Fraxinus excelsior L. Hedera helix L. Hydrangea arborescens L. Hydrangea paniculata Sieb. Mahonia aquifolium Nutt. Malus 'Van Eseltine' Partenocissus quinquefolia (L.) Planch. Partenocissus tricuspidata (Siebold & Zucc.) Planch. Physocarpa opulifolius Maxim. Pinus mugo Turra	Sæbø et al., 2012

TABLE 9.2 (Continued)
Name of plants with corresponding pollutants (outdoor)

Pollutant	Principle	Name of the Plants	References
		Pinus sylvestris L.	
		Popolus tremula L.	
		Prunus avium L.	
		Prunus laurocerasus L.	
		Prunus padus L.	
		Pyrus calleryana Decne.	
		Quercus robur L.	
		Robinia pseudoacacia L.	
		Rosa rugosa Thunb.	
		Rosa × × rugotida Boomkw.	
		Salix cinerea L.	
		Skimmia japonica Thunb.	
		Sorbus intermdia Ehrh.	
		Spirea × × cinerea Zab.	
		Spireae × × vanhuttei Zab.	
		Stephanandra incisa Thunb.	
		Symphoricarpus albus Blake.	
		Taxus baccata L.	
		Taxus × Rehder 'Hilli'	
		Tilia cordata Mill.	
		Tilia × × europaea (L.) 'Pallida'	
		Ulmus glabra Huds.	
Particulate matter	Air articles are captured, retained, trapped in the cuticle, and then expelled to the soil	Pinus pinea Cornus mas Acer pseudoplatanus	Terzaghi et al., 2013
Particulate matter	Accumulation on leaf surfaces	Pinus nigra var. maritima Cupressocyparis leylandii Acer campestre Sorbus intermedia Populus deltoides × trichocarpa 'Beaupre'	Beckett et al., 2000
Particulate matter	Settling upon leaf surfaces	Populus lomentosa Populus canadensis Salix matsudana Juglans regia Fraxinus chinensis Prunus persica Ailanthus altissima Sophora japonica Syringa oblata Euonymus japonicus Parthenocissus quinquefolia	Wang et al., 2006

time (Sedjo and Sohngen, 2012). The metabolisation of CO is highly variable for different plant species, depending upon whether they are going to be oxidised or reduced further, generating CO_2 and amino acids, respectively (Agrawal et al., 2019). To enhance the phytoremediation and carbon sequestration efficacy of plants, sustainable utilisation of microbes and industrial waste residue should be encouraged (Pandey and Bajpai, 2019). As reported by Bidwell and Bebee (1974), 35 woody plants were marked to have that efficiency of CO fixation (Bidwell and Bebee, 1974). Green plants are considered to be the best potential sink of carbon as they are able to carry out photosynthesis mainly via the removal/uptake of CO_2 from the atmosphere (Sinha and Singh, 2010; Weyens et al., 2015). Another potent secondary air contaminant is O_3, which after getting through stomata, dissolved into the apoplastic fluid and subsequently generated reactive oxygen species (ROS), thereby projecting harmful effect on plants (Wohlgemuth et al., 2002; Tripathi et al., 2011; Rai et al., 2012; Sarkar et al., 2015). Several reports suggest urban trees are an efficient scavenger of O_3 (Taha, 1996; Manes et al., 2012; Nowak et al., 2014). Since the ozone has antimicrobial properties, thereby the antioxidative bacteria are considered to be the better alternative for the phytoremediation of the potential oxidant O_3 (Wu et al., 2014; Van Sluys et al., 2002). Phytoremediation of SO_2 starts when it readily diffuses into the leaves through stomata openings. Thereafter, it leads to the generation of sulphur-containing amino acids by utilising the plant metabolic system of detoxification "reductive sulphur cycle". However, the plant detoxification system collapses if the concentration goes beyond the plant tolerance level (Gheorghe and Ion, 2011). The surface of vegetation acts as the efficient sink of NO_2 (Chaparro-Suarez et al., 2011). Some of the plants have the potential to incorporate NO_2 from ambient and thereafter use it as an alternative of fertiliser (Welburn, 1998). Phytoremediation of NO_2 mostly takes place via the surface (leaf and root) adsorption and stomatal uptake (Welburn, 1990). Additionally, the uptake of NO_2 especially depends on the stomatal aperture (Geßler et al., 2000). Mostly the NO_x gets metabolized via a nitrogen assimilation pathway in order to convert it to amino acids, which is again demonstrated by the ^{15}N isotope (Chaparro-Suarez et al., 2011). In accordance with Krupa (2003), plants can however intake nitrogen directly from the air in the form of NH_4^+. Though N_2 is a nutritive supplement for the plant, the prevalence of excessive N_2 in high concentrations is rather phytotoxic (Sinha and Singh, 2010). In another study, four woody plants (*Robinia pseudo-acacia, Sophora japonica, Populus nigra*, and *Prunus lannesiana*) are recommended as the most efficient remediators in terms of abatement of urban air pollution (Takahashi et al., 2005). As per the report of Morikawa et al. (1998,) *Magnolia kobus, Eucalyptus globulus, Eucalyptus grandis, P. nigra, Populus* sp., and *Erechtites hieracifolia* showed their efficacy in phytoremediation of NO_2.

9.6 Phytoremediation of Indoor Air Pollution

9.6.1 Indoor Air Pollution: Grave Alarm?

People have to breathe clean air in order to maintain good health. Indoor pollution projects adverse health impacts as deteriorating indoor air quality (IAQ) has become worse than ambient air in the present-day scenario (Boor et al., 2017; Du et al., 2018; Andrade and Dominski, 2018). To have a reality check, it should be marked that due to the depleting condition of indoor air pollution, approximately 3.8 millions deaths are attributed annually on a global scale. The accumulation, toxicity rate, exposure period, and concentration of indoor air pollution may show detrimental heath outcomes to mankind like lung pneumonia, heart ailments, lung cancer, chronic obstructive pulmonary disease, and even stroke depending upon individuals (Leung, 2015; Kumar et al., 2015; WHO, 2016; Banerjee et al., 2017; Amoatey et al., 2018; WHO, 2018) and also eventually contributing towards severe ailments like "sick building syndrome" and "building-related illness". Because the contemporary Western population is primarily exposed to the indoor rather than the outdoor environment, IAQ is perceiving unprecedented attention (Colbeck and Nasir, 2010; Molloy et al., 2012; Goldstein et al., 2020). The majority of people from developing countries, with special reference to India, mostly stay in the vicinity of both indoor and outdoor pollution because the pollution generally emerges from household and ambient air (India State-Level Disease Burden Initiative Air Pollution Collaborators, 2019). In India, 1.2 millions deaths have been reported in the year 2017 by means of both indoor and outdoor air pollution (Health Effect Institute, 2019). Indoor air is reported to contain two to four times greater amounts of pollutants than outdoor air (Jafari et al., 2015) and may be due to the variability in terms of ventilation level, humidity, or temperature that influence IAQ (Torpy et al., 2013; Zhang et al., 2015). Indoor PM pollution is 10–30% among total PM-generated diseases (Morawska et al., 2013). Keeping the major concern on people's adverse health impact for indoor air pollution, several reports have been analysed depending upon different conditions like poorly ventilated rooms, air-conditioned offices, hospitals with air-conditioning or without mechanical ventilation, school buildings with natural ventilation, and some restaurants with exhaust fans and chimneys (Morawska et al., 2017; Azuma et al., 2018; Mohammadyan et al., 2019; Taner et al., 2013).

9.6.2 Common Indoor Pollutants: Matters of Concern

Dependent upon the socioeconomic structure, the types, amounts, and availability of indoor pollutants are altered on a global scale (Colbeck and Nasir, 2010). The standard ventilation rate in construction is reduced by the recognised society, i.e., American Society of Heating, Refrigerating and Air-Conditioning Engineers (Burroughs and Hansen, 2004), in order to decrease building consumption energy (Seppänen, 2002). However, the utility of sources that might produce some indoor pollution (such as office equipment) is simultaneously increasing at the same time (Jafari et al., 2015) due to the ignorance of environment managers and policy makers for being inconsiderate of minor sources to cause indoor air pollution. VOCs, CO_2, PMs, NO_2, O_3, and other pollutants are considered to be prevalent in indoor air, mostly in the case of developed nations (Morawska and Salthammer, 2006; Wisthaler and Weschler, 2010; Lindgren, 2010; Wolkoff, 2013; Morawska et al., 2013; Bozkurt et al., 2015; Oliveira et al., 2019).

Particulate matters generally include a broad range of mobile particles with various chemical compositions. Increased mortality rate associated with respiratory, cardiovascular, and venous thromboembolic disorder have been found as an inevitable result of PM pollution (Bari et al., 2014). In fact, a minor elevation of $PM_{2.5}$ exposure is associated with a high mortality rate due to cardiovascular diseases from the USA (Wang et al., 2016) and Canada (Crouse et al., 2012). Similar trends are also reported from cities all over the world (Li et al., 2017). Indoor PM pollutants can be sourced to mechanical or natural ventilation from the outdoor environment. There is prominent evidence of health hazards caused due to PM exposure (Wyzga and Rohr, 2015; Junaid et al., 2018; Nezis et al., 2019); therefore, the presence of concerning levels of PMs have now become a significant matter of issue for indoor environments as well as in residential houses (He et al., 2004; Yassin et al., 2012), offices (Cheng, 2017), classrooms (Fromme et al., 2007), urban nurseries (Branco et al., 2014), commercial buildings (Goyal and Kumar, 2013), industrial communities (Tunno et al., 2015), and schools (Sánchez-Soberón et al., 2019). Not only through outside sourcing, PM pollution is also reported to be generated from indoor cooking, printing, smoking, cleaning, and increasing building occupancy (Long et al., 2000; He et al., 2007; Buonanno et al., 2009; Wheeler et al., 2011).

VOCs are of particular importance in terms of IAQ as the indoor environment is supposed to contain a mixture of several VOCs (Goodman et al., 2017) and has gained enormous research attention (Yu and Kim, 2010). VOCs are generally produced indoors by the influence of off-gassing from solvents, building materials, furnishings, plastics, cleaning aids, wall coverings, floors, electronics, and appliances (Zhang et al., 1996; de Gennaro et al., 2015; Schlink et al., 2010). Exposure of imperceptible lower concentrations of VOCs for shorter or longer time periods has been investigated to associate with several adverse health impacts like sick building syndrome (Goodman et al., 2017), asthma symptoms (Wichmann et al., 2009; Fuentes-Leonarte et al., 2009; McGwin Jr et al., 2010), and others related to renal, hepatic, and nervous problems (Sriprapat et al., 2014b). Nevertheless, exposure of few VOCs in low levels has been found to be carcinogenic (Khanchi et al., 2015). Among the benzene, toluene, ethylbenzene, and xylene (BTEX group), benzene has been reported to be the most toxic compound that is classified into group-I and proclaimed as carcinogenic by the International Agency for Research on Cancer (IARC). Benzene can even cause several other health hazards like haematological disease, acute and chronic lymphocytic leukemia, myeloma, and acute myeloid leukemia (Snyder, 2012; Collins et al., 2003). Apart from that, several other gasses are also responsible for degrading IAQ. CO_2 is one of the most predominant gasses in indoor environments that is primarily generated from respiratory emissions (Llewellyn and Dixon, 2011). A vicinity with excessive amounts of CO_2 is reported to cause the well-known sick building syndrome (Jafari et al., 2015); along with that, it is shown to decline workplace productivity and students' school attendance (Seppänen et al., 2006; Gaihre et al., 2014). The aforementioned reasons emphasise the critical need of removing contaminants from indoor air.

To ameliorate the indoor air pollution, several mechanical air purification techniques (i.e., photocatalysis, airfiltration, ionisation, UV pholysis, photocatalytic oxidation, ozonation, etc.) have been utilised that are able to remediate only a single form of pollutant. However, in the longer term, these energy extensive, potentially hazardous, complex techniques should be altered with a natural and sustainable approach of remediation. There is an urgent requirement to evolve the means of creating and sustaining inhabitable indoor environments, maintaining lower amounts of greenhouse gas emissions, and improving the utilisation of building energy in this increasingly energy-aware world. Thus, addressing and mitigating the entire issue by means of botanical, biological, and biotechnological methodologies might be a propitious choice of research (Luengas et al., 2015).

9.6.3 Phytoremediation of Various Indoor Air Pollutants

The phytoremediation technique has been reported as a way to lessen health concerns due to poor indoor air conditions. Plants have the potential to ameliorate indoor air contaminants in an organic and sustainable way. Through the efficacy of phytoremediation based on plants' species specificity and pollutants' characteristics (Papinchak et al., 2009), it may easily provide a low-cost, energy-efficient, self-regulating, potential green replacement (Irga et al., 2018; Luengas et al., 2015; Pettit et al., 2018a) either by active (green roof, green walls, bio-covers) or passive (potted plants and their association with microorganisms) approaches. There are several well-reported ornamental plants that are involved in abatement of particular air pollutions like *Dracaena sanderiana* (Treesubsuntorn and Thiravetyan, 2012), *Schefflera arboricola*, and *Spathiphyllum wallisii* (Parseh et al., 2018) for benzene; *Schefflera actinophylla* for xylene and toluene (Kim et al., 2016); *Dieffenbachia compacta* (Aydogan and Montoya, 2011) and *Chamaedorea elegans* (Teiri et al., 2020) for fomaldehyde; *Chlorophytum comosum, Dracaena deremensis,* and for trichloroethylene, benzene, and formaldehyde (Wolverton et al., 1989); and *Dypsis lutescens* for total volatile organic compounds, CO_2, and CO (Bhargava et al., 2021). The initial idea that green systems can effectively ameliorate high levels of indoor air pollutants was investigated from space science research in order to evolve a spacecraft that is self-sustaining by utilising a biological life support system (Salisbury et al., 1997; Andre and Chagvardieff, 1997).

9.6.3.1 Phytoremediation of Indoor Airborne Particulates

Green plants are well-established outdoor sinks of aerosolised particulate matter; however, only a few reports are available supporting the phytoremediation capacity of indoor plants to abate indoor PMs. A higher absorption of pollutant corresponds with the size of the leaf, texture of the surface, the amount of wax layer on leaf, pubescence, and complicated leaf morphology (Stapleton and Ruiz-Rudolph, 2018). Stapleton and Ruiz-Rudolph (2018) examined the potential of 11 indoor plants to combat against ultrafine particle (UFP) pollution. In this study, they found that except for *Dracena deremensis,* all of the plants were able to reduce the UFP significantly. Among them, *Juniperus chinensis* showed the maximum level of decrement of UFP pollution. Additionally, it was interesting to observe a strong intercorrelation between the reduction of UFP with the increased foliage density, which signifies that the most influential parameter would be plant surface area (dense yet variegated) in order to mitigate UFP pollution in a broader spectrum using vegetation. Another investigation on spider plants (*Chlorophytum comosum* L.) revealed that this potential plant is capable of accumulating both hydrophobic and hydrophilic and all the three size fractions mentioned of PMs from indoor air. Moreover, among the three size fractions of PMs, the largest one occupies a greater portion of the leaves and the accumulation forces/factors are more than only a simple gravitation force (Gawrońska and Bakera, 2015). The potted plants are reported to assimilate a major amount of PMs via foliar inception (Lohr and Pearson-Mims, 1996).

9.6.3.2 Phytoremediation of Indoor Volatile Organic Compounds

VOCs are the most predominant indoor air contaminant with a ubiquitous range source in comparison to the ambient environment. Plants are generally known to enhance indoor air quality by lowering the level of toxic VOCs in the air, making it better for human health (Dela Cruz et al., 2014a, b). Particular plants that are efficient in uptaking toxic VOCs from ambient play a crucial role as air purifiers, considering phytoremediation. Though there are limitations, the plant-based VOC removal technique has grown a lot from passive filtration utilising potted plants to a more advanced method, i.e., active botanical biofiltration (Kim et al., 2017). The wholesome idea of phytoremediation of indoor VOCs started when more than 300 different ranges of VOCs were found inside the closed interior of the Skylab space station, which attributed health hazards of the occupants associated with poor air quality, reported by The US National Aeronautics and Space Administration (NASA) (The National Aeronautics and Space Administration, 1974). That subsequently led to the investigation of those particular plants that have the active potency to eliminate VOCs from the air of indoor environments by the life science researchers of NASA. A pioneer study was carried out by Wolverton and McDonald in the year 1982, making use of *Scindapsus aureus* (Devil's ivy), *Syngonium podophyllum* (goosefoot plant), and *Ipomoea batatas* (sweet potato) to mostly exclude formaldehyde from indoor air. Again, Wolverton and colleagues investigated to find out the most efficient plants for the phytoremediation of air pollution from sealed chambers with *Chlorophytum elatum* (Ait.) (R.Br.) var.*vittatum* Hort.; *Syngonium aureus;* and *Syngonium podophyllum*. Their initial outcome was positive enough to highlight the potential of the experimental plants to reduce substantial concentrations of VOCs like trichloroethylene, benzene, and formaldehyde from a sealed chamber (Wolverton et al., 1984, 1989), which brought this work enough appreciation for its potential aspect inside building environments (AIRAH, 2016). There is a combination of strategies, i.e., direct mechanism via absorption and indirect mechanism via biotransformation, of the removal of VOCs from the air either by the root or shoot system, growing medium, or aided by the micro-flora present in the rhizosphre. The nature/characteristics of VOCs readily influence plants' absorption capacity as benzene may freely pass through the cuticle; however, the hydrophilic VOC formaldehyde can pass through stomatal openings (Dela Cruz et al., 2014a, b). Several early botanical air purifying research mostly reinforced the basic methodology of NASA through the passive use of potted plants in order to increase the quality of indoor air (Wolverton et al., 1984; Godish and Guindon, 1989; Wolverton and Wolverton, 1993; Wood et al., 2002, 2006; Orwell et al., 2004, 2006; Cruz et al., 2014; Chen et al., 2017; Teiri et al., 2018) and the experiments were carried out under high VOC concentrations within a small closed chamber (Llewellyn and Dixon, 2011).

9.6.3.2.1 Plant Aerial Part Mediated Phytoremediation
The potency of several plants in eliminating VOCs from indoor air corresponds with several biological, physical, and mechanical factors (Dela Cruz et al., 2014). Both the above- and below-ground parts differ in their efficiency, depending upon plant species and pollutant. The ornamental plants are reported to act as the most efficient supplement that positively reduce the VOC concentration from the indoor environment (Giese et al., 1994; Kim et al., 2009). After screening 28 different indoor ornamental plants, Yang and colleagues found variations in plant responses, especially in a few plants like *Hemigraphis alternate* (red ivy), *Hedera helix* (common ivy), *Hoya carnosa* (wax plant), and *Asparagus densiflorus* (asparagus fern), which showed immensely convincing results regarding the removal of VOCs from indoor air (Yang et al., 2009). In another experiment, among six different orchid plant species, only four, i.e., *Sedirea japonica* (Japan Sedirea), *Dendrobium phalaenopsis* (Cooktown orchid), and *Phalaeonopsis* sp. (moth orchids), showed a significant potential of removing formaldehyde from the indoor environment (Kim and Lee, 2008). Among 73 ornamental plants, only 10 were reported to have efficacy in eliminating benzene (Liu et al., 2007). According to the observation made by Kim and colleagues in their experiment, ferns are two to three times more capable of removing formaldehyde among all the other groups of plants, depending on the leaf area parameter (Kim et al., 2010). The potted plants enable the decomposition of VOCs and were reported to reduce benzene by 15% (Lim et al., 2009). Nonetheless, to observe the VOC removal capacity of aerial parts, in a few experiments the aboveground part of the plant was subjected to a physical barrier and isolated from the rest of the substrate, along with the root zone, which showed a positive response independently (Tani et al., 2007; Tani and Hewitt, 2009; Treesubsuntorn and Thiravetyan, 2012; Treesubsuntorn et al., 2013; Sriprapat and Thiravertyan, 2013; Sriprapat et al., 2014a, b). Aydogan and Montoya (2011) set an experiment to

observe the efficiency of the formaldehyde-removal capacity of plant root zones and aerial parts separately. As per the observation that there is concern both the above- and underground part of plants are capable of removing formaldehyde from indoor air however, interestingly, the root zone showed a faster rate of removal than the foliage part (Aydogan and Montoya, 2011). Although a plant's root zone is crucial for toluene and xylene degradation, the uptake and transfer of the pollutant to the root corresponds with the stem portion (Kim et al., 2016). The same trend of response was also observed when a potted *Chlorophytum comosum* was exposed to 500 ppm of benzene concentration with its root covered with aluminium foil, where removal of 68.77% of benzene had been reported within eight days (Setsungnern et al., 2017). Another experiment revealed that the soil and plant aerial surface has a similar capability in order to remove VOCs independently where the potted plants substrate was covered with foil and the aerial part showed the same potential of removing VOCs with that of potting-soil treatment (Hörmann et al., 2017). Toluene is effectively removed by the plants *Spathiphyllum* and *Dracaena deremensis* (Orwell et al., 2004). *Dracaena deremensis, Dracaena marginata, Schefflera actinophylla, Howea forsteriana, Spathiphyllum floribundum,* and *Schefflera floribundum* plants are potentialy enough to ameliorate benzene from indoor air (Orwell et al., 2004). Out of 94 investigated plants, *Chrysanthemum morifolium* and *Calathea rotundifolia* cv. *Fasciata* have been reported to remove the highest amounts of benzene and toluene (Yang and Liu, 2011). Alteration in water status and stomatal conductance are in correlation with the VOC removal capacity of that plant. Sometimes the selectivity of particular pollutants is plant specific. There is a perfect example of the *Kalanchoe blossfeldiana* plant that preferentially removes benzene over toluene (Cornejo et al., 1999). The above- and below-ground part of plants have different rates of VOC removal capacity that instantly suggest the potential of upper and below-ground plant tissue to have significant phytoremediation capacity (Kim et al., 2008). To achieve the high level of efficacy in terms of VOC removal, equal contribution is needed from both the root zone and aerial part to maintain mutual health (Wood et al., 2002; Kim et al., 2008; Wang et al., 2014). The plant and leaf age is another parameter over the efficiency of VOC removal as, simultaneously, the wax deposition increases in many species due to leaf age. As per the report of Ugrekhelidze et al. (1997), benzene and toluene intake capacity is much higher in the case of younger spinach, apple, and grape leaves than older ones (Ugrekhelidze et al., 1997). Moreover, a similar trend has been followed for the removal of formaldehyde in the case of the *Chlorophytum comosum* plant (Su and Liang, 2015). Because of the influence of stomatal conductance on VOCs' removal capacity of the plant, the type of photosynthesis in the respective plant is also thought to have more or less influence on the same. In an experiment, Sriprapat and colleagues investigated differences in xylene removal capacity in different combinations of plants having C3, C4, and CAM modes of photosynthesis where they found out the plants having CAM and C3 kind of metabolism were highly persistent in terms of xylene removal capacity (Sriprapat et al., 2014a). Similarly, another experiment with different species orchids with a C3 and CAM photosynthetic system showed C3 orchids to be more efficient formaldehyde removers than CAM orchids (Kim and Lee, 2008).

9.6.3.2.2 Microbial-Mediated Phytoremediation

The root system's associated microbes have been well recognised for their potential role in pollution remediation as they are able to metabolize various groups of organic compounds (Weyens et al., 2015). The dominant effect of microorganisms for the removal of pollutants has been well proved when sterilised soil loses its capacity for pollutant removal (Agrawal et al., 2019). Plant root systems have predominant efficiency for VOC removal either in direct ways or by increasing the bioavailability of contaminants for the microbial community (Wenzel, 2009). In an investigation by Torpy et al. (2013), it has been found out that comparing the benzene removal capacity of two potted plants, one of which has stimulated substrates for the microbial community, showed significantly better benzene removal capacity than the other untreated potted plant (Torpy et al., 2013). Similary, in another study by Sriprapat and Thiravetyan (2016), they identified a benzene degrading phyllospheric bacteria. Additionally, they observed that the potted plant inoculated with that particular microbe showed better benzene removal efficacy instead of the sterilised leaf containing ordinary potted plants (Sriprapat and Thiravetyan, 2016). In another experiment, an endophytic bacterium, *Bacillus cereus,* was introduced in two non-native host plants (*Zamioculas zamiifolia* and *Euphorbia milii*), and resulted in an increased level of tolerance against formaldehyde phytotoxicity (Khaksar et al., 2016a). Moreover, the same inoculation has been found to increase the seed germination of *Clitoria ternatea*, along with plant growth against formaldehyde (stress) phytotoxicity and consequently enhancement of formaldehyde removal capacity (Khaksar et al., 2016b). The bacterial species that have been well studied for their high efficiency for the degradation of BTEX are *Pseudomonas putida and Pseudomonas* spp. (Jindrova et al., 2002). There are reports of soil microbes with potential VOC removal capacity (Wolverton and McDonald, 1982; Wolverton et al., 1984; Wolverton and Wolverton, 1993; Yang et al., 2020). *Achromobacter* sp., which is an endophytic aromatic bacteria, is capable of degrading VOCs in plant roots, and is reported to be accommodated without interfering rhizospheric microflora (Yin-Ning Ho et al., 2013). According to the experiment conducted by De Kempeneer et al. (2004), the toluene degrading bacteria *Pseudomonas putida* is found to be inoculated in *Azalea indica*, and eventually increase the rate of toluene removal capacity from the air (De Kempeneer et al., 2004). Thus, for the prevalence of purified air in the indoor environment, the strategy of plant-microbe interactions is of great significance for enhancing the phytoremediation efficacy of indoor growing plants (Xu et al., 2010).

9.6.3.3 Phytoremediation of Other Indoor Air Pollutants

Indoor plants have the capacity to subside several other indoor pollutants other than PMs and VOCs by utilising several mechanisms. Indoor CO_2 removal can easily be facilitated by involving plants' normal physiological event, photosynthesis, where CO_2 is absorbed via stomata (Torpy et al., 2014). Additionally, Torpy et al. (2014) reported that the variation in indoor light level regulates the photosysnthetic capacity of plants, along with differences in CO_2 removal rate from indoor air and the removal rate goes to its maximum only when the light intensity increased more than the usual indoor light level (Torpy et al., 2014). SO_2 can be efficiently absorbed and metabolized by indoor plants (Lee and Sim, 1999). Coward et al. (1996) reported a 30% reduction in indoor NO_2 concentration by the activity of sun- and shade-loving plants (Coward et al., 1996). O_3, which is a phytotoxic air contaminant for indoor and outdoor environments, is reported to be ameliorated from the indoor air in rural areas sustainably by some common indoor plants (Smith, 2002). Heather et al. (2009) also reported some houseplants that are known to eliminate O_3 pollution from indoor environments (Heather et al., 2009). The concentration of NO_2 and CO is potentially reduced by the phytoremediation properties of common foliage plants like *Sansevieria trifasciata*, *Epipremnum aureum*, and *Chlorophytum comosum* (Wolverton et al., 1984). After the report of Selmar et al. (2015), *Mentha piperata* (peppermint plant) has the ability to absorb tobacco smoke from indoor air or from soil or plant debris (Selmar et al., 2015). As per another recent study by Bhargava et al. (2021), the Areca palm potted plant showed high indoor air pollution removal efficiency for improving indoor air quality. Among the four discrete naturally ventilated investigation sites (I, II, III, IV), Areca palm potted plants showed a 88.16% reduction of total VOCs in site no IV whereas in site no III CO_2 and CO found to be reduced by 52.33% and 95.70%, respectively (Bhargava et al., 2021).

9.6.4 Limitations of Potted Plant/Static Chamber Experiments

Most of the existing experiments and literature studies regarding the efficacy of potted plants to abate pollutants by utilising small-scale *in-vitro* chambers with high VOC content has been monitored over time; however, the result restricts the implementation of this experiment in real-world building settings (Soreanu, 2016; Torpy et al., 2018). Different constructions are much more complex systems, where the interior has the facility to exchange air from the exterior and there are several sources of VOCs from the interior as well which contrast to the small closed chambers having no such criteria (Thomas et al., 2015). The concentration of VOCs in chamber experiments is much higher (where VOCs are measured in parts per million/ppm scale) than that of a normal building indoor ambient (where VOCs are measured in parts per billion/ppb scale) (Torpy et al., 2018). Another matter of concern regards the sustainability of the microorganisms present in soil substrates of potted plants that are being subjected to fluctuating levels of VOC content (Guieysse et al., 2008). Additionally, the potted plant system constrains their potential by limiting the rhizospheric microflora to come in exposure to indoor environments directly (Waring, 2016). Therefore, there are limitations to adapt the green-based technologies in buildings for purifying the indoor air in recent days.

9.6.5 Active Botanical Biofiltration Technology with Functional Green Wall

To enhance the potential exposure of indoor air contaminants to the substrate and rhizosperic microflora, the active biofiltration system has been well introduced that utilises active air flow through the plant foliage across the substrate (Darlington et al., 2001; Pettit et al., 2017; Irga et al., 2017a, b). The active green wall system has a great approach by means of a vertical alignment, increased plant density within a small area, and high efficacy to remediate indoor air quality by which the contaminated air can go through the foliage as well as substrate aiding mechanical ventilation (Soreanu, 2016). Unlike the standard ventilation technique, there is a possibility that active green wall technology has the potential to maintain standard IAQ, utilising the unique approach of biofiltration of recirculating air (Chen et al., 2005). However, development and alteration of conventional strategies is needed in order to achieve the final goal.

9.6.5.1 Potential of Active Green Wall for PM Removal

Advancements in research and technology have revealed a possible way to remove suspended particular matter by the use of active green walls. The plant substrate present in active green systems attributes potentials like a porous filter matrix. The active green system has a suction capacity of the air through the plant growth medium that can effectively screen out the PM contaminants from that air flow, unlike the common potted plant system that is only restricted to PM deposition method on foliage (Irga et al., 2017b). The PM filtration efficacy of an active green wall system can be investigated by screening the single-pass particulate filtration efficiency of that system (Irga et al., 2017b). The single-pass pollutant removal efficacy of this green wall technology at its best air flow rate have been recorded as $53.35 \pm 9.73\%$, $53.51 \pm 15.99\%$, and $48.21 \pm 14.71\%$ for total suspended particles, PM10 and PM2.5 respectively, which have improved their potency by the subsequent development of technology (Irga et al., 2017b). Green wall technology using fern species recorded the highest rate of PM removal capacity for all particle sizes; however, single-pass removal performance varied amongst systems occupied by various plant species (Pettit et al., 2019). Nevertheless, various investigations revealed a number of different factors (plant type, humidity, root volume, plant morphology) that significantly influence air flow as well as biofilter potential of a green wall system (Abdo et al., 2016; Irga et al., 2017a; Pettit et al., 2017, 2019).

9.6.5.2 Potential of Active Green Wall for VOC Removal

Despite using plants, some of the filters use rhizospheric microbes and substrate filters for VOC removal as system's microbes can readily degrade and substrates can efficiently adsorb VOCs from indoor environments (Rene et al., 2018). The activity of removing pollutants by rhizospheric microbial communities is increased by the influence of the plants present in active green biofilters (Xu et al., 2010). The mechanism of VOC removal differs in a way that firstly, it needs to be transferred into a liquid phase before it get diffused into a microbial cell, followed by its degradation and substrate adsorption (Darlington et al., 2000, 2001; Aizpuru et al., 2003; Karanfil and Dastgheib, 2004; Vikrant et al., 2017). The VOCs should dissolve into the aqueous phase more efficiently in the case of active biofiltration as the active botanical biofilters induce a pressure drop over the substrate membrane (Wang and Zhang, 2011; Pettit et al., 2017; Irga et al., 2017b). Due to the evident propensity of VOCs to escape through the aqueous phase into the ambient, the absorbent should be selected that can ensure greater retention of VOCs inside of the substrate matrix. According to the investigation performed by Wang and Zhang (2011), their botanical biofilter showed single pass removal efficiency of 91.7% and 98.7% for toluene and formaldehyde, respectively, by the utilisation of activated carbon as an adsorbent in the substrate (Wang and Zhang, 2011). The investigation regarding the selection of a proper substrate for active green biofilters has become a crucial matter of concern as the matter still remains unexplored.

TABLE 9.3

List of investigations showing the capability of green wall plants for indoor air pollution removal

Plant Species	Substrate	Type of Experiment	Size of Green Wall	Pollutant	References
Chlorophytum comosum Epipremnum aureum Gibasis sp. Philodendron xanadu Peperomia obtusifolia	Organic substrates	Active green wall	60 m^2	Bio-aerosol	Fleck et al., 2020
Chamaedorea elegans Epipremnum aureum Ficus lyrata Neomarica gracillis Peperomia obtusifolia Spathiphyllum wallisii Syngonium podophyllum	Coir fibre	Passive green wall	1.5 m^2	PMs, TVOC	Pettit et al., 2018a
Epipremnum aureum Nephrolepis exaltata Peperomia obtusifolia Schefflera arboricola Spathiphyllum wallisii	coconut husk	Active green wall	1.5 m^2	PMs, TVOC	Pettit et al., 2018a
Chlorophytum comosum 'variegatum' Chlorophytum orchidastrum Ficus lyrata Nematanthus glabra Nephrolepis cordifolia 'duffii' Nephrolepis exaltata 'bostoniensis' Schefflera amate Schefflera arboricola		Active green wall	500 × 500 × 130 mm	PMs, TVOC, NO$_2$	Paull et al., 2018
Chlorophytum orchidastrum Ficus lyrata Nematanthus glabra Nephrolepis cordifolia 'duffii' Nephropelis exaltata 'bostoniensis' Schefflera amate Schefflera arboricola	Coconut husk	Active green wall	0.25 m^2	PMs	Pettit et al., 2017
Chlorophytum comosum 'variegatum' Schefflera arboricola	Coconut husks and fibre	Active green wall	500 × 500 × 130mm	Gaseous and particulate air pollutants	Abdo et al., 2019
Chlorophytum orchidastrum Nematanthus glabra Nephrolepis cordifolia 'duffii'	Coconut coir	Active green wall	500 × 500 × 130mm	VOCs	Irga et al., 2019
Nephrolepis exaltata bostoniensis	Coconut coir activated carbon	Active green wall	0.6 × 0.6 × 0.6 m; 216 L	PMs, VOCs	Pettit et al., 2018b
Neomarica sp. Philodendron xanadu Peperomia sp. Large Peperomia sp. Small Gibasis sp. Epipremnum aureum Chlorophytum comosum	Coconut fibre, fertiliser (NPK 18:3:10) and watering	Active green wall	500 × 500 × 130mm	CO$_2$	Torpy et al., 2017

(Continued)

TABLE 9.3 (Continued)

List of investigations showing the capability of green wall plants for indoor air pollution removal

Plant Species	Substrate	Type of Experiment	Size of Green Wall	Pollutant	References
Chlorophytum comosum *Epipremnum aureum*	Coconut fibre, watering and liquid fertiliser	Passive mode, no added cross-ventilation; and two active modes	500 × 500 × 130 mm; 0.25 m^2	PMs, VOCs	Torpy and Zavattaro, 2018
Spathiphyllum wallisii *Syngonium podophyllu*	Coconut husk	Active botanical biofilter	PVC pipe: 100 mm diameter 2.80 m length	NO$_x$, O$_3$	Pettit et al., 2019
Syngonium podophyllum	Coconut husk	Active botanical biofilter	0.5 m × 0.5 m	VOCs	Pettit et al., 2019
Indoor landscaping with various plants along with two mosses (*Plagiomnium cuspidatum* and *Taxiphyllum deplanatum*)	Hyroponics, wetted lava rock	Biofiltration	0.27 m^2	Toluene, ethylbenzene, xylene	Darlington et al., 2001
Chlorophytum comosum	Coconut husk	Botanical biofilter	0.25 m^2	PMs	Irga et al., 2017b
Philodendron scandens *Philodendron scandens* 'Brazil' *Asplenium antiquum* *Syngonium podophyllum*	Inorganic growth media	Active green wall	1.5 m^2	VOCs	Torpy et al., 2018
Epipremnum aureum	Activated carbon and porous shale pebbles	Dynamic bonanical filtration	1.08 m^2	Formaldehyde and toluene	Wang and Zhang, 2011
Epipremnum aureum	Activated carbon and porous shale pebbles	Dynamic bonanical filtration	1.08 m^2	Formaldehyde	Wang et al., 2014

9.6.5.3 Potential of Active Green Wall for CO$_2$ Removal

It is thought to be impractical to filter excess concentrations of CO$_2$ from a building efficiently only by potted plants, whilst a green wall technology manages to provide a better surface area in the wall for greater and denser occupancy of the plants to fulfil the criteria in that regard (Irga et al., 2013). Su and Lin in their study (Su and Lin, 2015) revealed that efficient amounts of CO$_2$ removal (2,000 ppm to 800 ppm) were achieved by using a 5.72 m^2 indoor green wall inside a 38.88 m^2 room within two hours only. However, the substrate matrix of the green wall was kept isolated with the help of aluminium foil to avoid the influence of substrate respiration, which is harmful for plant health in the long run. Further investigation for selecting suitable CO$_2$ sequestering plants for the active green wall is highly needed.

9.7 Concluding Remarks and Future Prospects

The variation in characteristics and sources of air pollutants project a challenge to formulate adequate methodologies to combat outdoor and indoor air pollution. Location-specific remediation of pollutants needs to be implemented as the predominance of particular pollutants varies in different microenvironments considerably, projecting it as a complex scenario. The interaction between soil-microbe-plant attributes is promising approach for improvement of air quality by metabolizing, sequestering, or degrading air contaminants. The varieties of plant specimens, along with microorganisms, reveal their potential to remediate both the indoor as well as outdoor environment. Though the general population along with the scientific communities is much aware of the fact those plantations of trees cause multiple advantages, phytoremediation is still an emerging concept. The capability of plants to abate indoor air pollution has been overlooked for a long time period. The specific identification of plant species in alignment with the type of pollutant prevailing and specificity of the environment carries an importance in terms of phytoremediation of air contaminants, instead of considering their aesthetic values only. The indoor plants are exposed to particular air pollutants that are in higher concentrations, whereas the outdoor plants correspond to another scenario where they have to combat against the synergistic effect of several air pollutants, stresses, and transform themselves to get adapted to varying climatic conditions, which indicate the requirement of selection of different soil-plant–microbe systems for each different environment. Despite having profound importance, the phytoremediation technique deals with a few limitations as well. Due to the slow pollution removal process, the contaminants get more time to adhere and accumulate over the plant surface and continued exposure of the contaminant could cause phytotoxicity or other damaging impacts to the plant. Therefore, the phytoremediation of air pollution provides more scope of research in this field. Several studies have produced promising findings that have estimated the air treatment capabilities of potted plants along with active botanical biofilters, but the need for further research is still there. Advanced technologies and superior tools are required to be developed in order to maintain a parallel sync between urban forestry and city planning to get a new version of ecoarchitechture and sustainable developing world (Table 9.3).

Acknowledgements

AS acknowledges the receipt of financial assistance in the form of a research project (Reference No.F.30-393/2017 (BSR), dated September 27, 2018) funded by the University Grants Commission, MHRD, Government of India. The research project helped to develop the idea and outcome of this chapter.

REFERENCES

Abdo, P., Huynh, B.P., Irga, P.J., & Torpy, F.R. (2019). Evaluation of air flow through an active green wall biofilter. *Urban Forestry & Urban Greening*, *41*, 75–84.

Abdo, P., Huynh, B.P., Avakian, V., Nguyen, T., Torpy, F.R., & Irga, P.J. (2016). Measurement of air flow through a green-wall module. In Paper presented at Australasian fluid mechanics conference, Perth, Australia.

Adnan, Z., Mir, S., & Habib, M. (2017). Exhaust gases depletion using non-thermal plasma (NTP). *Atmospheric Pollution Research*, *8*, 338–343.

Agrawal, P., Sarkar, M., Chakraborty, B., & Banerjee, T. (2019). Phytoremediation of air pollutants: prospects and challenges. In *Phytomanagement of Polluted Sites*, 221–241. Elsevier.

AIRAH Group (2016). Improving Australian housing envelope integrity: a net benefit case for post contstruction fan pressurisation testing. In *Australian institute of Refrigeration Air Conditioning and Heating (AIRAH)*. https://www.airah.orgau/Content_Files/Industryresearch/Improving-AustralianHousing-Envelope-Integrity-20-10-16.pdf. Accessed 27 Sep 2017

Aizpuru, A., Malhautier, L., Roux, J.C., & Fanlo, J.L. (2003). Biofiltration of a mixture of volatile organic compounds on granular activated carbon. *Biotechnology and Bioengineering*, *83*(4), 479–488.

Amoatey, P., Omidvarborna, H., Baawain, M.S., & Al-Mamun, A. (2018). Indoor air pollution and exposure assessment of the gulf cooperation council countries: A critical review. *Environment International*, *121*(1), 491–506.

Andrade, A., & Dominski, F.H. (2018). Indoor air quality of environments used for physical exercise and sports practice: systematic review. *Journal of Environmental Management*, *206*, 577–586.

Andre, M., & Chagvardieff, P. (1997). CELSS research: interaction between space and terrestrial approaches in plant science. In E. Goto, K. Kurata, M. Hayashi, Sase S. (Eds.), *Plant Production in Closed Ecosystems: The International Symposium on Plant Production in Closed Ecosystems Held in Narita, Japan, August 26–29, 1996*, 245–261. Dordrecht: Springer.

Atkinson, R.W., Carey, I.M., Kent, A.J., van Staa, T.P., Anderson, H.R., & Cook, D.G. (2013). Long-term exposure to outdoor air pollution and incidence of cardiovascular diseases. *Epidemiology*, 44–53.

Atkinson, R.W., Fuller, G.W., Anderson, H.R., Harrison, R.M., & Armstrong, B. (2010). Urban ambient particle metrics and health: a time-series analysis. *Epidemiology*, 501–511.

Awa, S.H., Hadibarata, T. (2020) Removal of heavy metals in contaminated soil by phytoremediation mechanism: a review. *Water, Air, & Soil Pollution*, *231*(47), 1–15.

Aydogan, A., & Montoya, L.D. (2011). Formaldehyde removal by common indoor plant species and various growing media. *Atmospheric Environment*, *45*(16), 2675–2682.

Azuma, K., Ikeda, K., Kagi, N., Yanagi, U., & Osawa, H. (2018). Physicochemical risk factors for building-related symptoms in air-conditioned office buildings: ambient particles and combined exposure to indoor air pollutants. *Science of the Total Environment*, *616*, 1649–1655.

Bae, S.W., Roh, S.A., & Kim, S.D. (2006). NO removal by reducing agents and additives in the selective non-catalytic reduction (SNCR) process. *Chemosphere*, *65*(1), 170–175.

Baik, J.H., Yim, S.D., Nam, I.S., Mok, Y.S., Lee, J.H., Cho, B.K., & Oh, S.H. (2004). Control of NO x emissions from diesel engine by selective catalytic reduction (SCR) with urea. *Topics in Catalysis*, *30*(1–4), 37–41.

Baker, A.J.M., & Brooks, R.R. (1989). Terrestrial higher plants which hyperaccumulate metallic elements – a review of their distribution, ecology and phytochemistry. *Biorecovery*, *1*, 81–126.

Banerjee, T., & Srivastava, R.K. (2010). Estimation of the current status of floral biodiversity at surroundings of Integrated Industrial Estate-Pantnagar, India. *International Journal of Environmental Research*, *4*(1), 41–48.

Banerjee, T., Kumar, M., Mall, R.K., & Singh, R.S. (2017). Airing 'clean air' in clean India mission. *Environmental Science and Pollution Research*, *24*(7), 6399–6413.

Banerjee, T., Murari, V., Kumar, M., & Raju, M.P. (2015). Source apportionment of airborne particulates through receptor modeling: Indian scenario. *Atmospheric Research*, *164*, 167–187.

Bañuelos, G.S., Zambrzuski, S., & Mackey, B. (2000). Phytoextraction of selenium from soils irrigated with selenium-laden effluent. *Plant and Soil*, *224*(2), 251–258.

Bari, M.A., MacNeill, M., Kindzierski, W.B., Wallace, L., Héroux, M. È., & Wheeler, A.J. (2014). Predictors of coarse particulate matter and associated endotoxin concentrations in residential environments. *Atmospheric Environment*, *92*, 221–230.

Baur, P., & Schönherr, J. (1995). Temperature dependence of the diffusion of organic compounds across plant cuticles. *Chemosphere*, *30*, 1331–1340.

Beckett, K.P., Freer-Smith, P., & Taylor, G. (2000). Particulate pollution capture by urban trees: effect of species and wind speed. *Global Change Biology*, *6*, 995–1003. doi: 10.1046/j.1365-2486.2000.00376.x.

Berger, A., & Tricot, C. (1992). The greenhouse effect. *Geophysical Survey*, *13*, 523–549.

Bhargava, B., Malhotra, S., Chandel, A., Rakwal, A., Kashwap, R.R., & Kumar, S. (2021). Mitigation of indoor air pollutants using Areca palm potted plants in real-life settings. *Environmental Science and Pollution Research*, *28*(7), 8898–8906.

Bidwell, R.G.S., & Bebee, G.P. (1974). Carbon monoxide fixation by plants. *Canadian Journal of Botany*, *52*(8), 1841–1847.

Blanusa, T., Fantozzi, F., Monaci, F., & Bargagli, R. (2015). Leaf trapping and retention of particles by holm oak and other common tree species in Mediterranean urban environments. *Urban Forestry & Urban Greening*, *14*(4), 1095–1101.

Bogodage, S.G., & Leung, A.Y. (2015). CFD simulation of cyclone separators to reduce air pollution. *Powder Technology*, *286*, 488–506.

Bolan, N.S., Park, J.H., Robinson, B., Naidu, R., & Huh, K.Y. (2011). Phytostabilization: a green approach to contaminant containment. In *Advances in Agronomy, 112*, 145–204. Academic Press.

Boor, B.E., Spilak, M.P., Laverge, J., Novoselac, A., & Xu, Y. (2017). Human exposure to indoor air pollutants in sleep microenvironments: a literature review. *Building and Environment, 125*, 528–555.

Boulanger, G., Redaelli, M. 2015. Association of carbon dioxide with indoor air pollutants and exceedance of health guideline values. *Building and Environment, 93*, 115–124.

Bozkurt, Z., Doğan, G., Arslanbaş, D., Pekey, B., Pekey, H., Dumanoğlu, Y., Bayram, A., & Tuncel, G. (2015). Determination of the personal, indoor and outdoor exposure levels of inorganic gaseous pollutants in different microenvironments in an industrial city. *Environmental Monitoring and Assessment, 187*(9), 590.

Brack, C.L. (2002). Pollution mitigation and carbon sequestration by an urban forest. *Environment Pollution, 116*, 195–200.

Branco, P., Alvim-Ferraz, M., Martins, F., & Sousa, S. (2014). Indoor air quality in urban nurseries at Porto city: Particulate matter assessment. *Atmospheric Environment, 84*, 133–143.

Brilli, F., Fares, S., Ghirardo, A., de Visser, P., Calatayud, V., Muñoz, A., Annesi-Maesano, I., Sebastiani, F., Alivernini, A., Varriale, V., & Menghini, F. (2018). Plants for sustainable improvement of indoor air quality. *Trends in Plant Science, 23*, 507–512.

Brockherhoff, M.P. (2000). An urbanizing world. *Population Bulletin, 55*, 3–44.

Brunekreef, B. & Holgate, S.T. (2002). Air Pollution and Health. *Lancet*, 1470-2045 9341 360 1233 1342

Buonanno, G., Morawska, L., & Stabile, L. (2009). Particle emission factors during cooking activities. *Atmospheric Environment, 43*(20), 3235–3242.

Burroughs, H.E., & Hansen, S.J. (2004). Managing indoor air quality (Vol. 1). CRC Press. By *Chlorophytum comosum* L. plants. *Air Quality, Atmosphere, and Health, 8*, 265-272.

Carrer, P., Maroni, M., Alcini, D., & Cavallo, D. (2001). Allergens in indoor air: environmental assessment and health effects. *Science of the Total Environment, 270*(1–3), 33–42.

Castaño-Vinyals, G., Cantor, K.P., Malats, N., Tardon, A., Garcia-Closas, R., Serra, C., ... & Kogevinas, M. (2008). Air pollution and risk of urinary bladder cancer in a case-control study in Spain. *Occupational and Environmental Medicine, 65*(1), 56–60.

Chaparro-Suarez, I.G., Meixner, F.X., & Kesselmeier, J. (2011). Nitrogen dioxide (NO_2) uptake by vegetation controlled by atmospheric concentrations and plant stomatal aperture. *Atmospheric Environment, 45*, 5742–5750.

Chen, L.Y., Lin, M.W., & Chuah, Y.K. (2017). Investigation of a potted plant (*Hedera helix*) with photoregulation to remove volatile formaldehyde for improving indoor air quality. *Aerosol and Air Quality Research, 17*(10), 2543–2554.

Chen, L., Liu, C., Zhang, L., Zou, R., & Zhang, Z. (2017). Variation in tree species ability to capture and retain airborne fine particulate matter (PM 2.5). *Scientific Reports, 7*(1), 1–11.

Chen, S.J., Tian, M., Zheng, J., Zhu, Z.C., Luo, Y., Luo, & Mai, B.X. (2014). Elevated levels of enantioselective biotransformation of chiral PCBs in plants. *Environmental Science & Technology, 48* (7), 38473855.

Chen, W., Zhang, J.S., & Zhang, Z. (2005). Performance of air cleaners for removing multiple volatile organic compounds in indoor air. *ASHRAE transactions, 111*(1), 1101–1114.

Cheng, Y.H. (2017). Measuring indoor particulate matter concentrations and size distributions at different time periods to identify potential sources in an office building in Taipei City. *Building and Environment, 123*, 446–457.

Cheung, K., Daher, N., Kam, W., Shafer, M.M., Ning, Z., Schauer, J.J., & Sioutas, C. (2011). Spatial and temporal variation of chemical composition and mass closure of ambient coarse particulate matter (PM10–2.5) in the Los Angeles area. *Atmospheric Environment, 45*(16), 2651–2662.

Cho, K., Shibato, J., Kubo, A., Kohno, Y., Satoh, K., Kikuchi, S., Sarkar, A.,.... & Rakwal, R. (2013). Comparative analysis of seed transcriptomes of ambient ozone-fumigated 2 different rice cultivars. *Plant Signaling & Behavior, 8*(11), e26300.

Cho, K., Tiwari, S., Agrawal, S.B., Torres, N.L., Agrawal, M., Sarkar, A., ... & Rakwal, R. (2011). Tropospheric ozone and plants: absorption, responses, and consequences. *Reviews of Environmental Contamination and Toxicology, 212*, 61–111.

Chu, W., & Windawi, H. (1996). Control VOCs via catalytic oxidation. *Chemical engineering Progress, 92*(3).

Cohen, A.J., Brauer, M., Burnett, R., Anderson, H.R., Frostad, J., Estep, K., ... & Forouzanfar, M.H. (2017). Estimates and 25-year trends of the global burden of disease attributable to ambient air pollution: an analysis of data from the Global Burden of Diseases Study 2015. *The Lancet*, *389*(10082), 1907–1918.

Colbeck, I., & Nasir, Z.A. (2010). Indoor air pollution. In *Human Exposure to Pollutants via Dermal Absorption and Inhalation*, 41–72. Dordrecht: Springer.

Collins, J.J., Ireland, B., Buckley, C.F., & Shepperly, D. (2003). Lymphohaematopoeitic cancer mortality among workers with benzene exposure. *Occupational and Environmental Medicine*, *60*(9), 676–679.

Cook, L.L., Inouye, R.S., & McGonigle, T.P. (2009). Evaluation of four grasses for use in phytoremediation of Cs contaminated arid land soil. *Plant and Soil*, *324*, 169–184.

Cornejo, J.J., Munoz, F.G., Ma, C.Y., & Stewart, A.J. (1999). Studies on the decontamination of air by plants. *Ecotoxicology*, *8*, 311e320.

Coury, J.R., Pisani, R., & Hung, Y.T. (2004). Cyclones. In *Air Pollution Control Engineering*, *1*, 97–151. Totowa, NJ: Humana Press.

Coward, M. et al. (1996). Pilot study to assess the impact of green plants on NO_2 levels in homes. In *Building Research Establishment Note N154/96*. Watford, UK.

Crouse, D.L., Peters, P.A., van Donkelaar, A., Goldberg, M.S., Villeneuve, P.J., Brion, O., ... & Burnett, R.T. (2012). Risk of nonaccidental and cardiovascular mortality in relation to long-term exposure to low concentrations of fine particulate matter: a Canadian national-level cohort study. *Environmental Health Perspectives*, *120*(5), 708–714.

Cruz, M.D., Christensen, J.H., Thomsen, J.D., & Müller, R. (2014). Can ornamental potted plants remove volatile organic compounds from indoor air? – a review. *Environmental Science and Pollution Research*, *21*(24), 13909–13928.

Cunningham, S.D., & Ow, D.W. (1996). Promises and prospects of phytoremediation. *Plant Physiology*, *110*(3), 715.

Cunningham, S.D., Berti, W.R., & Huang, J.W. (1995). Phytoremediation of contaminated soils. *Trends in Biotechnology*, *13*, 393–397. doi: 10.1016/S0167-7799(00)88987-8.

Darcovich, K., Jonasson, K.A., & Capes, C.E. (1997). Developments in the control of fine particulate air emissions. *Advanced Powder Technology*, *8*(3), 179–215.

Darlington, A., Chan, M., Malloch, D., Pilger, C., & Dixon, M. (2000). The biofiltration of indoor air: implications for air quality. *Indoor Air*, *10*(1), 39–46.

Darlington, A.B., Dat, J.F., & Dixon, M.A. (2001). The biofiltration of indoor air: air flux and temperature influences the removal of toluene, ethylbenzene, and xylene. *Environmental Science & Technology*, *35*(1), 240–246.

Das, J., Rene, E.R., Dupont, C., Dufourny, A., Blin, J., & van Hullebusch, E.D. (2019). Performance of a compost and biochar packed biofilter for gas-phase hydrogen sulfide removal. *Bioresource Technology*, *273*, 581–591.

Das, S., & Prasad, P. (2012). Particulate matter capturing ability of some plant species: implication for phytoremediation of particulate pollution around Rourkela steel plant, Rourkela, India. *Nature, Environment and Pollution Technology*, *11*(4), 657–665.

Das, S., Bharati, A.M., Kamilya, P., & Sarkar, A. (2020). Responses of some herbaceous road-side wild plant species against ambient particulate pollution (PM_{10}): A case study from Malda district, West Bengal. *Pleione*, *14*(2), 311–322.

Das, S., Biswas, A., & Sarkar, A. (2017). Air pollution toxicity and global agricultural perspective: a brief review. In P.K. Padhy, P.K. Patra, & U.K. Singh (Eds.), *Emerging Issues in Environmental Science: Concerns and Management*, chapter 7, 75–89. New Delhi: New Delhi Publishers. ISBN: 978-93-85503-77-1

Das, S., Kamilya, P., & Sarkar, A. (2018). Evaluation of plant responses under higher vehicular pollution alongside National Highway-34 within Malda District, West Bengal. In A. Chakravorty *Dynamic in Biology*, chapter-17, 152–161. New Delhi: Akinik publications.

Das, S., Pal, D., & Sarkar, A. (2021). Particulate Matter Pollution and Global Agricultural Productivity. *Sustainable Agriculture Reviews 50: Emerging Contaminants in Agriculture*, 79–107.

de Gennaro, G., Loiotile, A.D., Fracchiolla, R., Palmisani, J., Saracino, M.R., Tutino, M. (2015). Temporal variation of VOC emission from solvent and water based wood stains. *Atmospheric Environment*, *115*, 53–61.

De Kempeneer, L., Sercu, B., Vanbrabant, W., Van Langenhove, H., & Verstraete, W. (2004). Bioaugmentation of the phyllosphere for the removal of toluene from indoor air. *Applied Microbiology & Biotechnology*, *64*, 284–288.

Dela Cruz, M., Christensen, J.H., Thomsen, J.D., & Müller, R. (2014a). Can ornamental potted plants remove volatile organic compounds from indoor air? – a review. *Environmental Science and Pollution Research*, *21*, 13909–13928.

Dela Cruz, M., Müller, R., Svensmark, B., Pedersen, J.S., & Christensen, J.H. (2014b). Assessment of volatile organic compound removal by indoor plants – a novel experimental setup. *Environmental Science and Pollution Research*, *21*, 7838–7846.

Deng, N., He, H., Yan, J., Zhao, Y., Ticha, E.B., Liu, Y., ... & Cheng, B. (2019). One-step melt-blowing of multi-scale micro/nano fabric membrane for advanced air-filtration. *Polymer*, *165*, 174–179.

DiGiovanni, F., & Fellin, P. (2006). *Transboundary Air Pollution*. Oxford, UK: Eolss Publishers.

Donovan, R.P. (2020). *Fabric Filtration for Combustion Sources: Fundamentals and Basic Technology*. CRC Press.

Du, W., Li, X., Chen, Y., & Shen, G. (2018). Household air pollution and personal exposure to air pollutants in rural China – a review. *Environmental Pollution*, *237*, 625–638.

Duan K., Sun G., Zhang Y., Yahya K., Wang K.M., Madden J.M. (2017). Impact of air pollution induced climate change on water availability and ecosystem productivity in the conterminous United States. *Climate Change*, *140*, 259–272.

Dushenkov, V., Kumar, P.N., Motto, H., & Raskin, I. (1995). Rhizofiltration: the use of plants to remove heavy metals from aqueous streams. *Environmental Science & Technology*, *29*(5), 1239–1245.

Dzantor, E.K. (2007). Phytoremediation: the state of rhizosphere 'engineering' for accelerated rhizodegradation of xenobiotic contaminants. *Journal of Chemical Technology & Biotechnology: International Research in Process, Environmental & Clean Technology*, *82*(3), 228–232.

Dzierżanowski, K., & Gawroński, S.W. (2011). Use of trees for reducing particulate matter pollution in air. *Challenges of Modern Technology*, *2*(1).

Dzierzanowski, K., Popek, R., Gawronska, H., Saebo, A., & Gawronski, S.W. (2011). Deposition of particulate matter of different size fractions on leaf surface and waxes of urban forest species. *International Journal of Phytoremediation*, *13*, 1037–1046.

EPA. (1998). *A Citizen's Guide to Phytoremediation*. Available from: http://cluin.org/PRODUCTS/CITGUIDE/Phyto2.htm.

EPA. (2000). *Introduction to Phytoremediation*. National Risk Management Research Laboratory. US Environmental Protection Agency. Available from: http://www.clu-in.org/download/remed/introphyto.pdf.

Esworthy, R. (2013). Air quality: EPA's 2013 changes to the particulate matter (PM) standard. *Library of Congress, Congressional Research Service*.

Fang, M., Yi, N., Di, W., Wang, T., & Wang, Q. (2020). Emission and control of flue gas pollutants in CO2 chemical absorption system – a review. *International Journal of Greenhouse Gas Control*, *93*, 102904.

Faustini, A., Stafoggia, M., Colais, P., Berti, G., Bisanti, L., Cadum, E., ... & Forastiere, F. (2013). Air pollution and multiple acute respiratory outcomes. *European Respiratory Journal*, *42*(2), 304–313.

Favas, P.J., Pratas, J., Varun, M., D'Souza, R., & Paul, M.S. (2014). Phytoremediation of soils contaminated with metals and metalloids at mining areas: potential of native flora. *Environmental Risk Assessment of Soil Contamination*, *3*, 485–516.

Feng, Y., Ning, M., Lei, Y., Sun, Y., Liu, W., & Wang, J. (2019). Defending blue sky in China: effectiveness of the "air pollution prevention and control action plan" on air quality improvements from 2013 to 2017. *Journal of Environmental Management*, *252*, 109603.

Fleck, R., Gill, R.L., Pettit, T., Irga, P.J., Williams, N.L.R., Seymour, J.R., & Torpy, F.R. (2020). Characterisation of fungal and bacterial dynamics in an active green wall used for indoor air pollutant removal. *Building and Environment*, *179*, 106987.

Fowler, D., Cape, J.N., & Unsworth, M.H. (1989). Deposition of atmospheric pollutants on forests. *Philosophical Transactions of the Royal Society of London. B, Biological Sciences*, *324*(1223), 247–265.

Francke, K.P., Miessner, H., & Rudolph, R. (2000). Plasmacatalytic processes for environmental problems. *Catalysis Today*, *59*, 411–416.

Franklin, B.A., Brook, R., & Pope III, C.A. (2015). Air pollution and cardiovascular disease. *Current Problems in Cardiology*, *40*(5), 207–238.

Freer-Smith, P.H., El-Khatib, A.A., & Taylor, G. (2004). Capture of particulate pollution by trees: a comparison of species typical of semi-arid areas (*Ficus nitida* and *Eucalyptus globulus*) with European and North American species. *Water, Air, and Soil Pollution*, *155*(1), 173–187.

Fromme, H., Twardella, D., Dietrich, S., Heitmann, D., Schierl, R., Liebl, B., & Rüden, H. (2007). Particulate matter in the indoor air of classrooms – exploratory results from Munich and surrounding area. *Atmospheric Environment*, *41*(4), 854–866.

Fuentes-Leonarte, V., Tenías, J.M., & Ballester, F. (2009). Levels of pollutants in indoor air and respiratory health in preschool children: a systematic review. *Pediatric Pulmonology*, *44*(3), 231–243.

Fulekar, M.H., Singh, A., Thorat, V., Kaushik, C.P., & Eapen, S. (2010). Phytoremediation of 137 Cs from low level nuclear waste using *Catharanthus roseus*. *Indian Journal of Pure & Applied Physics*, *48*(07), 516–519.

Gaihre, S., Semple, S., Miller, J., Fielding, S., & Turner, S. (2014). Classroom carbon dioxide concentration, school attendance, and educational attainment. *Journal of School Health*, *84*(9), 569–574.

Gao, X., Liu, S., Zhang, Y., Luo, Z., Ni, M., & Cen, K. (2011). Adsorption and reduction of NO2 over activated carbon at low temperature. *Fuel Processing Technology*, *92*, 139–146.

Gawrońska, H., & Bakera, B.(2015). Phytoremediation of particulate matter from indoor air by *Chlorophytum comosum* L. plants. *Air Quality, Atmosphere & Health*, *8*(3), 265–272.

Gawronski, S.W., Gawronska, H., Lomnicki, S., Sæbo, A., & Vangronsveld, J. (2017). Chapter eight – plants in air phytoremediation. In A. Cuypers & J. Vangronsveld (Eds.), *Advances in Botanical Research*, 319–346. Academic Press.

Genc, S., Zadeoglulari, Z., Fuss, S.H., & Genc, K. (2012). The adverse effects of air pollution on the nervous system. *Journal of Toxicology*, *2012*.

Gerhardt, K.E., Huang, X.-D., Glick, B.R., Bruce, M., & Greenberg, B.M. (2009). Phytoremediation and rhizoremediation of organic soil contaminants: potential and challenges. *Plant Science*, *176*, 20–30.

Geßler, A., Rienks, M., & Rennenberg, H. (2000). NH$_3$ and NO$_2$ fluxes between beechtrees and the atmosphere ecorrelation with climatic and physiologicalparameters. *New Phytologist*, *147*, 539–560.

Gheorghe, I.F., & Ion, B. (2011). The effects of air pollutants on vegetation and the role of vegetation in reducing atmospheric pollution. *The Impact of Air Pollution on Health, Economy, Environment and Agricultural Sources*, 241–280.

Giese, M., Bauer-Doranth, U., Langebartels, C., & Sandermann Jr, H. (1994). Detoxification of formaldehyde by the spider plant (*Chlorophytum comosum* L.) and by soybean (*Glycine max* L.) cell-suspension cultures. *Plant Physiology*, *104*(4), 1301–1309.

Giles, L.V., Barn, P., Künzli, N., Romieu, I., Mittleman, M.A., van Eeden, S., ... & Brauer, M. (2011). From good intentions to proven interventions: effectiveness of actions to reduce the health impacts of air pollution. *Environmental Health Perspectives*, *119*(1), 29–36.

Godish, T., & Guindon, C. (1989). An assessment of botanical air purification as a formaldehyde mitigation measure under dynamic laboratory chamber conditions. *Environmental Pollution*, *62*, 13–20.

Goldstein, A.H., Nazaroff, W.W., Weschler, C.J., & Williams, J. (2020). How do indoor environments affect air pollution exposure? *Environmental Science & Technology*.

Goodman, N.B., Steinemann, A., Wheeler, A.J., Paevere, P.J., Cheng, M., & Brown, S.K. (2017). Volatile organic compounds within indoor environments in Australia. *Building and Environment*, *122*, 116–125.

Goyal, R., & Kumar, P. (2013). Indoor–outdoor concentrations of particulate matter in nine microenvironments of a mix-use commercial building in megacity Delhi. *Air Quality, Atmosphere & Health*, *6*(4), 747–757.

Guieysse, B., Hort, C., Platel, V., Munoz, R., Ondarts, M., & Revah, S. (2008). Biological treatment of indoor air for VOC removal: potential and challenges. *Biotechnology Advances*, *26*, 398–410.

Guo, Y., Li, Y., Zhu, T., & Ye, M. (2015). Investigation of SO$_2$ and NO adsorption species on activated carbon and the mechanism of NO promotion effect on SO$_2$. *Fuel*, *143*, 536–542.

Guo, Z., Xie, Y., Hong, I., & Kim, J. (2001). Catalytic oxidation of NO to NO$_2$ on activated carbon. *Energy Conversion and Management*, *42*, 2005–2018.

Gupta, V.K., & Verma, N. (2002). Removal of volatile organic compounds by cryogenic condensation followed by adsorption. *Chemical Engineering Science*, *57*(14), 2679–2696.

Gurjar, B.R., Ravindra, K., & Nagpure, A.S. (2016). Air pollution trends over India megacities and their local-to-global implications. *Atmospheric Environment*, *142*, 475–495.

He, C., Morawska, L., & Taplin, L. (2007). Particle emission characteristics of office printers. *Environmental Science & Technology*, *41*(17), 6039–6045.

He, C., Morawska, L., Hitchins, J., & Gilbert, D. (2004). Contribution from indoor sources to particle number and mass concentrations in residential houses. *Atmospheric Environment*, *38*(21), 3405–3415.

He, F., Gao, J., Pierce, E., Strong, P.J., Wang, H., & Liang, L. (2015). In situ remediation technologies for mercury-contaminated soil. *Environmental Science and Pollution Research*, *22*(11), 8124–8147.

Health Effects Institute. (2019). *State of Global Air 2019*.

Heather, L.P., Holcomb, E.J., Best, T.O., & Decoteau, D.R. (2009). Effectiveness of houseplants in reducing the indoor air pollutant ozone. *HortTechnology*, *19*(2), 286–290.

Heaton, A.C., Rugh, C.L., Wang, N.J., & Meagher, R.B. (1998). Phytoremediation of mercury-and methylmercury-polluted soils using genetically engineered plants. *Journal of Soil Contamination*, *7*(4), 497–509.

Ho, L.A., & Kuschner, W.G. (2012). Respiratory health in home and leisure pursuits. *Clinics in Chest Medicine*, *33*(4), 715–729.

Ho, Y.N., Hsieh, J.L., & Huang, C.C. (2013). Construction of a plant–microbe phytoremediation system: combination of vetiver grass with a functional endophytic bacterium, Achromobacter xylosoxidans F3B, for aromatic pollutants removal. *Bioresource Technology*, *145*, 43–47.

Hörmann, V., Brenske, K.R., & Ulrichs, C. (2017). Suitability of test chambers for analyzing air pollutant removal by plants and assessing potential indoor air purification. *Water, Air, & Soil Pollution*, *228*(10), 1–13.

Huang, R.J., Zhang, Y., Bozzetti, C., Ho, K.F., Cao, J.J., Han, Y., ... & Prévôt, A.S. (2014). High secondary aerosol contribution to particulate pollution during haze events in China. *Nature*, *514*(7521), 218–222.

Huang, Y., Ho, S.S.H., Lu, Y., Niu, R., Xu, L., Cao, J., & Lee, S. (2016). Removal of indoor volatile organic compounds via photocatalytic oxidation: a short review and prospect. *Molecules*, *21*(1), 56.

India State-Level Disease Burden Initiative Air Pollution Collaborators. (2019). The impact of air pollution on deaths, disease burden and life expectancy across the states of India: the Global Burden Of Disease Study 2017. *Lancet Planet Health*, *3*, e26–e39.

International Agency for Research on Cancer. (2013). *Outdoor Air Pollution a Leading Environmental Cause of Cancer Deaths, No. 221*. World Health Organization. Available online at: https://www.iarc.fr/en/media-centre/iarcnews/pdf/pr221_E.pdf.

Irga, P.J., Pettit, T.J., & Torpy, F.R. (2018). The phytoremediation of indoor air pollution: a review on the technology development from the potted plant through to functional green wall biofilters. *Reviews in Environmental Science and Bio/Technology*, *17*(2), 395–415.

Irga, P.J., Pettit, T., Irga, R.F., Paull, N.J., Douglas, A.N., & Torpy, F.R. (2019). Does plant species selection in functional active green walls influence VOC phytoremediation efficiency?. *Environmental Science and Pollution Research*, *26*(13), 12851–12858.

Irga, P., Torpy, F., & Burchett, M. (2013). Can hydroculture be used to enhance the performance of indoor plants for the removal of air pollutants? *Atmospheric Environment*, *77*, 267–271.

Irga, P.J., Abdo, P., Zavattaro, M., & Torpy, F.R. (2017a). An assessment of the potential fungal bioaerosol production from an active living wall. *Building and Environment*, *111*, 140–146.

Irga. P.J., Paull, N.J., Abdo, P., & Torpy, F.R. (2017b) An assessment of the atmospheric particle removal efficiency of an in-room botanical biofilter system. *Building and Environment*, *115*, 281–290.

Jafari, M.J., Khajevandi, A.A., Najarkola, S.A.M., Yekaninejad, M.S., Pourhoseingholi, M.A., Omidi, L., & Kalantary, S. (2015). Association of sick building syndrome with indoor air parameters. *Tanaffos*, *14*(1), 55.

Jafarinejad, S. (2017). 5 – control and treatment of air emissions. In S. Jafarinejad (Ed.), *Petroleum Waste Treatment and Pollution Control*, 149–183. Butterworth-Heinemann.

Jamrozik, E., & Musk, A.W. (2011). Respiratory Health Issues in the Asia-Pacific Region: An Overview. *Respirology*, 1440-1843 1 16 3 12

Javed, M.T., Irfan, N., & Gibbs, B.M. (2007). Control of combustion-generated nitrogen oxides by selective non-catalytic reduction. *Journal of Environmental Management*, *83*(3), 251–289.

Jin, Y., Andersson, H., & Zhang, S. (2016). Air pollution control policies in China: a retrospective and prospects. *International Journal of Environmental Research and Public Health*, *13*(12), 1219.

Jindrova, E., Chocova, M., Demnerova, K., & Brenner, V. (2002). Bacterial aerobic degradation of benzene, toluene, ethylbenzene and xylene. *Folia Microbiologica*, *47*(2), 83–93.

Johansson, C., Norman, M., & Gidhagen, L. (2007). Spatial & temporal variations of PM10 and particle number concentrations in urban air. *Environmental Monitoring and Assessment*, *127*(1), 477–487.

Juda-Rezler, K., Reizer, M., & Oudinet, J.P. (2011). Determination and analysis of PM10 source apportionment during episodes of air pollution in Central Eastern European urban areas: The case of wintertime 2006. *Atmospheric Environment*, *45*(36), 6557–6566.

Junaid, M., Syed, J.H., Abbasi, N.A., Hashmi, M.Z., Malik, R.N., & Pei, D.S. (2018). Status of indoor air pollution (IAP) through particulate matter (PM) emissions and associated health concerns in South Asia. *Chemosphere*, *191*, 651–663.

Juwarkar, A.A., Singh, S.K., & Mudhoo, A. (2010). A comprehensive overview of elements in bioremediation. *Reviews in Environmental Science and Bio/Technology*, *9*(3), 215–288.

Kampa, M., & Castanas, E. (2008). Human health effects of air pollution. *Environmental Pollution*, *151*, 362–367.

Kaplan, G.G., Tanyingoh, D., Dixon, E., Johnson, M., Wheeler, A.J., Myers, R.P., ... & Villeneuve, P.J. (2013). Ambient ozone concentrations and the risk of perforated and nonperforated appendicitis: a multicity case-crossover study. *Environmental Health Perspectives*, *121*(8), 939–943.

Karami, A., & Shamsuddin, Z.H. (2010). Phytoremediation of heavy metals with several efficiency enhancer methods. *African Journal of Biotechnology*, *9*(25), 3689–3698.

Karanfil, T., & Dastgheib, S.A. (2004). Trichloroethylene adsorption by fibrous and granular activated carbons: aqueous phase, gas phase, and water vapor adsorption studies. *Environmental Science & Technology*, *38*(22), 5834–5841.

Khaksar,G., Treesubsuntorn. C., & Thiravetyan, P. (2016a). Effect of endophytic Bacillus cereus ERBP inoculation into nonnative host: potentials and challenges for airborne formaldehyde removal. *Plant Physiology and Biochemistry*, *107*, 326–336

Khaksar, G., Treesubsuntorn, C., & Thiravetyan, P. (2016b). Endophytic *Bacillus cereus* ERBP-Clitoria ternatea interactions: potentials for the enhancement of gaseous formaldehyde removal. *Environtal Experimental Botany*, *126*, 10–20.

Khallaf, M. (Ed.). (2011). *The Impact of Air Pollution on Health, Economy, Environment and Agricultural Sources*. BoD – Books on Demand.

Khanchi, A., Hebbern, C.A., Zhu, J., & Cakmak, S.(2015). Exposure to volatile organic compounds and associated health risks in windsor, Canada. *Atmospheric Environment*, *120*, 152–159.

Kim, H., Kim, J., Kim, S., Kang, S.H., Kim, H.J., Kim, H., ... & Chae, I.H. (2017). Cardiovascular effects of long-term exposure to air pollution: a population-based study with 900 845 person-years of follow-up. *Journal of the American Heart Association*, *6*(11), e007170.

Kim, K.J., & Lee, D.W. (2008). Efficiency of volatile formaldehyde removal of orchids as affected by species and crassulacean acid metabolism (CAM) nature. *Horticulture Environmrnt and Biotechnology*, *49*(2), 132–137.

Kim, K.J., Jeong, M.I., Lee, D.W., Song, J.S., Kim, H.D., Yoo, E.H., ... & Kim, H.H. (2010). Variation in formaldehyde removal efficiency among indoor plant species. *HortScience*, *45*(10), 1489–1495.

Kim, K.J., Jung, H.H., Seo, H.W., Lee, J.A., & Kays, S.J. (2014). Volatile toluene and xylene removal efficiency of foliage plants as affected by top to root zone size. *HortScience*, *49*(2), 230–234.

Kim, K.J., Khalekuzzaman, M., Suh, J.N., Kim, H.J., Shagol, C., Kim, H.H., & Kim, H.J. (2018). Phytoremediation of volatile organic compounds by indoor plants: a review. *Horticulture, Environment, and Biotechnology*, *59*(2), 143–157.

Kim, K.J., Kil, M.J., Jeong, M.I., Kim, H.D., Yoo, E.H., Jeong, S.J., ... & Son, K.C. (2009). Determination of the efficiency of formaldehyde removal according to the percentage volume of pot plants occupying a room. *Korean Journal of Horticultural Science and Technology*, *27*, 305–311.

Kim, K.J., Kil, M.J., Song, J.S., Yoo, E.H., Son, K.C., & Kays, S.J. (2008). Efficiency of volatile formaldehyde removal by indoor plants: contribution of aerial plant parts versus the root zone. *Journal of the American Society for Horticultural Science*, *133*(4), 521–526.

Kim, K.J., Kim, H.J., Khalekuzzaman, M., Yoo, E.H., Jung, H.H., & Jang, H.S. (2016). Removal ratio of gaseous toluene and xylene transported from air to root zone via the stem by indoor plants. *Environmental Science and Pollution Research*, 23(7), 6149–6158.

Kosusko, M., & Nunez, C.M. (1990). Destruction of volatile organic compounds using catalytic oxidation. *Journal of the Air & Waste Management Association*, 40(2), 254–259.

Kumar, M., Singh, R.K., Murari, V., Singh, A.K., Singh, R.S., & Banerjee, T., (2016). Fireworks induced particle pollution: a spatio-temporal analysis. *Atmospheric Research*, 180, 78–91.

Kumar, M., Singh, R.S., & Banerjee, T. (2015). Associating airborne particulates and human health: exploring possibilities. *Environment International*, 84, 201–202.

Kumar, R., & Gupta, P. (2016). *Air Pollution Control Policies and Regulations* (pp. 133–149).

Kvesitadze, G., Khatisashvili, G., Sadunishvili, T., & Ramsden, J.J. (2006). *Biochemical Mechanisms of Detoxification in Higher Plants: Basis of Phytoremediation*. Springer Science & Business Media.

Lai, J.K., & Wachs, I.E. (2018). A perspective on the selective catalytic reduction (SCR) of NO with NH_3 by supported V2O5–WO3/TiO2 catalysts. *ACS Catalysis*, 8(7), 6537–6551.

Landrigan, P.J. (2017). Air pollution and health. *The Lancet Public Health*, 2(1), e4–e5.

Lasat, M. (2001). Phytoextraction of toxic metals: a review of biological mechanisms. *Journal of Environmental Quality*, 31, 109–120.

Lee, J.H., & Sim, W.K. (1999). Biological absorption of SO_2 by Korean native indoor species. In: M.D. Burchett et al. (Eds.), *Towards a New Millennium in People-Plant Relationships, Contributions from International People Plant Symposium*, 101–108. Sydney.

Lee, M., & Yang, M. (2010). Rhizofiltration using sunflower (*Helianthus annuus* L.) and bean (*Phaseolus vulgaris* L. var. *vulgaris*) to remediate uranium contaminated groundwater. *Journal of Hazardous Materials*, 173(1-3), 589–596.

Lemus, J., Martin-Martinez, M., Palomar, J., Gomez-Sainero, L., Gilarranz, M.A., & Rodriguez, J.J. (2012). Removal of chlorinated organic volatile compounds by gas phase adsorption with activated carbon. *Chemical Engineering Journal*, 211, 246–254.

Letter, C., & Jäger, G. (2020). Simulating the potential of trees to reduce particulate matter pollution in urban areas throughout the year. *Environment, Development and Sustainability*, 22(5), 4311–4321.

Leung, D.Y. (2015). Outdoor-indoor air pollution in urban environment: challenges and opportunity. *Frontiers in Environmental Science*, 2, 69.

Lewandowski, D.A. (1999). *Design of Thermal Oxidation Systems for Volatile Organic Compounds*. CRC Press.

Li, Z., Wen, Q., & Zhang, R. (2017). Sources, health effects and control strategies of indoor fine particulate matter (PM 2.5): A review. *Science of the Total Environment*.

Lim, Y.W., Kim, H.H., Yang, J.Y., Kim, K.J., & Lee, J.Y. (2009). Improvement of indoor air quality by house plants in new-built apartment buildings. *Journal of the Japanese Society for Horticultural*, 78(4), 456–462.

Limmer, M., & Burken, J. (2016). Phytovolatilization of organic contaminants. *Environmental Science & Technology*, 50(13), 6632–6643.

Lindgren, T. (2010). A case of indoor air pollution of ammonia emitted from concrete in a newly built office in Beijing. *Building and Environment*, 45(3), 596–600.

Liqun, G., & Yanqun, G. (2011). Study on building materials and indoor pollution. *Procedia Engineering*, 21, 789–794.

Liu, C., Hsu, P.C., Lee, H.W., Ye, M., Zheng, G., Liu, N., & Cui, Y. (2015). Transparent air filter for high-efficiency PM 2.5 capture. *Nature Communications*, 6(1), 1–9.

Liu, J., Cao, Z., Zou, S., Liu, H., Hai, X., Wang, S., ... & Jia, Z. (2018). An investigation of the leaf retention capacity, efficiency and mechanism for atmospheric particulate matter of five greening tree species in Beijing, China. *Science of the Total Environment*, 616, 417–426.

Liu, J., Mauzerall, D.L., Chen, Q., Zhang, Q., Song, Y., & Peng, W., et al. (2016). Air polltuant emissions from Chinese households: a major and underappreciated ambient pollutant source. *Proceedings of the National Academy of Sciences of the United States of America*, 113, 7756–7761.

Liu, Y.J., Mu, Y.J., Zhu, Y.G., Ding, H., & Arens, N.C. (2007). Which ornamental plant species effectively remove benzene from indoor air? *Atmospheric Environment*, 41(3), 650–654.

Liu, Y.-J., & Ding, H.U.I. (2008). Variation in air pollution tolerance index of plants near a steel factory: Implication for landscape-plant species selection for industrial areas. *WSEAS Transactions on Environment and Development*, 4(1), 24–32.

Llewellyn, D., & Dixon, M. (2011). 4.26 Can plants really improve indoor air quality. *Comprehensive Biotechnology* (2nd Edition), 331–338. Burlington: Academic Press.

Lohr, V.I., & Pearson-Mims, C.H. 1996. Particulate matter accumulation on horizontal surfaces in interiors: influence of foliage plants. *Atmospheric Environment*, 30(14), 2565–2568.

Lone, M.I., He, Z.L., Stoffella, P.J., & Yang, X.E. (2008). Phytoremediation of heavy metal polluted soils and water: progresses and perspectives. *Journal of Zhejiang University Science B*, 9(3), 210–220.

Long, C.M., Suh, H.H., & Koutrakis, P. (2000). Characterization of indoor particle sources using continuous mass and size monitors. *Journal of the Air & Waste Management Association*, 50(7), 1236–1250.

Luengas, A., Barona, A., Hort C., Gallastegui, G., Platel, V., & Elias, A. (2015). A review of indoor air treatment technologies. *Reviews in Environmental Science and Bio/Technology*, 14(3), 499–522.

Ma, Y., Oliveira, R.S., Freitas, H., & Zhang, C. (2016). Biochemical and molecular mechanisms of plant-microbe-metal interactions: revelance for phytoremediation. *Frontiers in Plant Science*, 7, 918. doi: 10.3389/fpls.2016.00918.

Macpherson, A.J., Simon, H., Langdon, R., & Misenheimer, D. (2017). A mixed integer programming model for National Ambient Air Quality Standards (NAAQS) attainment strategy analysis. *Environmental Modelling & Software*, 91, 13–27.

Madureira, J., Paciência, I., & Fernandes, E.D.O. (2012). Levels and indoor–outdoor relationships of size-specific particulate matter in naturally ventilated Portuguese schools. *Journal of Toxicology and Environmental Health, Part A*, 75(22-23), 1423–1436.

Mahendra, A., & Seto, K.C. (2019). *Upward and Outward Growth: Managing Urban Expansion for More Equitable Cities in the Global South.*

Maia, G.D., Gates, R.S., & Taraba, J.L. (2012). Ammonia biofiltration and nitrous oxide generation during the start-up of gas-phase compost biofilters. *Atmospheric Environment*, 46, 659–664.

Manes, F., Incerti, G., Salvatori, E., Vitale, M., Ricotta, C., & Costanza, R. (2012). Urban ecosystem services: tree diversity and stability of tropospheric ozone removal. *Ecological Applications*, 22(1), 349–360.

Manisalidis, I., Stavropoulou, E., Stavropoulos, A., & Bezirtzoglou, E. (2020). Environmental and health impacts of air pollution: a review. *Frontiers in Public Health*, 8.

Mate, A.R., & Deshmukh, R.R. (2015). To control effects of air pollution using roadside trees. *International Journal of Innovative Science Engineering and Technology*, 4(11), 1116711171.

McCormack, M.C., Breysse, P.N., Hansel, N.N., Matsui, E.C., Tonorezos, E.S., Curtin-Brosnan, J., ... & Diette, G.B. (2008). Common household activities are associated with elevated particulate matter concentrations in bedrooms of inner-city Baltimore pre-school children. *Environmental Research*, 106(2), 148–155.

McDonald, A. G., Bealey, W. J., Fowler, D., Dragosits, U., Skiba, U., Smith, R. I., & Nemitz, E. (2007). Quantifying the effect of urban tree planting on concentrations and depositions of PM10 in two UK conurbations. *Atmospheric Environment*, 41(38), 8455–8467.

McGwin, Jr, G., Lienert, J., & Kennedy, Jr, J.I. (2010). Formaldehyde exposure and asthma in children: a systematic review. *Environmental Health Perspectives*, 118(3), 313.

Mhawish, A., Banerjee, T., Broday, D.M., Misra, A., & Tripathi, S.N. (2017). Evaluation of MODIS Collection 6 aerosol retrieval algorithms over Indo-Gangetic Plain: Implications of aerosols types and mass loading. *Remote Sensing of Environment*, 201, 297–313.

Mhawish, A., Kumar, M., Mishra, A.K., Srivastava, P.K., & Banerjee, T. (2018). remote sensing of aerosols from space: retrieval of properties and applications. In: Islam, et al., (Eds.), *Remote Sensing of Aerosols, Clouds, and Precipitation*, 45–83. Amsterdam, The Netherland: Elsevier Inc. ISBN 978-0-12-810437-8.

Miller, K.A., Siscovick, D.S., Sheppard, L., Shepherd, K., Sullivan, J.H., Anderson, G.L., & Kaufman, J.D. (2007). Long-term exposure to air pollution and incidence of cardiovascular events in women. *New England Journal of Medicine*, 356(5), 447–458.

Misra, C., Geller, M.D., Shah, P., Sioutas, C., & Solomon, P.A. (2001). Development and Evaluation of a Continuous Coarse (PM10–PM25) Particle Monitor. *Journal of the Air & Waste Management Association*, 51(9), 1309–1317.

Moeckel, C., Thomas, G.O., Barber, J.L. & Jones, K.C. (2008). Uptake and storage of PCBs by plant cuticles. *Environmental Science & Technology*, *42*(1), 100105.

Mohammadyan, M., Keyvani, S., Bahrami, A., Yetilmezsoy, K., Heibati, B., & Pollitt, K.J.G. (2019). Assessment of indoor air pollution exposure in urban hospital microenvironments. *Air Quality, Atmosphere & Health*, *12*(2), 151–159.

Molloy, S.B., Cheng, M., Galbally, I.E., Keywood, M.D., Lawson, S.J., Powell, J.C., ... & Selleck, P.W. (2012). Indoor air quality in typical temperate zone Australian dwellings. *Atmospheric Environment*, *54*, 400–407.

Molnar, V.É., Simon, E., Tóthmérész, B., Ninsawat, S., & Szabó, S. (2020). Air pollution induced vegetation stress–the air pollution tolerance index as a quick tool for city health evaluation. *Ecological Indicators*, *113*, 106234.

Morawska, L., & Salthammer, T. (Eds.). (2006). *Indoor Environment: Airborne Particles and Settled Dust*. John Wiley & Sons.

Morawska, L., Afshari, A., Bae, G., Buonanno, G., Chao, C.Y.H., Hänninen, O., Hofmann, W., Isaxon,C., Jayaratne, E.R., & Pasanen, P. (2013). Indoor aerosols: from personal exposure to risk assessment. *Indoor Air*, *23*(6), 462–487.

Morawska, L., Ayoko, G.A., Bae, G.N., Buonanno, G., Chao, C.Y.H., Clifford, S., ... & Wierzbicka, A. (2017). Airborne particles in indoor environment of homes, schools, offices and aged care facilities: The main routes of exposure. *Environment International*, *108*, 75–83.

Mueller, J.G., Cerniglia, C.E., & Pritchard, P.H. (1996). Bioremediation of environments contaminated by polycyclic aromatic hydrocarbons. In E.L. Crawford, D.L. Crawford (Eds.), *Bioremediation: Principles and Applications*, 125–194. Cambridge: Cambridge University Press.

Newman, L.A., & Reynolds, C.M. (2004). Phytodegradation of organic compounds. *Current Opinion in Biotechnology*, *15*(3), 225–230.

Nezis, I., Biskos, G., Eleftheriadis, K., & Kalantzi, O.I. (2019). Particulate matter and health effects in offices – a review. *Building and Environment*, *156*, 62–73.

Nowak, D.J., Crane, D.E., & Stevens, J.C. (2006). Air pollution removal by urban trees and shrubs in the United States. *Urban Forestry & Urban Greening*, *4*, 115–123. doi: 10.1016/j.ufug.2006.01.007

Nowak, D.J., Crane, D.E., Stevens, J.C., Hoehn, R.E., Walton, J.T., & Bond, J. (2008). A ground-based method of assessing urban forest structure and ecosystem services. *Arboriculture & Urban Forestry*, *34*, 347e358.

Nowak, D.J., Hirabayashi, S., Bodine, A., & Greenfield, E. (2014). Tree and forest effects on air quality and human health in the United States. *Environmental Pollution*, *193*, 119–129.

Oliveira, M., Slezakova, K., Delerue-Matos, C., Pereira, M.C., & Morais, S. (2019). Children environmental exposure to particulate matter and polycyclic aromatic hydrocarbons and biomonitoring in school environments: a review on indoor and outdoor exposure levels, major sources and health impacts. *Environment International*, *124*, 180–204.

Orwell, R.L., Wood, R.A., Burchett, M.D., Tarran, J., & Torpy, F. (2006). The potted-plant microcosm substantially reduces indoor air VOC pollution: II. Laboratory study. *Water, Air, and Soil Pollution*, 1096, *177*(1–4), 59–80.

Orwell, R.L., Wood, R.L., Tarran, J., Torpy, F., & Burchett, M.D. (2004). Removal of benzene by the indoor plant/substrate microcosm and implications for air quality. *Water, Air, and Soil Pollution*, *157*(1–1099 4), 193–207.

Pandey, V.C., & Bajpai, O. (2019). Chapter 1 – phytoremediation: from theory toward practice. In C. Pandey & K. Bauddh (Eds.), V. *Phytomanagement of Polluted Sites*, 1–49. Elsevier.

Papinchak, H.L., Holcomb, E.J., Best, T.O., & Decoteau, D.R. (2009). Effectiveness of houseplants in reducing the indoor air pollutant ozone. *HortTechnology*, *19*(2), 286–290.

Parseh, I., Teiri, H., Hajizadeh, Y., & Ebrahimpour, K. (2018). Phytoremediation of benzene vapors from indoor air by Schefflera arboricola and Spathiphyllum wallisii plants. *Atmospheric Pollution Research*, *9*(6), 1083–1087.

Paull, N.J., Irga, P.J., & Torpy, F.R. (2018). Active green wall plant health tolerance to diesel smoke exposure. *Environmental Pollution*, *240*, 448–456.

Pei, J.J., & Zhang, J.S.S. (2011). Critical review of catalytic oxidization and chemisorption methods for indoor formaldehyde removal. *HVAC&R Research*, *17*, 476–503. doi: 10.1080/10789669.2011.587587.

Pérez-Padilla, R., Schilmann, A., & Riojas-Rodriguez, H. (2010). Respiratory health effects of indoor air pollution. *The International Journal of Tuberculosis and Lung Disease, 14*(9), 1079–1086.

Perrino, C. (2010). Atmospheric particulate matter. *Biophysics and Bioengineering Letters, 3*(1).

Pettit, T., Irga, P.J., & Torpy, F.R. (2018b). Functional green wall development for increasing air pollutant phytoremediation: substrate development with coconut coir and activated carbon. *Journal of Hazardous Materials, 360*, 594–603.

Pettit, T., Irga, P.J., & Torpy, F.R. (2018a). Towards practical indoor air phytoremediation: a review. *Chemosphere, 208*, 960–974.

Pettit, T., Irga, P.J., & Torpy, F.R. (2019). The in situ pilot-scale phytoremediation of airborne VOCs and particulate matter with an active green wall. *Air Quality, Atmosphere & Health, 12*(1), 33–44.

Pettit, T., Irga, P.J., Abdo, P., & Torpy, F.R. (2017). Do the plants in functional green walls contribute to their ability to filter particulate matter? *Building and Environment, 125*, 299–307.

Pilon-Smits, Elizabeth (2005-04-29). Phytoremediation. *Annual Review of Plant Biology, 56*(1), 15–39.

Pivetz, B.E. (2001). *Ground Water Issue: Phytoremediation of Contaminated Soil and Ground Water at Hazardous Waste Sites*, 1–36. Washington, D.C.: United States Environmental Protection Agency.

Popek, R., Gawrońska, H., Wrochna, M., Gawroński, S.W., & Sæbø, A. (2013). Particulate matter on foliage of 13 woody species: deposition on surfaces and phytostabilisation in waxes–a 3-year study. *International Journal of Phytoremediation, 15*(3), 245–256.

Popek, R., Łukowski, A., Bates, C., & Oleksyn, J. (2017). Accumulation of particulate matter, heavy metals, and polycyclic aromatic hydrocarbons on the leaves of Tilia cordata Mill. in five Polish cities with different levels of air pollution. *International Journal of Phytoremediation, 19*(12), 1134–1141.

Popescu, F., & Ionel, I. (2010). Anthropogenic air pollution sources. *Air Quality*, 1–22.

Porter, J.R. (1994). Toluene removal from air by *Dieffenbachia* in a closed environment. *Advances in Space Research, 14*(11), 99–103.

Prasad, M.N.V. (2011). *A State-of-the-Art Report on Bioremediation, Its Applications to Contaminated Sites in India*. Ministry of Environment & Forests, Government of India.

Pugh, T.A., MacKenzie, A.R., Whyatt, J.D., & Hewitt, C.N. (2012). Effectiveness of green infrastructure for improvement of air quality in urban street canyons. *Environmental Science & Technology, 46*, 7692–7699.

Qaroush, A.K., Assaf, K.I., Al-Khateeb, A.A., Alsoubani, F., Nabih, E., Troll, C., ... & Eftaiha, A.A.F. (2017). Pentaerythritol-Based Molecular Sorbent for CO_2 Capturing: A Highly Efficient Wet Scrubbing Agent Showing Proton Shuttling Phenomenon. *Energy & Fuels, 31*(8), 8407–8414.

Rai, R., Sarkar, A., Agrawal, S.B., & Agrawal, M. (2012). Evaluation of Tropospheric O3 Effects on Global Agriculture: A New Insight. *Improving Crop Productivity in Sustainable Agriculture*, 69–105.

Ramalho, O., Wyart, G., Mandin, C., Blondeau, P., Cabanes, P.-A., Leclerc, N., Mullot, J.-U., Rathinasabapathi, B., Ma, L.Q., & Srivastava, M. (2006). Arsenic hyperaccumulating ferns and their application to phytoremediation of arsenic contaminated sites. *Floriculture, Ornamental and Plant Biotechnology, 3*(32), 304–311.

Raziani, Y., & Raziani, S. (2021). The Effect of Air Pollution on Myocardial Infarction. *Journal of Chemical Reviews, 3*(1), 83–96.

Ren, Y., Ge, Y., Gu, B., Min, Y., Tani, A., & Chang, J. (2014). Role of management strategies and environmental factors in determining the emission of biogenic volatile organic compounds from urban green spaces. *Environmental Science & Technology, 48*, 6237–6246.

Rene, E.R., Sergienko, N., Goswami, T., López, M.E., Kumar, G., Saratale, G.D., ... & Swaminathan, T. (2018). Effects of concentration and gas flow rate on the removal of gas-phase toluene and xylene mixture in a compost biofilter. *Bioresource Technology, 248*, 28–35.

Ren-Jian, Z., Kin-Fai, H.O., & Zhen-Xing, S. (2012). The role of aerosol in climate change, the environment, and human health. *Atmospheric and Oceanic Science Letters, 5*(2), 156–161.

Richardson, G., Eick, S., & Jones, R. (2005). How is the indoor environment related to asthma?: literature review. *Journal of Advanced Nursing, 52*(3), 328–339.

Rizwan, S.A., Nongkynrih, B., & Gupta, S.K. (2013). "Air pollution in Delhi": its magnitude and effects on health. *Indian Journal of Community Medicine, 38*, 4–8. doi: 10.4103/0970-0218.106617.

Roland, U., Holzer, F., & Kopinke, F.D. (2002). Improved oxidation of air pollutants in a non-thermal plasma. *Catalysis Today, 73*, 315–323.

Sæbø, A., Popek, R., Nawrot, B., Hanslin, H.M., Gawronska, H., & Gawronski, S.W. (2012). Plant species differences in particulate matter accumulation on leaf surfaces. *Science of the Total Environment*, *427*, 347–354.

Sahu, L.K., Yadav, R., & Tripathi, N. (2020). Aromatic compounds in a semi-urban site of western India: seasonal variability and emission ratios. *Atmospheric Research*, *246*, 105114.

Sakakibara, M., Watanabe, A., Inoue, M., Sano, S., & Kaise, T. (2010, January). Phytoextraction and phytovolatilization of arsenic from As-contaminated soils by *Pteris vittata*. In *Proceedings of the Annual International Conference on Soils, Sediments, Water and Energy*, *12*(1), 26.

Salisbury F.B., Gitelson J.I., & Lisovsky G.M. (1997). Bios-3: Siberian experiments in bioregenerative life support. *Bioscience*, *47*, 575–585.

Salt, D.E., Blaylock, M., Kumer, N.P.B.A., Dushenkov, V., Ensley, B.D., Chet, I., et al. (1995). Phytoremediation: a novel strategy for the removal of toxic metals from the environment using plants. *Biotechnology*, *13*, 468–474. doi: 10.1038/nbt0595-468

Samarghandi, M.R., Babaee, S.A., Ahmadian, M., Asgari, G., Ghorbani Shahna, F., & Poormohammadi, A. (2014). Performance catalytic ozonation over the carbosieve in the removal of toluene from waste air stream. *Journal of Research in Health Sciences*, *14*(3), 227–232.

Sánchez-Soberón, F., Rovira, J., Sierra, J., Mari, M., Domingo, J.L., & Schuhmacher, M. (2019). Seasonal characterization and dosimetry-assisted risk assessment of indoor particulate matter (PM10-2.5, PM2.5-0.25, and PM0. 25) collected in different schools. *Environmental Research*, *175*, 287–296.

Sarkar, A., & Agrawal, S.B. (2010a). Elevated ozone and two modern wheat cultivars: an assessment of dose dependent sensitivity with respect to growth, reproductive and yield parameters. *Environmental and Experimental Botany*, *69*(3), 328–337.

Sarkar, A., & Agrawal, S.B. (2010b). Identification of ozone stress in Indian rice through foliar injury and differential protein profile. *Environmental Monitoring and Assessment*, *161*(1), 205–215.

Sarkar, A., & Agrawal, S.B. (2012). Evaluating the response of two high yielding Indian rice cultivars against ambient and elevated levels of ozone by using open top chambers. *Journal of Environmental Management*, *95*, S19–S24.

Sarkar, A., Agrawal, G.K., Shibato, J., Cho, K., & Rakwal, R. (2012a). Impacts of ozone (O_3) and carbon dioxide (CO2) environmental pollutants on crops: a Transcriptomics update. *INTECH Open Access Publisher*, 49–60.

Sarkar, A., Das, S., Srivastava, V., Singh, P., & Singh, R.P. (2018). Effect of wastewater irrigation on crop health in the indian agricultural scenario. *Emerging Trends of Plant Physiology for Sustainable Crop Production*, 357.

Sarkar, A., Islam, M.T., Zargar, S.M., Dogra, V., Kim, S.T., Gupta, R., ... & Rakwal, R. (2014). Proteomics potential and Its Contribution toward Sustainable agriculture. *Agroecology, Ecosystems, and Sustainability*, 151.

Sarkar, A., Rakwal, R., Bhushan Agrawal, S., Shibato, J., Ogawa, Y., Yoshida, Y.,....., & Agrawal, M. (2010). Investigating the impact of elevated levels of ozone on tropical wheat using integrated phenotypical, physiological, biochemical, and proteomics approaches. *Journal of Proteome Research*, *9*(9), 4565–4584.

Sarkar, A., Rakwal, R., Shibato, J., & Agrawal, G.K. (2012b). Toward Sustainable Agriculture through Integrated 'OMICS' Technologies: A Quest for Future Global Food Security. *Journal of Developments in Sustainable Agriculture*, *7*(1), 103–110.

Sarkar, A., Singh, A.A., Agrawal, S.B., Ahmad, A., & Rai, S.P. (2015). Cultivar specific variations in antioxidative defense system, genome and proteome of two tropical rice cultivars against ambient and elevated ozone. *Ecotoxicology and Environmental Safety*, *115*, 101–111.

Schlink, U., Thiem, A., Kohajda, T., Richter, M., & Strebel, K. (2010). Quantile regression of indoor air-concentrations of volatile organic compounds (VOC). *Science of the Total Environment*, *408*(18), 3840–3851.

Schröder, P., Harvey, P.J., & Schwitzgébel, J.P. (2002). Prospects for the phytoremediation of organic pollutants in Europe. *Environmental Science and Pollution Research*, *9*(1), 1–3. ISSN 0944-1344

Sedjo, R., & Sohngen, B. (2012). Carbon sequestration in forests and soils. *Annual Review of Resource Economics*, *4*(1), 127–144.

Selmar, D., Engelhardt, U., Hansel, S., Thrane, C., Nowak, M., & Kleiwachter, M. (2015). Nicotine uptake by peppermint plants as a possible source of nicotine in plantderived products. *Agronomy for Sustainable Development, 35*, 1185–1190.

Selmi, W., Weber, C., Rivière, E., Blond, N., Mehdi, L., & Nowak, D. (2016). Air pollution removal by trees in public green spaces in Strasbourg city, France. *Urban Forestry & Urban Greening, 17*, 192–201.

Seppänen, O. (2002). Ventilation, energy and indoor air quality. *Indoor Air*, 136–147.

Seppänen, O., Fisk, W.J., & Lei, Q. (2006). Ventilation and performance in office work. *Indoor Air, 16*(1), 28–36.

Setsungnern, A., Treesubsuntorn, C., & Thiravetyan, P. (2017). The influence of different light quality and benzene on gene expression and benzene degradation of *Chlorophytum comosum*. *Plant Physiology and Biochemistry, 120*, 95–102.

Sharma, J. (2018). *Introduction to Phytoremediation – A Green Clean Technology*. Available at SSRN 3177321.

Singh, S.N., & Verma, A. (2007). Phytoremediation of air pollutants: a review. In S.N. Singh & R.D. Tripathi (Eds.), *Environmental Bioremediation Technologies*, 293–314. Berlin: Springer Berlin Heidelberg.

Singh, S.N., & Tripathi, R.D. (2007). *Environmental Bioremediation Technologies*. Berlin, Germany: Springer. ISBN 9783540347934.

Sinha, R.K., & Singh, S. (2010). Plants combating air pollution. In R.K. Sinha (Ed.), *Green Plants and Pollution: Nature's Technology for Abating and Combating Environmental Pollution (Air, Water and Soil Pollution Science and Technology)*. Palo Alto, CA, USA: Nova Science Publishers.

Sitaras, I.E., & Siskos, P.A. (2008). The role of primary and secondary air pollutants in atmospheric pollution: Athens urban area as a case study. *Environmental Chemistry Letters, 6*, 59–69.

Smith, K.R. (2002). Indoor air pollution in developing countries: recommendations for research. *Indoor Air, 12*, 198–207.

Smith, P., Ashmore, M.R., Black, H.I., Burgess, P.J., Evans, C.D., Quine, T.A.,..., & Orr, H.G. (2013). The role of ecosystems and their management in regulating climate, and soil, water and air quality. *Journal of Applied Ecology, 50*(4), 812–829.

Snyder, R. (2012). Leukemia and benzene. *International Journal of Environmental Research and Public Health, 9*, 2875–2893.

Song, Y., Maher, B.A., Li, F., Wang, X., Sun, X., & Zhang, H. (2015). Particulate matter deposited on leaf of five evergreen species in Beijing, China: Source identification and size distribution. *Atmospheric Environment, 105*, 53–60.

Soreanu, G. (2016). Biotechnologies for improving indoor air quality. In *Start-Up Creation*, 301–328. Woodhead Publishing.

Soreanu, G., Dixon, M., & Darlington, A. (2013). Botanical biofiltration of indoor gaseous pollutants – a mini-review. *Chemical Engineering, 229*, 585–594.

Srimuruganandam, B., & Nagendra, S.S. (2012). Source characterization of PM10 and PM2.5 mass using a chemical mass balance model at urban roadside. *Science of the Total Environment, 433*, 8–19.

Sriprapat, W., & Thiravertyan, P. (2013). Phytoremediation of BTEX from indoor air by *Zamioculcas zamiifolia*. *Water, Air, & Soil Pollution, 224*, 1482.

Sriprapat, W., & Thiravetyan, P. (2016). Efficacy of ornamental plants for benzene removal from contaminated air and water: Effect of plant associated bacteria. *International Biodeterioration & Biodegradation, 113*, 262–268.

Sriprapat, W., Boraphech, P., & Thiravetyan, P. (2014a). Factors affecting xylene-contaminated air removal by the ornamental plant *Zamioculcas zamiifolia*. *Environmental Science and Pollution Research, 21*(4), 2603–2610.

Sriprapat, W., Suksabye, P., Areephak, S., Klantup, P., Waraha, A., Sawattan, A., & Thiravetyan, P. (2014b). Uptake of toluene and ethylbenzene by plants: removal of volatile indoor air contaminants. *Ecotoxicology and Environmental Safety, 102*, 147–151.

Srivastava, R.K., Jozewicz, W., & Singer, C. (2001). SO_2 scrubbing technologies: a review. *Environmental Progress, 20*(4), 219–228.

Srivastava, V., Sarkar, A., Singh, S., Singh, P., de Araujo, A.S., & Singh, R.P. (2017). Agroecological responses of heavy metal pollution with special emphasis on soil health and plant performances. *Frontiers in Environmental Science, 5*, 64.

Stapleton, E., & Ruiz-Rudolph, P. (2018). The potential for indoor ultrafine particle reduction using vegetation under laboratory conditions. *Indoor and Built Environment, 27*(1), 70–83.

Strömberg, A.M., & Karlsson, H.T. (1988). Limestone based spray dry scrubbing of SO2. In *Tenth International Symposium on Chemical Reaction Engineering*, 2095–2102. Pergamon.

Su, Y., & Liang, Y. (2015). Foliar uptake and translocation of formaldehyde with Bracket plants (*Chlorophytum comosum*). *Journal of Hazardous Materials, 291*, 120–128.

Su, Y.M., & Lin, C.H. (2015). Removal of indoor carbon dioxide and formaldehyde using green walls by bird nest fern. *The Horticulture Journal, 84*(1), 69–76.

Taha, H. (1996). Modeling impacts of increased urban vegetation on ozone air quality in the South Coast Air Basin. *Atmospheric Environment, 30*, 3423–3430.

Takahashi, M., Higaki, A., Nohno, M., Kamada, M., Okamura, Y., Matsui, K., ... & Morikawa, H. (2005). Differential assimilation of nitrogen dioxide by 70 taxa of roadside trees at an urban pollution level. *Chemosphere, 61*(5), 633–639.

Taner, S., Pekey, B., & Pekey, H. (2013). Fine particulate matter in the indoor air of barbeque restaurants: Elemental compositions, sources and health risks. *Science of the Total Environment, 454*, 79–87.

Tani, A., & Hewitt, C.N. (2009). Uptake of aldehydes and ketones at typical indoor concentrations by houseplants. *Environmental Science & Technology, 43*(21), 8338–8343.

Tani, A., Kato, S., Kajii, Y., Wilkinson, M., Owen, S., & Hewitt, N. (2007). A proton transfer reaction mass spectrometry based system for determining plant uptake of volatile organic compounds. *Atmospheric Environment, 41*(8), 1736–1746.

Teiri, H., Hajizadeh, Y., Samaei, M.R., Pourzamani, H., & Mohammadi, F. (2020). Modelling the phytoremediation of formaldehyde from indoor air by *Chamaedorea Elegans* using artificial intelligence, genetic algorithm and response surface methodology. *Journal of Environmental Chemical Engineering, 8*(4), 103985.

Teiri, H., Pourzamani, H., & Hajizadeh, Y. (2018). Phytoremediation of VOCs from indoor air by ornamental potted plants: a pilot study using a palm species under the controlled environment. *Chemosphere, 197*, 375–381.

Teper, E. (2009). Dust-particle migration around flotation tailings ponds:pine needles as passive samplers. *Environmental Monitoring and Assessment, 154*, 383–391. doi: 10.1007/s10661-008-0405-4.

Terzaghi, E., Wild, E., Zacchello, G., Cerabolini, B.E., Jones, K.C., & Di Guardo, A. (2013). Forest filter effect: role of leaves in capturing/releasing air particulate matter and its associated PAHs. *Atmospheric Environment, 74*, 378–384.

The National Aeronautics and Space Administration (1974). *NASA Technical Memorandium. The Proceedings of the Skylab Life Sciences Symposium*. Thomas CK, Kim KJ: Lyndon B. Johnson Space Center, 426.

Thomas, C.K., Kim, K.J., & Kays, S.J. (2015). Phytoremediation of indoor air. *HortScience, 50*, 765–768.

Tomé, F.V., Rodríguez, P.B., & Lozano, J.C. (2008). Elimination of natural uranium and 226Ra from contaminated waters by rhizofiltration using Helianthus annuus L. *Science of the Total Environment, 393*(2–3), 351–357.

Topsoe, N.Y., Topsoe, H., & Dumesic, J.A. (1995). Vanadia/titania catalysts for selective catalytic reduction (SCR) of nitric-oxide by ammonia: I. Combined temperature-programmed in-situ FTIR and on-line mass-spectroscopy studies. *Journal of Catalysis, 151*(1), 226–240.

Torpy, F.R., Zavattaro, M., & Irga, P.J. (2017). Green wall technology for the phytoremediation of indoor air: a system for the reduction of high CO_2 concentrations. *Air Quality, Atmosphere & Health, 10*(5), 575–585.

Torpy, F., & Zavattaro, M. (2018). Bench-study of green-wall plants for indoor air pollution reduction. *Journal of Living Architecture, 5*(1), 1–15.

Torpy, F., Clements, N., Pollinger, M., Dengel, A., Mulvihill, I., He, C., & Irga, P. (2018). Testing the single-pass VOC removal efficiency of an active green wall using methyl ethyl ketone (MEK). *Air Quality, Atmosphere & Health, 11*(2), 163–170.

Torpy, F., Irga, P., & Burchett, M. (2014). Profiling indoor plants for the amelioration of high CO_2 concentrations. *Urban Forestry & Urban Greening, 13*(2), 227–233

Torpy, F., Irga, P., Moldovan, D., Tarran, J., & Burchett, M. (2013). Characterization and biostimulation of benzene biodegradation in the potting-mix of indoor plants. *Journal of Applied Horticulture, 15*(1), 10–15.

Treesubsuntorn, C., Suksabye, P., Weangjun, S., Pawana, F., & Thiravetyan, P. (2013) Benzene adsorption by plant leaf materials: effect of quantity and composition of wax. *Water, Air, & Soil Pollution, 224*, 1736–1745.

Treesubsuntorn, C., & Thiravetyan, P. (2012). Removal of benzene from indoor air by *Dracaena sanderiana*: Effect of wax and stomata. *Atmospheric Environment, 57*, 317–321.

Tripathi, R., Sarkar, A., Pandey Rai, S., & Agrawal, S.B. (2011). Supplemental ultraviolet-B and ozone: impact on antioxidants, proteome and genome of linseed (Linum usitatissimum L. cv. Padmini). *Plant Biology, 13*(1), 93–104.

Tsai, Y.I., Yang, H.H., Wang, L.C., Huan, J.L., Young, L.H., Cheng, M.T., et al. (2011). The influences of diesel particulate filter installation on air pollutant emissions for used vehicles. *Aerosol and Air Quality Research, 11*, 578–583. doi: 10.4209/aaqr.2011.05.0066.

Tunno, B.J., Shields, K.N., Cambal, L., Tripathy, S., Holguin, F., Lioy, P., & Clougherty, J.E. (2015). Indoor air sampling for fine particulate matter and black carbon in industrial communities in Pittsburgh. *Science of the Total Environment, 536*, 108–115.

Ugrekhelidze, D., Korte, F., & Kvesitadze, G. (1997). Uptake and transformation of benzene and toluene by plant leaves. *Ecotoxicology and Environmental Safety, 37*(1), 24–29.

UNFPA. (2004). *State of World Population 2004: The Cairo Consensus at Ten: Population, Reproductive Health and the Global Effort to End Poverty: Chapter 4. Migration and Urbanization.* New York, NY: United Nations Population Fund.

United Nations. (2000). *World Urbanization Prospects (the 1999 Revision).* New York, NY: Population Division, Department of Economic and Social Affairs, United Nations.

United Nations. (2004). *World Urbanization Prospects (the 2003 Revision).* New York, NY: Population Division, Department of Economic and Social Affairs, United Nations.

Vallero, D.A. (2014). *Fundamentals of Air Pollution* (5th Edition). San Diego, CA: Elsevier.

Van Sluys, M.A., Monteiro-Vitorello, C.B., Camargo, L.E.A., Menck, C.F.M., Da Silva, A.C.R., Ferro, J.A.,..... & Simpson, A.J. (2002). Comparative genomic analysis of plant-associated bacteria. *Annual Review of Phytopathology, 40*(1), 169–189.

Verma, P., George, K.V., Singh, H.V., Singh, S.K., Juwarkar, A., & Singh, R.N. (2006). Modeling rhizofiltration: heavy-metal uptake by plant roots. *Environmental Modeling & Assessment, 11*(4), 387–394.

Verstraete, S., Hermia, J., & Vigneron, S. (1994). VOC separation on membranes: A review. *Studies in Environmental Science, 61*, 359–373.

Vigueras, G., Shirai, K., Martins, D., Franco, T.T., Fleuri, L.F., & Revah, S. (2008). Toluene gas phase biofiltration by Paecilomyces lilacinus and isolation and identification of a hydrophobin protein produced thereof. *Applied Microbiology and Biotechnology, 80*(1), 147.

Vikrant, K., Kim, K.H., Szulejko, J.E., Pandey, S.K., Singh, R.S., Giri, B.S., ... & Lee, S.H. (2017). Biofilters for the treatment of VOCs and odors-A review. *Asian Journal of Atmospheric Environment, 11*(3), 139–152.

Vineis, P., Hoek, G., Krzyzanowski, M., Vigna-Taglianti, F., Veglia, F., Airoldi, L., ... & Riboli, E. (2006). Air pollution and risk of lung cancer in a prospective study in Europe. *International Journal of Cancer, 119*(1), 169–174.

Vorholt, J.A. (2012). Microbial life in the phyllosphere. *Nature Reviews Microbiology, 10*(12), 828–840. Available from: 10.1038/nrmicro2910.

Wang, L.K., Lin, W., & Hung, Y.T. (2004d). Thermal oxidation. In *Air Pollution Control Engineering*, 347–367. Totowa, NJ: Humana Press.

Wang, L.K., Lin, W., & Hung, Y.T. (2004e). Catalytic oxidation. In *Air Pollution Control Engineering*, 369–394. Totowa, NJ: Humana Press.

Wang, L.K., Pereira, N.C., & Hung, Y.T. (Eds.). (2004f). *Air Pollution Control Engineering (Vol. 1).* Totowa, NJ: Humana press.

Wang, L.K., Taricska, J.R., Hung, Y.T., Eldridge, J.E., & Li, K.H. (2004b). Wet and dry scrubbing. In *Air Pollution Control Engineering*, 197–305. Totowa, NJ: Humana Press.

Wang, L.K., Taricska, J.R., Hung, Y.T., & Li, K.H. (2004g). Gas-Phase Activated Carbon Adsorption. In L.K. Wang, N.C. Pereira, & Y.T. Hung (Eds.), Air Pollution Control Engineering. Handbook of Environmental Engineering, vol 1. Totowa, NJ: Humana Press. doi:10.1007/978-1-59259-778-9_10

Wang, L.K., Williford, C., & Chen, W.Y. (2004a). Fabric filtration. In *Air Pollution Control Engineering*, 59–95. Totowa, NJ: Humana Press.

Wang, L.K., Williford, C., & Chen, W.Y. (2004c). Condensation. In *Air Pollution Control Engineering*, 307–328. Totowa, NJ: Humana Press.

Wang, Y.Q., Zhang, X.Y., & Arimoto, R. (2006). The contribution from distant dust sources to the atmospheric particulate matter loadings at XiAn, China during spring. *Science of the Total Environment, 368*(2-3), 875–883.

Wang, Y., Kloog, I., Coull, B.A., Kosheleva, A., Zanobetti, A., & Schwartz, J.D. (2016). Estimating causal effects of long-term PM 2. 5 exposure on mortality in New Jersey. *Environmental Health Perspectives, 124*(8), 1182.

Wang, Z., & Zhang, J.S. (2011). Characterization and performance evaluation of a full-scale activated carbon-based dynamic botanical air filtration system for improving indoor air quality. *Building and Environment, 46*(3), 758–768.

Wang, Z., Pei, J., & Zhang, J. (2013). Catalytic oxidization of indoor formaldehyde at room temperature – Effect of operation conditions. *Building and Environment, 65*, 49–57. doi: 10.1016/j.buildenv.2013.03.0

Wang, Z., Pei, J., & Zhang, J.S. (2014). Experimental investigation of the formaldehyde removal mechanisms in a dynamic botanical filtration system for indoor air purification. *Journal of Hazardous Materials, 280*, 235–243.

Waring, M.S. (2016). Bio-walls and indoor houseplants: Facts and fictions. In: *Microbiomes of the Built Environment: From Research to Application, Meeting #3*. Irvine: University of California.

Weber, F., Kowarik, I., & Säumel, I. (2014). Herbaceous plants as filters: Immobilization of particulates along urban street corridors. *Environmental Pollution, 2014*(186), 234–240.

Wei, X., Lyu, S., Yu, Y., Wang, Z., Liu, H., Pan, D., & Chen, J. (2017). Phylloremediation of air pollutants: exploiting the potential of plant leaves and leaf-associated microbes. *Frontiers in Plant Science, 8*, 1318.

Weichert, H.J. (1996). Study of air velocity and turbulence effects on organic compound emissions from building materials/furnishings using a new small test chamber. In *Characterizing Sources of Indoor Air Pollution and Related Sink Effects*. ASTM International.

Welburn, A. (1990). Why are atmospheric oxides of nitrogen usually phytotoxic and not alternative fertilizers? *New Phytologist, 115*, 395–429.

Welburn, A. (1998). Atmospheric nitrogenous compounds and ozone – Is NOx fixation by plants a possible solution? *New Phytologist, 139*, 5–9.

Wenzel, W.W. (2009). Rhizosphere processes and management in plant assisted bioremediation (phytoremediation) of soils. Plant and Soil, *321*, 385–408.

Weyens, N., Thijs, S., Popek, R., Witters, N., Przybysz, A., Espenshade, J., ... & Gawronski, S.W. (2015). The role of plant–microbe interactions and their exploitation for phytoremediation of air pollutants. *International Journal of Molecular Sciences, 16*(10), 25576–25604.

Weyens, N., Van Der Lelie, D., Artois, T., Smeets, K., Taghavi, S., Newman, L., ... & Vangronsveld, J. (2009a). Bioaugmentation with engineered endophytic bacteria improves contaminant fate in phytoremediation. *Environmental Science & Technology, 43*(24), 9413–9418.

Weyens, N., van der Lelie, D., Taghavi, S., & Vangronsveld, J. (2009b). Phytoremediation: plant–endophyte partnerships take the challenge. *Current Opinion in Biotechnology, 20*(2), 248–254.

Wheeler, A.J., Dobbin, N.A., Lyrette, N., Wallace, L., Foto, M., Mallick, R., Kearney, J., Van Ryswyk, K., Gilbert, N.L., & Harrison, I. (2011). Residential indoor and outdoor coarse particles and associated endotoxin exposures. *Atmospheric Environment, 45*(39), 7064–7071.

WHO. (2013). *Health Effects of Particulate Matter*. Available online at: http://www.euro.who.int/__data/assets/pdf_file/0006/189051/Health-effects-ofparticulate-matter-final-Eng.pdf.

WHO. (2018). *Burden of Disease from Household Air Pollution for 2016*. Geneva: World Health Organization.

WHO. (2016). Ambient air pollution: a global assessment of exposure and burden of diseases, 2016. In: *WHO Document Production Services*. Geneva, Switzerland. who.int/phe/health_topics/outdoorair/databases/en/(accessed 30 November 2016).

WHO. (2014a). Burden of Disease. World Health Organization. http://www.who.int/gho/phe/outdoor_air_pollution/burden_text/en/ (accessed 10 February 2016).

WHO. (2014b). *WHO Indoor Air Quality Guidelines:Household Fuel Combustion: Emissions of HealthDamaging Pollutants From Household Stoves*. World Health Organization. http://www.who.int/indoorair/guidelines/hhfc/en/ (accessed 22 March 2016).

Wichmann, F.A., Müller, A., Busi, L.E., Cianni, N., Massolo, L., Schlink, U., ... & Sly, P.D. (2009). Increased asthma and respiratory symptoms in children exposed to petrochemical pollution. *Journal of Allergy and Clinical Immunology*, *123*(3), 632–638.

Winter, K., & Holtum, J.A.M. (2014). Facultative crassulacean acid metabolism (CAM) plants: Powerful tools for unravelling the functional elements of CAM photosynthesis. *Journal of Experimental Botany*, *65*, 3425–3441.

Wisthaler, A., & Weschler, C.J. (2010). Reactions of ozone with human skin lipids: sources of carbonyls, dicarbonyls, and hydroxycarbonyls in indoor air. *Proceedings of the National Academy of Sciences*, *107*(15), 6568–6575.

Wohlgemuth, H., Mittelstrass, K., Kschieschan, S., Bender, J., Weigel, H.J., Overmyer, K., ... & Langebartels, C. (2002). Activation of an oxidative burst is a general feature of sensitive plants exposed to the air pollutant ozone. *Plant, Cell & Environment*, *25*(6), 717–726.

Wolkoff, P. (2013). Indoor air pollutants in office environments: assessment of comfort, health, and performance. *International Journal of Hygiene and Environmental Health*, *216*(4), 371–394.

Wolverton, B.C., McDonald, R.C., Watkins, E.A. Jr (1984). Foliage plants for removing indoor air pollutants from energy-efficient homes. *Economic Botany*, *38*, 224–228.

Wolverton, B.C. (1988). *Foliage Plants for Improving Indoor Air Quality*.

Wolverton, B.C., & McDonald, R.C. (1982). *Foliage Plants for Removing Formaldehyde from Contaminated Air Inside Energy-efficient Homes and Future Space Stations*. (TM-84674 NSTL 39529) NASA National Space Technology Labs, Bay St. Louis, Mississippi, USA

Wolverton, B.C., Johnson, A., & Bounds, K. (1989). *Interior Landscape Plants for Indoor Air Pollution Abatement*. National Aeronautics and Space Administration, John C. Stennis Space Center. MS 39529-6000. https://ntrs.nasa.gov/archive/nasa/casi.ntrs.nasa.gov/19930073077.pdf

Wolverton, B.C., & Wolverton, J.D. (1993). Plants and soil microorganisms: removal of formaldehyde, xylene, and ammonia from the indoor environment. *Mississippi Academy of Sciences*, *38*, 11–15.

Wood, R.A., Orwell, R.L., Tarran, J., Torpy, F., & Burchett, M. (2002). Potted-plant/growth media interactions and capacities for removal of volatiles from indoor air. *The Journal of Horticultural Science and Biotechnology*, *77*(1), 120–129.

Wood, R.A., Burchett, M.D., Alquezar, R., Orwell, R.L., Tarran, J., & Torpy, F. (2006). The potted-plant microcosm substantially reduces indoor air VOC pollution: I. Office field-study. *Water, Air, & Soil Pollution*, *175*(1), 163–180.

Wu, D., Sun, M.Z., Zhang, C., & Xin, Y. (2014). Antioxidant properties of *Lactobacillus* and its protecting effect to oxidative stress Caco-2 cells. *Journal of Animal &Plant Sciences*, *24*, 1766–1771.

Wu, Z., Zhang, X., Wu, X., Shen, G., Du, Q., & Mo, C. (2013). Uptake of di (2-ethylhexyl) phthalate (DEHP) by plants *Benincasa hispida* and its use for lowering DEHP content in intercropped vegetables. *Journal of Agricultural and Food Chemistry*, *6*(22), 5220–5225.

Wyzga, R.E., & Rohr, A. (2015). Long-term particulate matter exposure: Attributing health effects to individual PM components. *Journal of the Air & Waste Management Association*, *65*(5), 523–543.

Xu, Y., Xu, W., Mo, L., Heal, M.R., Xu, X., & Yu, X. (2018). Quantifying particulate matter accumulated on leaves by 17 species of urban trees in Beijing, China. *Environmental Science and Pollution Research*, *25*(13), 12545–12556.

Xu, Z., Qin, N., Wang, J., & Tong, H. (2010). Formaldehyde biofiltration as affected by spider plant. *Bioresource Technology*, *101*, 6930–6934.

Yadav, B.K., Siebel, M.A., & van Bruggen, J.J. (2011). Rhizofiltration of a heavy metal (lead) containing wastewater using the wetland plant *Carex pendula*. *CLEAN – Soil, Air, Water*, *39*(5), 467–474.

Yanai, J., Zhao, F.-J., McGrath, S.P., & Kosaki, T. (2006). Effect of soil characteristics on Cd uptake by the hyperaccumulator *Thlaspi caerulescens*. *Environmental Pollution*, *139*(1), 167–175.

Yang, D.S., Pennisi, S.V., Son, K.C., & Kays, S.J. (2009). Screening indoor plants for volatile organic pollutant removal efficiency. *HortScience*, *44*(5), 1377–1381.

Yang, H., & Liu, Y. (2011). In M. Khallaf (Ed.), *Phytoremediation on Air Pollution. The Impact of Air Pollution on Health, Economy, Environment and Agricultural Sources*, *1*, 281–294. Intech.

Yang, J., McBride, J., Zhou, J., & Sun, Z. (2005). The urban forest in Beijing and its role in air pollution reduction. *Urban Forestry & Urban Greening*, *3*(2), 65–78.

Yang, Y., Su, Y., & Zhao, S. (2020). An efficient plant–microbe phytoremediation method to remove formaldehyde from air. *Environmental Chemistry Letters, 18*(1), 197–206.

Yassin, M.F., AlThaqeb, B.E., & Al-Mutiri, E.A. (2012). Assessment of indoor PM2. 5 in different residential environments. *Atmospheric Environment, 56*, 65–68.

Yin, L., Niu, Z., Chen, X., Chen, J., Zhang, F., & Xu, L. (2014). Characteristics of water-soluble inorganic ions in PM 2.5 and PM 2.5–10 in the coastal urban agglomeration along the Western Taiwan Strait Region, China. *Environmental Science and Pollution Research, 21*(7), 5141–5156.

Yu, C.W.F., & Kim, J.T. (2010). Building pathology, investigation of sick buildings – VOC emissions. *Indoor and Built Environment, 19*(1), 30–39.

Yu, D., Deng, Q., Wang, J., Chang, X., Wang, S., Yang, R., ... & Yu, J. (2019). Air pollutants are associated with dry eye disease in urban ophthalmic outpatients: a prevalence study in China. *Journal of Translational Medicine, 17*(1), 1–9.

Yuan, C.S.J., & Shen, T.T. (2004). Electrostatistic Precipitation. In *Air Pollution Control Engineering*, 153–196. Totowa, NJ: Humana Press.

Zeng, Y., Xie, R., Cao, J., Chen, Z., Fan, Q., Liu, B., Lian, X., & Huang, H. (2020). Simultaneous removal of multiple indoorair pollutants using a combined process of electrostatic precipitation and catalytic decomposition. *Chemical Engineering Journal, 388*, 124219.

Zhang, H., Sun, Y., Xie, X., Kim, M.S., Dowd, S.E., & Paré, P.W. (2009). A soil bacterium regulates plant acquisition of iron via deficiency-inducible mechanisms. *The Plant Journal, 58*(4), 568–577.

Zhang, J.S., Shaw, C.Y., Kanabus-Kaminska, J.M., MacDonald, R.A., Magee, R.J., Lusztyk, E., Zhang, K., Huo, Q., Zhou, Y.Y., Wang, H.H., Li, G.P., Wang, Y.W., & Wang, Y.Y. (2019). Textiles/metal–organic frameworks composites as flexible air filters for efficient particulate matter removal. *ACS Applied Materials & Interfaces, 11*(19), 17368–17374.

Zhang, J. S., & Shaw, C. Y., Kanabus-Kaminska, J. M., MacDonald, R. A., Magee, R. J., Lusztyk, E., & Weichert, H. J. (1996). Study of air velocity and turbulence effects on organic compound emissions from building materials/furnishings using a new small test chamber. In *Characterizing Sources of Indoor Air Pollution and Related Sink Effects*. ASTM International.

Zhang, L., Weng, H.X., Gao, C.J., & Chen, H.L. (2002). Remove volatile organic compounds (VOCs) with membrane separation techniques. *Journal of Environmental Sciences, 14*(2), 181–187.

Zhang, X., Chen, B., & Fan, X. (2015). Different fuel types and heating approaches impact on the indoor air quality of rural houses in Northern China. *Procedia Engineering, 121*, 493–500.

Zhao, Y., & Wang, S. (2015). The relationship between urbanization, economic growth and energy consumption in China: an econometric perspective analysis. *Sustainability, 7*, 5609–5627. doi: 10.3390/su7055609.

10 Phytoremediation: Importance and General Mechanisms

Surbhi Singla, Aastha Baliyan, and Baljinder Singh
Department of Biotechnology, Panjab University, Chandigarh, India

CONTENTS

10.1	Introduction	197
10.2	Sources of Contamination	198
10.3	Consequences of Heavy Metal Contamination	199
10.4	Uptake and Translocation Mechanism	200
10.5	Phytoremediation Processes	200
	10.5.1 Phytoextraction	200
	10.5.2 Phytovolatilisation	201
	10.5.3 Phytodesalination	201
	10.5.4 Phytostabilisation	202
	10.5.5 Phytodegradation	202
	10.5.6 Rhizofiltration	202
	10.5.7 Phytostimulation	203
	10.5.8 Phytohydraulic Containment	203
10.6	Enhancing Phytoremediation Capability of Plants	203
10.7	Conclusion	203
References		204

10.1 Introduction

Increasing industrialisation and urbanisation have culminated in the accumulation of hazardous pollutants in the environment. The release of untreated contaminated water from resources like oil and gas industries, sewage sludge, metal mining, and smelting pollutes the water resources, which contaminates the soil when used for irrigation purposes. Excessive applications of fertilisers and pesticides to increase agricultural yield also results in environmental pollution. Among the contaminants released into the environment, heavy metals pose a serious threat to all life forms, which is due to their non-degradable nature and long half-life (>20 years) (Ashraf et al., 2019). Plants take up heavy metals present in the soil, which enter the food chain and get accumulated through biomagnification.

Elements with relatively high atomic weight and density, such as Cd, Hg, Pb, As, Zn, Cu, Ni, Fe, and Cr, are considered heavy metals (Yan et al., 2020). They are non-degradable and can change their oxidation state. Heavy metals have been categorised into essential and non-essential metals. Essential heavy metals (Cu, Fe, Mn, Ni, Zn) have physiological or biochemical roles. They are toxic only when their concentration increases, whereas non-essential heavy metals (Pb, Cd, As, Hg) have no known function and are highly toxic. Though the concentration of heavy metals ranges from 1 to 100,000 mg/kg, topsoil contains the heavy metals in the highest concentration because of the ability of organic horizons to bind heavy metals strongly. According to the U.S. EPA, more than 10 million people suffer from chronic diseases caused by heavy metals (Jan et al., 2015). It directly impacts the physiological and biochemical pathways in humans. It affects the plant's ability to take up essential nutrients and results in stunted growth and low yield.

Various techniques (such as physicochemical and biological) have been employed to remediate soil and groundwater. Physio-chemical techniques are ex-situ techniques that can be applied only to small regions that are highly contaminated. It includes soil incineration, soil washing, excavation, acid leaching, chemical treatment, electrokinetics, thermal or pyrometallurgical separation, activated carbon adsorption, microbes use, air stripping, and biochemical and biosorptive technologies (Hashim et al., 2011; Wuana and Okieimen, 2011). These techniques require high energy input, a long time, are costly, inefficient, and technically complex. It drastically affects the soil ecosystem and may introduce secondary pollutants. Biological techniques are in-situ techniques that include bioremediation and phytoremediation. Bioremediation employs microbes to immobilise the heavy metals and thereby, reducing their bioavailability. This method cannot degrade the heavy metal but can only convert it into other forms by altering its physical and chemical properties (Ayangbenro and Babalola, 2017). The limitations of all these techniques call for a method that is efficient, cheap, eco-friendly, and economically feasible. Phytoremediation is one such technology that has grabbed much attention. Phytoremediation is the removal of heavy metals from the soil by plants. It employs various strategies to detoxify heavy metals, including *phytoextraction, phytostabilization, phytodegradation, phytostimulation, phytovolatilization, rhizofiltration, phytohydraulic containment,* and *phytodesalination* (Yan et al., 2020). It has large-scale field applicability and also improves soil fertility. Despite all these advantages, phytoremediation has some limitations. Plant species selected for the purpose are slow growing or poorly adapted to a variety of environments. Hence, research is being done to improve plant species' accumulating ability and introduce new capabilities by genetically engineering plants. Microbes are also being used to increase the metal bioavailability and/or ameliorate soil health to enhance plant growth.

10.2 Sources of Contamination

Excessive anthropogenic activities along with natural processes, such as weathering of rocks, contaminate the environment. Discharging the effluents containing heavy metals from industries, mining sites, and smelting into water resources contaminates the water. Fuel manufacturing, agrochemical industry, and production of construction material release gases and dust in the atmosphere. Heavy metals exist in the aerosol form in the air, which gets precipitated and deposited in soil (Wei and Yang, 2010).

Nutrient deficiency in soil is a significant issue these days. Soil is being supplied with biosolids (sewage sludge, compost, livestock manure) and fertilisers to replenish the soil and increase its fertility. Pesticides and herbicides that are extensively being used in agriculture and horticulture are a common practice to increase crop yield. Application of these compounds leads to the buildup of harmful chemicals in soil. Pesticides contain heavy metals like copper and arsenate. With the increasing use of fertilisers, the amount of Pb and Cd is also increasing (Basta et al., 2005; Atafar et al., 2008). Besides this, ore processing, paint pigments, electroplating, medical and electrical waste, coal combustion, and fly ash also contribute to the increasing amount of heavy metals in the surroundings (Sumiahadi and Acar, 2018). From soil, heavy metals get leached to groundwater and further contaminate it.

Industrial products like wood preservatives; dye stuffs; agricultural products including insecticides, herbicides, and fungicides; and veterinary medicines to eradicate tapeworm are good sources of arsenic (Tchounwou et al., 1999). Manufacturing of batteries and synthesis of alloys and pigments increase Cd emission in the environment. The release of Pb in the atmosphere is growing with Pb-acid batteries, fuel combustion, mining, and manufacturing activities (Kocadal et al., 2020). Hg, which is highly toxic, is widely used in the electrical, pharmaceutical, wood, dentistry, and nuclear industries. In water, algae, and bacteria methylate Hg, which fish take up, and through consumption of contaminated fish, it enters the food chain and gets bioaccumulated (Bhan and Sarkar, 2005). Industries such as metal processing, tannery, chromate production, stainless steel welding, ferrochrome, and chrome pigment production result in Cr release in the environment (Cohen et al., 1993).

10.3 Consequences of Heavy Metal Contamination

Heavy metals affect both micro- and macro-organisms, including plants, animals, and humans. Metal-contaminated crops, such as vegetables, are unfit for human consumption as they cause carcinogenic and other chronic diseases. Heavy metals disrupt cellular processes through various mechanisms such as increased reactive oxygen species (ROS) production, inhibiting functional groups, and replacing basic metal ions. Elevated ROS levels cause oxidative stress, which damages DNA (Mittal and Mehta, 2008). Heavy metals interact with DNA and disrupt the DNA repair mechanisms, enhancing the rate of mutations and increasing the risk of carcinogenicity. They induce double-stranded breaks and favour its false repair (Morales et al., 2016). Ag, Hg, and Cu cause severe kidney, liver, and respiratory diseases. Cd, Ni, and Se are highly carcinogenic. Cd decreases DNA damage repair, upregulates cytokines and proto-oncogenes, and activates protein degradation (Hwua and Yang, 1998). Pb adversely affects the central nervous system, which leads to intellectual disability, memory loss, and impaired development. Pb penetrates the fetal tissue and causes neurodevelopmental deficiencies. Since Pb mimics calcium, it interacts with the skeleton similarly to calcium and makes the bones fragile (Flora et al., 2006). Some metals affect the hair and decrease fertility (Dixit et al., 2015). Factors such as the type of element, the form of existence, a person's individual susceptibility, duration, and route of exposure to heavy metal determine the extremity of the consequences of heavy metals. Exposure to heavy metals generally increases the mortality rate (Jan et al., 2015).

Since metals may have toxic effects on the plant, the plant must detoxify the heavy metals to remediate them. Plants adopt different strategies to lower the heavy metal concentration below the threshold level. As the first line of defense, plants try to limit the metal uptake and its movement into other tissues through various mechanisms such as root sorption, metal exclusion, and metal precipitation (Table 10.1).

TABLE 10.1

Common heavy metal contaminants, their sources, and toxic effects

Contaminant	Sources	Toxic Effects
Ag	Sewage sludge, metallurgical enterprises, pesticides, production of photo and electrical materials, the cement plants	Skin problems, breathing difficulties, lung and throat irritation, stomach pain
As	Pesticides and preservatives for wood	Important cellular activities, including oxidative phosphorylation and ATP production, are affected
Cd	Paint pigments, plastic stabilisers, phosphate fertilisers, electroplating	Carcinogenic, mutagenic, affects calcium regulation, fragile bones
Cr	Tanneries, steel mills, fly ash, and biosolids	Hair loss, carcinogenic
Cu	Pesticides, fertilisers	Multiple organ failure (brain damage, liver cirrhosis, kidney damage, stomach, and intestine inflammation)
Hg	Medical waste, biosolids, and emissions from Au-Ag mining and coal combustion	Autoimmune diseases, fatigue, hair loss, memory loss, disturbed vision, brain damage, tremors, lung and kidney failure
Ni	Automobile batteries, industrial effluents, household appliances, medical tools, steel alloys	Skin diseases, lung cancer, immunotoxic, neurotoxic, genotoxic, decreased fertility, hair loss
Pb	Lead petrol combustion, fertiliser, herbicides, insecticides, biosolids, and battery production	Reduced children's growth and intellect, short-term memory loss, and learning impairments
Se	Coal combustion; mining, smelting, and refining factories; sewage effluent	At high concentration, it is genotoxic and carcinogenic

10.4 Uptake and Translocation Mechanism

Heavy metals accumulate in plants through various mechanisms such as root uptake, xylem storing, root-to-shoot transport, compartmentalisation, and metal confinement. The majority of heavy metals exist in insoluble form and are therefore inaccessible to plants. Plants exude a variety of exudates that alter the pH of the rhizosphere and increase metal solubility and bioavailability (Dalvi 2013). Roots take up heavy metals either through the apoplastic or the symplastic pathway. Immobilisation of the metal ions in the extracellular space constitutes the apoplastic pathway, whereas immobilisation in the intracellular spaces, such as vacuoles, constitutes the symplastic pathway (Tangahu et al., 2011). Ions are transported to the shoots from the root symplasm, where these ions are sequestered in the apoplast or symplast and prevent its accumulation in cytosol.

Various metal ion transporters and cross-linking agents mediate the uptake and translocation of metal ions. Complexing agents, including organic acids, amino acids, metallothioneins, and phytochelatins, chelate heavy metal ions and form precipitate, which is taken up by plants (Hall, 2002). Some heavy metals induce the production of amino acids, which chelates heavy metals and detoxifies them. For example, Ni hyperaccumulation induces histidine production, and histidine acts as a chelator of Ni (Rai, 2002). These complexes are translocated from the cytosol into vacuoles, leaf petioles, and trichomes. These are inactive compartments where these metals cause less toxicity. Heavy metals are removed from the sites where vital cellular functions occur through sequestration and compartmentalisation, and thus, their detrimental effect is decreased.

Heavy metal ion transporters include H^+ coupled carriers, co- and anti-transporters, and channels that mediate specific metals' transport across the cellular membranes and from roots to shoots. Several families of metal transporters have been identified based on sequence homology, which includes ZRT-IRT-like proteins (ZIP family), metal transporter proteins (MTPs), heavy metal transporting ATPases (HMAs), and naturally resistant associated macrophage proteins (NRAMPs) (Yan et al., 2020). The function of ZIP family proteins is the translocation of certain heavy metals like Fe, Zn, and Mn from root to shoot. It has been observed in the overexpression of ZIP family transporters in the Zn hyperaccumulator plant *Arabidopsis halleri* (Guerinot, 2000). HMA3 is involved in vacuolar compartmentalisation and sequestration of Zn, Cd, Co, and Pb (Williams and Mills, 2005). The MTP family proteins regulate the translocation of metals into internal compartments and extracellular spaces. NRAMPs are localised in membranes and mediate Fe^{2+}, Mn^{2+}, Co^{2+}, Cu^{2+}, and Cd^{2+} ions (Cailliatte et al., 2010; Figure 10.1).

10.5 Phytoremediation Processes

10.5.1 Phytoextraction

Phytoextraction, also referred to as phytoaccumulation, is a process employed in which plants extract organic compounds and heavy metals present in the contaminated soil and water. In recent times, phytoextraction has become a cost-effective and environmentally friendly method to remediate contaminated soil (especially with heavy metals like Zn, Pb, Cr, Cd,Cu, and Ni) without any changes in soil fertility and texture (Kanwar et al., 2020). It is ideal for soils that contain low levels of heavy metals (Kanwar et al., 2020). The plants exhibiting a natural ability to phytoextract the contaminants are generally resistant to the pollutants they accumulate and are known as hyperaccumulators. Examples of hyperaccumulators include *Solanum lycopersicum, Triticum aestivum, Eichhornia crassipe, Linum usitatissimum, Sedum alfredii, Cannabis sativa, Nicotianna tobacum, Zea mays, Helianthus annus,* and several grass species. Phytoextraction aims to extract and remove the soil contaminants that can be harvested and reclaimed later from the plant; hence, it has a broad commercial application.

The process of phytoextraction includes mobilisation and absorption of the contaminants from soil or water by the roots of the hyperaccumulator plants, which are then translocated to the upper parts of the plants, thereby accumulating the contaminants into a harvestable plant biomass (Vamerali et al., 2010).

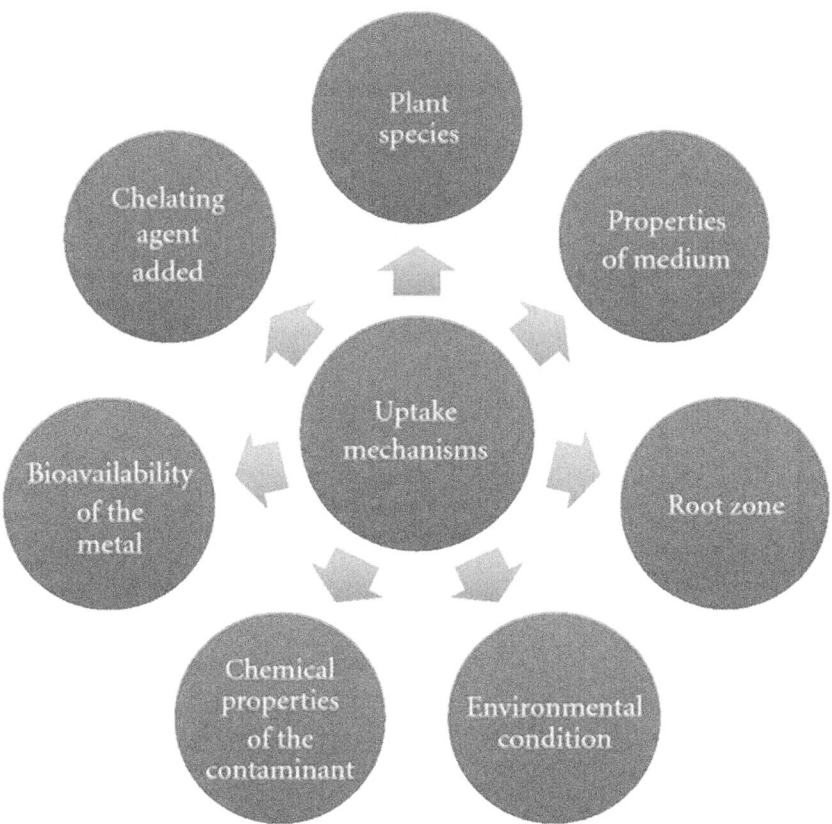

FIGURE 10.1 Factors affecting the contaminant uptake mechanisms in the plants.

Kanwar et al. (2020) categorised the phytoextraction process into (a) natural phytoextraction and (b) phytochelatin-assisted phytoextraction.

10.5.2 Phytovolatilisation

Phytovolatilisation is the process by which plants uptake the soil contaminants (like Se, Hg, tritium, Ar, ammonia, sulfur hexafluoride, methane, etc.) and convert them into less toxic, volatile substances, and release them into the atmosphere (Chaney et al., 1997). Examples of plant species that show natural phytovolatilisation are *Brassica juncea, Typha latifolia, Scirpus acutus, Fraxinus mandshurica, Alnus glutinosa, Phragmites australis, Scirpus lacustris,* etc.

There are two mechanisms for phytovolatilisation: direct and indirect. Direct phytovolatilisation is the mechanism by which the uptake of the soil contaminants, translocation, and eventual volatilisation from stem and leaves (Limmer and Burken, 2016). On the other hand, indirect phytovolatilisation is the mechanism involving an increased contaminant flux from root activities as the result of a lowered water table, increased soil permeability, chemical transport through hydraulic redistribution, etc. (Limmer and Burken, 2016).

10.5.3 Phytodesalination

Sodium and other salts can be accumulated in the soil due to either weathering of plant minerals and/or anthropogenic activities like improper resource management, resulting in saline-sodic soils (Qadir and Oster, 2004). Saline-sodic soil affects crop growth adversely due to decreased water availability to the plants (as a function of increased osmotic pressure) and structural changes in the soil affecting physical processes (Qadir and Oster, 2004). This can affect the movements of air and water, available water for plants, alter the water-holding capacity of the soil, decrease root penetration, seed emergence, etc., and such soil is more prone to erosion and tillage (Qadir and Oster, 2004). These problems are hurdles to sustainable agricultural practices and also decrease the yield substantially.

The process to remediate the calcareous and saline-sodic soil is known as vegetative bioremediation (Qadir and Oster, 2004). This method depends on the fact that some plants can naturally solubilise $CaCO_3$ because of root respiration and H^+ release. This decreases the soil salinity and sodity by reducing the soluble sodium content. The released Ca^{2+} from the solubilised $CaCO_3$ can substitute the Na^+ ions in the saline-sodic soil (Rabhi et al., 2010). Apart from sustainable agriculture, vegetative bioremediation is highly cost effective and, thereby, an attractive approach to bioremediate landfill leachate.

According to Qadir and Oster (2004), it depends on the following factors:

$$V_{Bio} = \sum R_{p(CO2)} + R_H + R_{Phy} + S_{Na+}$$

$RpCO_2$ is the partial pressure of CO_2 in the root zone, R.H. is the root proton (H+) released in N_2-fixing plants, RPhy is the physical effects of the roots in improving soil aggregation, and S_{Na+} is the amount of sodium removed by harvest.

10.5.4 Phytostabilisation

Usually, heavy metal contaminated sites contain more than one contaminant; however, most of the plants used for phytoremediation are specific for one contaminant. Hence, this poses a problem to commercially exploit plants to remediate soil contaminated with multiple metals (Van Nevel et al., 2007). Phytostabilisation is the most efficient method to solve this problem while also decreasing the risk of the contaminants entering the food chain and the surrounding environment by minimising the metal mobility and reducing their bioavailability (Yan et al., 2020).

Phytostabilisation is usually used as an alternative where phytoextraction may not be possible (like multiple contaminants and/or very high amounts of soil contaminants) (Van Nevel et al., 2007). In comparison to phytoextraction, phytostabilisation does not require the contaminant to be disposed off later. It works by plant root exudates interacting with the heavy metal contaminants, immobilising and adsorbing them onto the roots and/or precipitating them into the soil around the roots (Liu et al., 2020). The roots play a very crucial role in the whole process; the plant species used for phytostabilisation should have deep and dense root systems (Van Nevel et al., 2007); for example, *Festuca rubra, Agrostis capillaris, A. stolonifera, Lolium perenne, Trifolium repens,* and *Salix* spp.

10.5.5 Phytodegradation

Phytodegradation is the process to decompose organic contaminants like crude oil, petroleum hydrocarbons, SDS (sodium dodecyl sulfate), pesticides, antibiotics, petrochemical waste, azo dyes, and other organic compounds in the soil and wastewater and convert them into lesser toxic forms (Liu et al., 2020).

The phytodegradation process includes uptake of the contaminants from the rhizosphere and breaking down the pollutants through various internal metabolic processes (Abdullah et al., 2020). To a lesser extent, phytodegradation is also achieved by degrading the contaminants by the enzymes released by the plants into the rhizosphere (Abdullah et al., 2020). The plants that exhibit the phytodegradation ability include *Canna indica, Eichhornia crassipes, Spirodela polyrhiza, Phragmites australis,* etc.

10.5.6 Rhizofiltration

Rhizofiltration, also known as phytofiltration, means the filtration of toxic organic compounds and toxic heavy metals (like Pb, Cd, Cu, Ni, Cr, etc.) that are present in the soil through the roots (Kanwar et al., 2020; Chaudhry et al., 1998). The hyperaccumulating plant roots adsorb and absorb the pollutants from the soil and wastewater and are eliminated by the roots, both in-situ and ex-situ (Kanwar et al., 2020). Plants exhibiting the rhizofiltration capability are usually aquatic plants; however, many terrestrial plants also carry out this process. The examples of plants showing natural rhizofiltration ability are sunflower, rye, tobacco, spinach, Indian mustard, water hyacinth, Azola, duckweed, etc.

10.5.7 Phytostimulation

Phytostimulation, or rhizodegradation, is used to degrade organic contaminants present in the rhizosphere by employing enhanced microbial activity by inoculating endophytes to the rhizosphere (Ashraf et al., 2019). Various endophyte microbial species, called phytostimulators, can be added to the rhizosphere and increase plant growth directly. The phytostimulators generally secrete plant hormones like auxins, cytokinins, and gibberellins (Steenhoudt and Vanderleyden, 2000), stimulating root elongation and enhanced plant growth (Lugtenberg et al., 2002). *Azospirillum* is a commonly used phytostimulator that can enhance *Chlorella vulgaris*; hence, it can clean the contaminated water (Bloemberg and Lugtenberg, 2001).

10.5.8 Phytohydraulic Containment

According to Farmayan et al., phytohydraulic containment is the process where trees are used as natural pumps to create a depression zone in the saturation zone to restrict their migration to other zones and potentially remediate the groundwater from contaminants. In 1998, Burken and Schnoor demonstrated that poplar plants could be used to translocate and contain a wide variety of organic pollutants. In 2001, Farmayan et al. conducted research to assess the impact of deep-rooted poplar trees in hydraulically containing MTBE plume at the Houston site. It concluded that the poplar trees could locally depress the water table and create a capture zone for plume containment. In 1997, Gordon et al. showed that poplar trees removed about 95% of the groundwater contaminated with TCE. Barac et al. (2009) demonstrated that poplar trees and the microorganisms associated with them could reduce containment levels in groundwater and prevent contaminants from migrating off-site.

10.6 Enhancing Phytoremediation Capability of Plants

A significant bottleneck in the large-scale commercial application of phytoremediation is the inability of any given plant species to remediate a wide range of contaminants (which is usually the case in contaminated soil and water). It is also desirable that plants exhibit very high efficiency for removing contaminants present in very high concentrations at a high rate without adversely affecting plant health. This can be achieved by: (1) The selection of plant species that naturally exhibit greater efficiency for removing pollutants from affected areas and (2) selective of the plant genome manipulation through rational genetic engineering to increase the phytoremediation efficiency, i.e., either by manipulation of an existing gene in the plant genome or by introducing new genes in the genome of the plant to enable the plant to remediate a broader range of pollutants with higher efficiency and rate than naturally achievable—for example, introducing 2E1 (mammalian cytochrome P450) in tobacco plants to aid the oxidation of halogenated hydrocarbons like trichloroethylene (Chaudhry etal., 2002; Doty et al., 2000).

10.7 Conclusion

Environmental pollution has been accelerating tremendously nowadays, and newer, faster, and more efficient methods are required to solve this problem. Sustainable development is the key to moving forward with the available resources without further damaging the environment. Phytoremediation has come across as a cost-effective and environmentally friendly method to remediate various soil contaminants from the environment. As discussed in this chapter, these contaminants have serious health effects if they become part of the food chain. Apart from the impact on humans, these contaminants also pose severe and often lethal health effects to other species; hence, disrupting the food chains and the ecosystem.

In addition to the methods mentioned in this chapter, various methods are also employed to remediate the environment, including multiple microbes or physical methods to aid the process. Notably, these methods do not work alone and are not exclusive to only one type of contaminant. They are often used in combination to achieve the targeted result.

The natural ability of a plant to phytoremediate can be augmented either by combining with other methods or by genetically engineering the plant to optimise the phytoremediation capability of plants to use them for commercial purposes.

REFERENCES

Abdullah, S.R.S., Al-Baldawi, I.A., Almansoory, A.F., Purwanti, I.F., Al-Sbani, N.H., & Sharuddin, S.S.N. (2020). Plant-assisted remediation of hydrocarbons in water and soil: application, mechanisms, challenges, and opportunities. *Chemosphere*, 247. 10.1016/j.chemosphere.2020.125932.

Ashraf, S., Ali, Q., Zahir, Z.A., Ashraf, S., & Asghar, H.N. (2019). Phytoremediation: environmentally sustainable way for reclamation of heavy metal polluted soils. *Ecotoxicology and Environmental Safety*, 174. 10.1016/j.ecoenv.2019.02.068.

Atafar, Z., Mesdaghinia, A., Nouri, J., Homaee, M., Yunesian, M., Ahmadimoghaddam, M., & Mahvi, A.H. (2008). Effect of fertilizer application on soil heavy metal concentration. *Environmental Monitoring and Assessment*, 160(1), 83. 10.1007/s10661-008-0659-x.

Ayangbenro, A., & Babalola, O. (2017). A new strategy for heavy metal polluted environments: a review of microbial biosorbents. *International Journal of Environmental Research and Public Health*, 14(1), 94. 10.3390/ijerph14010094.

Barac, T., Weyens, N., Oeyen, L., Taghavi, S., Van Der Lelie, D., Dubin, D., Spliet, M., & Vangronsveld, J. (2009). Field note: hydraulic containment of a BTEX plume using poplar trees. *International Journal of Phytoremediation*, 11(5). 10.1080/15226510802655880.

Basta, N.T., Ryan, J.A., & Chaney, R.L. (2005). Trace element chemistry in residual-treated soil: key concepts and metal bioavailability. *Journal of Environmental Quality*, 34(1), 49–63. doi:10.2134/jeq2005.0049dup.

Bhan, A., & Sarkar, N.N. (2005). Mercury in the environment: effect on health and reproduction. *Reviews on Environmental Health*, 20(1), 39–56. 10.1515/reveh.2005.20.1.39.

Bloemberg, G.V., & Lugtenberg, B.J.J. (2001). Molecular basis of plant growth promotion and biocontrol by rhizobacteria. *Current Opinion in Plant Biology*, 4(4). 10.1016/S1369-5266(00)00183-7.

Cailliatte, R., Schikora, A., Briat, J.-F., Mari, S., & Curie, C. (2010). High-affinity manganese uptake by the metal transporter NRAMP1 is essential for *Arabidopsis* growth in low manganese conditions. *The Plant Cell*, 22(3), 904 LP–904917. 10.1105/tpc.109.073023

Chaney, R.L., Malik, M., Li, Y.M., Brown, S.L., Brewer, E.P., Angle, J.S., & Baker, A.J.M. (1997). Phytoremediation of soil metals. *Current Opinion in Biotechnology*, 8(3). 10.1016/S0958-1669(97)80004-3.

Chaudhry, Q., Schröder, P., Werck-Reichhart, D., Grajek, W., & Marecik, R. (2002). Prospects and limitations of phytoremediation for the removal of persistent pesticides in the environment. *Environmental Science and Pollution Research*, 9(1). 10.1007/BF02987313.

Chaudhry, T., Hayes, W., Khan, A., & Khoo, C. (1998). Phytoremediation – focusing on accumulator plants that remediate metal-contaminated soils. *Australasian Journal of Ecotoxicology*, 4(1), 37–51.

Cohen, M.D., Kargacin, B., Klein, C.B., & Costa, M. (1993). Mechanisms of chromium carcinogenicity and toxicity. *Critical Reviews in Toxicology*, 23(3), 255–281. 10.3109/10408449309105012.

Dalvi, A.A., & Bhalerao, S.A. (2013). Response of plants towards heavy metal toxicity: an overview of avoidance, tolerance and uptake mechanism. *Annals of Plant Sciences*, 2(09), 2013. https://www.annalsofplantsciences.com/index.php/aps/article/view/87.

Dixit, R., Wasiullah, Malaviya, D., Pandiyan, K., Singh, U.B., Sahu, A., Shukla, R., Singh, B.P., Rai, J.P., Sharma, P.K., Lade, H., & Paul, D. (2015). Bioremediation of heavy metals from soil and aquatic environment: an overview of principles and criteria of fundamental processes. *Sustainability (Switzerland)*, 7(2), 2189–2212. 10.3390/su7022189.

Doty, S.L., Shang, T.Q., Wilson, A.M., Tangen, J., Westergreen, A.D., Newman, L.A., Strand, S.E., & Gordon, M.P. (2000). Enhanced metabolism of halogenated hydrocarbons in transgenic plants containing mammalian cytochrome P450 2E1. *Proceedings of the National Academy of Sciences of the United States of America*, 97(12). 10.1073/pnas.97.12.6287.

Flora, S.J.S., Flora, G., & Saxena, G. (2006). Environmental occurrence, health effects and management of lead poisoning. In *Lead*, pp. 158–228. Elsevier Science BV.

Gordon, M., Choe, N., Duffy, J., Ekuan, G., Heilman, P., Muiznieks, I., Ruszaj, M., Shurtleff, B.B., Strand, S., Wilmoth, J., & Newman, L.A. (1998). Phytoremediation of trichloroethylene with hybrid poplars. *Environmental Health Perspectives*, 106(4). 10.1289/ehp.98106s41001.

Guerinot, M. Lou. (2000). The ZIP family of metal transporters. *Elsevier Science B.V. All*, 1465, 190–198.

Hall, J.L. (2002). Cellular mechanisms for heavy metal detoxification and tolerance. *Journal of Experimental Botany*, 53(366), 1–11. 10.1093/jexbot/53.366.1.

Hashim, M.A., Mukhopadhyay, S., Sahu, J.N., & Sengupta, B. (2011). Remediation technologies for heavy metal contaminated groundwater. *Journal of Environmental Management*, 92(10), 2355–2388. 10.1016/j.jenvman.2011.06.009.

Hong, M.S., Farmayan, W.F., Dortch, I.J., Chiang, C.Y., McMillan, S.K., & Schnoor, J.L. (2001). Phytoremediation of MTBE from a groundwater plume. *Environmental Science and Technology*, 35(6). 10.1021/es001911b.

Hwua, Y.S., & Yang, J.L. (1998). Effect of 3-aminotriazole on anchorage independence and mutagenicity in cadmium- and lead-treated diploid human fibroblasts. *Carcinogenesis*, 19(5), 881–888. 10.1093/carcin/19.5.881.

Jan, A.T., Azam, M., Siddiqui, K., Ali, A., Choi, I., & Haq, Q.M.R. (2015). Heavy metals and human health: mechanistic insight into toxicity and counter defense system of antioxidants. *International Journal of Molecular Sciences*, 16(12), 29592–29630. 10.3390/ijms161226183.

Kanwar, V.S., Sharma, A., Srivastav, A.L., & Rani, L. (2020). Phytoremediation of toxic metals present in soil and water environment: a critical review. *Environmental Science and Pollution Research*. 10.1007/s11356-020-10713-3.

Kocadal, K., Alkas, F.B., Battal, D., & Saygi, S. (2020). Cellular pathologies and genotoxic effects arising secondary to heavy metal exposure: a review. *Human and Experimental Toxicology*, 39(1), 3–13. 10.1177/0960327119874439.

Limmer, M., & Burken, J. (2016). Phytovolatilization of organic contaminants. *Environmental Science and Technology*, 50(13). 10.1021/acs.est.5b04113.

Liu, S., Yang, B., Liang, Y., Xiao, Y., & Fang, J. (2020). Prospect of phytoremediation combined with other approaches for remediation of heavy metal-polluted soils. *Environmental Science and Pollution Research*, 27(14). 10.1007/s11356-020-08282-6.

Lugtenberg, B.J.J., Chin-A-Woeng, T.F.C., & Bloemberg, G.V. (2002). Microbe-plant interactions: principles and mechanisms. *Antonie van Leeuwenhoek, International Journal of General and Molecular Microbiology*, 81(1–4). 10.1023/A:1020596903142.

Mittal, M., & Mehta, A. (2008). Flora SJS, Mittal M, Mehta A. Heavy metal induced oxidative stress & its possible reversal by chelation therapy. Indian J Med Res 128: 501–523. *The Indian Journal of Medical Research*, 128, 501–523.

Morales, M.E., Derbes, R.S., Ade, C.M., Ortego, J.C., Stark, J., Deininger, P.L., & Roy-Engel, A.M. (2016). Heavy metal exposure influences double strand break DNA repair outcomes. *PLoS One*, 11(3), e0151367. 10.1371/journal.pone.0151367.

Qadir, M., & Oster, J.D. (2004). Crop and irrigation management strategies for saline-sodic soils and waters aimed at environmentally sustainable agriculture. *Science of the Total Environment*, 323(1–3). 10.1016/j.scitotenv.2003.10.012.

Rabhi, M., Ferchichi, S., Jouini, J., Hamrouni, M.H., Koyro, H.W., Ranieri, A., Abdelly, C., & Smaoui, A. (2010). Phytodesalination of a salt-affected soil with the halophyte Sesuvium portulacastrum L. to arrange in advance the requirements for the successful growth of a glycophytic crop. *Bioresource Technology*, 101(17). 10.1016/j.biortech.2010.03.097.

Rai, V.K. (2002). Role of amino acids in plant responses to stresses. *Biologia Plantarum*, 45(4), 481–487. 10.1023/A:1022308229759.

Steenhoudt, O., & Vanderleyden, J. (2000). Azospirillum, a free-living nitrogen-fixing bacterium closely associated with grasses: genetic, biochemical and ecological aspects. *FEMS Microbiology Reviews*, 24(4). 10.1111/j.1574-6976.2000.tb00552.x.

Sumiahadi, A., & Acar, R. (2018). A review of phytoremediation technology: heavy metals uptake by plants. *IOP Conference Series: Earth and Environmental Science*, *142*(1). 10.1088/1755-1315/142/1/012023.

Tangahu, B.V., Sheikh Abdullah, S.R., Basri, H., Idris, M., Anuar, N., & Mukhlisin, M. (2011). A Review on heavy metals (As, Pb, and Hg) uptake by plants through phytoremediation. *International Journal of Chemical Engineering*, *2011*, 939161. 10.1155/2011/939161.

Tchounwou, P.B., Wilson, B., & Ishaque, A. (1999). Important considerations in the development of public health advisories for arsenic and arsenic-containing compounds in drinking water. *Reviews on Environmental Health*, *14*(4), 211–229. 10.1515/reveh.1999.14.4.211.

Vamerali, T., Bandiera, M., & Mosca, G. (2010). Field crops for phytoremediation of metal-contaminated land. A review. *Environmental Chemistry Letters*, *8*(1). 10.1007/s10311-009-0268-0.

Van Nevel, L., Mertens, J., Oorts, K., & Verheyen, K. (2007). Phytoextraction of metals from soils: how far from practice? *Environmental Pollution*, *150*(1). 10.1016/j.envpol.2007.05.024.

Wei, B., & Yang, L. (2010). A review of heavy metal contaminations in urban soils, urban road dusts and agricultural soils from China. *Microchemical Journal*, *94*(2), 99–107. 10.1016/j.microc.2009.09.014.

Williams, L.E., & Mills, R.F. (2005). P_{1B}-ATPases – an ancient family of transition metal pumps with diverse functions in plants. *Trends in Plant Science*, *10*(10), 491–502. 10.1016/j.tplants.2005.08.008.

Wuana, R.A., & Okieimen, F.E. (2011). Heavy metals in contaminated soils: a review of sources, chemistry, risks and best available strategies for remediation. *ISRN Ecology*, *2011*, 402647. 10.5402/2011/402647.

Yan, A., Wang, Y., Tan, S.N., Mohd Yusof, M.L., Ghosh, S., & Chen, Z. (2020). Phytoremediation: a promising approach for revegetation of heavy metal-polluted land. *Frontiers in Plant Science*, 11. 10.3389/fpls.2020.00359.

11

Phytoremediation Mechanisms of Heavy Metal Removal: A Step Towards a Green and Sustainable Environment

Nathaniel A. Nwogwu[1,2,3], Oluwaseyi A. Ajala[4,5], Fidelis O. Ajibade[2,3,6], Temitope F. Ajibade[3,6,7], Bashir Adelodun[8,9], Kayode H. Lasisi[3,6,7], Adamu Y. Ugya[10,11], Pankaj Kumar[12], Ifeoluwa F. Omotade[13], Toju E. Babalola[14], James R. Adewumi[6], and Christopher O. Akinbile[13]

[1] *Department of Agricultural and Bioresources Engineering, Federal University of Technology, Owerri, Nigeria*
[2] *Research Centre for Eco-Environmental Sciences, Chinese Academy of Sciences, Beijing, PR China*
[3] *University of Chinese Academy of Sciences, Beijing, PR China*
[4] *Department of Zoology and Environmental Sciences, Punjabi University, Patiala, Punjabi, India*
[5] *Department of Chemistry, Faculty of Science University of Ibadan, Ibadan, Nigeria*
[6] *Department of Civil and Environmental Engineering, Federal University of Technology, Akure, Nigeria*
[7] *Institute of Urban Environment, Chinese Academy of Sciences, Xiamen, PR China*
[8] *Department of Agricultural and Biosystems Engineering, University of Ilorin, Ilorin, Nigeria*
[9] *Department of Agricultural Civil Engineering, Kyungpook National University, Daegu, Korea*
[10] *Department of Environmental Management, Kaduna State University, Kaduna State, Nigeria*
[11] *Key Lab of Groundwater Resources and Environment of Ministry of Education, Key Lab of Water Resources and Aquatic Environment of Jilin Province, College of New Energy and Environment, Jilin University, Changchun, People's Republic of China*
[12] *Agro-ecology and Pollution Research Laboratory, Department of Zoology and Environmental Science, Gurukula Kangri (Deemed to be University), Haridwar, Uttarakhand, India*
[13] *Department of Agricultural and Environmental Engineering, Federal University of Technology, Akure, Nigeria*
[14] *Department of Water Resources Management & Agro-Meteorology, Federal University Oye-Ekiti, Oye, Ekiti State, Nigeria*

CONTENTS

11.1	Background	208
11.2	Phytoremediation: A Green Technology of Bioremediation	209
11.3	Techniques and Mechanisms Involved in Phytoremediation and Their Applications	210
	11.3.1 Phytoextraction	212

	11.3.2	Phytofiltration (Rhizofiltration)	213
	11.3.3	Phytostabilization	213
	11.3.4	Phytovolatilization	214
	11.3.5	Phytodegradation/Phytotransformation	214
11.4		Phytoremediation of Heavy Metals Using Aquatic Macrophytes	215
11.5		Factors Affecting the Mobility, Absorption, and Accumulation in Plants	216
11.6		Challenges Associated with Phytoremediation in Decontaminating Soil/Water	217
11.7		Significance of Plant Root Nature in Phytoremediation/Phytomining	218
11.8		Innovative Approaches for Optimizing Phytoremediation of Heavy Metals	219
	11.8.1	Chelators (Chemical-Assisted Phytoremediation)	219
	11.8.2	Biochar	219
	11.8.3	Genetic Engineering: Utilization of Transgenic Plants	220
	11.8.4	Microbes	221
	11.8.5	Exogenous Phytohormones	221
	118.6	Electro Kinetic-Enhanced Phytoremediation	221
	11.8.7	Nanoparticles	222
11.9		Phytoremediation as a Tool for Environmental Sustainability	222
11.10		Future Trends in Phytoremediation	222
11.11		Conclusion	224
References			225

11.1 Background

Environmental contamination has been an occurrence over time in history. It has always been noted to have developed as a result of environmental pollution that emanates from both the natural accumulation of these contaminants and some anthropogenic activities within the environment (Ajibade et al., 2021a). It is noted that environmental contamination has become more pronounced within the last two centuries, especially the 20th century. This upsurge is stated to be associated with the rapid population growth, urbanization, industrialization, and mining activities; agricultural practices; as well as wars and various conflicts that induce explosives and chemical warfare agents, as well as their manufacture, usage, and storage (Macek et al., 2004). Different types of organic or inorganic contaminants exist. Some of the organic contaminants include petroleum hydrocarbons (e.g., benzopyrene), linear halogenated hydrocarbons (e.g., trichloroethylene), chlorinated solvents (e.g., polychlorinated benzenes), explosives (e.g., trinitrotoluene), and volatile organic carbons while the inorganic compounds comprise phosphates, nitrates, and metals and metalloids such as chromium (Cr), nickel (Ni), mercury (Hg), lead (Pb), arsenic (As), cadmium (Cd), selenium (Se), silver (Ag), copper (Cu), and zinc (Zn), and nonradioactive or radioactive nuclides, like uranium (U), strontium (Sr), and cesium (Cs) (Chandra and Kumar, 2018). Despite the need to control environmental pollution and its associated health hazards, these wastes are still being dumped on land or discharged into water bodies (e.g., rivers, canals, lakes, streams, etc.). Metal accumulation in various compartments of the environment poses risks to living organisms including humans because living organisms accumulate these metal elements, causing biomagnification processes to occur. This biomagnification process happens when there is an increase in the concentration of the contaminants at successively higher levels in a food chain (Ali et al., 2013; Gomes et al., 2016). This continuous rise in the presence of these elements in the ecosystem results in the increased presence of metallic contaminants in the form of metallic nanoparticles in the environment has been of concern to researchers (Ebbs et al., 2016). Although there are several benefits associated with these chemicals when applied in different fields, such as pharmaceutics, cosmetics, energy, consumer products, and agriculture (Baker et al., 2014), their associated potential risks and negative effects as a result of their use and to the environment remain unclear (Ruffini-Castiglione and Cremonini, 2009).

Over the years, various physicochemical and biological treatment methods have been used as soil cleanup techniques. However, their costs and inefficiencies as well as the fact that sometimes they result in the transfer of the associated risks to other places, thereby causing the control of the contaminants to surround environmental compartments, has led to some research on the use of more effective biological remediation methods (Macek et al., 2004; Gomes et al., 2016). An effective option is the use of some plants that are capable of transferring, accumulating, and removing these pollutants from the environment or lessening their spread within the environment (Macek et al., 2000). This concept is known as phytoremediation. Phytoremediation is the utilization of plants in conjunction with their related microbes to extract, sequester, and detoxify various types of environmental and emerging contaminants from soils, sediments, air, and water (Rahman et al., 2016; Kumar and Chandra, 2017; Sarwar et al., 2017; Yadav et al., 2018; Patra et al., 2020). It is a plant-based technology that involves the discovery of hyperaccumulating plants (capable of accumulating, translocating, and concentrating metals in their harvestable biomass). It has received increasing attention in recent years (Ajibade et al., 2013; Ajibade and Adewumi, 2017; Patra et al., 2020). Furthermore, it has been proposed that this approach can remove inorganic and organic xenobiotics as well as contaminants found in soil, water, and air (Lasat, 2000; Ryslava et al., 2003). Examples of these pollutants are heavy metals, inorganic fertilizers, trace or radioactive elements, pesticides, and recalcitrant organic compounds like polychlorinated biphenyls (PCBs) or polyaromatic hydrocarbons (PAHs). In 1885, Baumann was the first to hypothesize plant accumulation of high amounts of metals for *Thlaspi caerulescens* and *Viola calaminaria* (Lasat, 2000). The use of metal-hyperaccumulating plants for selective removal and recycling of extreme soil metals was first made known in 1983, gained general acceptance in 1990, and it has been studied over time with suggestions that it is a more promising eco-friendly, practical, and cost-effective technology than the existing physicochemical techniques (Chaney et al., 1997). While there have been numerous review studies on phytoremediation (Ali et al., 2013; Ovečka and Takáč, 2014; Akinbile et al., 2016; Mahar et al., 2016; da Conceição Gomes et al., 2016; Antoniadis et al., 2017; Yadav et al., 2018; Akinbile et al., 2019; Omotade et al., 2019; Ugya et al., 2019; Patra et al., 2020), they somewhat provide scattered information.

Thus, this chapter presents a comprehensive synthesis of all aspects of phytoremediation, including numerous techniques (phytoextraction, phytofiltration/rhizofiltration, phytostabilization, phytovolatilisation, phytodegradation/phytotransformation). Additionally, the review highlights factors affecting the mobility, uptake, and accumulation of metals in plants, and challenges associated with phytoremediation in decontaminating soil and water. It further discusses innovative approaches for optimizing phytoremediation technology (chelators, biochar, transgenic plants, metal ions, microbes, etc.) and the use of phytoremediation techniques as tools for ensuring environmental sustainability. Lastly, future trends in phytoremediation were also addressed.

11.2 Phytoremediation: A Green Technology of Bioremediation

The term *phytoremediation* simply means "plant-based action". It is coined from a Greek word *phyto*, meaning "plant" and Latin word *remedium*, meaning "correct evil or to recover" or "restoring balance" (Erakhrumen and Agbontalor, 2007). The term *phytoremediation* was first mentioned in 1991 (Raskin et al., 1998). Many researchers have defined phytoremediation in various ways that collectively refer to a green technology involving the use of plants for environmental cleanup via pollutant removal. Some researchers define phytoremediation as follows:

> *Phytoremediation is defined as the use of green plants and their associated rhizospheric microorganisms, soil amendments, and agronomic techniques to remove, degrade, or detoxify harmful environmental pollutants. (Ouyang, 2002; Schwitzguébel, 2002)*

> *Phytoremediation is defined as the use of plants and their associated microbes to extract, sequester, and/or detoxify various kinds of environmental pollution or emerging contaminants*

from water, sediments, soils, and air (as magnetic particles lying in the particulate matter). (Rai and Singh, 2015; Rai and Chutia, 2016)

Phytotechnology (phytoremediation technology) has various advantages that make it a promising technology for remediation of hazardous sites. These advantages of phytoremediation can be summarized as follows.

Aesthetics: Phytoremediation is an aesthetically attractive technology (Tangahu et al., 2011). Its major plus is its effectiveness in the reduction of heavy metal ion concentration to a minimal level through its various techniques of degradation of contaminants (Rodriguez et al., 2005) while using low-cost biosorbent materials (Rakhshaee et al., 2009). According to the USEPA (2000), phytoremediation is the cleanest and least expensive approach for the treatment of designated hazardous areas.

Less disruptive: Phytoremediation is considered less disruptive than the conventional physicochemical techniques (Tangahu et al., 2011). In addition to providing on-site treatment, this system is environmentally friendly (e.g., phytoextraction) and conserves the topsoil; hence, leaving the topsoil in a utile state which may be recovered for agricultural use.

Effectiveness in contaminants reduction/removal: Phytoremediation is drawing attention as an innovative technology, basically due to its effectiveness and efficiency in the reduction and removal of contaminants (especially, toxic heavy metals). It is considered a cleanup method for contaminated air, water, and soil (Van Ginneken et al., 2007). Phytoremediation can be applied to treat diversified toxic metals and radionuclides, reduce environmental disorderliness, eliminate waterborne wastes, and secondary air, thus gaining more public acceptance (Liu et al., 2000a). Furthermore, phytoremediation can reduce organic contaminants to CO_2 and H_2O, thereby minimizing environmental toxicity (Mwegoha, 2008). When treating heavy metals in the soil and water, phytoremediation technology translocates them into recyclable metal-rich plant residue (Liu et al., 2000b).

Inexpensiveness: Phytoremediation is a cheap technology for the remediation of contaminated environments (Rakhshaee et al., 2009; Tangahu et al., 2011), mainly suitable for sites of comparatively low contamination (Van Ginneken et al., 2007). It is cheaper and less expensive (even up to 60–80% reduction) than conventional physicochemical methods, since it is solar-energy-driven, and at the same time neither expensive equipment nor highly skilled employees are required for management (Tangahu et al., 2011).

Wide range of application: Phytoremediation is suggested to be efficient in treating a wide range of both organic and inorganic pollutants in various environments such as water, soil, and air (Mwegoha, 2008; Van Ginneken et al., 2007). This equally makes the technology a very promising one compared to the conventional methods.

Environmental friendliness: Phytoremediation is better compared to harsher redressing technologies such as incineration, thermal vapourization, solvent washing, or other soil-washing techniques, which simultaneously destroy the biological components of the soil and equally alter its chemical and physical characteristics and consequently leave a comparatively nonviable solid waste (Tangahu et al., 2011). The utilization of phytoremediation generally improves the soil, resulting in a more functional ecosystem and at about one-tenth of the cost of today's technologies (Hinchman et al., 1995). Tangahu et al. (2011) stated that phytoremediation remains the most environmentally friendly cleanup technology for remediation of contaminated soils; hence, it is regarded as a green technology.

Other advantages and benefits of phytoremediation over conventional methods: requires fewer soil disposal sites, higher public acceptability, does not require excavation and transportation, and the possibility of much larger-scale remediation (Schnoor et al., 1995).

11.3 Techniques and Mechanisms Involved in Phytoremediation and Their Applications

Phytoremediation is an umbrella for many plant-related bioremediations. These include different techniques and mechanisms used by plants in detoxifying and decontaminating soil, sludge, groundwater, sediment, and wastewater. For effectiveness, phytoremediation technology is categorized into

Phytoremediation Mechanisms

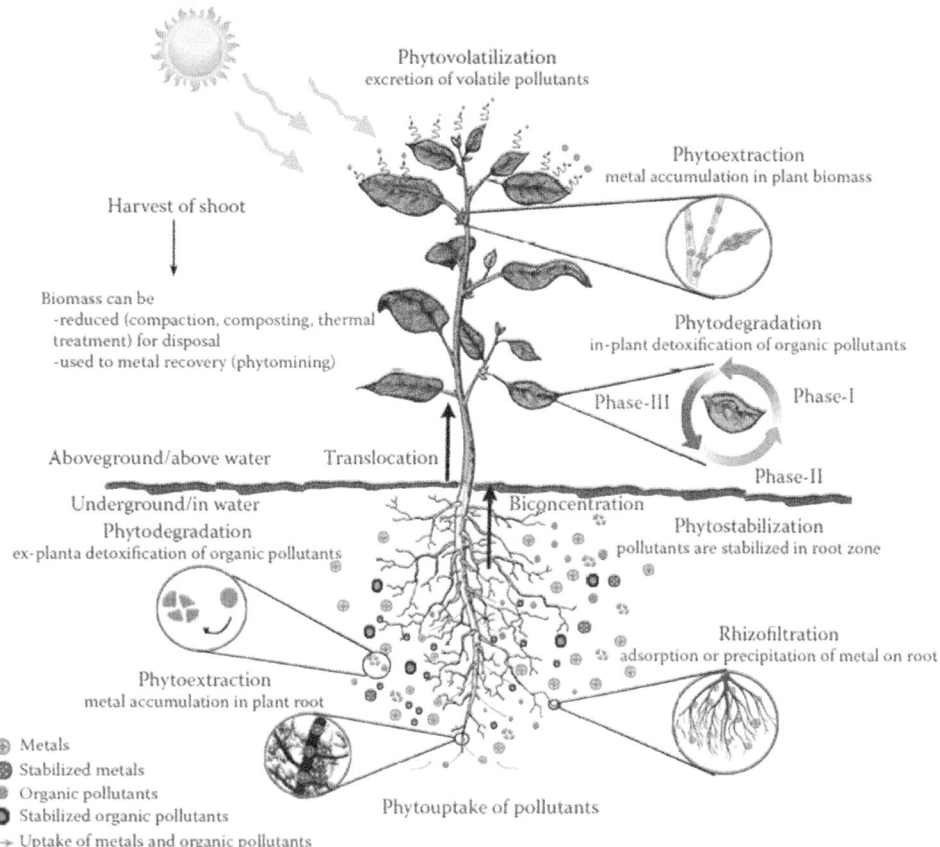

FIGURE 11.1 Graphical representation of various phytoremediation strategies (adapted from Chandra and Kumar, 2018).

phytoextraction, phytofiltration, phytostabilization, phytovolatilization, and phytodegradation, with each category having a different strategy or mechanism of action for redressing pollutants from contaminated soil, sludge, groundwater, surface water, sediment, and wastewater (Alkorta et al., 2004; Chandra and Kumar, 2018), as illustrated in Figure 11.1.

Moreover, the application of these phytoremediation technologies is dependent on some factors such as contaminants, site conditions, degree of cleanup required, and the plant type (Chandra and Kumar, 2018; Rai, 2018). Despite some specific features of plants used in each of these phytoremediation technologies, there are general characteristics that every plant selected for phytoremediation should possess (Sharma et al., 2014). These features are:

1. Ability to grow fast.
2. High tolerance to metals and other contaminants.
3. High resistance to diseases, pests, etc.
4. Easy to cultivate and harvest.
5. Extensive root and shoot system.
6. Unappealing to animals to minimize the transfer of metals and other toxic compounds to a higher trophic level of the terrestrial food chain (Bruce et al., 2003).

Furthermore, Rai (2018) stated that wetland plants could be good candidates for this purpose as they detoxicate soil and groundwater compartments of the environment in the following ways:

1. Altering the physicochemical properties of contaminated soils.
2. Release of root exudates, which increases organic carbon.
3. Improving aeration through the release of oxygen in the root zone, and enhancing porosity in the upper soil zone.
4. Interception and retardation of the mobility of the contaminants.
5. Plant enzymatic conversion of xenobiotics and recalcitrant chemicals, as well as performing co-metabolic microbial.
6. Reducing pollutants' vertical and lateral movement into groundwater by collecting available water and reversing the hydraulic gradient (Singh and Rai, 2016).

11.3.1 Phytoextraction

Phytoextraction is the process by which plants absorb toxic contaminants from soil, sediments, and/or sludge and store or concentrate them in their shoot tissues and harvestable parts of their roots (Garbiscu and Alkorta, 2001). It is also known as phytoaccumulation, phytoabsorption, or phytosequestration. Phytoextraction is the most recognized and used phytoremediation technology for the removal of harmful metal pollutants (Chandra, 2015; Gomes et al., 2016); however, its efficiency when in use is determined by factors like bioavailability of metals, properties of soil, and speciation of metal and plant species that are mainly on shoot metal concentration and biomass (Ali et al., 2013; Li et al., 2010). As illustrated in Figure 11.2, phytoextraction is classified into two categories: induced phytoextraction and continuous phytoextraction.

Induced phytoremediation is the use of high biomass metal-accumulating plants in which their metal accumulation capability is accelerated by the addition of chemicals like ethylenediaminetetraacetic acid (EDTA), ethylenediaminedisuccinic acid (EDDS), nitrilotriacetic acid (NTA), etc. (Saifullah et al., 2009). Continuous phytoextraction on the other hand involves the use of hyperaccumulators naturally for the decontamination of sites and accumulation of contaminating metals (McGrath et al., 2002).

Interestingly, some plants for phytoextraction have been described to possess a special characteristic known as phytoexcretion (Kadukova and Kavuličova, 2011). For instance, Mediterranean halophytic shrubs or trees, *Tamarix smyrnensis,* has been suggested to be able to use salt glands for the removal of excess salt from the soil. Similarly, in a situation where metals are available in the soil, the salt glands become the excretion point of these metals in the form of non-toxic crystals; this technique is referred to as phytoexcretion (Kadukova and Kalogerakis, 2007; Manousaki et al., 2008). According to UNEP (2016), approximately >400 plants have extreme affinity for metal absorption and some of these plants have been reported in recent studies for remediating heavy metals. They include *Jatropha curcas* (Marrugo-Negrete et al., 2015), *Stanleya pinnata* (Bañuelos et al., 2015), *Brassica juncea* (Kathal et al., 2016), *Pisum sativum* (Tariq and Ashraf, 2016), *Puccinellia frigida* (Rámila et al., 2016), *Brassica napus* (Dhiman et al., 2016), *Helianthus annuus* (Farid et al., 2017), *Salix viminalis* (Mleczek et al., 2018), *Coronopus didymus* (Sidhu et al., 2018), *Melilotus officinalis* and *Amaranthus retroflexus*

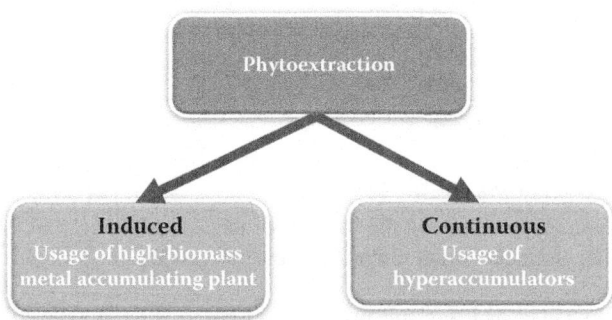

FIGURE 11.2 Phytoextraction classification.

Phytoremediation Mechanisms 213

(Ghazaryan et al., 2019), *Allium schoenoprasum* (Eisazadeh et al., 2019), *Amaranths hypochondriacus* (Sun et al., 2020), *Vetiveria zizanioides* (Pentyala and Eapen, 2020), and *Corchorus capsularis* (Saleem et al., 2020). However, in conjunction with the previously described mechanism, the special characteristics of plants used in phytoextraction include:

a. Presence of profuse root system.
b. Tolerance to high metal contents.
c. Exhibit rapid growth rate, (3) production of high biomass.
d. Have the ability to accumulate metals in harvestable parts.
e. Adaptation to extreme weather events (Ali et al., 2013; Yadav et al., 2018).

11.3.2 Phytofiltration (Rhizofiltration)

Rhizofiltration is a root zone phytoremediation strategy, also called phytofiltration, caulofiltration, and blastofiltration, in which terrestrial and aquatic plants are used for removing pollutants, mainly heavy metals in wastewater (Yadav et al., 2018). Contaminants are absorbed, adsorption, concentrated, and precipitated during this process by terrestrial and aquatic plants from polluted sites, usually aqueous sources such as wetlands and wastewater. It involves a series of physical and biochemical processes such as surface absorption of metal onto the plant roots, intracellular absorption, vacuole deposition, translocation to the shoot, or heavy metal precipitation by the action of plant exudates (Ajibade et al., 2013; Kumar and Chandra, 2017). Depending on the nature of the toxic metals, rhizofiltration can be accompanied by other phytoremediation processes such as phytoextraction, phytostabilization, and phytovolatilization. However, the effectiveness of rhizofiltration is dependent on the ability of plant roots to absorb, accumulate, and precipitate contaminants (Akinbile and Yusoff, 2012; Rahman et al., 2016; Bora and Sarma, 2020). Hence, the ideal plants to be used for rhizofiltration must: (1) Have the ability to grow rapidly; (2) produce a good quantity of roots; (3) have roots with large surface area; (4) ability to grow under submerged environments; and (5) can absorb, tolerate, and precipitate toxic metals over a lengthy duration of time (Rahman et al., 2016; Kumar and Chandra, 2017; Bora and Sarma, 2020). In addition, several terrestrial, aquatic, and wetland plants have been discovered and tested for their phytofiltration capacities to remediate heavy metals and some of these plant species include *Micranthemum umbrosum* and *Warnstorfia fluitans* (Islam et al., 2015), *Phragmites australis* and *Typha latifolia* (Kumari and Tripathi, 2015), *Typha domingensis* (Vymazal, 2016), *Salix matsudana* (Tang et al., 2017), *Micranthemum umbrosum* (Islam et al., 2017), *Pistia stratiotes* (Galal et al., 2018), *Warnstorfia fluitans* (Sandhi et al., 2018), *Potamogeton pusillus* (Griboff et al., 2018), *Nymphaea tetragona* (Li et al., 2019), and *Monosoleum tenerum* (Sut-Lohmann et al., 2020). Rhizofiltration is an effective and cost-efficient phytoremediation strategy; however, the use of aquatic plants has been reported to have limited capacity for rhizofiltration due to their small size, slow-growing, and high-water-content roots, which is responsible for their inability to dry, compost, and incinerate rapidly compared to terrestrial plants that have long and effective fibrous root systems and large root surface area for metal sorption (Dushenkov et al., 1995).

11.3.3 Phytostabilization

Phytostabilization, also called phytoimmobilization, is another phytoremediation strategy where contaminants such as heavy metals in the soil are immobilized to reduce their mobility and bioavailability in the environment by the use of certain plant species through absorption and accumulation by their roots, adsorption on their root surfaces, or precipitation within the root zone and physical stabilization of soil (Rahman et al., 2016). Unlike phytoextraction, phytostabilization deals with the sequestering of heavy metals within the rhizosphere (Figure 11.1), thus hindering their leaching into the groundwater and absorption by the plant tissues, therefore reducing metal risks to human health and the environment (Kumar and Chandra, 2017; Yadav et al., 2018). Phytostabilization is used mainly for immobilizing heavy metals such as Pb, As, Cd, Cr, Cu, and Zn. However, some organic contaminants and their metabolic by-products have the tendency to be attracted or attached to plant parts such as lignins, which can be related to phytostabilization in terms of phytolignification with a long history of application in removing organic contaminants. In addition, microorganisms such as bacteria and mycorrhiza present in the rhizosphere of these plants play an important role in overcoming phytotoxicity (Mastretta et al., 2006). One major difference between phytolignification and phytostabilization is that phytostabilization of heavy metals generally occurs in the soil, whereas phytolignification of organic contaminants occurs above the ground level (Kumar and Chandra, 2017). Consequently, the characteristic features of plants used in phytostabilization include: (1) must be tolerant to soil conditions, (2) have rapid growth rate, (3) have dense root systems, (4) easy to grow and maintain, and (5) have a relatively long life or able to self-propagate (Berti and Cunnigham, 2000). Metal-tolerant plants species commonly and recently used in the phytostabilization of heavy metals include *Brassica juncea, Typha domingensis, Phragmites australis, Lupinus luteus, Miscanthus sinensis, Rose plant, Epilobium dodonaei, Iris sibirica, Hordeum vulgare, Vicia villosa, Helichrysum microphyllum, Atriplex nummularia, Salix L., Agrostis* spp. and *Festuca* spp., and *Helianthus petiolaris* (Mahar et al., 2016; Ranđelović et al., 2016; Ma et al., 2017; Katoh et al., 2017, Bacchetta et al., 2018, Eissa 2019, Ilić et al., 2020; Saran et al., 2020). Macrophytes such as *Vossia cuspidata, Arundo donax, Zannichellia peltate, Apium nodiflorum, Nasturtium officinale, Typha domingensis,* and have also been recently identified and studied for phytostabilization of heavy metals in the aquatic environment (Galal et al., 2017; Bonanno and Vymazal, 2017).

11.3.4 Phytovolatilization

Phytovolatilization is the release of volatile contaminants via plant transpiration into the atmosphere (Figure 11.1). It is the process whereby organic contaminants are absorbed by plants from the soil and chemically converted into less toxic and volatile compounds and subsequent release into the atmosphere (Ali et al., 2013). Phytovolatilization is a complex phytoremediation strategy driven by biophysical mechanisms of mass transport of highly volatile contaminants through plants (Kumar and Chandra, 2017). Organic contaminants and toxic metals such as Hg, Se, and As are converted into non-toxic forms and released into the atmosphere via roots, shoots, and leaves of plants (Pivetz, 2001; Berken et al., 2002). For instance, the phytovolatilization of Se in plants occurs through a series of metabolic processes whereby several organic and inorganic species of Se such as selenate, selenite, and Se-Met are absorbed and converted into less toxic and volatile compounds (Berken et al., 2002). The volatilization of Se as methyl selenate is recommended as the main process of removal of Se from contaminated soil, while Indian mustard (*Brassica juncea*) is well reported in studies to be the most effective plant for accumulation and volatilization of Se coupled with its rapid growth rate (Berken et al., 2002; Banuelos et al., 2005). Limmer and Burken (2016) categorized phytovolatilization into two types: direct and indirect phytovolatilization. Direct phytovolatilization involves plants absorbing volatile pollutants from polluted locations and releasing them into the atmosphere through their stems, trunks, and leaves through transpiration, while indirect phytovolatilization involves volatile contaminant flux from the subsurface due to the activities of the plant roots (Limmer and Burken, 2016). The removal of heavy metals via ohytovolatilization is less effective because the contaminant must (1) be absorbed by plant roots, (2) move via the xylem to the leaves, (3) be metabolised into volatile forms, and (4) volatilize into the atmosphere (Mueller et al., 1999). Another serious disadvantage of phytovolatilization is the tendency of the volatilized contaminants to be precipitated back into the environment during rainfall (Hussein et al., 2007).

11.3.5 Phytodegradation/Phytotransformation

Phytodegradation, also termed *phytotransformation*, involves the degradation or breakdown of contaminants, mainly organic contaminants, through various plant metabolic processes or via the degrading effects of enzymes produced by the plants on the organic contaminants (Vishnoi and Srivastava, 2008; Ali et al., 2013). Plants are capable of producing enzymes that can facilitate and speed up the degradation of organic contaminants. Some of these enzymes include dehalogenase for degrading chlorinated compounds, peroxidase for transforming phenolic compounds, nitrilase for degrading cyanated aromatic compounds, nitroreductase for transforming explosives and nitrated compounds, and phosphatise for the degradation of organophosphate pesticides (Jabeen et al., 2009). Phytodegradation depends on the direct uptake of contaminants from the contaminated sites and the accumulation in the vegetation. This direct uptake of contaminants into the plant tissues through the root system depends on the uptake efficiency, transpiration rate, and the concentration of the contaminant in the contaminated sites. Also, the uptake efficiency depends on chemical speciation, physicochemical features, and plant properties, while the transpiration rate depends on the environmental conditions and plant characteristics such as nature, and the leaves' surface area (Kumar and Chandra, 2017). Recent studies have focused on the phytodegradation of organic compounds; for example, *Cyperus alternifolius* was employed for degrading and removal of ethanolamines from wastewater (Dolphen and Thiravetyan, 2015) and *Armoracia rusticana*'s ability to degrade benzophenone within its tissues has also been discovered (Chen et al., 2016).

11.4 Phytoremediation of Heavy Metals Using Aquatic Macrophytes

Heavy metals enter into the environment through natural sources such as volcanic eruption, erosion, mineralization (Ali et al., 2013), and man-made sources (Table 11.1). In contrast to organic contaminants, heavy metals and metalloids persist longer in the environment because of their non-biodegradability property (Ahmadpour et al., 2012). While some metals, namely Fe, Co, Mn, Mo, and Zo, are vital for human physiological and biochemical functions in small concentrations, metals like Pb, Cd, Hg, and Pu are detrimental even in their lowest concentrations (Table 11.2) (Tchounwou et al., 2012; Kumar et al., 2017). Several aquatic macrophytes have been suggested to be effective in phytoremediation of both inorganic and organic contaminants. In phytoremediation of inorganic contaminants, water hyacinth (*Eichhornia crassipes*) has been reported to be efficient in enhancing water quality parameters and it is affirmed as an adequate accumulator of Cd and Zinc (Zn) (Valadi et al., 2019; Ansari et al., 2020). Saha et al. (2017) suggested that plants can equally be employed in the remediation of water polluted with toxic hexavalent chromium (CrVI), while *Lemna minor* has been reported to be able to degrade soluble lead (Pb), nickel (Ni) (Axtell et al., 2003) as well as Cr (VI) (Uysal, 2013). Similarly, *Wolffia globosa* has also been found to have a significant concentration of trace metals in the roots and leaves of and *Najas marina* (Baldantoni et al., 2004). Bennicelli et al.

TABLE 11.1

Natural and anthropogenic sources of heavy metal(loid)s in recent studies

Sources	Test Medium	Heavy Metal(loid)s	References
Natural			
Lithogenic	Coal power soil	As, Pb, and Cd	Zhang et al., 2020a
	Agricultural soils	Fe, Cr, V, Cu, Ni, and Mn	Yuanan et al., 2020
Atmospheric deposition	Yangtze sediments	Cd	Han et al., 2017
Anthropogenic			
Coal mining	Coal power soil	Cu, Mn, and Zn	Zhang et al., 2020a
Vehicle emissions	Agricultural soils	Cd and Pb	Yuanan et al., 2020
Mining activities	Gold mining area	As, Cr, Pb, and Hg	Adewumi, 2020
Pesticides and fertilizers	Farm soil	As and Cd	Zhang et al., 2020b
Fertilizer application	Lake sediment	Cu and Cd	Dai et al., 2018
Shipping activities	Surface water	Cu and Cr	Ismail et al., 2016
Petroleum spills	Pipeline vandalised area	Ni and V	Ogunlaja et al., 2019
Wastewater irrigation	Cropland soils	As	Dong et al., 2019
Metal smelting	Pb/Zn smelter soil	Pb, Zn, Cd, and Cu	Fan et al., 2019

TABLE 11.2

Hazardous effect of heavy metals on human health

Metals	Effects	Reference
Cd	Nausea, muscular weakness, vomiting, dyspnea, and abdominal cramps; in extreme exposure, leads to pulmonary edema or death.	Duruibe et al., 2007; Sobha et al., 2007; Lalor, 2008
Zn	Diarrhea, vomiting, icterus, anemia, bloody urine, kidney or liver failure.	Duruibe et al., 2007; Jaiswal et al., 2018
Cu	Gastrointestinal and skin irritation, hepatic and renal damage, widespread capillary and central nervous system damage.	Argun et al., 2007; Stern et al., 2007; Chambers et al., 2010
Pb	Cardiovascular malfunction, gastrointestinal problems, haemoglobin synthesis inhibition, neurological disorders, and resulting in death.	Kazemipour et al., 2008; Odum, 2016
Cr	Strong oxidizing agent, soluble in alkaline and mildly acidic water, corrosive, and potentially carcinogenic and toxic to humans.	Shaffer et al., 2001; Jeyasingh and Philip, 2005; Huang et al., 2009
Hg	Gastrointestinal disorders like corrosive hematochezia and esophagitis, congenital malformation, and spontaneous abortion.	Duruibe et al., 2007
As	Causes protein coagulation, Guillian-Barre syndrome, and increased exposure lead to death.	Duruibe et al., 2007; Jaiswal et al., 2018

(2004), in their studies using *Azolla caroliniana* (a small water fern), stated that the plants can decontaminate waters contaminated by mercury (Hg) and Cr.

On the other hand, many free-floating macrophytes have also been reported to show reasonable phytoremediation potential for organic contaminants. For instance, Salvinia, Lemna, Eichhornia, and Pistia have been proposed to be effective in removing organic pollutants from water (Bhaskaran et al., 2013). *Chara vulgaris, Lemna minor, Myriophyllum spicatum,* and *Potamogetan perfoliatus* were also reported to remove alkylbenzene sulfonate (Liu and Wu, 2018; Zhou et al., 2018; Liu et al., 2019). Moreover, *Lemna minor* has been reported to have the capacity to enhance the quality of blue and textile dyes-polluted water (Ekperusi et al., 2019; Neag et al., 2018). It has been reported that *Lemna gibba* is used in phytoremediation of several biochemical processes: biochemical oxygen demand (BOD), total suspended solids, ammonium nitrate and phosphate, ammonia, and total nitrogen (Ekperusi et al., 2019).

11.5 Factors Affecting the Mobility, Absorption, and Accumulation in Plants

Several factors affect the solubility and bioavailability of heavy metals to plants and one of these determining factors is the soil physicochemical properties such as soil pH, electrical conductivity, soil texture, organic matters, cation exchange capacity, clay content, loading rate, and redox potential (Yadav et al., 2018; Patra et al., 2020). It is generally understood that heavy metals in soil with a high clay content, organic matter, or soil pH are less mobile and bioavailable to the plant. Another important factor responsible for the differences in metal mobility and uptake by plants is soil temperature. Plant roots in the soil absorb heavy metals after their solubility and mobilization. However, their mobility in the soil depends on the following: (1) Diffusion of heavy metals along the concentration gradient due to their uptake and depletion around the root vicinity, (2) root interception due to the displacement of soil volume by the roots, and (3) the bulk soil solution to water potential gradient flow of heavy metals (Marschner, 1995). Heavy metal absorption occurs by both roots and leaves and hence increases with increasing metal concentrations in the surrounding environment. However, there is no linear relationship between the increasing concentration of heavy metals in the external environment with the uptake of metals. This is because the metals bound in the plant tissue cause saturation, which is ruled by the metal uptake rate. Also, the efficiency of metal uptake by plants is highest at their low concentration in the external environment. When the accumulation of heavy metals occurs in the vascular system of plants, most of these metals are insoluble and restricted from moving freely, therefore forming sulphate, phosphate, or carbonate precipitates (Raskin et al., 1998). Generally, when metal ions are absorbed by root symplasm, their mobility to the xylem depends on three procedures: (1) metal sequestration into the root symplasm, (2) symplastic transport into the stele, and (3) release of metals into the xylem (Shah et al., 2010).

The membrane transport proteins mediate the movement of metal ions into the xylem. However, the competition between the transport of essential and non-essential metals using these transmembrane carriers exist. For instance, Cr (III) is taken up by plants in passive processes compared with Cr (VI) transport, which is mediated by sulphate carriers. Nevertheless, the affinity for Cr (VI) is low. In a barley seedlings experiment, inhibitors such as dinitrophenol and sodium azide inhibit the uptake of Cr (VI) as compared to Cr (III), which they did not (Skeffington et al., 1976). Shewry and Peterson (1974) reported group VI anions such as SO_4^{2-} as inhibitors to chromate uptake while Cr (VI) transport is stimulated by Ca^{2+}. Hence, the transportation of the inhibited chromate is due to the chemical similarities and competitiveness of these inhibitors, while the stimulated transport of Cr (VI) by Ca^{2+} occurred because of its important role in the uptake and transport of metals in the plants (Zayed and Terry, 2003; Montes-Holguin et al., 2006). Metals such as B, Cd, Mn, Mo, Se, and Zn are readily absorbed and transported within plant tissues, whereas other metals such as Al, Ag, Cr, Fe, Hg, and Pb are bound strongly to the soil components and plant roots, thus with less mobility (Chaney, 1983a, b). Nevertheless, at some concentrations, all these metals can be translocated within the plant system, even at a concentration gradient. For instance, essential metals such as Zn^{2+}, Cu^{2+}, and Ni^{2+} and nonessential Cd^{2+} have been shown using kinetic data as they compete for their transport through the same transmembrane carrier (Crowley et al., 1991). Other significant factors that influence the uptake mechanisms and efficiency of the various phytotechnologies employed in toxic metal removal from the soil, sludge, and wastewater include bioavailability of the metal, environmental condition, plant species, root zone, chemical properties of the contaminant, properties of the medium, and chelating agent added (Sheoran et al., 2016).

11.6 Challenges Associated with Phytoremediation in Decontaminating Soil/Water

Phytoremediation has been identified to be an efficient contaminant removal technology; however, some challenges or drawbacks have been reported to hinder its efficiency (Ramamurthy and Memarian, 2012; Mahar et al., 2016), especially in the removal of toxic metals from both soil and water environments. A few of these challenges include inadequate metal phytoavailability, biomass, low growth rate, and unable to target metal hyperaccumulators. As a result of these challenges, phytoremediation will take a long period to attain the desired remediation goal (Saxena et al., 2019); thus, leading to an increase in cost for treatment, safety, and accompanied liability of risks, because the longer time the more the cost when assessing the economic practicability at the field (Conesa et al., 2012; Mahar et al., 2016). The enhancement of the phytoremediation technology, such as the use of synthetic chelators for higher metal accumulation by plants, can also add challenges like an additional cost, unwanted environmental issues such as interruption of the physical and chemical properties of soils through mineral components dissolution, soil microorganisms and plant toxicity, and groundwater undesirable leaching, which makes chelate-assisted phytoremediation technology almost unfeasible (Saxena et al., 2019). Vangronsveld et al. (2009) also stated that most European countries have banned several of these metal mobilizing amendments (e.g., EDTA). Also, plants' growth when phytoremediation is taking place equally results in alteration in pH and increase in organic acids, thus, creating room for metals to be more bioaccessible to the food chain prior to their remediated process (Gerhardt et al., 2016), hence bringing up some environmental challenges that can negate some of the merits associated with phytoremediation (Saxena et al., 2019). Since these contaminants are accumulated in plants, there is the likelihood of consequent occurrence such that their toxic forms could be re-discharged or discharge into the soil through leaf debris. Furthermore, phytostabilized metals are not eliminated from the soil, and changing soil conditions may result in re-discharge of contaminants into the environment, necessitating long-term field monitoring to avoid future land-use changes (Gerhardt et al., 2016). Using genetically modified plants to improve phytoremediation may likewise lead to a higher cost of phytoremediation because of the huge maintenance required for contaminated sites, and government-required field application policies in terms of monitoring and dumping biomass (Stephenson and Black, 2014). Phytoremediation technology is still faced with the problem of inadequate research funding from both public and private sector agencies.

In addition to the explanation of the challenges of phytotechnology, Mahar et al. (2016), in their study, summarily stated the challenges of phytoremediation technology as follows:

i. A prolonged period is required for soil remediation.
ii. Reduced efficiency of phytoextraction of some hyperaccumulator as a result of degraded biomass and sulky growth rate.
iii. Proper and careful disposal of the polluted biomass is required as they are equally toxic after phytoextraction.
iv. On the occasion of attack from pests and diseases especially in tropical and sub-tropical regions, the efficiency and effectiveness of the hyperaccumulators are negatively influenced.
v. Difficulty in mobilizing tightly bound metal ions from the soil when there is limited bioavailability of the contaminants.
vi. Careful selection of plant species is required to avoid the introduction of hyperaccumulators that could be invasive to other indigenous floral diversity.
vii. Agronomic practices and soil amendments negatively influence the mobility of contaminants.
viii. Climate and weather condition influences.
ix. Due to unsustainable plant growth in soils of high contamination, phytoremediation applies primarily in areas of low to moderate metal pollution levels.
x. When biomass is mishandled, there could be a transfer of the accumulated metal contaminants into the food chain.

11.7 Significance of Plant Root Nature in Phytoremediation/Phytomining

Biotechnology is gradually transmuting global agriculture while adding new traits to crop plants at an immense level of advancement (Gleba et al., 1999). Besides the conventional benefits of plants such as the provision of food, fibre, medicines, and construction materials, biotechnology has disclosed peculiar uses of plants that have been gaining attention from the general public as well as the scientific community globally; these "value-added" uses are phytoremediation (Salt et al., 1998) and molecular farming (phyto-manufacturing), which is the use of plants for the production of valuable organic molecules and recombinant proteins (Nawrath et al., 1995; Franken et al., 1997).

Apart from root crops, plant roots are relatively less studied than the shoots (Gleba et al., 1999). However, this trend is taking a new paradigm due to the emerging biotechnologies (phytotechnologies) described that tap the ability of plants to transport important molecules into and out of their roots. These root-based technologies mainly include phytoextraction and rhizofiltration. Plant roots form large underground networks that serve as solar-energy-driven pumps that extract and concentrate essential elements and compounds from soil and water. For instance, in phytoextraction of metals, the roots of hyperaccumulators transport and concentrate the toxic metals from the soil into harvestable shoots above the ground (Kumar et al., 1995; Vassil et al., 1998). Hydroponically grown plant roots absorb, precipitate, and concentrate toxic metals from contaminated effluents through rhizofiltration (Dushenkov et al., 1995). Furthermore, Agostini et al. (2013) affirm that in organic contaminant removal, plant roots stimulate accelerated decontamination through the discharge of root exudates and oxido-reductive enzymes, such as peroxidases (Px) and laccases.

11.8 Innovative Approaches for Optimizing Phytoremediation of Heavy Metals

Phytoremediation has been noted to be a promising technology for environmental remediation and sustainability. Various traditional phytoremediation techniques as discussed in an earlier section have been suggested to be good in tackling the heavy metal problems in environments. However, some drawbacks associated with these traditional techniques have limited them from being applied on a large scale (Sarwar et al., 2017). These observed challenges associated with the traditional phytoremediation techniques prompted researchers to modify these techniques for improved results, reduced limitations, as well as promote large-scale application of phytoremediation. The improvement approaches in use for optimizing phytoremediation of heavy metals are explained.

11.8.1 Chelators (Chemical-Assisted Phytoremediation)

This is the strategy in which plants' heavy metal uptake potential is promoted and enhance the bioavailability of heavy metals by the addition of biodegradable physicochemical factors such as chelating agents and micronutrients, as well as by stimulating the heavy-metal-uptake capacity of the microbial community in and around the plant (Tangahu et al., 2011). This chelate-assisted phytoremediation technology is also referred to as induced phytoextraction. It is a difficult task to select hyperaccumulator plant species that are effective and efficient for the phytoremediation of heavy metals. However, Baker et al. (1994) suggested the adoption of the shoot-to-root metal concentration ratio to determine whether a plant can be considered as a hyperaccumulator. They concluded that if the ratio is greater than 1.0, it shows that the shoot stores more metal than the root of such plant species; hence, the plant species can be employed as a hyperaccumulator and as a matter of fact could be applied for phytoextraction. Non-hyperaccumulator plants that produce more aboveground biomass but have lower heavy metal extraction potential can equally be modified chemically to suit phytoremediation of heavy metals (Sarwar et al., 2017). Some synthetic chelating agents that have been used in phytoremediation include ethylene diamine tetra–acetic acid (EDTA), diethylene triamine penta–acetic acid (DTPA), and ethylene glycol tetra–acetic acid (AGTA) (Saifullah et al., 2009; Pereira et al., 2010), and nitrilotriacetic acid (NTA) (Tangahu et al., 2011). Furthermore, researchers noticed some inefficiency adaptation issues associated with heavy metal phytoremediation of some plant species, such as stunted growth and low biomass production. This led to the application of some chemicals that would help in improving the capacity of metal plants to work effectively in phytoremediation of heavy metals and at the same time cope with heavy metal stress (Sarwar et al., 2015; Popova et al., 2012). An example of such chemicals with a widespread phenolic compound and has been used is salicylic acid (He et al., 2010; Sarwar et al., 2017).

Besides the advantages of this approach, its most pronounced disadvantage is the high cost of these synthetic chelating agents. Hence, low-molecular-weight organic acids, examples of which are oxalic, citric, acetic, and malic, are used to serve as alternatives and can be effectively used as chelating agents of heavy metals. These organic acids can form metal complexes of low to moderate stability (Sarwar et al., 2017). Another advantage of using these organic acids is that they are more biodegradable in soil than synthetic chelating agents with minimum environmental contamination risks (Souza et al., 2013).

11.8.2 Biochar

Biochar is a carboniferous permeable material that could be produced from the pyrolysis of organic feedstocks such as plant materials, organic manures, and sludges (Paz-Ferreiro et al., 2014). Wood biochar (charcoal) is the most common biochar that has been in use. Biochar is suggested to be a resourceful material for improving traditional phytoremediation techniques due to its specific physicochemical properties (Ajibade et al., 2021c). These properties of biochar include high (alkaline) pH, large surface area for sorption of metals, high carbon content, and its ability to increase the bioavailability of metals in the rhizospheric region, which places it among good heavy metal remediators (Sarwar et al., 2017; Patra et al., 2020). Biochar is also resourceful in phytoremediation of plants as it enhances the biomass (Liu et al., 2013), and this increased plant biomass is possible as a result of high nutrients, water-holding capacity, and cation exchange capacity in biochar (Ahmad et al., 2016; Fellet et al., 2014; Liu et al., 2013). Elad et al. (2012) further affirm that biochar also helps in improving the activity of essential microbes in soil as well as suppressing the action of pathogenic microbes.

11.8.3 Genetic Engineering: Utilization of Transgenic Plants

As a way of enhancing the efficiency and effectiveness of the phytoremediation technology in the decontamination of heavy metals and other toxic contaminants, researchers have tried the application of some advanced technologies. One of these advances made is the utilization of transgenic plants. Transgenic plants in phytoremediation are plants that have been genetically modified for enhanced uptake and degradation of contaminants. The formation of transgenic plants is part of genetic strategies for phytoremediation which have improved the removal or detoxification of toxic metals in the environment due to their increased ability in metal uptake, accumulation, and tolerance to toxicity (Fasani et al., 2018). Currently, using transgenic plants is considered a promising option, despite several anatomical constraints that pose threats to the development of hybrid plants with increased phytoremediation capability (Yadav et al., 2018). Transgenic plants can be developed through a direct DNA method transfer of genes or simple dipping of *Agrobacterium tumefaciens* mediated transformation (Seth, 2012). An example is the inoculation of *Theobroma cacao* with an *Agrobacterium* suspension (Fister et al., 2016). Generally, the use of transgenic plants has been successfully used to phytoextract metals such as Cd, Pb, and Cu, and metalloids such as As and Se in the aboveground biomass (Sarwar et al., 2017; Yadav et al., 2018). It involves the use of metal transporters, improved production of sulphur-metabolizing enzymes such as phytochelations, and metal-detoxifying chelators-metallothioneins (Kotrba et al., 2009). Examples of transgenic plants developed for metal tolerance and phytoremediation are listed in Table 11.3.

Furthermore, Kärenlampi et al. (2000) found that transgenic plants with higher glutathione synthase or phytochelatins synthase show enhanced cadmium (Cd) accumulation. Transgenic plants bearing foreign genes for proteins translocating metals within membranes have equally been developed (Krämer and Chardonay, 2001). Though transgenic plants developed with genes, and huge resistance to certain heavy metals was a result of coding different forms of metallothionein (mammalian, yeast, insect and human) (Kärenlampi et al., 2000), hence the achievement of improved accumulation was not possible

TABLE 11.3

Transgenic plants developed for metal tolerance and phytoremediation

S/N	Transgenic Plants	Plant and Bacteria Source	DNA Method Transfer of Genes for Metals Tolerance and Phytoremediation	References
1	*V. sativa* homologue of Caffeoyl-CoA Omethyltransferase	*Vicia sativa*	sCCoAOMT gene transfer for Cd tolerance and accumulation	Xia et al., 2018
2	Metallothionein	*Sedum plumbizincicola*	SpMT1 gene transfer for Cd hyperaccumulation and hypertolerance	Peng et al., 2017
3	Ubiquitin (Ub)-conjugating enzyme	*Arabidopsis thaliana*	NtUBC1 gene transfer for Cd tolerance and accumulation	Bahmani et al., 2017
4	g-Glutamylcysteine synthetase-glutathione	*Streptococcus thermophils synthetase*	StGCS-GS gene transfer for Cd, Zn, and Cu accumulation	Liu et al., 2015
5	Glutathione synthase GS and γ-glutamylcysteine synthase ECS	*Escherichia coli*	γ-ECS gene transfer for Cd, Cr, As, Pb, and Zn tolerance and accumulation	Reisinger et al., 2008
6	Acyl-CoA-binding protein	*Arabidopsis thaliana*	ACBP1 gene transfer for Pb tolerance and accumulation	Xiao et al., 2008

(Liu et al., 2000b). Dietz and Schnoor (2001), however, stated that metals cannot be enzymatically degraded like organic contaminants; hence, they suggested a special improvement in genetic engineering for phytoremediation of heavy metals. Generally, using transgenic plants in phytoremediation has helped to reduce and address environmental contamination in recent years; however, there are still many challenges to overcome in this area (Yau and Stewart, 2013). One of the major challenges is that most of the research studies in the development and use of transgenic plants are carried out in controlled settings, instead of the physical environment (Abhilash et al., 2009; Bhargava et al., 2012). There is also a need for proper risk evaluation about the application of transgenic plants in phytoremediation of metals and their exposure to humans and wildlife to lead to a safer genetically modification of crops and wider opportunities for their usage (Wolfenbarger and Phifer, 2000; Davison, 2005; Sarwar et al., 2017).

11.8.4 Microbes

Soil microbes carry out some activities in the soil and within the rhizosperic region of plants that are beneficial to plants in their uptake of nutrients. This relationship between soil microorganisms and plants leads to its inclusion among the strategies that enhance the capability of plants in the phytoremediation of heavy metals. The relationship between some soil microbes and plant roots could be a symbiotic or free-living relationship (Sarwar et al., 2017). The primary component of living organism's component within the root zone has been recognized as mycorrhizal fungi (Sarwar et al., 2017), and they are found in many forms with most higher plants such as arbuscular mycorrhizas, ericaceous mycorrhizas, orchid mycorrhizas, and ectomycorrhizas with arbuscular mycorrhizal fungi (AMF) having the most frequent connections with terrestrial plant roots (Sheng and Xia, 2006). The relationship between fungi and plant roots are very important for enhancing the bioavailability of plant nutrients (Ca, N, P, Zn, Co, K, S, Ni, and Cu) via a wide range of hyphal network (Zaidi et al., 2006; Sheng and Xia, 2006). Similarly, these fungal interactions can change the chemical composition of root secretion and soil pH, hence increasing heavy metal bioavailability in the soil (Sarwar et al., 2017). For instance, an application of *Ixeris denticulate, Kummerowia striata*, and *Echinochloa crusgalli* mixed with AMF inoculums, showed an increased uptake and accumulation of Pb (Chen et al., 2003).

Some bacteria, such as plant-growth-promoting bacteria (PGPR) are another beneficial group of microorganisms for plants in tackling heavy metal contamination (Seth, 2012). They are similarly classified as symbiotic bacteria and free-living rhizobacteria (Liao et al., 2003). These bacteria enhance plant development through a variety of ways, including reduced ethylene synthesis under stress, nitrogen fixation, and particular enzyme activity (Glick et al., 1998). These organisms not only improve heavy metal absorption and accumulation, but they also reduce the impact of the toxicity and heavy metal stress in the plant (Sarwar et al., 2017; Farwell et al., 2007); hence, the application of the right microbial inoculum can help plant species effectively remove heavy metals from the soil.

11.8.5 Exogenous Phytohormones

Exogenous phytohormones are applied as supplementary enhancers to plants. They are primarily applied to enhance plant biomass, fitness, and stress tolerance for better phytoremediation efficiency (DalCorso et al., 2019). For instance, cytokinins have been suggested to be effective both in enhancing plant growth and increasing phytoextraction capacity (Aderholt et al., 2017); however, auxins showed contradictory results (Vamerali et al., 2011). Although the use of gibberellins or some commercial growth regulator mixes reduced metal buildup in plant tissues, the increase in plant biomass as a result of their application increased metal extraction per plant (Hadi et al., 2010).

118.6 Electro Kinetic-Enhanced Phytoremediation

This is a new strategy that has been proposed to overcome some of the limitations such as the long time and limited treatment depths that are associated with traditional phytoremediation (DalCorso et al., 2019). This technique applies the combination of plants and a physicochemical treatment, in other words, low-intensity electric fields, to the metal-polluted soil, favouring metal mobilization and bioavailability (Camesselle and Gouveia, 2019). Also, a reasonably high Pb, As, and Cs phytoextraction has been obtained by direct current (DC) electric field used with inert affordable graphite electrodes, because of the modification of soil pH and metal solubility from the electric field (Mao et al., 2016). This technology is fascinating as the application of a solar-cell-powered electric field has been tested in a real-scale field test for the phytoremediation of a metal-polluted electronic waste recycling centre by *Eucalyptus globulus*, and it showed an increased plant growth and metal accumulation (DalCorso et al., 2019). Even though the main power supply systems are more effective in metal transportation and containment, nevertheless the cost effectiveness of solar cells suggests that this technique is more sustainable (Luo et al., 2018).

11.8.7 Nanoparticles

Nanoparticles have equally been identified as effective in assisting materials for the phytoremediation of metals. Due to their unique characteristic, absorption capacity, or redox catalytic activity, many kinds of nanomaterials have been employed to remediate metal-polluted substrates (Mueller and Nowack, 2010). Liang et al. (2017) suggested that a combination of plants and nanoparticles can be used to improve the efficacy of phytostabilization by absorbing metal ions. Their combination can also be employed in phytoextraction to enhance plant fitness and stress tolerance as well as increase metal bioavailability (Moameri and Khalaki, 2017; Singh and Lee, 2016).

11.9 Phytoremediation as a Tool for Environmental Sustainability

The generation of environmental pollutants and toxic wastes has grown rapidly over the last two decades due to increasing global industrialization, urbanization, and population (Kumar and Chandra, 2017). These environmental pollutants and toxins include a wide range of organics such as dyes, phenolic compounds, humic substances, petroleum, pharmaceuticals, surfactants, and pesticides; and inorganic contaminants such as salts, heavy metals, and metalloids. These toxic wastes are generally discharged into water bodies or dumped on land with little or no treatment despite the need for pollution control and thus pose a great danger to public health, habitats, and the environment (Gupta et al., 2015; Yadav et al., 2017, 2018). Compared to other conventional methods for wastewater treatment (Table 11.4), phytoremediation has provided a more realistic, sustainable, cost-effective, and aesthetic approach for the remediation of toxic pollutants from the ecosystem to ensure environmental sustainability and preservation (Ajibade et al., 2021b; Rahbar et al., 2016; Vymazal and Březinová, 2016; Rezania et al., 2016). The use of phytoremediation as a tool for environmental sustainability can be dated to centuries ago, especially in the use of the plant for wastewater treatment (Raskin et al., 1998). The word *phytoremediation* was coined from two Greek words "phyto" which means "plant" and "remedium" which also means "to clear or restore". Given that phytoremediation is dependent on the usage of plants to extract, sequester, and detoxify contaminants, phytoremediation is globally known as an ecological, economical, and environmentally friendly alternative to conventional methods (Chandra et al., 2015). Several researchers have reported on the remediation capability of plants in different ecosystems i.e., terrestrial or aquatic environments (see section 3). Physically, plants have roots, stems, and leaves which, through different biophysical and biochemical processes including adsorption, translocation, mineralization, and transformation, can remediate contaminants.

11.10 Future Trends in Phytoremediation

Heavy metal contamination is a serious environmental and health problem due to its potential transfer via the food chain. However, phytoremediation strategies have proven to be a pertinent tool in decontaminating and removing of heavy metals from contaminated sites. Currently, phytoremediation is experiencing several new

TABLE 11.4

Conventional techniques for wastewater treatment

Techniques	Merits	Demerits	References
Electro-Chemical Treatments			
Electro-Coagulation (EC)	Small-scale, simple, and productive technology. Easy to operate and require less maintenance. No secondary pollution and can be used to remove a wide range of pollutants. Less sludge generation and generated sludges settle faster. Dewatering is very easy.	Poor systematic reactor design and electrode reliability. The use of electricity may be expensive. High conductivity of wastewater suspension. Loss of efficiency due to the formation of an impermeable oxide film on the electrode.	Emamjomeh and Sivakumar, 2009; Azimi et al., 2017; Syam Babu et al., 2019
Electro-Floatation (EF)	Simplicity in design, operation, and environmental compatibility. Low running cost. Good adaptability with small and compact units. Apply in a wide range of industries such as treatment of oil from oil-water emulsions, sewage and pit waters, groundwater disinfection, colloidal and suspended particles, juice and food processing effluents. Use for removing heavy metal pollutants at a very dilute solution (concentrations < 50 mgm^{-3}).	Power consumption may be high. Electrode dependency. Current density, pH, and electrode types affect EF operation.	Chen, 2004; Azimi et al., 2017; Kolesnikov et al., 2019
Electro-Deposition (ED)	Most efficient method among other electrochemical processes. No further reagents are needed. No sludge is produced. Highly selective and low cost.	Faces conventional aqueous solution problems such as the release of hydrogen gas, low thermal stability, and narrow electrochemical window. Physical parameters such as temperature, pH, and the presence of complexing and chelating agents affect the efficiency of ED.	Azimi et al., 2017
Physico-Chemical Treatments			
Chemical Precipitation	Simple, easily automated method. Wide application in removing different heavy metals.	High chemical consumption. Hight sludge production. Require constant monitoring and controlling, and generated effluent may cause further environmental pollution.	Azimi et al., 2017; Crini and Lichtfouse, 2019
Ion Exchange	Highly efficient, low cost, and less sludge production. Highly selective and excellent recovery for metal ions.	Physical parameters such as pH, anions, temperature, adsorbent and sorbate, and contact time affects its operation.	Crini and Lichtfouse, 2019; Shahedi et al., 2020
Adsorption	Simple and highly effective. Requires low treatment time.	Limited regeneration, costly, and efficiency depends on the adsorbent.	Crini and Lichtfouse, 2019
Membrane Filtration	Highly efficient and simple separation method. No chemical is required. No pollution loads and lower energy consumption compared with other	High cost of maintenance and operation. Low anti-compacting ability. Specific for some processes. Membrane clogging.	Azimi et al., 2017; Crini and Lichtfouse, 2019

(Continued)

TABLE 11.4 (Continued)
Conventional techniques for wastewater treatment

Techniques	Merits	Demerits	References
	conventional methods. Wide application in food and beverage industries, biotechnology, water filtration.		
Photocatalysis	Uses non-toxic materials as catalysts. Simple design, low-cost operation, high stability, and high removal efficiency. Wide apply in photosplitting and decomposition of water to produce hydrogen gas, disinfection of microorganisms, degradation of organic compounds, selective oxidation, manufacturing liquid chemical fuels and removing pollutants including heavy metals.	Recombination of electrode or hole. Production of unwanted by-product. Absorption of visible light.	Azimi et al., 2017

dimensions of research and applications in heavy metal removal and to ensure further understanding and additional gains in this area, further studies are needed in creating a new approach in the development of higher gene expression in plants and manipulation of metal transporters and their cellular organs to ensure safe partitioning of metals in areas where other cellular activities are not disturbed, development of plants with metal-tolerant capabilities and capable of repelling herbivores from their consumption to prevent the transfer of metals via the food web, and the development of transgenic plants with improved plant-microbe interactions or rhizosphere microbial functions. Also, most studies on transgenic plants in phytoremediation are limited to laboratory settings; hence, the need for open-field efficiency testing. There is a need for transgenic research in phytoremediation to focus on mixed contamination problems.

11.11 Conclusion

The environment is a great factor both as an aspect of the ecosystem and as an object of academic research. The significance of the environment and the need for its sustainability have led to the collaborative efforts of various experts such as scientists, engineers, environmentalists, health personnel, and even rural community groups in advocating for green engineering. The contamination of the environment by some polluting elements including heavy metals, which human activities are the major contributors for, in turn, made the environment unfavourable for both man and other living organisms. This chapter has expressly explicated the phytoremediation technology and its techniques and mechanisms (viz. phytoextraction, phytofiltration or rhizofiltration, phytostabilization, phytovolatilization, phytodegradation or phytotransformation) for the remediation of heavy-metal-contaminated environments. The advantages of the phytoremediation technology over traditional decontamination methods were explained, and its limitations were equally highlighted. Factors affecting the mobility, uptake, and accumulation of heavy metals in plants were discussed and some innovative approaches for improving heavy metal phytoremediation were equally discussed. Finally, possible future trends in phytoremediation for a more effective decontamination of heavy-metal-contaminated environments, by which the environmental and health problems associated with these heavy metals and other contaminants due to their potential transfer via the food chain, could be curbed were suggested. Hence, phytoremediation technology can be instrumental in the promotion of public health quality and environmental sustainability.

REFERENCES

Abhilash, P.C., Jamil, S., & Singh, N. (2009). Transgenic plants for enhanced biodegradation and phytoremediation of organic xenobiotics. *Biotechnology Advances*, 27(4), 474–488.

Aderholt, M., Vogelien, D.L., Koether, M., & Greipsson, S. (2017). Phytoextraction of contaminated urban soils by Panicum virgatum L. enhanced with application of a plant growth regulator (BAP) and citric acid. *Chemosphere*, 175, 85–96.

Adewumi, A.J. (2020). Contamination, sources and risk assessments of metals in media from Anka artisanal gold mining area, Northwest Nigeria. *Science of The Total Environment*, 718, 137235.

Agostini, E., Talano, M.A., González, P.S., Oller, A.L., & Medina, M.I. (2013). Application of hairy roots for phytoremediation: what makes them an interesting tool for this purpose? *Applied Microbiology and Biotechnology*, 97, 1017–1030. DOI: 10.1007/s00253-012-4658-z.

Ahmad, N., Imran, M., Marral, M.R., Mubashir, M., & Butt, B. (2016). Influence of biochar on soil quality and yield related attributes of wheat (Triticum aestivum L.). *Journal of Environmental and Agricultural Sciences*, 7, 68–72.

Ahmadpour, P., Ahmadpour, F., Mahmud, T.M.M., Abdu, A., Soleimani, M., & Tayefeh, F.H. (2012). Phytoremediation of heavy metals: a green technology. *African Journal of Biotechnology*, 11(76), 14036–14043.

Ajibade, F.O., Adelodun, B., Lasisi, K.H., Fadare, O.O., Ajibade, T.F., Nwogwu, N.A., Sulaymon, I.D., Ugya, A.Y., Wang, H.C., & Wang, A. (2021a). Environmental pollution and their socioeconomic impacts. In A. Kumar, V.K. Singh, P. Singh, & V.K. Mishra (Eds.), *Microbe Mediated Remediation of Environmental Contaminants*. Woodhead Publishing, Elsevier. 10.1016/B978-0-12-821199-1.00025.-0.

Ajibade, F.O., Nwogwu, N.A., Lasisi, K.H., Ajibade, T.F., Adelodun, B., Guadie, A., Ugya, A.Y., Adewumi, J.R., Wang, H.C., & Wang, A. (2021b). Removal of nitrogen oxyanion (nitrate) in constructed wetlands. In N.A. Oladoja & I.E. Unuabonah (Eds.), *Progress and Prospects in the Management of Oxoanion Polluted Aqua Systems*. The Netherlands: Springer Nature, Netherlands. 10.1007/978-3-030-70757-6_12.

Ajibade, F.O., Wang, H., Guadie, A.A., Ajibade, T.F., Fang, Y., Sharif, H.M.A., Liu, W., & Wang, A. (2021c). Total nitrogen removal in biochar amended non-aerated vertical flow constructed wetlands for secondary wastewater effluent with low C/N ratio: microbial community structure and dissolved organic carbon release conditions. *Bioresource Technology*, 124430. 10.1016/j.biortech.2020.124430.

Ajibade, F.O., & Adewumi, J.R. (2017). Performance evaluation of aquatic macrophytes in a constructed wetland for municipal wastewater treatment. *FUTA Journal of Engineering and Engineering Technology*, 11(1), 1–11.

Ajibade, F.O., Adeniran, K.A., & Egbuna, C.K. (2013). Phytoremediation efficiencies of water hyacinth in removing heavy metals in domestic sewage (A Case Study of University of Ilorin, Nigeria). *The International Journal of Engineering and Science*, 2(12), 16–27.

Akinbile, C.O., Ikuomola, B.T., Olanrewaju, O.O., & Babalola, T.E. (2019). Assessing the efficacy of Azolla pinnata in four different wastewater treatment for agricultural re-use: a case history. *Sustainable Water Resources Management*, 5(3), 1009–1015.

Akinbile, C.O., Ogunrinde, T.A., Che bt Man, H., & Aziz, H.A. (2016). Phytoremediation of domestic wastewaters in free water surface constructed wetlands using Azolla pinnata. *International Journal of Phytoremediation*, 18(1), 54–61.

Akinbile, C.O., Yusoff, M.S., & Shian, L.M. (2012). Leachate characterization and phytoremediation using water hyacinth (Eichorrnia crassipes) in Pulau Burung, Malaysia. *Bioremediation Journal*, 16(1), 9–18.

Akinbile, C.O., & Yusoff, M.S. (2012). Assessing water hyacinth (Eichhornia crassipes) and lettuce (Pistia stratiotes) effectiveness in aquaculture wastewater treatment. *International Journal of Phytoremediation*, 14(3), 201–211.

Ali, H., Khan, E., & Sajad, M.A. (2013). Phytoremediation of heavy metals – concepts and applications. *Chemosphere*, 91(7), 869–881. 10.1016/j.chemosphere.2013.01.075.

Alkorta, I., Hernández-Allica, J., Becerril, J.M., Amezaga, I., Albizu, I., & Garbisu, C. (2004). Recent findings on the phytoremediation of soils contaminated with environmentally toxic heavy metals and metalloids such as zinc, cadmium, lead, and arsenic. *Reviews in Environmental Science and Biotechnology*, 3(1), 71–90.

Ansari, A.A., Naeem, M., Gill, S.S., & AlZuaibr, F.M. (2020). Phytoremediation of contaminated waters: an eco-friendly technology based on aquatic macrophytes application. *The Egyptian Journal of Aquatic Research*, 10.1016/j.ejar.2020.03.002.

Antoniadis, V., Levizou, E., Shaheen, S.M., Ok, Y.S., Sebastian, A., Baum, C.,... & Rinklebe, J. (2017). Trace elements in the soil-plant interface: phytoavailability, translocation, and phytoremediation – a review. *Earth-Science Reviews*, *171*, 621–645.

Argun, M.E., Dursun, S., Ozdemir, C., & Karatas, M. (2007). Heavy metal adsorption by modified oak sawdust: thermodynamics and kinetics. *Journal of Hazardous Materials*, *141*(1), 77–85.

Axtell, N.R., Sternberg, S.P., & Claussen, K. (2003). Lead and nickel removal using Microspora and Lemna minor. *Bioresource Technology*, *89*, 41–48. 10.1016/s0960-8524(03)00034-8.

Azimi, A., Azari, A., Rezakazemi, M., & Ansarpour, M. (2017). Removal of heavy metals from industrial wastewaters: a review. *ChemBioEng Reviews*, *4*(1), 37–59.

Bacchetta, G., Boi, M.E., Cappai, G., De Giudici, G., Piredda, M., & Porceddu, M. (2018). Metal tolerance capability of Helichrysum microphyllum Cambess. subsp. tyrrhenicum Bacch., Brullo & Giusso: a candidate for phytostabilization in abandoned mine sites. *Bulletin of Environmental Contamination and Toxicology*, *101*(6), 758–765.

Bahmani, R., Kim, D., Lee, B.D., & Hwang, S. (2017). Over-expression of tobacco UBC1 encoding a ubiquitin-conjugating enzyme increases cadmium tolerance by activating the 20S/26S proteasome and by decreasing Cd accumulation and oxidative stress in tobacco (Nicotiana tabacum). *Plant Molecular Biology*, *94*(4-5), 433–451.

Baker, T.J., Tyler, C.R., & Galloway, T.S. (2014). Impacts of metal and metaloxidenanoparticles on marine organisms. *Environmental Pollution*, *186*, 257–271. DOI: 10.1016/j.envpol.2013.11.014.

Baker, A.M., Reeves, R.D., & Hajar, A.M. (1994). Heavy metal accumulation and tolerance in british populations of the metallophyte Thlaspicaerulescens J and C Presl (Brassicaceae). *New Phytologist*, *127*, 61–68.

Baldantoni, D., Alfani, A., Di Tommasi, P., Bartoli, G., & De Santo, A.V. (2004). Assessment of macro and microelement accumulation capability of two aquatic plants. *Environmental Pollution*, *130*(2), 149–156. doi:10.1016/j.envpol.2003.12.015.

Bañuelos, G.S., Arroyo, I., Pickering, I.J., Yang, S.I., & Freeman, J.L. (2015). Selenium biofortification of broccoli and carrots grown in soil amended with Se-enriched hyperaccumulator Stanleya pinnata. *Food Chemistry*, *166*, 603–608.

Bañuelos, G.S., Lin, Z.-Q., Arroyo, I., & Terry, N. (2005). Selenium volatilization in vegetated agricultural drainage sediment from the San Luis Drain, Central California. Chemosphere, *60*(9), 1203–1213. doi:10.1016/j.chemosphere.2005.02.033.

Bennicelli, R., Stępniewska, Z., Banach, A., Szajnocha, K., & Ostrowski, J. (2004). The ability of Azolla caroliniana to remove heavy metals (Hg(II), Cr(III), Cr(VI)) from municipal waste water. *Chemosphere*, *55*(1), 141–146. doi:10.1016/j.chemosphere.2003.11.01.

Berken, A., Mulholland, M. M., LeDuc, D. L., & Terry, N. (2002). Genetic Engineering of Plants to Enhance Selenium Phytoremediation. *Critical Reviews in Plant Sciences*, *21*(6), 567–582. doi: 10.1080/0735-260291044368.

Berti, W.R., & Cunnigham, S.D. (2000). Phytostabilization of metals. In I. Raskin & B.D. Ensley (Eds.), *Phytoremediation of Toxic Metals: Using Plants to Clean Up the Environment*, 71–88. New York: Wiley.

Bhargava, A., Carmona, F.F., Bhargava, M., & Srivastava, S. (2012). Approaches for enhanced phytoextraction of heavy metals. *Journal of Environmental Management*, *105*, 103–120.

Bhaskaran, K., Nadaraja, A.V., Tumbath, S., Shah, L.B., & Veetil, P.G. (2013). Phytoremediation of perchlorate by free floating macrophytes. *Journal of Hazardous Materials*, *260*, 901–906. 10.1016/j.jhazmat.2013.06.008.

Bonanno, G., & Vymazal, J. (2017). Compartmentalization of potentially hazardous elements in macrophytes: insights into capacity and efficiency of accumulation. *Journal of Geochemical Exploration*, *181*, 22–30.

Boonyapookana, B., Upatham, E.S., Kruatrachue, M., Pokethitiyook, P., & Singhakaew, S. (2002). Phytoaccumulation and Phytotoxicity of Cadmium and Chromium in Duckweed Wolffia globosa. *International Journal of Phytoremediation*, *4*(2), 87–100.

Bora, M.S., & Sarma, K.P. (2020). Phytoremediation of heavy metals/metalloids by native herbaceous macrophytes of wetlands: current research and perspectives. *Emerging Issues in the Water Environment During Anthropocene*, 261–284. Singapore: Springer.

Bruce, S.L., Noller, B.N., Grigg, A.H., Mullen, B.F., Mulligan, D.R., Ritchie, P.J.,... & Ng, J.C. (2003). A field study conducted at Kidston gold mine, to evaluate the impact of arsenic and zinc from mine tailing to grazing cattle. *Toxicology Letters*, *137*, 23–34.

Cameselle, C., & Gouveia, S. (2019). Phytoremediation of mixed contaminated soil enhanced with electric current. *Journal of Hazardous Materials*, *361*, 95–102.

Chambers, A., Krewski, D., Birkett, N., Plunkett, L., Hertzberg, R., Danzeisen, R.,... & Jones, P. (2010). An exposure-response curve for copper excess and deficiency. *Journal of Toxicology and Environmental Health, Part B*, *13*(7-8), 546–578.

Chandra, R. (Ed.). (2015). *Advances in Biodegradation and Bioremediation of Industrial Waste*. CRC Press.

Chandra, R., & Kumar, V. (2018). Phytoremediation: a green sustainable technology for industrial waste management. In R. Chandra, N.K. Dubey, & V. Kumar (Eds.), *Phytoremediation of Environmental Pollutants*, 1–42. Boca Raton: Taylor & Francis Group. 10.1201/9781315161549.

Chandra, R., Saxena, G., & Kumar, V. 2015. Phytoremediation of environmental pollutants: an ecosustainable green technology to environmental management. R. Chandra (Ed.), *Advances in Biodegradation and Bioremediation of Industrial Waste*, 1–29. Boca Raton, FL: CRC Press.

Chaney, R.L. (1983a). Potential effects of waste constituents on the food chain. J.F. Parr, P.B. Marsh & J.M. Kla (Eds.), *Land Treatment of Hazardous Wastes*, 152–240. Park Ridge, New Jersey, USA: Noyes Data Corp.

Chaney, R.L. (1983b). Plant uptake of inorganic waste. J.F. Parr, P.B. Marsh and J.M. Kla. (Eds.), *Land Treatment of Hazardous Wastes*, 50–76. Park Ridge, New Jersey, USA: Noyes Data Corp.

Chaney, R. L., Malik, M., Li, Y. M., Brown, S. L., Brewer, E. P., Angle, J. S., & Baker, A. J. (1997). Phytoremediation of soil metals. *Current Opinion in Biotechnology*, *8*(3), 279–284. doi: 10.1016/S0958-1669(97)80004-3.

Chen, G. (2004). Electrochemical technologies in wastewater treatment. *Separation and Purification Technology*, *38*(1), 11–41.

Chen, J., Yang, L., Yan, X., Liu, Y., Wang, R., Fan, T.,... & Cao, S. (2016). Zinc-finger transcription factor ZAT6 positively regulates cadmium tolerance through the glutathione-dependent pathway in Arabidopsis. *Plant Physiology*, *171*(1), 707–719.

Chen, B.D., Li, X.L., Tao, H.Q., Christie, P., & Wong, M.H. (2003). The role of arbuscular mycorrhizal in zinc uptake by red clover growing in a calcareous soil spiked with various quantities of zinc. *Chemosphere*, *50*, 839–846.

Conesa, M.H., Evangelou, W.H., Robinson, H.B., & Schulin, R. (2012). A critical view of the current state of phytotechnologies to remediate soils: still a promising tool? *Scientific World Journal*, doi: 10.1100/2012/173829.

Crini, G., & Lichtfouse, E. (2019). Advantages and disadvantages of techniques used for wastewater treatment. *Environmental Chemistry Letters*, *17*(1), 145–155.

Crowley, D.E., Wang, Y.C., Reid, C.P.P., & Szaniszlo, P.J. (1991). Mechanisms of iron acquisition from siderophores by microorganisms and plants. In *Iron Nutrition and Interactions in Plants*, 213–232. Dordrecht: Springer.

da Conceição Gomes, M.A., Hauser-Davis, R.A., de Souza, A.N., & Vitória, A.P. (2016). Metal phytoremediation: general strategies, genetically modified plants and applications in metal nanoparticle contamination. *Ecotoxicology and Environmental Safety*, *134*, 133–147.

Dai, L., Wang, L., Li, L., Liang, T., Zhang, Y., Ma, C., & Xing, B. (2018). Multivariate geostatistical analysis and source identification of heavy metals in the sediment of Poyang Lake in China. *Science of the Total Environment*, *621*, 1433–1444.

DalCorso, G., Fasani, E., Manara, A., Visioli, G., & Furini, A. (2019). Heavy metal pollutions: state of the art and innovation in phytoremediation. *International Journal of Molecular Sciences*, *20*, 3412. doi:10.3390/ijms20143412.

Davison, J. (2005). Risk mitigation of genetically modified bacteria and plants designed for bioremediation. *Journal of Industrial Microbiology and Biotechnology*, *32*(11-12), 639–650.

Dhiman, S.S., Selvaraj, C., Li, J., Singh, R., Zhao, X., Kim, D.,... & Lee, J.K. (2016). Phytoremediation of metal-contaminated soils by the hyperaccumulator canola (Brassica napus L.) and the use of its biomass for ethanol production. *Fuel, 183,* 107–114.

Dietz, A.C., & Schnoor, J.L. (2001). Advances in Phytoremediation. *Environmental Health Perspectives, 109*(1), 163–168.

Dolphen, R., & Thiravetyan, P. (2015). Phytodegradation of Ethanolamines by Cyperus alternifolius: effect of Molecular Size. *International Journal of Phytoremediation, 17*(7), 686–692.

Dong, B., Zhang, R., Gan, Y., Cai, L., Freidenreich, A., Wang, K.,... & Wang, H. (2019). Multiple methods for the identification of heavy metal sources in cropland soils from a resource-based region. *Science of the Total Environment, 651,* 3127–3138.

Duruibe, J.O., Ogwuegbu, M.O.C., & Egwurugwu, J.N. (2007). Heavy metal pollution and human biotoxic effects. *International Journal of Physical Sciences, 2*(5), 112–118.

Dushenkov, V., Kumar, P.A., Motto, H., & Raskin, I. (1995). Rhizofiltration: the use of plants to remove heavy metals from aqueous streams. *Environmental Science and Technology, 1239–1245,* 1239–1245. 10.1021/es00005a015.

Ebbs, S.D., Bradfield, S.J., Kumar, P., White, J.C., Musante, C., & Ma, X. (2016). Accumulation of zinc, copper, or cerium in carrot (Daucus carota) exposed to metal oxide nanoparticles and metal ions. *Environmental Science: Nano, 3*(1), 114–126. doi:10.1039/C5EN00161G.

Eisazadeh, S., Kapourchal, S.A., Homaee, M., Noorhosseini, S.A., & Damalas, C.A. (2019). Chive (Allium schoenoprasum L.) response as a phytoextraction plant in cadmium-contaminated soils. *Environmental Science and Pollution Research, 26*(1), 152–160.

Eissa, M.A. (2019). Effect of Compost and Biochar on Heavy Metals Phytostabilization by the Halophytic Plant Old Man Saltbush [Atriplex Nummularia Lindl]. *Soil and Sediment Contamination: An International Journal, 28*(2), 135–147.

Ekperusi, A.O., Sikoki, F.D., & Nwachukwu, E.O. (2019). Application of common duckweed (Lemna minor) in phytoremediation of chemicals in the environment: state and future perspective. *Chemosphere.* doi: 10.1016/j.chemosphere.02.025.

Elad, Y., Cytryn, E., Harel, Y.M., Lew, B., & Graber, E.R. (2012). The biochar effect: plant resistance to biotic stresses. *Phytopathologia Mediterranea, 50,* 335–349.

Emamjomeh, M.M., & Sivakumar, M. (2009). Review of pollutants removed by electrocoagulation and electrocoagulation/flotation processes. *Journal of Environmental Management, 90*(5), 1663–1679.

Erakhrumen, A., & Agbontalor, A. (2007). Review phytoremediation: an environmentally sound technology for pollution prevention, control and remediation in developing countries. *Educational Research and Reviews, 2*(7), 151–156.

Fan, S., Wang, X., Lei, J., Ran, Q., Ren, Y., & Zhou, J. (2019). Spatial distribution and source identification of heavy metals in a typical Pb/Zn smelter in an arid area of northwest China. *Human and Ecological Risk Assessment: An International Journal, 25*(7), 1661–1687.

Farid, M., Ali, S., Rizwan, M., Ali, Q., Abbas, F., Bukhari, S.A.H.,... & Wu, L. (2017). Citric acid assisted phytoextraction of chromium by sunflower; morpho-physiological and biochemical alterations in plants. *Ecotoxicology and Environmental Safety, 145,* 90–102.

Farwell, A.J., Vesely, S., Nero, V., Rodriguez, H., McCormack, K., Shah, S.,... Glick, B.R. (2007). Tolerance of transgenic canola plants (Brassica napus) amended with plant growth promoting bacteria to flooding stress at a metal contaminated field site. *Environmental Pollution, 147,* 540–545.

Fasani, E., Manara, A., Martini, F., Furini, A., & DalCorso, G. (2018). The potential of genetic engineering of plants for the remediation of soils contaminated with heavy metals. *Plant, Cell & Environment, 41*(5), 1201–1232. doi: 10.1111/pce.12963.

Fässler, E., Evangelou, M.W., Robinson, B.H., & Schulin, R. (2010). Effects of indole-3-acetic acid (IAA) on sunflower growth and heavy metal uptake in combination with ethylene diamine disuccinic acid (EDDS). *Chemosphere, 80,* 901–907.

Fellet, G., Marmiroli, M., & Marchiol, L. (2014). Elements uptake by metal accumulator species grown on mine tailings amended with three types of biochar. *Science of the Total Environment, 468,* 598–608.

Fister, A.S., Shi, Z., Zhang, Y., Helliwell, E.E., Maximova, S.N., & Guiltinan, M.J. (2016). Protocol: transient expression system for functional genomics in the tropical tree Theobroma cacao L. *Plant Methods, 12*(1), 1–13. doi: 10.1186/s13007-016-0119-5.

Franken, E., Teuschel, U., & Hain, R. (1997). Recombinant proteins from transgenic plants. *Current Opinion in Biotechnology*, 8, 411–416.

Galal, T.M., Eid, E.M., Dakhil, M.A., & Hassan, L.M. (2018). Bioaccumulation and rhizofiltration potential of Pistia stratiotes L. for mitigating water pollution in the Egyptian wetlands. *International Journal of Phytoremediation*, 20(5), 440–447.

Galal, T.M., Gharib, F.A., Ghazi, S.M., & Mansour, K.H. (2017). Phytostabilization of heavy metals by the emergent macrophyte Vossia cuspidata (Roxb.) Griff.: a phytoremediation approach. *International Journal of Phytoremediation*, 19(11), 992–999.

Garbiscu, C., & Alkorta, I. (2001). Phytoextraction: a cost effective plant based technology for the removal of metals from the environment. *Bioresource Technology*, 77, 229–236.

Gerhardt, K.E., Gerwing, P.D., & Greenberg, B.M. (2016). Opinion: taking phytoremediation from proven technology to accepted practice. *Plant Science*, 256, 170–185. doi: 10.1016/j.plantsci.2016.11.016.

Ghazaryan, K.A., Movsesyan, H.S., Minkina, T.M., Sushkova, S.N., & Rajput, V.D. (2019). The identification of phytoextraction potential of Melilotus officinalis and Amaranthus retroflexus growing on copper-and molybdenum-polluted soils. *Environmental Geochemistry and Health*, 1–9.

Gleba, D., Borisjuk, N.V., Borisjuk, L.G., Kneer, R., Poulev, A., Skarzhinskaya, M.,... & Raskin, I. (1999). Use of plant roots for phytoremediation and molecular farming. *Proceedings of the National Academy of Sciences*, 96, 5973–5977.

Glick, B.R., Penrose, D.M., & Li, J. (1998). A model for the lowering of plant ethylene concentrations by plant growth promoting bacteria. *Journal of Theoretical Biology*, 190, 63–68.

Gomes, M.C., Hauser-Davis, R.A., de Souza, A.N., & Vitória, A.P. (2016). Metal phytoremediation: general strategies, genetically modified plants and applications in metal nanoparticle contamination. *Ecotoxicology and Environmental Safety*, 134, 133–147. doi:10.1016/j.ecoenv.2016.08.024.

Griboff, J., Wunderlin, D.A., & Monferran, M.V. (2018). Phytofiltration of As3+, As5+, and Hg by the aquatic macrophyte Potamogeton pusillus L, and its potential use in the treatment of wastewater. *International Journal of Phytoremediation*, 20(9), 914–921.

Gupta, N., Yadav, K.K., & Kumar, V. (2015). A review on current status of municipal solid waste management in India. *Journal of Environmental Sciences*, 37, 206–217.

Hadi, F., Bano, A., & Fuller, M.P. (2010). The improved phytoextraction of lead (Pb) and the growth of maize (Zea mays L.): the role of plant growth regulators (GA3 and IAA) and EDTA alone and in combinations. *Chemosphere*, 80, 457–462.

Hamon, R., Wundke, J., McLaughlin, M., & Naidu, R. (1997). Availability of zinc and cadmium to different plant species. *Soil Research*, 35(6), 1267–1278.

Han, D., Cheng, J., Hu, X., Jiang, Z., Mo, L., Xu, H.,... & Wang, H. (2017). Spatial distribution, risk assessment and source identification of heavy metals in sediments of the Yangtze River Estuary, China. *Marine Pollution Bulletin*, 115(1-2), 141–148.

He, J., Ren, Y., Pan, X., Yan, Y., Zhu, C., & Jiang, D. (2010). Salicylic acid alleviates the toxicity effect of cadmium on germination, seedling growth, and amylase activity of rice. *Journal of Plant Nutrition and Soil Science*, 173, 300–305.

Hinchman, R.R., Negri, M.C., & Gatliff, E.G. (1995). *Phytoremediation: using green plants to clean up contaminated soil, groundwater, and wastewater*. Retrieved from Argonne National Laboratory Hinchman, Applied Natural Sciences, Inc.: http://www.treemediation.com/Technical/Phytoremediation_1998.pdf.

Huang, S.H., Bing, P.E.N.G., Yang, Z.H., Chai, L.Y., & Zhou, L.C. (2009). Chromium accumulation, microorganism population and enzyme activities in soils around chromium-containing slag heap of steel alloy factory. *Transactions of Nonferrous Metals Society of China*, 19(1), 241–248.

Hussein, H.S., Ruiz, O.N., Terry, N., & Daniell, H. (2007). Phytoremediation of mercury and organomercurials in chloroplast transgenic plants: enhanced root uptake, translocation to shoots, and volatilization. *Environmental Science & Technology*, 41(24), 8439–8446.

Ilić, Z.H., Pajević, S., Borišev, M., & Luković, J. (2020). Assessment of phytostabilization potential of two Salix L. clones based on the effects of heavy metals on the root anatomical traits. *Environmental Science and Pollution Research*, 27, 29361–29383. doi: 10.1007/s11356-020-09228-8.

Islam, M.S., Saito, T., & Kurasaki, M. (2015). Phytofiltration of arsenic and cadmium by using an aquatic plant, Micranthemum umbrosum: phytotoxicity, uptake kinetics, and mechanism. *Ecotoxicology and Environmental Safety*, *112*, 193–200.

Islam, M.S., Sikder, M.T., & Kurasaki, M. (2017). Potential of Micranthemum umbrosum for phytofiltration of organic arsenic species from oxic water environment. *International Journal of Environmental Science and Technology*, *14*(2), 285–290.

Ismail, A., Toriman, M.E., Juahir, H., Zain, S.M., Habir, N.L.A., Retnam, A.,… & Azid, A. (2016). Spatial assessment and source identification of heavy metals pollution in surface water using several chemometric techniques. *Marine Pollution Bulletin*, *106*(1-2), 292–300.

Jabeen, R., Ahmad, A., & Iqbal, M. (2009). Phytoremediation of heavy metals: physiological and molecular mechanisms. *The Botanical Review*, *75*(4), 339–364.

Jaiswal, A., Verma, A., & Jaiswal, P. (2018). Detrimental effects of heavy metals in soil, plants, and aquatic ecosystems and in humans. *Journal of Environmental Pathology, Toxicology and Oncology*, *37*(3), 183–197.

Jeyasingh, J., & Philip, L. (2005). Bioremediation of chromium contaminated soil: optimization of operating parameters under laboratory conditions. *Journal of Hazardous Materials*, *118*(1-3), 113–120.

Kadukova, J., & Kalogerakis, N. (2007). Lead accumulation from non-saline and saline environment by Tamarix smyrnesis Bunge. *European Journal of Soil Biology*, *43*, 216–223.

Kadukova, J., & Kavuličova, J. (2011). Phytoremediation of heavy metal contaminated soils – plant stress assessment. In I.A. Golubev (Ed.), *Handbook of Phytoremediation*, 185–222. New York: Nova Science Publishers, Inc.

Kärenlampi, S., Schat, H., Vangronsveld, J., Verkleij, J.A., van der Lelie, D., Mergeay, M., & Tervahauta, A.I. (2000). Genetic engineering in the improvement of plants for phytoremediation of metal polluted soils. *Environmental Pollution*, *107*, 225–231.

Kathal, R., Malhotra, P., Kumar, L., & Uniyal, P.L. (2016). Phytoextraction of Pb and Ni from the Polluted Soil by Brassica juncea L. *Journal of Environmental and Analytical Toxicology*, *6*, 394.

Katoh, M., Risky, E., & Sato, T. (2017). Immobilization of lead migrating from contaminated soil in rhizosphere soil of barley (Hordeum vulgare L.) and hairy vetch (Vicia villosa) using hydroxyapatite. *International Journal of Environmental Research and Public Health*, *14*(10), 1273.

Kazemipour, M., Ansari, M., Tajrobehkar, S., Majdzadeh, M., & Kermani, H.R. (2008). Removal of lead, cadmium, zinc, and copper from industrial wastewater by carbon developed from walnut, hazelnut, almond, pistachio shell, and apricot stone. *Journal of Hazardous Materials*, *150*(2), 322–327.

Kolesnikov, V.A., Il'in, V.I., & Kolesnikov, A.V. (2019). Electroflotation in wastewater treatment from oil products, dyes, surfactants, ligands, and biological pollutants: a review. *Theoretical Foundations of Chemical Engineering*, *53*(2), 251–273.

Kotrba, P., Najmanova, J., Macek, T., Ruml, T., & Mackova, M. (2009). Genetically modified plants in phytoremediation of heavy metal and metalloid soil and sediment pollution. *Biotechnology Advances*, *27*(6), 799–810.

Krämer, U., & Chardonay, A. (2001). The use of transgenic plants in the bioremediation of soils contaminated with trace elements. *Applied Microbiology and Biotechnology*, *55*, 661–672.

Kumar, B., Smita, K., & Flores, L.C. (2017). Plant mediated detoxification of mercury and lead. *Arabian Journal of Chemistry*, *10*, S2335–S2342.

Kumar, P.A., Dushenkov, V., Motto, H., & Raskin, I. (1995). Phytoextraction: the use of plants to remove heavy metals from soils. *Environmental Science and Technology*, *29*, 1232–1238. doi:10.1021/es00005 a014.

Kumar, V., & Chandra, R. (2017). Phytoremediation: a green sustainable technology for industrial waste management. In *Phytoremediation of Environmental Pollutants*, 15–56. Boca Raton: CRC Press.

Kumari, M., & Tripathi, B.D. (2015). Efficiency of Phragmites australis and Typha latifolia for heavy metal removal from wastewater. *Ecotoxicology and Environmental Safety*, *112*, 80–86.

Lalor, G.C. (2008). Review of cadmium transfers from soil to humans and its health effects in the Jamaican environment. *Science of the Total Environment*, *400*(1-3), 162–172.

Lasat, M.M. (2000). Phytoextraction of toxic metals: a review of biological mechanisms. *Journal of Environmental Quality*, *31*, 109–125.

Li, C., Wang, M., & Luo, X. (2019). Uptake of uranium from aqueous solution by Nymphaea tetragona Georgi: the effect of the accompanying heavy metals. *Applied Radiation and Isotopes*, *150*, 157–163.

Li, J.-T., Liao, B., Lan, C.Y., Ye, Z.H., Baker, A.J., & Shu, W.S. (2010). Cadmium accumulation in a high-biomass tropical tree (Averrhoa carambola) and its potential for phytoextraction. *Journal of Environmental Quality, 39*(4), 1262–1268.

Liang, S.X., Jin, Y., Liu, W., Li, X., Shen, S.G., & Ding, L. (2017). Feasibility of Pb phytoextraction using nano-materials assisted ryegrass: results of a one-year field-scale experiment. *Journal of Environmental Management, 190*, 170–175.

Liao, J.P., Lin, X.G., Cao, Z.H., Shi, Y.Q., & Wong, M.H. (2003). Interactions between arbuscular mycorrhizae and heavy metals under a sand culture experiment. *Chemosphere, 50*, 847–853.

Limmer, M., & Burken, J. (2016). Phytovolatilization of organic contaminants. *Environmental Science & Technology, 50*(13), 6632–6643.

Liu, N., & Wu, Z. (2018). Toxic effects of linear alkylbenzenesulfonate on Chara vulgaris L. *Environmental Science and Pollution Research*, 4934–4941.

Liu, Y., Liu, N., Zhou, Y., Wang, F., Zhang, Y., & Wu, Z. (2019). Growth and physiological responses in Myriophyllum spicatum L. exposed to linear alkylbenzenesulfonate. *Environmental Toxicology and Chemistry, 38*(9), 2073–2081. doi: 10.1002/etc.4475.

Liu, D., An, Z., Mao, Z., Ma, L., & Lu, Z. (2015). Enhanced heavy metal tolerance and accumulation by transgenic sugar beets expressing Streptococcus thermophilus StGCS-GS in the presence of Cd, Zn and Cu alone or in combination. *PLoS One, 10*(6), e0128824.

Liu, X., Zhang, A., Ji, C., Joseph, S., Bian, R., Li, L., & Paz Ferreiro, J. (2013). Biochar's effect on crop productivity and the dependence on experimental conditions a meta-analysis of literature data. *Plant Soil, 373*, 583–594.

Liu, D., Jiang, W., Liu, C., Xin, C., & Hou, W. (2000a). Uptake and accumulation of lead by roots, hypocotyls and shoots of Indian mustard [Brassica juncea (L.)]. *Bioresource Technology, 71*(3), 273–277.

Liu, J.R., Suh, M.C., & Choi, D. (2000b). Phytoremediation of cadmium contamination: overexpression of metallothionein in transgenic tobacco plants. *Burtdesgesundheitsbl Gesundheitsforsch-Gesundheitsschutz, 2*, 126–130.

Luo, J., Yang, D., Qi, S., Wu, J., & Gu, X.S. (2018). Using solar cell to phytoremediate field-scale metal polluted soil assisted by electric field. *Ecotoxicology and Environmental Safety, 165*, 404–410.

Ma, N., Wang, W., Gao, J., & Chen, J. (2017). Removal of cadmium in subsurface vertical flow constructed wetlands planted with Iris sibirica in the low-temperature season. *Ecological Engineering, 109*, 48–56.

Macek, T., Pavlikova, D., & Mackova, M. (2004). Phytoremediation metals and Inorganic Pollutants. In A. Singh & O.P. Ward (Eds.), *Applied Bioremediation and Phytoremediation (Vol. 1)*, 135–157. Berlin, Heidelberg, New York: Springer-Verlag.

Macek, T., Mackova, M., & Kas, J. (2000). Exploitation of plants for the removal of organics in environmental remediation. *Biotechnology Advances, 18*, 23–35.

Mahar, A., Wang, P., Ali, A., Awasthi, M.K., Lahori, A.H., Wang, Q.,… & Zhang, Z. (2016). Challenges and opportunities in the phytoremediation of heavy metals contaminated soils: a review. *Ecotoxicology and Environmental Safety, 126*, 111–121.

Manousaki, E., Kadukova, J., Papadantonakis, N., & Kalogerakis, N. (2008). Phytoextraction and phytoexcretion of Cd by the leaves of Tamarix smyrnensis growing on contaminated non saline and saline soils. *Environmental Research, 106*(3), 326–332.

Mao, X., Han, F.X., Shao, X., Guo, K., McComb, J., Arslan, Z., & Zhang, Z. (2016). Electro-kinetic remediation coupled with phytoremediation to remove lead, arsenic and cesium from contaminated paddy soil. *Ecotoxicology and Environmental Safety, 125*, 16–24.

Marrugo-Negrete, J., Durango-Hernández, J., Pinedo-Hernández, J., Olivero-Verbel, J., & Díez, S. (2015). Phytoremediation of mercury-contaminated soils by Jatropha curcas. *Chemosphere, 127*, 58–63.

Marschner H. (1995). *Mineral Nutrition of Higher Plants*. Cambridge: Academic Press.

Mastretta, C., Barac, T., Vangronsveld, J., Newman, L., Taghavi, S., & Lelie, D.V.D. (2006). Endophytic bacteria and their potential application to improve the phytoremediation of contaminated environments. *Biotechnology and Genetic Engineering Reviews, 23*(1), 175–188.

McGrath, S.P., Zhao, F.J., & Lombi, E. (2002). Phytoremediation of metals, metalloids, and radionuclides. *Advances in Agronomy, 75*, 1–56.

Mleczek, M., Gąsecka, M., Waliszewska, B., Magdziak, Z., Szostek, M., Rutkowski, P.,… & Niedzielski, P. (2018). Salix viminalis L.-A highly effective plant in phytoextraction of elements. *Chemosphere, 212*, 67–78.

Moameri, M., & Khalaki, M.A. (2017). Capability of Secale montanum trusted for phytoremediation of lead and cadmium in soils amended with nano-silica and municipal solid waste compost. *Environmental Science and Pollution Research*, 1–8.

Montes-Holguin, M.O., Peralta-Videa, J.R., Meitzner, G., Martinez-Martinez, A., de la Rosa, G., Castillo-Michel, H.A., & Gardea-Torresdey, J.L. (2006). Biochemical and spectroscopic studies of the response of Convolvulus arvensis L. to chromium (III) and chromium (VI) stress. *Environmental Toxicology and Chemistry: An International Journal*, 25(1), 220–226.

Mueller, N.C., & Nowack, B. (2010). Nanoparticles for remediation: solving big problems with little particles. *Elements*, 6, 395–400.

Mueller, B., Rock, S., Gowswami, D., & Ensley, D. (1999). Phytoremediation decision tree. *Prepared by- Interstate Technology and Regulatory Cooperation Work Group*, 1–36.

Mwegoha, W.J. (2008). The use of phytoremediation technology for abatement soil and groundwater pollution in Tanzania: opportunities and challenges. *Journal of Sustainable Development in Africa*, 10(1), 140–156.

Nawrath, C., Poirier, Y., & Somerville, C. (1995). Plant polymers for biodegradable plastics: cellulose, starch and polyhydroxyalkanoates. *Molecular Breeding*, 1, 105–122.

Neag, E., Malschi, D., & Măicăneanu, A. (2018). Isotherm and kinetic modelling of Toluidine Blue (TB) removal from aqueous solution using Lemna minor. *International Journal of Phytoremediation*, 20, 1049–1054. doi: 10.1080/15226514.2018.1460304.

Odum, H.T. (Ed.). (2016). *Heavy Metals in the Environment: Using Wetlands for their Removal*. 19–344. Boca Raton: CRC Press. doi: 10.1201/9781420032840.

Ogunlaja, A., Ogunlaja, O.O., Okewole, D.M., & Morenikeji, O.A. (2019). Risk assessment and source identification of heavy metal contamination by multivariate and hazard index analyses of a pipeline vandalised area in Lagos State, Nigeria. *Science of the Total Environment*, 651, 2943–2952.

Omotade, I.F., Alatise, M.O., Olanrewaju, O.O. Recycling of aquaculture wastewater using charcoal based constructed wetlands. *International Journal of Phytoremediation*, 21 (2019), 399–404. doi: 10.1080/15226514.2018.1537247.

Ouyang, Y. (2002). Phytoremediation: modelling plant uptake and contaminant transport in the soil-plant-atmosphere continuum. *Journal of Hydrology*, 266, 66–82.

Ovečka, M., & Takáč, T. (2014). Managing heavy metal toxicity stress in plants: biological and biotechnological tools. *Biotechnology Advances*, 32(1), 73–86.

Patra, D.K., Pradhan, C., & Patra, H.K. (2020). Toxic metal decontamination by phytoremediation approach: concept, challenges, opportunities and future perspectives. *Environmental Technology & Innovation*, 18, 100672.

Paz-Ferreiro, J., Lu, H., Fu, S., Méndez, A., & Gascó, G. (2014). Use of phytoremediation and biochar to remediate heavy metal polluted soils: a review. *Solid Earth*, 5, 65–75.

Peng, J.S., Ding, G., Meng, S., Yi, H.Y., & Gong, J.M. (2017). Enhanced metal tolerance correlates with heterotypic variation in SpMTL, a metallothionein-like protein from the hyperaccumulator Sedum plumbizincicola. *Plant, Cell & Environment*, 40(8), 1368–1378.

Pentyala, V.B., & Eapen, S. (2020). High-efficiency phytoextraction of uranium using Vetiveria zizanioides L. Nash. *International Journal of Phytoremediation*, 1–10.

Pereira, B.F., De–Abreu, C.A., Herpin, U., De–Abreu, M.F., & Berton, R.S. (2010). Phytoremediation of lead by jack beans on a rhodic hapludox amended with EDTA. *Scientia Agricola*, 67, 308–318.

Pilon-Smits, E.A., Hwang, S., Lytle, C.M., Zhu, Y., Tai, J.C., Bravo, R.C.,… & Terry, N. (1999). Overexpression of ATP sulfurylase in Indian mustard leads to increased selenate uptake, reduction, and tolerance. *Plant Physiology*, 119(1), 123–132.

Pivetz, B.E. (2001). *Phytoremediation of Contaminated Soil and Ground Water at Hazardous Waste Sites*. US Environmental Protection Agency, Office of Research and Development, Office of Solid Waste and Emergency Response.

Popova, L.P., Maslenkova, L.T., Ivanova, A., & Stoinova, Z. (2012). Role of salicylic acid in alleviating heavy metal stress. In *Environmental Adaptations and Stress Tolerance of Plants in the Era of Climate Change*, 447–466. New York: Springer.

Rahbar, A., Farjadfard, S., Leili, M., Kafaei, R., Haghshenas, V., & Ramavandi, B. (2016). Experimental data of biomaterial derived from Malva sylvestris and charcoal tablet powder for Hg2+ removal from aqueous solutions. *Data in Brief, 8*, 132–135.

Rahman, M.A., Reichman, S.M., De Filippis, L., Sany, S.B.T., & Hasegawa, H. (2016). Phytoremediation of toxic metals in soils and wetlands: concepts and applications. In *Environmental Remediation Technologies for Metal-contaminated Soils*, 161–195. Tokyo: Springer.

Rai, P.K. (2018). *Phytoremediation of Emerging Contaminants in Wetlands*. Boca Raton: Taylor & Francis Group.

Rai, P.K., & Chutia, B. (2016). Biomagnetic monitoring through Lantana leaves in an Indo-Burma hot spot region. *Environmental Skeptics and Critics, 5*(1), 1–11.

Rai, P.K., & Singh, M.M. (2015). Lantana camara invasion in urban forests of an Indo-Burma hotspot region and its ecosustainable management implication through biomonitoring of particulate matter. *Journal of Asia-Pacific Biodiversity, 8*, 375–381.

Rakhshaee, R., Giahi, M., & Pourahmad, A. (2009). Studying the effect of cell wall's carboxyl-carboxylate ratio change of Lemna minor to remove heavy metals from aqueous solution. *Journal of Hazardous Materials, 163*(1), 165–173.

Ramamurthy, A.S., & Memarian, R. (2012). Phytoremediation of mixed soil contaminants. *Water, Air, & Soil Pollution, 223*, 511–518.

Rámila, C.D.P., Contreras, S.A., Domenico, C.D., Montenegroc, M.A.M., Vega, A., Handford, M., Bonilla, C.A., & Pizarroa, G.E. (2016). Boron stress response and accumulation potential of the extremely tolerant species Puccinellia frigid. *Journal of Hazardous Materials, 317*, 476–484.

Ranđelović, D., Gajić, G., Mutić, J., Pavlović, P., Mihailović, N., & Jovanović, S. (2016). Ecological potential of Epilobium dodonaei Vill. for restoration of metalliferous mine wastes. *Ecological Engineering, 95*, 800–810.

Raskin, I., Salt, D.E., & Smith, R.D. (1998). Phytoremediation. *Plant Molecular Biology, 49*, 643–668.

Reisinger, S., Schiavon, M., Terry, N., & Pilon-Smits, E.A. (2008). Heavy metal tolerance and accumulation in Indian mustard (*Brassica juncea L.*) expressing bacterial γ-glutamylcysteine synthetase or glutathione synthetase. *International Journal of Phytoremediation, 10*(5), 440–454.

Rezania, S., Taib, S.M., Din, M.F.M., Dahalan, F.A., & Kamyab, H. (2016). Comprehensive review on phytotechnology: heavy metals removal by diverse aquatic plants species from wastewater. *Journal of Hazardous Materials, 318*, 587–599.

Rodriguez, L., Lopez-Bellido, F.J., Carnicer, A., Recreo, F., Tallos, A., & Monteagudo, J.M. (2005). Mercury recovery from soils by phytoremediation. In *Book of Environmental Chemistry*, 197–204. Berlin, Germany: Springer.

Ruffini-Castiglione, M., & Cremonini, R. (2009). Nanoparticles and higher plants. *Caryologia, 62*, 161–165.

Ryslava, E., Krejcik, Z., Macek, T., Novakova, H., Demnerova, K., & Mackova, M. (2003). Study of PCB degradation in real contaminated soil. *Fresenius Environmental Bulletin, 12*, 296–301.

Saha, P., Shinde, O., & Sarkar, S. (2017). Phytoremediation of industrial mines wastewater using water hyacinth. *International Journal of Phytoremediation, 19*, 87–96.

Saifullah, M.E., Qadir, M., de Caritat, P., Tack, F.M., Du Laing, G., & Zia, H. (2009). EDTA-assisted Pb phytoextraction. *Chemosphere, 74*, 1279–1291.

Saleem, M.H., Fahad, S., Adnan, M., Ali, M., Rana, M.S., Kamran, M.,... & Hussain, R.M. (2020). Foliar application of gibberellic acid endorsed phytoextraction of copper and alleviates oxidative stress in jute (Corchorus capsularis L.) plant grown in highly copper-contaminated soil of China. *Environmental Science and Pollution Research*, 1–13.

Salt, D. E., Smith, R.D., & Raskin, I. (1998). Phytoremediation. *Annual Review of Plant Biology, 49*(1), 643–668.

Sandhi, A., Landberg, T., & Greger, M. (2018). Phytofiltration of arsenic by aquatic moss (Warnstorfia fluitans). *Environmental Pollution, 237*, 1098–1105.

Saran, A., Fernandez, L., Cora, F., Savio, M., Thijs, S., Vangronsveld, J., & Merini, L.J. (2020). Phytostabilization of Pb and Cd polluted soils using Helianthus petiolaris as pioneer aromatic plant species. *International Journal of Phytoremediation, 22*(5), 459–467.

Sarwar, N., Imran, M., Shaheen, M.R., Ishaque, W., Kamran, M.A., Matloob, A.,... & Hussain, S. (2017). Phytoremediation strategies for soils contaminated with heavy metals: modifications and future perspectives. *Chemosphere, 171*, 710–721. doi:10.1016/j.chemosphere.2016.12.116.

Sarwar, N., Ishaq, W., Farid, G., Shaheen, M.R., Imran, M., Geng, M., & Hussain, S. (2015). Zinc–cadmium interactions: impact on wheat physiology and mineral acquisition. *Ecotoxicology and Environmental Safety*, *122*, 528–536.

Saxena, G., Purchase, D., Mulla, S.I., Saratale, G.D., & Bharagava, R.N. (2019). Phytoremediation of heavy metal-contaminated sites: eco-environmental concerns, field studies, sustainability issues and future prospects. *Reviews of Environmental Contamination and Toxicology*, doi: 10.1007/398_2019_24.

Schnoor, J.L., Licht, L.A., McCutcheon, S.C., Wolfe, N.L., & Carreira, L.H. (1995). Phytoremediation of organic contaminants. *Environmental Science & Technology*, *29*, 318–323.

Schwitzguébel, J.-P. (2002). Hype or hope: the potential of phytoremediation as an emerging green technology. *Federal Facilities Environmental Journal*, 109–125.

Seth, C.S. (2012). A review on mechanisms of plant tolerance and role of transgenic plants in environmental clean-up. *The Botanical Review*, *78*(1), 32–62.

Shahedi, A., Darban, A.K., Taghipour, F., & Jamshidi-Zanjani, A. (2020). A review on industrial wastewater treatment via electrocoagulation processes. *Current Opinion in Electrochemistry*, *22*,154–169. doi: 10.1016/j.coelec.2020.05.009.

Shaffer, R.E., Cross, J.O., Rose-Pehrsson, S.L., & Elam, W.T. (2001). Speciation of chromium in simulated soil samples using X-ray absorption spectroscopy and multivariate calibration. *Analytica Chimica Acta*, *442*(2), 295–304.

Shah, F.U.R., Ahmad, N., Masood, K.R., & Peralta-Videa, J.R. (2010). Heavy metal toxicity in plants. *In-Plant Adaptation and Phytoremediation*, 71–97. Dordrecht: Springer.

Sharma, S., Singh, S., & Manchanda, V.K. (2014). Phytoremediation: role of terrestrial plants and aquatic plants and aquatic macrophytes in the remediation of radionuclides and heavy metal contaminated soil and water. *Environmental Science and Pollution Research*, doi: 10.1007/s11356-014-3635-8.

Sheng, X.F., & Xia, J.J. (2006). Improvement of rape (Brassica napus) plant growth and cadmium uptake by cadmium–resistant bacteria. *Chemosphere*, *64*, 1036–1042.

Sheoran, V., Sheoran, A.S., & Poonia, P. (2016). Factors affecting phytoextraction: a review. *Pedosphere*, *26*(2), 148–166.

Shewry, P.R., & Peterson, P.J. (1974). The uptake and transport of chromium by barley seedlings (Hordeum vulgare L.). *Journal of Experimental Botany*, *25*(4), 785–797.

Sidhu, G.P.S., Bali, A.S., Singh, H.P., Batish, D.R., & Kohli, R.K. (2018). Ethylenediamine disuccinic acid-enhanced phytoextraction of nickel from contaminated soils using Coronopus didymus (L.) Sm. *Chemosphere*, *205*, 234–243.

Singh, J., & Lee, B.K. (2016). Influence of nano-TiO_2 particles on the bioaccumulation of Cd in soybean plants (Glycine max): a possible mechanism for the removal of Cd from the contaminated soil. *Journal of Environmental Management*, *170*, 88–96.

Singh, M.M., & Rai, P.K. (2016). Microcosm investigation of Fe (iron) removal using macrophytes of Ramsar Lake: a phytoremediation approach. *International Journal of Phytoremediation*, *18*(12), 1231–1236.

Skeffington, R.A., Shewry, P.R., & Peterson, P.J. (1976). Chromium uptake and transport in barley seedlings (Hordeum vulgare L.). *Planta*, *132*(3), 209–214.

Sobha, K., Poornima, A., Harini, P., & Veeraiah, K. (2007). A study on biochemical changes in the fresh water fish, Catla catla (Hamilton) exposed to the heavy metal toxicant cadmium chloride. *Kathmandu University Journal of Science, Engineering and Technology*, *3*(2), 1–11.

Souza, L.A., Piotto, F.A., Nogueirol, R.C., & Azevedo, R.A. (2013). Use of non–hyperaccumulator plant species for the phytoextraction of heavy metals using chelating agents. Scientia Agricola, *70*, 290–295.

Stephenson, C., & Black, C.R. (2014). One step forward, two steps back: the evolution of phytoremediation into commercial technologies. *Bioscience Horizons*, *7*, 1–15.

Stern, B.R., Solioz, M., Krewski, D., Aggett, P., Aw, T.C., Baker, S.,... & Keen, C. (2007). Copper and human health: biochemistry, genetics, and strategies for modeling dose-response relationships. *Journal of Toxicology and Environmental Health, Part B*, *10*(3), 157–222.

Sun, S., Zhou, X., Cui, X., Liu, C., Fan, Y., McBride, M.B.,... & Zhuang, P. (2020). Exogenous plant growth regulators improved phytoextraction efficiency by Amaranths hypochondriacus L. in cadmium contaminated soil. *Plant Growth Regulation*, *90*(1), 29–40.

Sut-Lohmann, M., Jonczak, J., & Raab, T. (2020). Phytofiltration of chosen metals by aquarium liverwort (Monosoleum tenerum). *Ecotoxicology and Environmental Safety*, *188*, 109844.

Syam Babu, D., Anantha Singh, T.S., Nidheesh, P.V., & Suresh Kumar, M. (2019). Industrial wastewater treatment by electrocoagulation process. *Separation Science and Technology*, 1–33.

Tang, C., Song, J., Hu, X., Hu, X., Zhao, Y., Li, B.,... & Peng, L. (2017). Exogenous spermidine enhanced Pb tolerance in Salix matsudana by promoting Pb accumulation in roots and spermidine, nitric oxide, and antioxidant system levels in leaves. *Ecological Engineering*, *107*, 41–48.

Tangahu, B.V., Abdullah, S.R., Basri, H., Idris, M., Anuar, N., & Mukhlisin, M. (2011). A Review on Heavy Metals (As, Pb, and Hg) Uptake by Plants through Phytoremediation. *International Journal of Chemical Engineering*. doi:10.1155/2011/939161.

Tariq, S.R., & Ashraf, A. (2016). Comparative evaluation of phytoremediation of metal contaminated soil of firing range by four different plant species. *Arabian Journal of Chemistry*, *9*(6), 806–814.

Tchounwou, P.B., Yedjou, C.G., Patlolla, A.K., & Sutton, D.J. (2012). Heavy metal toxicity and the environment. In *Molecular, Clinical and Environmental Toxicology*, 133–164. Basel: Springer.

Ugya, A.Y., Hua, X., & Ma, J. (2019). Phytoremediation as a tool for the remediation of wastewater resulting from dyeing activities. *Applied Ecology and Environmental Research*, *17*(2), 3723–3735.

UNEP. (2016). United Nations Environment Programme. Phytoreme*diation: An Environmentally Sound Technology for Pollution Prevention, Control and Remediation, Newsletter and Technical Publications, Freshwater Management Series No. 2.*

USEPA (Environmental Protection Agency). (2000). *Introduction to Phytoremediation*. Retrieved from National Risk Management Research Laboratory, EPA/600/R-99/107: http://www.clu-in.org/download/remed/introphyto.pdf

Uysal, Y. (2013). Removal of chromium ions from wastewater by duckweed, Lemna minor L. by using a pilot system with continuous flow. *Journal of Hazardous Materials*, *263*, 486–492. doi: 10.1016/j.jhazmat.2013.10.006.

Valadi, A.S., Hatamzadeh, A., & Sedaghathoor, S. (2019). Study of the accumulation of contaminants by Cyperus alternifolius, Lemna minor, Eichhornia crassipes, and Canna generalis in some contaminated aquatic environments. *Environmental Science and Pollution Research*, *26*, 21340–21350. doi: 10.1007/s11356-019-05203-0.

Vamerali, T., Bandiera, M., Hartley, W., Carletti, P., & Mosca, G. (2011). Assisted phytoremediation of mixed metal (loid)-polluted pyrite waste: effects of foliar and substrate IBA application on fodder radish. *Chemosphere*, *84*, 213–219.

Van Ginneken, L., Meers, E., Guisson, R., Ruttens, A., Elst, K., Tack, F.M.,... & Dejonghe, W. (2007). Phytoremediation for heavy metal-contaminated soils combined with bioenergy production. *Journal of Environmental Engineering and Landscape Management*, *15*(4), 227–236.

Vangronsveld, J., Herzig, R., Weyens, N., Boulet, J., Adriaensen, K., Ruttens, A.,... & Mench, M. (2009). Phytoremediation of contaminated soils and groundwater: lessons from the field. *Environmental Science and Pollution Research*, *16*, 765–794.

Vassil, A.D., Kapulnik, Y., Raskin, I., & Salt, D.E. (1998). The Role of EDTA in Lead Transport and Accumulation by Indian Mustard. *Plant Physiology*, *117*, 447–453. DOI: 10.1104/pp.117.2.447.

Vishnoi, S.R., & Srivastava, P.N. (2008). Phytoremediation–green for environmental clean. In *Proceedings of Taal2007: the 12th World Lake Conference* (*Vol. 1016*), 1021.

Vymazal, J. (2016). Concentration is not enough to evaluate accumulation of heavy metals and nutrients in plants. *Science of the Total Environment*, *544*, 495–498.

Vymazal, J., & Březinová, T. (2016). Accumulation of heavy metals in aboveground biomass of Phragmites australis in horizontal flow constructed wetlands for wastewater treatment: a review. *Chemical Engineering Journal*, *290*, 232–242.

Wolfenbarger, L.L., & Phifer, P.R. (2000). The ecological risks and benefits of genetically engineered plants. *Science*, *290*(5499), 2088–2093.

Xia, Y., Liu, J., Wang, Y., Zhang, X., Shen, Z., & Hu, Z. (2018). Ectopic expression of Vicia sativa Caffeoyl-CoA O-methyltransferase (VsCCoAOMT) increases the uptake and tolerance of cadmium in Arabidopsis. *Environmental and Experimental Botany*, *145*, 47–53.

Xiao, S., Gao, W., Chen, Q.F., Ramalingam, S., & Chye, M.L. (2008). Overexpression of membrane-associated acyl-CoA-binding protein ACBP1 enhances lead tolerance in Arabidopsis. *The Plant Journal*, *54*(1), 141–151.

Yadav, K.K., Gupta, N., Kumar, A., Reece, L.M., Singh, N., Rezania, S., & Khan, S.A. (2018). Mechanistic understanding and holistic approach of phytoremediation: a review on application and future prospects. *Ecological Engineering, 120*, 274–298.

Yadav, K.K., Gupta, N., Kumar, V., & Singh, J.K. (2017). Bioremediation of heavy metals from contaminated sites using potential species: a review. *Indian Journal of Environmental Protection, 37*(1), 65.

Yaseen, D.A., & Scholz, M. (2017). Comparison of experimental ponds for the treatment of dye wastewater under controlled and semi-natural conditions. *Environmental Science and Pollution Research International, 24*, 16031–16040.

Yau, Y.Y., & Stewart, C.N. (2013). Less is more: strategies to remove marker genes from transgenic plants. *BMc Biotechnology, 13*(1), 36.

Yuanan, H., He, K., Sun, Z., Chen, G., & Cheng, H. (2020). Quantitative source apportionment of heavy metal (loid) s in the agricultural soils of an industrializing region and associated model uncertainty. *Journal of Hazardous Materials, 391*, 122244.

Zaidi, S.S., Usmani, B.R., & Singh, M.J. (2006). Significance of Bacillus subtilis strain SJ–101 as a bioinoculant for concurrent plant growth promotion and nickel accumulation in Brassica juncea. *Chemosphere, 64*, 991–997.

Zayed, A.M., & Terry, N. (2003). Chromium in the environment: factors affecting biological remediation. *Plant and Soil, 249*(1), 139–156.

Zhang, Y., Wu, D., Wang, C., Fu, X., & Wu, G. (2020a). Impact of coal power generation on the characteristics and risk of heavy metal pollution in nearby soil. *Ecosystem Health and Sustainability, 6*(1), 1787092.

Zhang, M., Wang, X., Liu, C., Lu, J., Qin, Y., Mo, Y.,… & Liu, Y. (2020b). Identification of the heavy metal pollution sources in the rhizosphere soil of farmland irrigated by the Yellow River using PMF analysis combined with multiple analysis methods – using Zhongwei city, Ningxia, as an example. *Environmental Science and Pollution Research*, 1–12.

Zhou, J., Wu, Z., Yu, D., Pang, Y., Cai, H., & Liu, Y. (2018). Toxicity of linear alkylbenzene sulfonate to aquatic plant Potamogeton perfoliatus L. *Environmental Science and Pollution Research, 25*, 32303–32311.

12

Phytoremediation: A Sustainable Technology for Pollution Control and Environmental Cleanup

Poonam Yadav[1], Anwesha Chakraborty[2], Sudhakar Srivastava[1], Shalini Sahani[3], and Pardeep Singh[4]
[1]Institute of Environment and Sustainable Development, Banaras Hindu University, India
[2]Department of Life Sciences, Presidency University, Kolkata, WB, India
[3]Department of Material Science & Engineering, Gachon University, 1342 Seongnamdaero, Seongnam, Korea
[4]Department of Environmental Science, PGDAV College University of Delhi, New Delhi, India

CONTENTS

12.1	Introduction	237
12.2	Phytodetoxification Mechanism	239
	12.2.1 Avoidance	239
	12.2.2 Tolerance	239
12.3	Phytoremediation Strategies	240
	12.3.1 Phytoextraction and Accumulation	242
	12.3.2 Phytodegradation	242
	12.3.3 Phytostabilisation	242
	12.3.4 Phytovolatilisation	242
	12.3.5 Phytofiltration	242
12.4	Case Studies	243
11.5	Phytoremediation Technique: Advantages and Limitations	244
11.6	Conclusion and Future Prospects	245
References		245

12.1 Introduction

Industrialisation and rapid urbanisation have resulted in the release of large amounts of heavy metal(loids) in our surroundings, along with other life-threatening pollutants in the environment. That has reached an alarming situation and is of serious concerns among the scientific community throughout the world (Kanwar et al., 2020). These heavy metal(loids) come into the environment from natural and anthropogenic sources, such as the use of fertilisers and pesticides in agriculture (Hamzah et al., 2016; Rani et al., 2020); sewage discharge, mining, and smelting processes (Chen et al., 2016); fossil fuel burning and electroplating, etc. Hence, they concentrate in soil and water, and pose a great threat to the environment because heavy metals are non-degradable. Therefore, they are persistent in the environment and can reside in the soil for prolonged periods. They make their way from the roots of plants to our food chain via accumulation in crops. Thus, they accumulate in our body through the process of biomagnification and affect our health, leading to some serious illnesses (Rehman et al., 2017). So, it is becoming necessary to decontaminate the soil and water, and stop the entry of the contaminants into our environment and food chain (Gerhardt et al., 2017). The extraction of these toxic pollutants from contaminated soil and water by various physical, chemical, and biological methods is the current interest. So far, many remediation and decontamination approaches have been proposed that have also evolved with time due to continued research and advanced techniques. Traditional methods of remediation such as incineration, flushing, electrokinetics, landfilling, and vitrification are very expensive and result in damage to the soil properties. One of the most popular, attractive, and sustainable techniques is phytoremediation of contaminated soil or water (Kanwar et al., 2020).

Phytoremediation is a technology that makes use of plants (hence the name "phyto") and microbial flora associated with it to accumulate or degrade pollutants of organic and inorganic origin present in the system. This is a novel and efficient cleanup strategy that utilises biological entities (Pilon-Smits, 2005). So, phytoremediation is the utilisation of living plants for degradation, accumulation (assimilation in plants' tissue), and containment of pollutants from water, soil, and other sources within its tissues (Iori et al., 2015) and rendering toxic environmental pollutants harmless. It can also be described by its potential for immobilisation, sequestration, extraction, removal, destruction, remediation, uptake, and stabilisation of soil/water contaminants (Kanwar et al., 2020).

Green plant-based remediation and management of contaminated soil or water is not a new concept. Floating and submerged plant systems, reed beds, and constructed wetlands have been in common use for the management of some wastewaters for many years. But, with time, the concept has evolved and developed. In literature, many plant species have been reported as hyper-accumulators, i.e., accumulation beyond a certain threshold limit. There are a large number of diverse sets of plants available (terrestrial and aquatic both) that have been reported to show hyper-accumulation potential for different metal (loids); for example, Ni, Zn, As, Cd, etc. *Alyssum lesbiacum, Alyssum bertolonii,* and *Thlaspi goesingense* are known to hyper-accumulate Ni from contaminated sources (Küpper et al., 2001); *Pteris vitata* is known for As hyper-accumulation (Ma et al., 2001); and *Arabidopsis halleri* is reported for Zn hyper-accumulation (Zhao et al., 2000). There are other plants species also, like alpine pennycress, hemp, mustard plants, and pigweed, that have proven to be good and result-oriented hyper-accumulators for toxic contaminants at waste sites. Some of the aquatic macrophytes, free-floating, submerged, and rooted emergent plants, are also found to have very effective phytoremediation potential; for example, *Hydrilla verticillata, Lemna minor, Ceratoplhyllum, Typha latifolia, Pistia* sp., *Spirodela* sp., and *Eichornia crassipes* (Valipour and Ahn, 2017; Poonam et al., 2017; Srivastava et al., 2007; Yadav and Srivastava, 2020). Other hyper-accumulator plants, highly studied and reported by the scientific community, are *Sedum alfredii* sp., *Thlaspi* sp., *Arabidopsis* sp., and *Thlaspi* species were found to hyper-accumulate many metals (*T. caerulescens* for Pb, Ni, Cd, and Zn; *T. ochroleucum* and *T. goesingense* for Ni and Zn; and *T. rotundifolium* for Ni, Pb, and Zn) (Vara Prasad and de Oliveira Freitas, 2003). *Thlaspi caerulescens* is the most studied plant and has received much attention as a potential candidate for phytoremediation of soils rich in Cd and Zn. The desired condition for phytoremediation is that plants used should be native, tolerant, and adaptive to change. They should have high biomass production, easy propagation, high accumulation potential for the contaminant, and easily facilitate the breakdown of pollutants in the soil (Freitas et al., 2004). One major criterium for deciding on a plant as a hyper-accumulator is that it should accumulate the minimum threshold concentration in its tissues (for example, 0.1% for arsenic in dry weight of plants) (Wang et al., 2007). Here, it can be noted that every phytoremediation plant is specific to a particular metal(loids) in terms of response and tolerance. A single plant species cannot be used for all the contaminants, organic or inorganic. Thus, the selection of plant species for dominating contaminants present in the source is a crucial step that decides the success of the process.

Plants that produce high-biomass and high-valued products are also economically feasible for phytoremediation (Jiang et al., 2015). Some of the commercial flowering plants like marigolds and sunflowers can be proposed as ameliorating plants for contaminated soil, such as arsenic-contaminated soils. It will help in the management of contaminated soil and provide financial support to the local farmers as well (Poonam and Srivastava, 2019). The phytoremediation technique is an easy and applicable method, ex-situ and in-situ both, depending upon the contaminated media. Static water is required for the application of phytoremediation techniques (phyto-filtration), whereas any type of soil can be remediated using suitable plant species. Further, for the contaminated groundwater in some specific areas, rooted plants or trees with the capability to tap groundwater can be used to remediate the soil. Various contaminant sites have been remediated under phytoremediation science projects; for example, landfill leachates, pesticides, polyaromatic hydrocarbon derivatives, explosives, industrial effluents, mine areas, and crude oil, etc. (Section 4). Thus, identification of a fast-growing plant with a larger affinity for nutrients and tolerant hyper-accumulating genotypes are very important aspects. In this respect, genetically modified and engineered plants are also a beneficial step for the success of the technique. Some studies developed GM plants and advocated the use of genetically modified plants for more efficient phytoextraction of Cd, Hg, Pb, Se, and As (Zhang et al., 2013; Hunt et al., 2014).

12.2 Phytodetoxification Mechanism

The main purpose of the phytoremediation technique is the heavy-metal detoxification of soil and water systems. In the presence of any metal(loids) stress, plants respond in two different ways by utilising their defense mechanisms to fight the toxicity, i.e., avoidance and tolerance. By adopting these two strategies, the plant controls the entry of toxic metal(loids) into the plant cells for better management of stress (Hall, 2002; Yan et al., 2020).

12.2.1 Avoidance

In the avoidance mechanism, plants check the uptake of heavy metal in its roots and, thus, restrict further entry into the plant tissue (Dalvi and Bhalerao, 2013; Yan et al., 2020). Toxic metal(loids) are immobilised by the root through root sorption, metal ion precipitation, and exclusion. In these strategies, root exudates play an important role as metal ligands to form immobilised and stable metal ion complexes. This activity also involves changes in the pH of the rhizosphere, which helps in the precipitation of metal and, hence, limits the bioavailability of the toxic metal to the root, whereas in metal exclusion, a biological barrier plays a role between the root cells and shoot system and thus metal remains in the root tissue/cells only. In that way, the aerial shoot part remains stress-free. Moreover, arbuscular mycorrhizas utilise the absorption, adsorption, and chelation of the toxic metal in the root rhizosphere. Thus, mycorrhiza too may restrict the entry of metal into the root systems of the plants. The plant also embeds the metal(loids) in their cell wall as a result of the avoidance mechanism (Memon and Schröder, 2009). Negatively charged organic compounds (carboxylic groups of polygalacturonic acids in the pectin) of the cell wall also help in binding the metal. This is called the cation exchange mechanism and acts as barrier for the entry of heavy metals inside the root cells (Ernst et al., 1992; Yan et al., 2020).

12.2.2 Tolerance

Tolerance is another highly effective defense machinery of the plants with several defensive ways to fight the stress (abiotic and biotic) that have evolved with time (Yadav et al., 2020). In the presence of any toxic element [metal(loids)] in the cytosol, the plant starts to change its energetics (electron transport chain, photosynthesis, etc.) as a tolerance strategy to minimise the stress level. Various changes via different mechanistic pathways take place, like chelation, inactivation, and compartmentalisation of toxic metal ions at the intracellular level inside the plants' cells (Dalvi and Bhalerao, 2013). These processes come under the detoxification mechanism of plants when excess toxic metal ions get stored inside the cells (Manara, 2012). Chelation helps in reducing the concentrations of the toxic form of free metal ions to relatively low levels. Amino acids, metallothioneins, phytochelatins, organic acids, and other organic compounds present in the cell walls (proteins, polyphenols, pectins) of the plants help in the chelation (Gupta et al., 2013; Yadav et al., 2020). Thus, the availability of toxic metal ions to plants gets reduced. Chelation of Ni in the leaves of *T. goesingense* is mediated by citrate (Krämer et al., 2000), whereas in the leaves of *Solanum nigrum*, chelation of Cd is achieved via acetic and citric acids (Sun et al., 2006), and chelation of Zn in *A. halleri* is done by malate (Sarret et al., 2002). Besides, metal toxicity also induces the increased production of amino acids in plants under toxic metal stress. Cd and

As metal(loids) toxicity is found to induce increased production of cysteine in *Arabidopsis thaliana* and rice plants, respectively (Domínguez-Solís et al., 2004; Yadav et al., 2020), Ni increases the histidine accumulation and proline production is induced by Zn, Cd, Pb, As, and Cu stress (Roy and Bera, 2002).

Among all these mentioned detoxification strategies, vacuolar sequestration and compartmentalisation are more effective defense strategies against any damaging effect of a toxic metal ion. It helps in the protection of plant cell metabolism and other functions. These are the special defense machinery of some plant species that makes them potential and sustainable hyper-accumulators or tolerant to any specific metal(loids) ion toxicity. For vacuolar sequestration of most of the metal(loids), plants rely on the complexation with glutathione (GSH) and phytochelatins (PCs), and the complexed metal(loid)s are exported to vacuoles with the help of specific transporters (Song et al., 2014). In addition, plants may have specific transporters for a metal(loid) that can sequester it without the need of complexation, such as Arsenic Compounds Resistance 3 (ACR3) in *P. vittata* (Indriolo et al., 2010). Apart from effective detoxification via complexation, metal(loid)-tolerant plants also possess an efficient antioxidant defense system to tackle toxic impacts of metal(loid)s (Srivastava et al., 2007).

12.3 Phytoremediation Strategies

Plant species that are generally used for phytoremediation approaches are either tolerant or are potent hyper-accumulators of one or more contaminating element (Shukla and Srivastava, 2017). The hyper-accumulator plants are peculiar in their way of accumulating metals or metalloids in different tissues of the root or above-ground part at a much higher concentration compared to other plants which are non-hyper-accumulators and survive on similar soil compositions (Van der Ent et al., 2013). Thus, phytoremediation through plants includes various metal(loids)-accumulating ferns, weeds, and many aquatic plants (*Hydrilla, Ceratophyllum, Lemna,* etc.) that can easily flourish in contaminated soil and water (Figure 12.1), and are able to complete their life cycles (Srivastava et al., 2007; Poonam et al., 2017; Yadav and Srivastava, 2020). A few examples of hyper-accumulator plants have been mentioned in Table 12.1. Plants that are capable of accumulating Ni concentrations of >1,000 mg g^{-1} in their tissues are referred to as Ni-hyper-accumulators (Van der Ent et al., 2013). Many plant species have been identified with higher accumulation properties such as *Hybanthus austrocaledonicus* (family Violaceae), endemic to New Caledonia, that has been proved to accumulate high Ni concentrations in its phloem (Paul et al., 2020). Sabah (Malaysia), on the island of Borneo, has remarkable species like *Phyllanthus balgooyi, Phyllanthus rufuschaneyi,* and *Actephila alanbakeri* (Phyllanthaceae) that are known for their Ni hyper-accumulation roles (Mesjasz-Przybylowicz et al., 2016; Bouman et al., 2018).

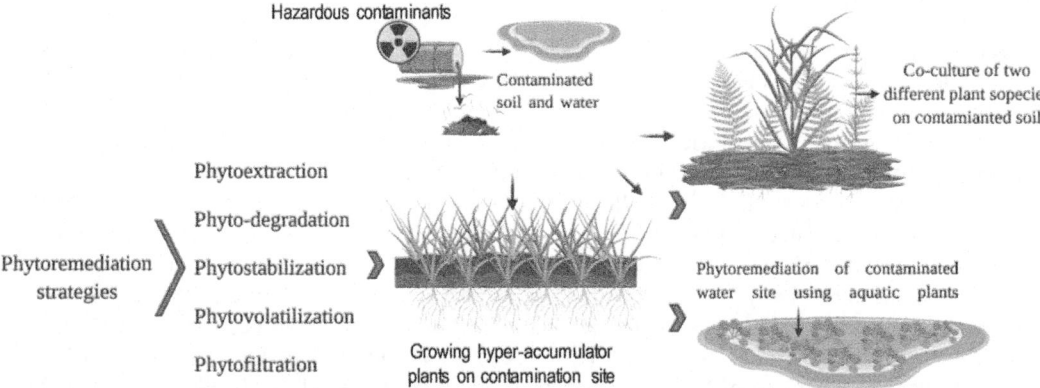

FIGURE 12.1 Illustration of phytoremediation strategies for contaminated soil and water.
Source: Image created with BioRender.com.

TABLE 12.1

List of hyper-accumulator plants for various metal(loids) contaminations in soil and water

Plant's Name	Pollutants Removed	Source	References
Streptanthus polygaloides	Nickel, zinc, copper, cobalt, manganese, and lead	Serpentine soil	Boyd and Davis, 2001
Thlaspi caerulescens	Zinc, cadmium, and nickel	Metalliferous soils	Milner and Kochian, 2008
Ceratophyllum demersum	Chromium and lead	Aqueous solutions (lake)	Abdallah, 2012
Lemna gibba	Chromium and lead	Aqueous solutions (lake)	Abdallah, 2012
Corydalis davidii	Zinc	Artisanal zinc smelting area (soil)	Lin et al., 2012
Juncus acutus	Zinc	Zinc contaminated lands/soil	Mateos-Naranjo et al., 2014
Psychotria gabriellae	Nickel	Soil	Merlot et al., 2014
Actephila alanbakeri	Nickel	Ultramafic soils	Van der Ent et al., 2015
Arabidopsis halleri	Cadmium, zinc	Contaminated soil	Claire-Lise and Nathalie, 2012; Tlustoš et al., 2016
Noccaea praecox	Cadmium, zinc	Contaminated soil	Tlustoš et al., 2016
Noccaea caerulescens	Cadmium, zinc	Contaminated soil	Tlustoš et al., 2016
Phyllanthus balgooyi	Nickel	Soil	Mesjasz-Przybylowicz et al., 2016
Catharanthus roseus (L.)	Lead, nickel, zinc, cadmium, and chromium	Soil	Subhashini and Swamy, 2017
Dichapetalum gelonioides	Zinc	Soil	Nkrumah et al., 2018
Phyllanthus rufuschaneyi	Nickel	Soil	Bouman et al., 2018
Senecio coronatus	Nickel	Soil	Meier et al., 2018
Pteris vitatta	Arsenic	Soil	Yan et al., 2019, Ma et al., 2001
Hydrilla verticillata	Copper, phenol, arsenic	Aquatic system (water body)	Venkateswarlu et al., 2019; Chang et al., 2019; Yadav and Srivastava, 2020.
Pistia stratiotes	Copper	Aquatic system (water body)	Venkateswarlu et al., 2019
Hybanthus austrocaledonicus	Nickel	Soil	Paul et al., 2020

Thlaspi caerulescens is one of the most widely studied species that has a unique behaviour of accumulating not only Ni but also Zn and Cd from the soil in its shoots (Milner and Kochian, 2008). Similarly, *Streptanthus polygaloides*, a member of Brassicaceae, can collectively hyper-accumulate six metals: zinc, nickel, copper, cobalt, manganese, and lead (Boyd and Davis, 2001). Two other species of family Brassicaceae viz., *Noccaea caerulescens* and *Arabidopsis halleri*, are known as potent Cd and Zn hyper-accumulators (Tlustoš et al., 2016). *Catharanthus roseus* has been identified as a hyper-accumulator of five metals viz., Pb, Ni, Zn, Cd, and Cr. Thus, it has been recommended for use in phytoextraction of these metals from contaminated soil (Subhashini and Swamy, 2017). The genes responsible for Ni and other metal(loid)s hyper-accumulation have been identified. This has paved a way to utilise gene sequencing technologies to study the transcriptomes of various transporters of metals during the accumulation process (Merlot et al., 2014). RNA-Seq analysis can be used as a potential technique to trace the evolutionary lineage of any particular species; an example has been done by Meier et al. (2018) in *Senecio coronatus*. The ability of hyper-accumulation and thus phytoremediation is not only restricted to land plants, but aquatic macrophytes like *Ceratophyllum demersum, Hydrilla verticillata,* and *Lemna gibba* have been found very effective in removing the toxic metals lead, chromium (Abdallah, 2012), and As (Srivastava et al., 2007).

12.3.1 Phytoextraction and Accumulation

Phytoextraction is the process of utilising the accumulation potential of the harvestable parts of plants to extract the pollutants from the medium. It is found to be a green, sustainable, and environmentally friendly way to cleanup of contaminated soils and waters (Kanwar et al., 2020). Thus, in this process, plant parts take up the metal(loids) and concentrate them in roots or shoots. Here, the accumulation is defined in terms of efficiency. The efficiency is calculated by acquiring the translocation and bio-concentration factor (BCF) from the phytoremediation plants and their multiplication product is called phytoextraction efficiency. Thus, the accumulation efficiency of remediating plants is highly dependent on the BCF (Ladislas et al., 2012) and high biomass production, whereas there are plant species with low biomass but they have high hyper-accumulating properties; for example, Thlaspi and Arabidopsis. Studies showed that phytoextraction for managing the polluted site can be a feasible approach for cleaning purposes (Robinson et al., 2015). Apart from cleaning, phytoextraction is found to be valuable in extracting precious metals, such as platinum, nickel, and gold. That can be further recovered by phytomining for commercial purposes. Phytoaccumulation plants accumulate contaminants in their tissue in higher concentration because of their high translocation factor from root to shoot. For the phytoaccumulation, mostly non-edible plants are recommended for use due to human exposure and safety reasons.

12.3.2 Phytodegradation

Phytodegradation is the process of utilising the plant- and root-associated microorganisms to degrade the organic pollutants present in the soil (Alkorta and Garbisu, 2001). The same plants may utilise the many functions as per the requirements or any one of the functions of phytoremediation strategies (functions related to adaptation and tolerance).

12.3.3 Phytostabilisation

Plants' roots absorb the toxic metal(loids) present in the contaminated soil and try to keep it in the rhizosphere zone. Thus, it prevents them from leaching and making them harmless. This function of phytoremediation by plants is termed phytostabilisation. This process can also be defined as in-situ confinement of pollutants by the plants (Wan et al., 2017; Yadav and Srivastava, 2020).

12.3.4 Phytovolatilisation

Phytovolatilisation is another strategy adapted by the plants to clean up the polluted soil and water. It involves the use of aerial parts of the plant to volatilise contaminants, e.g., Se, As, and Hg. Plants' roots takes up the metal contaminants from the soil/water through the root and metabolizes them in gaseous form in the aerial part of the plants. These volatile forms are released back into the atmosphere. Dushenkov (2003) removed the radioactive tritium, an isotope of hydrogen (3H), from the medium by exploiting the volatilisation potential of the plant.

12.3.5 Phytofiltration

Phytofiltration or rhizofiltration uses the process of adsorption or precipitation for pollutants present in the root zone of the plants. Phytofiltration involves the plants' roots for the removal/filtration of metals/excess nutrients from aqueous wastes like wastewater or sewage water. Phytofiltration technology is an emerging and eco-friendly technology that employs various aquatic/semi-aquatic plants like *Micranthemum umbrosum, Hydrilla vericillata,* and *ceratophyllum* etc. to remove/filtrate the contaminants from water or

Phytoremediation: A Sustainable Technology 243

aqueous solutions. To some extent, the phytofiltration process is similar to the rhizofiltration. The only difference is that rhizofiltration is the main component of the whole process of phytofiltration, as it happens in the rhizosphere zone (Kanwar et al., 2020).

12.4 Case Studies

Agricultural soils near mining sites are reported to have high concentrations of toxic heavy metal(loids) such as lead, Cd, Mg, As, Sb, etc., that pose a high risk to soil and human health (Fan et al., 2017). A notable procedure to make these lands reusable and resourceful is phytoremediation. Many plant species are reported there (Table 12.1) that can be grown in contaminated land stretches to refine them. Again, a combination of more than two plants can be uptaken as a prospective strategy to maximise the heavy metal accumulation to clear up contaminated areas in less time. Studies suggest that a combination of aquatic plant species *Ceratophyllum demersum, Lemna minor,* and *Hydrilla verticillate* were found very effective in As removal at a raised rate from the contaminated medium as compared to individual plants. Thus, more than one set of plants can be undertaken for phytoremediation strategies (Srivastava et al., 2014; Poonam et al., 2017). So, besides these aquatic plants, many more potential hyper-accumulators are reported and can also be used for removing heavy metals from industrial and municipal wastewaters, e.g., *Phragmites australis, Typha latifolia,* and *Scirpus* (Bulrush). A field study was performed by Kertulis-Tartar et al. (2006) using Chinese brake fern (*Pteris vittata* L.) for phytoremediation of land contaminated with chromated copper arsenate (CCA). *Pteris vittata* is a fast-growing plant that preferably grows in alkaline soil and is responsible for translocating and accumulating arsenic in tissues above the ground. It is the first reported arsenic hyper-accumulating plant amongst many others known to date. The live fronds of *Pteris* accumulate a greater amount of arsenic compared to the senesced fronds of the fern. The study conducted by Ma et al. (2001) showed that a significantly higher amount of arsenic was removed from the soil by concerned ferns, proving the fact that *Pteris vittata* can be used as a potential arsenic hyper-accumulator in treating arsenic-contaminated soil. In a similar field study, arsenic hyper-accumulating capacity of two ferns, *Pityrogramma calomelanos var. austroamericana* and *Pteris vittata,* were analyzed on arsenic-contaminated cattle-dip site located in Wollongbar, NSW, Australia (Niazi et al., 2012). Their study spanned around 27 months, after which it was observed that *P. calomelanos var. austroamericana* removed around 2.65 times higher arsenic (8,053 mg arsenic) compared to *Pteris vittata* (3,042 mg arsenic). Their data revealed that the frond tissues of *P. calomelanos var. austroamericana* extracted about 1.7–3.9% and *Pteris vittata* removed 0.53–1.5% of total arsenic from the soil of their study area. With these results, it can be concluded that two ferns have enough potential compared to other reported plant species.

Aquatic plant species *Hydrilla verticillate* is also one of the potential candidates for an arsenic hyper-accumulator in arsenic-contaminated water. The study shows that approximately 72% of the total arsenic supplied (8,546 g of the total) was removed from the test contaminated water by using *Hydrilla verticillate* (Srivastava et al., 2011). *Hydrilla* is an easily available plant, grows fast, and requires less management; hence its suitability for phytoremediation increases. Co-culture of *Hydrilla* with that of rice was also proposed as a low-cost and sustainable technique for ameliorating As stress in rice plants growing in arsenic-contaminated fields (Yadav and Srivastava, 2020). The concept of co-culture was found both feasible and sustainable. It has been proved effective to decrease arsenic accumulation in rice (Yadav and Srivastava, 2020). In another study, phytoremediation capacity of *Triticale, Helianthus annuus,* and *Brassica juncea* was examined on a large heavy metal/radionuclide contaminated area from an old uranium mining site in East Germany (Willscher et al., 2013). In the study, the effect and interaction of fungi and bacteria with the aforementioned plants was analyzed with increasing pH, organic matter, etc., whereas Chang et al. (2019) proposed that corn plants can also be used for phytoremediation purposes. Their study used combined circulation-enhanced electrokinetics (CEEK) and corn plants to decontaminate the agricultural land polluted by Pb in Taiwan. The study showed that lead concentration in soil was reduced by around 63% when CEEK and phytoremediation were performed back to back for three stages, i.e., CEEK + phytoremediation using corn + CEEK. Similarly, *Azolla caroliniana* (water fern) was also reported an effective plant with a high bio-accumulation factor for the management of metal-enriched fly ash pond (Pandey, 2012). *Azolla caroliniana* showed a BCF from 1.7 to 18.6 in roots and 1.8 to 11.0 in the fronds.

Literature also suggests the use of some algae for the remediation of the contaminated medium. Algae belonging to the divisions *Chlorophyta, Cyanophyta, Euglenophyta,* and *Heterokontophyta* have been proven to play an important role in controlling metal concentration in aquatic environments like lakes and oceans. Mitra et al. (2012) have suggested the utilisation of algal biomass for phytoremediation of arsenic and boron (B) present in three springs from the Sang-E-Noghreh area in Iran. They calculated the bio-concentration and bio-accumulation factors of algae for As and B. Bio-accumulation factors of both B and As were higher than ten in algae collected from three sites of study. Some other plant species such as *Leersia hexandra, Juncus effusus,* and *Equisetum ramosisti* were also found to accumulate toxic heavy metals like Pb, Zn, Cu, and Cd in their roots (Deng et al., 2004). These plants were grown on the wetlands of China and it was inferred that plants adapted an exclusion strategy for metal tolerance. In a similar wetland-based study of the Erh-Chung wetlands of Taiwan, the ability of water hyacinths (*Eichhornia crassipes* Mart. Solms.) was analyzed for hyper-accumulation and translocation factors of toxic heavy metals like Cd, Pb, Cu, Zn, and Ni (Liao and Chang, 2004). The study showed that the concentration of these five elements was 3–15 times higher in the roots compared to the shoots. In the root tissues, the order in which the aforementioned metals were accumulated was Cu > Zn > Ni > Pb > Cd. Thus, it can be concluded that *E. crassipes* can be a potential candidate of phytoremediation of wastewater and wetland. The above-mentioned studies are limited to some plant species only. It is necessary to explore more plant species from the large plant diversity worldwide.

11.5 Phytoremediation Technique: Advantages and Limitations

The term *phytoremediation* itself is self-advocating for sustainable and eco-friendly technology. This technology has many advantages over other techniques. If we compare the conventional methods of phytoremediation, ex-situ and in-situ phytoremediation are found to be less expensive and feasible (Pilon-Smits and Freeman, 2006). Applications of phytoremediation techniques in-situ are very easy and can be sustained with low maintenance costs. If we analyze the various case study findings (Table 12.1), it can be easily concluded that phytoremediation is a natural and solar-energy-based mitigation strategy for moderately contaminated soils and sites over large areas. It has huge potential to evolve as the most accepted green technology for remediation, provided that selection of hyper-accumulator plants have been done carefully and agronomic practices are utilised for maximum phyto-availability of contaminants (Schwitzguébel, 2017). This can also improve the soil texture and nutrient quality as plants and microbial metabolic processes will add the nutrients and oxygen to the soil via the in-situ phytoremediation technique (Wiszniewska et al., 2016). Apart from this, remediating plants will help in checking soil erosion by water and wind by their ground cover and capacity of their roots to bind soil at its place. Talking about the phytoremediation applicability, we can apply it to any site and any geographical location of the world by using the plants that can be grown in that area. Furthermore, selection of the suitable plant species, varieties or cultivar, and ecotype according to the type of contamination and medium should be done carefully and exploited at full potential for better results of the techniques. These positive characteristics of phytoremediation make phytoremediation a promising and attractive technique for remediation adapted by the public and industries. However, there are some technical and built-in limitations of the techniques, such as contaminates should be in the range of the roots of the plants (Schwitzguébel, 2017). The availability of contaminants in the root zone will pose serious threat and stress to plants by affecting various factors such as availability of water, nutrient, depth, and other factors essential for plant development and biomass production. Secondly, the area of the contaminated site is also a limiting factor, as correct application of various agronomic applications of farming techniques is not possible in the small area of land or soil. Further, due to being dependent on plant growth, the process is slow and proceeds as the plant grows. The need to have fast-growing plants compromises their biomass generation, while trees with a large biomass and metal(loid) removal take years to grow. So, there are certain concerns associated with phytoremediation. However, considering the pros and cons of physico-chemical and phytoremediation methods, phytoremediation is justifiably far superior, sustainable, lower cost, and widely applicable option to take further. However, this would require supplementing phytoremediation plants with microorganisms, nanoparticles, or other physical and chemical methods. The need of the hour is to devise a mixed approach to achieve desirable results in a short time to relieve the stress on the environment and develop a clean, healthy environment.

11.6 Conclusion and Future Prospects

With the increasing global population, urbanisation, and industrialisation, pollution load into the air, water, and soil has increased over the years and this has emerged as a great threat to the environment. Thus, sustainable and economic technologies are required to reduce the pollution and remediate the polluted environmental matrices. Many technological developments have been achieved with time for remediation of the soil and water and, among these, phytoremediation, being sustainable and eco-friendly, is the most suitable to be adopted to clean soil and water systems. The success of these techniques entirely depends on the feasibility of plants used and their detoxification mechanism and potential. Plants used for remediation should be hyper-accumulators or hyper-tolerant for their suitability as bioremediation. Moreover, phytoremediation processes and techniques do not require any highly specialised and expensive equipment or any sophisticated mechanisms. Therefore, it is easy to implement and manage compared to other methods. As there are huge numbers of plants unexplored for their remediation potential, more study is needed to explore such plants that can be much better as hyper-accumulators than the existing known ones. Additionally, we should also think about saving the environment by changing our attitude towards resource use and its exploitation. We should focus on adapting the principles of sustainable development in every aspect of life, either technology-driven or where resource use is concerned.

REFERENCES

Abdallah, M.A.M. (2012). Phytoremediation of heavy metals from aqueous solutions by two aquatic macrophytes. *Ceratophyllum demersum* and *Lemna gibba* L. *Environmental Technology*, *33*(14), 1609–1614.

Afrous, A., Manshourisup, M., Liaghatsup, A., Pazirasup, E., & Sedghisup, H. (2011). Mercury and arsenic accumulation by three species of aquatic plants in Dezful, Iran. *African Journal of Agricultural Research*, *6*(24), 5391–5397.

Alkorta, I., & Garbisu, C. (2001). Phytoremediation of organic contaminants in soils. *Bioresource Technology*, *79*(3), 273–276.

Alkorta, I., Hernández-Allica, J., Becerril, J.M., Amezaga, I., Albizu, I., & Garbisu, C. (2004). Recent findings on the phytoremediation of soils contaminated with environmentally toxic heavy metals and metalloids such as zinc, cadmium, lead, and arsenic. *Reviews in Environmental Science and Biotechnology*, *3*(1), 71–90.

Ashraf, S., Ali, Q., Zahir, Z.A., Ashraf, S., & Asghar, H.N. (2019). Phytoremediation: environmentally sustainable way for reclamation of heavy metal polluted soils. *Ecotoxicology and Environmental Safety*, *174*, 714–727.

Bouman, R., van Welzen, P., Sumail, S., Echevarria, G., Erskine, P.D., & van der Ent, A. (2018). *Phyllanthus rufuschaneyi*: a new nickel hyperaccumulator from Sabah (Borneo Island) with potential for tropical agromining. *Botanical Studies*, *59*(1), 9.

Boyd, R.S., & Davis, M.A. (2001). Metal tolerance and accumulation ability of the Ni hyperaccumulator *Streptanthus polygaloides* Gray (Brassicaceae). *International Journal of Phytoremediation*, *3*(4), 353–367.

Chang, J.H., Dong, C.D., & Shen, S.Y. (2019). The lead contaminated land treated by the circulation-enhanced electrokinetics and phytoremediation in field scale. *Journal of Hazardous Materials*, *368*, 894–898.

Chen, H., Chen, R., Teng, Y., & Wu, J. (2016). Contamination characteristics, ecological risk and source identification of trace metals in sediments of the Le'an River (China). *Ecotoxicology and Environmental Safety*, *125*, 85–92.

Claire-Lise, M., & Nathalie, V. (2012). The use of the model species Arabidopsis halleri towards phytoextraction of cadmium polluted soils. *New Biotechnology*, *30*(1), 9–14.

Dalvi, A.A., & Bhalerao, S.A. (2013). Response of plants towards heavy metal toxicity: an overview of avoidance, tolerance and uptake mechanism. *Annals of Plant Sciences*, *2*, 362–368.

Deng, H., Ye, Z.H., & Wong, M.H. (2004). Accumulation of lead, zinc, copper and cadmium by 12 wetland plant species thriving in metal-contaminated sites in China. *Environmental Pollution*, *132*(1), 29–40.

Domínguez-Solís, J.R., López-Martín, M.C., Ager, F.J., Ynsa, M.D., Romero, L.C., & Gotor, C. (2004). Increased cysteine availability is essential for cadmium tolerance and accumulation in *Arabidopsis thaliana*. *Plant Biotechnology Journal*, *2*, 469–476.

Dushenkov D. (2003). Trends in phytoremediation of radionuclides. Plant and Soil, *249*, 167–175.

Ernst, W.H., Verkleij, J., & Schat, H. (1992). Metal tolerance in plants. *Acta Botanica Neerlandica*, *41*, 229–248.

Fan, Y., Zhu, T., Li, M., He, J., & Huang, R. (2017). Heavy metal contamination in soil and brown rice and human health risk assessment near three mining areas in central China. *Journal of Healthcare Engineering*, *2017*.

Freitas, H., Prasad, M.N.V., & Pratas, J. (2004). Plant community tolerant to trace elements growing on the degraded soils of Sao Domingos mine in the south east of Portugal: environmental implications. *Environment International*, *30*(1), 65–72.

Gerhardt, K.E., Gerwing, P.D., & Greenberg, B.M. (2017). Opinion: taking phytoremediation from proven technology to accepted practice. *Plant Science*, *256*, 170–185.

Gupta, D.K., Vandenhove, H., & Inouhe, M. (2013). Role of phytochelatins in heavy metal stress and detoxification mechanisms in plants. In D.K. Gupta, F.J. Corpas, & J.M. Palma (Eds.), *Heavy Metal Stress in Plants*, 73–94. Berlin: Springer.

Hall, J. (2002). Cellular mechanisms for heavy metal detoxification and tolerance. *Journal of Experimental Botany*, *53*, 1–11.

Hamzah, A., Hapsari, R.I., & Wisnubroto, E.I. (2016). Phytoremediation of cadmium-contaminated agricultural land using indigenous plants. *International Journal of Environmental & Agriculture Research*, *2*(1), 8–14.

Indriolo, E., Na, G., Ellis, D., Salt, D.E., & Banks, J.A. (2010). A vacuolar arsenite transporter necessary for arsenic tolerance in the arsenic hyperaccumulating fern Pteris vittata is missing in flowering plants. *Plant Cell*, *22*, 2045–2057.

Hunt, A.J., Anderson, C.W., Bruce, N., García, A.M., Graedel, T.E., Hodson, M., Meech, J.A., Nassar, N.T., Parker, H.L., Rylott, E.L., & Sotiriou, K. (2014). Phytoextraction as a tool for green chemistry. *Green Processing and Synthesis*, *3*(1), 3–22.

Iori V., Pietrini F., Massacci A., & Zacchini M. (2015). Morphophysiological responses, heavy metal accumulation and phytoremoval ability in four willow clones exposed to cadmium under hydroponics. In A.A. Ansari, S.S. Gill, R. Gill, G.R. Lanza, & L. Newman (Eds.), *Phytoremediation: Management of Environmental Contaminant*, 87–98. Cham: Springer.

Jiang, Y., Lei, M., Duan, L., & Longhurst, P. (2015). Integrating phytoremediation with biomass valorisation and critical element recovery: a UK contaminated land perspective. *Biomass and Bioenergy*, *83*, 328–339.

Kanwar, V.S., Sharma, A., Srivastav, A.L., & Rani, L. (2020). Phytoremediation of toxic metals present in soil and water environment: a critical review. *Environmental Science and Pollution Research*, 1–26.

Kertulis-Tartar, G.M., Ma, L.Q., Tu, C., & Chirenje, T. (2006). Phytoremediation of an arsenic-contaminated site using *Pteris vittata* L.: a two-year study. *International Journal of Phytoremediation*, *8*(4), 311–322.

Krämer, U., Pickering, I.J., Prince, R.C., Raskin, I., & Salt, D.E. (2000). Subcellular localization and speciation of nickel in hyperaccumulator and non-accumulator *Thlaspi* species. *Plant Physiology*, *122*, 1343–1354.

Küpper, H., Lombi, E., Zhao, F.J., Wieshammer, G., & McGrath, S.P. (2001). Cellular compartmentation of nickel in the hyperaccumulators *Alyssum lesbiacum*, *Alyssum bertolonii* and *Thlaspi goesingense*. *Journal of Experimental Botany*, *52*(365), 2291–2300.

Ladislas, S., El-Mufleh, A., Gerente, C., Chazarenc, F., Andres, Y., & Bechet, B. (2012). Potential of aquatic macrophytes as bio-indicators of heavy metal pollution in urban storm water runoff. *Water Air & Soil Pollution*, 223, 877–888.

Liao, S., & Chang, W.L. (2004). Heavy metal phytoremediation by water hyacinth at constructed wetlands in Taiwan. *Photogrammetric Engineering and Remote Sensing*, 54, 177–185.

Lin, W., Xiao, T., Wu, Y., Ao, Z., & Ning, Z. (2012). Hyperaccumulation of zinc by Corydalis davidii in Zn-polluted soils. *Chemosphere*, 86(8), 837–842.

Ma, L.Q., Komar, K.M., Tu, C., Zhang, W., Cai, Y., & Kennelley, E.D. (2001). A fern that hyperaccumulates arsenic. *Nature*, 409(6820), 579–579.

Manara, A. (2012). Plant responses to heavy metal toxicity, Furini, A. (Ed.) *Plants and Heavy Metals*, 27–53, Dordrecht: Springer. doi:10.1007/978-94-007-4441-7_2.

Mateos-Naranjo, E., Castellanos, E.M., & Perez-Martin, A. (2014). Zinc tolerance and accumulation in the halophytic species *Juncus acutus*. *Environmental and Experimental Botany*, 100, 114–121.

Meier, S.K., Adams, N., Wolf, M., Balkwill, K., Muasya, A.M., Gehring, C.A., ... & Ingle, R.A. (2018). Comparative RNA-seq analysis of nickel hyperaccumulating and non-accumulating populations of *Senecio coronatus* (Asteraceae). *The Plant Journal*, 95(6), 1023–1038.

Memon, A.R., & Schröder, P. (2009) Implications of metal accumulation mechanisms to phytoremediation. *Environmental Science and Pollution Research*, 16, 162–175.

Merlot, S., Hannibal, L., Martins, S., Martinelli, L., Amir, H., Lebrun, M., & Thomine, S. (2014) The metal transporter PgIREG1 from the hyperaccumulator *Psychotria gabriellae* is a candidate gene for nickel tolerance and accumulation. *Journal of Experimental Botany*, 65(6), 1551–1564.

Mesjasz-Przybyłowicz, J., Nakonieczny, M., Migula, P., Augustyniak, M., Tarnawska, M., & Reimold, U., et al. (2004) Uptake of cadmium, lead nickel and zinc from soil and water solutions by the nickel hyperaccumulator *Berkheya coddii*. *Acta Biologica Cracoviensia Series Botanica*, 46, 75–85.

Mesjasz-Przybylowicz, J., Przybylowicz, W., Barnabas, A., & Van Der Ent, A. (2016). Extreme nickel hyperaccumulation in the vascular tracts of the tree *Phyllanthus balgooyi* from Borneo. *New Phytologist*, 209(4), 1513–1526.

Milner, M.J., & Kochian, L.V. (2008). Investigating heavy-metal hyperaccumulation using *Thlaspi caerulescens* as a model system. *Annals of Botany*, 102(1), 3–13.

Mitra, N., Rezvan, Z., Ahmad, M.S., & Hosein, M.G.M. (2012). Studies of water arsenic and boron pollutants and algae phytoremediation in three springs, Iran. *International Journal of Ecosystem*, 2(3), 32–37.

Nejatzadeh-Barandozi, F., & Gholami-Borujeni, F. (2014). Effectiveness of phytoremediation technologies to clean up of metalloids using three plant species in Iran. *Water Environment Research*, 86(1), 43–47.

Niazi, N.K., Singh, B., Van Zwieten, L., & Kachenko, A.G. (2012). Phytoremediation of an arsenic-contaminated site using *Pteris vittata* L. and *Pityrogramma calomelanos* var. *austroamericana*: a long-term study. *Environmental Science and Pollution Research*, 19(8), 3506–3515.

Nkrumah, P.N., Echevarria, G., Erskine, P.D., & van der Ent, A. (2018). Contrasting nickel and zinc hyperaccumulation in subspecies of *Dichapetalum gelonioides* from Southeast Asia. *Scientific Reports*, 8(1), 1–15.

Pandey, V.C. (2012). Phytoremediation of heavy metals from fly ash pond by Azolla caroliniana. *Ecotoxicology and Environmental Safety*, 82, 8–12.

Paul, A.L., Gei, V., Isnard, S., Fogliani, B., Echevarria, G., Erskine, P.D., ... & van der Ent, A. (2020). Nickel hyperaccumulation in New Caledonian *Hybanthus* (Violaceae) and occurrence of nickel-rich phloem in *Hybanthus austrocaledonicus*. *Annals of Botany*, 126(5), 905–914.

Pilon-Smits, E. (2005). Phytoremediation. *Annual Review of Plant Biology*, 56, 15–39.

Pilon-Smits, E.A., & Freeman, J.L. (2006). Environmental cleanup using plants: biotechnological advances and ecological considerations. *Frontiers in Ecology and the Environment*, 4(4), 203–210.

Poonam, & Srivastava, S. (2019). Assessing the phytoremediation potential of a flowering plant Zinnia angustifolia for arsenic contaminated soil. In *Environmental Arsenic in a Changing World*, 547–548. London: CRC Press, Taylor & Francis Group.

Poonam, Upadhyay, M.K., Gautam, A., Mallick, S., & Srivastava, S. (2017). A successive application approach for effective utilization of three aquatic plants in arsenic removal. *Water, Air, & Soil Pollution*, 228(2), 54.

Rani, L., Thapa, K., Kanojia, N., Sharma, N., Singh, S., Grewal, A.S., Srivastav, A.L., & Kaushal, J. (2020). An extensive review on the consequences of chemical pesticides on human health and environment. *Journal of Cleaner Production*, 124657.

Rehman, Z.U., Khan, S., Brusseau, M.L., & Shah, M.T. (2017). Lead and cadmium contamination and exposure risk assessment via consumption of vegetables grown in agricultural soils of five-selected regions of Pakistan. *Chemosphere*, *168*, 1589–1596.

Robinson, B.H., Anderson, C.W.N., & Dickinson, N.M. (2015). Phytoextraction: where's the action? *Journal of Geochemical Exploration*, *151*, 34–40.

Roy, S.B., & Bera, A. (2002) Individual and combined effect of mercury and manganese on phenol and proline content in leaf and stem of mungbean seedlings. *Journal of Environmental Biology*, *23*, 433–435.

Sarret, G., Saumitou-Laprade, P., Bert, V., Proux, O., Hazemann, J.-L., Traverse, A., et al. (2002). Forms of zinc accumulated in the hyperaccumulator *Arabidopsis halleri*. *Plant Physiology*, *130*, 1815–1826.

Schwitzguébel, J.P. (2017). Phytoremediation of soils contaminated by organic compounds: hype, hope and facts. *Journal of Soils and Sediments*, *17*(5), 1492–1502.

Shukla, A., & Srivastava, S. (2017). Emerging aspects of bioremediation of arsenic. In *Green Technologies and Environmental Sustainability*, 395–407. Cham: Springer.

Song W.Y., Yamaki T., Yamaji N., Ko D., Jung K.H., Fujii-Kashino M., An G., Martinoia E., Lee Y. Ma J.F. (2014) A rice ABC transporter, OsABCC1, reduces arsenic accumulation in the grain. *Proceedings of the National Academy of Sciences of the United States of America*, *111*, 15699–15704.

Srivastava S., Mishra S., Tripathi R.D., Dwivedi S., Trivedi P.K., Tandon P.K. (2007). Phytochelatins and antioxidant systems respond differentially during arsenate and arsenite stress in *Hydrilla verticillata* (L.f.) Royle. *Environmental Science and Technology*, *41*, 2930–2936.

Srivastava, S., Shrivastava, M., Suprasanna, P., & D'souza, S.F. (2011). Phytofiltration of arsenic from simulated contaminated water using *Hydrilla verticillata* in field conditions. *Ecological Engineering*, *37*(11), 1937–1941.

Srivastava, S., Sounderajan, S., Udas, A., & Suprasanna, P. (2014). Effect of combinations of aquatic plants (*Hydrilla, Ceratophyllum, Eichhornia, Lemna* and *Wolffia*) on arsenic removal in field conditions. *Ecological Engineering*, *73*, 297–301.

Subhashini, V., & Swamy, A.V.V.S. (2017). Potential of Catharanthus roseus (L.) in Phytoremediation of Heavy Metals. In *Catharanthus Roseus*, 349–364. Cham: Springer.

Sun, R.-L., Zhou, Q.-X., & Jin, C.-X. (2006). Cadmium accumulation in relation to organic acids in leaves of *Solanum nigrum* L. as a newly found cadmium hyperaccumulator. *Plant and Soil*, *285*, 125–134.

Tlustoš, P., Břendová, K., Száková, J., Najmanová, J., & Koubová, K. (2016). The long-term variation of Cd and Zn hyperaccumulation by *Noccaea spp* and *Arabidopsis halleri* plants in both pot and field conditions. *International Journal of Phytoremediation*, *18*(2), 110–115.

Valipour, A., & Ahn, Y.H. (2017). A Review and Perspective of Constructed Wetlands as a Green Technology in Decentralization Practices. In *Green Technologies and Environmental Sustainability*, 1–43. Cham: Springer.

Van der Ent, A., Baker, A.J., Reeves, R.D., Pollard, A.J., & Schat, H. (2013). Hyperaccumulators of metal and metalloid trace elements: facts and fiction. *Plant and Soil*, *362*(1–2), 319–334.

van der Ent, A., van Balgooy, M., & van Welzen, P. (2015). *Actephila alanbakeri* (Phyllanthaceae): a new nickel hyperaccumulating plant species from localised ultramafic outcrops in Sabah (Malaysia). *Botanical Studies*, *57*(1), 1–8.

Vara Prasad, M.N., & de Oliveira Freitas, H.M. (2003). Metal hyperaccumulation in plants: biodiversity prospecting for phytoremediation technology. *Electronic Journal of Biotechnology*, *6*(3), 285–321.

Venkateswarlu, V., Venkatrayulu, C.H., & Bai, T.J.L. (2019). Phytoremediation of heavy metal Copper (II) from aqueous environment by using aquatic macrophytes *Hydrilla verticillata* and *Pistia stratiotes*. *International Journal of Fisheries and Aquatic Studies*, *7*(4), 390–393.

Wan, X., Lei, M., Chen, T., & Yang, J. (2017). Intercropped *Pteris vittata* L. and *Morus alba* L. presents a safe utilization mode for arsenic-contaminated soil. *Science of the Total Environment*, *579*, 1467–1475.

Wang, H.B., Wong, M.H., Lan, C.Y., Baker, A.J.M., Qin, Y.R., Shu, W.S., Chen, G.Z., & Ye, Z.H. (2007). Uptake and accumulation of arsenic by 11 *Pteris* taxa from southern China. *Environmental Pollution*, *145*(1), 225–233.

Willscher, S., Mirgorodsky, D., Jablonski, L., Ollivier, D., Merten, D., Büchel, G., ... & Werner, P. (2013). Field scale phytoremediation experiments on a heavy metal and uranium contaminated site, and further utilization of the plant residues. *Hydrometallurgy*, *131*, 46–53.

Wiszniewska, A., Hanus-Fajerska, E., Muszyńska, E., & Ciarkowska, K. (2016). Natural organic amendments for improved phytoremediation of polluted soils: a review of recent progress. *Pedosphere*, *26*(1), 1–12.

Yadav, P., & Srivastava, S. (2020) Co-culturing *Hydrilla verticillata* with rice (*Oryza sativa*) plants ameliorates arsenic toxicity and reduces arsenic accumulation in rice. *Environmental Technology & Innovation*, 100722.

Yadav, P., Srivastva, S., Patil, T., Raghuvanshi, R., Srivastava, A.K., & Suprasanna, P. (2020). Tracking the time-dependent and tissue-specific processes of arsenic accumulation and stress responses in rice (*Oryza sativa* L.). *Journal of Hazardous Materials*, 124307.

Yan, A., Wang, Y., Tan, S.N., Yusof, M.L.M., Ghosh, S., & Chen, Z. (2020). Phytoremediation: a promising approach for revegetation of heavy metal-polluted land. *Frontiers in Plant Science*, 11.

Yan, H., Gao, Y., Wu, L., Wang, L., Zhang, T., Dai, C., ... & He, Z. (2019). Potential use of the Pteris vittata arsenic hyperaccumulation-regulation network for phytoremediation. *Journal of Hazardous Materials*, *368*, 386–396.

Zhang, Y., Liu, J., Zhou, Y., Gong, T., Wang, J., & Ge, Y. (2013). Enhanced phytoremediation of mixed heavy metal (mercury)-organic pollutants (trichloroethylene) with transgenic alfalfa co-expressing glutathione S-transferase and human P450 2E1. *Journal of Hazardous Materials*, *260*, 1100–1107.

Zhao, F.J., Lombi, E., & Breedon, T.M.S.P. (2000). Zinc hyperaccumulation and cellular distribution in *Arabidopsis halleri*. *Plant, Cell & Environment*, *23*(5), 507–514.

13 Nanophytoremediation: A Promising Strategy for the Management of Environmental Contaminants

Nair G. Sarath[1], P. Pravisya[2], A.M. Shackira[3], and Jos T. Puthur[1]
[1]Plant Physiology and Biochemistry Division, Department of Botany, University of Calicut, Kerala, India
[2]Department of Botany, Malabar Christian College, Calicut, Kerala, India
[3]Department of Botany, Sir Syed College, Taliparamba, Kannur, Kerala, India

CONTENTS

13.1 Introduction 251
    13.1.1 Environmental Contamination 251
    13.1.2 Major Source of Contamination 252
        13.1.2.1 Natural Sources 252
        13.1.2.2 Agricultural Wastes 252
        13.1.2.3 Industrial Wastes 253
    13.1.3 Sustainable Management of Environmental Contaminants 253
    13.1.4 Nanotechnology in Phytoremediation 254
13.2 Nanophytoremediation 255
    13.2.1 Phytoremediation Strategies Used by Plants 255
    13.2.2 Remediation of Organic Contaminants 257
    13.2.3 Remediation of Toxic Heavy Metals 258
13.3 Conclusion 258
Acknowledgements 259
References 259

13.1 Introduction

13.1.1 Environmental Contamination

Environmental pollution is a serious issue faced by our planet, and it has become faster in the 21st century. Rapid increase in population growth accompanied with industrialisation accelerates the rate of pollution. Any kind of unwanted transformation in abiotic characteristics of the environment constituents like air, water, and soil can result in a negative impact on the biotic component. Based on the origin of pollutant, environmental contaminants are of two types: natural and anthropogenic. Major environmental pollutions by natural factors include earthquakes, floods, cyclones, volcanoes, etc. In a few cases, human activities strengthen the impact of natural pollution factors such as earthquakes and floods. Unmanaged urbanisation, industrial development, deforestation, etc. are the main anthropogenic activities that directly cause destruction of the environment and also accelerate the negative impact of natural pollution factors. Different kinds of pollutants (majorly due to human activities) are responsible for the deterioration of nature and its affects on the characteristics of land, air, and water. Pollutants like pesticides, insecticides, fertilisers, petroleum compounds, high salinity, nutrients, radionuclides, heavy metals, etc. cause temporary or persistent effects on the environment (Sarath and Puthur 2020). According to the Federal Environmental Protection Agency (FEPA) Act of 1990, water, air, and land are considered major environmental components; any interference in these elements threatens the existence of all living components on earth. So, among various environmental pollutions, the pollution occurring in water, on land, and in the air are considered primary pollutions that badly affect the living organisms, including the human race.

Presently, the combined effect of pollution and climate change due to anthropogenic activities adversely affect the flora and fauna of the biotic system and will result in the total imbalance of the biome. The overpollution load on water, air, and land leads to the modification in its natural characteristics and thereby cause effects from microorganisms to higher animals (Yang et al., 2018). The deposition of municipal solid waste into the land or landfill management has more pronounced effects on public health. The noxious and toxic compounds in gases or solids persistently survive in the environment and these potent chemicals also leach out into the ground and contaminate the water column (Han et al., 2019). The pollution due to toxic metals is hastened day by day because its production through human activities is higher than natural ones (D'amore et al., 2005). The most commonly occurring lethal metal detected in contaminated areas is lead and the other metals are arranged in the order Cromium > Arsenic > Zinc > Cadmium > Copper depending upon their presence (United States Environmental Protection Agency 1996). Due to the immortality of lethal metals, once they coordinate with soil, water, or air, they can take many decades to detoxify. Bioaccumulation and biomagnification capacity of toxic metals lead to a reduction in agriculture production which badly affect the economic growth of poorly developed countries like India.

13.1.2 Major Source of Contamination

13.1.2.1 Natural Sources

Weathering of rock is considered the most important contributor of toxic metals; the speed of this process depends on the nature of the rock, climate, and heavy metal constituent in rock (Abdu et al., 2011). Metals like Mn, Cr, Co, Cu, Ni, Zn, Sn, Cd, Hg, and Pb are added to the soil by this process in higher concentrations (Srivastava et al., 2017). Volcanic eruptions cause air pollution by the emission of toxic gases containing the following metals: Al, Zn, Mn, Pb, Ni, Cu, Hg, Fe, Cr, and Pb (Seaward and Richardson, 1989; Nagajyoti et al., 2010). Volatile lethal metals like selenium and mercury are emitted to the atmosphere by forest fires (Ross, 1994; Naidu et al., 1997; Nagajyoti et al., 2010).

13.1.2.2 Agricultural Wastes

Irrational application of chemicals in the farming system results in the accumulation of heavy metals all over the world, particularly in India. Fungicides, phosphate fertilisers, and inorganic fertilisers act as major sources of different lethal metals and other compounds like DDT, trichloroethylene, endosulfan, and trinitrotoluene, etc. (Kelepertzis, 2014; Tóth et al., 2018). Accumulation of Cd, Fe, Hg, and Pb increase as a result of unscientific use of phosphorus fertilisers. Nowadays, wastewater from municipal waste lines, industries, and effluents are commonly used for irrigation in many countries (Khalid et al., 2018). The studies conducted by Jiménez (2006) and Bjuhr (2007) reported that approximately about 20 million hectares of crop area are watered with heavy metals containing water and also 50% of horticulture land in Asian and African countries are affected by contaminated water. Usually farmers do not bother with the negative impact of this because lower doses of heavy metals result in plant growth and yield promotion and they get dual benefits while using wastewater for irrigation. But studies revealed that long-term wastewater use in agricultural land leads to the addition of toxic metals in soil and thereby into the crops.

The application of biowaste as a fertiliser is a common practice among farmers and this enhances the amassing of toxic metals like As, Cd, Cr, Cu, Pb, Hg, Ni, Se, Mo, Zn, Tl, and Sb (Sharma et al., 2017). The excreta of pig and cattle are usually used as fertiliser but metals like Cu and Zn, which are added in

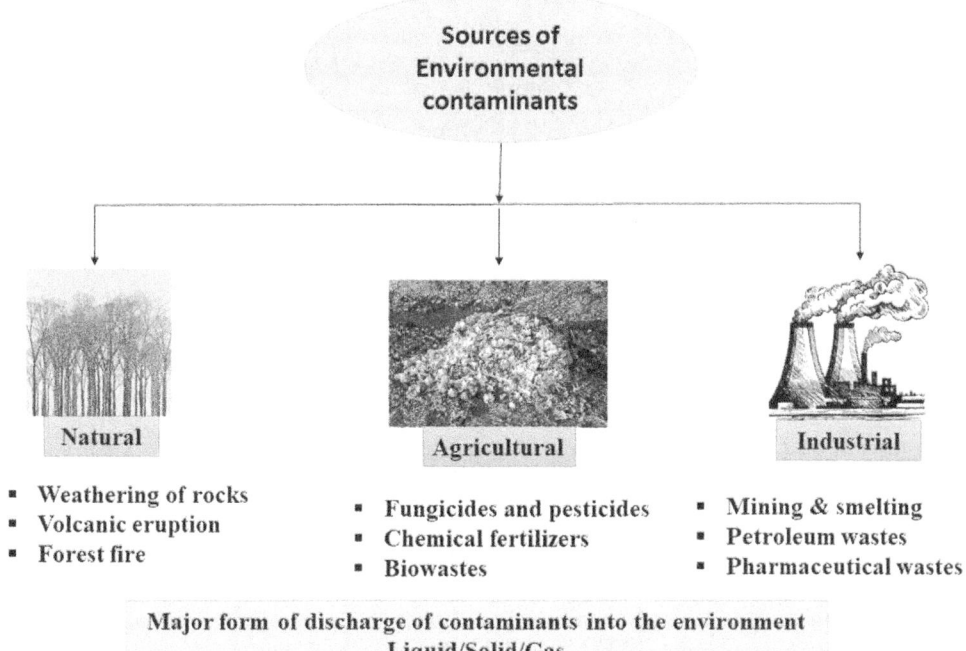

FIGURE 13.1 Various sources of environmental contaminants.

their diets, accumulate in the soil through this (Provolo et al., 2018). Sewage sludge is organic solid products, produced as a by-product of wastewater. Toxic metals like Pb, Ni, Cd, Cr, Cu, and Zn are accumulating in soil as a consequence of biosolids application (Silveira et al., 2003). Leaching results in groundwater contamination and finally gets biomagnified in humans (Wuana and Okieimen, 2011) (Figure 13.1).

13.1.2.3 Industrial Wastes

The industries related to mining and milling of metal ores are some of the main reasons behind soil metal contamination. They eliminate metal waste, pollute soil, and thereby contaminate the entire ecosystem. Over Pb and Zn ore mining and smelting results in the increase of soil heavy metal contents and it causes serious issues in biota. Many remediation strategies are applied in these regions but they all are tedious, high cost, and even may not revitalise soil fertility. Other materials generated by a variety of industries such as textile, tanning, petrochemicals from accidental oil spills, or utilisation of petroleum-based products, pesticides, and pharmaceuticals are highly variable in composition. Although some of them are disposed in soil, a few have benefits in agriculture and forestry, but many are potentially dangerous because of heavy metal contents (Cr, Pb, and Zn) or toxic organic compounds. Others are very low in plant nutrient content or have no soil conditioning properties (Sumner, 2000). The large amount of untreated waste released from the industries such as textiles, oil refineries, and chemical industries including nuclear plants, rapidly contaminate the environment and the toxicity of the pollutant causes a high health hazard to ecosystems (Singh et al., 2017). Various sources of environmental contaminants are depicted in Figure 13.2.

13.1.3 Sustainable Management of Environmental Contaminants

This ever-increasing environmental problem is faced in many countries around the world and has attracted more researchers for finding ways and means for sustainable reclamation methods (Zeng et al.,

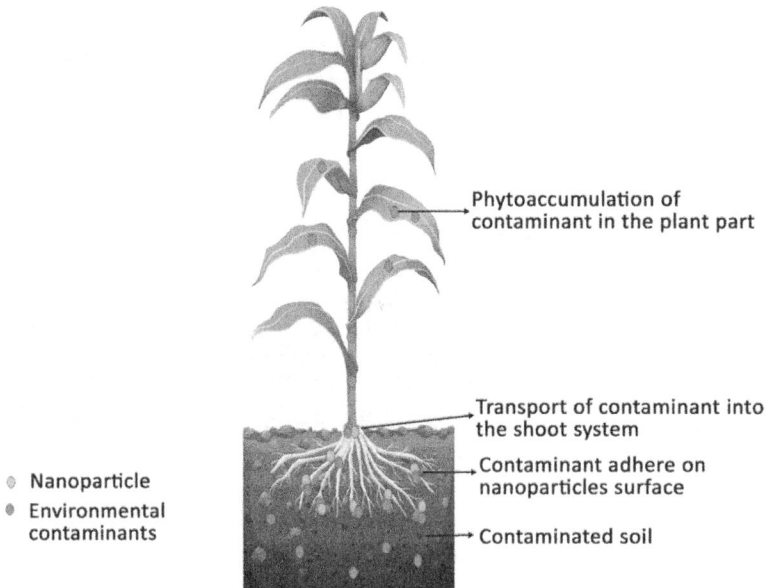

FIGURE 13.2 The enhancement of phytoremediation using nanoparticles for the remediation of environmental contaminants.

2020). Remediation strategies that were conventionally practiced have several disadvantages like high cost, harmful effect to other organisms, complex implementation strategies, less efficiency, etc. Hence, planning and implementation of a cost-effective technique for the proper and sustainable management of environmental contaminants and their recuperation by natural means are of more importance nowadays. Different agents are used in the sustainable management of environmental contaminants. For this purpose, various agents such as microbes, algae (phycoremediation), fungi (mycoremediation), and higher plants (phytoremediation) are used and these are regarded as effective bioremediating agents (Sarath et al., 2020). Among these, plants have better mechanisms for the operative elimination of pollutants from soil. They utilise different methods for the decontamination of soil. The co-application of different agents such as bacteria, fungi, and other materials like nanoparticles with higher plants enhance the remediation potential of these plants and are more effective than the independent treatments. The potential of the hyper-accumulating plants can be enhanced with the symbiotic association of arbuscular mycorrhiza. It results in the enhanced root growth and morphology, improved free radicle scavenging mechanisms, and biomass of the hyper-accumulators and thereby easily flourishes and tolerates hyper-metal concentrations in the growing habitats (Janeeshma and Puthur, 2020). Etesami (2018) reported the use of PGPRs (plant growth-promoting rhizobacteria) in metal tolerance of plants and this association led to the enhancement in plant biomass in polluted soil (especially heavy metals) and also helped to produce safe agronomic products from the crop plant with less toxic metals in their tissues. The application of nanoparticles in phycoremdiation act as an efficient way of environmental remediation with a potential in different areas like production of energy sources like bioethanol, biomethanol, biodiesel, etc. (Bansal et al., 2018). So, sustainable management of environmental pollutants has its own significance. Phytoremediation and nanoparticle-assisted plant-based remediation (nanophytoremediation) are successfil mechanisms in remediation aspects and their operational ease, low cost, and in-situ applications help to decontaminate the pollutants.

13.1.4 Nanotechnology in Phytoremediation

Nanotechnology has emerged as a potent technology with several applications. Recently, the application of nanotechnology in phytoremediation is reveled but still is in the nascent stage and has great promise in the environmental cleanup processes (Kharisov et al., 2014). Nanoparticles have a crucial role in the remediation by directly removing the pollutants from the environment (due to their large surface area to volume ratio), by increasing the phytoavailability of pollutants and also by indirectly promoting plant growth. This technique of remediation, with the help of nanoparticles, is generally termed *nanoremediation*. However, nanophytoremediation has proven to be much more promising when compared with the two operating independently, and nowadays this combined technique is applied for the decontamination of various pollutants. The two major goals of nanophytoremediation include the synthesis of nanoparticles for the specific purpose of biosphere health or by the direct application of nanoparticles to remediate pollutants from the environment. The second and most important goal of this green nanotechnology is to accelerate the potential of microbes, plants, or algae for removal of environmental contaminants (McKenzie and Hutchison, 2004). Depending upon the co-partner of nanoparticles used for environment cleanup, it is divided into four types: nanophycoremediation (with algae), nanozooremediation (with animal), nano-microbial remediation (microbe), and nanophytoremediation (plant) (Rajan et al., 2015; Gil-Díaz et al., 2016; Wang et al., 2016).

13.2 Nanophytoremediation

The emerging method, nanophytoremediation, is the combination of nanotechnology and phytoremediation utilised for the purpose of toxicant removal from polluted soil. The removal of the toxicant using plants has a disadvantage of taking more time (often several years) to complete the process of remediation. But, nowadays, the combined action of nanotechnology and phytoremediation has revealed success for remediation of environmental contaminants (Song et al., 2019). The nano-sized particles or nanoparticles or ultrafine particles have a particle matter between 1 and 100 nanometres. They have higher applications in many fields (Tang et al., 2014). Based on their composition, nanoparticles are of two types: organic and inorganic nanoparticles (Yadav et al., 2017; Ramezani et al., 2021). Carbon-containing nanoparticles (fullerenes) come under organic nanoparticles while inorganic nanoparticles consist of magnetic nanoparticles, noble metal nanoparticles like gold and silver, and semiconductor nanoparticles like TiO_2 and ZnO.

The potential of nanoparticles for elimination/stabilisation/degradation of environmental contaminants and their synergic effects with plants has developed as an important strategy for bioremediation. Therefore, nanoparticle-based phytoremediation has emerged as a potent technique for the remediation of toxicants from contaminated sites. This combined technology is ecologically mingled, costs less, and has emerged as the best alternative strategy to existing techniques (Yadav et al., 2017). The application of nanoparticles in phytoremediation augments the plants' growth and enhances the phytoavailability of the environmental contaminant. As a result, the plant absorption gets increased and it makes plants an effective extraction tool from soil (Figure 13.2). Besides this, nowadays, nanomaterials are used as better candidates for the abstraction of pollutants like toxic metals and organic contaminants, etc., but reports state that co-application of nanomaterials with various living organisms is more effective than pure form application. According to Srivastav et al. (2018), nanoparticles used for remediation should possess some important features:

- It must be non-toxic for plant.
- Stimulate seed germination, growth, and biomass.
- Enhance the antioxidant machinery in plants.
- Stimulate the synthesis of phytohormones.
- High efficiency to bind with contaminants.

13.2.1 Phytoremediation Strategies Used by Plants

The exact mechanism of nanophytoremediation has not been revealed. However, there are some reports stating that nanoparticles increase the phytoavailability of the environmental contaminant and increase

TABLE 13.1

Different methods adopted by plants for the remediation of environmental contaminants from soil

Methods Used by Plants	Mechanism	Example	References
Phytostabilisation	Stabilisation of the contaminant in plant root system and reduces its toxicity to nearby regions	Phytostabilisation of copper by *Bruguiera cylindrica*	Sruthi and Puthur (2019)
Phytovolatilisation	Conversion of the pollutant into volatile products	Phytovolatilisation arsenic by *Arundo donax*	Guarino et al. (2020)
Phytoextraction	Extracting the toxicant from the soil and accumulate in the plant biomass	Phytoextraction of copper by *Linum usitatissimum*	Saleem et al. (2020)
Phytodegradation	Degradation of contaminant by plants using their enzymatic machinery	Phytodegradation of petroleum hydrocarbon by	Al-Baldawi et al. (2015)

the phytoremediation potential of a phytoremediation candidate. Sruthi et al. (2015) state that the plants with phytoremediation potential have mechanisms to modulate their metabolic pathways todetoxify or volatise the environmental contaminants like xenobiotics and heavy metals. In addition to this, some plants have specific strategies to tackle the pollutants in their biomass or may reduce the toxicity or volatise without any more deleterious effect on the plant. Plants utilise different strategies of phytoremediation, such as phytostabilisation, phytoextraction, phytovolatilisation, and phytodegradation. These mechanisms are shortlisted in Table 13.1.

The plants with phytostabilisation potential have the capacity to reduce the mobility of the toxicant and thereby the pollutant stabilised (Trivedi and Ansari, 2015). It helps to sequester (binding and sorption) and immobilise the pollutant so that it is changed to unavailable forms. The plant *Acanthus ilicifolius* has a phytostabilisation potential for Cd and Zn and it immobilises the toxic metal in the root (Shackira et al., 2017; Shackira and Puthur, 2019). In phytoextraction, plants absorb the pollutant from the soil and transport it into the shoot system where they accumulate. In the remediation techniques, after attaining the proper biomass, these plants can be harvested and result in the permanent elimination of toxicants from the polluted area. The success of the phytoextraction is dependent upon several characteristics, such as the ability of accumulation of the toxicant in the plant body, easy establishment, higher growth rate under contaminated soil conditions, and capacity to adapt to other harsh conditions like alkalinity, sodicity, pH, and water availability (Blaylock and Huang, 2000). *Sesuvium portulacastrum* tolerates heavy metal stress using its own inherent phytoextraction capability. It accumulates Cd in vacuoles and salt glands and reduces the toxicity of the pollutant by compartmentalisation. It leads to the easy establishment of the plant in the polluted soil conditions (Ayyappan et al., 2016). Studies of da Silva Teófilo et al. (2020) state that the plants *Canavalia ensiformes* and *Stilizobium aterrimum* are excellent candidates for phytoextraction of highly toxic herbicides such as diuron and hexazinone. They recommended these plants species for remediation purposes. Phytovolatilisation is another mechanism in plants where they convert the toxic compound into volatile forms. The potent hazardous metals like Hg, Se, and As are absorbed and transformed into less poisonous volatile forms through phytovolatilisation (Cristaldi et al., 2017). The plants showed indirect and direct phytovolatilisation. In the former one, through roots, and the latter through aerial parts like leaves. Phytodegradation or phytotransformation are the mechanisms in plants in which the pollutant is broke down into less toxic compounds through the specific cellular process of plants. Zazouli et al. (2014) reported the phytodegradation potential of the aquatic plant *Azolla filiculoides*. It remediates the highly toxic environmental pollutant bisphenolA.

13.2.2 Remediation of Organic Contaminants

Toxic organic environmental contaminants released as a result of anthropogenic activities affect the healthy ecosystem. The contaminants' persistency in the ecosystem causes durable impairments (Lasserre et al., 2009; Kanissery and Sims, 2011; Sánchez-Sánchez et al., 2013). Due to their persistence, they have the capacity to move long distances in the environment, bioaccumulate in animals, and finally get biomagnified in food chains. Because of this reason, these contaminants were banned in many countries (Sánchez-Sánchez et al., 2013). In 2004, the Stockholm Convention devoted about 90 signatory countries to wipe out the use of persistent organic chemicals (Jepson and Law, 2016).

Biomagnification of organic pollutants has a negative impact on human health. It includes endocrine disorders, reproductive and immune impairments, neurodegenerative diseases, and uncontrolled cell proliferation (commonly, cancer on breast) (Goff et al., 2005). Detoxification of organic pollutants occurs through i) biodegradation, ii) chemical degradation, and iii) photodegradation. Microorganisms use organic pollutants as the primary resource of carbon, cell-building material, electrons, and energy. However, phytoremediation is considered an effective method for recycling of the degraded environment, especially soil pollutants like pesticides, polycyclic aromatic hydrocarbons, petroleum, and explosives (Kang, 2014). The co-application of plants with nanomaterials enhanced the remediation potential of organic pollutants like trichloroethylene, endosulfan, and trinitrotoluene (Ma and Wang, 2010; Jiamjitrpanich, et al., 2012; Pillai and Kottekottil, 2016). Engineered nanoparticles of iron oxides, manganese oxides, cerium oxides, titanium oxides, or zinc oxides act as detoxification agents. This greater affinity is directly related to their number of active surface sites and their significant surface area (Martínez-Fernández et al., 2017). Cottonwood treated with carbon nanoparticles increases the uptake of trichloroethylene (Ma and Wang, 2010). Jiamjitrpanich et al. (2012) observed that priming of *Panicum maximum* with nanoscale zero-valent iron (nZVI) particles has improved its trinitrotoluene scavenging potential. Pillai and Kottekottil (2016) reported the nZVI-assisted phytoremediation in *Alpinia calcarata*, *Ocimum sanctum*, and *Cymbopogon citrates* for the removal of endosulfan showed improved elimination of endosulfan from the polluted soil. According to Gao and Zhou (2013), iron nanoparticles play a key role in degradation of toxicants because of their reducing properties as electron donors. nZVI improves the biomass of *Impatiens balsamina* by 30% in pollution-free soil and 54% in soil contaminated with pollutants. Waraporn et al. (2012) suggested that zero valent iron-treated *Panicum maximum* showed a higher potential for the breakdown of trinitrotoluene (TNT). It is used in the remediation of TNT from contaminated soil. Similarly, Waraporn et al. (2012) reported that nZVI treatment in *Panicum maximum* and *Helianthus annuus* detoxify TNT contaminants in soil. Also, plants treated with nZVI showed significantly higher polychlorinated biphenyls (PCBs) detoxification efficiency compared with untreated plants. These types of nanophytoaccumulation may be used to improve the texture of degraded soil.

Like soil purification, nanomaterials are also used to scavenge pollutants in water. Activated carbons with high binding affinity have been employed to adsorb residues of undesirable organic toxicants including PPCPs from water (Yong et al., 2002; White et al., 2009). Carbon-based nanomaterials, principally single-wall/multi-wall carbon nanotubes, have been utilised to adsorb PPCPs and (endocrine-disrupting compounds) EDCs in water, such as carbamazepine, ciprofloxacin, doxycycline hydrochloride, ibuprofen, macrolides, norfloxacin, quinolones, sulfonamides, oxytetracycline, tetracycline, triclosan, and bisphenol A (BPA) tetrabromobisphenol (HeBeveridge and Murray, 1980; Watson et al., 1999; Windt et al., 2005; He et al., 2007; Shi et al., 2007; Peng 2010; Majumber, 2010; Singh et al., 2011; Pattrick et al., 2012; Murty et al., 2013). Chlorfenapyr is an important broad-spectrum insecticide/acaricide and it is under the hazardous insecticide category of the WHO and cause health risks in birds and also cause ecological impacts. The application of nanophytoremdiation with *Plantago major* and green nanoparticles (F–Fe0, Ip–Ag0, and Br–Ag0) e) reduces the chlorfenapyr content in water and soil (Romeh and Saber, 2020). *Salvinia molesta* is used in the remediation of effluents of pulp and paper mills. The application of silver nanoparticles in phytoremediation is shown to be more effective than independent treatments (Bhardwaj et al., 2018).

13.2.3 Remediation of Toxic Heavy Metals

Nanoparticles have huge applications in the omission of toxic contaminants like toxic metals from soil. The unique features, such as enormous surface area and high surface energy of nanoparticles, help them to absorb more pollutants from the contaminated site and it also increases the beakdown rate of contaminants (Poorva et al., 2013). The nanoparticles are easily also used for the purpose of the remediation of toxicants from underground because of their ultra-small size and also they are easily transported with the flow of water. The application of nanoparticles into soil is also very easy with any mode of excavation. The long durability and potential for retention of their activity in the contaminated site makes their functionality the best remediation agent (Zhang, 2003; Karn et al., 2009). Nanomaterials based on carbon, nanomaterials based on metals, and engineered nanoparticles are used widely in remediation purposes (Chen et al., 2017).

Nanoscale zero-valent iron (nZVI) is highly used in remediation strategies. nZVI has less phytotoxicity and high reactivity, making it extra efficient in plants and leads to better remediation of environmental contaminants from the soil (Terzi et al., 2016; Gil-Díaz and Lobo 2018; Radziemska et al. 2021). Iron nanoparticles are important nanoparticles and have many applications in the remediation of environmental contaminants, including toxic metals (Nwadinigwe and Ugwu, 2018). Its strong reducing properties help in the remediation processes (Zhang 2003). Gao and Zhou (2013) studied the potential use of nZVI with the plant *Impatiens balsamina*. The results from this study revealed that the combination (plant with nZVI) makes it more efficient in the remediation of e-waste-contaminated soil than without nZVI.

Liang et al. (2017) reported that nano-hydroxyapatite improves the remediation capacity of *Lolium perenne* (ryegrass) towards lead. The 1-month-old plantlets of *Lolium perenne* can remediate about 47% of the toxic lead from the soil with the help of nano-hydroxyapatites compared to the non-nanoparticle-treated plant. Hussain et al. (2019) reported the potential of iron oxide nanoparticles (Fe NPs) for the mitigation of cadmium stress in wheat plants. Iron oxide nanoparticles (20 ppm) diminish the effects of Cd toxicity by antioxidant pathways and result in improved growth, higher efficiency in photosynthesis, and decrease the Cd absorption. Similarly, Cd(II) and Cr(VI) toxicity in wheat plants was reduced by the application of citrate-coated magnetite nanoparticles (López-Luna et al., 2016).

The applied aspects of nanotechnology in the field of phytoremediation and biotechnology are helpful in the sustainable agriculture and crop production in the contaminated land (Abhilash and Dubey, 2015). Nanoparticle-mediated agriculture systems will help to produce consumption of safe agronomical products from polluted land by different means, including decreasing the toxicity of heavy metals, immobilising the pollutant in soil, and degradation of organic toxicants. It reduces the entry of toxicants into plant biomasses (Abhilash et al., 2016). Similar to this, Cai et al. (2017) found that a decrease in the accumulation of Pb in rice plants when they were treated with titanium dioxide (TiO_2) nanoparticles. It has a potential importance in the revitalisation of contaminated lands and production of safe phytoprodutcs. Moameri and Khalaki (2019) carried out phytoremediation experiments with the help of the plant *Secale montanum* in the soil collected from areas of a national lead and zinc factory in Zanjan, Iran. These soils contained high amounts of metals like lead and zinc. They concluded that the phytoremediation capacity of the *S. montanum* can be improved with the help of nano-silica (NS500). The NS500 treated *S. montanum* showed a higher translocation factor, bioconcentration factor, and remediation factor. A list of nanoparticles, along with plant species for the remediation of environmental contaminants, are listed in Table 13.2.

13.3 Conclusion

Overexploitation of natural resources and higher anthropogenic activities lead to higher pollution rates, badly affecting the natural ecosystem. The conventional technologies for removing toxic contaminants from the environment are expensive and bear a limitation to produce secondary products. Therefore, the use of biological resources for the sustainable management of pollutants have been explored worldwide. The efficacy of the bioremediating agents can be accelerated with some other biological agents or with

TABLE 13.2

List of nanoparticles along with plant species for the environmental contaminants' remediation

Nanoparticle	Plant	Contaminant	Reference
Iron nanoparticle (nZVI)	*Panicum maximum*	Trinitrotoluene	Jiamjitrpanich et al. (2012)
Iron nanoparticle (nZVI)	*Alpinia calcarata, Ocimum sanctum,* and *Cymbopogon citrates*	Endosulfan	Pillai and Kottekottil (2016)
Iron nanoparticle (nZVI)	*Impatiens balsamina*	e-waste	Gao and Zhou (2013
Iron nanoparticle (nZVI)	Panicum maximum and Helianthus annuus	Detoxify TNT	Jiamjitrpanich et al. (2012)
Fe nanoparticle	Wheat	Cd	Hussain et al. (2019)
TiO2 nanoparticle	Rice	Pb	Cai et al. (2017)
Citrate-coated Magnetite nanoparticle	Wheat, oat, sorghum	Cr	López-Luna et al. (2016)
Ni/Fe bimetallic nanoparticle	Chinese cabbage	Polybrominated diphenyl ethers	Wu et al. (2016)
Salicylic acid nanoparticle	*Isatis cappadocica*	Arsenic	Souri et al. (2017)

other materials. Nanoparticles are promising candidates for enhancing the potential of good bioremediating agents. Nanoparticles augment the growth of plants with phytoremediation potential, improve better intake of contaminants, and are eco-friendly in nature. Moreover, this technique is cost effective and can be practiced without many complications.

Acknowledgements

NGS and JTP acknowledge the financial assistance provided by the Kerala State Council for Science, Technology and Environment in the form of KSCSTE Research Grant (KSCSTE/5179/2017-SRSLS). The authors extend their sincere thanks to the Department of Science & Technology (DST), Government of India, for granting funds under Fund for Improvement of S&T Infrastructure (FIST) programme SR/FST/LSI-532/2012.

REFERENCES

Abdu, N., Agbenin, J.O., & Buerkert, A. (2011). Geochemical assessment, distribution, and dynamics of trace elements in urban agricultural soils under long-term wastewater irrigation in Kano, northern Nigeria. *Journal of Plant Nutrition and Soil Science, 174*(3), 447–458.

Abhilash, P.C., & Dubey, R.K. (2015). Root system engineering: prospects and promises. *Trends in Plant Science, 20*, 1360–1385.

Abhilash, P.C., Tripathi, V., Edrisi, S.A., Dubey, R.K., Bakshi, M., Dubey, P.K., Singh, H.B., & Ebbs, S.D. (2016). Sustainability of crop production from polluted lands. *Energy, Ecology and Environment, 1*(1), 54–65.

Al-Baldawi, I.A., Abdullah, S.R.S., Anuar, N., Suja, F., & Mushrifah, I. (2015). Phytodegradation of total petroleum hydrocarbon (TPH) in diesel-contaminated water using Scirpus grossus. *Ecological Engineering, 74*, 463–473.

Ayyappan, D., Sathiyaraj, G., & Ravindran, K.C. (2016). Phytoextraction of heavy metals by Sesuvium portulacastrum l. a salt marsh halophyte from tannery effluent. *International Journal of Phytoremediation, 18*(5), 453–459.

Bansal, A., Shinde, O., & Sarkar, S. (2018). Industrial wastewater treatment using phycoremediation technologies and co-production of value-added products. *Journal of Bioremediation and Biodegradation, 9*(1), 1–10.

Bhardwaj, P., Kaushal, J., Naithani, V., & Singh, A.P. (2018). Phytoremediation of pulp and paper mill effluents by Salvinia molesta and its comparison with nanoparticles inclusive phytoremediation. *Egyptian Journal of Basic and Applied Sciences*, 8(8), 338–356.

Blaylock, M.J. (2000). Phytoextraction of metals. In Raskin, I., Ensley, B. D., (Eds.). *Phytoremediation of Toxic Metals: Using Plants to Clean Up the Environment*, 53–70. New York: Wiley.

Blaylock, M., & Huang, J. (2000). Phytoextraction of metals, In Raskin, I., & Ensley, B. D. (Eds.). *Phytoremediation of Toxic Metals: Using Plants to Clean-up the Environment*, 303, New York, NY: John Wiley & Sons, Inc.

Bjuhr, J. (2007). Trace metals in soils irrigated with waste water in a periurban area downstream Hanoi City, Vietnam. Seminar Paper, Institution for Narkvetenskap, Sveriges Lantbruks Universitet (SLU), Uppsala, Sweden.

Cai, F., Wu, X., Zhang, H., Shen, X., Zhang, M., Chen, W., Gao, Q., White, J.C., Tao, S., & Wang, X. (2017). Impact of TiO2 nanoparticles on lead uptake and bioaccumulation in rice (Oryza sativa L.). *NanoImpact*, 5, 101–108.

Chen, B., Bi, H., Ma, Q., Tan, C., Cheng, H., Chen, Y., He, X., Sun, L., Lim, T.T., Huang, L., & Zhang, H. (2017). Preparation of graphene-MoS 2 hybrid aerogels as multifunctional sorbents for water remediation. *Science China Materials*, 60(11), 1102–1108.

Cristaldi, A., Conti, G.O., Jho, E.H., Zuccarello, P., Grasso, A., Copat, C., & Ferrante, M. (2017). Phytoremediation of contaminated soils by heavy metals and PAHs. A brief review. *Environmental Technology & Innovation*, 8, 309–326.

D'amore, J.J., Al-Abed, S.R., Scheckel, K.G., & Ryan, J.A. (2005). Methods for speciation of metals in soils: a review. *Journal of Environmental Quality*, 34(5), 1707–1745.

da Silva Teófilo, T.M., Mendes, K.F., Fernandes, B.C.C., de Oliveira, F.S., Silva, T.S., Takeshita, V., de Freitas Souza, M., Tornisielo, V.L., & Silva, D.V. (2020). Phytoextraction of diuron, hexazinone, and sulfometuron-methyl from the soil by green manure species. *Chemosphere*, 127059.

De La Torre-Roche, R., Hawthorne, J., Deng, Y., Xing, B., Cai, W., Newman, L.A., Wang, Q., Ma, X., Hamdi, H., & White, J.C. (2013). Multiwalled carbon nanotubes and C60 fullerenes differentially impact the accumulation of weathered pesticides in four agricultural plants. *Environmental Science & Technology*, 47(21), 12539–12547.

Edelstein, M., & Ben-Hur, M. (2018). Heavy metals and metalloids: sources, risks and strategies to reduce their accumulation in horticultural crops. *Scientia Horticulturae*, 234, 431–444.

Etesami, H. (2018). Bacterial mediated alleviation of heavy metal stress and decreased accumulation of metals in plant tissues: mechanisms and future prospects. *Ecotoxicology and Environmental Safety*, 147, 175–191.

Gao, Y.Y., & Zhou, Q.X. (2013). Application of nanoscale zero valent iron combined with Impatiens balsamina to remediation of e-waste contaminated soils. *Advanced Materials Research*, 790, 73–76.

Gil-Díaz, M., & Lobo, M.C. (2018). Phytotoxicity of nanoscale zerovalent iron (nZVI) in remediation strategies. In *Phytotoxicity of Nanoparticles*, 301–333. Cham: Springer.

Gil-Díaz, M., Diez-Pascual, S., González, A., Alonso, J., Rodríguez-Valdés, E., Gallego, J.R., & Lobo, M.C. (2016). A nanoremediation strategy for the recovery of an As-polluted soil. *Chemosphere*, 149, 137–145.

Goff, K.F., Hull, B.E., & Grasman, K.A. (2005). Effects of PCB 126 on primary immune organs and thymocyte apoptosis in chicken embryos. *Journal of Toxicology and Environmental Health, Part A*, 68(6), 485–500.

Guarino, F., Miranda, A., Castiglione, S., & Cicatelli, A. (2020). Arsenic phytovolatilization and epigenetic modifications in Arundo donax L. assisted by a PGPR consortium. *Chemosphere*, 251, 126310.

Hamdi, H., De La Torre-Roche, R., Hawthorne, J., & White, J.C. (2015). Impact of non-functionalized and amino-functionalized multiwall carbon nanotubes on pesticide uptake by lettuce (Lactuca sativa L.). *Nanotoxicology*, 9(2), 172–180.

Han, Z., Zeng, D., Li, Q., Cheng, C., Shi, G., & Mou, Z. (2019). Public willingness to pay and participate in domestic waste management in rural areas of China. *Resources, Conservation and Recycling*, 140, 166–174.

He, S., Guo, Z., Zhang, Y., Zhang, S., Wang, J., & Gu, N. (2007). Biosynthesis of gold nanoparticles using the bacteria *Rhodopseudomonas capsulata*. *Materials Letters*, 61(18), 3984–3987.

HeBeveridge, T.J., & Murray, R.G. (1980). Sites of metal deposition in the cell wall of *Bacillus subtilis*. *Journal of Bacteriology*, *141*(2), 876–887.

Hussain, A., Ali, S., Rizwan, M., Ur Rehman, M.Z., Qayyum, M.F., Wang, H., & Rinklebe, J. (2019). Responses of wheat (Triticum aestivum) plants grown in a Cd contaminated soil to the application of iron oxide nanoparticles. *Ecotoxicology and Environmental Safety*, *173*, 156–164.

Janeeshma, E., & Puthur, J.T. (2020). Direct and indirect influence of arbuscular mycorrhizae on enhancing metal tolerance of plants. *Archives of Microbiology*, *202*(1), 1–16.

Jepson, P.D., & Law, R.J. (2016). Persistent pollutants, persistent threats. *Science*, *352*(6292), 1388–1389.

Jiamjitrpanich, W., Parkpian, P., Polprasert, C., & Kosanlavit, R. (2012, June). Enhanced phytoremediation efficiency of TNT-contaminated soil by nanoscale zero valent iron. In *2nd International Conference on Environment and Industrial Innovation IPCBEE* (Vol. 35), 82–86.

Jiménez, B. (2006). Irrigation in developing countries using wastewater. *International Review for Environmental Strategies*, *6*(2), 229–250.

Kang, J.W. (2014). Removing environmental organic pollutants with bioremediation and phytoremediation. *Biotechnology Letters*, *36*(6), 1129–1139.

Kanissery, R.G., & Sims, G.K. (2011). Biostimulation for the enhanced degradation of herbicides in soil. *Applied and Environmental Soil Science*, *2011*, 1–10.

Karn, B., Kuiken, T., & Otto, M. (2009). Nanotechnology and in situ remediation: a review of the benefits and potential risks. *Environmental Health Perspectives*, *117*(12), 1813–1831.

Kelepertzis, E. (2014). Accumulation of heavy metals in agricultural soils of Mediterranean: insights from Argolida basin, Peloponnese, Greece. *Geoderma*, *221*, 82–90.

Khalid, S., Shahid, M., Bibi, I., Sarwar, T., Shah, A.H., & Niazi, N.K. (2018). A review of environmental contamination and health risk assessment of wastewater use for crop irrigation with a focus on low and high-income countries. *International Journal of Environmental Research and Public Health*, *15*(5), 895.

Kharisov, B.I., Dias, H.R., & Kharissova, O.V. (2014). Nanotechnology-based remediation of petroleum impurities from water. *Journal of Petroleum Science and Engineering*, *122*, 705–718.

Lasserre, J.P., Fack, F., Revets, D., Planchon, S., Renaut, J., Hoffmann, L., Gutleb, A.C., Muller, C.P., & Bohn, T. (2009). Effects of the endocrine disruptors atrazine and PCB 153 on the protein expression of MCF-7 human cells. *Journal of Proteome Research*, *8*(12), 5485–5496.

Liang, S., Jin, Y., Liu, W., Li, X., Shen, S., & Ding, L. (2017). Feasibility of Pb phytoextraction using nanomaterials assisted ryegrass: Results of a one-year field-scale experiment. *Journal of Environmental Management*, *190*, 170–175.

López-Luna, J., Silva-Silva, M.J., Martinez-Vargas, S., Mijangos-Ricardez, O.F., González-Chávez, M.C., Solís-Domínguez, F.A., & Cuevas-Díaz, M.C. (2016). Magnetite nanoparticle (NP) uptake by wheat plants and its effect on cadmium and chromium toxicological behavior. *Science of the Total Environment*, *565*, 941–950.

Ma, X., & Wang, C. (2010). Fullerene nanoparticles affect the fate and uptake of trichloroethylene in phytoremediation systems. *Environmental Engineering Science*, *27*(11), 989–992.

Martínez-Fernández, D., Vítková, M., Michálková, Z., & Komárek, M. (2017). Engineered nanomaterials for phytoremediation of metal/metalloid-contaminated soils: implications for plant physiology. In *Phytoremediation*, 369–403. Cham: Springer.

McKenzie, L.C., & Hutchison, J.E. (2004). Green nanoscience. *Chimica oggi*, *22*(9), 30–33.

Mehndiratta, P., Jain, A., Srivastava, S., & Gupta, N. (2013). Environmental pollution and nanotechnology. *Environment and Pollution*, *2*(2), 49.

Moameri, M., & Khalaki, M.A. (2019). Capability of Secale montanum trusted for phytoremediation of lead and cadmium in soils amended with nano-silica and municipal solid waste compost. *Environmental Science and Pollution Research*, *26*(24), 24315–24322.

Murty, B.S., Shankar, P., Raj, B., Rath, B.B., & Murday, J. (2013). *Textbook of Nanoscience and Nanotechnology*. Berlin, Heidelberg: Springer Science & Business Media.

Nagajyoti, P.C., Lee, K.D., & Sreekanth, T.V.M. (2010). Heavy metals, occurrence and toxicity for plants: a review. *Environmental Chemistry Letters*, *8*(3), 199–216.

Naidu, R., Kookana, R.S., Sumner, M.E., Harter, R.D., & Tiller, K.G. (1997). Cadmium sorption and transport in variable charge soils: a review. *Journal of Environmental Quality*, *26*(3), 602–617.

Nwadinigwe, A.O., & Ugwu, E.C. (2018). Overview of nano-phytoremediation applications. In *Phytoremediation*, 377–382. Cham: Springer.

Pattrick, R.A., Coker, V.S., Pearce, C.I., Telling, N.D., van der Laan, G., & Lloyd, J.R. (2012). Extracellular bacterial production of doped magnetite nanoparticles. *Nanoscience: Nanostructures Through Chemistry*, *1*, 102–111.

Pillai, H.P., & Kottekottil, J. (2016). Nano-phytotechnological remediation of endosulfan using zero valent iron nanoparticles. *Journal of Environmental Protection*, *7*(05), 734.

Provolo, G., Manuli, G., Finzi, A., Lucchini, G., Riva, E., & Sacchi, G.A. (2018). Effect of pig and cattle slurry application on heavy metal composition of maize grown on different soils. *Sustainability*, *10*(8), 2684.

Radziemska, M., Gusi-Atin, Z.M., Holatko, J., Ham-Merschmiedt, T., Głuchowski, A., Mizerski, A., Jaskulska, I., Baltazar, T., Kintl, A., Jaskulski, D., & Brtnicky, M. (2021). Nano Zero Valent Iron (nZVI) as an Amendment for Phytostabili-zation of Highly Multi-PTE Con-taminated Soil. *Materials 2021*(14), 2559.

Ramezani, M., Rad, F.A., Ghahari, S., Ghahari, S., & Ramezani, M. (2021). Nano-Bioremediation Application for Environment Contamination by Microorganism. In *Microbial Rejuvenation of Polluted Environment*, 349–378. Singapore: Springer.

Rajan, R., Chandran, K., Harper, S.L., Yun, S.I., & Kalaichelvan, P.T. (2015). Plant extract synthesized silver nanoparticles: an ongoing source of novel biocompatible materials. *Industrial Crops and Products*, *70*, 356–373.

Romeh, A.A., & Saber, R.A.I. (2020). Green nano-phytoremediation and solubility improving agents for the remediation of chlorfenapyr contaminated soil and water. *Journal of Environmental Management*, *260*, 110104.

Ross, S.M. (1994). *Toxic Metals in Soil–Plant Systems*, 4. Chichester: Wiley.

Sánchez-Sánchez, R., Ahuatzi-Chacon, D., Galíndez-Mayer, J., Ruiz-Ordaz, N., & Salmerón-Alcocer, A. (2013). Removal of triazine herbicides from aqueous systems by a biofilm reactor continuously or intermittently operated. *Journal of Environmental Management*, *128*, 421–426.

Sarath, N.G., Puthur, J.T. (2020). Heavy metal pollution assessment in a mangrove ecosystem scheduled as a community reserve. *Wetlands Ecology and Management*, 1–12.

Sarath, N.G., Sruthi, P., Shackira, A.M., & Puthur, J.T. (2020). Heavy metal remediation in wetlands: mangroves as potential candidates. In M.N. Grigore (Ed.), Handbook of Halophytes: From Molecules to Ecosystems Towards Biosaline Agriculture, 1–27. Switzerland: Springer International Publishing.

Seaward, M.R.D., & Richardson, D.H.S. (1989). Atmospheric sources of metal pollution and effects on vegetation. *Heavy Metal Tolerance in Plants: Evolutionary Aspects*, 75–92.

Saleem, M.H., Kamran, M., Zhou, Y., Parveen, A., Rehman, M., Ahmar, S., Malik, Z., Mustafa, A., Anjum, R.M.A., Wang, B., & Liu, L. (2020). Appraising growth, oxidative stress and copper phytoextraction potential of flax (Linum usitatissimum L.) grown in soil differentially spiked with copper. *Journal of Environmental Management*, *257*, 109994.

Shackira, A. M., Puthur, Jos T., & Nabeesa Salim, E. (2017). Acanthus ilicifolius L. a promising candidate for phytostabilization of zinc. *Environmental Monitoring and Assessment*, *189*(282), 1–13.

Shackira, A.M., & Puthur, Jos T. (2019). Cd2+ influences metabolism and elemental distribution in roots of Acanthus ilicifolius L. *International Journal of Phytoremediation*, *21*(9), 866–877.

Sharma, B., Sarkar, A., Singh, P., & Singh, R.P. (2017). Agricultural utilization of biosolids: a review on potential effects on soil and plant grown. *Waste Manage*, *64*, 117–132. doi: 10.1016/j.wasman.2017.03.002.

Shi, L., Squier, T.C., Zachara, J.M., & Fredrickson, J.K. (2007). Respiration of metal (hydr) oxides by Shewanella and Geobacter: a key role for multihaem c-type cytochromes. *Molecular Microbiology*, *65*(1), 12–20.

Silveira, M.L.A., Alleoni, L.R.F., & Guilherme, L.R.G. (2003). Biosolids and heavy metals in soils. *Scientia Agricola*, *60*(4), 793–806.

Singh, S., Barick, K.C., & Bahadur, D. (2011). Surface engineered magnetic nanoparticles for removal of toxic metal ions and bacterial pathogens. *Journal of Hazardous Materials*, *192*(3), 1539–1547.

Song, B., Xu, P., Chen, M., Tang, W., Zeng, G., Gong, J., Zhang, P., & Ye, S. (2019). Using nanomaterials to facilitate the phytoremediation of contaminated soil. *Critical Reviews in Environmental Science and Technology*, *49*(9), 791–824.

Souri, Z., Karimi, N., Sarmadi, M., & Rostami, E. (2017). Salicylic acid nanoparticles (SANPs) improve growth and phytoremediation efficiency of Isatis cappadocica Desv., under As stress. *IET Nanobiotechnology, 11*(6), 650–655.

Srivastav, A., Yadav, K.K., Yadav, S., Gupta, N., Singh, J.K., Katiyar, R., & Kumar, V. (2018). Nanophytoremediation of pollutants from contaminated soil environment: current scenario and future prospects. In *Phytoremediation*, 383–401. Cham: Springer.

Srivastava, V., Sarkar, A., Singh, S., Singh, P., de Araujo, A.S., & Singh, R.P. (2017). Agroecological responses of heavy metal pollution with special emphasis on soil health and plant performances. *Frontiers in Environmental Science, 5*, 64.

Sruthi, P., & Puthur, J.T. (2019). Characterization of physiochemical and anatomical features associated with enhanced phytostabilization of copper in Bruguiera cylindrica (L.) Blume. *International Journal of Phytoremediation, 21*(14), 1423–1441.

Sruthi, P., Shackira, A. M., & Puthur, Jos T. (2016). Heavy metal detoxification mechanisms in halophytes: an overview. *Wetlands Ecology and Management, 25*(2), 129–148.

Sumner, M.E. (2000). Beneficial use of effluents, wastes, and biosolids. *Communications in Soil Science and Plant Analysis, 31*(11-14), 1701–1715.

Tang, W.W., Zeng, G.M., Gong, J.L., Liang, J., Xu, P., Zhang, C., & Huang, B.B. (2014). Impact of humic/fulvic acid on the removal of heavy metals from aqueous solutions using nanomaterials: a review. *Science of the Total Environment, 468*, 1014–1027.

Terzi, K., Sikinioti-Lock, A., Gkelios, A., Tzavara, D., Skouras, A., Aggelopoulos, C., Klepetsanis, P., Antimisiaris, S., & Tsakiroglou, C.D. (2016). Mobility of zero valent iron nanoparticles and liposomes in porous media. *Colloids and Surfaces A: Physicochemical and Engineering Aspects, 506*, 711–722.

Tóth, G., Hermann, T., da Silva, M.R., & Montanarella, L. (2018). Monitoring soil for sustainable development and land degradation neutrality. *Environmental Monitoring and Assessment, 190*(2), 57.

Trivedi, S., Ansari, A.A., (2015). Molecular mechanisms in the phytoremediation of heavy metals from coastal waters. In Ansari, A.A. et al. (Eds.). *Phytoremediation: Management of Environmental Contaminants*, 2, 219–231, Switzerland: Springer International Publishing.

Wang, P., Lombi, E., Zhao, F.J., & Kopittke, P.M. (2016). Nanotechnology: a new opportunity in plant sciences. *Trends in Plant Science, 21*(8), 699–712.

Waraporn, J., Preeda, P., Chongrak, P., & Rachain, K. (2012). Enhance Phytoremediation efficiency of TNT contaminated soil by nanoscale zerovalent iron. In *Second International Conference on Environmental and Industrial Innovation*. Volume 35. Singapore: IACSIT Press, 82–86.

Watson, J.H.P., Ellwood, D.C., Soper, A.K., & Charnock, J. (1999). Nanosized strongly-magnetic bacterially-produced iron sulfide materials. *Journal of Magnetism and Magnetic Materials, 203*(1-3), 69–72.

White, B.R., Stackhouse, B.T., & Holcombe, J.A. (2009). Magnetic γ-Fe2O3 nanoparticles coated with poly-l-cysteine for chelation of As (III), Cu (II), Cd (II), Ni (II), Pb (II) and Zn (II). *Journal of Hazardous Materials, 161*(2-3), 848–853.

Windt, W.D., Aelterman, P., & Verstraete, W. (2005). Bioreductive deposition of palladium (0) nanoparticles on Shewanella oneidensis with catalytic activity towards reductive dechlorination of polychlorinated biphenyls. *Environmental Microbiology, 7*(3), 314–325.

Wu, J., Xie, Y., Fang, Z., Cheng, W., & Tsang, P.E. (2016). Effects of Ni/Fe bimetallic nanoparticles on phytotoxicity and translocation of polybrominated diphenyl ethers in contaminated soil. *Chemosphere, 162*, 235–242.

Wuana, R.A., & Okieimen, F.E. (2011). Heavy metals in contaminated soils: a review of sources, chemistry, risks and best available strategies for remediation. *Isrn Ecology, 2011*.

Yadav, K.K., Singh, J.K., Gupta, N., & Kumar, V.J.J.M.E.S. (2017). A review of nanobioremediation technologies for environmental cleanup: a novel biological approach. *Journal of Materials and Environmental Science, 8*(2), 740–757.

Yang, X., Warren, R., He, Y., Ye, J., Li, Q., & Wang, G. (2018). Impacts of climate change on TN load and its control in a River Basin with complex pollution sources. *Science of the Total Environment, 615*, 1155–1163.

Yong, P., Rowson, N.A., Farr, J.P.G., Harris, I.R., & Macaskie, L.E. (2002). Bioreduction and biocrystallization of palladium by Desulfovibrio desulfuricans NCIMB 8307. *Biotechnology and Bioengineering, 80*(4), 369–379.

Zazouli, M.A., Mahdavi, Y., Bazrafshan, E., & Balarak, D. (2014). Phytodegradation potential of bisphenolA from aqueous solution by Azolla Filiculoides. *Journal of Environmental Health Science and Engineering*, *12*(1), 66.

Zeng, J., Liu, L., Li, J., Dong, J., & Cheng, Z. (2020). Properties of nanofibril produced from wet ball milling after enzymatic treatment Vs mechanical grinding of bleached softwood craft fibres. *Bioresources*, *15*, 3809–3820.

Zhang, W.X. (2003). Nanoscale iron particles for environmental remediation: an overview. *Journal of Nanoparticle Research*, *5*(3-4), 323–332.

14

Approaches of Overproduction and Purification of Pleurotus Laccase for the Treatment of Sugarcane Vinasse

Joberson Alves Junior[1], Débora da Silva Vilar[1], Carlos Eduardo Maynard Santana[1], Ram Naresh Bharagava[2], Muhammad Bilal[3], Sikandar I. Mulla[4], Pankaj Kumar Arora[5], Álvaro Silva Lima[1,6], Ranyere Lucena de Souza[1,6], and Luiz Fernando Romanholo Ferreira[1,6]

[1]*Graduated Program in Process Engineering, Tiradentes University (UNIT), Av. Murilo Dantas, Farolândia, Aracaju-Sergipe, Brazil*
[2]*Laboratory for Bioremediation and Metagenomics Research (LBMR), Department of Environmental Microbiology (EM), Babasaheb Bhimrao Ambedkar University (A Central University), Vidya Vihar, Raebareli Road, Lucknow, U.P., India*
[3]*School of Life Science and Food Engineering, Huaiyin Institute of Technology, Huaian, China*
[4]*Department of Biochemistry, School of Applied Sciences, REVA University, Bangalore, India*
[5]*Department of Environmental Microbiology, Babasaheb Bhimrao Ambedkar University, Lucknow, India*
[6]*Institute of Technology and Research (ITP), Tiradentes University (UNIT), Av. Murilo Dantas, Farolândia, Aracaju, Sergipe, Brazil*

CONTENTS

14.1 Introduction 265
14.2 Badiomycetes – *Pleurotus* Genus 267
14.3 Sugarcane Residue 267
14.4 Lignin-Modifying Enzyme 269
    14.4.1 Laccase 269
    14.4.2 Laccase Production 270
    14.4.3 Laccase Purification 272
    14.4.4 Laccase Application 273
14.5 Conclusion and Future Perspectives 274
References 274

14.1 Introduction

In recent years, there has been a significant increase in the demand for biofuels, also known as "green fuels", as an alternative to replacing fossil fuels (Rosen, 2018). From this, the production of different crops, such as sugar cane and others, were intensified. Brazil is the world's largest producer of sugarcane, with around 5 million hectares of cultivated area, and 50% of this production is destined for ethanol. According to the National Supply Company (CONAB), in 2018, approximately 32.31 million m^3 of ethanol was produced in the country, with a forecast to continue with high production in the coming years. However, a traditional production plant generates for each liter of ethanol 9–15 L an effluent, known as vinasse (España-Gamboa et al., 2017; CONAB, 2019).

Water, inorganic minerals, suspended solids, and organic pollutants such as phenolic compounds and melanoidin are part of vinasse composition. It has a brown color, a corrosive effect due to its low pH (3.5–5.0), high COD (70–150 g/L), and high BOD values (35–50 g/L), making it a complex residue that is difficult to degrade (Parsaee et al., 2019). It is used as a fertilizer in crop irrigation; however, it can cause soil effects such as saturation and pH elevation (España-Gamboa et al., 2017). It causes the oxygen dissolved in water to be consumed faster in watercourses, endangering aquatic biota (Vilar et al., 2018). The water streams can also be contaminated by vinasse impurities due to the deep percolation of irrigation water, which infiltrates and reaches underground aquifers (González and Mejía, 2015). Therefore, the vinasse must undergo treatment before being discarded or reused.

The degradation of this effluent from the sugarcane industry by microorganisms proved to be an efficient form of treatment, mainly using the basidiomycete of the genus *Pleurotus* spp. (Vilar et al., 2018; Aragão et al., 2019; Romanholo Ferreira et al., 2020b). The biodegradation alternative is also a method with biotechnological purposes, as in the fermentation process the fungus produces and excretes enzymes of great commercial interest, such as laccase (EC 1.10.3.2) (Romanholo Ferreira et al., 2020b). This enzyme is primarily produced by fungi and catalyzes reactions with various chemical substrates. It has a high potential for biotechnological processes, including the food processing industry (Mayolo-Deloisa et al., 2020), in the delignification of cellulosic compounds (Malhotra and Suman, 2021), paper bleaching (Gupta et al., 2020), degradation of synthetic dyes (Xu et al., 2020), wastewater treatment (Unuofin et al., 2019), formulation of biosensors (Debnath and Saha, 2020), and the degradation of pesticides (Bilal et al., 2019).

With so many applications, laccase production has attracted a lot of attention and can be considerably increased by the use of substrate inducers (Fokina et al., 2016; Liu et al., 2016; Zhuo et al., 2017). Aromatic compounds like pyrogallol and ferulic acid have been shown to stimulate laccase production in *Pleurotus* spp. (Zucca et al., 2011). However, the majority of these compounds are toxic to humans or have high costs, preventing their use in industrial purposes. According to some authors, different alcohols may be more suitable and have a cost advantage in stimulating enzyme production (Manavalan et al., 2013; Hernández et al., 2015). Copper is another well-known inducer, as it increases laccase production by fungi due to the existence of copper atoms in their catalytic domain (Khammuang et al., 2013; Fokina et al., 2016).

Based on the application and the intention of commercial production, the laccase must be purified. Filtration, followed by membrane ultrafiltration, is one of the most common methods used to purify this enzyme (Zucca et al., 2011), followed by precipitation dialysis (Halaburgi et al., 2011), freezing at −20°C, and thawing followed by centrifugation (Jordaan et al., 2004) and chromatographic techniques (Schwienheer et al., 2015). Using these techniques, several purification steps are necessary to obtain the pure enzyme, which causes loss of enzymatic activity and high cost (Nadar et al., 2017).

Thus, an alternative that has been used for the purification and separation of molecules, such as enzymes, proteins, and nucleic acids, is the liquid-liquid extraction using aqueous biphasic systems (ABSs), which has low cost and efficiency, such as purity and activity enzymatic (Rajagopalu et al., 2016; Sánchez-Trasviña et al., 2017). As a result, purification processes have gained widespread acceptance as a promising alternative in biochemical processes, with the difference that they can eliminate or minimize the level of impurities that could inhibit the enzyme, allowing the production of degradable products while also meeting the requirements of a given application (Lemes et al., 2019).

In this scenario, the proposal of this work addresses the biotechnological potential of the fungus genus *Pleurotus* spp. in organic compound degradation and provides an overview of the structural properties of its laccase enzyme, its mechanism of action, and therefore its role in biorefinery. Furthermore, it focuses on the development of high value-added products from the use of agroindustrial substrates (sugarcane vinasse), which simultaneously allows the treatment of these residues, in addition to elucidating the main enzyme purification techniques. Thus, this study motivates the development of new biocatalytic systems that allow the development of industrial sustainability with ecologically renewable inputs, as well as describes future perspectives that can create opportunities in the field of biocatalysis, as well as its technical-economic and environmental issues.

14.2 Badiomycetes – *Pleurotus* Genus

Pleurotus spp. is a genus of white-rot basidiomycete fungi. They are used in a variety of mycoremediation processes, primarily in biotechnology, due to their low cost, great capacity for degradation of recalcitrant pollutants, and acquired resistance in harsh environmental conditions (Zhuo et al., 2017). Due to its nutritional characteristics rich in proteins, fibers, minerals, and vitamins, this genus ranks third in the world for commercial mushroom production, with a total production of 9 Mt/year. (Chanakya et al., 2015; Han et al., 2016). Furthermore, they have therapeutic and bioactive properties that include high antioxidant, antimicrobial, antiproliferative, immunomodulatory, antidiabetic, antihypertensive, and anti-inflammatory potential (Knop et al., 2014; Lavelli et al., 2018).

Since then, this genus has retained the ability to grow rapidly due to chemostatic induction through organic carbon sources, which causes hyphae branching and elongation and thus allows colonization of contaminated areas (Przystaś et al., 2018). It is estimated that about 200 species of this genus have been identified; however, so far only a few have been used for food applications, such as *P. pulmonarius, P. eryngii, P. ostreatus, P. sapidus, P. cystidiosus,* and *P. sajor-caju.*

According to Voběrková et al. (2018), these microorganisms can still quickly adapt their metabolism to different substrates in order to promote a great amount of extracellular ligninolytic enzymes and degrade and/or mineralize harmful chemicals. In addition, this fungi consortium produces and excretes only the enzymes peroxidase, manganese peroxidase (MnP), and laccase (Lac), which are nonspecific and oxidative in the disruption of the plant cell wall of lignin (phenolic compounds). Moreover, they can convert bioavailable polysaccharides in the plant wall structure (cellulose and hemicellulose) into easily assimilated sugars (Saha et al., 2016; Fasanella et al., 2018). This delignification process, however, varies depending on the fungal strain. In this regard, this genus can still make use of a variety of agroindustrial by-products as growth substrates (Rouches et al., 2016), as shown in Table 14.1.

14.3 Sugarcane Residue

Among the various agroindustrial residues, sugarcane residue has gained prominence in the sugar-energy sector and the production of ligninolytic enzymes, as it has nutritional and functional properties, with the presence of carbon and nitrogen, phenolic compounds, and soluble sugars (Romanholo Ferreira

TABLE 14.1

Different agrowastes used in the cultivation of *Pleurotus* spp.

References	Genus *Pleurotus* spp.	Agroindustrial Residues
(Dias et al., 2003)	*P. sajor-caju*	Bean straw and corn straw
(Reddy et al., 2003)	*P. sajor-caju e, P. ostreatus*	Banana residue
(Alexandrino et al., 2007)	*P. ostreatus*	Orange residue
(Gregori et al., 2008)	*P. ostreatus*	Wheat bran and beech sawdust
(Aguiar et al., 2010)	*P. sajor-caju, P. ostreatus e P. ostreatoroseus*	Sugarcane bagasse
(Bernardi et al., 2013)	*P. sajor-caju e, P. ostreatus*	Elephant grass, castor bean residues, and rice straw
(Corrêa et al., 2016)	*Pleurotus* spp.	Cotton residues, wheat straw, coffee husks, pea stalks, and peanuts
(Ozcirak Ergun and Ozturk Urek, 2017)	*P. ostreatus*	Potato peel
(Silva, Da et al., 2017)	*Pleurotus pulmonarius*	Pineapple residue
(Cruz et al., 2018)	*P. sajor-caju*	Pulp wash
(Vilar et al., 2018)	*P. sajor-caju*	Vinasse

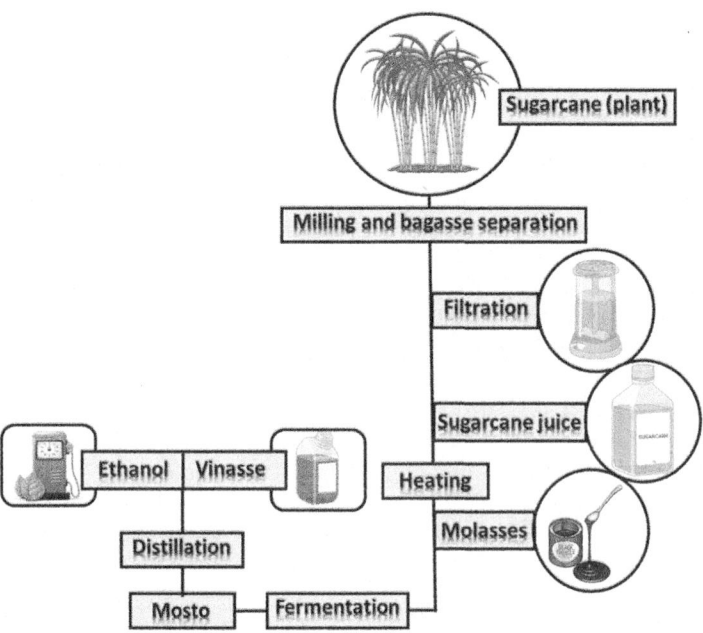

FIGURE 14.1 Outlook ethanol production process and sugarcane vinasse.

et al., 2020b). Liquid waste (vinasse) and solid waste (bagasse) are the main by-products generated by the sugar-energy industry. The first results from the manufacture of alcohol from sugarcane (Figure 14.1) after the process of must fermentation and wine distillation (Aguıar et al., 2018; Vilar et al., 2018; Romanholo Ferreira et al., 2020b).

España-Gamboa et al. (2017) reported that the chemical formation of vinasse varies depending on the origin of the raw material, the procedure used to prepare the must, the selected fermentation technique, the species of yeast used, and the operating conditions of the distillery and facilities used. About 2–6% of the constituents of vinasse are solid, basically formed by organic matter, based on organic acids and, to a lesser extent, are mostly potassium (K) and also phosphorus (P), magnesium (Mg), calcium (Ca), manganese (Mn), and organic nitrogen (N) (Christofoletti et al., 2013; Godoi De et al., 2019).

It is estimated that to produce a liter of alcohol it takes about 10–18 liters of vinasse. Its use can be performed in several ways: by tank trucks, by infiltration furrows, by spraying, and currently, there is the possibility of dispersing the diluted vinasse through center pivots (España-Gamboa et al., 2017). Characterized by Romanholo Ferreira et al. (2020a), the vinasse is of great importance for sugarcane crops, due to the rich presence of mineral elements such as potassium, chemical oxygen demand (COD), and biochemical oxygen demand (BOD). In addition to polluting properties as the strong acidity pH (3.7–5.0), high corrosivity due to the free sulfuric acid disposed of in the fermentation vats, it also has a dark color and a characteristic smell, due to the presence of melanoidins, which are polymers with high molecular weight and difficult to degrade. These are generated from a Maillard reaction with proteins (Sadeghi et al., 2016).

For these reasons, this residue has high power as a soil fertilizer and environmental pollutant. Its pollutant potential is about 100 times greater than that of domestic sewage, in addition, it has high levels of BOD and COD, which range from 35,000–50,000 mg L^{-1} and 70,000–150,000 mg L^{-1}, respectively. With these data we can see that vinasse has a large amount of organic matter, which leads to the eutrophication of effluents where this waste is dumped, resulting in substances that can compromise water quality (Robles-González et al., 2012).

From a biological standpoint, vinasse should also be considered a harmful agent, responsible for the increase of microbial activity in the soil, due to being a recalcitrant compound, having phytotoxic, antibacterial substances and antioxidant properties, which cause changes in the soil's physical-chemical properties. Thus, if the residue is not treated before the destination or disposal in the environment, and used incorrectly, it will cause damage to the fauna and flora, polluting rivers and seas, as well as the organisms residing in these ecosystems (España-Gamboa et al., 2011).

As a result, industries are searching for innovative ways to dispose of and reuse vinasse, as this residue emits a high level of contaminants into the environment (España-Gamboa et al., 2017). The reuse of vinasse appears in several forms: sugarcane washing processes, with water reuse, biogas generation (anaerobic digestion), recycling in fermentation, distribution in sacrifice zones (non-cultivable soils), incineration, combustion, generation of yeasts, crop irrigation through fertigation, and other activities intrinsic to the industrial process (Godoi De et al., 2019).

This compound, which was previously released in nature into water bodies, is now released into the ground, most often by fertigation, highlighting the use of this practice, as it does not require large investments. Furthermore, it has a low maintenance cost, does not involve the use of complex technology, enables a rapid elimination of large quantities of this material, improves soil fertility, and acts as a reducing agent for the factors that cause pollution of water resources (Godoi De et al., 2019). However, the use of vinasse in fertigation cannot be excessive, since it can harm the environment, due to the accumulation of contaminants, and its leaching can contaminate underground water sources and salinize the soils, and can also promote rapid microbial degradation of organic matter, which contributes to the proliferation of flies and makes the environment unhealthy and unstable. At the same time, it became necessary to adopt decrees in the various areas of action of the environmental inspection bodies, to maintain the ecological balance and minimize the negative environmental impacts caused when the vinasse is dumped into water bodies without proper treatment (Sousa et al., 2019).

Therefore, with this polluting effect of vinasse and the large existing alcohol production, this residue must be reused or treated before being discarded in the environment. Its degradation by fungi is an alternative treatment for biotechnological purposes, since in the fermentation process the fungus excretes enzymes like laccase (Junior et al., 2020).

14.4 Lignin-Modifying Enzyme

14.4.1 Laccase

The laccase enzymes (benzenediol: oxygen oxidoreductase, EC 1.10.3.2) belong to the family of polyphenol oxidases, widely found in plants, insects, bacteria, and filamentous fungi (Agrawal et al., 2018). They have a molecular mass ranging from 50 to 140 KDa and are produced by most basidiomycetes, as well as four copper (Cu) ions present in three binding sites, each of which plays a significant role in the catalytic reaction that occurs during phenolic substrate oxidation, as molecular oxygen is reduced to water (Agrawal et al., 2018).

Figure 14.2 illustrates the three-dimensional structure of Lac, which is comprised of four copper molecules linked together by three different types of nuclear bonds (T1, T2, and T3). Due to its strong interaction with T2, mononuclear copper (T1) is responsible for removing electrons from the substrate and transferring them to T2 and T3, via the tripeptide sequence (His-Cys-His), as well as promoting the reduction of molecular oxygen (O_2) in water. For such reasons, its catalytic mechanism has piqued the interest of several researchers, primarily because it is considered a "green process," as it requires O_2 as the sole co-substrate for catalytic reactions, excluding the participation of H_2O_2, and thus offers the advantage of degrading lignin with non-toxic derivatives (Kumar and Chandra, 2020).

These fungal laccases are classified according to their redox potential as low (430 mV), medium (430–710 mV), and high (790 mV), which allows them to convert toxic substances into metabolic ones via oxidation (Piscitelli et al., 2011). However, these enzymes have difficulties penetrating the substrate (biomass) and degrade lignin, due to their high molecular weight. Furthermore, they can degrade phenolic lignin compounds only on the substrate surface. On the other hand, they are unable to oxidize non-phenolic lignin compounds present on the surface and have a high redox potential, as these enzymes generally have low redox potential (Singh and Arya, 2019). With the use of chemical mediators, which are low molecular weight compounds that allow oxidized radicals to chemically react with high

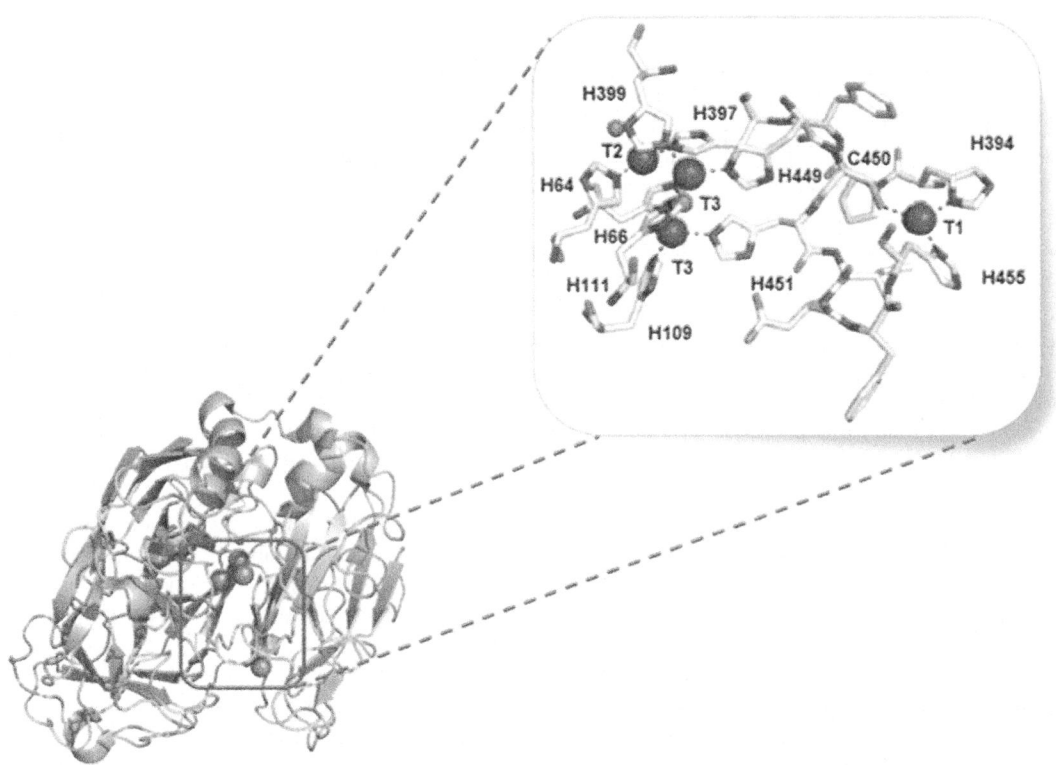

FIGURE 14.2 Three-dimensional structure of the active site of laccase (Lac) belonging to the genus *Pleurotus*, which describes the orientation of copper atoms and its distance (blue, interatomic distances in Å), being integrated by four copper atoms distributed in three catalytic centers: Cu (T1) and the tri-nuclear Cu group (T2–T3) (adapted from Rodgers et al., 2010; Kumar and Chandra, 2020).

redox potential target substrates, these limitations can be overcome, since these mediators act as intermediate substrates for laccases (Shraddha et al., 2011).

Most of these enzymes are inducible, so inducers such as aromatic or phenolic compounds related to lignin or lignin derivatives are responsible for increasing the production of this enzyme through fermentation processes. The same inducer can either increase the production of laccase for a particular species of fungus or not affect another species. Therefore, the optimal inducer is not common to all fungi, and its choice will depend on the studied fungus (Mann et al., 2015).

The production of laccase is affected by physiological differences that exist in the cultivation conditions and between fungi cultures, as well as in the low-cost procedures, which make the application of these enzymes viable (Kocyigit et al., 2012). The ideal temperature of its activity and its stability vary according to the different sources of enzymes, normally more stable in acidic pH (Morsi et al., 2020).

14.4.2 Laccase Production

Pleurotus laccase is extracellular and is secreted into the medium by the fungus mycelium. Several parameters affect the production of laccase during its cultivation, which include presence or absence of inducers or inhibitors, pH, temperature, incubation time, agitation, type of culture, and the species of microorganism used (Morsi et al., 2020). The pH and temperature of enzyme production differ greatly, but most authors agree that the initial pH (before inoculation) should be between 4.5 and 6 and the optimal incubation temperature should be between 25 and 30°C; cultures with temperatures above 30°C have reduced enzymatic activity (Aragão et al., 2019; Romanholo Ferreira et al., 2020a).

Fermentation can take place in a submerged or solid state. Submerged fermentation involves the cultivation of the fungus in a liquid medium, containing necessary nutrients in the substrate for its growth. The formation of mycelia during fungal cell growth can limit oxygen and mass transfer, and some strategies have already been employed to deal with these limitations, such as cell immobilization (Corrêa et al., 2016; Ozcirak Ergun and Ozturk Urek, 2017). The production of laccase in a liquid medium is influenced by agitation. The gentle agitation favours the enzyme production, while the excessive agitation compromises its production due to the mechanical stress caused and the damage that the agitation can cause to the mycelia (Couto and Toca-Herrera, 2007). Hess et al. (2002) found that laccase production by *Trameter multicolor* decreased considerably when the fungus was cultivated under agitation, while Galhaup et al. (2002) found that with shaking the production of laccase from *Trameter pubescens* increased twice, and the enzyme activity was also higher.

The fungus species also influences laccase production (Sartori et al., 2015); for example, *Phlebia fascicularia* and *Dichomitus squalens* are better laccase producers than *Trametes versicolor*, under culture conditions for 20 days in 10 mL of MSB (medium composed of mineral salts broth) distributed in 100 mL Erlenmeyer flasks and inoculated with two mycelial discs of 7 mm in diameter, obtained from the cultivation of the fungi in Petri dishes with YGA medium (yeast-glucose agar) and incubated at 25 ± 1°C (Arora and Gill, 2000).

The promotion of laccase production has attracted a lot of attention due to its potential application in environmental and industrial areas (Zhuo et al., 2017). Production can be considerably increased by adding various supplements to the substrate. These supplements are known as inducers, as they induce greater enzyme production, which can be the enzyme substrate itself or a structurally analogous compound. Among the factors that influence the greater production of laccase are the chemical nature of the inducer, the amount added, and the moment in which its addition influences the greater production of laccase (Manavalan et al., 2013; Hernández et al., 2015).

Laccase activity is increased and induced by the addition of organic contaminants such as lignin, xylidine, and veratryl alcohol (Shraddha et al., 2011). According to one study, the addition of cellobiose can stimulate laccase activity in some *Trametes* species (Brijwani et al., 2010). Copper also exhibited the effect as inducer over laccase activity (Passarini et al., 2015; Fonseca et al., 2016). Ethanol is indicated as a cheap and low-toxic inducer for the production of laccase (Hernández-López et al., 2015). Using ethanol, Chen et al. (2016) induced laccase from *T. versicolor*, Manavalan et al. (2013) induced the production of laccase by *Ganoderma lucidum*, using concentrations of 1, 2, 3, 4, and 5% (v/v), where a higher enzymatic activity was obtained with 3% ethanol after 15 days, being 6 or 5 times larger than the control sample.

Several agricultural residues work as a substrate for the fungus' growth, and, in addition, some of these residues are also described as inducers for greater enzymatic activity, such as corn, rice, and wheat bran, orange peel, grapefruit and tamarind, bagasse, and straw of wheat (Vilar et al., 2019). Table 14.2 shows some inducers used to increase laccase production in a certain fungal species.

TABLE 14.2

Inductors used in laccase production

Indutor	Fungo	References
ABTS	*Trametes versicolor*	(Abhijit M, 2015)
Ácido ferúlico	*Pleurotus sajor-caju*	(Patrick et al., 2011)
Cobre	*Dichomitus squalens; Trametes* sp.	(Kannaiyan et al., 2012; Fonseca et al., 2016)
Etanol	*Trametes versicolor; Pycnoporus cinnabarinus*	(Lee et al., 1999; Hernández et al., 2015)
Guaiacol	*Trametes versicolor*	(Kuhar and Papinutti, 2014)
Resíduos agrícolas	*Pleurotus sajor-caju*	(Junior et al., 2020)
Seringaldazina	*Trametes versicolor*	(Abhijit M, 2015)
Xilidina	*Pleurotus sajor-caju*	(Zucca et al., 2011)

The extracellular production of laccase dispenses with the use of intracellular extraction techniques, which generally lead to loss of yield. The enzyme extraction from liquid medium cultivation is easier and allows the total extraction of extracellular enzymes produced, while in solid-state cultivation, additional steps are necessary to obtain the enzymatic extract (Aguiar et al., 2018).

14.4.3 Laccase Purification

The purification process consists of removing or reducing most of the impurities present in the enzymes, through a combination of steps that make it possible to isolate, concentrate, and stabilize the target protein. Furthermore, this process is important in the field of biotechnology because it provides benefits such as increased production economy, the ability to optimize, and also promoting a high degree of protein purity and minimizing the use of chemical additives. This optimization involves several variables inherent to the enzyme (characterization and purification), and to the fermentation process (pH, temperature, rotation, cultivation time, microorganism, substrate, and reactor type) (Patel et al., 2017; Lemes et al., 2019).

In general, the enzymatic purification process is regarded as an excellent replacement for conventional chemical processes because it reduces the undesirable residues production and usually requires fewer steps to achieve the ultimate product. Furthermore, the purification process of the natural enzyme favours high yields at low cost, contrary to what normally occurs with commercial enzymes, which cost around $2,400/kg, and in this case, a greater number of process steps are used (Lemes et al., 2019).

Regarding LMEs, most are extracellular and specific, in particular, Lac and MnP, which facilitate the purification process and, consequently, allow the obtainment of degradable products and reduce the amount of waste generated. Currently, the main techniques used in this process were developed to reduce the number of contaminants that can inhibit the activity of proteins; they are ultrafiltration, ion exchange, molecular exclusion, precipitation with salts, centrifugation, and liquid-liquid extraction in aqueous systems biphasic (ABS) (Table 14.3) (Guzman et al., 2018; Hidayah et al., 2018). In general, this last method is the most effective because of its outstanding properties, which include high yield and bio-compatibility with solutes, as well as fast and continuous processing, which allows biomolecules to participate in one of the system's phases (Campos-Pinto et al., 2017).

These immiscible (top and bottom) phases are formed when two hydrophilic polymers or when a polymer and an inorganic salt (Figure 14.3) dissolve in an aqueous solution above their critical concentrations (Torres-Acosta et al., 2018). The most-used polymers are polyethylene glycol (PEG) and dextran, and the salt is potassium phosphate. In addition, this method consists primarily of water (65–90%), which favours the stability of proteins during separation, when compared to traditional systems composed of organic solvents (González-Amado et al., 2017). Characteristics of high selectivity of enzymes, ease of operation, and low cost make this system promising and of great interest to the chemical and related industries.

TABLE 14.3

Various purification techniques performed in laccase

Purification Techniques	References
Filtration followed by membrane ultrafiltration	(Zucca et al., 2011)
Precipitation with ammonium sulfate or acetone, followed by dialysis to remove salts	(Halaburgi et al., 2011; Younes Ben and Sayadi, 2011)
Freezing at −20 °C and thawing followed by centrifugation	(Jordaan et al., 2004)
Ion exchange chromatography	(Saito et al., 2003)
Partition chromatography	(Schwienheer et al., 2015)
Affinity chromatography	(Jaiswal et al., 2014)
Gel permeation chromatography	(Marques De Souza and Peralta, 2003)

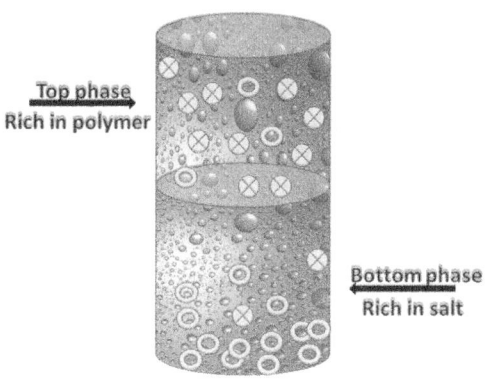

FIGURE 14.3 Biphasic aqueous system composed of polymer and salt.

However, in addition to its wide versatility, this process is based on the principle of "green chemistry", which serves the synthesis of less hazardous products, the development of safe chemicals that degrade into non-toxic products, and that allow recovery and/or recycling of its components (Ruiz et al., 2017). Studies suggest that the stability and activation of the enzyme, when added to the SAB, are highly dependent on some factors, such as size and hydrophobic nature of the biomolecule, molecular weight and polymer and salt concentration, temperature, and pH. Thus, it is essential to more specifically assess the parameters that influence the behaviour of enzymes in the presence of water (Silva et al., 2016; Đorđević and Antov, 2017).

14.4.4 Laccase Application

Laccase is widely studied to this day due to the high number of applications in different areas of biotechnology (Fokina et al., 2016). Laccase can be applied in several areas of the food industry. It is used in the production of bread to increase specific volume while decreasing crumb and chewing hardness (Renzetti et al., 2010). In wine stabilization, the enzyme aims to remove phenols from the must, facilitating regeneration and subsequent reuse, in addition to inhibiting oxidants and preventing discoloration (Riebel et al., 2017). In the wine industry, it can be used in the preparation of cork stoppers, reducing the taste of the stopper that can occur in bottled wine (Nunes and Kunamneni, 2018). Furthermore, it is used to clarify fruit juice, reducing phenols that cause strong turbidity and affect the quality of the juice (Lettera et al., 2016).

In the paper industry, laccase is of fundamental importance for the removal of lignin, allowing it to obtain white paper and improving its flexibility (Rodríguez-Couto, 2019). It can also be used to lighten eucalyptus pulp, without the use of chlorine, improving the properties of the pulp (Sharma et al., 2014) and in the discoloration of fungal pigments on aged paper and parchment (Abd El Monssef et al., 2016). The use of laccase in the textile industry promotes benefits such as reduced use of water and energy, the use of smaller amounts of chemicals, and faster and more sustainable production processes (Polak et al., 2016). This enzyme is also used to discolor dyes and bio bleach tissues (Zhuo et al., 2017; Bagewadi et al., 2018). This promotes the reduction of cotton bleaching costs if combined with hydrogen peroxide, producing higher levels of whiteness than conventional methods (Gonçalves et al., 2014).

The enzyme is also used in bioremediation processes, being used to reduce the environmental impact caused by industrial effluents, such as in the removal of chemical endocrine disruptors, herbicides, xenobiotics, among others, which commonly pollute waste and agricultural water (Kumar et al., 2014; Daâssi et al., 2016; Yousefi-Ahmadipour et al., 2016). It can be applied as a biosensor to assess industrial contamination, determining the presence of pesticides in the environment, based on enzymatic inhibition (Taylor et al., 2011). In the area of medicine, it is used in detection protocols for other enzymes or as an enzymatic marker for immunochemical, histochemical, cytochemical, and nucleic acid detection tests (Shraddha et al., 2011). It can act as a bactericide and fungicide, being used in combination with a protein or peptide capable of destroying microbial cells (Sampaio et al., 2016). They are used as catalysts for the synthesis of important products in the pharmaceutical industry, such as anticancer drugs, antioxidants, anti-inflammatory drugs, analgesics, and antiproliferative (Kudanga et al., 2011; Morsi et al., 2020).

Laccase also has other applications, such as in the production of bioethanol, oxidizing compounds that inhibit fermentation (Kacem et al., 2016; Zabed et al., 2017). With the aid of laccase, it is possible to improve the production of biogas and eliminate toxic compounds released in the combustion process, being a system of great application potential for in-situ removal of toxic substances in industries (Hom-Diaz et al., 2016). Laccase can be used in fuel cells, enabling the conversion of chemical energy into electrical energy (Arrocha et al., 2014; Aquino Neto et al., 2016).

14.5 Conclusion and Future Perspectives

Pleurotus spp. has many peculiarities, which contribute to its biodiversity, taxonomic characteristics, and biochemical composition. The main biodegradation mechanism employed by this genus is the lignin degradation system by extracellular LMEs, specifically laccase. Because it is non-specific and oxidative, it can mineralize a huge spectrum of organic pollutants that are similar in structure to lignin. As a result, in current history, there has been a significant increase in interest in using these basidiomycetes to degrade a variety of lignocellulosic and xenobiotic materials. Although these lignocellulosic residues, specifically sugarcane vinasse, have a high polluting potential, it is still considered promising in fermentation processes, with the difference that it has low cost and nutritional properties that induce the production of enzymes. From this, these oxidoreductase enzymes can be further purified by liquid-liquid extraction using SAB and facilitate the extraction yield, sequentially improving the separation conditions, leaving the purest strata free of contaminants. As a result, the biotechnological interest in these white-rot fungi and their produced enzymes appears to be promising for replacing traditional chemical processes in a variety of industries. It is critical to advance our understanding of laccase bioprospecting to improve its productivity and degradation mechanism. In this sense, more studies are required to investigate its application in industrial waste bioconversion, as well as to develop new biotechnological processes that are proficient in producing laccase commercially on a large scale. Selection of white-rot fungal strains and optimization of growing conditions can also enhance enzyme production. Furthermore, the conventional improvement and/or molecular techniques of laccase are promising in the field of future research for its application in biorefineries based on lignocellulosic raw material.

REFERENCES

Abd El Monssef, R.A., Hassan, E.A., & Ramadan, E.M. (2016). Production of laccase enzyme for their potential application to decolorize fungal pigments on aging paper and parchment. *Annals of Agricultural Sciences, 61*(1), 145–154.

Abhijit M, A.C. (2015). Screening and isolation of laccase producers, determination of optimal condition for growth, laccase production and choose the best strain. *Journal of Bioremediation & Biodegradation, 06*(04), 1–8.

Agrawal, K., Chaturvedi, V., & Verma, P. (2018). Fungal laccase discovered but yet undiscovered. *Bioresources and Bioprocessing, 5*(1), 1–12. Springer Berlin Heidelberg. Disponível em: https://bioresourcesbioprocessing.springeropen.com/articles/10.1186/s40643-018-0190-z.

Aguiar, M., Lfr, F., & Rtr, M. (2010). Use of vinasse and sugarcane bagasse for the production of enzymes by lignocellulolytic fungi. *Brazilian Archives of Biology and Technology, 53*, 1245–1254.

Aguiar, M.M., Pietrobon, V.C., Salles Pupo, M.M.D.E., et al. (2018). Evaluation of commercial cellulolytic enzymes for sugarcane bagasse hydrolysis. *Cellulose Chemistry and Technology, 52*, 695-699.

Alexandrino, A.M., De Faria, H.G., De Souza, C.G.M., & Peralta, R.M. (2007). Reutilisation of orange waste for production of lignocellulolytic enzymes by Pleurotus ostreatus (Jack:Fr). *Ciencia e Tecnologia de Alimentos, 27*(2), 364–368.

Aquino Neto, S., Zimbardi, A.L.R.L., Cardoso, F.P., et al. (2016). Potential application of laccase from Pycnoporus sanguineus in methanol/O2 biofuel cells. *Journal of Electroanalytical Chemistry, 765*, 2–7.

Aragão, M.S., Menezes, D.B., Ramos, L.C., et al. (2019). Mycoremediation of vinasse by surface response methodology and preliminary studies in air-lift bioreactors. *Chemosphere*, 244, 125–432. Elsevier Ltd. Disponível em: 10.1016/j.chemosphere.2019.125432.

Arora, D.S., & Gill, P.K. (2000). Laccase production by some white rot fungi under different nutritional conditions. *Bioresource Technology*, 73(3), 283–285.

Arrocha, A.A., Cano-Castillo, U., Aguila, S.A., & Vazquez-Duhalt, R. (2014). Enzyme orientation for direct electron transfer in an enzymatic fuel cell with alcohol oxidase and laccase electrodes. *Biosensors and Bioelectronics*, 61, 569–574.

Bagewadi, Z.K., Mulla, S.I., & Ninnekar, H.Z. (2018). Optimization of endoglucanase production from Trichoderma harzianum strain HZN11 by central composite design under response surface methodology. *Biomass Conversion and Biorefinery*, 8(2), 305–316.

Bernardi, E., Minotto, E., & Do Nascimento, J.S. (2013). Evaluation of growth and production of Pleurotus sp. in sterilized substrates. *Arquivos do Instituto Biológico*, 80(3), 318–324.

Bilal, M., Iqbal, H.M.N., & Barceló, D. (2019). Persistence of pesticides-based contaminants in the environment and their effective degradation using laccase-assisted biocatalytic systems. *Science of the Total Environment*, 695, 133896.

Brijwani, K., Rigdon, A., & Vadlani, P.V. (2010). Fungal Laccases: Production, Function, and Applications in Food Processing. *Enzyme Research*, 2010, 149-748.

Campos-Pinto, I., Espitia-Saloma, E., Rosa, S.A.S.L., Rito-Palomares, M., & Aguilar, O. (2017). Integration of cell harvest with affinity-enhanced purification of monoclonal antibodies using aqueous two-phase systems with a dual tag ligand. *Separation and Purification Technology*, 173, 129–134. Disponível em: 10.1016/j.seppur.2016.09.017.

Chanakya, H.N., Malayil, S., & Vijayalakshmi, C. (2015). Cultivation of Pleurotus spp. on a combination of anaerobically digested plant material and various agro-residues. *Energy for Sustainable Development*, 27, 84–92. Disponível em: 10.1016/j.esd.2015.04.007.

Chen, L., Yi, X., Deng, F., et al. (2016). A novel ethanol-tolerant laccase, Tvlac, from Trametes versicolor. *Biotechnology Letters*, 38(3), 471–476.

Christofoletti, C.A., Escher, J.P., Correia, J.E., Marinho, J.F.U., & Fontanetti, C.S. (2013). Sugarcane vinasse: Environmental implications of its use. *Waste Management*, 33(12), 2752–2761.

Corrêa, R.C.G., Brugnari, T., Bracht, A., Peralta, R.M., & Ferreira, I.C.F.R. (2016) Biotechnological, nutritional and therapeutic uses of Pleurotus spp. (Oyster mushroom) related with its chemical composition: A review on the past decade findings. *Trends in Food Science and Technology*, 50, 103–117.

Couto, S.R., & Toca-Herrera, J.L. (2007). Laccase production at reactor scale by filamentous fungi. *Biotechnology Advances*, 25(6), 558–569.

Cruz, Y.W.G., Vieira, Y.A., Vilar, D.S., et al. (2018). Pulp wash: a new source for production of ligninolytic enzymes and biomass and its toxicological evaluation after biological treatment. *Environmental Technology*, 41(14), 1–28. Taylor & Francis. Disponível em: 10.1080/09593330.2018.1551428.

Daâssi, D., Zouari-Mechichi, H., Frikha, F., et al. (2016). Sawdust waste as a low-cost support-substrate for laccases production and adsorbent for azo dyes decolorization. *Journal of Environmental Health Science and Engineering*, 14(1).

Debnath, R., & Saha, T. (2020). An insight into the production strategies and applications of the ligninolytic enzyme laccase from bacteria and fungi. *Biocatalysis and Agricultural Biotechnology*, 26.

Dias, E.S., Koshikumo, É.M.S., Schwan, R.F., & Silva, R.D.A. (2003). Cultivo do cogumelo Pleurotus sajor-caju em diferentes resíduos agrícolas. *Ciência e Agrotecnologia*, 27(6), 1363–1369.

Đorđević, T., & Antov, M. (2017). Ultrasound assisted extraction in aqueous two-phase system for the integrated extraction and separation of antioxidants from wheat chaff. *Separation and Purification Technology*. Disponível em: 10.1016/j.seppur.2017.03.025.

España-Gamboa, E., Mijangos-Cortes, J., Barahona-Perez, L., & Dominguez-Maldonado, J. (2011). Vinasses: characterization and treatments. *Waste Management & Research*, 29, 1235–1250.

España-Gamboa, E., Vicent, T., Font, X., Dominguez-Maldonado, J., & Canto-Canché, B. (2017). Pretreatment of vinasse from the sugar refinery industry under non-sterile conditions by Trametes versicolor in a fluidized bed bioreactor and its effect when coupled to an UASB reactor. *Journal of Biological Engineering*, 1–11. Disponível em: 10.1186/s13036-016-0042-3.

Fasanella, C.C., Aguiar, M.M., Ferreira, L.F.R., Pupo, M.M.S., & Salazar-Banda, G.R. (2018). Microscopic analysis of sugarcane bagasse following chemical and fungal treatment. *Cellulose Chemistry and Technology*, 52(1–2).

Fokina, O., Eipper, J., Kerzenmacher, S., & Fischer, R. (2016). Selective natural induction of laccases in Pleurotus sajor-caju, suitable for application at a biofuel cell cathode at neutral pH. *Bioresource Technology*. Disponível em: 10.1016/j.biortech.2016.06.126.

Fonseca, M.I., Tejerina, M.R., Sawostjanik-Afanasiuk, S.S., et al. (2016). Preliminary studies of new strains of Trametes sp. From Argentina for laccase production ability. *Brazilian Journal of Microbiology*, 47(2), 287–297.

Galhaup, C., Wagner, H., Hinterstoisser, B., & Haltrich, D. (2002). Increased production of laccase by the wood-degrading basidiomycete Trametes pubescens. *Enzyme and Microbial Technology*, 30(4), 529–536.

Godoi De, L.A.G., Camiloti, P.R., Bernardes, A.N., et al. (2019). Seasonal variation of the organic and inorganic composition of sugarcane vinasse: main implications for its environmental uses. *Environmental Science and Pollution Research*, 26(28), 29267–29282.

Gonçalves, I., Martins, M., Loureiro, A., et al. Sonochemical and hydrodynamic cavitation reactors for laccase/hydrogen peroxide cotton bleaching. *Ultrasonics Sonochemistry*, 21(2), 774–781, (2014).

González-Amado, M., Rodil, E., Arce, A., Soto, A., & Rodríguez, O. (2017). Polyethylene glycol (1500 or 600) – Potassium tartrate Aqueous Two-Phase Systems. *Fluid Phase Equilibria*. Disponível em: 10.1016/j.fluid.2017.11.009.

González, L., & Mejía, M. (2015). Impact of Ferti-Irrigation with Vinasse on Groundwater Quality. *Irrigation and Drainage*, 64(3), 400–407.

Gregori, A., Švagelj, M., Pahor, B., Berovič, M., & Pohleven, F. (2008). The use of spent brewery grains for Pleurotus ostreatus cultivation and enzyme production. *New Biotechnology*, 25(2–3), 157–161.

Gupta, G.K., Kapoor, R.K., & Shukla, P. (2020). Advanced Techniques for Enzymatic and Chemical Bleaching for Pulp and Paper Industries. *Microbial Enzymes and Biotechniques*, 43–56. Singapore: Springer. Disponível em: https://link.springer.com/chapter/10.1007/978-981-15-6895-4_3. Acesso em: 10/6/2021.

Guzman, G.Y.F., Hurtado, G.B., & Ospina, S.A. (2018). New dextransucrase purification process of the enzyme produced by Leuconostoc mesenteroides IBUN 91. 2. 98 based on binding product and dextranase hydrolysis. *Journal of Biotechnology*, 265(July 2017), 8–14. Disponível em: 10.1016/j.jbiotec.2 017.10.019.

Halaburgi, V.M., Sharma, S., Sinha, M., Singh, T.P., & Karegoudar, T.B. (2011). Purification and characterization of a thermostable laccase from the ascomycetes Cladosporium cladosporioides and its applications. *Process Biochemistry*, 46(5), 1146–1152. Elsevier Ltd. Disponível em: 10.1016/j.procbio.2011.02.002.

Han, N.S., Wan Ahmad, W.A.N., & Wan Ishak, W.R. (2016). Quality characteristics of Pleurotus sajor-caju powder: Study on nutritional compositions, functional properties and storage stability. *Sains Malaysiana*, 45(11), 1617–1623.

Hernández-López, E.L., Ayala, M., & Vazquez-Duhalt, R. (2015). Microbial and Enzymatic Biotransformations of Asphaltenes. *Petroleum Science and Technology*, 33(9), 1017–1029.

Hernández, C.A., Sandoval, N., & Mallerman, J., et al. (2015). Electronic Journal of Biotechnology Ethanol induction of laccase depends on nitrogen conditions of Pycnoporus sanguineus. *EJBT*, 18(4), 327–332. Elsevier B.V. Disponível em: 10.1016/j.ejbt.2015.05.008.

Hess, J., Leitner, C., Galhaup, C., et al. (2002). Enhanced formation of extracellular laccase activity by the white-rot fungus Trametes multicolor. *Applied Biochemistry and Biotechnology – Part A Enzyme Engineering and Biotechnology. Anais....* 98–100, 229–241.

Hidayah, N., Sukohidayat, E., Zarei, M., Manap, M.Y., & Baharin, B.S. (2018). Purification and Characterization of Lipase Produced by Leuconostoc mesenteroides Subsp. mesenteroides ATCC 8293 Using an Aqueous Two-Phase System (ATPS) Composed of Triton X-100 and Maltitol. *Molecules*, 23, 1800.

Hom-Diaz, A., Passos, F., Ferrer, I., Vicent, T., & Blánquez, P. (2016). Enzymatic pretreatment of microalgae using fungal broth from Trametes versicolor and commercial laccase for improved biogas production. *Algal Research*, 19, 184–188.

Jaiswal, N., Pandey, V.P., & Dwivedi, U.N. (2014). Purification of a thermostable alkaline laccase from papaya (Carica papaya) using affinity chromatography. *International Journal of Biological Macromolecules, 72,* 326–332.

Jordaan, J., Pletschke, B.I., & Leukes, W.D. (2004). Purification and partial characterization of a thermostable laccase from an unidentified basidiomycete. *Enzyme and Microbial Technology, 34,* 635–641.

Junior, J.A., Vieira, Y.A., Cruz, I.A., et al. (2020). Sequential degradation of raw vinasse by a laccase enzyme producing fungus Pleurotus sajor-caju and its ATPS purification. *Biotechnology Reports, 25,* 4–11.

Kacem, I., Koubaa, M., Maktouf, S., et al. (2016). Multistage process for the production of bioethanol from almond shell. *Bioresource Technology, 211,* 154–163.

Kannaiyan, R., Mahinpey, N., Mani, T., Martinuzzi, R.J., & Kostenko, V. (2012). Enhancement of Dichomitus squalens tolerance to copper and copper-associated laccase activity by carbon and nitrogen sources. *Biochemical Engineering Journal, 67,* 140–147.

Khammuang, S., Yuwa-Amornpitak, T., & Svasti, J. (2013). Biocatalysis and Agricultural Biotechnology Copper induction of laccases by Lentinus polychrous under liquid-state fermentation. *Biocatalysis and Agricultural Biotechnology, 2*(4), 357–362. Elsevier. Disponível em: 10.1016/j.bcab.2013.05.004.

Knop, D., Yarden, O., & Hadar, Y. (2014). The ligninolytic peroxidases in the genus Pleurotus: divergence in activities, expression, and potential applications. *Applied Microbiology and Biotechnology, 99*(3), 1025–1038.

Kocyigit, A., Pazarbasi, M.B., Yasa, I., Ozdemir, G., & Karaboz, I. (2012). Production of laccase from Trametes trogii TEM H2: A newly isolated white-rot fungus by air sampling. *Journal of Basic Microbiology, 52*(6), 661–669.

Kudanga, T., Nyanhongo, G.S., Guebitz, G.M., & Burton, S. (2011). Potential applications of laccase-mediated coupling and grafting reactions: A review. *Enzyme and Microbial Technology, 48(3),* 195–208.

Kuhar, F., & Papinutti, L. (2014). Optimization of laccase production by two strains of Ganoderma lucidum using phenolic and metallic inducers. *Revista Argentina de Microbiología, 46*(2), 144–149.

Kumar, A., & Chandra, R. (2020). Ligninolytic enzymes and its mechanisms for degradation of lignocellulosic waste in environment. *Heliyon, 6*(2), e03170. Elsevier Ltd. Disponível em: 10.1016/j.heliyon.2020.e03170.

Kumar, V.V., Sivanesan, S., & Cabana, H. (2014). Magnetic cross-linked laccase aggregates – bioremediation tool for decolorization of distinct classes of recalcitrant dyes. *Science of the Total Environment, 487*(1), 830–839.

Lavelli, V., Proserpio, C., Gallotti, F., Laureati, M., & Pagliarini, E. (2018). Circular reuse of bio-resources: The role of: Pleurotus spp. in the development of functional foods. *Food and Function, 9*(3), 1353–1372.

Lee, I.Y., Jung, K.H., Lee, C.H., & Park, Y.H. (1999). Enhanced production of laccase in Trametes vesicolor by the addition of ethanol. *Biotechnology Letters, 21*(11), 965–968.

Lemes, A.C., Silvério, S.C., Rodrigues, S., & Rodrigues, L.R. (2019). Separation and Puri fi cation Technology Integrated strategy for puri fi cation of esterase from Aureobasidium pullulans. *Separation and Purification Technology, 209*(June 2018), 409–418. Disponível em: 10.1016/j.seppur.2018.07.062.

Lettera, V., Pezzella, C., Cicatiello, P., et al. (2016). Efficient immobilization of a fungal laccase and its exploitation in fruit juice clarification. *Food Chemistry, 196,* 1272–1278.

Liu, J., Yu, Z., Liao, X., et al. (2016). Scalable production, fast purification, and spray drying of native Pycnoporus laccase and circular dichroism characterization. *Journal of Cleaner Production, 127,* 600–609.

Malhotra, M., & Suman, S.K. (2021). Laccase-mediated delignification and detoxification of lignocellulosic biomass: removing obstacles in energy generation. *Environmental Science and Pollution Research, 1,* 1–16.

Manavalan, T., Manavalan, A., Thangavelu, K.P., & Heese, K. (2013). Characterization of optimized production, purification and application of laccase from Ganoderma lucidum. *Biochemical Engineering Journal, 70,* 106–114. Elsevier B.V. Disponível em: 10.1016/j.bej.2012.10.007.

Mann, J., Markham, J.L., Peiris, P., et al. (2015). Use of olive mill wastewater as a suitable substrate for the production of laccase by Cerrena consors. *International Biodeterioration and Biodegradation.*

Marques De Souza, C.G., & Peralta, R.M. (2003). Purification and characterization of the main laccase produced by the white-rot fungus Pleurotus pulmonarius on wheat bran solid state medium. *Journal of Basic Microbiology, 43*(4), 278–286.

Mayolo-Deloisa, K., González-González, M., & Rito-Palomares, M. (2020). Laccases in Food Industry: Bioprocessing, Potential Industrial and Biotechnological Applications. *Frontiers in Bioengineering and Biotechnology, 8*, 222.

Morsi, R., Bilal, M., Iqbal, H.M.N., & Ashraf, S.S. (2020). Laccases and peroxidases: The smart, greener and futuristic biocatalytic tools to mitigate recalcitrant emerging pollutants. *Science of the Total Environment, 714,* 136572.

Nadar, S.S., Pawar, R.G., & Rathod, V.K. (2017). International Journal of Biological Macromolecules Recent advances in enzyme extraction strategies: A comprehensive review. *International Journal of Biological Macromolecules, 101,* 931–957. Elsevier B.V. Disponível em: 10.1016/j.ijbiomac.2017.03.055.

National Supply Company —CONAB. (2019). Monitoring of the Brazilian sugarcane crop, 2018/2019 crop—third survey. Brasilia.

Nunes, C.S., & Kunamneni, A. (2018). Laccases-properties and applications. *Enzymes in Human and Animal Nutrition: Principles and Perspectives,* 133–161.

Ozcirak Ergun, S., & Ozturk Urek, R. (2017). Production of ligninolytic enzymes by solid state fermentation using Pleurotus ostreatus. *Annals of Agrarian Science, 15*(2), 273–277.

Parsaee, M., Kiani Deh Kiani, M., & Karimi, K. (2019). A review of biogas production from sugarcane vinasse. *Biomass and Bioenergy, 122,* 117–125.

Passarini, M.R.Z., Ottoni, C.A., Santos, C., Lima, N., & Sette, L.D. (2015). Induction, expression and characterisation of laccase genes from the marine-derived fungal strains Nigrospora sp. CBMAI 1328 and Arthopyrenia sp. CBMAI 1330. *AMB Express, 5*(1).

Patel, A.K., Singhania, R.R., & Pandey, A. (2017). *Production, Purification, and Application of Microbial Enzymes, 1,* 13–41.

Patrick, F., Mtui, G., Mshandete, A.M., & Kivaisi, A. (2011). Optimization of laccase and manganese peroxidase\rproduction in submerged culture of Pleurotus sajorcaju\r. *African Journal of Biotechnology, 10*(50), 10166–10177. Disponível em: http://www.academicjournals.org/AJB.

Piscitelli, A., Giardina, P., & Lettera, V., et al. (2011). Induction and Transcriptional Regulation of Laccases in Fungi. *Current Genomics, 12*(2), 104–112.

Polak, J., Jarosz-Wilkolazka, A., Szuster-Ciesielska, A., et al. (2016). Toxicity and dyeing properties of dyes obtained through laccase-mediated synthesis. *Journal of Cleaner Production, 112,* 4265–4272.

Przystaś, W., Zabłocka-Godlewska, E., & Grabińska-Sota, E. (2018). Efficiency of decolorization of different dyes using fungal biomass immobilized on different solid supports. *Brazilian Journal of Microbiology, 49*(2), 285–295.

Rajagopalu, D., Show, P.L., Tan, Y.S., et al. (2016), Recovery of laccase from processed Hericium erinaceus (Bull.: Fr) Pers. fruiting bodies in aqueous two-phase system. *Journal of Bioscience and Bioengineering, xx*(xx), 6–11. Elsevier Ltd. Disponível em: 10.1016/j.jbiosc.2016.01.016.

Reddy, G.V., Ravindra Babu, P., Komaraiah, P., Roy, K.R.R.M., & Kothari, I.L. (2003). Utilization of banana waste for the production of lignolytic and cellulolytic enzymes by solid substrate fermentation using two Pleurotus species (P. ostreatus and P. sajor-caju). *Process Biochemistry, 38*(10), 1457–1462.

Renzetti, S., Courtin, C.M., Delcour, J.A., & Arendt, E.K. (2010). Oxidative and proteolytic enzyme preparations as promising improvers for oat bread formulations: Rheological, biochemical and microstructural background. *Food Chemistry, 119*(4), 1465–1473.

Riebel, M., Sabel, A., Claus, H., et al. (2017). Antioxidant capacity of phenolic compounds on human cell lines as affected by grape-tyrosinase and Botrytis-laccase oxidation. *Food Chemistry, 229,* 779–789.

Robles-González, V., Galíndez-Mayer, J., & Rinderknecht-Seijas, N. (2012). Treatment of mezcal vinasses: A review. *Journal of biotechnology, 157,* 524–546.

Rodgers, C.J., Blanford, C.F., Giddens, S.R., et al. (2010). Designer laccases: a vogue for high-potential fungal enzymes? *Trends in Biotechnology, 28*(2), 63–72.

Rodríguez-Couto, S. (2019). *Fungal Laccase: A Versatile Enzyme for Biotechnological Applications* 429–457.

Romanholo Ferreira, L.F., Torres, N.H., Armas, R.D.D.E., et al. (2020a). Fungal lignin-modifying enzymes induced by vinasse mycodegradation and its relationship with oxidative stress. *Biocatalysis and Agricultural Biotechnology, 27,* 101–691.

Romanholo Ferreira, L.F., Torres, N.H., Armas, R.D.D.E., et al. (2020b). Fungal lignin-modifying enzymes induced by vinasse mycodegradation and its relationship with oxidative stress. *Biocatalysis and Agricultural Biotechnology*, 27, 101–691. Elsevier Ltd.

Rosen, M.A. (2018). Environmental sustainability tools in the biofuel industry. *Biofuel Research Journal*, 5(1), 751–752.

Rouches, E., Herpoël-Gimbert, I., Steyer, J.P., & Carrere, H. (2016). Improvement of anaerobic degradation by white-rot fungi pretreatment of lignocellulosic biomass: A review. *Renewable and Sustainable Energy Reviews*, 59, 179–198. Disponível em: 10.1016/j.rser.2015.12.317.

Ruiz, C.A.S., Van Den Berg, C., Wijffels, R.H., & Eppink, M.H.M. (2017). Rubisco separation using biocompatible aqueous two-phase systems. *Separation and Purification Technology*. Disponível em: 10.1016/j.seppur.2017.05.001.

Sadeghi, S.H., Hazbavi, Z., & Harchegani, M.K. (2016). Controllability of runoff and soil loss from small plots treated by vinasse-produced biochar. *Science of the Total Environment*, 541, 483–490.

Saha, B.C., Qureshi, N., Kennedy, G.J., & Cotta, M.A. (2016). Biological pretreatment of corn stover with white-rot fungus for improved enzymatic hydrolysis. *International Biodeterioration and Biodegradation*, 109, 29–35. Disponível em: 10.1016/j.ibiod.2015.12.020.

Saito, T., Hong, P., Kato, K., et al. (2003). Purification and characterization of an extracellular laccase of a fungus (family Chaetomiaceae) isolated from soil. *Enzyme and Microbial Technology. Anais….* 33, 520–526.

Sampaio, L.M.P., Padrão, J., Faria, J., et al. (2016). Laccase immobilization on bacterial nanocellulose membranes: Antimicrobial, kinetic and stability properties. *Carbohydrate Polymers*, 145, 1–12.

Sánchez-Trasviña, C., Mayolo-Deloisa, K., González-Valdez, J., & Rito-Palomares, M. (2017). Refolding of laccase from Trametes versicolor using aqueous two phase systems: Effect of different additives. *Journal of Chromatography A*. Elsevier B.V. Disponível em: 10.1016/j.chroma.2017.05.023.

Sartori, S.B., Ferreira, L.F.R., Messias, T.G., et al. (2015). Pleurotus biomass production on vinasse and its potential use for aquaculture feed. *Mycology*, 6(1), 28–34.

Schwienheer, C., Prinz, A., Zeiner, T., Merz, J. (2015). Separation of active laccases from Pleurotus sapidus culture supernatant using aqueous two-phase systems in centrifugal partition chromatography. *Journal of Chromatography B*, 1002, 1–7. Elsevier B.V. Disponível em: 10.1016/j.jchromb.2015.07.050.

Sharma, A., Thakur, V.V., Shrivastava, A., et al. (2014). Xylanase and laccase based enzymatic kraft pulp bleaching reduces adsorbable organic halogen (AOX) in bleach effluents: A pilot scale study. *Bioresource Technology*, 169, 96–102.

Shraddha, Shekher, R., Sehgal, S., Kamthania, M., & Kumar, A. (2011). Laccase: Microbial sources, production, purification, and potential biotechnological applications. *Enzyme Research*, 1.

Silva Da, B.P., Gomes Correa, R.C., & Kato, C.G. (2017). Characterization of a Solvent-tolerant Manganese Peroxidase from Pleurotus pulmonarius and its Application in Dye Decolorization. *Current Biotechnology*, 6(4). Disponível em: http://www.eurekaselect.com/139753/article.

Silva, D.F.C., Azevedo, A.M., Fernandes, P., Chu, V., & Conde, J.P. (2016). Determination of partition coefficients of biomolecules ina microfluidic aqueous two phase system platform using fluorescence microscopy. *Journal of Chromatography A*. Disponível em: 10.1016/j.chroma.2016.12.036.

Singh, G., & Arya, S.K. (2019). Utility of laccase in pulp and paper industry: A progressive step towards the green technology. *International Journal of Biological Macromolecules*, 134, 1070–1084.

Sousa, R.M.O.F., Amaral, C., Fernandes, J.M.C., et al. (2019). Hazardous impact of vinasse from distilled winemaking by-products in terrestrial plants and aquatic organisms. *Ecotoxicology and Environmental Safety*, 183(July), 109493. Elsevier Inc. Disponível em: 10.1016/j.ecoenv.2019.109493.

Taylor, P., Strong, P.J., & Claus, H. (2011). Critical Reviews in Environmental Science and Technology Laccase: A Review of Its Past and Its Future in Bioremediation Laccase: A Review of Its Past and Its Future in Bioremediation. *Critical Reviews in Environmental Science and Technology*, 41(March 2013), 37–41.

Torres-Acosta, M.A., Pereira, J.F.B., Freire, M.G., et al. (2018). Economic evaluation of the primary recovery of Tetracycline with traditional and novel aqueous two- phase systems. *Separation and Purification Technology*. Disponível em: 10.1016/j.seppur.2018.04.041.

Unuofin, J.O., Okoh, A.I., & Nwodo, U.U. (2019). Aptitude of oxidative enzymes for treatment of wastewater pollutants: A laccase perspective. *Molecules*, *24(11)*, 2064.

Vilar, D.S., Carvalho, G.O., Pupo, M.M.S., et al. (2018). Vinasse degradation using Pleurotus sajor-caju in a combined biological – Electrochemical oxidation treatment. *Separation and Purification Technology*, *192*(January), 287–296. Elsevier.

Vilar, D.S., Cruz, I.C., Torres, N.H., et al. (2019). Agro-industrial Wastes: Environmental Toxicology, Risks, and Biological Treatment Approaches. *Environmental Contaminants: Ecological Implications and Management*, 1–23. Singapore: Springer.

Voběrková, S., Solčány, V., Vršanská, M., & Adam, V. (2018). Immobilization of ligninolytic enzymes from white-rot fungi in cross-linked aggregates. *Chemosphere*, *202*, 694–707. Disponível em: 10.1016/j.chemosphere.2018.03.088.

Xu, L., Sun, K., Wang, F., et al. (2020). Laccase production by Trametes versicolor in solid-state fermentation using tea residues as substrate and its application in dye decolorization. *Journal of Environmental Management*, *270*.

Younes Ben, S., & Sayadi, S. (2011). Purification and characterization of a novel trimeric and thermotolerant laccase produced from the ascomycete Scytalidium thermophilum strain. *Journal of Molecular Catalysis B: Enzymatic*, *73*(1–4), 35–42.

Yousefi-Ahmadipour, A., Bozorgi-Koshalshahi, M., Mogharabi, M., et al. (2016). Laccase-catalyzed treatment of ketoconazole, identification of biotransformed metabolites, determination of kinetic parameters, and evaluation of micro-toxicity. *Journal of Molecular Catalysis B: Enzymatic*, *133*, 77–84.

Zabed, H., Sahu, J.N., Suely, A., Boyce, A.N., & Faruq, G. (2017). Bioethanol production from renewable sources: Current perspectives and technological progress. *Renewable and Sustainable Energy Reviews*.

Zhuo, R., Yuan, P., Yang, Y., Zhang, S., & Ma, F. (2017). Induction of laccase by metal ions and aromatic compounds in Pleurotus ostreatus HAUCC 162 and decolorization of different synthetic dyes by the extracellular laccase. *Biochemical Engineering Journal*, *117*, 62–72. Disponível em: 10.1016/j.bej.2016.09.016.

Zucca, P., Rescigno, A., Olianas, A., et al. (2011). Induction, purification, and characterization of a laccase isozyme from Pleurotus sajor-caju and the potential in decolorization of textile dyes. *Journal of Molecular Catalysis B, Enzymatic*, *68*(2), 216–222. Elsevier B.V. Disponível em: 10.1016/j.molcatb.2010.11.008.

15 Enhanced CO_2 Assimilation by *Engineered* Escherichia coli (E. coli)

Navamallika Gogoi[1], Moharana Choudhury[2], Anwesha Gohain[3], and Anu Sharma[4]
[1]Department of Chemistry, Arunachal University of Studies, Namsai, A.P., India
[2]Voice of Environment (VoE), Guwahati, Assam, India
[3]Department of Botany, Arunachal University of Studies, Namsai, A.P., India
[4]Govt. Degree College, Bhaderwah, Doda, University of Jammu, Union Territory of J&K, India

CONTENTS

15.1 Introduction 281
15.2 Natural Carbon Biosequestration 282
15.3 Designed C Biosequestration 284
15.4 Limitation of Natural C Biosequestration Pathway 285
15.5 CO_2-Utilising Microbes 286
15.6 Characteristics of *Escherichia coli* 287
15.7 Engineered *Escherichia coli* for Enhancing CO_2 Biomitigation 288
15.8 Limitations, Challenges 289
15.9 Future Research Perspectives 291
15.10 Conclusion 291
References 292

15.1 Introduction

In present days, the biggest challenge of mankind is to slow down carbon dioxide emissions as well as to reverse global warming. In this context, we need innovative ideas for capturing carbon dioxide and bio-based technologies will play a key role where microbial CO_2 sequestration is one of the innovative ideas among them. Different types of chemical, physical, and biological process have been reported for CO_2 sequestration (Orsini and Marrone, 2019). Various physical and chemical CO_2 sequestration strategies (cryogenic distillation, absorption, adsorption, chemical looping combustion, oxyfuel combustion, hydrate-based CO_2 separation) have been applied to reduce CO_2 from the environment but these approaches are expensive, have low efficiency, require intensive energy, are unsustainable, and complex (Salehizadeh et al., 2020), whereas in biological processes CO_2 sequestration is a clean, eco-friendly, and prominent approach (Salmon et al., 2009). Biological CO_2 sequestration is basically a natural process and performed by living organisms although few engineered strategies are also employed. Despite the fact biosequestration is a natural carbon fixation approach, it is not able to be utilised by CO_2 into value-added products from a commercial point of view. In addition, natural biosequestration is influenced by the lack of efficiency of the CO_2-fixing enzymes and pathways under favourable environments. Engineered or synthesised biosequestration mainly focuses on designing and relocating natural CO_2-fixing pathways by enhancing CO_2 supply, developing and enhancing CO_2-fixing enzyme activity and stability to achieve successful CO_2-fixing. Also, conversion of CO_2 into different types of value-added products with the help of engineered biochemical pathway plays a significant role in synthetic biology.

Microbial CO_2 fixation has attracted the attention in order to turn CO_2 into a high-quality commodity because it uses organic organisms and biocatalysts to encourage an eco-friendly behaviour (Mohan et al., 2016). The microbial conversion for biodynamic chemical processing was very much noted for the mild reactions and sustainability conditions (Buschke et al., 2013). *Rhodobacter sphaeroides*, *Acetabacterium woodii*, *Ralstonia eutropha*, *Clostridium kluyveri*, *Chlorella vulgaris*, *Rhodococci*, *Synechococcus elongatus*, *Clostridium ijungdahlii*, and *Clostridium aceticum* are different types of microbes that help to sequester CO_2 (Bhatia et al., 2019).

Various biological mechanisms such as bicarbonate hydration by carbonic anhydrase (CA) hydrogenation of formate, adaptation of CO_2 into methanol, RubisCo enzyme reduction into methane are employed through the transformation of carbon dioxide into different value-added products. Different C assimilation pathways used by microbes are summarised in Table 15.1. Phosphoribulokinase (PRK) and ribulose-1, 5-bisphosphate carboxylase (RubisCo) enzyme of C-B-B cycle have been initiates into *Escherichia coli* and *Saccharomyces cerevisiae* to assist CO_2 cycling as well as boost ethanol synthesis.

Basically, there are two types of CO_2-fixing microorganisms and they are autotrophic and heterotrophic CO_2-fixing microorganisms. Autotrophic microorganisms are the principal producers of food which produce their food with the help of sunlight, carbon dioxide, and chlorophyll present in the plants (as green pigment) through the process of photosynthesis; heterotrophs cannot synthesise their own food and rely on other organisms. The basic carbon fixation efficiency in autotrophic species is relatively poor and limited energy is available (Gong et al., 2018). Heterotropic species typically have the advantage that they normally have higher growth and development rates than the autotrophic life cycle. Therefore, researchers attempted to reproduce the entire carbon fixation route in the heterotrophs. The challenge of improving efficient development of bio-based CO_2 products is up to genetic engineering today (Salehizadeh et al., 2020).

The first microbial carbon fixation of *Propioibacterium pentosaceum* was documented by Wood and Werkman in 1940. Later on, CO_2 fixation among several heterotrophic microorganisms like *Escherichia coli* was shown to be widespread. Recent studies have shown that integrating a natural carbon assimilation pathway in several steps into a heterotrophic organism could establish a bypass CO_2-fixation route that allows the host to absorb CO_2 at the cost of carbohydrates. *Escherichia coli* is the ideal host in strict heterotrophics. Due to its rapid development, well-studied physiology and various genetic resources, *Escherichia coli* is considered one of the most significant biotechnological platforms. Many value added chemicals have been synthesised successfully using *Escherichia coli* as well as biodegradable polymers (Chen et al., 2013). Reconstruction of the *tetrahydroflote* cycle and reverse glycine cleavage pathway construct an modified *Escherichia coli* strain that can sustain growth on CO_2 and formic acid. As *Escherichia coli* grow in the absence of oxygen, it can produce a metal containing enzyme called FHL that can convert gaseous CO_2 to liquid formic acid. Engineered *Escherichia coli* cells to express on the membrane of CA and used a foam bioreactor to assimilate carbon dioxide.

Researchers quantify that the CO_2-fixing capacity of engineered heterotrophic *Escherichia coli* match with autotrophic cyanobacteria and algae (Gong et al., 2015). The *Escherichia coli* distinctive CO_2-fixing concentrations are superior to most of the autotrophs due to their extreme development rate and rich genetic platform. Because *Escherichia coli* can reach a high cell density in a bioreactor when correctly managed, this heterotroph may serve as an alternate pathway for carbon dioxide fixation, with great potential. Expressed of type I RuBisCO has a notable impact on the metabolism of *Escherichia coli* under diverse fermentation conditions. Apart from this homologous expression of Park, CA, and FDH in *Escherichia coli*, it was also reported that *Escherichia coli* had CA and FDH.

Compared to a standard solution, engineered biosequestration produces better efficacies. Biologically active carbon-fixation technologies are rapidly developing, focusing on the investigation of *E. coli* as a research topic.

15.2 Natural Carbon Biosequestration

Biological sequestration requires specifically the sequestration of biotic oceans and terrestrial sequestration. Ocean fertilisation is basically involving *phytoplanktonic* photosynthesis process. It improved the growth and production of *phytoplanktons* by maximising the concentration of nutrients and

TABLE 15.1
Natural carbon fixation pathways

Pathway	CBB Cycle	rTCA Cycle	WL Pathway	3HP	Di-4HB Cycle	3HP-4HB Cycle
Energy	Light	Light and Sulfur	Hydrogen	Bicycle light	Hydrogen & sulfur	Hydrogen & sulfur
Key reactions	$3CO_2 + 6NAD(P)H + 9ATP + 5 H_2O \rightarrow G3P + 9ADP + 6NAD(P)^+ + 8Pi$	$2CO_2 + 2ATP + 2NAD(P)H + FADH + Fd_{red} + CoASH \rightarrow$ acetyl-CoA $+ 2ADP + 2Pi + 2NAD(P)^+ + FAD + + Fd_{ox}$	$2CO_2 + 2NAD(P)H + ATP + + 2Fd_{red} + HSCoA \rightarrow$ acetyl-CoA $+ ADP + Pi + 2NADP + + 2Fd_{ox}$	$3HCO_3^- + 5ATP + + 5NAD(P)H \rightarrow$ Pyruvate $+ 3ADP + 2AMP + + 2PPi + 5NAD(P)^+ + 3Pi$	$CO_2 + HCO_3^- + NAD(P)H + 3 ATP + Fd_{red} + 4MV_{red} + HSCoA \rightarrow$ acetyl-CoA $+ 2ADP + AMP +$ diphosphate $+ 2PPi + NAD(P)^+ + Fd_{ox} + 4MV_{ox}$	$2HCO_3^- + 4ATP + CoASH + 4NAD(P)H \rightarrow$ acetyl-CoA $+ 3ADP + 3Pi + AMP + PPi + 4NADP^+$
CO_2-fixing enzymes	RuBisCo, *Phosphoribulokinase*	2-Oxoglutarate synthase, Phosphoenolpyruvate carboxylase Isocitrate dehydrogenase	Formate dehydrogenase & COdehydrogenate/ AcetylCoA synthase	Acetyl-CoA-propionyl/ CoAcarboxylase, Malyl-CoA sLyas	Pyruvate synthase & Phosphoenolpyruvate carboxylase	Acetyl-CoA carboxylase & Propionyl-CoA carboxylase
Discoverd	1948	1966	1972	1993	2007	2008

CBB: Calvin-Benson-Bassham cycle, rTCA: Reductive citric acid cycle, WL pathway: Wood-Ljundahl pathway, 3HP:3-hydroxypropionate bicycle, Di/4HB:Dicarboxylate/4-hydroxbutyrate cycle, 3HP/4HB:3 Hydroxypropionate/4-hydroxybutyrate cycle.

eventually promotes carbon sequestration (Nogia et al., 2013). *Phytoplankton* photosynthesis requires large amount of micronutrient and macronutrients and also light and CO_2. Wolff et al. reported that iron fertilisation enhances the growth of *phytoplanktons* in low chlorophyll marine water (Wolff et al., 2011).

Oceans will take up about 2 billion metric tonnes of CO_2 each year, the amount of carbon that would double the atmospheric load. In recent years, plans for ocean fertilisation have been contentious because of potential risks and benefits related to the possible implementation of the large-scale carbon sequestration process (Galaz, 2012) because ocean fertilisation has some technical complications, is very expensive, and unaffordable (Aumont and Bopp, 2006). Nevertheless ocean fertilisation could be stored and decompose organic matter.

To contribute to climate change mitigation, terrestrial C sequestration, a mechanism by which plants uptake atmospheric CO_2 by photosynthesis and then store it in biomass and soils, might considerably reduce CO_2 in the atmosphere.

While much of the earth's carbon is contained by the oceans, soils store 75% of the total carbon pool on land, which is three times higher than all the vegetation and also two times more than that present in the atmosphere (Batjes and Sombroek, 1997). Therefore, a significant role is required to be played in ensuring a balanced global carbon cycle. Soils contain up to 100 cm depth 1,500 Pg C and up to 200 cm depth 2,500 Pg C (1 Pg = 1 × 1,015 g = 1 giga tonne) compared to 650 Pg C in vegetation and 750 Pg C in the atmosphere. The difference between the input from dead plant materials (mainly leaves and roots) and the output from decomposition and mineralisation of soil organic carbon to the atmosphere is known as carbon storage in soils. It has been shown that much of the C in the soil is labile in nature if the conditions are aerobic and is thus likely to go back to the atmosphere via the processes of CO_2 efflux or soil respiration. It has been shown that only 1.0% (0.4 Pg yr^{-1}) of carbon (55 Pg yr^{-1}) entering the soil from different sources eventually accumulates in stable fractions and is maintained for a long period of residence in the soil. This is the sum of soil carbon sequestered in the soil that can be claimed to be, and it is part of the global carbon balance. The changes in soil organic C due to agriculture and related activities are also found to influence the overall stock of terrestrial C and are an important factor to be considered for the associated global changes.

Another terrestrial sequestration of atmospheric CO_2 capture is phytosequestration. It has been reported that the contribution of vegetation and microbial biomass to 99.9% carbon biota. In a properly maintained and undisturbed plantation, phytosequestration could serve as a suitable carbon reservoir for many centuries. In reality, deforestation is one of the primary contributors to climate change and also a significant effect on the enhancement of atmospheric CO_2, so it might reverse the process by eliminating the source. Photosynthesis is a natural mechanism by which inorganic carbon is converted into organic carbon. So, in the presence of solar power, CO_2 goes into sugar and thus atmospheric carbon is distributed to different parts of the plant. It was expected that atmospheric carbon sequestration would increase by improving photosynthetic efficiencies, either by using natural or synthetic perspectives.

15.3 Designed C Biosequestration

Appropriate and financially feasible methods can be built using synthetic biology (Jajesniak et al., 2014). By modifying active biochemical pathways and introducing heterologous gene expression into organisms with desirable traits, synthetic biology has developed as a novel method for living organisms.

Heterotrophic model organisms are, in addition to natural pathways, an attractive medium for testing and building synthetic pathways. Genetic engineering approaches in carbon-reducing perspectives are basically a combined effect of biological and non-biological methods. Numerous genetically modified technologies are employed for enhancing photosynthetic activity of C3 plant. In the case CO_2 fixation in C3 photosynthesis (3-phophoglycerate), RuBisCO enzymes performance needs to be improved to increase the photosynthesis rate. So RuBisCO should be modified so that its specificity decreases towards CO_2, in order to reduce photorespiration, and to enhance the catalytic rate such that CO_2 assimilation is improved.

Another strategy is using cyanobacterial bicarbonate transporters on the chloroplast membrane of C3 plants. Cyanobacterial cells have many CO_2 absorption transporters, whereas C3 plants only have stomata for gas exchange. By having an efficient transporter system in C3 plants, the photosynthetic yield as well as greater CO_2 sequestration could be increased.

Enhanced CO₂ Assimilation

On the basis of four principles, synthetic CO_2-fixing pathways can be built: They should consist of few enzymes, independently linked to central metabolism, favourable thermodynamics and superior kinetics, and capable of working under both aerobic and anaerobic circumstances (Bar-Even et al., 2013).

Heterotrophic model organisms are in addition to natural pathways that are effective and trendy tools in analyzing and building artificial conduits. In order to achieve artificial CO_2 fixation and also additional carbon molecules such as formate, several such designs were suggested. Attempts to boost the biological CO_2 fixation have been focused on the evolution and specificity of RuBisCO in non-autotrophic organisms such as *Escherichia coli*, developing effective photorespiration and transplantation of natural pathways of CO_2 fixation. The implementation of synthetic pathways therefore involves a new strategy, which first allows for more detailed testing and optimisation conditions (Schwander et al., 2016). In living organisms, the complex association of various enzymes is under the threat of synthetic pathways.

15.4 Limitation of Natural C Biosequestration Pathway

Naturally occurring species used in carbon sequestration are often hampered by their adverse features, such as weak yields, costly conditions of cultivation, and slow growth. A comparative analysis is provided in Table 15.2 for the different biological carbon sequestration strategies. Some significant limitations of natural C sequestration are summarised:

a. Because of low efficiency and limited genetic modification capacity, industrial demand cannot be met by natural carbon fixation approaches.
b. There are multiple reaction steps to natural C biosequestration caused by engineered biology, discovering new natural carbon-fixation methods, and developing new artificial carbon assimilation pathway.
c. Natural CO_2 biosequestration primarily provides biomass, not a dedicated commodity. Moreover, under optimum conditions, the inefficiency of CO_2-fixing enzymes and pathways also directly affects biological CO_2 fixation; for instance, RuBisCO (ribulose1, 5bisphosphate carboxylase/oxygenase). The CBB cycle's carboxylase is a sluggish enzyme with a strong side reaction with oxygen, resulting in a process known as photorespiration, which results in a 30% deficient of fixed carbon and hence photosynthetic energy (Walker et al., 2016).

TABLE 15.2

Advantage and limitation of biological carbon sequestration and synthetic carbon sequestration

Biological C Sequestration	Advantages	Limitations
Ocean fertilisation	Improving *phytoplanktonic* CO_2 fixation with carbon sequestration. Could decompose organic matter.	Possible risks on large-scale carbon sequestration process, impact on ocean ecosystem, very expensive, and unaffordable.
Terrestrial		
a. Soil carbon sequestration	Improved soil structure, reduce soil erosion and degradation	Considerably slow process. Only reflect CO_2 emissions, not other global warming gas.
b. Phyto-sequestration	Huge storage capability, natural photosynthesis process contributes to cost-effective carbon sequestration	Photosynthetic production and planting must be improved globally.
Engineering C sequestration	Contributes indirectly to carbon sequestration enhancing natural photosynthetic yield	If efficient, performance will be tremendous, but more research is required.

TABLE 15.3

CO_2-fixing microorganisms

Bacterial	Archaeal	
• Cyanobacteria:	• Euryarchaeota: (Methanococcales, Methanobacteriales, Methanopyrales, Methanocellales, Methanomicrobiales, Archaeoglobales)	• Crenarchaeota: (Sulfolobale, Desulfurococcales, Cenarchaeales, Thermoproteales)
• Aquificae		
• Actinobacteria: (Acidimicrobiales, Actinomycetales)		
• Chloroflexi: (Chloroflexales)		
• Proteobacteria: (Desulfobacterales, Desulfovibrionales, Burkholderiales, Nitrosomonadales, Hydrogenophilales, Campylobacterales, Nautiliales Campylobacterales, Thiotrichales, Acidithiobacillales Pseudomonadales Pasteurellales Rhizobiales, Rhodobacterales Rhodospirillales)		
• Chlorobia: (Chlorobiales)		
• Firmicutes: (Clostridiales, Lactobacillales, Bacillales)		
• Thermodesulfobacteria: (Thermodesulfobacteriales)		

15.5 CO_2-Utilising Microbes

There are numerous microorganisms that can carry out CO_2 fixation through different pathways. Nutrients, pH, temperature, buffer capacity, bioreactors, etc. are influences on microbial CO_2 fixation capacity (Khandavalli et al., 2018). Researchers have been accounted for various engineering pathways in autotrophic microorganisms to improve genetic, autotrophic capacities in heterotrophic microbes (e.g., *E. coli, S. cerevisiae*) (Claassens et al., 2016, Antonovsky et al., 2016). Microorganisms that carry out carbon dioxide fixation can be categorised into photosynthetic and non-photosynthetic. The taxonomy of CO_2-fixing microorganisms are summarised in Table 15.3. While Hu et al. separated CO_2-fixing microbes into autotrophs and heterotrophs, the autotrophic microbes can propagate, utilising carbon dioxide as a single source of carbon; a large portion of this has a place with bacterial and archaeal community (e.g., *Cyano bacteria, Crenarchaeota, Betaproteo bacteria*).

Photosynthetic microorganisms, particularly cyanobacteria and microalgae, are widely used for the study of CO_2 fixation. Biosequestration by cyanobacteria in photobioreactors is a promising technique; *Synechococcus* has accomplished CO_2 assimilation of 0.6 g/l-day (Nouha et al., 2015). *Synechocystis* sp. PCC 6803, *S. elongatus* PCC 7942, and *Anabaena* sp. are various model strains of cyanobacteria (Angermayr et al., 2015). Microalgae are among the quickest developing microorganisms. In addition, it has high photosynthetic efficiencies and is appropriate to the carbon reduction strategy (Kurano et al., 1996). Various models of microalgae are *C. reinhardtii, T. pseudonana*, and *P. tricornutum*. Acetogenic microorganisms are anaerobic chemolitho autotrophs and have considerably improved CO_2-fixing abilities (Liu et al., 2017). *Clostridium ljungdahlii, Clostridium autoethanogenum, C. Carboxydivorans, E. limosum,* and *Acetobacterium woodii* are the most generally used microbes for fixing CO_2. Apart from this, *Rhodobacter sphaeroides*, an anoxygenic photoautotroph, has also been used in synthetic biology (Gonzales et al., 2019), while most of the *Euryarchaeota* are methanogens (*Archaeoglobales, Methanomicrobiales, Methanobacteriales, Methanococcales, Methanocellales, Methanopyrals, Methanopyrales*), i.e., they produce methane as a metabolic by-product to remediate carbon dioxide sequestration.

CO₂ fixation through microbial means, on a marketable level, is currently inadequate in applications. Micro-algae, producing up to 5,000 tonnes of biomass dry algal biomes each year, are the key strains for fixing CO_2 used in the industry (Borowitzka, 2013). A large number of additional valuable goods received by CO_2-fixing strains could be manufactured on a commercial scale in the future through the advancement of CO_2 sequestration technologies. Either terrestrial plants or microbes biologically assimilate carbon dioxide. Both physical and chemical approaches have some storage limits, while biological means continuously turn them into an organic biomass. CO_2-fixing microorganisms are essentially identical to plants in the carbon cycle. They may behave as principal or self-producing organisms well within the great depths of the seas or oceanic thermal-vent arrangement due to their chemoautotrophic existence. These species thrive on the hydrothermal solutions and gases dissolved therein from hydrothermal winds in these uninhabited habitats. Carbon dioxide is the sole source of cell carbon; therefore, they possess the capability to be used and applied in bringing forth a variety of know-hows to extract carbon dioxide from different production units. We know factories are the biggest producers of CO_2; hence, the know-hows and techniques involved here would undoubtedly work against the increasing concentration of CO_2 by adding them to the source. In addition to offering global warming solutions, biomass manufactured by microorganisms that fix CO_2 can also be used in the development of bioenergy, biodiesel, and biohydrogen. Such processes may act as substitutes to traditional sources of energy coal, petroleum, and natural gas. Thus, they alleviate to a certain degree the burden of fossil fuel consumption. *Relostonia eutropha* H16 has been used lately in decomposable plastic products by poly[R-(–)-3-hydroxybutyrate] (Pohlmann et al., 2006). Thus, these microorganisms can solve many environmental problems.

15.6 Characteristics of *Escherichia coli*

In 1885, Theodor Escherich isolated *Escherichia coli* from human newborns for the first time (Shulman et al., 2007). Gram-negative short rods facultative anaerobic gamma proteobacterium *Escherichia coli* DNA are binding the proteins. (HU, H1) are found in *Escherichia coli* and aid in DNA replication (Feng et al., 2002). *Escherichia coli*, a common bacterium found in mammalian intestines, are some of the well-studied organisms and play a significant role in biology, medicine, and industry.

Because of their rapid growth in chemically defined conditions and extensive molecular tools available for diverse uses, *Escherichia coli* were termed the "workshop" of molecular biology. Cracking the genetic code, revealing the essence of DNA replication, making dramatic utilisation in gene organisation and regulation, and creating genetically modified organisms are just a few examples. Due to the speedy development, relaxed settings in the cultivation of genetic engineering goods, metabolic flexibility of biological and chemical as well as physiological abundance of information, *E. coli* is also one of the finest metabolic and synthetic biological host organisms. Used normally, *E. coli* strains are commonly known as inoffensive. *E. coli* is by no means the most industrially flexible organism. The choice of *E. coli* is depends on two factors. The fact that *E. coli* has developed quickly in chemically established growth media, and the cells don't clump, it was an experimental organism. Most genetic examination relies on the populace derived (cloned) from a single cell. When cells do not cluster up and form a thick mass, cloning gets simplified. An extra justification to choose *E. coli* as the host to a number of widely studied molecular biological viruses (Cairns et al., 1966) which delivered a number of apparatuses for genetic as well as biotechnological manipulation of this bacterium. Facility of biochemical *E. coli* research popularised this organism. Analysts can easily increase big cell volumes and protein from the cells was easily removed. *E. coli* has 55% protein, 25% nucleic acids, 9% lipids, 6% cell wall, 2.5% glycogen, and 3% other metabolites. Furthermore, due to its ease of cultures and rapid growth, the breathing of genetic equipment, and a deep biochemical and physiological point of view, *E. coli* is possibly the best-studied organism in laboratories (Pontrelli et al., 2018). *E. coli* is a model microbe that biologists frequently employ. *E. coli*, in particular, may be easily genetically manipulated to increase the capacity to study. Their physiology produces new phenotypes; *E. coli* has been used to develop various molecular cloning techniques and genetic equipment. Chemical composition of *E. coli* is composed of about 70% H_2O. Proteins account for 55% of the remaining weight. The other major components are RNA and phosphatides, which account for 21% and 9% of the dry weight. DNA makes up only 3% of the dry weight of the human body. There are around 800 distinct metabolites and ions in a normal *E. coli* cell. Under all conditions, the pyridine nucleotides (NAD+, NADH, NADP+, and NADPH) and biotin are required. There are around 400 million pyridine nucleotide molecules in each cell that participate in reversible oxidation and reduce different metabolism electrons, but only a few hundred biotin molecules that act as the primary step in fatty acid production. Furthermore, the intracellular concentration of metabolites varies substantially depending on the cell's medium and development mode. For example, acetyl-CoA's key metabolic intermediary is 20 times higher in glucose-producing cells than in acetate-producing cells. The total chemical arrangement of cellular nature molecules varies adversely depending on the growth settings (Neidhardt et al., 1990).

Several microorganisms have been developed to produce compounds from renewable resources, with *Escherichia coli* being one of the most studied. Engineering *E. coli* for producing fuels and chemicals from various renewable resources took a lot of time and effort. Engineered *E. coli* can be used an alternative source for the production of a vast array of fuel and chemicals from renewal resources (Zhao et al., 2019). The choice of *E. coli* as a host for metabolic engineering was one of the most important. In addition to serving as a proof-of-concept model organism, *E. coli* has been employed widely as an agricultural producer. The development of lysine, 1, 3-propanediol (PDO), and 1, 4-butanediol are three prominent and influential examples (Pontrelli et al., 2018). *E. coli* has been used to metabolise tryptophan, phenylalanine, threonine, lysine, nucleotides, succinic acid, methylidenesuccinic acid, polyhydroxybutyrate (PHB), and 1, 3-PDO.

15.7 Engineered *Escherichia coli* for Enhancing CO_2 Biomitigation

Genetic engineering enhances value-added products from CO_2 fixation by utilising a photosynthesis pathway engineered in microorganisms. Engineered approaches mainly focus on improving and maximising the effectiveness of the RubisCo enzyme, minimisation of photorespiration, driving pathway design by blocking competition and storage pathway and improving the expression of enzymes to enhance the target pathways (Salehizadeh et al., 2020). Because of the relatively high development rate, simple cultivation, and vast reservoir of genetic resources, the gram-negative *Escherichia coli* bacteria alone attracts the attention of scientists and becomes the pioneer in developing new technologies for carbon dioxide reduction and utilisation. Heterotrophic *Escherichia coli* are able to fix carbon dioxide, which is equivalent to autotrophic microbes (Gong et al., 2015). Notably, *Escherichia coli*'s unusual concentrations of CO_2 fixation were superior to other autotrophic microbes. Transformation of heterotrophs into autotrophs is technically not so easy. Gleizer et al. (2019) are researching the conversion of the *Escherichia coli* microorganism into an autotroph that creates its entire CO_2 biomass (Gleizer et al., 2019). In an anaerobic circumstance, *E. coli* strains synthesise significant quantities of acetate, lactate, format, and ethanol. NADH formed by glycolysis is unable to produce NAD+ via the respiratory chain, so pyruvate or acetate are utilised as electron acceptors to produce lactate dehydrogenase (ldhA) and alcohol dehydrogenase (adhE) and most notably these two by-products consume carbon sources. Hence, *Escherichia coli* metabolic engineering must focus on blocking sub-products of biosynthesis.

According to current research, an *E. coli* strain can synthesise all CO_2-derived sugar biomass elements using a non-native CBB cycle (Antonovsky et al., 2016). CO_2 hydrogenation into a format was documented using a variety of biochemical mechanisms. The capacity of genetically modified *E. coli* to produce exogenous formate as a carbon source was achieved through genetic engineering (Yishai et al., 2016). On CO_2 and FA (formic acid), *E. coli* strains quickly generate a higher cell density. Many researchers have observed high levels of formic acid synthesis from CO_2 and H_2 (Nielsen et al., 2019; Roger et al., 2018). Bang et al. announced in 2020 that *Escherichia coli* was designed to develop on CO_2 and FA (formic acid) alone by introducing a synthetic CO_2 and formic acid-fixing pathway, fine-tuning metabolic fluxes, and optimising cytochrome bo3 and bd-I ubiquinol oxidase levels.

CO_2 reduction into format has also been reported with entire cell catalysis and immobilised enzymes. A dehydrogenase H format of *Escherichia coli* (EcFDH-H) model organism has been described with correct electrocatalytic CO_2 reduction (Bassegoda et al., 2014). FDH-H from the *Escherichia coli* may decrease CO_2 in bioelectrochemical settings without the need for electron mediators (Yuan et al., 2018). Roger et al. produced the hydrocarbon-dependent CO_2 reduction engineered strain RT1 (DhyaB, DhybC, DpflA, DfdhE). Siegel et al. built a pathway for formulas that can convert CO_2 to format, formaldehyde, and dihydroxyacetone phosphate (DHAP), formal-CoA; therefore, binding to glycolysis through the initiation conduit in *E. coli*.

A pure CA prokaryotic (*Escherichia coli* bacteria) was identified for the first gene sequence from 1992–1993, the metalloenzyme being the first β-type CA. Furthermore, there are numerous reports on *E. coli* expressing heterologous CA. About 253–267 mg $CaCO_3$ mg^{-1} CA was registered with *E. coli* for the highest CO_2 capabilities (Oviya et al., 2013). In *Bacillus mucilaginous*, a gram-negative bacterium, and cloning of CA to *E. coli*, quantitative PCR gene control (CA) and mineral (calcite) formation may be investigated in real time.

Xiao et al. (2014) performed the experiment and found the up-regulation of CA genes accelerating calcite dissolution. On the other hand, Jo et al. (2013) successfully demonstrated the effective fixation of gaseous CO_2 as calcium carbonate ($CaCO_3$).

CA on their membrane was expressed by *E. coli* cells through a foam bioreactor that provided a broad CO_2 gas-fluid absorption interface and turned CA into bicarbonate (Watson et al., 2016). RuBisCO and phosphoribulokinase (PRK) have been expressed heterologically in *Escherichia coli* and have decreased CO_2 under different fermentation setting. *E. coli* was grown in L-arabinose supplemented media and approximately 15% reduction in carbon dioxide emissions was observed throughout fermentation. This could account for the extraordinary CO_2-fixing efficiency of Rubisco-based *E. coli* compared to microalgae and cyanobacteria.

As far as fixing carbon dioxide is concerned, mixotrophic capacity in *E. coli* allows it to serve as a potential microbial production bio-based yield. Rubisco-based CO_2 recycling is efficient and viable in which around 67 mg-CO_2 mole-arabinose^{-1} L^{-1} h^{-1} can be repaired (Zhuang and Li, 2013). In periplasmic expression of *E. coli*, entire cell biocatalyst for CO_2 hydration is applied if carbonic anhydrate (CA) is produced from methanosarcina thermophile and methanobacterium thermoautotrophicum (Patel et al., 2013). In 2013, Jo et al. developed a whole-cell biocatalyst *Neisseria gonorrheae* (ngCA) in *Escherichia coli*'s periplasm. *E. coli* periplasm expression of ngCA considerably stimulates calcium carbonate ($CaCO_3$) and has affected the $CaCO_3$ maximum produced in moderately low pH (8.5). In *E. coli*, there are different reports of expression of heterologous carbohydrase (CA). Over-expression in the *E. coli* strain BA002, L. lactis pyruvate carboxylase(pyc) genome with phosphoenolpyruvate carboxylase (ppc) elimination improving the pitching of 8.13 to 62.56 mg L^{-1} h^{-1} CO_2 (Liu et al., 2013). The use of partial cyanobial BBC cycles and carbon concentration produces an *E. coli* strain capable of fixing CO_2 at the amount of 19.6 mg CO_2 L^{-1} h^{-1} or specific rate of 22.5 mg CO_2 g DCW^{-1} h^{-1} (Gong et al., 2015). Because of its oxygen calamity, significant CO_2 fastening levels, and novel intermediates, this 3-HPA bicycle is an attractive target for metabolic engineering. A bioplastic precursor was regarded as the eponymous substance 3-HPA (Meng et al., 2012), showing potential in the field of *E. coli* engineering. Mattozzi et al. (2013) expressed in subpathways and determined practical exposition the novel 3-HPA bicycle enzymes of *E. coli* by genetic testing and phänotyping. Table 15.4 shows significant developments of engineered CO_2 assimilation by *E. coli*.

15.8 Limitations, Challenges

However, if *Escherichia coli* are exploited as industrial hosts, some issues remain. Despite the numerous advantages associated with *Escherichia coli* as a host for metabolic engineering, there are a few issues and disadvantages to employing this bacterium:

- Has minimal capacity for generating glycosylated materials, hard-to-assemble proteins (Mueller et al., 2018), or proteins with multiple disulfide bonds.
- It is not considered ideal for cultivation under elevated temperatures or alkaline condition (Pontrelli et al., 2018).

TABLE 15.4

Significant advancement of engineered CO_2 fixation by *E. coli*

Area of Focus	Reference
1. Light-driven ATP production using the protorhodopsin photosystem from aquatic picoplankton produced in *E. coli*.	Martinez et al., 2007
2. To free CO_2 into the chloroplast, a bypass photorespiration route address is entered into the *E. coli* glycolate metabolic pathway.	Kebeish et al., 2007
3. In a mineral salt medium, a self-induction agitator was used to stir a two-stage culture of *Escherichia coli* NZN111 for succinic acid synthesis. CO_2 has cycled inside this reactor, and a suitable CO_2 transfer rate was controlled by removing CO_2 lost due to ventilation. This integrated system demonstrates that succinate and bioethanol may be generating efficiently while CO_2 emissions from ethanol fermentation are significantly decreased.	Wu et al., 2012
4. Fixing carbon dioxide in *E. coli* by using a cyanobacterial carboxysome has been engineered.	(Bonacci et al., 2012)
5. CA is periplasmically expressed in *Escherichia coli* and used to catalyze the hydration of carbon dioxide in a biocatalytic process. As a result, the cell biocatalyst's ability to resist heat enhanced, reducing expenses, and enhancing performance. This type of catalyst might be significant for various carbon collection and utilisation technologies.	Patel et al., 2013
6. Regarding *E. coli*, the individual and combined expression of Rubisco (ribulose-1,5-bisphosphate carboxylase/oxygenase) and phosphoribulokinase (PrkA) were conducted under different fermentation conditions. Reduction in CO_2 emissions mean a 15% decrease in carbon dioxide emissions, producing 38% of theoretical carbon dioxide reduction. CO_2 that has been recycled has been used to feed a modified *E. coli*, a small subset of the cyanobacteria Calvin cycle.	Zhuang and Li, 2013
7. *E. coli* bacteria produce an FHL enzyme that hydrolyzes formate to formate hydrogen. FHL acts as an effective hydrogen-dependent carbon dioxide reductase when hydrogen and carbon dioxide are put under increasing pressure.	Roger et al., 2018
8. To synthesise the enzyme 3-hydroxypropanoate-3-hydroxypropanoic acid (3-HPA), all of the unique enzymes found in the 3-HPA bicycle are heterologous. The appropriate grouping of these enzymes will enable them to use the substrates and produce products exported and imported by *E. coli's* endogenous metabolism. The collection of enzymes utilised in this experiment shows that they can create a wide range of new metabolites.	Mattozzi et al., 2013
9. Although ngCA was found to be abundantly expressed in the periplasm of *E. coli*, the fact that it was designed into a cell with periplasmic ngCA as a biocatalyst demonstrated that a cell with periplasmic ngCA can successfully serve as an efficient biocatalyst for CO_2 sequestration.	Jo et al., 2013
10. The partial cyanobacterial Calvin cycle and carbon concentrating method were combined to enhance CO2 fixation in *E. coli*. This strain has shown similar CO_2 fixation rates to autotrophic cyanobacteria and algae, highlighting its considerable prospects for heterotrophic CO_2 assimilation.	Gong et al., 2015
11. Ethylene oxide off-gas is treated with succinate-producing *E. coli* to produce a platform C4 chemical, which is carbon dioxide from ethylene oxide off-gas.	Wu et al., 2017
12. Non-native CO_2 fixation cycle where all the intermediate steps and yield are generated purely from CO_2.	Antonovsky et al., 2016
13. Conduct carbon fixation simulations in silico to increase carbon fixation ability by applying the reductive tricarboxylic acid (TCA) and CBB cycles.	Cheng et al., 2019
14. Mostly RBEC-based *E. coli* was the first to discover and accurately measured the ability of CO_2 fixers such as fermenters and bubblers to maintain stable carbon dioxide levels on the basis of the mass balance in three devices: Flask-based incubator (FIC), two-layered device (TLD), and CO_2 bubbling device (CBD).	Tan and Ng, 2020

- In addition, *E. coli* is not as resistant as yeast to different inhibitors. While applying in the industries in different processes, *E. coli* cells are unable to be used to minimise overall process costs as feed additives such as yeast.
- Furthermore, by using *E. coli*, safety issues are posed as a host. So *E. coli* can probably be one of the most advanced metabolic cell manufacturing units. As a host, *E. coli* cannot produce biopharmaceutical glycosylated products, proteins requiring complex assemblies, and proteins with high disulfide bonds (Meyer and Schmidhalter, 2012).
- For example, huge configurations (like cellulosomes) cannot be formed, complex proteins expressed (such as methane monooxygenase), or photosynthetic machinery developed in *E. coli*.
- Especially, *E. coli* also cannot use biomass or methane cellulosic and are unable to reap the light of the sun to repair CO_2.
- In addition, *E. coli* cannot grow at temperatures greater than 45°C when put to harsh circumstances including low or high pH and high-salt environments. These phenotypes can be useful for particular industries. In combination with logical metabolic architecture and evolution, the rapid development of genome editing and system analysis methods doesn't impractically lead to a drastic rises in the phänotypical range exposures of *E. coli*.

15.9 Future Research Perspectives

Researchers have created many innovative engineered biological CO_2 reduction and use strategies in recent years. The fundamental approaches are designing and planning carbon fastenings pathways, parallel transfer of natural carbon fasteners, and engineering forms of energy delivery. Synthetic biologists favour the invention and construction of new and competent carbon-fixation enzymes and modules focusing on natural enzymes and tracts. In the provision of ATP and NAD (P) H from various energy sources and multiple modules can also be installed in new carbon fixation systems and synthetic energy modules are necessary. Novel progress in the field of synthetic biology might increase regulation of alternative CO_2-fixing pathways, such as the computer-assisted design of new synthetic CO_2-fixing pathways. The cheap resource is the radiant energy from the sun when it comes to energy supply. Although re-creating complicated biological photosynthesis processes is challenging, it's pretty difficult to do. Multiple energy resources can be employed to fix carbon dioxide based on preliminary results from microbial electrical use and photovoltaic-growing technologies. The industry will be able to produce carbon dioxide and chemicals by using synthetic modules or microbes shortly.

Sub-paths express heterological enough chloroflexus carbon fixing to supplement mutations of the host. Therefore, this may create a synthesis of industrial bioproducts, biofuels, and biopharmaceuticals utilising an artificially autotrophic host cell.

In addition, future conclusions could focus on microbial CO_2 sequestration directions, such as the development of innovative and new autotrophic CO_2-fixing chassis (ii). The design of highly efficient autotrophic hosts' genetic systems, (iii) the creation of new chemicals for synthesised chemical products, combining carbon-efficient pathways (i.e., NOG) with CO requirements, too, and (iv) reduce ATP requirements.

15.10 Conclusion

Even though CO_2 emissions increase by 13% and remain at that level until 2050, profound climate change will ensue. Consequently, reducing CO_2 is a vital solution for climate change mitigation. The future for biotechnology-engineered *Escherichia coli* is regarded as a workshop of genetic resources.

The rapid advancement of synthetic biology over recent years has brought many novel strategies for the fixation of biocarbon, such as the horizontal transfer nature of design carbon pathways and the synthetic carbon assimilation pathways. Synthetic biologists need to organise and construct new and more efficient enzymes and routes on the basis of the spectrum of natural enzymes and pathways. Well-studied species include *Escherichia coli*, which have thoroughly explored metabolic and genetic networks. They are considerably easier steps for prospective approaches to harvesting new CO_2 fastening cycles in a preexisting metabolism. Due to comparatively rapid growth and performance, *Escherichia coli* developed good bacteria to moderate CO_2 fastening and use. The large-scale *Escherichia coli* cannot nevertheless be generated to express the complex protein.

Furthermore, *Escherichia coli* are prevented from using cellulosic biomass or methane and cannot recover sunlight in CO_2. In addition, *Escherichia coli* cannot develop at high or low pH or in high-alkaline environments or at temperatures above 45°C under extreme circumstances. Adapting all current and potential biotechnological weaknesses efficiently will make the destiny of our environment more stable.

REFERENCES

Angermayr, S.A., Rovira, A.G., & Hellingwerf, K.L. (2015). Metabolic engineering of cyanobacteria for the synthesis of commodity products. *Trends in Biotechnology, 33*, 352–361.

Antonovsky, N., Gleizer, S., Noor, E., Zohar, Y., Herz, E., Barenholz, U., Zelcbuch, L., Amram, S., Wides, A., Tepper, N., et al. (2016). Sugar synthesis from CO_2 in *Escherichia coli*. *Cell, 166*, 115–125.

Antonovsky, N., Gleizer, S., & Milo, R. (2017). Engineering carbon fixation in E. coli: from heterologous RuBisCO expression to the Calvin–Benson–Bassham cycle. *Current Opinion in Biotechnology, 47*, 83–91.

Aumont, O., & Bopp, L. (2006). Globalizing results from ocean in situ iron fertilization studies. *Global Biogeochemical Cycles, 20*, GB2017.

Bar-Even, A., Noor, E., Flamholz, A., & Milo, R. (2013). Design and analysis of metabolic pathways supporting formatotrophic growth for electricity- dependent cultivation of microbes. *Biochimica et Biophysica Acta, 1827*, 1039–1047.

Bassegoda, A., Madden, C., Wakerley, D.W., Reisner, E., & Hirst, J. (2014). Reversible Interconversion of CO_2 and Formate by a Molybdenum Containing Formate Dehydrogenase. *Journal of the American Chemical Society, 136*, 15473–15476.

Batjes, N.H., & Sombroek, W.G. (1997). Possibilities for carbon sequestration in tropical and subtropical soils. *Global Change Biology, 3*, 161–173.

Bhatia, S.K., Bhatia, R.K., Jeon, J.M., Kumar, G., & Yang, Y.H. (2019). Carbon dioxide capture and bioenergy production using biological system – a review. *Renewable & Sustainable Energy Reviews, 110*, 143–158.

Bonacci, W., Teng, P.K., Afonso, B., Niederholtmeyer, H., Grob, P., Silver, P.A., & Savage, D.F. (2012). Modularity of a carbon-fixing protein organelle. *Proceedings of the National Academy of Sciences of the United States of America, 109*, 478–483.

Borowitzka, M.A. (2013). High-value products from microalgae-their development and commercialisation. *Journal of Applied Phycology, 25*, 743–756.

Buschke, N., Schäfer, R., Becker, J., & Wittmann, C. (2013). Metabolic engineering of industrial platform microorganisms for biorefinery applications – optimization of substrate spectrum and process robustness by rational and evaluative strategies. *Bioresource Technology, 135*, 544–554.

Cairns, J., Stent, G.S., & Watson, J.D. (1966) *Phage and the Origins of Molecular Biology*. Cold Spring Harbor, NY: Cold Spring Harbor Laboratory of Quantitative Biology.

Chen, X., Zhou, L., Tian, K., Kumar, A., Singh, S., Prior, B.A., & Wang, Z. (2013). Metabolic engineering of *Escherichia coli*: a sustainable industrial platform for bio-based chemical production. *Biotechnology Advances, 31*, 1200–1223.

Cheng, H.T.Y., Lob, S.C. Huang, C.C., Hoa, T.Y., & Yang, Y.T. (2019). Detailed profiling of carbon fixation of in silico synthetic autotrophy with reductive tricarboxylic acid cycle and Calvin-Benson-Bassham cycle in *Esherichia coli* using hydrogen as an energy source. *Synthetic and Systems Biotechnology, 4*, 165–172.

Claassens, N.J., Sousa, D.Z., Martins dos Santos, V.A.P., de Vos, W.M., & van der Oost, J. (2016). Harnessing power of microbial autotrophy. *Nature Reviews Microbiology, 14*, 692–706.

Feng, P., Weagant, S., Grant, M., & Burkhardt, W. (2002). Bacteriological analytical manual: enumeration of Escherichia coli and the Coliform bacteria. *Bacteriological Analytical Manual, 6*, 1–13.

Galaz, V. (2012). Geo-engineering, governance and socio-ecological systems: critical issues and joint research needs. *Ecology and Society, 17*, 24.

Gong, F., Liu, G., Zhai, X., Zhou, J., Cai, Z., & Li, Y. (2015). Quantitative analysis of an engineered CO_2-fixing Escherichia coli reveals great potential of heterotrophic CO_2 fixation. *Biotechnology for Biofuels*, 8, 86.

Gleizer, S., Ben-Nissan, R., Bar-On, Y.M., Antonovsky, N., Zohar, Y., Jona, G., Krieger, E., Noor, E., Shamshoum, M., Bar-Even, A., et al. (2019). Conversion of Escherichia coli to generate all biomass carbon from CO_2. *Cell*, 179, 1255–1263.

Gong, F., Zhu, H., Zhang, Y., & Li, Y. (2018). Biological carbon fixation: from natural to synthetic. *Journal of CO_2 Utilization*, 28, 221–227.

Gonzales, J.N., Matson, M.M., & Atsumi, S. (2019). Nonphotosynthetic biological CO_2 reduction. *Biochemistry*, 58, 1470–1477.

Hu, G., Li, Y., Ye, Y., Liu, L., & Chen, X. (2016). Engineering Microorganisms for Enhanced CO_2 Sequestration. *Trends in Biotechnology*, 37, 532–547.

Jajesniak, P., Ali, H.E.M.O., & Wong, T.S. (2014). Carbon capture and utilization using biological systems: opportunities and challenges. *Journal of Bioprocessing & Biotechniques*, 4, 155.

Jo, B.H., Kim, I.G., Seo, J.H., Kang, D.G., & Cha, H.J. (2013). Engineered Escherichia coli with periplasmic carbonic anhydrase as a biocatalyst for CO_2 sequestration. *Applied and Environmental Microbiology*, 79, 6697–6705.

Kebeish, R., Niessen, M., Thiruveedhi, K., Bari, R., Hirsch, H.J., Rosenkranz, R., Stabler, N., Schönfeld, B., Kreuzaler, F., & Peterhansel, C. (2007). Chloroplastic photorespiratory bypass increases photosynthesis and biomass production in Arabidopsis thaliana. *Nature Biotechnology*, 25, 593–599.

Khandavalli, L.V.N.S., Lodha, T., Abdullah, M., Guruprasad, L., Chintalapati, S., & Chintalapati, V.R. (2018). Insights into the carbonic anhydrases and autotrophic carbon dioxide fixation pathways of high CO_2 tolerant *Rhodovulum viride JA756*. *Microbiology Research*, 215, 130–140.

Kurano, N., Ikemoto, H., Miyashita, H., Hasegawa, T., Hata, H., & Miyachi, S. (1996). Fixation and utilization of carbon dioxide by microalgal photosynthesis. *Energy Conversion and Management*, 36, 689–692.

Liu, H., Shi, J., Zhan, X., Zhang, L., Fu, B., & Liu, H. (2017). Selective acetate production with CO_2 sequestration through acetogen-enriched sludge inoculums in anaerobic digestion. *Biochemical Engineering Journal*, 121, 163–170.

Liu, R.M., Liang, L.Y., Wu, M.K., Chen, K.Q., Jiang, M., Ma, J.F., Wei, P., & Ouyang, P.K. (2013). CO2 fixation for succinic acid production by engineered Escherichia coli co-expressing pyruvate carboxylase and nicotinic acid phosphoribosyltransferase. *Biochemical Engineering Journal*, 79, 77–83.

Martinez, A., Bradley, A.S., & Waldbauer, J.R., et al. (2007) Proteorhodopsin photosystem gene expression enables photophosphorylation in a heterologous host. *Proceedings of the National Academy of Sciences of the United States of America*, 104, 5590–5595.

Mattozzi, M.D., Ziesack, M., Voges, M.J., Silver, P.A., & Way, J.C. (2013). Expression of the sub-pathways of the Chloroflexus aurantiacus 3-hydroxypropionate carbon fixation bicycle in E. coli: toward horizontal transfer of autotrophic growth. *Metabolic Engineering*, 16, 130–139.

Meng, D.C., Shi, Z.Y., Wu, L.P., Zhou, Q., Wu, Q., Chen, J.C., & Chen, G.Q. (2012). Production and characterization of poly (3-hydroxypropionate-co-4-hydroxybutyrate) with fully controllable structures by recombinant Escherichia coli containing an engineered pathway. *Metabolic Engineering*, 14, 317–324.

Meyer, H.P., & Schmidhalter, D.R. (2012). Microbial expression systems and manufacturing from a market and economic perspective. Innov. Biotechnol, 211–250.

Mohan, S.V., Modestra, J.A., Amulyak, K., Butti, S.K., & Velvizhi, G. (2016). A circular bioeconomy with biobased products from CO_2. *Trends in Biotechnology*, 34, 506–519.

Mourato, C., Martins, M., Da Silva, S.M., & Pereira, I.A.C. (2017). A continuous system for biocatalytic hydrogenation of CO_2 to formate. *Bioresource Technology*, 235, 149–156.

Mueller, P., Gauttam, R., Raab, N., Handrick, R., Wahl, C., Leptihn, S., Zorn, M., Kussmaul, M., Scheffold, M., & Eikmanns, B. (2018). High level in vivo mucin-type glycosylation in Escherichia coli. *Microbial Cell Factories*, 17, 168.

Neidhardt, F., Ingraham, J.L., & Schaechter, M. (1990). *Physiology of the Bacterial Cell. A Molecular Approach*. Sunderland, MA: Sinauer Associates.

Nielsen, C.F., Lange, L., & Meyer, A.S. (2019). Classification and enzyme kinetics of formate dehydrogenases for biomanufacturing via CO_2 utilization. *Biotechnology Advances*, 37, 107408.

Nogia, P., Sidhu, G.K., Mehrotra, R., & Mehrotra, S. (2013). Capturing atmospheric carbon: biological and nonbiological methods. *International Journal of Low-Carbon Technologies, 11*, 266–274.

Nouha, K., John, R. P., Yan, S., Tyagi, R. , Surampalli, R. Y., & Zhang, T. C. (2015). Carbon capture and sequestration: Biological technologies, Surampalli R. Y., (Ed.). *Carbon Capture and Storage. Physical, Chemical, and Biological Methods* (pp. 65–111), Reston: American Society of Civil Engineers. 10.1061/9780784413678.ch04.

Orsini, F., & Marrone, P. (2019). Approaches for a low-carbon production of building materials: a review. *Journal of Cleaner Production, 241*, 118380.

Oviya, M., Sukumaran, V., & Giri, S.S. (2013). Immobilization and characterization of carbonic anhydrase purified from E. coli MO1 and its influence on CO_2 sequestration. *World Journal of Microbiology & Biotechnology, 29*, 1813–1820.

Patel, T.N., Park, A.H., & Banta, S. (2013). Periplasmic expression of carbonic anhydrase in Escherichia coli: a new biocatalyst for CO2 hydration. *Biotechnology & Bioengineering, 110*, 1865–1873.

Pohlmann, A., Fricke, W.F., Reinecke, F., Kusian, B., Liesegang, H., Cramm, R., et al. (2006). Genome sequence of the bioplastic-producing "Knallgas" bacterium *Ralstonia eutropha H16*. *Nature Biotechnology, 24*, 1257–1262.

Pontrelli, S., Chiu, T.-Y., Lan, E.I., Chen, F.Y., Chang, P.C., & Liao, J.C. (2018). Escherichia coli as a host for metabolic engineering. *Metabolic Engineering, 50*, 16–46.

Roger, M., Brown, F., Gabrielli, W., & Sargent, F. (2018). Efficient hydrogen-dependent carbon dioxide reduction by Escherichia coli. *Current Biology, 28*, 140–145.

Salehizadeh, H., Yan, N., & Farnood, Ramin. (2020). Recent advances in microbial CO_2 fixation and conversion to value-added products. *Chemical Engineering Journal, 390*, 124584.

Salmon, S., Saunders, P., & Borchert, M. (2009). Enzyme technology for carbon dioxide separation from mixed gases. *IOP Conference Series: Earth and Environmental Science, 6*, 172018.

Schwander, T., von Borzyskowski, L.S., Burgener, S., Cortina, N.S., & Erb, T.J. (2016). A synthetic pathway for the fixation of carbon dioxide in vitro. *Science, 354*, 900–904.

Shulman, S.T., Friedmann, H.C., & Sims, R.H. (2007). Theodor Escherich: the first pediatric infectious diseases physician? *Clinical Infectious Diseases, 45*, 1025–1029.

Siegel, J.B., et al. (2015). Computational protein design enables a novel one-carbon assimilation pathway. *Proceedings of the National Academy of Sciences of the United States of America, 112*, 3704–3709.

Tan, S.I., & Ng, I.S. (2020). Design and optimization of bioreactor to boost carbon dioxide assimilation in RuBisCo-equipped Escherichia coli. *Bioresource Technology, 314*, 123785.

Watson, S.K., Han, Z., Su, W.W., Deshusses, M.A., & Kan, E. (2016). Carbon dioxide capture using Escherichia coli expressing carbonic anhydrase in a foam bioreactor. *Environmental Technology, 37*, 3186–3192.

Wolff, G.A., Billett, D.S.M., Bett, B.J., et al. (2011). The effects of natural iron fertilisation on deep-sea ecology: the crozct plateau, southern Indian ocean. *PLoS One, 6*, e20697.

Wu, H., Li, Q., Li, Z.M., & Ye, Q. (2012). Succinic acid production and CO_2 fixation using a metabolically engineered Escherichia coli in a bioreactor equipped with a self-inducing agitator. *Bioresource Technology, 107*, 376–384.

Wu, M., Zhang, W., Ji,Y., Yi, X., Ma, J., Wu, H., & Jiang, M. (2017). Coupled CO2 fixation from ethylene oxide off-gas with bio-based succinic acid production by engineered recombinant Escherichia coli. *Biochemical Engineering Journal, 117*, 1–6.

Xiao, L., Hao, J., Wang, W., Lian, B., Shang, G., Yang, Y., et al. (2014). The up-regulation of carbonic anhydrase genes of Bacillus mucilaginosus under soluble Ca^{2+} deficiency and the heterologously expressed enzyme promotes calcite dissolution. *Geomicrobiology Journal, 31*, 632–641.

Yishai, O., Lindner, S.N., Gonzalez de la Cruz, J., Tenenboim, H., & Bar-Even, A. (2016). The formate bioeconomy. *Current Opinion in Chemical Biology, 35*, 1–9.

Yuan, M., Sahin, S., Cai, R., Abdellaoui, S., Hickey, D. P., Minteer, S. D., & Milton, R. D. (2018). Creating a low-potential redox polymer for efficient electroenzymatic CO2 reduction. *Angewandte Chemie International Edition, 57*, 6582–6586. 10.1002/anie.201803397.

Zhao, C., Zhang, Y., & Lia, Y. (2019). Production of fuels and chemicals from renewable resources using engineered Escherichia coli. *Biotechnology Advances, 37*, 107402.

Zhuang, Z.Y., & Li, S.Y. (2013). Rubisco-based engineered Escherichia coli for in situ carbon dioxide recycling. *Bioresource Technology, 150*, 79–88.

16 Role of Biopolymers in Development of Sustainable Remediation Technologies

Monika Yadav[1], Manita Das[2], Sonal Thakore[2], and R.N. Jadeja[1]
[1]Department of Envrionmental Studies, Faculty of Science, The Maharaja Sayajirao University of Baroda, Vadodara, India
[2]Department of Chemistry, Faculty of Science, The Maharaja Sayajirao University of Baroda, Vadodara, India

CONTENTS

16.1	Introduction	296
	16.1.1 Cellulose	297
	16.1.2 Chitin	297
	16.1.3 Chitosan	297
	16.1.4 Alginate	297
	16.1.5 Dextran	298
	16.1.6 Cyclodextrin	298
	16.1.7 Starch	298
16.2	Characterisation of Biopolymers	298
	16.2.1 SEM	299
	16.2.2 XRD	299
	16.2.3 FTIR	300
	16.2.4 NMR	301
	16.2.5 TGA	301
	16.2.6 Optical Microscopy	302
16.3	Processing of Biopolymers	302
	16.3.1 Aerogels	302
	16.3.2 Hydrogels	303
	16.3.3 Nano Biopolymers/Nanofibrilation	305
	16.3.4 Hybrid Systems of Biopolymers	306
16.4	Applications of Biopolymer-Based Materials in Adsorption	307
	16.4.1 Removal of Dyes	307
	16.4.2 Removal of Metals	308
	16.4.3 Removal of Micropollutants	309
	16.4.4 Gas Adsorption	310
16.5	Application of Biopolymer-Based Materials in Photocatalysis	311
16.6	Biopolymers from Food Waste	311
18.7	Conclusions	313
References		314

16.1 Introduction

In present industrial economies and biological world, polymers play a crucial role. Polymers synthesised by living organisms are biopolymers. Repeating monomeric units are covalently linked to form long-chain biomolecules, i.e., polymeric biomolecules (Dassanayake R et al., 2019). Various renewable resources like bacteria, algae, fungi, animals, plants, or chemically generated natural products like sugars, starch, natural fats, and oils, etc. synthesise a wide range of biopolymers like deoxyribonucleic acid (DNA), ribonucleic acid (RNA), proteins, starch, cellulose, and chitin (Azeem et al., 2017). The stable synthetic polymers are resistant to physical and chemical degradation and also pose challenges to solid waste treatment plants deteriorating the environmental quality (Kyrikou and Briassoulis, 2007). Excess exploitation and utilisation of non-renewable resources like coal, fossil fuel, etc. and several environmental challenges have attracted researchers' attention toward biodegradable polymers instead of synthetic polymers (Taboada et al., 2010). Biopolymers are eco-friendly, biodegradable, and easy to be sterilised, which does not express any toxic response inordinate to their useful effects (Nilani et al., 2010). Universal occurrences of three main classes of biopolymers include polynucleotides (13 or more nucleotide monomers constituting long polymers like DNA, RNA), polypeptides (amino acids are monomeric units constituting short polymers via amide bond), and polysaccharides (composed of monomeric sugars via O glycosidic linkage) (Klemm et al., 2005; Numata, 2015). Starch, pectin, xylan, chitosan, lignin, galactoglucomannan, etc. are plant-based polymers extracted from cellulose, corn starch, plant oil, weeds, potato starch, hemp, sugarcane, etc. (Reddy et al., 2013). Monomeric units of biopolymers, their chain length, linkage type, etc. govern the variation in solubility, gelling potential, mechanical properties, and interfacial and surface properties. Different forms of bio-based materials like fibres, food casings, membranes, films, hydrogels, aerogels, sponges, etc. have a wide range of applications in food, pharmaceuticals, biomedical, adsorption, etc. (Wang et al., 2016). Material properties of biopolymers have attracted interest in academics and industries; Figure 16.1 shows different types of

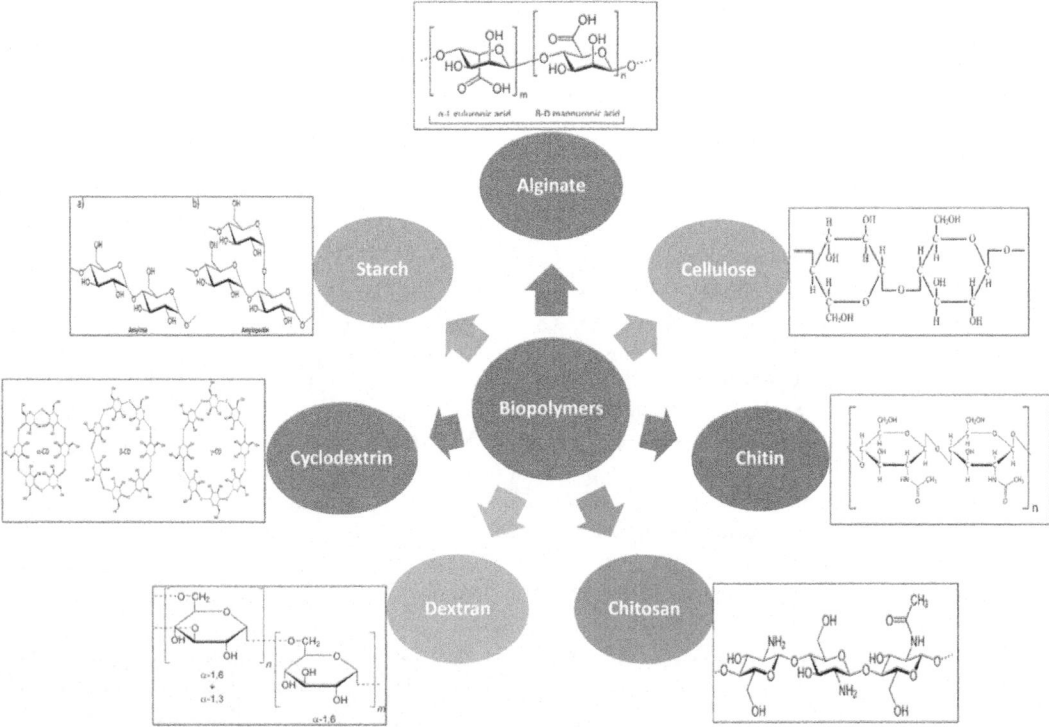

FIGURE 16.1 Different types of biopolymers.

biopolymers. This book chapter is focused on biopolymers like cellulose, dextran, chitin, alginate, chitosan, cyclodextrin, starch, etc. and their utilisation in sustainable remediation methods.

16.1.1 Cellulose

Cellulose has a flat-ribbon-like conformation due to a linear chain of glucose molecules. In the case of plant cell walls, it is considered a major structural component that is a most common organic polymer and abundantly available renewable resource on the planet with 1.5×10^{12} tons of total annual biomass production. It is one of the inexhaustible raw materials to meet the enhancing demand of eco-friendly, sustainable, and biocompatible technologies (Klemm et al., 2005; Moon et al., 2011). Cellulose consists of two anhydroglucose rings joined via oxygen and covalently linked to carbon 1 of one glucose ring and carbon 4 of the adjacent glucose ring (1-4 linkage), termed a β1-4 glycosidic bond (Ahmed et al., 2005). The degree of polymerisation in cellulose differs depending upon its source and treatments applied during the extraction process (George, 2015). Cellulose is a polymer with a higher molecular weight and extremely crystalline nature. It is tough, hydrophilic, and fibrous in nature but not soluble in water. Apart from plant-derived cellulose, there is cellulose derived from bacteria that has a thin nanofibrous network structure and high hydrophilicity and moldability (Khan and Ahmad, 2013).

16.1.2 Chitin

Chitin is an important natural polymer on earth; second in terms of abundance after cellulose. Chitin is derived from "chiton", a Greek word meaning "coat of nail". It is major component of exoskeletons of crustaceans, insects, and invertebrates; fungi and yeast cell walls are composed of chitin (Tan et al., 1996). Chitin's structural unit is made up of a linear polymer of 2-acetamido-2-deoxy-D-glucopyranose binded by β(1-4) glycosidic bond (Tanaka et al., 2014). The biosynthesis of chitin occurs in chitin synthase, which is a membrane-bound protein complex. Approximately 90% degree of acetylation is observed in chitin after extracting and purifying from source materials. Extensive hydrogen bonding among the polymer chains results in an arrangement of crystalline microfibrils in chitin. Chitin is harder to process than cellulose; it is tough, hydrophobic, and insoluble in water and other common solvents (Chen et al., 2015; Velde et al., 2004).

16.1.3 Chitosan

Chitosan is a poly-β-1,4-linked glucosamine, synthesised by alkaline deacetylation of chitin from crustacean sources under alkaline conditions in commercial production. It has less than a 50% degree of acetylation (Kumar, 2000). It is found in various morphological forms like primary, unorganised structure, crystalline, and semicrystalline forms. It posses high biological and mechanical properties and is biorenewable, biodegradable, and bio-functional; thus, it is useful in terms of environmental remediation. It has been copolymerised to convert into various forms like nanoparticles, hydrogels, aerogels, fibers, membranes, and porous scaffolds (Alhwaige et al., 2020). It can be soluble in weak acids; its solubility is facilitated by the protonation of free glucosamine in acidic condition (Kurita, 2001).

16.1.4 Alginate

Alginate is a linear anionic polysaccharide derived from brown seaweed like *Laminaria, Macrocystis,* and *Ascophyllum* species or from soil bacteria commercially (McHugh, 2003). They are unbranched, linear polysaccharides containing different amounts of (1→4')-linked β-d-mannuronic acid and α-l-guluronic acid residues. Alginate can be considered a true block copolymer, consisting of a homopolymeric region of guluronic acid and mannuronic acid. It is biodegradable and has controllable porosity; it can be linked/encapsulated with other bioactive molecules. It has hemostatic properties and thus used in different types of gels and sponges for wound treatment. Calcium alginate has the properties of adhesion and proliferation. It forms a weak hydrophilic gel that can absorb water and fluids more than 20 times its weight (Thomas, 2000).

16.1.5 Dextran

Dextran is the most versatile and biocompatible exopolysaccharide that is synthesised by *Leuconostoc mesenteroides*, i.e., lactic acid bacteria (Demirbilek and Dinc, 2016). It is generally synthesised by sucrose-containing products such as sugar beets, molasses, and sugarcane. The process parameter generally affects the yield and molecular weight of dextran. Dextran is broadly constructed by α-1,6-glycosidic linkage and some branched linkages of α-1,2; α-1,3; and α-1,4. Under optimised conditions, a medium having 15% sucrose produced high molecular weight dextran, i.e., 2 million Dalton (Sarwat et al., 2008). It helps in improvement of the dispersion of components in cosmetic formulations acting as binding agents and also improves antiaging activity. It also acts as a bulking agent due to its water-binding capacity and increases the bulk of formation. Sulphonation of dextran produces dextran sulfate, whereas dextran salt, along with amphoteric or anionic surfactants, produces cationic dextran (Sajna et al., 2015). Dextran can provide a surface to conjugate with several molecules like aromatic rings, aliphatic or cyclic hydrocarbons, etc. and can be utilised in several applications like drug delivery, environmental remediation, cosmetics, and many others (Maiti and Kumari, 2016).

16.1.6 Cyclodextrin

Cyclodextrins are the cyclic organic compound derived from starch due to the activity of cyclodextrin glycosyltransferase enzyme. Alpha, beta, and gamma cyclodextrins are the predominate cyclodextrins formed by the enzymatic activity having 6, 7 and 8 (1→4)-linked α-D-glucopyranosyl units, respectively. *Bacillus macerans, Bacillus megaterium, Klebsiella pneumoniae, Thermoanaerobacter* sp., etc. are some bacteria that produce the enzyme used for the production of cyclodextrins (Hedges, 2009). The condition and duration of reaction decides the final ratio of alpha, beta, and gamma cyclodextrins. Cyclodextrin molecules appear like rings, doughnuts, or truncated cones. The inner wall of cyclodextrin is hydrophobic, whereas the outer wall is hydrophilic, which can help in formation of inclusion complexes with organic molecules via host guest interactions (Yadav et al., 2019). In comparison to linear dextrins, cyclodextrins are more resistant to non-enzymatic hydrolysis. Cyclodextrins are stable in the solid state and can be stored for several years at room temperature without any degradation (Szejtli, 1998).

16.1.7 Starch

The chief storage carbohydrate of green plants is starch, an attractive raw material for polymer application. It is generally insoluble and a semi-crystalline granules present in storage tissues. D-glucose, i.e., amylose (unbranched) and amylopectin (highly branched), are the two polymers responsible for starch synthesis. Approximately 10% of the moisture content is present in native starch. Approximatley, 98–99% dry weight of native granules of starch is composed of amylase and amylopectin, whereas the remainder has less amounts of minerals, lipids, and phosphorous. The size of starch granules vary from 1 to 100 mm diameter and their shape might be polygonal, lenticular, spherical, etc. (Copeland et al., 2009). They can vary on the basis of degree of crystallinity, structure, content, and organisation of amylase and amylopectin molecules (Tyler, 2004). Flour enriched with starch is one of the cheapest materials and its utilisation in the application of wastewater treatment can be economically viable and eco-friendly (Gimbert et al., 2008).

16.2 Characterisation of Biopolymers

The most used characterisation techniques for biopolymers are Fourier Transform Infrared Spectroscopy (FTIR), X-ray Diffraction (XRD), Thermogravimetric Analysis (TGA), Scanning Electron Microscopy (SEM), and Nuclear Magnetic Resonance (NMR) spectroscopy.

Role of Biopolymers 299

FIGURE 16.2 SEM images of dried magnetic beads: (a) low magnification and (b) high magnification. Reproduced from Asadi et al. (2018) with permission from the American Chemical Society.

16.2.1 SEM

SEM is used to determine external morphology on the basis of a nanometer (nm) to micrometer (μm) scale of heterogeneous organic and inorganic materials. It gives magnified three-dimensional images of their surface (Gnanavel et al., 2010). The morphology of magnetic composite beads and calcium alginate was visualised through dried magnetic bead SEM images in the study conducted by Asadi et al. (2018). Figure 16.2 demonstrates the iron oxide nanoparticles covered with a fine layer of calcium alginate (Asadi et al., 2018).

16.2.2 XRD

X-ray powder diffraction plays a crucial role in material research, development, and their application. X-rays are diffracted by crystals and quantitatively analyze the concentrations of each phase in multiphase samples. In polycrystalline materials, it also observes unit cell metrics and microstructure analysis. The "reflected" diffracted beam from a plane passes through crystal lattice points in a way that form this crystal-lattice planes similar to mirrors, so that the angle of incidence is equal to the angle of reflection (Wang et al., 2005). The XRD pattern revealed that the ZnO sample shows a hexagonal

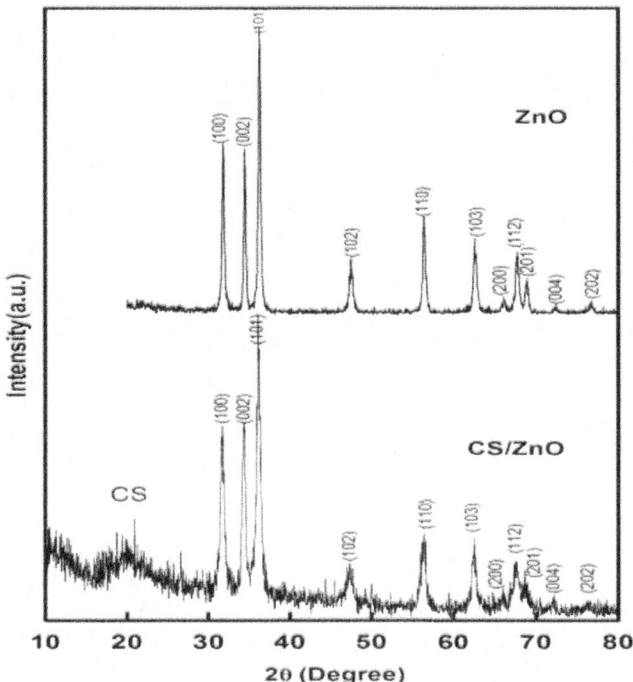

FIGURE 16.3 XRD pattern of ZnO and CS-ZnO. Reproduced from reference Dhanavel et al. (2014) with permission from the International Journal of ChemTech Research.

structure. Determination of lattice parameters is a = 3.257 Å, b = 3.257 Å, and c = 5.204 Å. Scherrer's formula was utilised for calculation of average crystallite size of the ZnO nanoparticles observed as 23 nm. Figure 16.3 shows the XRD pattern of the CS/ZnO nanocomposite. The formation of the composite was confirmed as the peaks of both chitosan and ZnO were observed in XRD (Dhanavel et al., 2014).

16.2.3 FTIR

FTIR is one of the identification methods of undefined substances and their chemical structure. A substance or material FTIR spectrum is considered a fingerprint of that material. The absence or presence of few groups could be corresponded with the presence or absence of specific wavelength absorption when it is impossible to identify or detect any material from the spectrum. FTIR spectroscopy is an interferometric method. In IR spectra plot of absorption is basically represented as a function of wavelength (cm^{-1}). The spectrum of biopolymers is recorded within the range of 700–4,000/cm, having 2/cm of resolution (Bayarı and Severcan, 2005). The pure β-CD FTIR spectra in pregel condition was represented in Figure 16.4 and studied by Kundu et al. (2019); 3,450 cm^{-1} represents the O–H stretching broad peak and C–H stretching at 2,922 cm^{-1} has been obtained. The –CH_2 peak bending is observed at 1,456 cm^{-1} and at 1,338 cm^{-1}, skeletal vibrations could be observed; 1,159 cm^{-1} represents glycosidic linkage, which is due to the stretching vibration of α-(1 → 4), and 937 cm^{-1} shows vibration of the glycosidic linkage; 1,082 cm^{-1} is associated with the bending of C–C, whereas 1,028 cm^{-1} shows stretching of C–O, bonded by two pyranose rings. The ring vibration representing glycosidic deformation of C–H has been observed at 856 cm^{-1} (Kundu et al., 2019).

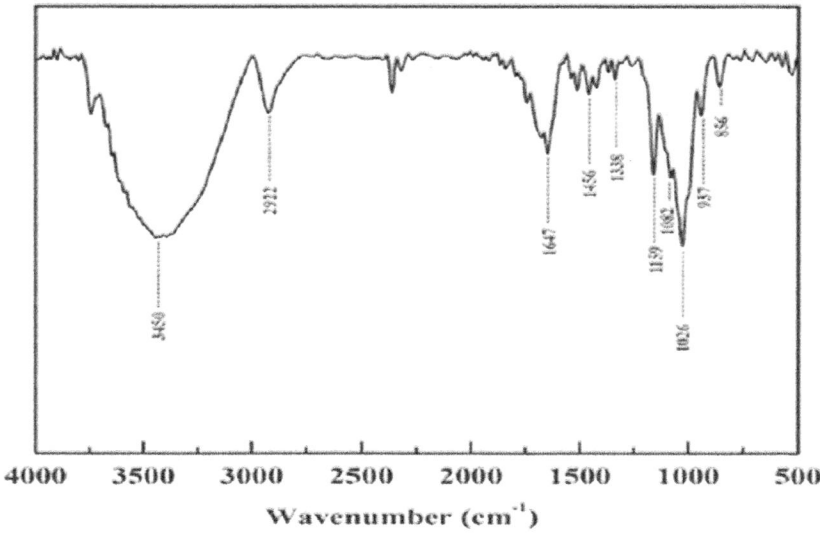

FIGURE 16.4 FTIR spectra of pure βCD. Reproduced from Kundu et al. (2019) from American Chemical Society.

16.2.4 NMR

The movement of nuclear spin represents NMR spectroscopy to detect nuclei and estimate them quantitatively. The magnetic field associated with spinning charge is equivalent to the conductor-carrying current. There are two broadly studied nuclei for NMR, i.e., ^{1}H and ^{13}C. The uniform magnetic field influence tends to align these nuclei with the field or opposite the field. Magnetisation induces current to receiver due to spin inversion; the signal obtained has been registered, amplified, and then plotted to give the spectrum. The solid-state ^{13}C NMR spectrum of Chitosan and Cs-PMA (Chitosan-Polymethacrylic acid) is investigated by Samit Kumar Ray et al. (2018). The methl group peaks in its acetamido moiety is observed at 24.714 ppm in chitosan. In the case of chitosan, 106.067, 58.298, 83.934, and 61.887 ppm corresponds to C1, C2, C4, and C6 ring signals, respectively, while the peaks of the C3 and C5 ring are observed at 76.072 ppm. Additionally, the signal of carbonyl group (C=O) is observed at 175.201 ppm due to the acetamido moiety. The Cs-PMA NMR spectrum represents the main carbon backbone by large peak at 46.334, whereas the methyl group carbon atom and carboxylic group is represented by the peaks at 17.194 and 183.319 ppm, respectively. Maity and Ray (2018)^{1}H NMR spectrum of a solubilised chitosan nanocrystal (CsNWs, 50% NaOH, 50°C, 48 h) has been shown in Figure 16.5 (Pereira et al., 2015).

16.2.5 TGA

In thermal gravimetric analysis (TGA), measurement samples were scanned in 25°C to 400°C temperature range at 10°C/min heating rate under a nitrogen condition with a flow rate of 20 mL/min to avoid oxidation of the sample. Five mg samples were taken for TGA analysis (Al et al., 2018). The curves of TGA Mag-Ben, Mag-Ben/CCS/Alg, Alg, and CCS 324.3%, 10.1%, 6.7%, and 14.8% represent the weight loss percent at ambient temperatures (0–150°C) of Mag-Ben, Mag-Ben/CCS/Alg, Alg, and CCS, respectively, due to the evaporation of moisture from the sample. The thermal stability of final materials and each component at T20% (the temperature representing 20% sample weight) was investigated by Xiao-kun Ouyang et al. (2019). Good thermal stability of Mag-Ben was observed due to the stability of Fe3O4 and Ben. The T20% values for Alg, CCS, and Mag-Ben/CCS/Alg was observed at 212°C, 286°C, and 307°C, respectively. The polymer constituent's thermal decomposition at higher temperatures represents the next stage of degradation. The observed final Mag-Ben weight is still 86%, mainly due to magnetic bentonite/carboxymethyl chitosan/sodium alginate (Mag-Ben/CCS/Alg) having

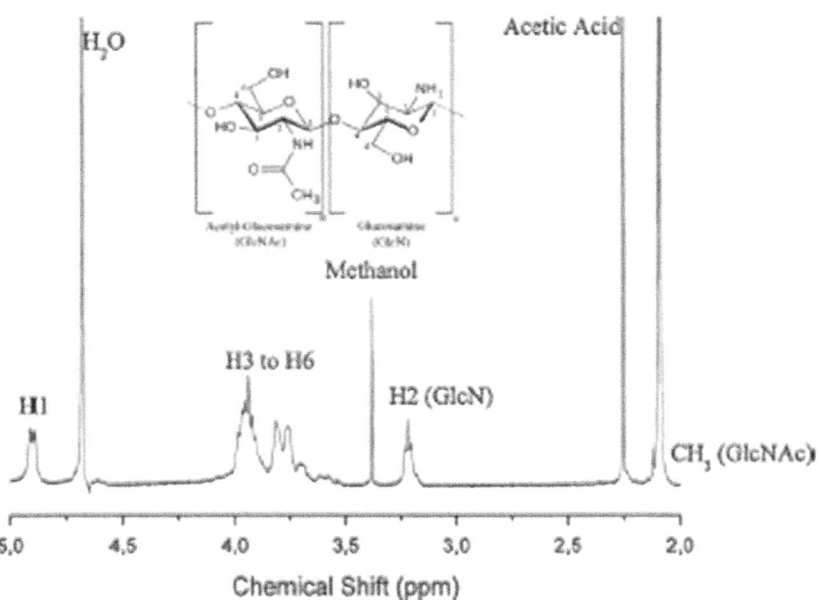

FIGURE 16.5 ^{1}H NMR spectrum of solubilised chitosan nanocrystal (CsNWs, 50% NaOH, 50°C, 48 h). Reproduced from reference Pereira et al. (2015) from Carbohydrate Polymers.

48% weight. At higher temperatures, the degradation of biopolymers CCS and Alg are approximately 20%, Alg degradation is faster than CCS but at 800°C CCS degrades thoroughly (Zhang et al., 2019). Thermogravimetric traces of virgin polymers (solid line) and recycled polymers (dashed line) has been shown in Figure 16.6 (Shojaeiarani et al., 2019).

16.2.6 Optical Microscopy

The hydrogels morphologies could be visualised with the help of optical microscopy. The hydrogels βCD-MCC, βCD-xylan, and βCD-CMC are shown in optical microscopy as glossy structures (a–d) investigated by Kundu et al. (2019) represented in Figure 16.7. βCD-MCC gel is a form of interconnected gel, whereas others are shown as secluded assemblies. The metal-ion adsorbed hydrogel represents fluorescent microscopy images in the figure (e-l) and concentrated metal-ion spots are also visible in the images. These may be due to the entrapment of Cd(II) and Ni(II) ions inside the network (Kundu et al., 2019).

16.3 Processing of Biopolymers

16.3.1 Aerogels

Aerogels are highly porous sol-gel-derived materials and they can be classified on the basis of their appearance, porosity, or composition (Zhao et al., 2018). Widely, they are divided into three classes, i.e., organic aerogels, inorganic aerogels and organic-inorganic hybrid aerogels (Stergar, 2016). Removal of the liquid component from the hydrogel produces an aerogel. This process involves freeze-drying or critical point drying to evade actual gel micro-structure destruction that has been synthesised using a solution of biopolymers. Aerogels show significant applications in adsorption, biomedicine, catalyst, separation, and photo-electricity because of their high specific surface area, low density, and high porosity (Wang et al., 2016). Bio-based materials like cellulose, alginate, etc. are more useful in preparation of aerogels compared to synthetic polymers. Crosslinking of polymers increases the density of

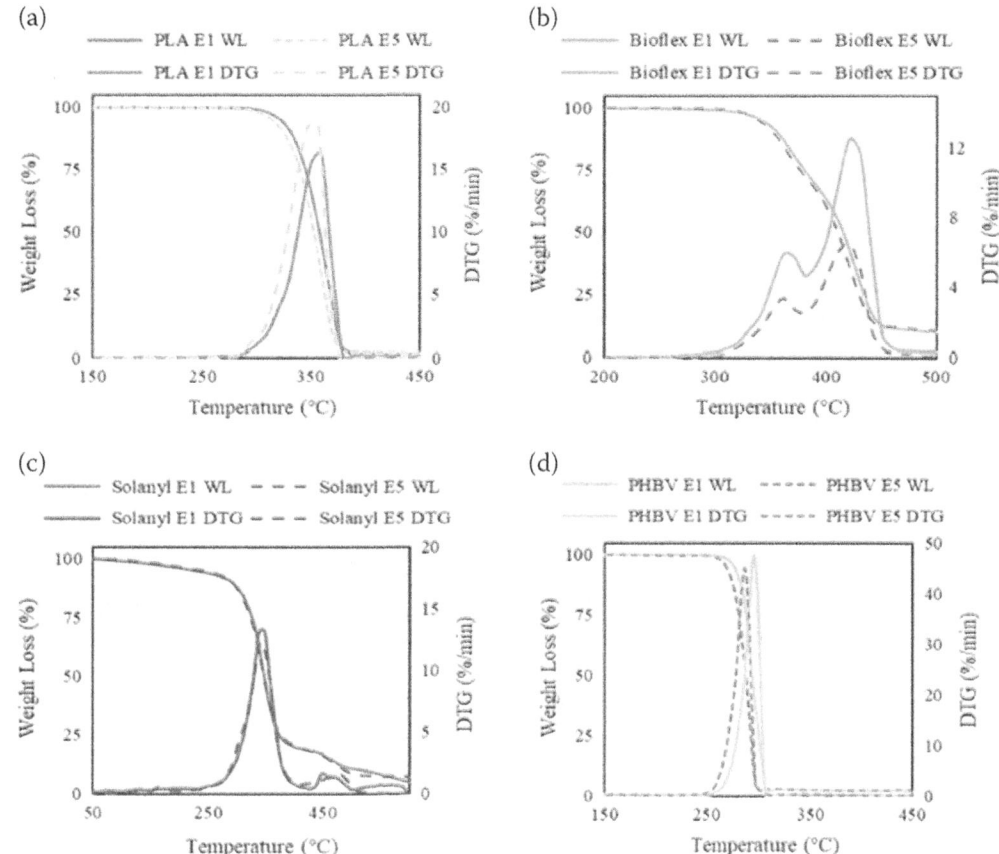

FIGURE 16.6 Thermogravimetric traces of virgin polymers (solid line) and recycled polymers (dashed line): (a) PLA, (b) Bioflex, (c) Solanyl, and (d) PHBV. Reproduced from Shojaeiarani et al. (2019) from Polymers.

porosity and elevates the mechanical integrity of the aerogels (Klemm et al., 2018). Aerogels based on cellulose have been broadly utilised as a support system for metal nanoparticles because of their good stability, biodegradability, high mechanical strength, non-toxicity, low density, and web-shaped network (Cai et al., 2009). Clearly defined morphologies of pore and precise sorption properties of aerogels gathered a variety of interest for sorbents, insulating materials, and act as templates for various functional materials (De France et al., 2017). The adsorption of ketoprofen and griseofulvin was studied for oral administration using hydrophilic silica aerogels (Smirnova et al., 2004). Salzano de Lunaa et al. (2019) represent the formation of chitosan aerogels via freeze drying for dye adsorption. A fibrous cellulose aerogel has been shown in Figure 16.8 that has been fabricated by a simple approach of disintegrating electrospun cellulose fibers (ECFs) (Jiang and Hsieh, 2018).

16.3.2 Hydrogels

Hydrogels are composed of hydrophilic homopolymers or a copolymer network and they show swelling behaviour in the presence of water. A hydrogel is a kind of solid yet jelly-like polymer with a three dimensional-network structure. Hydrogels are broadly categorised as chemical or physical hydrogels depending on cross-linking types (Hoare and Kohane, 2008). Hydrogels are widely used in drug delivery, tissue engineering, agriculture, and healthcare applications because of their extremely moist, porous, three-dimensional structure that imitates biological tissue and high water absorption capacity. Hydrogels have a polyfunctional structure; thus, they can also be used as adsorbents for environmental remediation

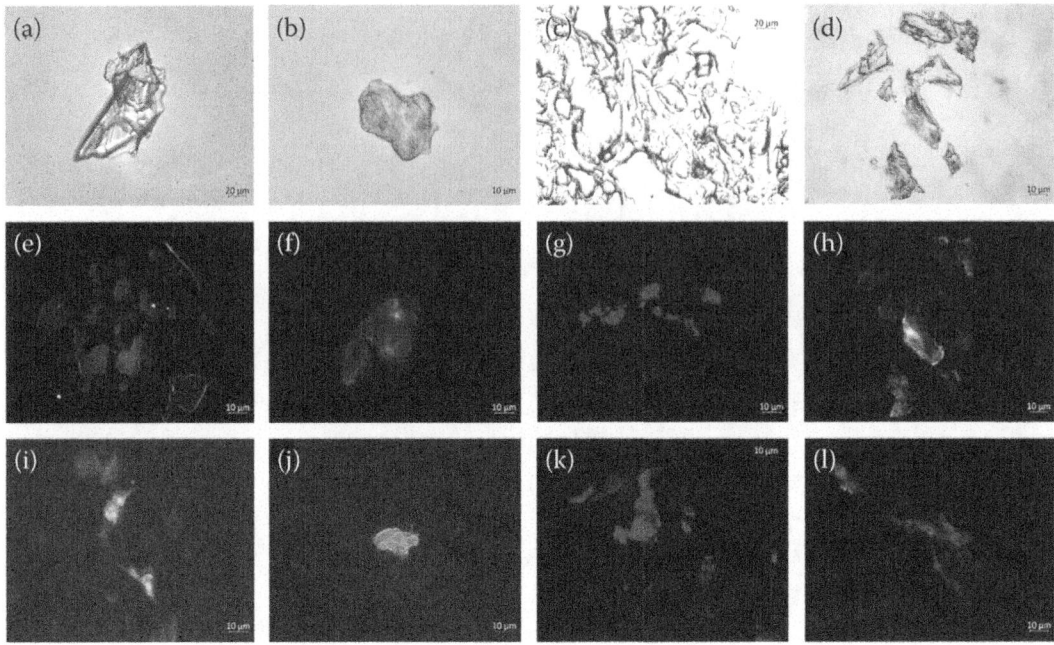

FIGURE 16.7 Optical microscopy images of hydrogels (no metal-ion adsorption): (a) βCD-xylan, (b) βCD-CMC, (c) βCD-MCC, and (d) βCD gel. Fluorescent microscopy images of Cd(II)-loaded hydrogels (e–h) and Ni(II)-loaded hydrogels (i–l). (e, i) βCD-xylan, (f, j) βCD-CMC, (g–k) βCD-MCC, and (h–l) βCD gel. Reproduced from Kundu et al. (2019) from American Chemical Society.

FIGURE 16.8 Cellulose fibrous aerogels disintegrated electrospun cellulose fibers (ECFs). Reproduced from Jiang and Hsieh (2018) from ACS Omega.

technologies. The presence of hydrophilic groups like – NH_2, –COOH, –OH, –$CONH_2$, –CONH–, and –SO_3H is responsible for the hydrophilic nature of a hydrogel network (Pandey et al., 2020). The polymeric hydrogels possessing naturally derived polysaccharides have gathered significant interest as useful adsorbent systems for the elimination of dyes and heavy metals (Jiang et al., 2019). A bacterial

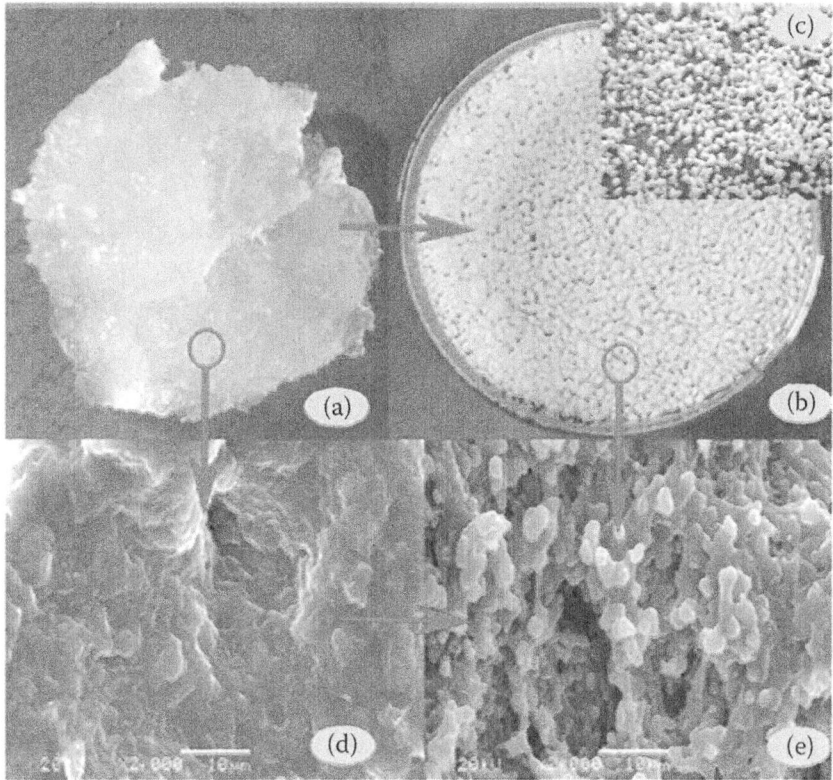

FIGURE 16.9 Digital images of (a) wet state of SA-g-PAA hydrogel, (b) wet state of SA-g-PAA/PVP/GE (SA2) hydrogel, and (c) dry state of SA-g-PAA/PVP/GE (SA2) hydrogel. SEM micrographs of (d) SA-g-PAA and (e) SA-g-PAA/PVP/GE (SA2). Reproduced from Wang et al. (2013) from Journal of Polymer Research.

cellulose-hydrogel was prepared by utilising gamma radiation polymerisation, which is a pH-responsive hydrogel. The acrylic acid (AA) monomer was grafted onto the bacterial cellulose at two different of ratios, as investigated by Hakam et al. (2015) and Pewarna et al. (2015). A novel chitosan/REC/cellulose composite hydrogel was synthesised via chemical crosslinking with high strength and highly cost effective for dye adsorption by Tu et al. (2017). The knotting and twisting nature of chitosan/cellulose hydrogels without any fracture represents high strength, resilience, and good elasticity due to which they restore quickly after compression. The increase in efficacy of hydrogel composites and thermal stability is due to the incorporation of rectorite into hydrogels (Tu et al., 2017). Alginate-based hydrogels were synthesised via crosslinking and grafting between acrylic acid, sodium alginate, polyvinylpyrrolidone, and gelatin (Wang et al., 2013; Figure 16.9).

16.3.3 Nano Biopolymers/Nanofibrilation

Biomolecules and biopolymers impart great effects on particles, clusters, and crystals like nano objects of the surroundings due to water solubility and biocompatibility (El Hankari et al., 2019). Nanocellulose, nanochitin, nanochitosan, etc. are extracted from their particular biopolymers with a minimum one dimension in the range of nanometer. Destroying the original structure of hierarchy of the biopolymers prepares nanoparticles. In hydrogels, nanocomposite cellulose nanocrystals act as potential fillers because of their rigidity and mechanical strength. Shearing, electrospinning, and strong acid hydrolysis treatment are used to convert biopolymers into their nanofibers and nanocrystals (Ifuku, 2014). Hydrochloric acid and sulfuric acid are generally used to generate nanocrystals' needle-shaped structure. Drying from the suspension, the nanoparticles of biopolymers

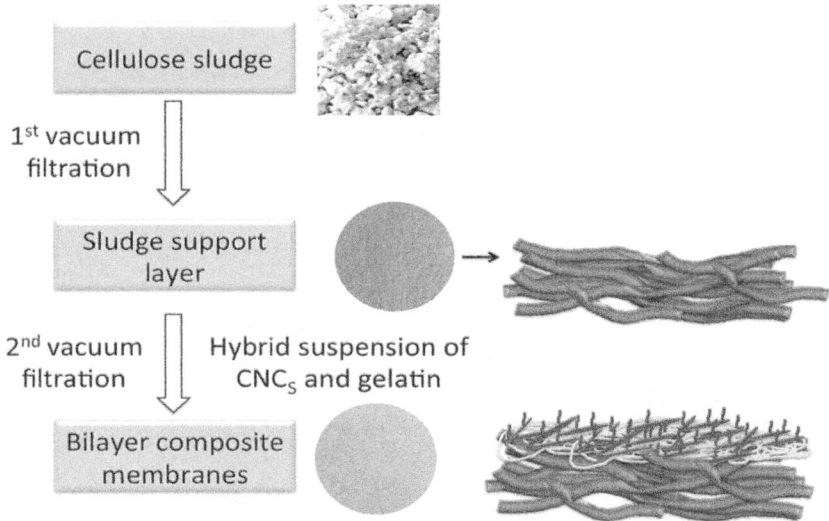

FIGURE 16.10 Synthesis of layered cellulose nanocomposites. Reproduced from Karim et al. (2016) from Royal Society of Chemistry.

could be directly prepared into aerogels and films. A simple and fast method of thin-layered nanocomposite membranes of cellulose was preparation via vacuum filtration as reported by Karim et al. (2016). The slender functional layer was fabricated by further nanocellulose filtration (cellulose nanocrystals$_{BE}$, cellulose nanocrystals$_{SL}$, and phosphate cellulose nanocrystals$_{SL}$)–gelatin suspension. The aluminium plates were used to press the synthesised membranes after drying for 12 h at room temperature in a compression-molding machine with a load of 60–70 kN (Fontune Presses, Elastocon, Sweden) and heating (80°C) to get membranes, as shown in Figure 16.10 (Karim et al., 2016).

16.3.4 Hybrid Systems of Biopolymers

Hybrid materials are synthesised by several methods, utilising polysaccharides like cellulose, chitosan, dextran, etc. and inorganic materials that have been studied for different applications. Polyvinyl chloride, magnetite, montmorillonite, bentonite, oil palm ash, polyvinyl alcohol, perlite, polyurethane, silica, activated clay, kaolinite, etc. are some systems that could form composites with biopolymers (Budnyak et al., 2016). The polymerisation of aniline via the chemical oxidative method in the presence of dispersed nanocomposites of starch-montmorillonite synthesises starch-montmorillonite/polyaniline (St-MMT/PANI) (Olad and Azhar, 2014). Specific and complementary adsorption properties were observed in a hybrid adsorbent system of activated carbon, clay, and chitosan. Hybrid material may possess a single remediation ability for different types of pollutants; thus, it might handle aromatic, anionic, and cationic species simultaneously. Higher adsorption efficiency containing new hybrid adsorbents may be utilised for accomplishing "at source" treatment, evading the pollutants mixing in order to decrease the difficulties of treatment (Bouyahmed et al., 2018). A new cellulose–clay hydrogel with excellent mechanical performance, superabsorbent properties, and better removal efficiency for dye was synthesised by Peng et al. (2016), as shown in Figure 16.11. The chemically cross-linked carboxymethyl cellulose (CMC), cellulose, and the interspersed NaOH/urea aqueous solution in clay ("green" cellulose solvent) results in the formation of superabsorbent hydrogels (Peng et al., 2016). The hydrogels' adsorption capacity for methylene blue (MB) dye was investigated and MB adsorption on the hydrogels' mechanism was discussed in the paper.

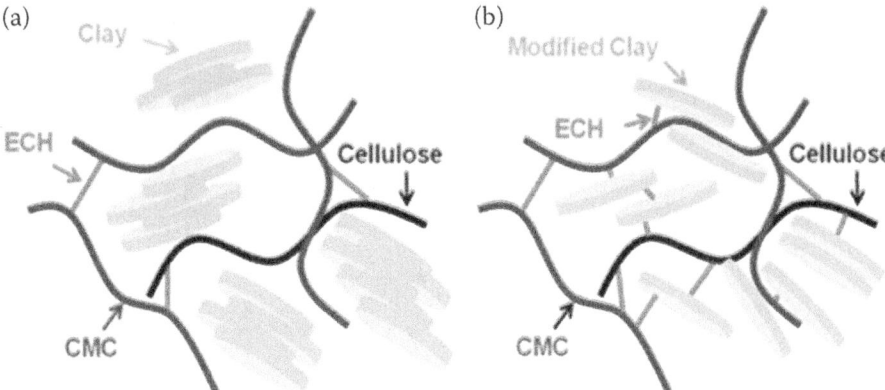

FIGURE 16.11 Synthesis schemes of cellulose–clay nanocomposite hydrogels. (a) Hydrogel network containing unmodified clay. (b) Epoxidised clay (modified clay) cross-linked in the hydrogel networks. Reproduced from Peng et al. (2016) from American Chemical Society.

16.4 Applications of Biopolymer-Based Materials in Adsorption

16.4.1 Removal of Dyes

Dyes are extensively utilised in several industries like textiles, printing, painting, dyeing, and different chemical industries, which cause severe water contamination if they are not eliminated before being discharged into the natural enviornment because of their highly toxic, non-biodegradable, carcinogenic, and mutagenic properties (Dong et al., 2018). Presences of dyes are distinctly noticeable in water even in trace amounts. Therefore, demand of sustainable remediation methods to treat water contaminated with organic dyes has gained attention in the recent era (Narula and Rao, 2019). Several processes have been utilised to eliminate dyes from polluted water such as biological methods, physical-chemical methods, adsorption, electrochemical methods, oxidation, flocculation, coagulation, and membrane separation, etc. (Junfeng Li et al., 2017). Adsorption seems to be superior to others in terms of its cost effectiveness, ease of operation, and renewable capability.

The multi-walled carbon nanotube composite fiber using calcium alginate (CA/MWCNTs) was prepared by Kunyan Sui et al. in which $CaCl_2$ was used as a cross-linking agent by following wet spinning. This system adsorbs heavy metals and dyes effectively without breaking off its composition. Therefore, further formation of micropollutants in water is restricted (Sui et al., 2012).

The low cost, non-toxic, and biocompatible magnetic hydrogel of calcium alginate was synthesised as a potential sorbent to eliminate methyl violet from an aqueous solution, represented in Figure 16.12; 713 and 889 mg/g was the maximum adsorption capacities of magnetic hydrogel beads and calcium alginate hydrogel beads for the removal of MV (Asadi et al., 2018).

A novel, low-cost hydrogel was synthesised using locust bean gum [LBG-cl-Poly(DMAAm)] for fast and highly effective removal of BG dye, showing a higher adsorption capacity than other reported systems. The maximum adsorption capacity obtained was 142.85 mg/g and 97.7% of BG dye with an initial concentration of 50 mg/L (Pandey et al., 2020).

ZSM-5 zeolite was modified by the addition of chitin (biopolymer/ZSM-5 zeolite) to generate a novel material for cationic dye adsorption from aqueous solutions, such as CV, BF, and MB dyes with higher adsorption capacities (Brião et al., 2018).

Synthesis of a dextran-based hydrogel was done via a glycidyl methacrylate (GMA) reaction to generate modified dextran (Dex-MA) of double-bond structure. Further, modified dextran and acrylic acid (AA) has been copolymerised to produce Dex-MA/PAA hydrogel. Dex-MA/PAA hydrogel shows removal efficiency 86.4% and 93.9% for crystal violet (CV) and methylene blue (MB) within 1 minute of having an initial concentration of 50 ppm. The 1,994 mg/g adsorption capacity for MB and for CV is 2,390 mg/g (Yuan et al., 2019).

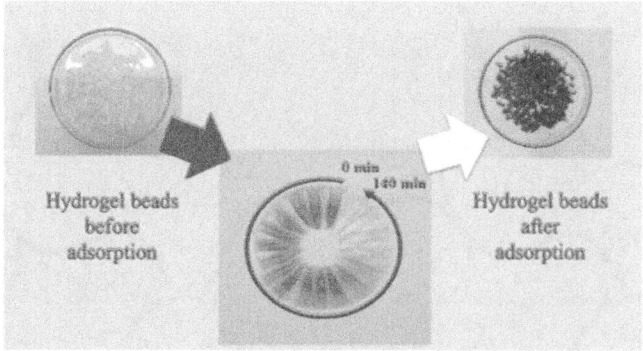

FIGURE 16.12 Hydrogel of calcium alginate for adsorption of methyl violet. Reproduced from Asadi et al. (2018) from American Chemical Society.

The basic parameter for adsorption performance optimisation of chitosan aerogel is a cross-linking step. The aerogel nanocomposite CS/GO shows enhanced mechanical strength and also exhibits higher adsorption capacity, removing both cationic and anionic pollutants from an aqueous solution, i.e., indigo carbine and methylene blue (Gimbert et al., 2008).

16.4.2 Removal of Metals

Metallurgical processes and mining activities repeatedly introduce major amounts of heavy metals to the adjacent environment; mainly in the nearby water bodies and groundwater systems (Pour and Ghaemy, 2015). Metal ions like iron, manganese, zinc, copper, etc. are important for living organisms like plants and animals but become virulent in higher concentrations. Highly considerable contaminating metal ions like mercury, cadmium, lead, and arsenic have huge lethality for living organisms (Mahmoud, 2013). Heavy metals can precisely bind with nucleic acid, metabolites, and proteins in living organisms, suppressing their functionalities. Heavy metals can accumulate in organisms (bioaccumulation) and be consumed by other organisms along the food chain (Ip et al., 2005).

The hydrogel membranes based on graphene oxide/alginate (GAHMs) were prepared by thorough mixing of graphene oxide, sodium alginate, urea, and deionised (DI) water, further cross-linking with the calcium chloride solution. GAHMs are used for the adsorption of HMIs, Pb(II), and Cr(III). The maximum adsorption capacity for Cr(III) was lower than Pb(II) (118.6 and 327.9 mg/g, respectively) (Bai et al., 2019).

Polydopamine and a chitosan cross-linked graphene oxide (GO/CS/PDA) composite aerogel having a porous three-dimensional network structure and strong mechanical strength was synthesised for efficient removal of chromium through adsorption. It is a dense cross-linked network structure providing multiple active sites for removal of Cr(VI) ions with 312.05 mg/g maximum adsorption capacity at 298 K (Li et al., 2019).

Abundance and renewable nature of cellulosic biofibers make them a better option in the wastewater remediation field. Four heavy-metal ions, zinc(II), copper(II), lead(II), and cadmium(II), were adsorbed in a batchwise manner on NaOH-treated cellulosic fibers of *Abelmoschus esculentus*. The maximum adsorption capacities are 19.21, 16.85, 44.42, and 67.24 mg/g for Cu^{2+}, Zn^{2+}, Cd^{2+}, and Pb^{2+} metal ions, respectively, at 25°C (Singha and Guleria, 2014).

Polymer hydrogel blends of chitosan and poly(vinyl alcohol) (PVA/CS) were synthesised and adsorption of Cu(II) was studied (Jamnongkan and Singcharoen, 2016).

Spherical-shaped chitosan was prepared for reduction of agglomeration in an acidic medium and to enhance the active binding sites of metal ions. This system was used for eliminating Zn(II) and Cr(VI) ions from their aqueous solution. The maximum adsorption capacity of chitosan beads was 79.56 mg/g in case Cr(VI) and 109.18 mg/g in case of Zn(II) having an initial concentration of 1,000 mg/L (Salih and Ghosh, 2017).

An ultrasonic-assisted method is used to prepare hydrogels using glucan and chitosan as the main raw materials. GL/CS hydrogel was used to investigate the adsorption of Cu^{2+}, Co^{2+}, Ni^{2+}, Cd^{2+}, and Pb^{2+}, showing maximum adsorption capacities at 342, 232, 184, 269, 395 mg/g, respectively, at pH 7.0 and 0.01 g adsorbent quantity (Jiang et al., 2019).

Methylene blue (MB) and (Pb(II)) were precisely adsorbed by using the composites of poly(acrylamide and reduced graphene oxide (RGO/PAM), the adsorption capacity of Pb(II) was investigated as 1,000 mg/g, as shown in Figure 16.13. The adsorption kinetics of RGO/PAM and Pb(II) fit well into the pseudo-second-order model (Yang et al., 2013).

16.4.3 Removal of Micropollutants

Pharmaceuticals, pesticides, hormones, and industrial chemicals present in less concentrations are termed micropollutants. They can cause significant toxicological effects on aquatic and natural ecosystems and consequently on human beings (Guzzella et al., 2002). A large number of micropollutants are found in water resources, independent of their source. Since the micropollutants are non-biodegradable and obstinate in nature, they reside in soil for a longer time prior reaching into the groundwater (Hijosa-Valsero et al., 2010; Qi et al., 2015). Even though they are present in relative low levels in drinking water like part per billion or parts per trillion, they can cause adverse health effects following chronic exposure (Huerta-Fontela et al., 2011).

The magnetic nanoparticles coated by polymer polyvinylpyrrolidone (PVP) were synthesised to remove emerging micropollutants from an aqueous solution (Bisphenol-A, Estriol, Metolachlor, Tonalide, Ketoprofen and Triclosan) (Alizadeh Fard et al., 2017).

Bio-based trifunctional adsorbent using chitosan-EDTA-β-cyclodextrin (CS-ED-CD) has been synthesised using EDTA as a cross-linking agent through simple and green one-pot synthesis approach for the adsorption of organic micropollutants and harmful metal ion wastewater. The novel adsorbent displayed maximum adsorption capacity of 1.258, 0.803 mmol/g, for Cd(II) and Pb(II), respectively,

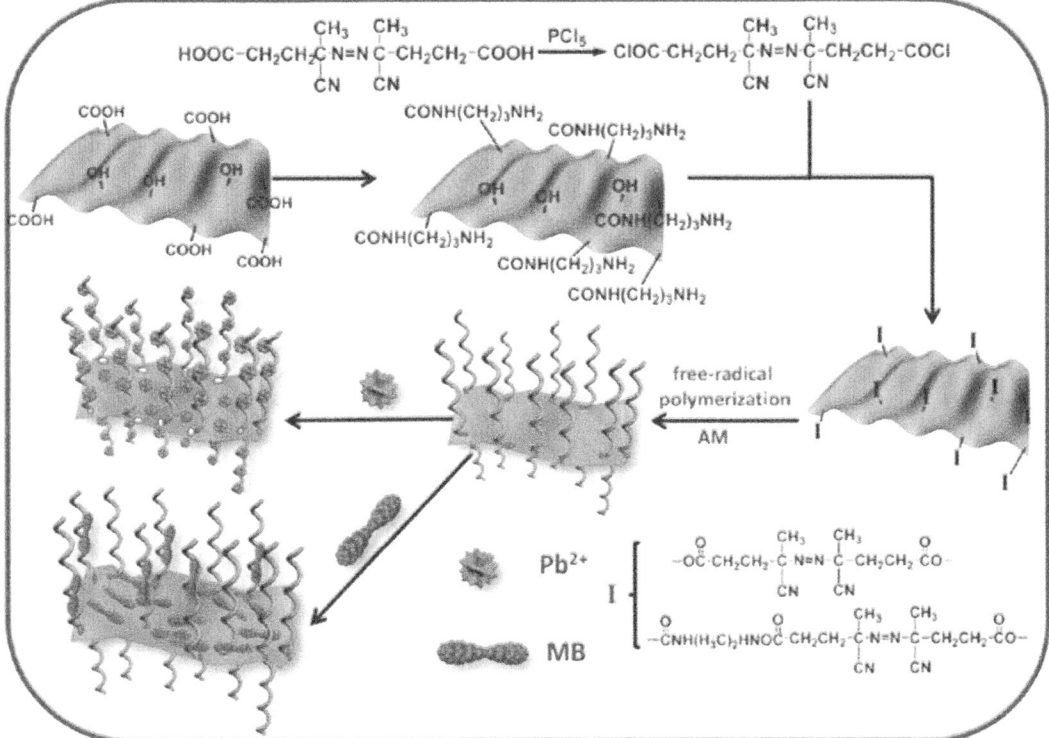

FIGURE 16.13 Synthesis of PAM chains on RGO sheets by free radical polymerisation and adsorption of Pb (II) and MB. Reproduced from Yang et al. (2013) from American Chemical Society.

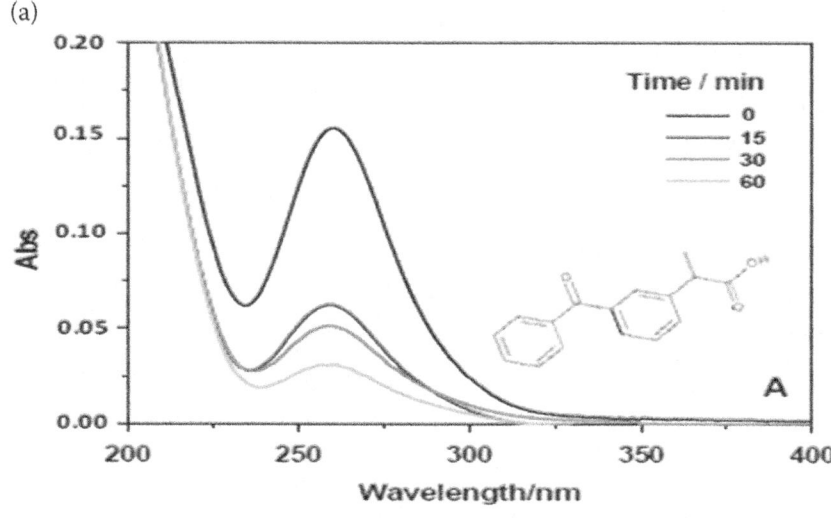
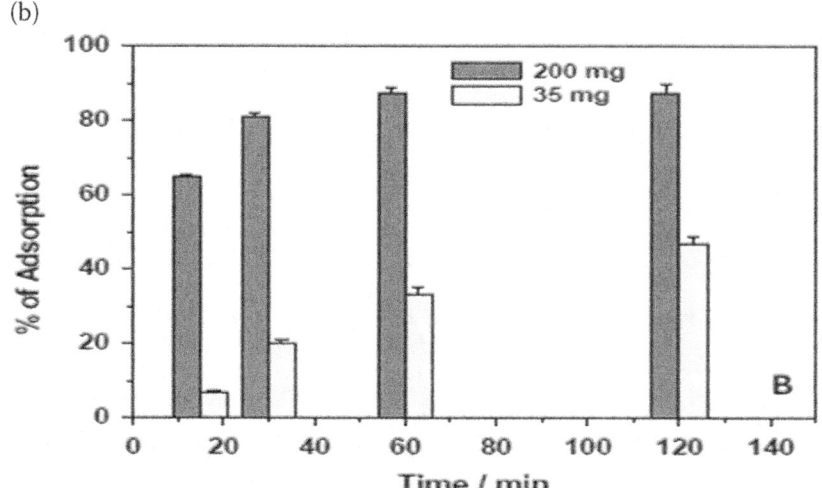

FIGURE 16.14 (a) Ultraviolet–visible light (UV–Vis) spectra of ketoprofen (Kp) solution, pH 5 in the presence of 150 mg of adsorbent; (b) Kp adsorption percentage calculated at different contact times in the presence of 200 mg and 35 mg of adsorbent. Reproduced from Rizzi et al. (2019) from Materials.

whereas for ciprofloxacin, procaine, bisphenol-S, and imipramine a heterogeneous adsorption capacity of 0.142, 0.203, 0.177, and 0.149 mmol/g, respectively, following Langmuir adsorption isotherm (Zhao et al., 2017). Figure 16.14 represents shell-waste-derived chitosan-based solid film for removal of the micropollutant Ketoprofen (Rizzi et al., 2019).

Low cost and environmentally friendly, biochar is a promising way to eliminate pharmaceuticals and other micropollutants from wastewater. Surface area and porosity of biochar enhances via an activation process that ultimately increases adsorption capacity. A new magnetic nature cross-linked chitosan/activated biochar (CMCAB) was prepared to remove emerging micropollutants: ibuprofen (IBP), naproxen (NPX), and diclofenac (DCF) from the aqueous solution (Mojiri et al., 2019).

16.4.4 Gas Adsorption

Glucose/graphene-based aerogels (G/GAs), three-dimensional structures, are prepared utilising the process of CO_2 activation method as well as hydrothermal reduction. A tapering mesopore-sized G/GAs exhibited a higher surface area (763 m²/g). It acts as an encouraging adsorbent system for elimination of CH_4 (16.8 mg/g at 298 K), H_2 g at 77 K), and CO/g at 77 K), and CO_2 (76.5 mg/g at 298 K) (Liu et al., 2018).

16.5 Application of Biopolymer-Based Materials in Photocatalysis

Generally electron hole pairs were generated by a light absorption photocatalytic mechanism that results in the production of superoxide anion and hydroxyl radicals, which leads to degradation of organic pollutants. The condition when some materials are supported by photocatalysts having functionality, porosity, and high surface area, their performance is enhanced due to higher dispersion and pollutant interactions (Kumar et al., 2019).

In sodium alginate, immobilisation of ZnO form spheres of calcium alginate so that mineralisation and degradation of 10 mg /L of TCS in aqueous solutions could be obtained. Degradation occurs through heterogeneous photocatalysis utilising natural and artificial sources of radiation (Kosera et al., 2017).

Photocatalysts synthesised with a biopolymer-metal complex wool-Pd/CdS show excellent photocatalytic activity and the photodegradation mechanism of model Rh B was obtained (Wang et al., 2013).

The ionic complexation results in the formation of hydrogels between a poly(acrylic acid) (PAA) as the polyanion and cationic side chain containing conjugated polymers as the polycation. In a hydrogel photocatalyst, the enhanced photocatalytic activity of the hydrogel was observed. The organic dye photodegradation and the synthesis of enzyme cofactors, i.e., nicotinamide adenine dinucleotide (NAD^+) with the help of photo-oxidation in water, were studied as shown in Figure 16.15 (Byun et al., 2019).

The synthesis of bionanocomposite hydrogel (TGB-hydrogel) with the help of TiO_2 nanorods and a functionalised gum ghatti (Gg) have been investigated. It is used to eliminate a toxic dye brilliant green (BG). The dye-loaded TGB-hydrogel was further processed for 3 hours at 550°C and again used for the photocatalytic degradation of the antibiotic ciprofloxacin (CIP). The reinstated spent photocatalyst completed the cycle after the BG dye adsorption. The photocatalyst shows a high photocatalytic efficiency of 88.7% and CIP degradation within 180 min (Kumar et al., 2018).

Nano-hetero-assembly $ZnSe-WO_3$ is supported by gum ghatti for the elimination of bisphenol S (BPA), an endocrine disruptor using solar power. The 99.5% removal was achieved in just 45 min by synergistic effects of photocatalysis-adsorption-ozonation (Kumar et al., 2017).

16.6 Biopolymers from Food Waste

The rich organic matter biowastes have changed the scenario of disposing waste to exploiting and converting them into value-added products. Unique properties of microbes utilise the bioorganic matter of food processing wastes in order to attain energy for their growth and development (Ranganathan et al., 2020). Microbial biotransformation produces several functional foods, biopolymers, biofuel, pharmaceutical preparations, bioactive compounds, and different other biotechnological applications (Yang et al., 2017). Fruit

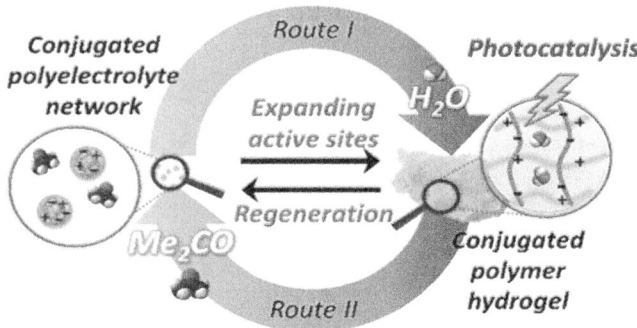

FIGURE 16.15 Schematic illustration of the conjugated polymer hydrogel photocatalyst with expandable active sites in water. The swelling of hydrogel photocatalyst was reversible by a solvent exchange. Reproduced from Byun et al. (2019) from American Chemical Society.

processing industries wastes are usually composed of cores, seeds, soft tissues, peels, and stems. Nutrients, polysaccharides, proteins, phenolic compounds, flavours, dietary fibers, phytochemicals, peptides, proteins, etc. are the high-value components extracted from the wastes of these industries (Lee et al., 2019).

A vegetable and fruit waste mixed stock having equal quantities of carrot, potato, apple, lettuce, and banana contains hemicellulose, cellulose, and lignin fractions of 19.2%, 3.9%, and 1.0%, respectively (Mäkipää, 2015). Recently, fruit peels have gained the attention of researchers and are used either in native or in modified form for the removal of environmental pollutants. Several citrus fruits, bananas, watermelons, and mixed fruit peels have been investigated. The methylene blue dye removal from aqueous solutions was investigated by Pavan et al. (2008) using yellow passion fruit peels (Passiflora edulis Sims. f. flavicarpa Degener), a powdered alternative low-cost adsorbent for solid waste (Cristina and Gushikem, 2008). The adsorption of malachite green using chemically modified HCHO (PP) and H_2SO_4 (APP) treated potato peels were investigated. The natural, methylated, and activated carbon taken from banana peels was used for the decolorisation of biologically treated pal oil mill (Anastopoulos and Kyzas, 2014). To enhance the active adsorption sites on biochar, the Fe_3O_4/GO/citrus peel-derived magnetic bio-nanocomposite (mGOCP) was synthesised by a facile one-pot hydrothermal approach to eliminate fluoroquinolone antibiotics ciprofloxacin (CIP) and sparfloxacin (SPA) efficiently from an aqueous solution, as shown in Figure 16.16. In this system, the Freundlich

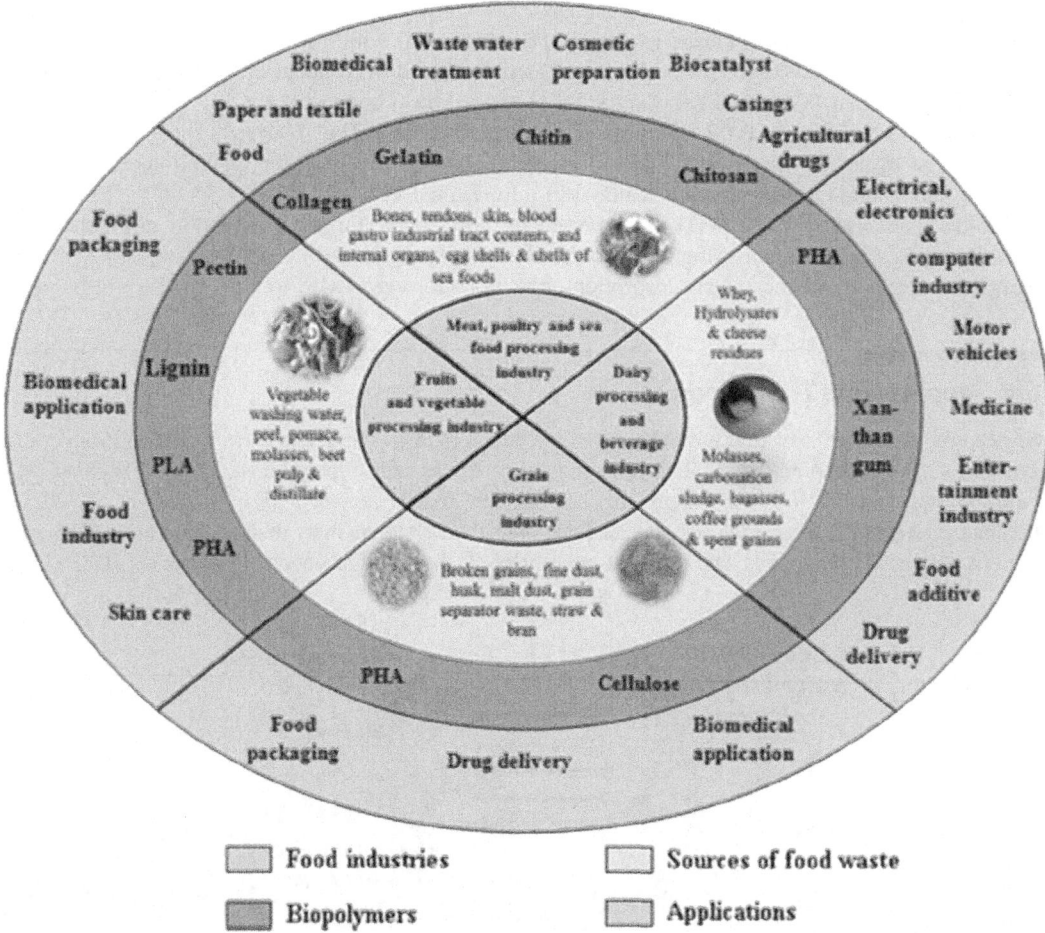

FIGURE 16.16 Applications of biopolymers produced from various waste streams. Reproduced from Ranganathan et al. (2020) from Heliyon.

TABLE 16.1

Several biopolymer-based materials utilised for removal of pollutants

S. No.	Biopolymer	Biopolymer-Based Materials	Pollutants	References
1	Cellulose	Cationic Cellulose Nanocrystals	Ag^+, Cu^{2+}, Fe^{2+}, Fe^{3+}	(Karim et al., 2016)
2		Xylan-type hemicellulosesg-acrylic acid (AA) ionic hydrogel	Pd^{2+}, Cd^{2+}, Zn^{2+}	(Peng et al., 2012)
3		Bio-based nanofibers	Cu^{2+}, Ni^{2+}, Cr^{3+}, Zn^{2+}	(Sehaqui et al., 2014)
4		Olysaccharide-epichlorohydrin (PS–EPI) copolymers	p-nitrophenol	(Dehabadi and Wilson, 2014)
5	Alginate	Superparamagnetic sodium alginate-coated Fe3O4 nanoparticles (Alg–Fe3O4)	Malachite Green	(Mohammadi et al., 2014)
6		Calcium alginate beads	Uranium	(Gok and Aytas, 2009)
7		Sodium alginate (SA) beads	Methylene Blue, Pb^{2+}, Cu^{2+}, Ni^{2+},	(Shao et al., 2018)
8	Chitosan	Bifunctionalised chitosan adsorbent	Cu^{2+}, Cr^{6+}	(Moreira et al., 2018)
9		Composite polymer adsorbent (Chitosan/polymethacrylic acid/halloysite nanotube	Pb^{2+}, Cd^{2+}	(Maity and Ray, 2018)
10		EGTA-modified chitosan	Pb^{2+}, Cd^{2+}	(Zhao et al., 2013)
11	Starch	Cross-linked amphoteric starch	Acid Light Yellow 2G, Acid Red G, Methyl Green, Methyl Violet	(Xu et al., 2006)
12	Cyclodextrin	γ-cyclodextrin (γ-CD) polymer	Polychlorobiphenyls	(Kawano et al., 2014)
13		Magnetic β-cyclodextrin-graphene oxide nanocomposites (Fe3O4/β-CD/GO)	Malachite Green	(Wang et al., 2015)
14	Dextran	N-benzyltriazole derivatised dextran	Methyl Violet	(Cho et al., 2015)
15		Cationic amphiphilic dextran hydrogels	Methyl Orange and Rose Bengal	(Stanciu and Nichifor, 2018)

model ($R^2 = 0.993$–0.996) was more suitable for antibiotic (CIP and SPA) adsorption in comparison to Langmuir and Tempkin models, which reveals that the adsorption process was surface multilayer and heterogeneous (Zhou et al., 2019; Table 16.1).

18.7 Conclusions

The excellent sorption capacities of biopolymers and bio-based materials are studied widely, considering their applicability in sustainable remediation. They are very easy to operate, biodegradable, low cost, and present abundantly in nature, which can be utilised before and after modification. In this chapter, several biopolymers like cellulose, chitosan, chitin, dextran, cyclodextrin, starch, alginate, etc. are discussed with their properties, characterisation, modifications, and utilisation in environmental remediation technologies. Recent developments of biomaterial-based systems in organic and inorganic pollutant removal including dyes, metals, gas adsorption, micropollutants, etc. have gathered the attention of industries and academics and are also discussed extensively.

REFERENCES

Ahmed, M., Azizi, S., Alloin, F., & Dufresne, A. (2005). Review of recent research into cellulosic whiskers, their properties and their application in nanocomposite field. *Biomacromolecules, 6*, 612–626.

Al, G., Aydemir, D., Kaygin, B., Ayrilmis, N., & Gunduz, G. (2018). Preparation and characterization of biopolymer nanocomposites from cellulose nanofibrils and nanoclays. *Journal of Composite Materials 52* (5), 689–700. 10.1177/0021998317713589.

Alhwaige, A.A., Ishida, H., & Qutubuddin, S. (2020). Chitosan/polybenzoxazine/clay mixed matrix composite aerogels: preparation, physical properties, and water absorbency. *Applied Clay Science, 184*(July 2019), 105403. 10.1016/j.clay.2019.105403.

Alizadeh Fard, M., Vosoogh, A., Barkdoll, B., & Aminzadeh, B. (2017). Using polymer coated nanoparticles for adsorption of micropollutants from water. *Colloids and Surfaces A: Physicochemical and Engineering Aspects, 531*, 189–197. 10.1016/j.colsurfa.2017.08.008.

Anastopoulos, I., & Kyzas, G.Z. (2014). Agricultural peels for dye adsorption: a review of recent literature. *Journal of Molecular Liquids.* 10.1016/j.molliq.2014.11.006.

Asadi, S., Eris, S., & Azizian, S. (2018). Alginate-based hydrogel beads as a biocompatible and efficient adsorbent for dye removal from aqueous solutions. *ACS Omega, 3*(11), 15140–15148. 10.1021/acsomega.8b02498.

Azeem, M., Batool, F., Iqbal, N., & Ikram-ul-Haq. (2017). *Algal-Based Biopolymers.* Elsevier. 10.1016/B978-0-12-812360-7.00001-X.

Bai, C., Wang, L., & Zhu, Z. (2019). Adsorption of Cr(III) and Pb(II) by graphene oxide/alginate hydrogel membrane: characterization, adsorption kinetics, isotherm and thermodynamics studies. *International Journal of Biological Macromolecules, Iii.* 10.1016/j.ijbiomac.2019.09.249.

Bayarı, S., & Severcan, F. (2005). FTIR study of biodegradable biopolymers: P (3HB), P (3HB-Co-4HB) and P (3HB-Co-3HV). *Journal of Molecular Structure, 747*, 529–534. 10.1016/j.molstruc.2004.12.029.

Bouyahmed, F., Cai, M., Reinert, L., Duclaux, L., Dey, R., Youcef, H., Lahcini, M., Muller, F., & Delpeux-Ouldriane, S. (2018). A wide adsorption range hybrid material based on chitosan, activated carbon and montmorillonite for water treatment. *C, 4*(2), 35. 10.3390/c4020035.

Brião, G.V., Jahn, S.L., Foletto, E.L., & Dotto, G.L. (2018). Highly efficient and reusable mesoporous zeolite synthetized from a biopolymer for cationic dyes adsorption. *Colloids and Surfaces A: Physicochemical and Engineering Aspects, 556*(August), 43–50. 10.1016/j.colsurfa.2018.08.019.

Budnyak, T.M., Yanovska, E.S., Kichkiruk, O.Y., Sternik, D., & Tertykh, V.A. (2016). Natural minerals coated by biopolymer chitosan: synthesis, physicochemical, and adsorption properties. *Nanoscale Research Letters 11* (1). 10.1186/s11671-016-1696-y.

Byun, J., Landfester, K., & Zhang, K.A.I. (2019). Conjugated Polymer Hydrogel Photocatalysts with Expandable Photoactive Sites in Water. *Chemistry of Materials, 31*(9), 3381–3387. 10.1021/acs.chemmater.9b00544.

Cai, J., Kimura, S., Wada, M., & Kuga, S. (2009). *Nanoporous Cellulose as Metal Nanoparticles Support, Biomacromolecules, 10*, 87–94. 10.1021/bm800919e.

Chen, X., Gao, Y., Wang, L., Chen, H., & Yan, N. (2015). Effect of treatment methods on chitin structure and its transformation into nitrogen-containing chemicals. *Chempluschem, 80*, 1565–1572. 10.1002/cplu.201500326.

Cho, E., Tahir, M.N., Kim, H., Yu, J.H., & Jung, S. (2015). Removal of Methyl Violet Dye by Adsorption onto N-Benzyltriazole Derivatized Dextran. *RSC Advances, 5*(43), 34327–34334. 10.1039/c5ra03317a.

Copeland, L., Blazek, J., Salman, H., & Tang, M.C. (2009). Form and functionality of starch. *Food Hydrocolloids, 23*(6), 1527–1534. 10.1016/j.foodhyd.2008.09.016.

Cristina, A., & Gushikem, Y. (2008). Removal of methylene blue dye from aqueous solutions by adsorption using yellow passion fruit peel as adsorbent. *Bioresource Technology, 99*, 3162–3165. 10.1016/j.biortech.2007.05.067.

De France, K.J., Hoare, T., & Cranston, E.D. (2017). Review of Hydrogels and Aerogels Containing Nanocellulose. *Chemistry of Materials, 29*(11), 4609–4631. 10.1021/acs.chemmater.7b00531.

Dehabadi, L., & Wilson, L.D. (2014). Polysaccharide-based materials and their adsorption properties in aqueous solution. *Carbohydrate Polymers, 113*, 471–479. 10.1016/j.carbpol.2014.06.083.

Demirbilek, C., & Dinc, C.O. (2016). Diethylaminoethyl dextran/epichlorohydrin (DEAE-D/ECH) hydrogel as adsorbent for murexide. *Desalination and Water Treatment, 57*, 6884–6893. 10.1080/19443994.2015.1013506.

Dhanavel, S., Nivethaa, E.A.K., Narayanan, V., & Stephen, A. (2014). Photocatalytic activity of chitosan/ZnO nanocomposites for degrading methylene blue. *International Journal of ChemTech Research, 6*(3), 1880–1882.

Dong, C., Lu, J., Qiu, B., Shen, B., Xing, M., & Zhang, J. (2018). Developing stretchable and graphene-oxide-based hydrogel for the removal of organic pollutants and metal ions. *Applied Catalysis B: Environmental, 222*(October 2017), 146–156. 10.1016/j.apcatb.2017.10.011.

El Hankari, S., Bousmina, M., & El Kadib, A. (2019). Biopolymer@metal-organic framework hybrid materials: a critical survey. *Progress in Materials Science, 106*(May), 100579. 10.1016/j.pmatsci.2019.100579.

George, J. (2015). Cellulose nanocrystals: synthesis, functional properties, and applications. *Nanotechnology, Science and Applications, 8*, 45–54.

Gimbert, F., Morin-Crini, N., Renault, F., Badot, P.M., & Crini, G. (2008). Adsorption isotherm models for dye removal by cationized starch-based material in a single component system: error analysis. *Journal of Hazardous Materials, 157*(1), 34–46. 10.1016/j.jhazmat.2007.12.072.

Gnanavel, P., Poongodi, S., & Ananthakrishnan, T. (2010). Characterization techniques. *Man-Made Textiles in India, 53*(2), 52–57. 10.1201/9781315229263-4.

Gok, C., & Aytas, S. (2009). Biosorption of uranium(VI) from aqueous solution using calcium alginate beads. *Journal of Hazardous Materials, 168*(1), 369–375. 10.1016/j.jhazmat.2009.02.063.

Guzzella, L., Feretti, D., & Monarca, S. (2002). Advanced oxidation and adsorption technologies for organic micropollutant removal from lake water used as drinking-water supply. *Water Research, 36*(17), 4307–4318. 10.1016/S0043-1354(02)00145-8.

Hakam, A., Rahman, I.A., Jamil, M. S., Othaman, R., Amin, M.C.I.M., & Lazim, A.M. (2015). Removal of methylene blue dye in aqueous solution by sorption on a bacterial-g-poly-(acrylic acid) polymer network hydrogel. *Sains Malaysiana, 44*(6), 827–834. 10.17576/jsm-2015-4406-08.

Hedges, A. (2009). Cyclodextrins: properties and applications. *Starch*, 833–851. 10.1016/B978-0-12-746275-2.00022-7.

Hijosa-Valsero, M., Matamoros, V., Sidrach-Cardona, R., Martín-Villacorta, J., Bécares, E., & Bayona, J.M. (2010). Comprehensive assessment of the design configuration of constructed wetlands for the removal of pharmaceuticals and personal care products from urban wastewaters. *Water Research, 44*(12), 3669–3678. 10.1016/j.watres.2010.04.022.

Hoare, T.R., & Kohane, D.S. (2008). Hydrogels in drug delivery: progress and challenges*. Polymers with aligned carbon nanotubes: active composite materials. *Polymer, 49*(8), 1993–2007. 10.1016/j.polymer.2008.01.027.

Huerta-Fontela, M., Galceran, M.T., & Ventura, F. (2011). Occurrence and removal of pharmaceuticals and hormones through drinking water treatment. *Water Research, 45*(3), 1432–1442. 10.1016/j.watres.2010.10.036.

Ifuku, S. (2014). Chitin and chitosan nanofibers: preparation and chemical modifications. *Molecules, 19*(11), 18367–18380. 10.3390/molecules191118367.

Ip, C.C.M., Li, X.D., Zhang, G., Wong, C.S.C., & Zhang, W.L. (2005). Heavy metal and pb isotopic compositions of aquatic organisms in the Pearl River Estuary, South China. *Environmental Pollution, 138*, 494–504. 10.1016/j.envpol.2005.04.016.

Jamnongkan, T., & Singcharoen, K. (2016). Towards novel adsorbents: the ratio of PVA/chitosan blended hydrogels on the copper (II) ion adsorption. *Energy Procedia, 89*(Ii), 299–306. 10.1016/j.egypro.2016.05.038.

Jiang, F., & Hsieh, Y. (2018). Dual wet and dry resilient cellulose II fibrous aerogel for hydrocarbon – water separation and energy storage applications. *ACS Omega, 3*, 3530–3539. 10.1021/acsomega.8b00144.

Jiang, C., Wang, X., Wang, G., Hao, C., Li, X., & Li, T. (2019). Adsorption performance of a polysaccharide composite hydrogel based on crosslinked glucan/chitosan for heavy metal ions. *Composites Part B: Engineering, 169*(April), 45–54. 10.1016/j.compositesb.2019.03.082.

Junfeng, L., et al. (2017). Biomass based hydrogel as an adsorbent for the fast removal of heavy metal ions from aqueous solutions. *Journal of Materials Chemistry A*, 1–13. 10.1039/C6TA10513K.

Karim, Z., Mathew, A.P., Kokol, V., Wei, J., & Grahn, M. (2016). High-flux affinity membranes based on cellulose nanocomposites for removal of heavy metal ions from industrial effluents. *RSC Advances, 6*(25), 20644–20653. 10.1039/c5ra27059f.

Kawano, S., Kida, T., Miyawaki, K., Noguchi, Y., Kato, E., Nakano, T., & Akashi, M. (2014). Cyclodextrin polymers as highly effective adsorbents for removal and recovery of polychlorobiphenyl (PCB) contaminants in insulating oil. *Environmental Science and Technology*, *48*(14), 8094–8100. 10.1021/es5 01243v.

Khan, F., & Ahmad, S.R. (2013). Polysaccharides and their derivatives for versatile tissue engineering application. *Macromolecular Bioscience*, *13*, 395–421. 10.1002/mabi.201200409.

Klemm, D., Heublein, B., Fink, H., & Bohn, A. (2005). Polymer science cellulose: fascinating biopolymer and sustainable raw material. *Angewandte*, 3358–3393. 10.1002/anie.200460587.

Klemm, D., Cranston, E.D., Fischer, D., Gama, M., Kedzior, S.A., Kralisch, D., Kramer, F., Kondo, T., Lindström, T., Nietzsche, S., et al. (2018). Nanocellulose as a natural source for groundbreaking applications in materials science: today's state. *Materials Today*, *xxx*(xx). 10.1016/j.mattod.2018.02.001.

Kosera, V.S., Cruz, T.M., Chaves, E.S., & Tiburtius, E.R.L. (2017). Triclosan degradation by heterogeneous photocatalysis using ZnO immobilized in biopolymer as catalyst. *Journal of Photochemistry and Photobiology A: Chemistry*, *344*, 184–191. 10.1016/j.jphotochem.2017.05.014.

Kumar, M.N.V.R. (2000). A review of chitin and chitosan applications. *Reactive and Functional Polymers*, *46*, 1–27.

Kumar, N., Mittal, H., Alhassan, S.M., & Ray, S.S. (2018). Bionanocomposite hydrogel for the adsorption of dye and reusability of generated waste for the photodegradation of ciprofloxacin: a demonstration of the circularity concept for water purification. *ACS Sustainable Chemistry & Engineering*, *6*(12), 17011–17025. 10.1021/acssuschemeng.8b04347.

Kumar, A., Naushad, M., Rana, A., Inamuddin, Preeti Sharma, G., Ghfar, A.A., Stadler, F.J., & Khan, M.R. (2017). ZnSe-WO3 nano-hetero-assembly stacked on gum ghatti for photo-degradative removal of bisphenol a: symbiose of adsorption and photocatalysis. *International Journal of Biological Macromolecules*, *104*, 1172–1184. 10.1016/j.ijbiomac.2017.06.116.

Kumar, A., Sharma, G., Naushad, M., Al-Muhtaseb, A.H., García-Peñas, A., Mola, G.T., Si, C., & Stadler, F.J. (2019). Bio-inspired and biomaterials-based hybrid photocatalysts for environmental detoxification: a review. *Chemical Engineering Journal*, 122937. 10.1016/j.cej.2019.122937.

Kundu, D., Mondal, S.K., & Banerjee, T. (2019). Development of β-cyclodextrin-cellulose/hemicellulose-based hydrogels for the removal of Cd (II) and Ni (II): synthesis, kinetics, and adsorption aspects. *Journal of Chemical & Engineering Data*, *64*, 2601–2617. 10.1021/acs.jced.9b00088.

Kurita, K. (2001). Controlled functionalization of the polysaccharide chitin. *Progress in Polymer Science*, *26*, 1921–1971.

Kyrikou, I., & Briassoulis, Æ.D. (2007). Biodegradation of agricultural plastic films: a critical review biodegradation of agricultural plastic films: a critical review. *Journal of Polymers and the Environment*, *15*, 125–150. 10.1007/s10924-007-0053-8.

Lee, J.-K., Patel, S.K.S., Sung, B.H., & Kalia, V.C. (2019). Biomolecules from municipal and food industry wastes: an overview. *Bioresource Technology*, *122346*. 10.1016/j.biortech.2019.122346.

Li, L., Wei, Z., Liu, X., Yang, Y., Deng, C., Yu, Z., Guo, Z., Shi, J., Zhu, C., Guo, W., et al. (2019). Biomaterials cross-linkedgraphene oxide composite aerogel with a macro–nanoporous network structure forefficient Cr (VI) removal. *International Journal of Biological Macromolecules*, *156*, 1337–1346. 10.1016/j.apcatb.2019.118214.

Liu, K.K., Jin, B., & Meng, L.Y. (2018). Glucose/graphene-based aerogels for gas adsorption and electric double layer capacitors. *Polymers (Basel)*. *11*(1). 10.3390/polym11010040.

Mahmoud, G.A. (2013). Adsorption of copper(II), lead(II), and cadmium(II) Ions from aqueous solution by using hydrogel with magnetic properties. *Monatshefte fur Chemie*, *144*(8), 1097–1106. 10.1007/s00706-013-0957-z.

Maiti, S., & Kumari, L. (2016). *Smart Nanopolysaccharides for the Delivery of Bioactives*. Elsevier Inc. 10.1 016/B978-0-323-47347-7.00003-3.

Maity, J., & Ray, S.K. (2018). Chitosan based nano composite adsorbent – synthesis, characterization and application for adsorption of binary mixtures of Pb(II) and Cd(II) from water. *Carbohydrate Polymers*, *182*, 159–171. 10.1016/j.carbpol.2017.10.086.

Mäkipää, J. (2015). Food waste conversion into biopolymers and other high value-added products in Hong Kong. *Feasibility Study*, November.

McHugh, D.J. (2003). *A Guide to the Seaweed Industry*. Food and Agriculture Organization of the United Nations.

Mohammadi, A., Daemi, H., & Barikani, M. (2014). Fast removal of malachite green dye using novel superparamagnetic sodium alginate-coated Fe_3O_4 nanoparticles. *International Journal of Biological Macromolecules, 69*, 447–455. 10.1016/j.ijbiomac.2014.05.042.

Mojiri, A., Kazeroon, R.A., & Gholami, A. (2019). Cross-linked magnetic chitosan/activated biochar for removal of emerging micropollutants from water: optimization by the artificial neural network. *Water (Switzerland), 11*(3), 1–18. 10.3390/w11030551.

Moon, R.J., Martini, A., Nairn, J., Youngblood, J., Martini, A., & Nairn, J. (2011). Cellulose nanomaterials review: structure, properties and nanocomposites. *Chemical Society Reviews, 40*, 3941–3994. 10.1039/c0cs00108b.

Moreira, A.L. da S.L., Pereira, A. de S., Speziali, M.G., Novack, K.M., Gurgel, L.V.A., & Gil, L.F. (2018). Bifunctionalized chitosan: a versatile adsorbent for removal of Cu(II) and Cr(VI) from aqueous solution. *Carbohydrate Polymers, 201*, 218–227. 10.1016/j.carbpol.2018.08.055.

Narula, A., & Rao, C.P. (2019). Hydrogel of the supramolecular complex of graphene oxide and sulfonatocalix[4]arene as reusable material for the degradation of organic dyes: demonstration of adsorption and degradation by spectroscopy and microscopy. *ACS Omega, 4*(3), 5731–5740. 10.1021/acsomega.9b00545.

Nilani, P., Raveesha, P., Rahul Nandkumar, B., Kasthuribai, N., Duraisamy, B., Dhamodaran, P., & Elango, K. (2010). Formulation and evaluation of polysaccharide based biopolymer – an ecofriendly alternative for synthetic polymer. *Journal of Pharmaceutical Sciences and Research, 2*(3), 178–184.

Numata, K. (2015). Poly (amino acid) s/polypeptides as potential functional and structural materials. *Polymer Journal*, 1–9. 10.1038/pj.2015.35.

Olad, A., & Azhar, F.F. (2014). Eco-friendly biopolymer/clay/conducting polymer nanocomposite: characterization and its application in reactive dye removal. *Fibers and Polymers, 15*(6), 1321–1329. 10.1007/s12221-014-1321-6.

Pandey, S., Do, J.Y., Kim, J., & Kang, M. (2020). Fast and highly efficient removal of dye from aqueous solution using natural locust bean gum based hydrogels as adsorbent. *International Journal of Biological Macromolecules, 143*, 60–75. 10.1016/j.ijbiomac.2019.12.002.

Pavan, F. A., Mazzocato, A. C., & Gushikem, Y. (2008). Removal of methylene blue dye from aqueous solutions by adsorption using yellow passion fruit peel as adsorbent. *Bioresource Technology, 99*, 3162–3165. 10.1016/j.biortech.2007.05.067.

Peng, X., Zhong, L., Ren, J., & Sun, R. (2012). Highly effective adsorption of heavy metal ions from aqueous solutions by macroporous xylan-rich hemicelluloses-based hydrogel. *Journal of Agricultural and Food Chemistry*. 10.1021/jf300387q.

Peng, N., Hu, D., Zeng, J., Li, Y., Liang, L., & Chang, C. (2016). Superabsorbent cellulose – clay nanocomposite hydrogels for highly efficient removal of dye in water. *ACS Sustainable Chemistry & Engineering, 4*, 7217–7224. 10.1021/acssuschemeng.6b02178.

Pereira, A.G.B., Muniz, E.C., & Hsieh, Y. (2015). 1H NMR and 1H–13C HSQC surface characterization of chitosan – chitin sheath-core nanowhiskers. *Carbohydrate Polymers, 123*, 46–52. 10.1016/j.carbpol.2015.01.017.

Pewarna, P., Biru, M., & Akueus, L. (2015). Removal of methylene blue dye in aqueous solution by sorption on a bacterial-g-poly-(acrylic acid). *Polymer Network Hydrogel, 44*(6), 827–834.

Pour, Z.S., & Ghaemy, M. (2015). Removal of dyes and heavy metal ions from water by magnetic hydrogel beads based on polyvinyl. *RSC Advances*, 64106–64118. 10.1039/c5ra08025h.

Qi, W., Singer, H., Berg, M., Müller, B., Pernet-Coudrier, B., Liu, H., & Qu, J. (2015). Elimination of polar micropollutants and anthropogenic markers by wastewater treatment in Beijing, China. *Chemosphere, 119*, 1054–1061. 10.1016/j.chemosphere.2014.09.027.

Ranganathan, S., Dutta, S., Moses, J.A., & Anandharamakrishnan, C. (2020). Utilization of food waste streams for the production of biopolymers. *Heliyon, 6*, e04891. 10.1016/j.heliyon.2020.e04891.

Reddy, R.L., Reddy, V.S., & Gupta, G.A. (2013). Study of bio-plastics as green & sustainable alternative to plastics. *International Journal of Emerging Technology and Advanced Engineering, 3*(5), 82–89.

Rizzi, V., Gubitosa, J., Fini, P., Romita, R., Nuzzo, S., & Cosma, P. (2019). Chitosan biopolymer from crab shell as recyclable film to remove/recover in batch ketoprofen from water: understanding the factors affecting the adsorption process. *Materials (Basel), 12*(23), 3810.

S Dassanayake, R., Acharya, S., & Abidi, N. (2019). Biopolymer-based materials from polysaccharides: properties, processing, characterization and sorption applications. *Advanced Sorption Process Applications*, 1–24. 10.5772/intechopen.80898.

Sajna, K.V., Gottumukkala, L.D., Sukumaran, R.K., & Pandey, A. (2015). *White Biotechnology in Cosmetics*. Elsevier B.V. 10.1016/B978-0-444-63453-5.00020-3.

Salih, S.S., & Ghosh, T.K. (2017). Preparation and characterization of bioadsorbent beads for chromium and zinc ions adsorption. *Cogent Environmental Science*, *3*(1), 1–14. 10.1080/23311843.2017.1401577.

Salzano de Luna, M., Ascione, C., Santillo, C., Verdolotti, L., Lavorgna, M., Buonocore, G.G., Castaldo, R., Filippone, G., Xia, H., & Ambrosio, L. (2019). Optimization of dye adsorption capacity and mechanical strength of chitosan aerogels through crosslinking strategy and graphene oxide addition. *Carbohydrate Polymers*, *211*, 195–203. 10.1016/j.carbpol.2019.02.002.

Sarwat, F., Qader, S.A.U., Aman, A., & Ahmed, N. (2008). Production & characterization of a unique dextran from an indigenous leuconostoc mesenteroides CMG713. *International Journal of Biological Sciences*, *4*(6), 379–386. 10.7150/ijbs.4.379.

Sehaqui, H., de Larraya, U.P., Liu, P., Pfenninger, N., Mathew, A.P., Zimmermann, T., & Tingaut, P. (2014). Enhancing adsorption of heavy metal ions onto biobased nanofibers from waste pulp residues for application in wastewater treatment. *Cellulose*, *21*(4), 2831–2844. 10.1007/s10570-014-0310-7.

Shao, Z.j., Huang, X.l., Yang, F., Zhao, W.f., Zhou, X.z., & Zhao, C.s. (2018). Engineering sodium alginate-based cross-linked beads with high removal ability of toxic metal ions and cationic dyes. *Carbohydrate Polymers*, *187*(November 2017), 85–93. 10.1016/j.carbpol.2018.01.092.

Shojaeiarani, J., Bajwa, D.S., Rehovsky, C., Bajwa, S.G., & Vahidi, G. (2019). Deterioration in the physico-mechanical and thermal properties of biopolymers due to reprocessing. *Polymers (Basel)*, *11*, 1–17. 10.3390/polym11010058.

Singha, A.S., & Guleria, A. (2014). Use of low cost cellulosic biopolymer based adsorbent for the removal of toxic metal ions from the aqueous solution. *Separation Science and Technology*, *49*(16), 2557–2567. 10.1080/01496395.2014.929146.

Smirnova, I., Suttiruengwong, S., Seiler, M., & Arlt, W. (2004). Dissolution rate enhancement by adsorption of poorly soluble drugs on hydrophilic silica aerogels. *Pharmaceutical Development and Technology*, *9*(4), 443–452. 10.1081/PDT-200035804.

Stanciu, M.C., & Nichifor, M. (2018). Influence of dextran hydrogel characteristics on adsorption capacity for anionic dyes. *Carbohydrate Polymers*, *199*, 75–83. 10.1016/j.carbpol.2018.07.011.

Stergar, J. (2016). Review of aerogel-based materials in biomedical applications. 10.1007/s10971-016-3968-5.

Sui, K., Li, Y., Liu, R., Zhang, Y., Zhao, X., Liang, H., & Xia, Y. (2012). Biocomposite fiber of calcium alginate/multi-walled carbon nanotubes with enhanced adsorption properties for ionic dyes. *Carbohydrate Polymers*, *90*(1), 399–406. 10.1016/j.carbpol.2012.05.057.

Szejtli, J. (1998). Introduction and general overview of cyclodextrin chemistry. *Chemical Reviews*, *98*(5), 1743–1753. 10.1021/cr970022c.

Taboada, C., Millán, R., & Míguez, I. (2010). Composition, nutritional aspects and effect on serum parameters of marine algae Ulva rigida. *Journal of the Science of Food and Agriculture*, *90*(3), 445–449. 10.1002/jsfa.3836.

Tan, S.C., Tan, T.K., Wong, S.M., & Khorb, E. (1996). The chitosan yield of zygomycetes at their optimum harvesting time. *Carbohydrate Polymers*, *30*, 239–242.

Tanaka, K., Yamamoto, K., & Kadokawa, J. (2014). Facile nanofibrillation of chitin derivatives by gas bubbling and ultrasonic treatments in water. *Carbohydrate Research*, *398*, 25–30. 10.1016/j.carres.2014.08.008.

Thomas, S. (2000). Alginate dressings in surgery and wound management: part 2. *Journal of Wound Care*, *9*(3), 1–5.

Tu, H., Yu, Y., Chen, J., Shi, X., Zhou, J., Deng, H., & Du, Y. (2017). Highly cost-effective and high-strength hydrogels as dye adsorbents from natural polymers chitosan and cellulose. *Polymer Chemistry*, *8* (19), 2913–2921. 10.1039/c7py00223h.

Tyler, R.T. (2004). Analytical, biochemical and physicochemical aspects of starch granule size, with emphasis on small granule starches. A review. *Starch*, *56*, 89–99. 10.1002/star.200300218.

Velde, K.V.D., & Kiekens, P. (2004). Structure analysis and degree of substitution of chitin, chitosan and dibutyrylchitin by FT-IR spectroscopy and solid state 13 C NMR. *Carbohydrate Polymers*, *58*, 409–416. 10.1016/j.carbpol.2004.08.004.

Wang, W., Kang, Y., & Wang, A. (2013). One-step fabrication in aqueous solution of a granular alginate-based hydrogel for fast and efficient removal of heavy metal ions. *Journal of Polymer Research, 20*, 101. 10.1007/s10965-013-0101-0.

Wang, S., Lu, A., & Zhang, L. (2016). Progress in polymer science recent advances in regenerated cellulose materials. *Progress in Polymer Science, 53*, 169–206. 10.1016/j.progpolymsci.2015.07.003.

Wang, D., Liu, L., Jiang, X., Yu, J., & Chen, X. (2015). Adsorption and removal of malachite green from aqueous solution using magnetic β-cyclodextrin-graphene oxide nanocomposites as adsorbents. *Colloids and Surfaces A: Physicochemical and Engineering Aspects, 466*, 166–173. 10.1016/j.colsurfa.2014.11.021.

Wang, S.F., Shen, L., Tong, Y.J., Chen, L., Phang, I.Y., Lim, P.Q., & Liu, T.X. (2005). Biopolymer chitosan/montmorillonite nanocomposites: preparation and characterization. *Polymer Degradation and Stability, 90*, 123–131. 10.1016/j.polymdegradstab.2005.03.001.

Wang, Q., Li, J., Bai, Y., Lu, X., Ding, Y., Yin, S., Huang, H., Ma, H., Wang, F., & Su, B. (2013). Photodegradation of textile dye rhodamine B over a novel biopolymer–metal complex wool-Pd/CdS photocatalysts under visible light irradiation. *Journal of Photochemistry and Photobiology B: Biology, 126*, 47–54. 10.1016/j.jphotobiol.2013.07.007.

Xu, S., Wang, J., Wu, R., Wang, J., & Li, H. (2006). Adsorption behaviors of acid and basic dyes on crosslinked amphoteric starch. *Chemical Engineering Journal, 117*(2), 161–167. 10.1016/j.cej.2005.12.012.

Yadav, M., Das, M., Savani, C., Thakore, S., & Jadeja, R. (2019). Maleic anhydride cross-linked β-cyclodextrin-conjugated magnetic nanoadsorbent: an ecofriendly approach for simultaneous adsorption of hydrophilic and hydrophobic dyes. *ACS Omega, 4*, 11993–12003. 10.1021/acsomega.9b00881.

Yang, Y.Y., Ma, S., Wang, X.X., & Zheng, X.L. (2017). Modification and application of dietary fiber in foods. *Journal of Chemistry, 2017*. 10.1155/2017/9340427.

Yang, Y., Xie, Y., Pang, L., Li, M., Song, X., Wen, J., & Zhao, H. (2013). Preparation of reduced graphene oxide/poly (acrylamide) nanocomposite and its adsorption of Pb (II) and methylene blue. *Langmuir, 29*, 10727–10736.

Yuan, Z., Wang, J., Wang, Y., Liu, Q., Zhong, Y., Wang, Y., Li, L., Lincoln, S.F., & Guo, X. (2019). Preparation of a poly(acrylic acid) based hydrogel with fast adsorption rate and high adsorption capacity for the removal of cationic dyes. *RSC Advances, 9*(37), 21075–21085. 10.1039/c9ra03077h.

Zhang, H., Omer, A.M., Hu, Z., Yang, L.Y., Ji, C., & Ouyang, X.k. (2019). Fabrication of magnetic bentonite/carboxymethyl chitosan/sodium alginate hydrogel beads for Cu (II) adsorption. *International Journal of Biological Macromolecules, 135*, 490–500. 10.1016/j.ijbiomac.2019.05.185.

Zhao, F., Repo, E., Yin, D., & Sillanpää, M.E.T. (2013). Adsorption of Cd(II) and Pb(II) by a novel EGTA-modified chitosan material: kinetics and isotherms. *Journal of Colloid and Interface Science, 409*, 174–182. 10.1016/j.jcis.2013.07.062.

Zhao, F., Repo, E., Yin, D., Chen, L., Kalliola, S., Tang, J., Iakovleva, E., Tam, K.C., & Sillanpää, M. (2017). One-pot synthesis of trifunctional chitosan-EDTA-β-cyclodextrin polymer for simultaneous removal of metals and organic micropollutants. *Scientific Reports, 7*(1), 1–14. 10.1038/s41598-017-16222-7.

Zhao, S., Malfait, W. J., Guerrero-Alburquerque, N., Koebel, M. M., & Nyström, G. (2018). Biopolymer aerogels and foams: Chemistry, properties, and applications. *Angewandte Chemie International Edition, 57*(26), 7580–7608. 10.1002/anie.201709014.

Zhou, Y., Cao, S., Xi, C., Li, X., Zhang, L., Wang, G., & Chen, Z. (2019). A novel Fe_3O_4/graphene oxide/citrus peel-derived bio-char based nanocomposite with enhanced adsorption affinity and sensitivity of ciprofloxacin and sparfloxacin. *Bioresource Technology, 292*, 121951. 10.1016/j.biortech.2019.121951.

17 Nanocatalyst Synthesis by the Green Route: Mechanism and Application

Naresh K Sethy, Zeenat Arif, PK Mishra, P Kumar, and Rajesh Saha
Department of Chemical Engineering, IIT (BHU), Varanasi, UP, India

CONTENTS

- 17.1 Introduction 321
  - 17.1.1 Nanoparticle Synthesis by Different Approaches 323
  - 17.1.2 Why We Go for Green Synthesis (Plants/Microorganisms) over Chemical Synthesis 324
  - 17.1.3 Green Synthesis via Plants and Microorganisms 325
- 17.2 Nanoparticle Synthesis Using Microorganisms 325
  - 17.2.1 Mechanism Behind the Nanoparticle Synthesis by Microorganisms 326
    - 17.2.1.1 Oxidoreductase Enzymes for Bioreduction 327
    - 17.2.1.2 Mediated Reduction by NADH-Dependent Reductase 329
    - 17.2.1.3 Mediated Reduction by Nitrate/Nitrite Reductase 329
- 17.3 Metallic Nanoparticles Synthesis via Different Parts of Plants 329
  - 17.3.1 Mechanism Behind Nanoparticle Synthesis by Different Parts of Plants 331
- 17.4 Application of Nanoparticles Synthesised Through Microorganisms and Plant Extracts 332
- 17.5 Conclusion 332
- References 337

17.1 Introduction

Nanotechnology is one of the rapidly emerging fields of 21st century and covers a wide range of technologies that operate at the nanoscale to cater to various industrial needs. The behaviour of materials at the nanoscale is often found to be desirable in comparison to macro-scale due to size confinement and high surface-to-volume ratios (Singh and Abdullah, 2016). These unique properties of nanostructured materials, nanoparticles, and other related nanotechnologies provide them good catalytic activity, increased strength, improved photo activity, and many other interesting characteristics with applications in numerous fields such as chemical industry, synthetic biology, electronics, drug-gene delivery (Pissuwan et al., 2011), photo catalysis (Amal and Low, 1999), water treatment (Xu et al., 2012), and photo-electrochemical applications (Chen et al., 2006). Various applications of nanoparticles in different fields are illustrated in Figure 17.1.

In this chapter, a primary concern is laid upon the applications of nanoparticles as bio-catalysts for environmental remediation.

Catalysts are chemical species that enable a ramp up in reaction speeds in a targeted chemical reaction, wherein photocatalysts are employed for ramping up the chemical reactions in the presence of UV light (Singh and Abdullah, 2016). Photocatalytic systems must possess proper band gap, good morphology, more exposed surface area, high stability, and better reusability.

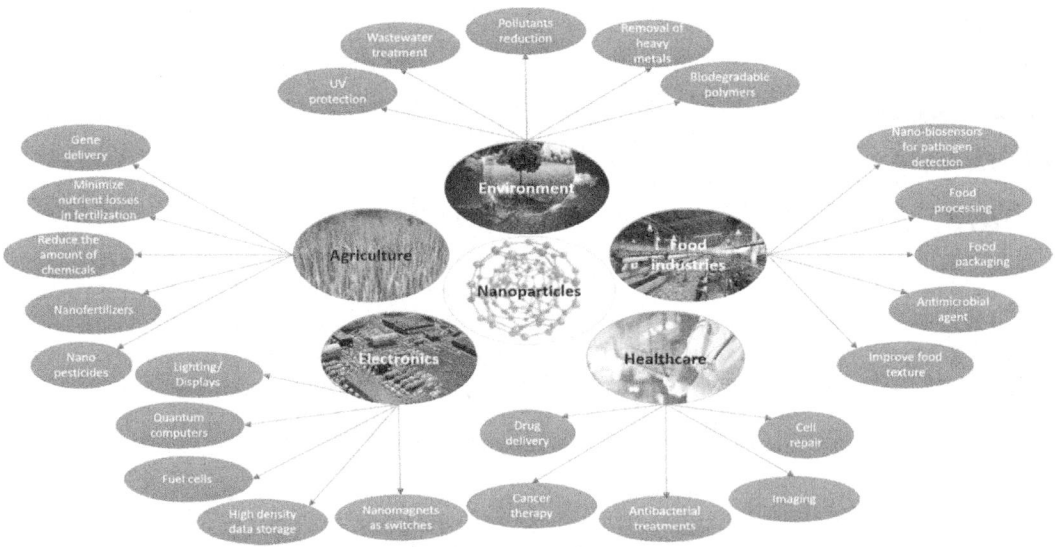

FIGURE 17.1 Applications of nanoparticles.

An ideal photocatalyst, besides being highly photoactive, must also be stable, inexpensive, and nontoxic. Another prominent criterion in applications, especially for the degradation of organic compounds, is that the redox potential of the $H_2O/•OH$ couple ($OH^- \rightarrow •OH + e^-$; $E^0 = -2.8$ V) must lie within the band gap of the semiconductor. Choice of semiconductor material based on the targeted application stands important and there are several nanoparticles such as TiO_2, ZnO, SiO_2, α-Fe_2O_3, Ag, Au, CuO, SnO_2, etc. with band gap energies sufficient for catalysing a wide range of chemical reactions (Amal and Low, 1999).

Photocatalysis refers to the redox reactions occurring on the surfaces of photocatalyst material initiated by the absorption of UV light radiation. The phenomenon is commenced when a photon is absorbed by a semiconductor having energy ≥ band gap of the semiconductor material. This absorption enables the excitation of an electron (e−) from the valence band (VB) to the conduction band (CB) of the semiconductor, thus generating vacant positive charge holes (h+) in the valence band. However, the recombination process of the electron and the hole must be suppressed as it affects the efficiency of the photocatalytic process because those excitations are not involved in chemical reactions. Holes react with water to generate hydroxyl radicals (HO•) and forms hydrogen gas, while negatively charged electrons react with organic species or react with the molecular oxygen (O_2), reducing it to superoxide radical anion ($O_2^{-•}$). Hydroxyl radicals (HO•) oxidize organic pollutants into non-toxic materials and also disinfect few bacteria and viruses (Singh and Abdullah, 2016). This phenomenon is described in Figure 17.2.

Photocatalysis plays a significant role in conversion of toxic and hazardous organic compounds to non-toxic products like CO_2 and H_2O through degradation and mineralisation. Decomposition of the air pollutants such as NO_2 and CO, destruction of all the waterborne microorganisms, and green synthesis of industrially important chemicals are done by the photocatalysis method (Singh and Abdullah, 2016).

It is found that metal nanoparticles function as excellent photocatalysts and are used in photocatalytic degradation of several dyes, organic compounds, and pollutants present in wastewater.

Many processes such as adsorption, advanced oxidation, air stripping, etc., which are currently in use to treat various toxins in wastewaters, targets removal by concentrating the chemicals present to facilitate removal but do not convert them into non-toxic forms. In this regard, photocatalytic processes attract the application with no requirement of secondary disposal methods.

Another advantage of this process is that it also does not demand the use of expensive oxidising chemicals (O_3, H_2O_2, etc.) compared to other advanced oxidation technologies, as ambient oxygen acts

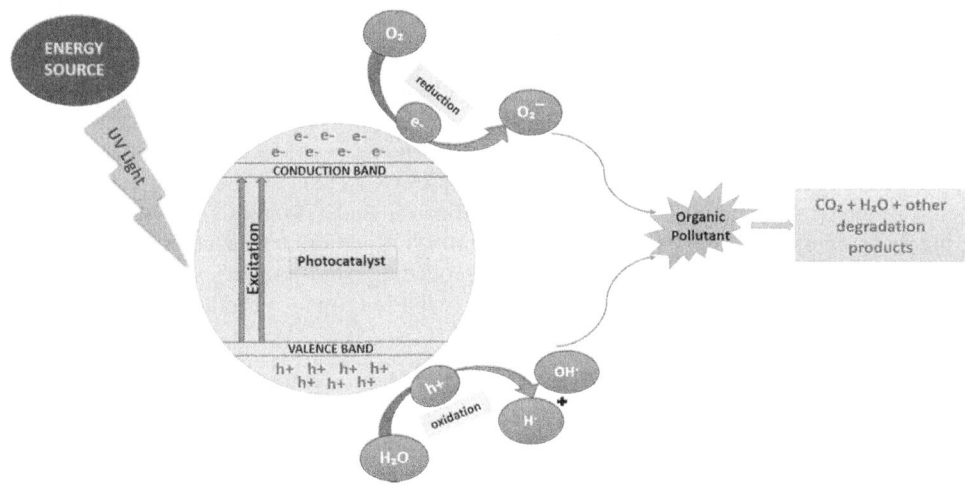

FIGURE 17.2 Phenomenon of photocatalytic degradation.

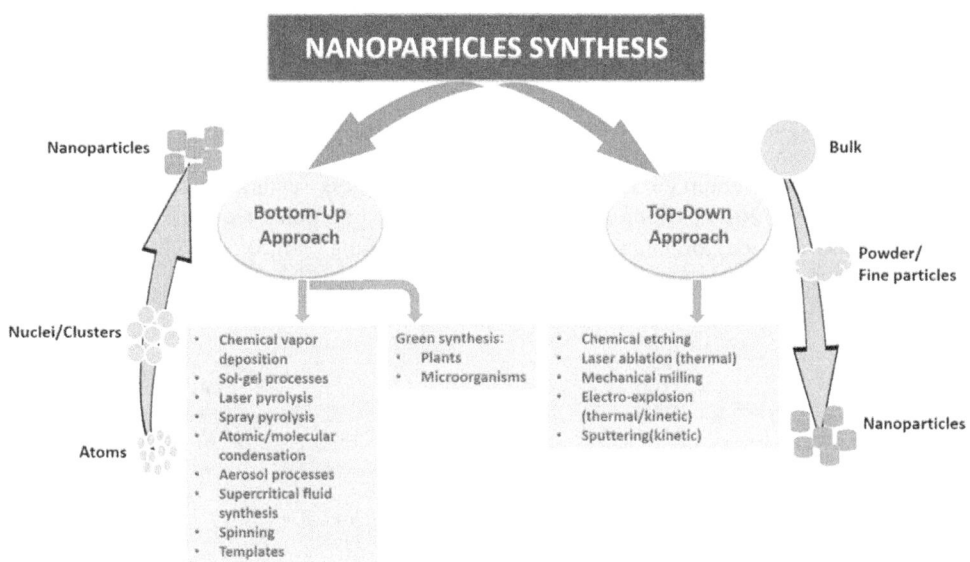

FIGURE 17.3 Top-down approach and bottom-up approach.

as the oxidant. Also, self-regenerated reuse and recycling is one added advantage of photocatalysts. Methods such as solar photocatalysis also bring advantage of use of renewable source of energy, making the process more viable for targeted applications (Amal and Low, 1999).

17.1.1 Nanoparticle Synthesis by Different Approaches

Generally, the synthesis of nanoparticles follows two methods, either a "top-down" approach or "bottom-up" approach (Iqbal et al., 2012), as shown in Figure 17.3. The top-down approach is a process of breaking down bulk materials (macro-crystalline) into nanosized particles. This approach includes methods like chemical etching (Hu et al., 2012), laser ablation (Ayyub et al., 2001), mechanical milling (Indris et al., 2005), electro-explosion (Indris et al., 2005), and sputtering (Ayyub et al., 2001). The bottom-up approach involves the building of nanomaterials from the atomic scale. This approach is further subdivided into two types: chemical approach and green approach. Methods like chemical vapour deposition (Swihart, 2003), sol-gel processes (Mackenzie and Bescher, 2007), laser pyrolysis (Figgemeier et al., 2007), aerosol processes (Xia et al., 2001), and spinning are said to be chemical approaches. In contrast, the green approach includes synthesis of metal nanoparticles using plants or microorganisms. The bottom-up approach is more advantageous than the top-down approach because the former has a better chance of producing nanostructures with less defects, more homogeneous chemical composition, and better short- and long-range ordering.

17.1.2 Why We Go for Green Synthesis (Plants/Microorganisms) over Chemical Synthesis

The synthesis of metal nanoparticles can be done by using various physical and chemical methods like sol-gel processes, laser pyrolysis, spray pyrolysis, chemical etching, mechanical milling, etc., as discussed previously. These processes require immense labour, toxic chemicals, operate in high temperature and high pressure, and ethical capital. Moreover, the accumulation of chemical waste released pollutes the environment because it cannot be recycled or reused. To counter these limitations, it becomes an obvious need to switch over to an alternate approach for the synthesis of metal nanoparticles that is cost effective, requires minimal/no use of harsh and toxic chemicals, doesn't require high temperature or high pressure, and is environmentally friendly. These requirements are fulfilled by an innovative, eco-friendly synthesis technique called "green synthesis". Contrary to the chemically synthesised nanoparticles, the biosynthesized or green nanoparticles pose less hazards to the environment and avoid the synthesis of unwanted or harmful by-products due to buildup of stable compounds, making the entire process sustainable and eco-friendly for synthesis (Singh et al., 2018). Green synthesis of metal nanoparticles requires either plants (leaf, flower, fruit, root, or seed) or microorganisms (bacteria, fungi, algae, or yeast) as the starting material, which act as reducing agents as well as capping agents. Figure 17.4 briefly illustrates the advantages of green synthesis.

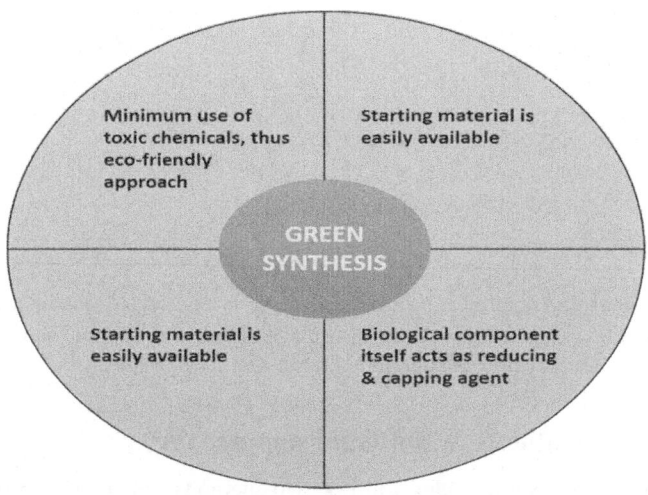

FIGURE 17.4 Advantages of green synthesis.

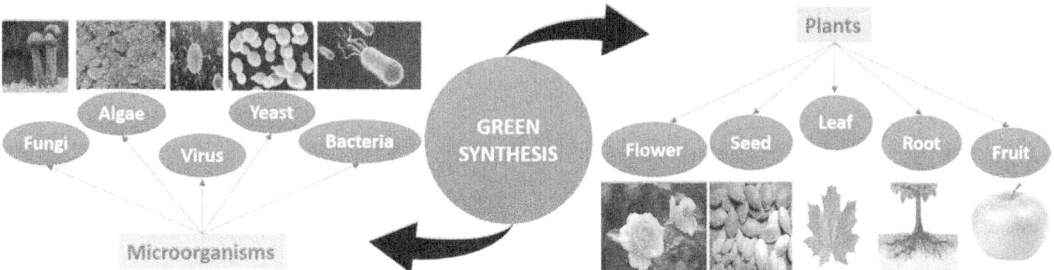

FIGURE 17.5 Green synthesis via different plant-derived extracts and microorganisms.

17.1.3 Green Synthesis via Plants and Microorganisms

The green synthesis of metal/metal oxide nanoparticles is carried out using plants or microorganisms, as shown in Figure 17.5. Among various options mentioned in the table, use of plant extracts is comparatively an amicable process to produce nanoparticles at larger scales. Also, the diversity within the plant, due to presence of various phytochemicals like ketones, aldehydes, flavones, amides, carboxylic acids, phenols, terpenoids, and ascorbic acids in leaves, stems, roots, extracts, etc., play a prominent role.

Most of the plants comprise phenolic and nitrogen compounds, vitamins, reducing sugars, and other metabolites that are rich in antioxidant activity and act as free radical scavengers. Also, plants used to synthesise nanoparticles are rich in antioxidants and polyols (Mohamad et al., 2014). These components are capable of reducing metal salts into metal nanoparticles.

In the case of microorganisms, nanoparticles are biosynthesised when the microorganisms grab target ions from their environment and then convert the metal ions into the element metal through enzymes generated by the cell activities.

Such nanomaterials are investigated for their use in biomedical diagnostics, molecular sensing, antimicrobials, optical imaging, and catalysis applications. Biosynthesis using plants is more favoured due to the faster rate of biosynthesis in plants than in microorganisms and the nanoparticles thus obtained are of various shapes and sizes (Iravani, 2011). The synthesis using microorganisms requires a stringent and careful control of cell structure, thus making it less favoured.

17.2 Nanoparticle Synthesis Using Microorganisms

Over the past few years, synthesis of metal nanoparticles by use of microorganisms such as bacteria (such as actinomycetes), fungi, viruses, and yeasts, have been extensively studied. Extra- and intracellular modes of synthesis using microorganisms are elucidated in Figure 17.6.

An array of biological protocols for nanoparticle synthesis has been reported using bacterial biomass, supernatant, and derived components. Among the extra- and intracellular synthesis, the former has received much attention due to its benefits in the downstream processing steps that do not require steps like sonication, centrifugation, washing, and purification as in intracellular ones. Moreover, the presence of proteins, enzymes, genes, peptides, cofactors, and other organic materials act as reducing agents, which benefits the synthesis of nanoparticles by preventing the aggregation and helps them to remain stable, thus providing capping action with additional stability (Figure 17.7) (Singh et al., 2016).

A few salient features of microbial synthesis of nanoparticles are tabulated in Table 17.1 (Prasad et al., 2016).

A brief procedure of nanoparticle synthesis using microorganisms is shown in Figure 17.8.

Table 17.2 gives some nanoparticles synthesised using microorganisms, their morphology, size, and the type of mechanism, i.e., intracellular or extracellular.

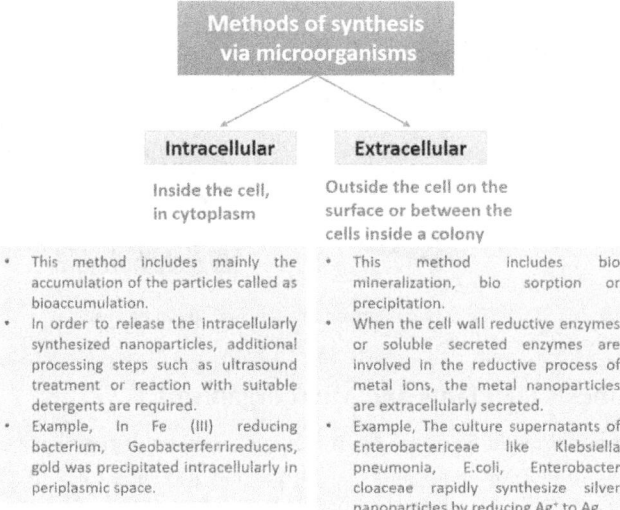

FIGURE 17.6 Intracellular and extracellular synthesis.

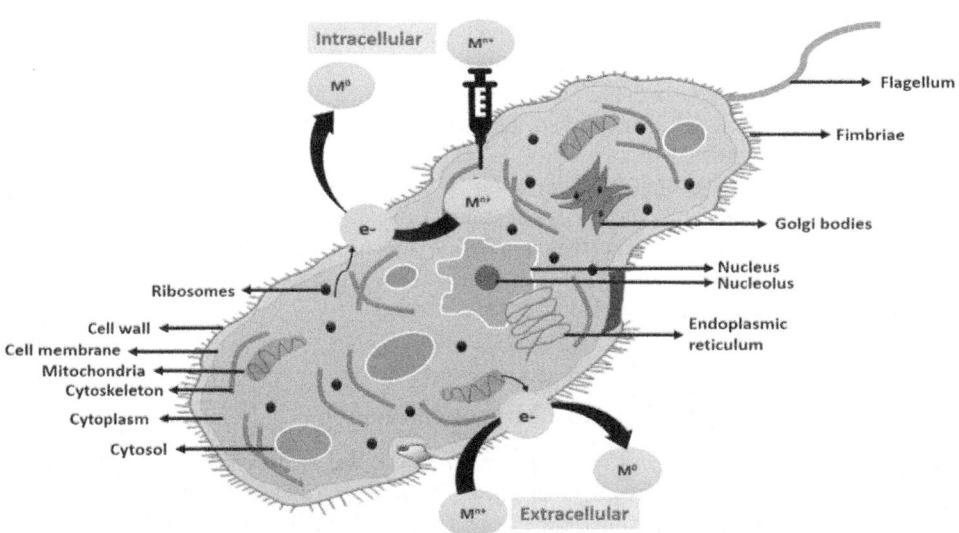

FIGURE 17.7 Intracellular and extracellular mechanism in a eukaryotic cell.

17.2.1 Mechanism Behind the Nanoparticle Synthesis by Microorganisms

The intracellular strategy involves explicit particle transportation into the microbial cell for nanoparticle development within the sight of cell proteins. Here, the cell mass of these organisms plays a key job as a channel. Ensuing to catalyst intervened amalgamation, the nanoparticles are shipped through dispersion through the cell divider. Extracellular amalgamation, on the other hand, is chiefly founded on nitrate reductase mediated, helps in reduction of the caught metal particles on the cell surface, with a couple of other announced biosorption and complexation measures. Notwithstanding the significant unthinking standards examined in the past segment, here we elucidate a portion of the vital strides in intracellular and extracellular nanoparticle creation. By the way, the mechanism and key steps behind the

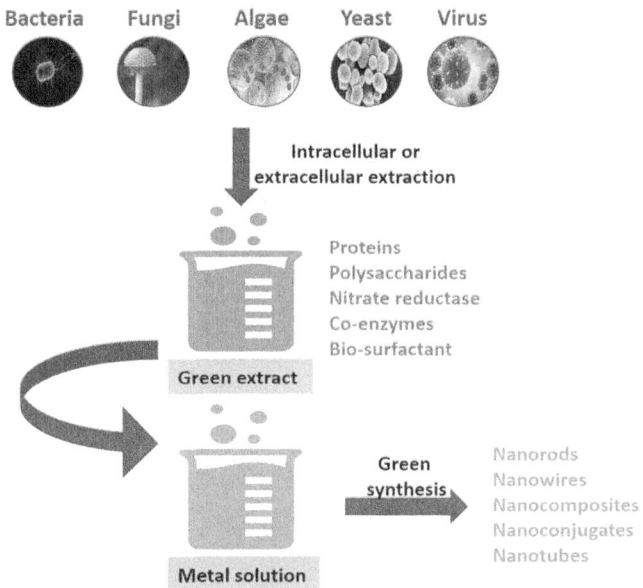

FIGURE 17.8 Green synthesis using microbial extract.

TABLE 17.1

Advantages and disadvantages involved for microbial synthesis of nanoparticles

Microorganism	Advantages	Disadvantages
Prokaryotes	• Growth rate is high • Genetically very easy to manipulate • Cost-effective cultivation	• Sophisticated instruments are used to get clear filtrate from colloidal solution. • Difficult to identify bio-capping agents.
Eukaryotes	• Ease of handling and process scale up • Higher surface areas and suitable growth of the mycelia enhance higher production rates • Easy downstream processing	• Transfer of gene to fungus is rigorous work • Growth rate of cell is low • It is difficult to get pure nanoparticles without capping agents and other biomolecules
Viruses	• Nanoparticles are nondispersive in nature • Robust and stable synthesis • Transfer of gene in viral platforms is regularly performed	• Process yet to be investigated further • For virus expression, a host microorganism is needed • Requires broad exploration for mass-scale application

intracellular and extracellular synthesis is primarily based on three enzymes and their different processes are described as follows (Ahmad et al., 2016).

17.2.1.1 Oxidoreductase Enzymes for Bioreduction

Mainly, a large number of proteins, carbohydrates, and biomembranes are involved for the bioreduction process in microorganisms. Nanoparticles are formed on the surfaces of the cell wall by passing through the following steps. Initially, the metal ions are trapped on the surface due to the electrostatic interaction between the metal ions and negatively charged groups in enzymes present at the cell wall. This is followed by enzymatic reduction for the formation of nanoparticles. NADH (nicotinamide adenine

TABLE 17.2

Metal nanoparticles synthesized using different microorganisms

Microorganism	Nanoparticle	Morphology	Size (nm)	Intracellular (I)/ Extracellular (E)	References
Actinomycetes					
Thermomonospora sp.	Au	Spherical	8	E	(Ahmad et al., 2016; Sastry et al., 2003)
Bacteria					
Escherichia coli	CdS	Spherical	2–51	I	(Sweeney et al., 2004)
Pseudomonas aeruginosa	Au	Spherical	15–30	E	(Husseiny et al., 2007)
Pseudomonas stutzeri	Ag	Various shapes	Up to 200	I	(Belliveau et al., 1987)
Fungus					
Aspergillus flavus	Ag	Spherical	8–10	I	(Vigneshwaran et al., 2007)
Colletotrichum sp.	Au	Spherical	20–40	E	(Shankar et al., 2003)
Fusarium oxysporum	Au	Spherical, triangular	20–40	E	(Mukherjee and Senapati, 2002)
Volvariella volvacea	Au & Ag	Spherical, hexagonal	20–150	E	(Philip, 2009)
Algae					
Chlorella vulgaris	Au	Spheroid, polyhedral	40–60	I	(Luangpipat et al., 2011)
Sargassum wightii	Au & Ag	Spheroid		E	(Singaravelu et al., 2007)
Virus					
M13 bacteriophage	ZnS	Hexagonal wurtzite structure	3–5	E	(Mao et al., 2003)
M13 bacteriophage	Hydroxyapatite (HAP)	Hydroxyapatite fibrils	-	E	(He et al., 2010; Wang et al., 2012)
Bacteriophage	Ca	Fibrils	-	E	(Wang et al., 2010; Xu et al., 2011)
Tobacco mosaic virus (TMV)	Silica	Various shapes	-	E	(Fernandes et al., 2014; Royston et al., 2009)
Tobacco mosaic virus (TMV)	SiO_2, CdS, PbS, Fe_2O_3	-	-	E	(Lee et al., 2002; Shenton et al., 1999)
Yeast					
Candida glabrata	CdS	Spherical	2	I	(Dameron et al., 1989)
Saccharomycetes cerevisiae	Sb_2O_3	Spherical	3–10	I	(Journal et al., 2016)
Yeast strain MKY3	Ag	Hexagonal	2–5	E	(Husseiny et al., 2007)
Schizosaccharomyces pombe	CdS	Hexagonal	1–2	I	(Kowshik et al., 2002)
Torulopsis sp.	PbS	Spherical	2–5	I	(Synthesis and Nanocrystallites, 2002)

dinucleotide) is a specific reducing enzyme used by microorganisms for bioreduction (Mukhopadhyay and Sarkar, 2006). The oxidoreductases enzymes are pH sensitive.

17.2.1.2 Mediated Reduction by NADH-Dependent Reductase

This is another way for the intracellular and extracellular synthesis of nanoparticles. Here, it has been found that NADH-subordinate reductase is the principal protein liable for nanoparticle biosynthesis for mediated reduction process. This reductase acquires electrons from NADH and oxidizes it to NAD^+. The catalyst is then oxidized by the synchronous reduction of metal particles. It has been noticed that after NADH-subordinate reductase, quinine subordinates of naphthoquinones and anthraquinonesare also help in the reduction of nanoparticles. Expectedly, nanoparticles created in an extracellular way were settled by the protein and diminishing specialists emitted by the parasite. This protein interceded bioreduction measure has moreover been noticed for microorganisms. Microorganisms secret the cofactor NADH and NADH-subordinate proteins, which give off an impression of being answerable for the bioreduction of nanoparticles and the ensuing development of nanoparticles (Durán et al., 2005).

17.2.1.3 Mediated Reduction by Nitrate/Nitrite Reductase

In this process, the catalyst is first formed with an electron donor, and then changes it to an elemental form through reduction. These responses are commonly led under anaerobic conditions and furthermore within the sight of NADPH, phytochelatin, and 4-hydroxyquinoline. These compounds act as cofactors, stabilizers of nanoparticles, and electron carriers, respectively, for this reduction process (Republic, 2010).

17.3 Metallic Nanoparticles Synthesis via Different Parts of Plants

Out of various sources under the green synthesis heading, plants and plant parts can be used for massive and substantial production of metallic nanoparticles due to their low cost of production of synthesis than by microorganisms and the nanoparticles thus obtained are also of various shapes and sizes while the synthesis using microorganisms requires a very strict and careful control of cell structure, thus making it less favoured. Plant components are more favoured due to the faster rate (Shah et al., 2015). This is illustrated in Figure 17.9.

Green synthesis from plants involves the use of various plant parts like roots, stems, leaves, fruits, flowers, etc. The biomolecules present in them help to synthesise the nanoparticles through reduction and stabilisation mechanisms. Table 17.3 summarises the different plant extracts used for the synthesis

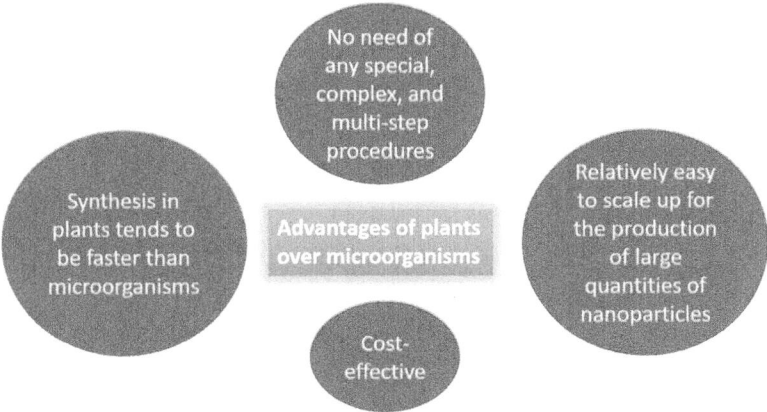

FIGURE 17.9 Advantages of green synthesis using plants over microorganisms.

TABLE 17.3

Metal nanoparticles synthesized using different plant extracts

Plant Extract	Nanoparticle	Precursor Used	Morphology	Size (nm)	References
Leaves					
Dalbergia spinosa	Ag	$AgNO_3$	Spherical	18	(Muniyappan and Nagarajan, 2014)
Thymbra spicata	Ag	$AgNO_3$	Spherical	7	(Veisi et al., 2018)
Ginkgo biloba Linn	Cu	$CuCl_2 \cdot 2H_2O$	Spherical	15–20	(Nasrollahzadeh and Sajadi, 2015)
Camellia sinensis	$\alpha\text{-}Fe_2O_3$	$Fe(NO_3)_3 \cdot 9H_2O$	Spherical	60	(Ahmmad et al., 2013)
Jatropha curcas	TiO_2	$TiCl_4$	Spherical	13	(Pratap et al., 2018)
Phyllanthus niruri	ZnO	Zinc nitrate	Quasi-spherical	25.61	(Anbuvannan et al., 2015)
Calotropis procera	ZnO	$Zn(NO_3)_2 \cdot 6H_2O$	Spherical	15–25	(Gavade and Babar, 2017)
Azadirachta indica	ZnO	$[Zn(CH_3COO)_2] \cdot 2H_2O$	Spherical	9.6–25.5	(Bhuyan et al., 2015)
Roots					
Catunaregam spinosa	Ag	$AgNO_3$	Spherical	33	(Haritha et al., 2017)
Rheum palmatum L.	CuO	$CuCl_2$	Spherical	10–20	(Bordbar and Shari, 2017)
Thymus vulgaris L.	Cu	$CuSO_4 \cdot 5H_2O$	Spherical	56	(Issaabadi et al., 2017)
Seeds					
Bunium persicum	Ag	$AgNO_3$	Spherical	20–50	(Rostami-vartooni et al., 2016)
Trigonella foenum-graecum	Ag	$AgNO_3$	Spherical	17	(Vidhu and Philip, 2014)
Trigonella foenum-graecum	Au	$HAuCl_4 \cdot 3H_2O$	Spherical	15–25	(Aromal and Philip, 2012)
P. granatum	Cu	Copper chloride	Semi-spherical	40–80	(Nazar et al., 2018)
Persia Americana	SnO_2	Stannous chloride	Very fine flakes with tiny agglomerates	4	(Elango et al., 2015)
Garcinia gummi-gutta	ZnO	$Zn(NO_3)_2 \cdot 6H_2O$	Irregular	10–20	(Raghavendra et al., 2017)
Peel					
Moringa oleifera	CeO_2	CAN	Spherical	45	(Surendra and Roopan, 2016)
Banana peel	CuO	$Cu(NO_3)_2 \cdot 3H_2O$	Spherical	60	(Aminuzzaman et al., 2017)
	ZnO	$Zn(NO_3)_2 \cdot 6H_2O$	Spherical		

TABLE 17.3 (Continued)

Metal nanoparticles synthesized using different plant extracts

Plant Extract	Nanoparticle	Precursor Used	Morphology	Size (nm)	References
Nephelium lappaceum L. Fruit	ZnO	Zn(NO$_3$)$_2$·6H$_2$O	Spherical	25–40	(Karnan et al., 2016)
Piper pedicellatum	Ag	AgNO$_3$	Spherical	30	(Tamuly et al., 2013)
Cynometra ramiflora	Fe$_2$O$_3$	FeCl$_3$:FeCl$_2$ = 1:2	Spherical	68.17	(Bishnoi et al., 2018)
Gardenia jasminoides Ellis	Pd	PdCl$_2$	Spherical, Rod, 3D polyhedra	3–5	(Jia et al., 2009)
Artocarpus gomezianus sp.	ZnO	Zn(NO$_3$)$_2$·6H$_2$O	Spherical	11.53	(Suresh et al., 2015)
Garcinia xanthochymus	ZnO	Zn(NO$_3$)$_2$·6H$_2$O	Spherical	20–30	(Nethravathi et al., 2015)

of various metal nanoparticles, the precursor used, morphology and size of the synthesised nanoparticle, and its characterisation and application.

17.3.1 Mechanism Behind Nanoparticle Synthesis by Different Parts of Plants

Herein, a brief and comprehensive study is carried out to understand the mechanism involved for the formation of nanoparticles using different parts of plants and to identify the functional groups present in them that are responsible for the green synthesis of nanoparticles (Kulkarni and Muddapur, 2014). Due to these interesting properties, plants have been considered a more environmentally friendly route for biologically synthesising metallic nanoparticles and for detoxification applications (Love et al., 2014; Suhanovsky et al., 2013). Constituents of plant extract act both as reducing and capping agents in nanoparticle synthesis. The plant extracts contain a variety of naturally occurring phytomolecules such as alkaloids, water-soluble flavonoids, and several other phenolic compounds, which are widely classified as polyphenols. These polyphenol-based phytomolecules are associated with strong reducing properties and have a strong tendency to adsorb on the nanoparticles surface and act as stabilisers. The antioxidant properties of phenolic phytomolecules are principally due to their reducing abilities, which enable them to act as reductants and singlet oxygen quenchers. The process begins by mixing a sample of plant extract with a metal salt solution, known as a metal precursor, at room temperature. Biochemical reduction of the salts starts immediately and the formation of nanoparticles is indicated by a change in the colour of the reaction mixture. During synthesis, there is an initial activation period when processed metal ions are converted from their mono or divalent oxidation states to zero-valent states and nucleation of the reduced metal atoms takes place. This is immediately followed by a period of growth when smaller neighbouring particles coalesce to form larger nanoparticles that are thermodynamically more stable while further biological reduction of metal ions takes place. As growth progresses, nanoparticles aggregate to form a variety of morphologies such as cubes, spheres, triangles, hexagons, pentagons, rods, and wires. Nanoparticle size and shape depend

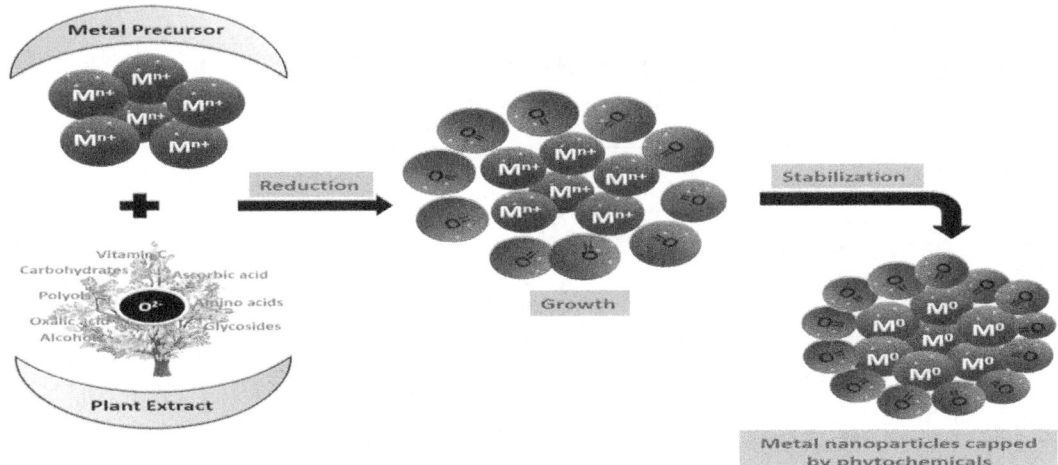

FIGURE 17.10 A general mechanism of synthesis of metal nanoparticles using plant extracts.

on the variation in composition and concentration of these active biomolecules between different plant-derived extracts and their subsequent interaction with aqueous metal ions (Akhtar et al., 2013). In the final stage of synthesis, the plant extracts' ability to stabilise the nanoparticle ultimately determines its most energetically favourable and stable morphology. Under the suitable conditions (plants extract concentration, metal salt concentration, reaction time, pH of reaction solution, and temperature) significantly influence the quality, size, and morphology of the synthesized nanoparticles (Mittal et al., 2013; Dwivedi and Gopal, 2010). A general mechanism of synthesis of metal nanoparticles using plant extracts is illustrated in Figure 17.10.

17.4 Application of Nanoparticles Synthesised Through Microorganisms and Plant Extracts

In the previous sections of this chapter we have seen the mechanism and advantages of nanoparticles synthesised via plant and microbial extracts. This section elucidates various applications of usage of metallic and non-metallic nanoparticles synthesised. Individual representation of synthesised nanoparticles from plant and microbial extracts along with their subsequent application of usage are tabulated in Tables 17.4 and 17.5, respectively. From the tabulated information it can be observed that a large variety of nanoparticles such as Ag, Au, Cu, Fe, Ni, Ti, some of their oxides, etc., are synthesised and these nanoparticles have a major role in applications relevant to environmental remediation with the majority working towards degradation of toxic substances such as dyes, heavy metals, and phenol compounds from wastewater and industrial effluents.

17.5 Conclusion

In this book chapter, brief information is given on the green synthesis of nanoparticles and their applications in various fields as well as detailed mechanisms is discussed thoroughly to give an insight on the formation of nanoparticles. The method of synthesis of nanoparticles plays a significant role in deciding their properties. Different approaches used for synthesising nanoparticles can be grouped under two broad groups as the top-down approach and bottom-up approach. Compared to the top-down approach, the bottom-up approach is simple, needs normal conditions of pressure and temperature, and is

TABLE 17.4

Applications of metal nanoparticles synthesized using plant extracts.

Plant Extract	Precursor	Size (nm)	Application
Silver nanoparticles			
Vaccinium macrocarpon (fruit)	AgNO$_3$	30	Degradation of methyl orange (MO), methylene blue (MB), rhodamine B (RhB), congo red (CR) • MO, MB, RhB, and CR reductions using Ag NPs were complete within 138, 40, 90, and 270 s, respectively. • The catalyst could be recycled for at least six times in the reduction of MO, MB, RhB, and CR with almost no loss of catalytic activity (Khodadadi et al., 2017).
Ficus benjamina (leaves)	AgNO$_3$	60–105	Removal of cadmium • Cd (II) removal percentage recorded the highest values at 40 min. • The percentage removal of Cd (II) removal increased with increase in pH from 1 to 6 and the sharpest increase in Cd (II) uptake was obtained at pH of 6. • Cd (II) removal percentage diminishes with an increase in the metal concentrations, whereas the practical amount of adsorbed Cd (II) ions increases. • When the agitation speed increases from 100 to 500 rpm, Cd (II) removal percentage by Ag NPs increases and then the removal percentage decreases after 500 rpm (Al-qahtani, 2017).
Gold nanoparticles			
Alpinianigra sp. (leaf)	HAuCl$_4$·3H$_2$O	21.52	Degradation of methyl orange (MO) and rhodamine B (RhB) • Au NPs in the presence of sunlight catalysed the degradation of the anthropogenic pollutant dyes MO and RhB with percent degradation of 83.25% and 87.64%, repetively. • The kinetics of the degradation reaction is pseudo first order (Baruah et al., 2018).
Copper oxide nanoparticles			
Psidium guajava (leaf)	Copper acetate monohydrate	11.07	Degradation of nile blue (NB) • Degradation efficiency was found to be 97% under sunlight. • Complete degradation of NB dye was seen in 100 minutes. • Lower molecular weight of NB facilitated its rapid diffusion towards photocatalytic degradation. • The synthesized CuO NPs exhibited significant photocatalytic activity even after five cycles of NB degradation. Degradation of reactive yellow 160 (RY160) • Degradation efficiency was found to be 80% under sunlight. • Complete degradation of RY160 dye was seen in 120 minutes. • Slower photocatalytic degradation and lower degradation efficiency was attributed to its higher molecular weight (Singh et al., 2019).
Catharanthus roseus (leaf)	CuSO$_4$	1	Chromium (VI) removal • The Cr (VI) rejection at 1 bar applied pressure was 85.37% and increased upto 88.08% at 3 bar pressure. • Maximum Cr rejections by negatively charged CuO coated membrane was observed at pH of 6.8 of the feed solution (Choudhury et al., 2018).
Fe$_2$O$_3$@SiO$_2$ nanoparticles			
Musa balbisiana (peel)	FeCl$_3$·6H$_2$O	9.2	Degradation of methyl red (MR) dye • Rate constant was highest (0.0775 min^{-1}) when 0.5 mol% Fe$_2$O$_3$ @ SiO$_2$ nanoparticles was used as photocatalysts.

(Continued)

TABLE 17.4 (Continued)

Applications of metal nanoparticles synthesized using plant extracts.

Plant Extract	Precursor	Size (nm)	Application
			• The rate constant was found almost similar after fifth cycle of photocatalytic reaction (Hazarika et al., 2016).
Ni@Fe$_3$O$_4$ and CuO nanoparticles			
Euphorbia maculata (aerial parts)	NiCl$_2$·6H$_2$O and CuSO$_4$·5H$_2$O	30 and 18	Degradation of organic dye pollutants • Higher photocatalytic activity was observed for CuO NPs compared to Ni@Fe$_3$O$_4$ NPs. • The analyses showed a reduction in the intensity of the peak in absorbance vs wavelength graph with time, but at the same time, the declining degree of the dyes followed the order MB > RhB > CR. • Reduction of solution pH from 6 to 12 increased the CR dye degradation percentage of Ni @ Fe$_3$O$_4$ and CuO NPs photocatalysts from 53% to 78% and 50% to 85%, respectively. • Effective photocatalytic degradation between 80% and 90% was observed at a pH of 10 for RhB. • The MB photodecolorisation efficiency was improved at higher pH values and the best efficiency of decolorization (94%) was observed at a pH of 10 (Pakzad et al., 2019).
Silver nanoparticles			
Carpobrotus acinaciformis (leaf)	AgNO$_3$	20–50	Degradation of methyl orange (MO) and congo red (CR) • Complete degradation of 20 ppm MO and CR occurred in 15.5 and 73 minutes, respectively. • Photocatalytic activity does not significantly change after four cycles (Rostami-vartooni et al., 2016).
Titanium dioxide nanoparticles			
Ageratina altissima (leaf)	TiO(OH)$_4$	60–100	Degradation of dyes • The percent of dye degradation recorded was 86.79%, 76.32%, 77.59%, and 69.06% for methylene blue, alizarin red, crystal violet, and methyl orange, respectively (Ganesan et al., 2016).
Jatropha curcas (leaf)	TiCl$_4$	13	Degradation of tannery wastewater • After the solar photocatalytic treatment of TWW with TiO$_2$ NPs, the removal efficiency was found to be 82.26%. • 76.48% removal of Cr from TWW was achieved (Pratap et al., 2018).
Ferrous oxide nanoparticles			
Ruellia tuberosa (leaf)	FeSO$_4$	52.78	Degradation of crystal violet dye • The photocatalytic ability of the synthesized FeO NPs was demonstrated by degrading crystal violet dye under solar irradiation upto 80%. • At the initial period of 0 minutes, absorbance of the dye was around 1.65 which was reduced to 0.35 at 150 minutes (Vasantharaj et al., 2019).
Stannic oxide nanoparticles			
Cyphomandra betacea (fruits)	SnCl$_2$	21	Degradation of methylene blue dye • MB degradation was 0.0952 per min and exactly at 70 minutes, it was completely degraded. • As time increases, the percentage rate of degradation also increases (Elango and Roopan, 2016).

TABLE 17.5

Applications of metal nanoparticles synthesised using microorganisms

Microorganism	Precursor	Size (nm)	Application
Silver nanoparticles			
Leuconostoc lactis (bacteria)	AgNO$_3$	35	Degradation of MO & CR • The rate of degradation increases with the amount of catalyst added up to 5 mg, and further increases in the amount of catalyst did not contribute to the significant enhancement in the degradation rate. Hence, 5 mg of the nanocatalyst has been chosen as the optimum dosage for efficient catalytic degradation of MO and CR dyes. • Complete degradation was achieved in 240 minutes (Saravanan et al., 2017).
Rhodococcus sp. NCIM 2891 (bacteria)	AgNO$_3$	5	Reduction of 4-nitrophenol to 4-aminophenol • The time required for the complete reduction of the 4-NP by homogeneous catalytic reaction was about 8 minutes, and the rate of reaction obtained was found to be 0.39 min^{-1}. • For the heterogeneous catalytic activity, the time required was 11 minutes, and rate of reaction was about 0.30 min^{-1}. • The availability of the nanoparticles for the catalytic reaction is more in the homogeneous catalytic reaction than that of the heterogeneous catalytic reaction, so the rate of reaction is more in case of homogeneous catalytic reactions (Otari et al., 2014).
Ulva lactuca (sea weed)	AgNO$_3$	48.59	Degradation of methyl orange • The adsorption of Ag NPs onto the methyl orange solution was initially low and further increased with constant increase in time (Kumar et al., 2013).
Palladium nanoparticles			
Saccharomyces cerevisiae (yeast)	Palladium acetate	32	Degradation of direct blue 71 dye • The yeast synthesised Pd NPs degraded 98% of direct blue 71 dye photochemically within 60 minutes under UV light. • The rate of the reaction was found to be 0.04579 min^{-1} under UV light (Sriramulu and Sumathi, 2018).
Gold nanoparticles			
Escherichia coli K12 (bacteria)	HAuCl$_4$	50	Degradation of 4-nitrophenol • Aqueous 4-NP shows maximum UV-vis absorbance at 317 nm. When NaBH$_4$ (pH > 12) was added to reduce 4-NP, an intense yellow colour appeared due to formation of 4-nitrophenolate ion red-shifting the absorption peak to 400 nm. • The kinetic reaction rate constant under the given set of reaction conditions was estimated to be 1.24×10^{-2} min^{-1} (Srivastava et al., 2013).
Cladosporium oxysponum AJP03 (soil fungus)		72.32	Degradation of Rhodamine B

(Continued)

TABLE 17.5 (Continued)

Applications of metal nanoparticles synthesised using microorganisms

Microorganism	Precursor	Size (nm)	Application
	Gold (III) chloride trihydrate ($H_7AuCl_4O_3$)		• The reaction rate was found to be higher for biologically synthesized Au NPs (12.7×10^{-3} s^{-1}) as compared to chemically synthesized Au NPs (1.3×10^{-3} s^{-1}) in reducing 2.5×10^{-5} M of RhB (Bhargava et al., 2016).
Bacillus marisflavi YCISMN 5 (bacteria)	$HAuCl_4$	14	Degradation of CR & MB • The reduction reaction follows pseudo first-order kinetics with a reaction rate constant of 0.2192 and 0.2484 minute^{-1} for CR and MB, respectively. • The biosynthesized Au NPs showed 98% CR degradation in the presence of $NaBH_4$ within 20 minutes. • The biosynthesized Au NPs showed 88% MB degradation in the presence of $NaBH_4$ within 10 minutes, and complete catalytic degradation occurred within the next 1 minute (Nadaf and Kanase, 2016).
Stannic oxide nanoparticles			
Erwinia herbicola (bacteria)	$SnCl_2.2H_2O$	28.89	Degradation of dyes • Approximately 95.3, 97.8, and 94.0% degradation of methylene blue, erichrome black T, and methyl orange was observed, respectively. • The increase in pH provides higher degradation efficiency of EBT dye (Srivastava and Mukhopadhyay, 2014).
Fe_3O_4/bacterial cellulose nanocomposites			
Gluconacetobacter xylinium (bacteria)	$FeCl_3 + FeCl_2$	15	Adsorbents for heavy metal ions • When Pb^{2+} concentration ranges from 0 to 100 mg/mL, the adsorption quantity of Pb^{2+} is proportional to its concentration and the adsorption capacity is higher than 90%. • However, the adsorption quantity approaches a constant at 52 mg/g, when Pb^{2+} concentration ranges from 100 to 200 mg/mL and the adsorption capacity is decreased from 90% to 65%. • When the concentration of Mn^{2+} and Cr^{3+} is above 60 mg/mL, the adsorption capacity for Mn^{2+} decreases from 46% to 33% and from 43% to 25% for Cr^{3+}. • The adsorption capacity of these three ions at the same ion concentration is sequenced as $Pb^{2+} > Mn^{2+} > Cr^{3+}$ (Zhu et al., 2011).

less energy intensive and eco-friendly. In the green route, plant extracts or microorganisms are used to prepare nanoparticles by the biological components present in the extract or microbial cells that act as reducing, capping, and stabilising agents in the synthesis process. Green synthesis is a quick and environmentally friendly way for the preparation of nanoparticles with good photocatalytic properties, which are helpful for the environment.

REFERENCES

Ahmad, A., Senapati, S., Khan, M.I., Kumar, R., & Sastry, M. (2016). Extracellular biosynthesis of monodisperse gold nanoparticles by a novel extremophilic actinomycete, *Thermomonospora* Sp. *Journal of Nanobiotechnology, 18*(2), 1–28.

Ahmmad, B., Leonard, K., Islam, S., & Kurawaki, J. (2013). Photocatalytic activity. *Advanced Powder Technology, 24*(1), 160–167. 10.1016/j.apt.2012.04.005.

Akhtar, M.S., Panwar, J., & Yun, Y.S. (2013). Biogenic synthesis of metallic nanoparticles by plant extracts. *ACS Sustainable Chemistry & Engineering*, 1, 591–602.

Al-Qahtani, K.M. (2017). Cadmium removal from aqueous solution by green synthesis zero valent silver nanoparticles with benjamina leaves extract. *Egyptian Journal of Aquatic Research, 43*(4), 269–274. 10.1016/j.ejar.2017.10.003.

Amal, R., & Low, G. (1999). Role of nanoparticles in photocatalysis role of nanoparticles in photocatalysis. *Journal of Nanoparticles Research*, 439–458. 10.1023/A.

Aminuzzaman, M., Kei, L.M., & Liang, W.H. (2017). Green synthesis of copper oxide (CuO) nanoparticles using banana peel extract and their photocatalytic activities. *AIP Conference Proceedings*, 10.1063/1.4979387.

Anbuvannan, M., Ramesh, M., Viruthagiri, G., Shanmugam, N., & Kannadasan, N. (2015). Synthesis, characterization and photocatalytic activity of ZnO nanoparticles prepared by biological method. *Spectrochimica Acta Part A: Molecular and Biomolecular Spectroscopy*, 143, 304–308. 10.1016/j.saa.2015.01.124.

Aromal, S.A., & Philip, D. (2012). Green synthesis of gold nanoparticles using trigonella foenum-graecum and its size-dependent catalytic activity. *Spectrochimica Acta Part A: Molecular and Biomolecular Spectroscopy*, 97, 1–5. 10.1016/j.saa.2012.05.083.

Ayyub, P., Chandra, R., Taneja, P., Sharma, A.K., & Pinto, R. (2001). Synthesis of nanocrystalline material by sputtering and laser ablation at low temperatures. *Applied Physics A: Materials Science and Processing, 73*(1), 67–73. 10.1007/s003390100833.

Baruah, D., Goswami, M., Narayan, R., Yadav, S., & Yadav, A. (2018). Biogenic synthesis of gold nanoparticles and their application in photocatalytic degradation of toxic dyes. *Journal of Photochemistry & Photobiology, B: Biology, 186*(July), 51–58. 10.1016/j.jphotobiol.2018.07.002.

Belliveau, B.H., Starodub, M.E., Cotter, C., & Trevors, J.T. (1987). Metal resistance and accumulation in bacteria. *Nanobiotechnology*, 5, 101–127.

Bhargava, A., Jain, N., Khan, M.A., Pareek, V., Dilip, R.V., & Panwar, J. (2016). Utilizing metal tolerance potential of soil fungus for efficient synthesis of gold nanoparticles with superior catalytic activity for degradation of Rhodamine B. *Journal of Environmental Management*, 183, 22–32. 10.1016/j.jenvman.2016.08.021.

Bhuyan, T., Mishra, K., Khanuja, M., & Prasad, R. (2015). Biosynthesis of zinc oxide nanoparticles from Azadirachta Indica for antibacterial and photocatalytic applications. *Materials Science in Semiconductor Processing*, 32, 55–61. 10.1016/j.mssp.2014.12.053.

Bishnoi, S., Kumar, A., & Selvaraj, R. (2018). Facile synthesis of magnetic iron oxide nanoparticles using inedible Cynometra Rami Fl Ora fruit extract waste and their photocatalytic degradation of methylene blue dye. *Materials Research Bulletin, 97*(March 2017), 121–127. 10.1016/j.materresbull.2017.08.040.

Bordbar, M., & Shari, Z. (2017). Green synthesis of copper oxide nanoparticles/clinoptilolite using *Rheum Palmatum* L. root extract: high catalytic activity for reduction of 4-nitro phenol, Rhodamine B, and Methylene Blue. *Journal of Nanoscience and Nanotechnology*, 724–733. 10.1007/s10971-016-4239-1.

Chen, S., Paulose, M., Ruan, C., Mor, G.K., Varghese, O.K., Kouzoudis, D., & Grimes, C.A. (2006). Electrochemically synthesized CdS nanoparticle-modified TiO_2 nanotube-array photoelectrodes: preparation, characterization, and application to photoelectrochemical cells. *Journal of Photochemistry and Photobiology A: Chemistry, 177*(2–3), 177–184. 10.1016/j.jphotochem.2005.05.023.

Choudhury, P., Mondal, P., Majumdar, S., Saha, S., & Sahoo, G.C. (2018). Preparation of ceramic ultra filtration membrane using green synthesized CuO nanoparticles for chromium (VI) removal and optimization by response surface methodology. *Journal of Cleaner Production*, 203, 511–520. 10.1016/j.jclepro.2018.08.289.

Dameron, C.T., Mehra, K., Carrollt, J., Brust, L.E., & Winge, D.R. (1989). Biosynthesis of cadmium sulphide quantum semiconductor crystallites. *Journal of Biological Chemistry, 338*(April), 596–597.

Durán, N., Marcato, P.D., Alves, O.L., De, G.I.H., & Esposito, E. (2005). Mechanistic aspects of biosynthesis of silver nanoparticles by several fusarium oxysporum strains. *Journal of Nanobiotechnology*, 7, 1–7. 10.1186/1477-3155-3-8.

Dwivedi, A.D., & Gopal, K. (2010). Aspects biosynthesis of silver and gold nanoparticles using chenopodium album leaf extract. *Colloids and Surfaces A: Physicochemical and Engineering, 369*(1–3), 27–33. 10.1016/j.colsurfa.2010.07.020.

Elango, G., & Roopan, S.M. (2016). Efficacy of SnO2 nanoparticles toward photocatalytic degradation of methylene blue dye. *Journal of Photochemistry & Photobiology, B: Biology, 155,* 34–38. 10.1016/j.jphotobiol.2015.12.010.

Elango, G., Manoj, S., Santhosh, S., Muthuraja, S., & Mohana, S. (2015). Molecular and biomolecular spectroscopy green synthesis of SnO_2 nanoparticles and its photocatalytic activity of phenolsulfonphthalein dye. *Spectrochimica Acta Part A, 145,* 176–180. 10.1016/j.saa.2015.03.033.

Fernandes, F.M., Coradin, T., & Aimé, C. (2014). *Silica Materials,* 792–812. 10.3390/nano4030792.

Figgemeier, E., Kylberg, W., Constable, E., Scarisoreanu, M., Alexandrescu, R., Morjan, I., Soare, I., Birjega, R., Popovici, E., Fleaca, C., et al. (2007) Titanium dioxide nanoparticles prepared by laser pyrolysis: synthesis and photocatalytic properties. *Applied Surface Science, 254*(4), 1037–1041. 10.1016/j.apsusc.2007.08.036.

Ganesan, S., Babu, I.G., Mahendran, D., Arulselvi, P.I., Elangovan, N., Geetha, N., & Venkatachalam, P. (2016). Green engineering of titanium dioxide nanoparticles using *Ageratina altissima* (L.) King & H.E. Robines. Medicinal plant aqueous leaf extracts for enhanced photocatalytic activity. *Annals of Phytomedicine, 5*(2), 69–75. 10.21276/ap.2016.5.2.8.

Gavade, V.V.G.N.L., & Babar, H.M.S.S.B. (2017). Green synthesis of ZnO nanoparticles by using calotropis procera leaves for the photodegradation of methyl orange. *Journal of Materials Science: Materials in Electronics, 28*(18), 14033–14039. 10.1007/s10854-017-7254-2.

Haritha, E., Mohana, S., Madhavi, G., Elango, G., & Arunachalam, P. (2017). Catunaregum Spinosa capped Ag NPs and its photocatalytic application against amaranth toxic azo dye. *Journal of Molecular Liquids, 225,* 531–535. 10.1016/j.molliq.2016.11.120.

Hazarika, M., Saikia, I., Das, J., & Tamuly, C. (2016). Biosynthesis of Fe2O3@SiO2 nanoparticles and its photocatalytic activity. *Material Letters, 164,* 480–483. 10.1016/j.matlet.2015.11.042.

He, T., Abbineni, G., Cao, B., & Mao, C. (2010). Nanofibrous Bio-Inorganic hybrid structures formed through self-assembly and oriented mineralization of genetically engineered phage nanofibers. *Small,* 2230–2235. 10.1002/smll.201001108.

Hu, M., Furukawa, S., Ohtani, R., Sukegawa, H., Nemoto, Y., Reboul, J., Kitagawa, S., & Yamauchi, Y. (2012). Synthesis of Prussian blue nanoparticles with a hollow interior by controlled chemical etching. *Angewandte Chemie International Edition, 51*(4), 984–988. 10.1002/anie.201105190.

Husseiny, M.I., El-Aziz, M.A., Badr, Y., & Mahmoud, M.A. (2007). Biosynthesis of gold nanoparticles using *Pseudomonas aeruginosa. Spectrochim Acta A, 67,* 1003–1006. 10.1016/j.saa.2006.09.028.

Husseiny, M.I., El-Aziz, M.A., Badr, Y., & Mahmoud, M.A. (2007). *Biosynthesis of Gold Nanoparticles Using Pseudomonas Aeruginosa, 67,* 1003–1006. 10.1016/j.saa.2006.09.028.

Indris, S., Amade, R., Heitjans, P., Finger, M., Haeger, A., Hesse, D., Grünert, W., Börger, A., & Becker, K.D. (2005). Preparation by high-energy milling, characterization, and catalytic properties of nanocrystalline TiO_2. *Journal of Physical Chemistry B, 109*(49), 23274–23278. 10.1021/jp054586t.

Iqbal, P., Preece, J.A., & Mendes, P.M. (2012). Nanotechnology: the "Top-Down" and "Bottom-Up" approaches. *Supramolecular Chemistry.* 10.1002/9780470661345.smc195.

Iravani, S. (2011). Green synthesis of metal nanoparticles using plants. *Green Chemistry, 13*(10), 2638–2650. 10.1039/c1gc15386b.

Issaabadi, Z., Nasrollahzadeh, M., & Sajadi, S.M. (2017). Green synthesis of the copper nanoparticles supported on bentonite and investigation of its catalytic activity. *Journal of Cleaner Production, 142,* 3584–3591. 10.1016/j.jclepro.2016.10.109.

Jia, L., Zhang, Q., Li, Q., & Song, H. (2009). The biosynthesis of palladium nanoparticles by antioxidants in Gardenia Jasminoides Ellis: long lifetime nanocatalysts for p-nitrotoluene hydrogenation. *Surface Biology,* 10.1088/0957-4484/20/38/385601.

Journal, A.I., Korbekandi, H., Mohseni, S., Jouneghani, R.M., Iravani, S., Korbekandi, H., Mohseni, S., Jouneghani, R.M., Pourhossein, M., & Iravani, S. (2016). Biosynthesis of silver nanoparticles using *Saccharomyces cerevisiae. Artificial Cells Nanomedicine Biotechnology, 1401.* 10.3109/21691401.2014.937870.

Karnan, T., Arul, S., & Selvakumar, S. (2016). Biosynthesis of ZnO nanoparticles using rambutan (*Nephelium lappaceum* L.) peel extract and their photocatalytic activity on methyl orange dye. *Journal of Molecular Structure, 1125,* 358–365. 10.1016/j.molstruc.2016.07.029.

Khodadadi, B., Bordbar, M., & Yeganeh-Faal, A. (2017). Green synthesis of Ag nanoparticles/clinoptilolite using vaccinium macrocarpon fruit extract and its excellent catalytic activity for reduction of organic dyes. *Journal of Alloys and Compounds, 719,* 82–88. 10.1016/j.jallcom.2017.05.135.

Kowshik, M., Deshmukh, N., Vogel, W., Urban, J., Kulkarni, S.K., & Paknikar, K.M. (2002). Microbial synthesis of semiconductor CdS nanoparticles, their characterization, and their use in the fabrication of an ideal diode. *Biotechnology and Bioengineering,* 10.1002/bit.1023.

Kulkarni, N., & Muddapur, U. (2014). Biosynthesis of metal nanoparticles: a review. *Journal of Nanotechnology, 2014,* 1–8. 10.1155/2014/510246.

Kumar, P., Govindaraju, M., Senthamilselvi, S., & Premkumar, K. (2013). Photocatalytic degradation of methyl orange dye using silver (Ag) nanoparticles synthesized from Ulva Lactuca. *Colloids Surfaces B Biointerfaces, 103,* 658–661. 10.1016/j.colsurfb.2012.11.022.

Lee, S., Mao, C., Flynn, C.E., & Belcher, A.M. (2002). Ordering of quantum dots using genetically engineered viruses. *Journal of Materials Chemistry, 296*(May), 892–896.

Love, A.J., Makarov, V., Yaminsky, I., Kalinina, N.O., & Taliansky, M.E. (2014). The use of tobacco mosaic virus and cowpea mosaic virus for the production of novel metal nanomaterials. *Virology, 449,* 133–139. 10.1016/j.virol.2013.11.002.

Luangpipat, T., Beattie, I.R., Chisti, Y., & Haverkamp, R.G. (2011). Gold nanoparticles produced in a microalga. *Journal of Nanoparticle Research, 6439*–6445. 10.1007/s11051-011-0397-9.

Mackenzie, J.D., & Bescher, E.P. (2007). Chemical routes in the synthesis of nanomaterials using the sol-gel process. *Accounts of Chemical Research, 40*(9), 810–818. 10.1021/ar7000149.

Mao, C., Flynn, C.E., Hayhurst, A., Sweeney, R., Qi, J., Georgiou, G., Iverson, B., & Belcher, A.M. (2003). Viral assembly of oriented quantum dot nanowires. *Proceedings of the National Academy of Sciences, 100*(12), 6946–6951.

Mittal, A.K., Chisti, Y., & Banerjee U.C. (2013). Synthesis of metallic nanoparticles using plant extracts. *Biotechnology Advances,* 31, 346–356.

Mohamad, N.A.N., Arham, N.A., Jai, J., & Hadi, A. (2014). Plant extract as reducing agent in synthesis of metallic nanoparticles: a review. *Advanced Materials Research, 832*(September 2015), 350–355. 10.4028/www.scientific.net/AMR.832.350.

Mukherjee, P., & Senapati, S. (2002). Extracellular synthesis of gold nanoparticles by the fungus fusarium oxysporum. *ChemBioChem, 3*(5), 461–463.

Mukhopadhyay, D., & Sarkar, G. (2006). The use of microorganisms for the formation of metal nanoparticles and their application. *Applied Microbiology and Biotechnology,* 485–492. 10.1007/s00253-005-0179-3.

Muniyappan, N., & Nagarajan, N.S. (2014). Green synthesis of silver nanoparticles with Dalbergia Spinosa leaves and their applications in biological and catalytic activities. *Process Biochemistry, 49*(6), 1054–1061. 10.1016/j.procbio.2014.03.015.

Nadaf, N.Y., & Kanase, S.S. (2016). Biosynthesis of gold nanoparticles by *Bacillus marisflavi* and its potential in catalytic dye degradation. *Arabian Journal of Chemistry.* 10.1016/j.arabjc.2016.09.020.

Nasrollahzadeh, M., & Sajadi, S.M. (2015). Green synthesis of copper nanoparticles using Ginkgo Biloba L. leaf extract and their catalytic activity for the Huisgen Cycloaddition of azides and alkynes at room temperature. *Journal of Colloid and Interface Science, 457,* 141–147. 10.1016/j.jcis.2015.07.004.

Nazar, N., Bibi, I., Kamal, S., Iqbal, M., Nouren, S., Jilani, K., & Ata, S. (2018). Cu nanoparticles synthesis using biological molecule of *P. granatum* seeds extract as reducing and capping agent: Growth mechanism and photo-catalytic activity. *International Journal of Biological Macromolecules, 106,* 1203–1210. 10.1016/j.ijbiomac.2017.08.126

Nethravathi, P.C., Shruthi, G.S., Suresh, D., Nagabhushana, H., & Sharma, S.C. (2015). Garcinia Xanthochymus mediated green synthesis of ZnO nanoparticles: photoluminescence, photocatalytic and antioxidant activity studies. *Ceramics International, 41*(7), 8680–8687. 10.1016/j.ceramint.2015.03.084.

Otari, S.V., Patil, R.M., Nadaf, N.H., Ghosh, S.J., & Pawar, S.H. (2014). Green synthesis of silver nanoparticles by microorganism using organic pollutant: its antimicrobial and catalytic application. *Environmental Science and Pollution Research,* 1503–1513. 10.1007/s11356-013-1764-0.

Pakzad, K., Alinezhad, H., & Nasrollahzadeh, M. (2019). Green synthesis of Ni@Fe3O4 and CuO nanoparticles using euphorbia maculata extract as photocatalysts for the degradation of organic pollutants under UV-irradiation. *Ceramic International, 45*(14), 17173–17182. 10.1016/j.ceramint.2019.05.272.

Philip, D. (2009). Spectrochimica Acta Part A: Molecular and Biomolecular Spectroscopy biosynthesis of Au, Ag and Au–Ag nanoparticles using edible mushroom extract. *Molecular & Biomolecular Spectroscopy, 73*, 374–381. 10.1016/j.saa.2009.02.037.

Pissuwan, D., Niidome, T., & Cortie, M.B. (2011). The forthcoming applications of gold nanoparticles in drug and gene delivery systems. *Journal of Controlled Release, 149*(1), 65–71. 10.1016/j.jconrel.2009.12.006.

Prasad, R., Pandey, R., & Barman, I. (2016). Engineering Tailored Nanoparticles with Microbes: Quo Vadis?, *8* (April). 10.1002/wnan.1363.

Pratap, S., Saxena, G., Singh, V., & Kumar, A. (2018). Green synthesis of TiO2 nanoparticles using leaf extract of *Jatropha curcas* L. for photocatalytic degradation of tannery wastewater. *Chemical Engineering Journal*, 336 (December 2017), 386–396. 10.1016/j.cej.2017.12.029.

Raghavendra, M., Yatish, K.V., & Lalithamba, H.S. (2017). Plant-mediated green synthesis of ZnO nanoparticles using garcinia gummi-gutta seed extract: photoluminescence, screening of their catalytic activity in antioxidant, formylation and biodiesel production. *Science.gov*. 10.1140/epjp/i2017-11627-1.

Republic, C. (2010). Mechanistic aspects of biosynthesis of nanoparticles by several. *B Biointerfaces, 81*, 430–433.

Rostami-Vartooni, A., Nasrollahzadeh, M., & Salavati-Niasari, M. (2016). Photocatalytic degradation of azo dyes by titanium dioxide supported silver nanoparticles prepared by a green method using carpobrotus acinaciformis extract. *Journal of Alloys and Compounds, 689*, 15–20. 10.1016/j.jallcom.2016.07.253.

Rostami-Vartooni, A., Nasrollahzadeh, M., & Alizadeh, M. (2016). Green synthesis of seashell supported silver nanoparticles using bunium persicum seeds extract: application of the particles for catalytic reduction of organic dyes. *Journal of Colloid and Interface Science, 470*, 268–275. 10.1016/j.jcis.2016.02.060.

Royston, E.S., Brown, A.D., Harris, M.T., & Culver, J.N. (2009). Preparation of silica stabilized tobacco mosaic virus templates for the production of metal and layered nanoparticles. *Journal of Colloid Interface Science, 332*(2), 402–407. 10.1016/j.jcis.2008.12.064.

Saravanan, C., Rajesh, R., Kaviarasan, T., Muthukumar, K., Kavitake, D., & Shetty, P.H. (2017). Synthesis of silver nanoparticles using bacterial exopolysaccharide and its application for degradation of azo-dyes. *Biotechnology Reports*, 10.1016/j.btre.2017.02.006.

Sastry, M., Ahmad, A., Islam Khan, M., & Kumar, R. (2003). Biosynthesis of metal nanoparticles using fungi and actinomycete. *Current Science, 85*(2), 162–170.

Shah, M., Fawcett, D., Sharma, S., & Tripathy, S.K. (2015). Green synthesis of metallic nanoparticles via biological entities. *Biotechnology*, 10.3390/ma8115377.

Shankar, S.S., Ahmad, A., & Sastry, M. (2003). Bioreduction of chloroaurate ions by geranium leaves and its endophytic fungus yields gold nanoparticles of different shapes. *Journal of Materials Chemistry*, 10.1 039/b303808b.

Shenton, B.W., Douglas, T., Young, M., Stubbs, G., & Mann, S. (1999). Inorganic and organic nanotube composites from template mineralization of tobacco mosaic virus. *Advanced Materials, 3*, 253–256.

Singaravelu, G., Arockiamary, J.S., Kumar, V.G., & Govindaraju, K. (2007). A novel extracellular synthesis of monodisperse gold nanoparticles using marine alga, Sargassum Wightii Greville. *Colloids Surface Biointerfaces, 57*, 97–101. 10.1016/j.colsurfb.2007.01.010.

Singh, P., & Abdullah, M.M. (2016). Role of nanomaterials and their applications as photo-catalyst and senors: A review abstract. *IMed. Pub. J, 2*, 1–10.

Singh, P., Kim, Y., & Zhang, D. (2016). Biological synthesis of nanoparticles from plants and microorganisms. *Trends in Biotechnology, 34*(7), 588–599. 10.1016/j.tibtech.2016.02.006.

Singh, J., Kumar, V., Kim, K., & Rawat, M. (2019). Biogenic synthesis of copper oxide nanoparticles using plant extract and its prodigious potential for photocatalytic degradation of dyes. *Environmental Research, 177*(July), 108569. 10.1016/j.envres.2019.108569.

Singh, J., Dutta, T., Kim, K.H., Rawat, M., Samddar, P., & Kumar, P. (2018). "Green" synthesis of metals and their oxide nanoparticles: applications for environmental remediation. *Journal of Nanobiotechnology, 16*(1), 1–24. 10.1186/s12951-018-0408-4.

Sriramulu, M., & Sumathi, S. (2018). Biosynthesis of palladium nanoparticles using *Saccharomyces cerevisiae* extract and its photocatalytic degradation behaviour. *Advances in Natural Sciences: Nanoscience and Nanotechnology, 9*, 025018. 10.1088/2043-6254/aac506.

Srivastava, N., & Mukhopadhyay, M. (2014). Biosynthesis of SnO_2 nanoparticles using bacterium Erwinia Herbicola and their photocatalytic activity for degradation of dyes. *Industrial & Engineering Chemistry Research*, 10.1021/ie5020052.

Srivastava, S. K., Yamada, R., Ogino, C., & Kondo, A. (2013). Biogenic synthesis and characterization of gold nanoparticles by Escherichia coli K12 and its heterogeneous catalysis in degradation of 4-nitrophenol. *Nanoscale Research Letters, 8*, 70.

Suhanovsky, M.M., Kale, A., Bao, Y., Zhou, Z., & Prevelige, P.E. (2013). Directed self-assembly of CdS quantum dots on bacteriophage P22 coat protein templates. *Biomacromolecules*, 10.1088/0957-4484/24/4/045603.

Surendra, T.V., & Roopan, S.M. (2016). Photocatalytic and antibacterial properties of phytosynthesized CeO2 NPs using moringa oleifera peel extract. *Journal of Photochemistry & Photobiology, B: Biology, 161*, 122–128. 10.1016/j.jphotobiol.2016.05.019.

Suresh, D., Shobharani, R.M., Nethravathi, P.C., Kumar, M.A.P., Nagabhushana, H., & Sharma, S.C. (2015). Artocarpus gomezianus aided green synthesis of ZnO nanoparticles: luminescence, photocatalytic and antioxidant properties intensity (Au). *Spectrochimica Acta Part A: Molecular and Biomolecular Spectroscopy, 141*, 128–134. 10.1016/j.saa.2015.01.048.

Sweeney, R.Y., Mao, C., Gao, X., Burt, J.L., Belcher, A.M., Georgiou, G., & Iverson, B.L. (2004). Of Cadmium Sulfide Nanocrystals. *11*, 1553–1559. 10.1016/j.

Swihart, T.M. (2003). Vapour-phase synthesis of nanoparticles. *COCIS*, 8, 127–133.

Synthesis, M., & Nanocrystallites, S.P. (2002). Microbial synthesis of semiconductor PbS. *Advanced Materials*, 11, 815–818.

Tamuly, C., Hazarika, M., Bordoloi, M., & Das, M.R. (2013). Photocatalytic activity of Ag nanoparticles synthesized by using piper Pedicellatum C. DC fruits. *Materials Letters, 102–103*, 1–4. 10.1016/j.matlet.2013.03.090.

Vasantharaj, S., Sathiyavimal, S., Senthilkumar, P., & Lewisoscar, F. (2019). Biosynthesis of iron oxide nanoparticles using leaf extract of *Ruellia tuberosa*: antimicrobial properties and their applications in photocatalytic degradation. *Journal of Photochemistry & Photobiology, B: Biology, 192*(December 2018), 74–82. 10.1016/j.jphotobiol.2018.12.025.

Veisi, H., Azizi, S., & Mohammadi, P. (2018). Green synthesis of the silver nanoparticles mediated by Thymbra Spicata extract and its application as a heterogeneous and recyclable nanocatalyst for catalytic reduction of a variety of dyes in water. *Journal of Cleaner Production, 170*, 1536–1543. 10.1016/j.jclepro.2017.09.265.

Vidhu, V.K., & Philip, D. (2014). Catalytic degradation of organic dyes using biosynthesized silver nanoparticles. *Micron, 56*, 54–62. 10.1016/j.micron.2013.10.006.

Vigneshwaran, N., Ashtaputre, N.M., Varadarajan, P.V., Nachane, R.P., Paralikar, K.M., & Balasubramanya, R.H. (2007). Biological synthesis of silver nanoparticles using the fungus aspergillus flavus. *Materials Letters, 61*, 1413–1418. 10.1016/j.matlet.2006.07.042.

Wang, F., Cao, B., & Mao, C. (2010). Bacteriophage bundles with prealigned Ca 2þ initiate the oriented nucleation and growth of hydroxylapatite. *Journal of Materials Chemistry*, (9), 3630–3636. 10.1021/cm902727s.

Wang, F., Nimmo, S.L., Cao, B., & Mao, C. (2012). Oxide formation on biological nanostructures via a structure-directing agent: towards an understanding of precise structural transcription. *Chemical Science*, 2639–2645. 10.1039/c2sc00583b.

Xia, B. Bin; Lenggoro, W., & Okuyama, K. (2001). Novel route to nanoparticle synthesis. *Advanced Materials, 13*(20), 1579–1582. 10.1002/1521-4095(200110)13:20<1579::AID-ADMA1579>3.0.CO;2-G.

Xu, H., Cao, B., George, A., & Mao, C. (2011). Self-assembly and mineralization of genetically modifiable biological nanofibers driven by β-structure formation. *Advanced Materials*, 2193–2199. 10.1021/bm200274r.

Xu, P., Zeng, G.M., Huang, D.L., Feng, C.L., Hu, S., Zhao, M.H., Lai, C., Wei, Z., Huang, C., Xie, G.X., et al. (2012). Use of iron oxide nanomaterials in wastewater treatment: a review. *Science of the Total Environment, 424*, 1–10. 10.1016/j.scitotenv.2012.02.023.

Zhu, H., Jia, S., Wan, T., Jia, Y., Yang, H., Li, J., Yan, L., & Zhong, C. (2011). Biosynthesis of spherical Fe3O4/bacterial cellulose nanocomposites as adsorbents for heavy metal ions. *Carbohydrate Polymers, 86*(4), 1558–1564. 10.1016/j.carbpol.2011.06.061.

18 Bio-Based Polymeric Material for Environmental Remediation

Zeenat Arif, Naresh K Sethy, Pradeep Kumar Mishra, Pradeep Kumar, and Pratiksha Pandey
Indian Institute of Technology (BHU), Varanasi, India

CONTENTS

- 18.1 Introduction 343
  - 18.1.1 Properties of Polymers 344
  - 18.1.2 Factors Affecting Properties of Polymers (Lagaron et al., 2004) 344
- 18.2 Major Issues Due to Synthetic Conventional Polymers 344
- 18.3 Bio-Based Polymer 344
  - 18.3.1 Why Bio-Based Polymers 345
- 18.4 Classification of Bio-Based Polymers 346
- 18.5 Biodegradable Polymers 347
- 18.6 Different Types of Bio-Based/Biodegradable Polymers and Their Applications 347
- 18.7 Major Application of Biodegradable Polymeric Material to Treat Wastewater ... 347
- 18.8 Technique and Biopolymer Used for Treatment of Wastewater 349
  - 18.8.1 Chitosan Biopolymers 349
  - 18.8.2 Bio-nanocomposites (BNCs) 349
  - 18.8.3 Cellulose Nanofibers (CNFs) 350
  - 18.8.4 Polyhydroxyalkanoates (PHAs) 350
  - 18.8.5 Foam Membranes 351
  - 18.8.6 Nano crystalline cellulose (NCC) 351
  - 18.8.7 Different Types of Pollutant Removal Using Bio-polymers 351
- 18.9 Mechanism of Biodegradation 353
  - 18.9.1 Degradation of Biodegradable Polymers 353
  - 18.9.2 Hydrolytic Degradation 353
  - 18.9.3 Acid-Catalysed Hydrolysis 353
  - 18.9.4 Alkali-Catalysed Polyester Hydrolysis 353
  - 18.9.5 Enzymatic Degradation 353
- 18.10 Challenges and Limitations 354
- 18.11 Future Prospective of Bio-Polymers 354
- 18.12 Conclusion 355
- References 355

18.1 Introduction

Polymers are found in nature as part of plant and animal structures like cellulose, starch, chitin, etc. and also synthesised by engineers, like plastic.

Polymeric material or polymers, the term coined from the Greek "poly" means many and "mers" means unit or part. These are long chains of repeating units known as monomers. And the molecular weight of polymers varyies from one to another depending on type and number of monomeric units used (Shen and Bever, 1972).

18.1.1 Properties of Polymers

Due to their physical and chemical properties such as high tensile strength, comparatively stiff and strong, corrosive resistive material, and light weight are a few of the important features of polymers that make it differ from other metallic or ceramic materials.

18.1.2 Factors Affecting Properties of Polymers (Lagaron et al., 2004)

- Molecular weight, structural aspects, crystallinity
- Monomeric nature signifies the family of polymers
- Molecular weight is dependent on the type of monomeric unit used and its distribution
- Number of branches and crosslinking in monomeric units
- Relative positions of functional groups, i.e., tactility
- Ordering in the position of the chain branches (crystallinity)

18.2 Major Issues Due to Synthetic Conventional Polymers

Synthetic polymers exist in a variety of forms, such as plastics, nylon, or the surface of a non-stick pan, and these synthetic polymers degrade ecosystems. Conventional polymers like polyethylene and polypropylene persist for many years after disposal; that is, they have a long durability but they are non-biodegradable (Gross and Kalra, 2002). This led to the death of millions of dying seabird species as they have ingested the polymers, considering it to be their food. Also, leaching of harmful chemicals and accumulation of waste in landfills are a few problems linked with the application of synthetic polymers. Around only 9% of plastic waste is recycled from the total plastic used, whereas 12% has been incinerated, and the remaining 79% is accumulated in landfills or are being dumped in the natural environment.

The other major issue associated with synthetic polymers is utilisation of non-renewable resources for their production. More than 99% of plastics are produced from chemicals derived from oil, natural gas, and coal, which are non-renewable resources. Due to this reason and environmental concern in recent years, the worldwide interest is shifting towards bio-based and biodegradable polymers. Figure 18.1 compares the production capacity of bio-based polymers presently and years 2018 and 2023, respectively.

18.3 Bio-Based Polymer

Bio-based polymers are produced by using biological sources or renewable resources. This can be synthesised directly in polymeric form within the biological sources like microorganisms and others can be formed by using bio-based monomers. Bio-based polymers can be classified as biodegradable or non-biodegradable based on their chemical structure and properties. Bio-based and biodegradable polymers are two terms that are used interchangeably in literature, but both terms have a key difference (Amass et al., 1998). Biodegradable polymers are defined as the material that possesses polymeric physical and chemical properties but undergoes complete degradation into carbon dioxide, water, and methane (anaerobic processes) when exposed to anaerobic processes (microbes) or anaerobic processes, whereas bio-based polymers are obtained from biological sources as stated previously. It can be biodegradable like polylactide (formed by polymerisation of lactic acid monomer extracted from lactobacillus), and

Bio-Based Polymeric Material 345

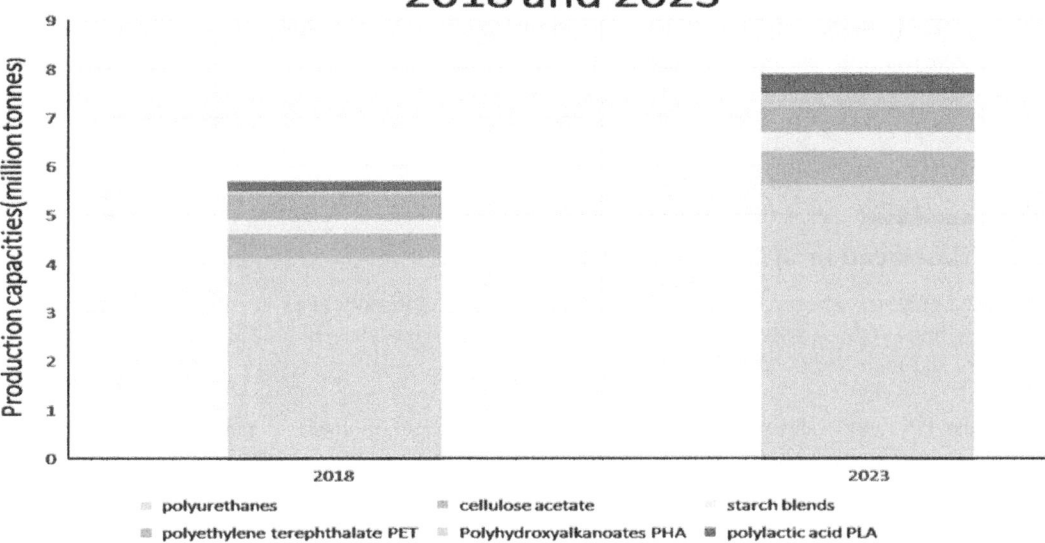

FIGURE 18.1 Production capacity of a few bio-polymers in a current and future scenario.

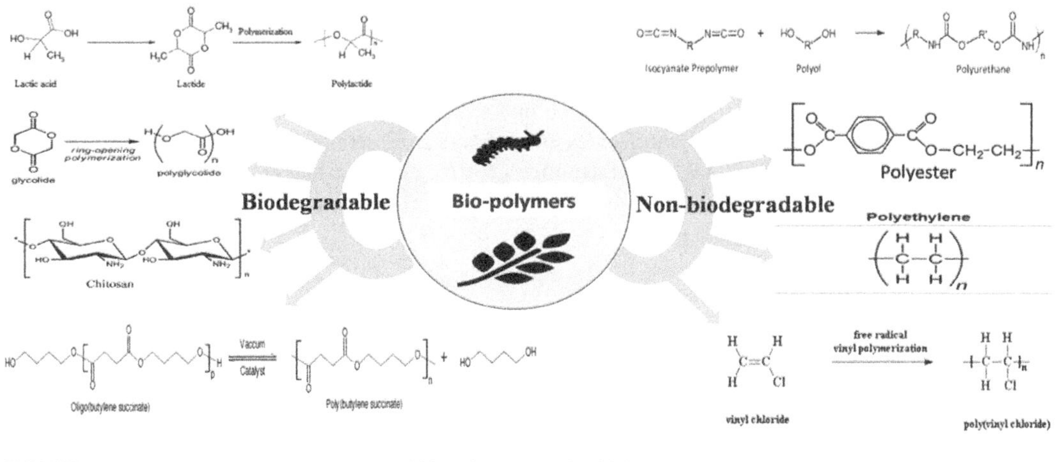

FIGURE 18.2 Examples of different types of biopolymers used widely.

non-degradable like biopolyethylene. Hence, we conclude that many bio-based polymers are biodegradable (e.g., starch and polyhydroxyalkanoate); not all biodegradable polymers are bio-based (polycaprolactone) (Amass et al., 1998). Figure 18.2 shows a few examples of biopolymers classified into biodegradable and non-biodegradable polymers.

18.3.1 Why Bio-Based Polymers

- **Reducing the dependency on non-renewable resources or fossil fuel–based polymers like petroleum oil based.**

    Dependency on non-renewable resources is reducing rapidly due to the high rate of their exploitation. On the other hand, this will help in reducing carbon dioxide emissions, which is the measure of issues of fossil fuel combustion. At the same time, worldwide there is high demand for replacing petroleum-based raw materials to renewable-based raw materials for the synthesis of polymers. Hence, producing polymers from agricultural feed such as corn, potatoes, and carbohydrate feedstock, etc. are gaining interest in the research field (Amass et al., 1998).

- **Biodegradable property:** Most of the biopolymers are biodegradable, as stated previously; hence, they solve the environmental problems of soil pollution, water pollution, and air pollution. Biodegradable plastics are now easily degrading into simple carbon dioxide and water within a few days or weeks. That somehow reduces the solid waste problems.

18.4 Classification of Bio-Based Polymers

Bio-based polymers are broadly classified into three classes: naturally derived polymers, bioengineered polymers, and synthetic polymers (Nakajima et al., 2017). Figure 18.3 shows different classifications of polymers and their details are discussed below:

Class-I Naturally derived polymers: Direct application of biomass as polymeric material, e.g., cellulose, cellulose acetate, starches, chitin, etc.

Class-II Bioengineered polymers: Polymers synthesized from microorganisms and plants, e.g., poly (hydroxy alkenoates), poly (glutamic acid).

Class-III Synthetic polymers: Polylactide, bio-polyolefins, bio-poly (ethylene terephthalic acid) (bio-PET) etc.

There are three principal ways to produce biopolymers from renewable resources (Amass et al., 1998):

- *Agro-based or biomass resources:* Extraction and separation of raw material from biomass to synthesise polymers. Example: polysaccharides and lipids (starch, cellulose), proteins. From oil products such as polycaprolactones (PCL), polyesteramides (PEA), aliphatic, and aromatic copolyester.
- *Microorganisms source:* The monomeric unit is formed by fermentation of yeast bacteria, fungi, etc. (e.g., polyhydroxlalkanoates)
- *Biotechnology by conventional synthesis:* e.g., polylactide, PBS, PBS-PE, etc.

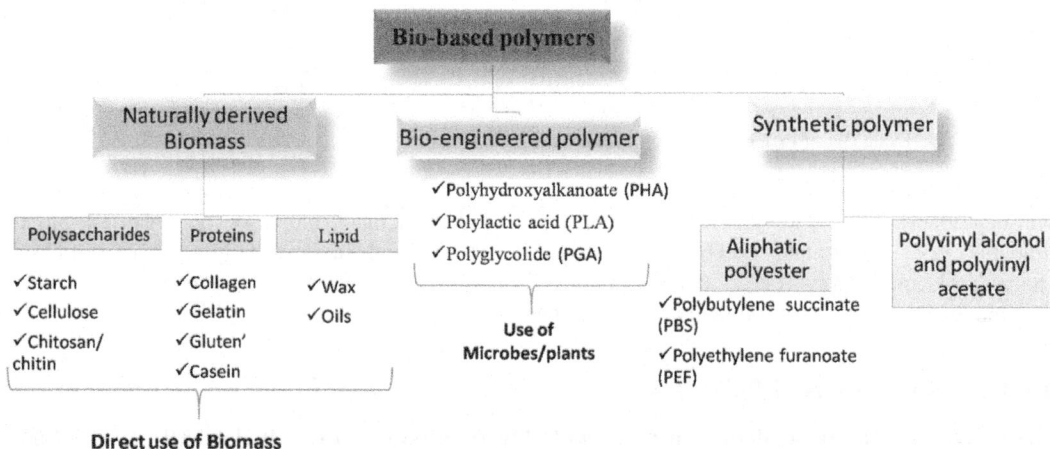

FIGURE 18.3 Schematic representation of classification of biopolymers.

18.5 Biodegradable Polymers

In its simplest form, biodegradable is said to be "capable of undergoing decomposition into simpler compounds such as carbon dioxide, methane, water, inorganic compounds and biomass". The biodegradable property of the bio-based polymers is gaining interest for its production. Now there are many biodegradable polymers launched and commercialised in markets that can be treated as solutions to the waste problems existing in the field of agriculture, marine fishery, and other goods industries. Mostly, developed polymers are polylactides (PLA) and poly(hydroxyalkanoates) (PHAs) are sauccinate-derived polymers. This way of enhancement in production of biodegradable polymers combats environmental issues and is considered the most successful and demanding innovation in the polymer industry (Díaz et al., 2014).

Biodegradable polymers consist of mainly amides, esters, and ether bonds, which are hydrolysable in nature. Biodegradability depends not only on the origin of the polymer but also on its chemical structure and the environmental degrading condition.

Characteristics of biodegradable polymers:

- Non-toxicity
- Bio-compatibility
- Tensile strength
- Mechanical strength
- Controlled rates of degradation

18.6 Different Types of Bio-Based/Biodegradable Polymers and Their Applications

1. Synthetic bio-based/biodegradable polymers

These synthetic polymers are having carbon backbones along with hydrolysable groups, such as ester, amide, and urethane, or polymers. It includes aliphatic polyesters (polyglycolide, polylactide, polycaprolactone, etc.), aromatic copolyesters poly(butylene adipate-co-terephthalate), polyamides, poly(ester-amide), polyurethanes, polyanhydrides, vinyl polymers[poly (ortho-ester), poly(propylene fumarate)]. Some of these are explained in Table 18.1 with applications.

2. Natural bio-based/biodegradable polymers

This is very important step in polymer industries which from global environmental point of view. These group includes naturally occurring polymers e.g., starch, cellulose, chitin, polysaccharides, proteins, etc. These polymers have a wide range of application in environmental remediation and manufacturing processes (Maharana et al., 2009). Some of these are discussed in Table 18.2.

18.7 Major Application of Biodegradable Polymeric Material to Treat Wastewater

Water is very essential for all living organisms on earth; freshwater is unequally distributed on the earth. Agriculture, energy production, industrial, and human consumption are the main demanding sources for water. Figure 18.3 depicts different sources that create water pollution. It was interpreted that by 2050 the global demand for water is expected to increase by 70%. This will lead to the condition of water scarcity. As a result, vast research is being carried out in the field of water treatment in order to reuse the

TABLE 18.1

List of synthetic biopolymers, their properties, and applications (Vroman and Tighzert, 2009; Lux, 2016; Babu et al., 2013; Maharana et al., 2009)

Biopolymers/Biodegradable Polymers	Properties	Application
Polylactic acid (PLA)	• Polycondensation of D- or L-lactic acid or polymerisation of lactide produces PLA. • Hydrophobic polymer. • Resistive to to hydrolysis • Glass transition temperature (Tg) 63.8°C • Melting temperature (Tm) 178°C	• Food packaging due to glossy appearance • Fabrication of products such as trays, tablewares, and cutleries. • Manufacturing of mulching films or delayed release material for spraying of pesticides and fertilizers • In the medical industry as implants, stents, and bone support splints.
Polybutylene succinate (PBS)	• Polycondensation of gylcols gives PBS • Glass transition temperature −45 to −10°C • Good processability and toxicity • Aliphatic polyester	• In a packaging industry where priority for oxygen permeability is less and biodegradibilty is important criteria. • Maunfacturing of compostable waste such as bags, film and compostable pouches, plates, and cutlery. • Manufacturing of mulching films, nets, trappers to be used as a deployment in agriculture and forestry industry. • High potential application in medical industry.
Polyhydroxyalkanoate (PHA)	• Use of bacteria and renewable waste feedstock such as lignocellulosics gas, vegetable oils, and fatty acids to produce PHA. • Poor thermo-mechanical properties • Good resistance to moisture, aroma barrier properties • Low melting point 40 to 180°C	• Fabrication of compostable films and bags • High potential application in medical and pharmaceutical industries including sutures, bone plates grafts, artificial oesophagus, drug delivery, skin regeneration, food additives
Polyethylenefuranoate (PEF)	• Condensation polymerization between monoethylene glycol and 2.3-furandicarboxylic acid produces PEF • Possess water and O_2 barrier (ten times better than PET and twice of PET) • Barrier properties make strong contender for the packaging market • Melting point is 230°C • Attractive potential energy savings and good tensile strength	• The barrier properties and relatively easy processing give PEF a great potential in the packaging industry. • Good tensile strength makes it a potential application in the textile sector.
Polyglycolide (PGA)	• It is a linear aliphatic polyester. • Ring polymerisation of a cyclic lactone, glycolide produces PGA. • Highly crystalline. • High melting point 220–225°C • Tg is 35–40°C.	• Fixed installations such as bone plates, bone screws, surgical sutures, spinning. • In oil field chemical process (time release agent for corrosion inhibitors, dispersants), temporary plugging agent. • Used in multilayer PET bottles for carbonated drinks • Fused for surgical sutures in medical, drug delivery
Polycaprolactone (PCL)	• It is a semi-crystalline linear polymer.	

TABLE 18.1 (Continued)

List of synthetic biopolymers, their properties, and applications (Vroman and Tighzert, 2009; Lux, 2016; Babu et al., 2013; Maharana et al., 2009)

Biopolymers/Biodegradable Polymers	Properties	Application
	• Ring-opening polymerization of caprolactone forms PCL • Catalyst used is tin octate catalyst • Good solubility in different solvents • Tg 60 °C. • Melting point is 60–65 °C. • Easy biodegraded by enzymes and fungi	• In controlled drug delivery systems for e.g., release of pesticides, herbicides, and fertilizers • Substitute to traditional plaster, dental impressions, and oncology immobilization systems • Used for making films and laminates • Chemical compound in coating materials or modifiers for plastics
Poly(ethylene terephthalate) (PET)	• Bio-based PET is obtained from ethylene glycol and terephthalic acid.	• Containers for packaging food products, drinks (soft and alcoholic beverages), detergents, cosmetics, and pharmaceutical products. • Making of films, for injection molding application.

wastewater discharged from different sources by different techniques within the process wherever possible (Sidek et al., 2019). Figure 18.4 shows different sources causing water pollution.

18.8 Technique and Biopolymer Used for Treatment of Wastewater

Conventional wastewater treatment processes: conventional wastewater treatment consists of a combination of physical, chemical, and biological processes. However, less efficiency, large volume of sludge production, and large consumption of chemicals leads to the development of advanced technologies with improved efficiency that includes emergence of membrane treatment technology, desalination technologies, etc. Different agents are used for the treatment of wastewater that can be inorganic (aluminium sulphate, ferric chloride) or synthetic polymers (polyacrylamides). Their utilisation results in the improvement of water quality and also maintains public health environmental quality. Environmental concern leads to the development of biopolymers that are used in membrane synthesis for the treatment of wastewater. The various biopolymers used in the treatment of water are discussed as follows.

18.8.1 Chitosan Biopolymers

A chitosan biopolymer is a naturally occurring biopolymer composed of chitin and derived from recycled crustacean cells and fungi. It is utilised as a flocculant in water treatment processes. In the Great Lakes' region, chitosan is used for the treatment of stormwater.

Chitosan can chelate the dissolved metals in water due to its unique property of chelation. Chelation is defined as a process where the multiple binding sites associated with the polymer chain readily bind with the metal to eliminate it from a solution. These advanced properties of chitosan make it a potential application in treating difficult industrial stormwater [Ziemer, 2021].

18.8.2 Bio-nanocomposites (BNCs)

The doping of species (inorganic or organic nanoparticles) with natural polymers is termed bio-nanocomposites (BNCs), or green composites. The presence of inorganic or biological moieties enhances its biocompatibility and functionality. Lignin, hemicellulose, cellulose, E1204/glucan, protein, PHA, and linear

TABLE 18.2

List of synthetic biopolymers, their properties, and general applications (Maharana et al., 2009)

Natural Bio-Based Polymers	Example	Application
Starch	• Starch/ethylene vinyl alcohol • Starch/polycaprolactone blends • Starch/cellulose acetate blends with methylmethacrylate and acrylic acid • Modified starch/its derivatives • Thermoplastic starch	• Orthopaedic implant devices as bone filler • Bone replacement/fixation • Implants/orthopaedic application • Bone cements • Food packaging application and drug delivery • Fabrication of containers, as textile sizing agents and adhesives
Cellulose	• Cellulose esters • Carboxylated methylcellulose/cellulose acetate fibres • Modified cellulose/microfibrous cellulose • Cellulose nanofibers and particles	• Used for the synthesis of membranes for separation • As a binder for drugs, film-coating agent, ointment base, wound dressings • In food packaging to prevent it from moisture • Textile application and chromatography application, chiral separation
Alginates	• Linear polysaccharide	• Utilized as viscosifiers, stabilisers, and water binding agents. • Potential applications in textile printing. • Manufacturing of ceramics to fabricate welding rods • In pharmaceutical and food applications as geling agent. • Due to high solubility in cold water and thermostable gel characteristics, the polymer is used in custard cream production. • Utilised as stabiliser and thickener in beverages, ice cream, and sauces. • Also used in wound dressing, controlled drug delivery, dental impression, etc.
Protein polymers	• Collagen, gelatine, glutenin, soy protein	• Wide application as film-forming agent to produce flexible and edible films
Chitin	• Chitosan	• Used as flocculant in wastewater treatment • Removing heavy metals from wastewater • Biomedical application • Used in cosmetic and wound treatment

polyester are some biopolymers, including both synthetic and natural, that are used for the synthesis of BNC. These are exploited for biomineralisation processes and other environmental remediation (Dogre et al., 2019).

18.8.3 Cellulose Nanofibers (CNFs)

Cellulose nanofibers have a wide range of applications, such as dialyzer-based haemodialysis, distillation, and micro, ultra, and nanofiltration. The membrane applications of cellulose are due to its porous, cost effective, semi-permeable hydrophilic nature (Richards et al., 2011; Li et al., 2015).

18.8.4 Polyhydroxyalkanoates (PHAs)

Polyhydroxy butyrate is a microbial polymer produced biotechnologically by specific bacteria. Because of its biodegradability and hydrophobicity, it is widely used for water filtration and decontamination. PHA-based polymers are used to remove both gram-positive and gram-negative bacteria and yeasts. After use, the filter material can be composted, which completely degrades within several days or weeks (Marova et al., 2015).

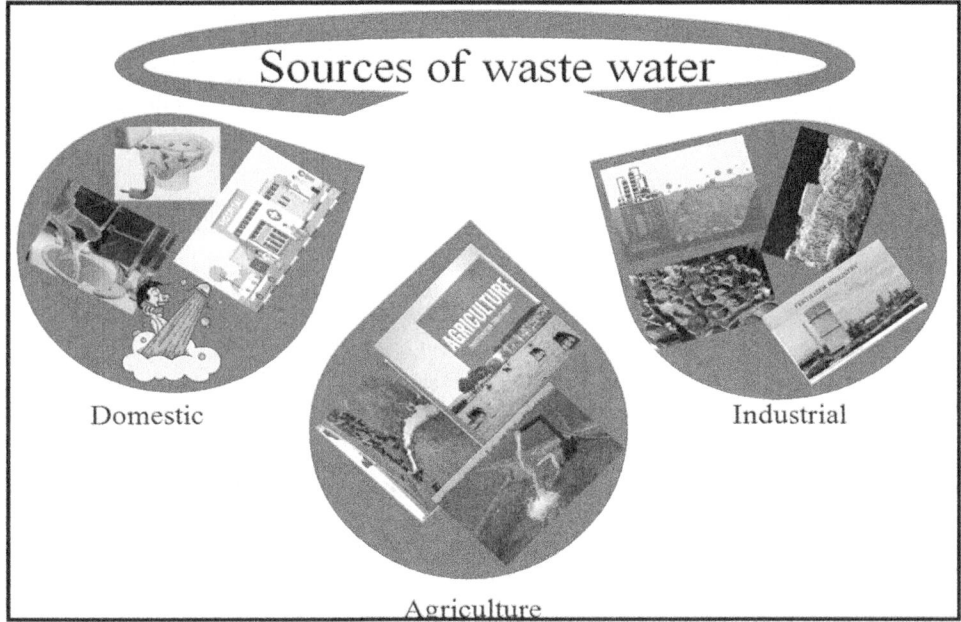

FIGURE 18.4 Different sources of water pollution.

18.8.5 Foam Membranes

Foam membranes have a unique capillary microstructure (10–45 μm) membrane derived from agarose and gelatine with non-toxic fruit extract and natural crosslinker, genipin; it is synthesized by the lyophilisation process to permit selective permeation of water. Foam membranes are used for the oil-water separation and have easy membrane cleaning capability after use, which provides retaining ability for long performance (Chaudhary et al., 2014).

18.8.6 Nano crystalline cellulose (NCC)

NCC is a biopolymer-based material derived from renewable and sustainable resources such as cotton and wool, which have a large specific surface associated with plenty of hydroxyl groups and anionic sulphate ester groups attached on the surface; this makes it suitable for developing a membrane for wastewater treatment and a perfect substrate to synthesize a composite absorbing and/catalytic materials. It has the ability to remove organic pollutants and oil pollutants effectively in water. The synthesis of a biomembrane by combining NCC crystals with micropores with diameters ranging from 10 to 13 nm on chitosan can be used efficiently for water treatment. The maximum removal rate of Victoria Blue 2B, Methyl Violet 2B, and Rh6G reached 98%, 84%, and 98%, respectively (Liang and Hu, 2016).

Apart from the above-mentioned polymer, there are other bio-based materials that are also utilizsd for the removal of oil spills. Biomass and conventional sorbents, surfactants, and separators are the material used for oil removal, dispersion, biodegradation, and oil recovery. A few bio-based polymers (Doshi et al., 2018) and their roles are summarized in Table 18.3.

18.8.7 Different Types of Pollutant Removal Using Bio-polymers

- **Heavy metal ion removal:** Due to the environmental friendliness, various types of adsorbents are synthesized using bio-based polymers (Zhan et al., 2018; Wu et al., 2018). Cellulose acetate (CA) nonwoven biopolymer, prepared via electro spinning used for the synthesis of membranes for the absorption of Cu^{2+}, Hg^{2+}, and Cd^{2+} (heavy metal ions) from wastewater and these ions can be

TABLE 18.3

Use of different types of biopolymers for treatment of oil spills using different techniques

Bio-Based Polymer		Method/Solvent	Separation of Oil
Sorbent Electrospun polyvinyl chloride/polystyrene fibers(nano-fibers)		Electrospinning	Motor oil, peanut oil, diesel
Surfactant based material		**Method**	Removal of crude oil, crude oil isooctane, waxy crude 35 oil and fresh asphaltenic oil, light or heavy oil kerosene
Dispersant	**Surfactant**		
Modified eggshell	Hexadecyl-trimethyl-ammonium-bromide (HDTMA-Br)	Mixing shell with HDTMA-Br followed by washing and drying	
Food-grade amphiphiles	Lecithin and Tween 80	Surfactants blending in ethanol	
Microbial biosurfactants	Surfactin fatty acyl-glutamate (FA-Glu)	Modified strains of *Bacillus subtillis* on glucose (2%) + mineral salts	
Octyl-carboxymethyl chitosan	Octyl aldehyde	Carboxylation acylation reduction	
Membrane/Separator		**Method**	Removal of toluene, isooctane, n-hexane, chloroform, crude oil, bio-diesel, oil-spill wastewater
Stainless steel + Residue of waste potato + polyurethane		Spraying	
Membrane composed of nylon-fibre covered by tunicate cellulose nancrystal		Facile dip-coating	
Foam membrane from agarose, gelatin, genipin		Blending + lypholization	
Aerogel membrane synthesized from chitosan, agarose, genipin		Blending + lypholization	

easily de-adsorbed after the treatment process, thereby increasing the reusability of membrane and increasing the life of the membrane. In addition to it, natural cellulose from jute and coconut was also used to synthesise an activated carbon for heavy metal ion removal using an absorption technique because of their high absorption capacity due to high specific surface area and pore volume (Phan et al., 2006).

- **Photocatalytic degradation of organic pollutants:** The photocatalytic activity of cellulose polymer was further enhanced by doping different semiconductor materials. The catalytic activity of nanocellulose–metal oxide was reported to be higher than the individual materials. Table 18.4 lists some of the nanocellulose–metal oxides as photocatalysts to remove different organic pollutants from wastewater.

TABLE 18.4

List of nanocellulose–metal oxides as photocatalysts to remove organic pollutants (Gao et al., 2014; Chen et al., 2017; Dong et al., 2018)

Type of Catalyst	Pollutants	Photocatalytic Degradation Efficiency (%)
ZnO/cellulose nanocomposite	Methylene blue	79%,
Nanocrystalline TiO_2/cellulose fabric	99.995% CO_2, flow rate- 300 mL/min for 1 h	Removal of 194.0 ppm/g and 50.8 ppm/g CO and CH_4, respectively.
Cellulose fibre supported zinc phthalocyanine	Basic green 1	98%
Cellulose acetate/Ag@AgCl membrane	Methyl orange	73%
Hydrogel of nanocomposite (Alginate/carboxymethyl cellulose and nano size TiO_2)	Congo red	91.5%

18.9 Mechanism of Biodegradation

Degradation of polymers happens mainly by changing properties like tensile strength, colour, shape, etc. under the influence of heat, light, or chemicals (one or more environmental factors). The biodegradation of polymers occurs by microbial action, photodegradation, or chemical degradation. Enzymatic and chemical action associated with the living organism results in the general biodegradation of material (Lucas et al., 2008). It is a two-step process, as described below:

Step 1 Abiotic (oxidation, photodegradation, or hydrolysis) or biotic reaction (degradation by microbes) results in the fermentation of polymeric material into lower mass substances.

Step 2 Bioassimilation of the polymer fragments by microorganisms and mineralization.

18.9.1 Degradation of Biodegradable Polymers

Degradation of biodegradable polymers is carried out either by hydrolysis (without the use of enzyme catalysis) or by enzymatic mechanism. Although degradation by hydrolysis is the main mechanism, they may undergo partial enzymatic degradation depending on the polymer structure.

18.9.2 Hydrolytic Degradation

The term "hydrolytic degradation" signifies "in the presence of water, hydrolytically breakdown of unstable polymer backbone". The lead to penetration of water molecules into the polymer bulk is followed by random cleavage of the chemical bonds. The result of breakdown tends to convert long polymeric chains into shorter chains with a decrease in the polymer molecular weight. At the same time, polymer physical and mechanical properties remain unchanged by keeping the structure together by the crystalline region. On further degradation, the chemical bonds break in the crystalline regions, thereby altering the physical and mechanical properties. In the last stage, in vivo metabolisation of fragments takes place by the enzymes. Hydrolysis can be also catalysed with an acid, base, or enzyme (Rydz et al., 2015).

18.9.3 Acid-Catalysed Hydrolysis

Protonation of the carbonyl oxygen of the ester group in polyesters under acidic conditions by a hydronium ion induces more electrophilicity to carbonyl carbon because of the positive charge followed by an attack on the carbonyl carbon by water molecules. This generates a tetrahedral intermediate, which further decomposes to a smaller compound such as carboxylic acid and alcohol.

18.9.4 Alkali-Catalysed Polyester Hydrolysis

Under alkaline conditions, the ester group of the carbonyl compound is attacked by the hydroxide anion and generates a tetrahedral intermediate. It is a reversible step and the ability of the leaving alcohol (ROH) to stabilise a negative charge determines the preference of the tetrahedral intermediate. The degradation rates in an alkaline medium are very high when compared to acidic conditions.

18.9.5 Enzymatic Degradation

In enzymatic degradation, enzymes (specific protein) cause polymer degradation. It is a complex mechanism and is strongly influenced by the composition of polymers. Some polymers degrade in biological environments and also in the presence of microorganisms. The degradation steps include (Pathak, 2017):

- Biodegradation: results in changes in the physical and chemical properties
- Biofragmentation: enzymatic action results in breaking the complex polymer structure into simpler forms
- Assimilation: molecule uptake by microorganisms
- Mineralization: after degradation, formation of CO_2, CH_4, H_2O products.

18.10 Challenges and Limitations

Undoubtedly, biopolymers prove to be an excellent material for environmental remediation; still, many challenges are faced during application. A few of the limitations are listed below (Shak et al., 2018):

- To promote commercialisation and marketability, it becomes important that the material should possess easy scalability with low cost, and should produce valuable end products.
- Low yield of biocomposite polymers make it difficult for large-scale production.
- The economic and regulatory hindrances are the obstacles for successful application and marketability.
- Low mechanical properties in a humid environment and fast degradation rate sometimes limit the application.
- Difficult to process in existing processing equipment.
- In order to make these materials viable, significant factors affecting logistics are biomass feedstocks, new microbial strains/enzymes, and efficient processing techniques to recover bio-based products.

18.11 Future Prospective of Bio-Polymers

Over the past two decades, research and development in the bio-based polymer industry is growing up with the fossil-fuel-based chemical industry (Zia et al., 2015). Being renewable, sustainable, biodegradable, compostable, and carbon impartial, there is an increasing demand for bio-based polymers such as bio-PE, bio-PET, and PLA and the demand is growing swiftly in the product market. The packaging market is considered a prime application area for biopolymers. Customers are looking for market-based eco-solutions and consumer behaviour studies show that consumers are willing to pay more for environmentally friendly goods. In an emerging sector, the bottle industry is quickly increasing the charge over the forecast duration. Even the authorities' interventions in green procurement rules are increasing the penetration of biopolymers in the dominating software market.

Cellulose regenerated from wood or plants has received significant attention in recent years in the research field and this trend is likely to continue. Bio-based coatings and/or combining bio-based and inorganic layers together can become an effective solution in the future.

In the future, it was also predicted that the cost of bio-based materials will decrease due to increased raw material supply and higher capacity and efficiency in the production. Bio-based industries are focussing on monomer and polymer bioconversions and nano reinforcement is a promising path for many applications in the production of new biomaterials.

Bioplastics have the potential to replace conventional plastics in the market and has a wide range of applications, such as fabrication of plastic plates, cups, cutlery, food packaging industry, and can also be used in storage bags/containers or other plastic or composite material items and therefore helps in making the environment sustainable. Due to its sustainability, there will be an increase in the future market for bioplastics. The bio-based materials have a major potential for being compostable. Developing technology, innovation, and global support are important methods to commercialise bioplastics. At the same time, in order to increase the material sustainability and processes throughout their lifetime, bioplastics must be environmentally friendly, and they should not be competing with traditional sources (Sidek et al., 2019). The use of biopolymers and bioplastics has the potential to reduce the petroleum consumption by 15–20% by 2025. Also, improved technical properties and innovations create an opportunity to open new markets with higher profit potentials in the fields of automotive, medicine, and electronics.

Polyesters and hydrophilic natural polymer composites are receiving significant interest as novel biodegradable polyesters with properties suitable for extraordinary biomedical applications. PCL-CS/PEO-GEL crosslinked GTA are used widely for tissue engineering, whereas PLLA/chitosan nanofibers are used as drug carriers. Composites of PECE/collagen/n-HA hydrogel have great potential in guided bone regeneration. Composite materials of POC-HA are used for the bone healing process and hybrid hydrogel composed of Arg-UPEA/GMA-chitosan is used in wound healing (Amulya et al., 2016).

18.12 Conclusion

Conventional polymers, for their adverse effects, are frequently being replaced by bio-based polymers. The discovery of biopolymers in the area of materials science and engineering has achieved a milestone. The environmentally friendly nature of biopolymers with low toxicity and functionality makes them potential candidates for environmental applications such as treatment of wastewater, carbon sequestration, etc. They have proven to be advantageous when compared with non-biodegradable material for their applications in the removal of heavy metal by adsorption technique and dye removal using the photocatalytic method with improved removal efficiency. However, modification is still required to improve its functionality in order to remove a wide range of target pollutants without losing its strength.

REFERENCES

Amass, W., Amass, A., & Tighe, B. (1998). A review of biodegradable polymers: uses, current developments in the synthesis and characterization of biodegradable polyesters, blends of biodegradable polymers and recent advances in biodegradation studies. *Polymer International, 47,* 89–144.

Amulya, K., Dahiya, S., & Venkata Mohan, S. (2016). Building a bio-based economy through waste remediation: innovation towards sustainable future. *Bioremediation and Bioeconomy,* 497–521.

Babu, R.P., Connor, K., & Seeram, R. (2013). Current progress on bio-based polymers and their future trends. *Progress in Biomaterials, 2,* 8–24.

Chaudhary, J.P., Nataraj, S.K., Gogda, A., & Meena, R. (2014). Bio-based superhydrophilic foam membranes for sustainable oil–water separation. *Journal Green Chemistry, 16,* 4552–4558.

Chen, P., Liu, X., Jin, R., Nie, W., & Zhou, Y. (2017). Dye adsorption and photo-induced recycling of hydroxypropyl cellulose/molybdenum disulphide composite hydrogels. *Carbohydrate Polymer, 167,* 36–43.

Díaz, A., Katsarava, R., & Puiggalí, J. (2014). Synthesis, properties and applications of biodegradable polymers derived from diols and dicarboxylic acids: from polyesters to poly(ester amide)s. *International Journal of Molecular Science, 15,* 7064–7123.

Dogre, R.S., Deshmukh, K.K., Mehta, A., Basu, S., Meshram, J.S., Maadeed, M.A.A.A., & Karim, A. (2019). Natural polymer based composite membranes for water purification: a review. *Polymer-Plastics Technology and Engineering, 58,* 1295–1310.

Dong, P., Cheng, X., Huang, Z., Chen, Y., Zhang, Y., Nie, X., & Zhang, X. (2018). *In-situ* and phase controllable synthesis of nanocrystalline TiO_2 on flexible cellulose fabrics via a simple hydrothermal method. *Materials Research Bulletin, 97,* 89–95.

Doshi, B., Sillanp, M., & Kalliola, S. (2018). A review of bio-based materials for oil spill treatment. *Water Research, 135,* 262–277.

Gao, M., Li, N., Lu, W., & Chen, W. (2014). Role of cellulose fibers in enhancing photosensitized oxidation of basic green 1 with massive dyeing auxiliaries. *Applied catalysis B: Environmental, 147,* 805–812.

Gross, R.A., & Kalra, B. (2002). Biodegradable polymers for the environment. *Science, 297*(5582), 803–807.

Lagaron, J.M., Catalá, R., & Gavara, R. (2004). Structural characteristics defining high barrier properties in polymeric materials, *Material Science and Technology, 20,* 1–7.

Li, R., Zhang, L., & Wang, P. (2015). Rational design of nanomaterials for water treatment. *Nanoscale, 7*, 17167–17194.

Liang, H., & Hu, X. (2016). A quick review of the applications of nano crystalline cellulose in wastewater treatment. *Journal of Bioresources and Bioproducts, 1*(4), 199–204.

Lucas, N., Bienaime, C., Belloy, C., Queneudec, M., Silvestre, F., & Nava-Saucedo, J.E. (2008). Polymer biodegradation: mechanisms and estimation techniques. *Chemosphere, 73*, 429–442.

Lux, A.M. (2016). Introduction to bio-based polymers. *Multilayer Flexible Packaging*, 47–52.

Maharana, T., Mohanty, B., & Negi, Y.S. (2009). Melt-solid polycondensation of lactic acid and its biodegradability. *Progress in Polymer Science, 34*, 99–124.

Marova, I., Kundrat, V., Benesova, P., Matouskova, P., & Obruca, S. (2015). Use of biodegradable PHA-based nanofibers to removing microorganisms from water, 2015, IEEE 15th International Conference on Nanotechnology (IEEE-NANO), 10.1109/NANO.2015.7388958

Nakajima, H., Dijkstra, P., & Loos, K. (2017). The recent developments in bio-based polymers toward general and engineering applications: polymers that are upgraded from biodegradable. *Polymers, Analogous to Petroleum-Derived Polymers, and Newly Developed. Polymers, 9*(10), 523.

Niaounakis, M. (2015). *Biopolymers: Applications and Trends*. Elsevier Science, 1–90 (ISBN No- 978-0-323-35399-1)

Pathak, V.M. (2017). Navneet, Review on the current status of polymer degradation: a microbial approach. *Bioresource and Bioprocessing, 4*, 15–46.

Phan, N.H., Rio, S., Faur, C., Le Coq, L., Le Cloirec, P., & Nguyen, T.H. (2006). Production of fibrous activated carbons from natural cellulose (jute, coconut) fibers for water treatment applications. *Carbon, 44*, 2569–2577.

Richards, L.A., Richards, B.S., & Schafer, A.I. (2011). Renewable energy powered membrane technology: salt and inorganic contaminant removal by nanofiltration/reverse osmosis. *Journal of Membrane Science, 369*, 188–195.

Rydz, J., Sikorska, W., Kyulavska, M., & Christova, D. (2015). Polyester-based (bio)degradable polymers as environmentally friendly materials for sustainable development. *International Journal of Molecular Science, 16*, 564–596.

Shak, K.P.Y., Pang, Y.L., & Mah, S.K. (2018). Nanocellulose: recent advances and its prospects in environmental remediation. *Beilstein Journal of Nanotechnology, 9*, 2479–2498.

Shen, M., & Bever, M.B. (1972). Gradients in polymeric materials, *Materials Science, 7*, 741–746.

Sidek, I.S., Draman, S.F.S., Abdullah, S.R.S., & Anuar, N. (2019). Current development on bioplastics and its future prospects: an introductory review. *i TECH MAG, 1*, 03–08.

Vroman, I., & Tighzert, L. (2009). Biodegradable polymers. *Materials, 2*, 307–344.

Wu, J., Qu, Y., Yu, Q., & Chen, H. (2018). Gold nanoparticle layer: a versatile nanostructured platform for biomedical applications. *Materials Chemistry Frontiers, 2*, 2175–2190.

Zhan, W., Qu, Y., Wei, T., Hu, C., Pan, Y., Yu, Q., & Chen, H. (2018). Sweet switch: sugar-responsive bioactive surfaces based on dynamic covalent bonding. *ACS Applied Material & Interfaces, 10*, 10647–10655.

Zia, K.M., Noreen, A., Zuber, M., Tabasum, S., & Mujahid, M. (2015). Recent developments and future prospects on bio-based polyesters derived from renewable resources: a review, *International Journal of Biological Macromolecules*, 10.1016/j.ijbiomac.2015.10.040.

Ziemer, J. Biopolymer Offers Biodegradable Alternative in Water Water treatment chemicals play an important role in maintaining public health and environmental quality. (Accessed online: https://www.waterworld.com/home/article/16205531/biopolymer-offers-biodegradable-alternative-in-water-treatment).

19 Cyanide: Sources, Health Issues, and Remediation Methods

Priyanka Yadav, Manisha Verma, and Vishal Mishra
School of Biochemical Engineering, IIT (BHU), Varanasi, India

CONTENTS

19.1 Introduction 357
19.2 Compounds Where Toxic Cyanide Exposure Occurs in Human Beings 359
19.3 Physiochemical Analysis of Hydrogen Cyanide 360
19.4 Fates of Cyanide in the Body 361
19.5 Sources of Cyanide 363
19.6 Industrial Utilisation of Cyanide 363
19.7 Cyanide Contamination in Water 363
19.8 Cyanide Contamination in Air 363
19.9 Cyanide Contamination in Soil 365
19.11 Methods of Cyanide Removal 366
      19.11.1 Remediation by Physical Methods 368
            19.11.1.1 Dilution Method 368
            19.11.1.2 Photolysis 368
            19.11.1.3 Membranes 369
            19.11.1.4 Electrowinning 369
            19.11.1.5 Hydrolysis Followed by Distillation 369
      19.11.2 Chemical Methods of Cyanide Removal 369
            19.11.2.1 Acidification Along with Volatilisation 369
            19.11.2.2 Solvent Extraction 370
            19.11.2.3 Adsorption 370
      19.11.3 Biological Method 370
      19.11.4 Kinetics and Isotherms Used 371
      19.11.5 Cyanide Removal from Soil, Air, and Water 371
19.12 Conclusion 371
References 372

19.1 Introduction

Cyanide is considered to be one of the most hazardous chemicals originating from various industrial activities like metal finishing, electroplating, steel tempering, mining (metal extraction like silver and gold), pharmaceuticals, automobile parts manufacturing, coal processing, and metal finishing. Through these industrial activities, cyanide enters into the water stream, which in turn becomes toxic for human beings as well as animals, due to binding of cyanide with *cytochrome oxidase* (iron-containing enzymes), which is meant for cells to respire aerobically. So, before discharge of cyanide in the environment, it is favourable to treat cyanide-containing wastewater (Dwivedi et al., 2016).

Cyanide can be found in wide variety of inorganic as well as organic compounds because cyanide is a carbon–nitrogen radical. Hydrogen cyanide is a very common form that is a colorless gas as well as a liquid with a faint, almond-like odour and bitter in nature. Due to the contamination of cyanide with other organic compounds and metals, formation of salts (complex and simple) and compounds take place and in most cases sodium cyanide and hydrogen cyanide as well as potassium cyanide is formed (Dash et al., 2009). Metal cyanide is commonly applicable in pesticide and electroplating industries and is considered to be highly toxic in nature. There are lot of techniques that are used in the treatment of cyanide from the aqueous solutions like sulfur oxidation, ion-exchange, microbiological degradation, photochemical destruction, peroxide oxidation, alkaline chlorination, etc. but on the other side there few economical and technical limitations with all the above-mentioned methods. Alkaline chlorination is a commonly used method and but it is not beneficial in the removal of iron cyanides. So, there is no recovery of cyanide and cannot be further reused. The free chlorine and chloroamines remain in the solution, which can further cause secondary contaminants. On the other side, microbial degradation is useful in the treatment of iron cyanide and it is useful in the treatment of wastewater that contains higher concentrations of cyanides. High cyanide treatment is a method called gas-filled microporous membranes that consist of microphobic structures useful in the separation of two aqueous streams and the resultant comes in the form of immobilized gas-liquid interface at both membrane surfaces. Membrane pores were air filled and when a volatile species like prussic acid is available in one aqueous stream then vaporization takes place from the feed solution, which further passes into gas-filled pores. The gas passes through pores of membranes and adsorption into strip solution takes place. The continuation of this process can take place until the volatile species activities attain equilibrium in both aqueous solutions (Shen et al., 2006). Only a small amount of cyanide is naturally produced and at present cyanogenic compounds may be present in more than 3,000 species of microbes, animal, and fungi as well as plants (Razanamahandry et al., 2017). Many studies have been done on cyanide poisoning and, due to its toxic nature, the experiments are performed on animals (Salkowski and Penney, 1994).

Figure 19.1 illustrates four forms of cyanide, i.e., free cyanide (contains sum of molecular HCN as well as the cyanide anions), simple cyanide, organic cyanide (contains nitriles), as well as complex cyanide (contains metallocyanides).

Figure 19.2 illustrates the toxic effects of cyanide on the human body like tachycardia, hypotension, paralysis, etc. The term *clinical manifestation* of poisoning caused by cyanide is mainly a reflection of intracellular hypoxia. In the case of poisoning caused by acute cyanide, skin must give a normal appearance or slightly ashen despite tissue hypoxia and arterial oxygen saturation remaining normal. An early respiratory sign of poisoning caused by cyanide contains deep respirations (Borron et al., 2007). Cyanide is directly combined with myoglobin, nitrate reductase, cytochrome oxidase, ribulose diphosphate carboxylase, as well as catalase. Binding with cytochrome oxidase, a mitochondrial enzyme useful for cellular respiration, results in aerobic metabolism inhibition as a consequent state of histotoxic anoxia. Unconsciousness, dyspnea, as well as cyanosis were the three most frequently reported signs of

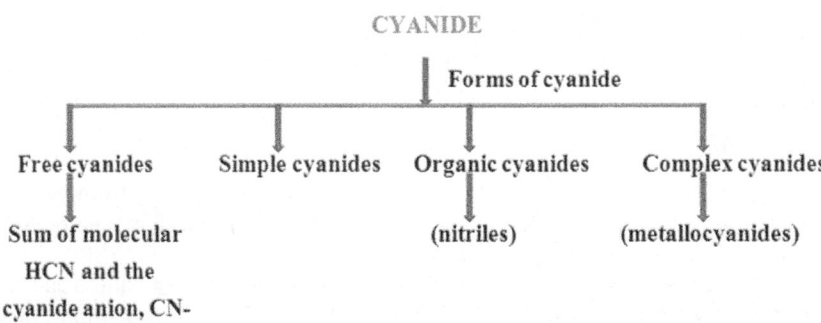

FIGURE 19.1 Different forms of cyanide.

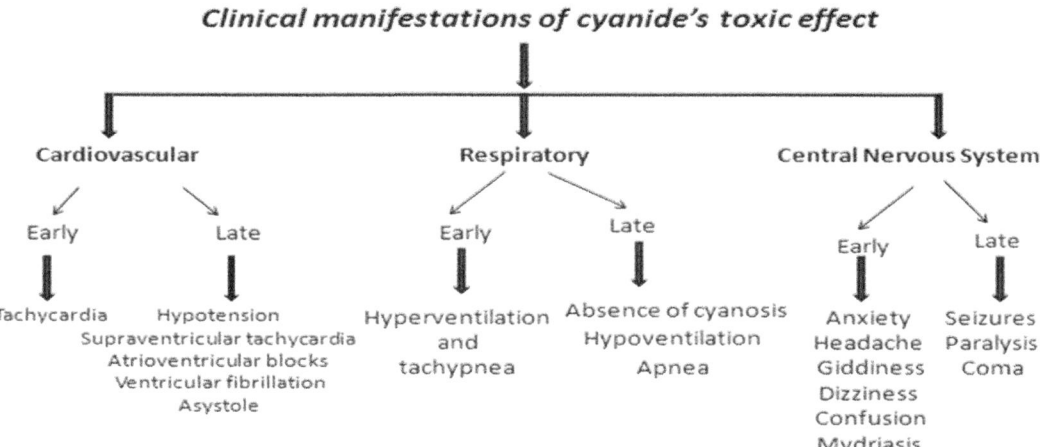

FIGURE 19.2 Toxic effects of cyanide.

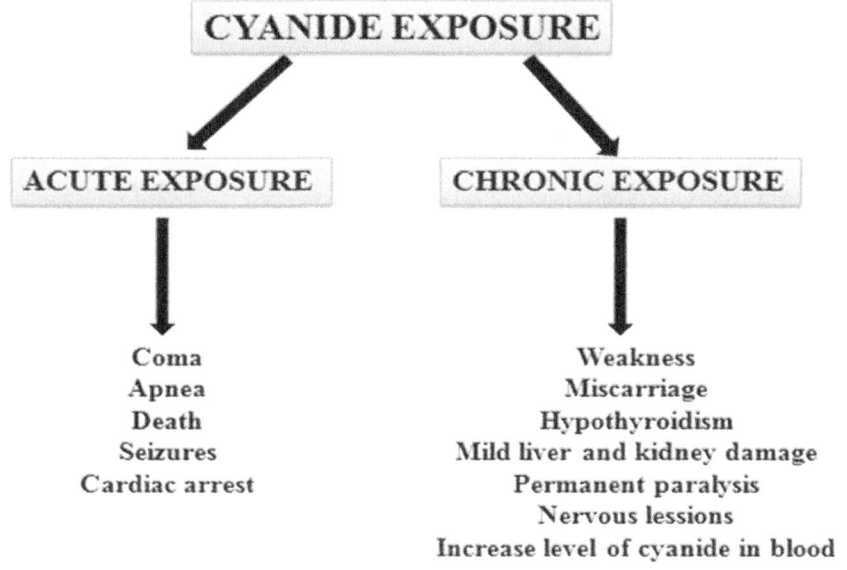

FIGURE 19.3 Cyanide exposure.

cyanide poisoning (Ruangkanchanasetr et al., 1999). Figure 19.3 illustrates the exposure of cyanide-like acute and chronic exposure that causes many diseases like coma, apnea, seizures, cardiac arrest, weakness, nervous lesions, as well as even death.

19.2 Compounds Where Toxic Cyanide Exposure Occurs in Human Beings

Cyanide exposure usually occurs in the following: cyanide salts and hydrogen cyanide gas and house fire smoke (Anseeuw et al., 2013) and from metal polishing workshops (Holland and Kozlowski, 1986), insecticides used in house and agricultural fields and nitroprusside medication (Egekeze and Oehme,

TABLE 19.1

Symptoms on basis of various blood level in µg/mL

Symptoms	Blood Level in µg/mL
Normal	<0.03
Hyperventilation	0.5–1
Decrease mental status	1–3
Death	>3

1980), in seeds of some fruits like apricots and apples (Martin et al., 1969). Cyanide in a fluid form has the capability to absorb within the skin (Martin and Patel, 1969). Cyanide ions get involved in respiration at the cellular level and cause cells to be unable to utilise oxygen (Reade et al., 2012). Diagnosis is the most difficult condition, as cyanide poisoning is possibly suspected in an individual when he is trapped in a workplace/house fire and get unconscious by high blood lactate and lower blood pressure (Hall et al., 1989). Blood analysis for cyanide is available but it takes a long time (Graham et al., 1977) and the analysis of these cyanides are as follows: mild levels – 0.5–1 mg/L; moderate – 1–2 mg/L; severe – 3 mg/L; lethal - more than 3 mg/L (Graham et al., 1977).

Table 19.1 illustrates symptoms of cyanide on a basis of various blood levels in µg/mL. As for the normal, it is <0.03, for hyperventilation it is 0.5–1, for decreased mental status it is 1–3, and for death it is >3.

In case cyanide exposure is found:

- Individual should be taken away from the place/source of cyanide exposure and decontaminated (Hamel, 2011)
- Provide them 100% oxygen (Hamel, 2011)
- As an antidote, hydroxocobalamin is found useful (Thompson and Marrs, 2012)
- Sodium thiosulphate may also be given (Anseeuw et al., 2013)

Cyanide shows great affinity towards metals like trivalent iron and cobalt and sulfate derivatives like sodium thiosulfate that have a sulfur–sulfur bond. In severe exposure, cyanide binds to them in present in cytochrome a3, which inhibits the electron transport chain by breaking the oxidative phosphorylation and formation of adenosine triphosphate (ATP) molecule. So, utilisation of oxygen within the cells ceases. Now, cells start anaerobic respiration that accumulates lactic acid and causes metabolic acidosis/acid–base imbalances (US Department of Health and Human Services, 1984). In lower concentration, rhodanase and hepatic enzymes convert cyanide into thiocyanate, and then it is excreted through urine. A smaller concentration of cyanide can be metabolised into carbon dioxide and passed out from the body by exhalation. Several times cyanide in the body reacts to hydroxocobalamin and produces vitamin B12. Normally, less exposure of cyanide leaves the human body in a single day (US Department of Health and Human Services, 1984).

19.3 Physiochemical Analysis of Hydrogen Cyanide

The physicochemical properties of HCN make it a good chemical warfare agent. Pure anhydrous HCN is a colourless liquid containing a peculiar odour of bitter almonds, which is not possible to inhale by humans (Figure 19.4). Slow hydrolysis of CN compounds takes place in water with subsequent gradual loss of toxicity as they may be oxidised by only strong oxidants like those of potassium permanganate (Raza and Jaiswal, 1994).

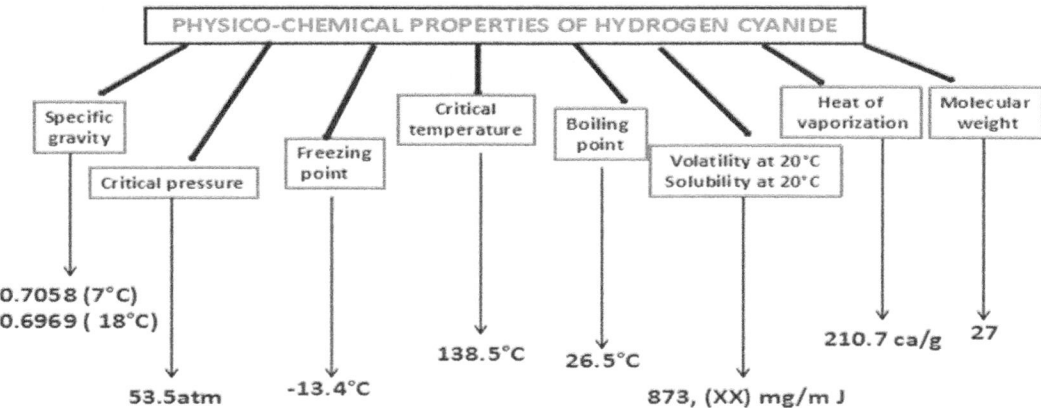

FIGURE 19.4 Physiochemical analysis of cyanide.

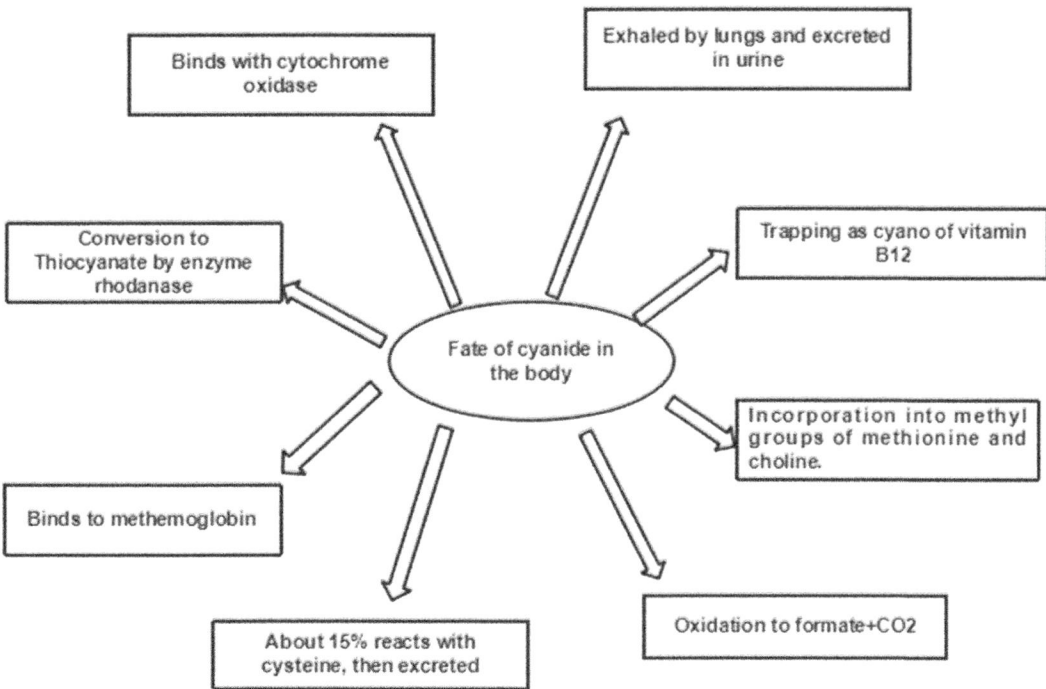

FIGURE 19.5 Different fates of cyanide in body.

19.4 Fates of Cyanide in the Body

There are number of forms of cyanide that affect human metabolism after exposure. Cyanide poisoning symptoms are as follows: headache, dizziness, lack of breath, fast heartbeat with vomiting. Symptoms usually occur just after minutes of exposure; if any individual survives from a lethal effect of cyanide, some neurological disorder remains for the long term (Ostrowski et al., 1999) (Figure 19.5 and Figure 19.6).

Figure 19.7 illustrates that the cyanide distribution in the human body can be done after inhalation of high doses of cyanide, which may distributed in the blood, kidney, lungs, brain, heart, as well as liver of the deceased human by cyanide exposure (Newhouse and Chiu, 2016).

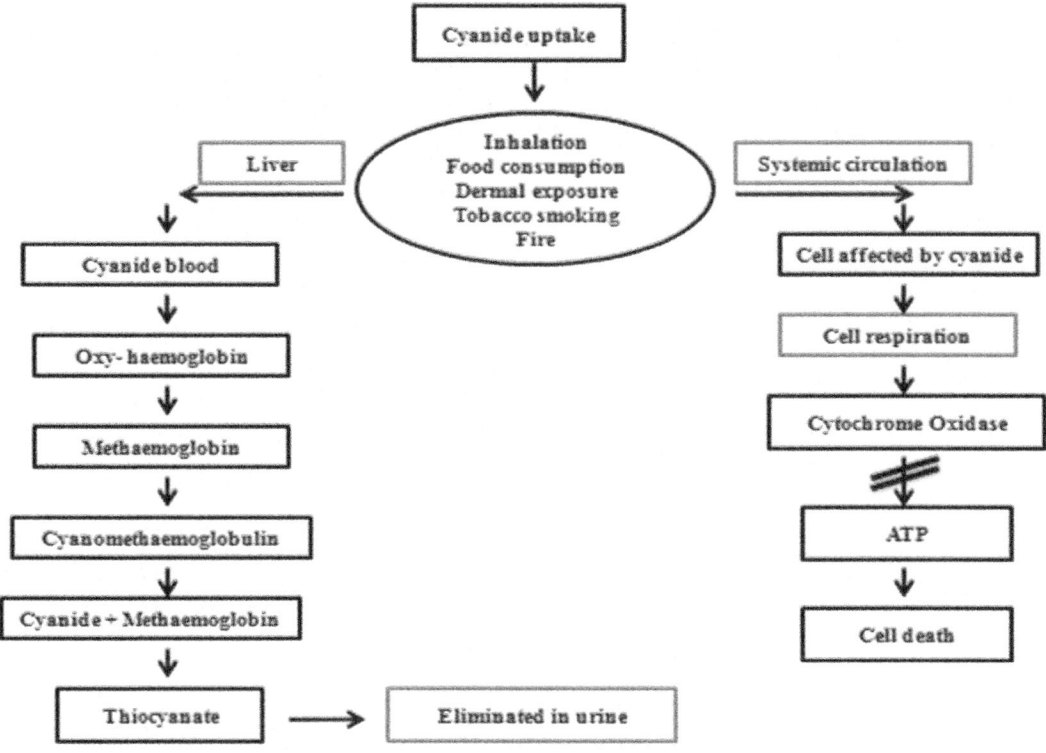

FIGURE 19.6 Cyanide distribution in liver and its elimination.

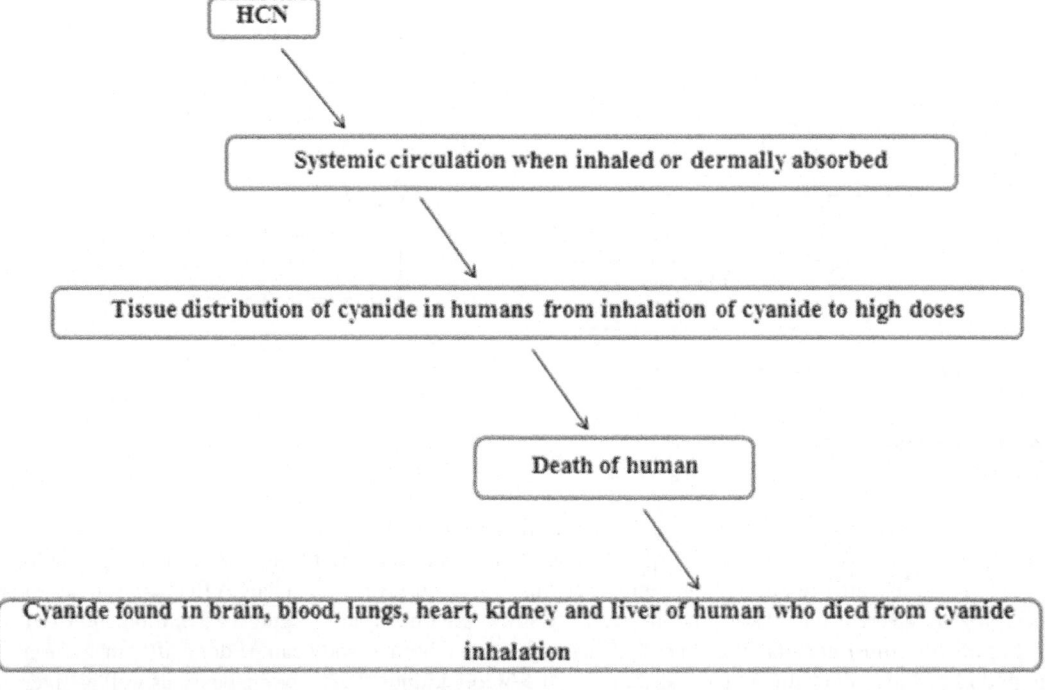

FIGURE 19.7 Cyanide distribution in several parts of the body.

19.5 Sources of Cyanide

The cyanides naturally occur in plants and processed foods and the cyanide ions are naturally obtained from the cyanogenic glycosides (found in apricot kernels, bamboo shoots, cassava roots) and that of hydrogen cyanide used in mining of gold and silver, plastic production of all types of dyes, and in chemical laboratories (Jaszczak et al., 2017). Cyanide is the compound that may be produced naturally by living organisms including algae, fungi, bacteria, as well as plants as a mechanism of defense against predation (Gebresemati et al., 2017). Cyanide as well as the compounds of cyanide were found in soil, air, food, as well as in water because of both anthropogenic and natural sources. On the other side, plants and other living organisms may form minute quantities of cyanide (Jennings, 2012). Cyanide was also investigated in insecticides, which are useful in commercial mass fumigation (Borron et al., 2007).

19.6 Industrial Utilisation of Cyanide

There are several industrial uses of cyanide like case-hardening for metals, electroplating, floatation of base metals, gasification of coal, the fumigation of ships, railway tracks, vehicles, civil constructions, flour mills, extraction (cyanidation) of silver, gold, etc. from their respective ores (Ma and Dasgupta, 2010). NaCN is widely used for adding CN into various organic compounds; particularly it is a reaction of halogen compounds (organic) to obtain nitriles. These nitriles are used to obtain various kinds of amides, carboxylic acids, amines, and esters. KCN is used as an electrolyte component for copper, silver, and gold extraction from their respective ores and electrolytic refining of platinum and for metal coloring (Monosson, 2004). The complex of cyanides with cadmium, copper, and zinc are widely utilised in electroplating, principally used in the plating of steel, iron as well as zinc, while cyanide salts are beneficial for chelation in some reactions as chelating agents (Cummings, 2004). For fumigation purposes, mostly calcium cyanide is used, as it generates hydrogen cyanide in the presence of air; in agriculture as fertilizer, as herbicide and rodenticide; cement stabilizer; and in manufacturing of stainless steel (Eisnor and Morgan, 2019). In silver plating, potassium silver cyanide is used. It is also used as a bactericide. Massive scale cyanide pollution is associated with industries, which give rise to the occurrence of cyanide in water resources, specifically in surface water. Lower concentrations of cyanide are easy to remove by some specific treatment methods (de Rosa, 2005).

19.7 Cyanide Contamination in Water

Cyanides are mainly detected at very low concentrations in drinking water sources. Both potassium and sodium cyanide are soluble in water; these aqueous solutions are strongly alkaline as well as rapidly decompose. Both form hydrogen cyanide on contact with acids as well as acid salts. Calcium cyanide is applied as a fumigant chiefly, as it releases hydrogen cyanide readily with exposure to air; as defoliant, fertilizer, herbicide as well as rodenticide; as a stabilizer for cement; as well as in manufacturing of stainless steel (Mousavi et al., 2013) (Figures 19.8).

19.8 Cyanide Contamination in Air

Mainly cyanide is found in gaseous form as gaseous hydrogen cyanide and tiny dust particles in air. The removal of cyanide particles is done by snow and rain from air, but the removal of gaseous hydrogen cyanide is not removed by snow or rain. In atmosphere hydrogen cyanide has half life time of about 1–3 years (Razanamahandry et al., 2017). The most common source of the cyanide exposure is inhalation of smoke from industrial and residential fires (Reade et al., 2012) (Figure 19.9).

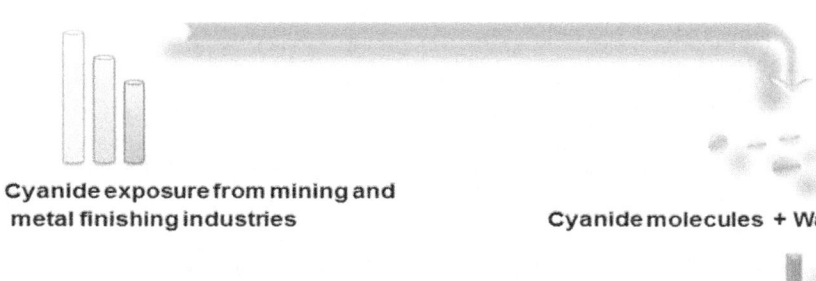

FIGURE 19.8 Flow diagram of cyanide contamination in water.

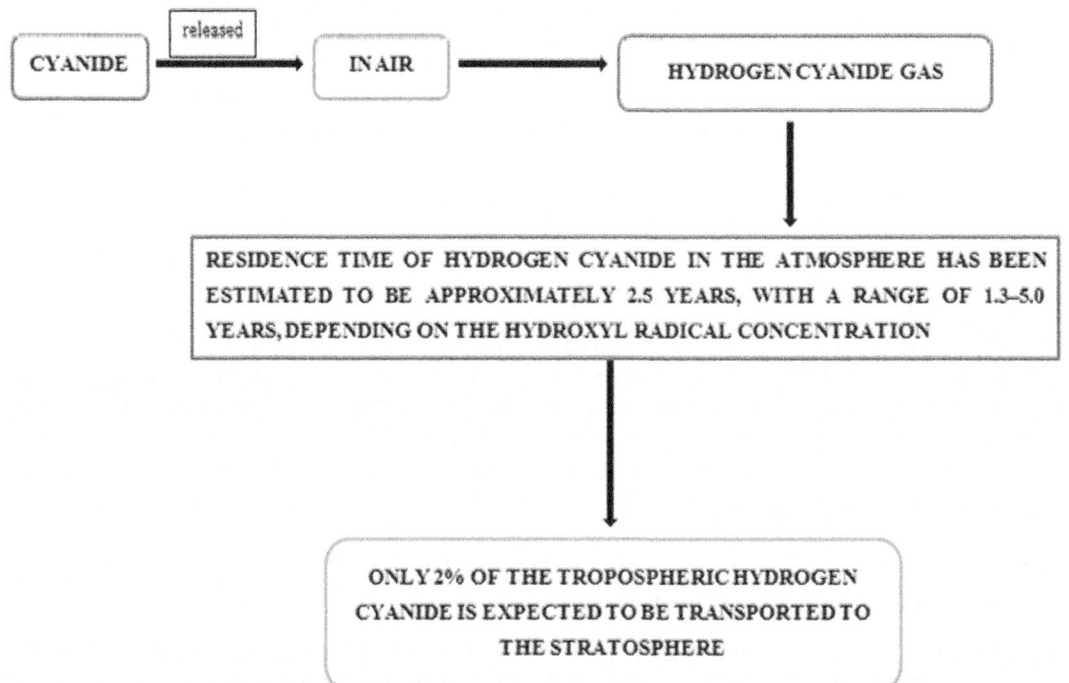

FIGURE 19.9 Contamination of cyanide in air.

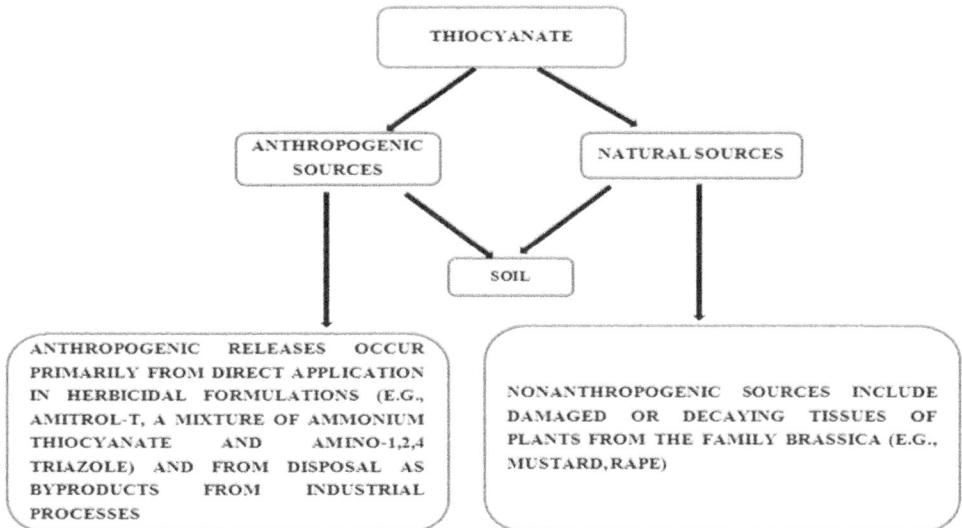

FIGURE 19.10 Contamination of cyanide in soil.

19.9 Cyanide Contamination in Soil

It was reported that mobility of cyanide compounds in soil depend on the dissociation characteristics as well as on the stability of compounds, soil chemistry, and permeability together with presence of anaerobic and aerobic microorganisms. A soil proves the major potential pathway for cyanide contamination of groundwater (Jennings, 2012). The contaminants' behavior in soils are mainly ruled by interactions between contaminants (dissolved) and soil. In the case of soil contamination with cyanide in the form of iron cyanide complexes, this interaction is possibly dominated by the term "precipitation of prussian blue" (Kjeldsen, 1999) (Figure 19.10).

The exposure of cyanide on golden hamsters by subcutaneous infusion finds to excrete only a relatively lower percentage, like 10–15% of dose, as the thiocyanate in urine (Sharma and Bhattacharya, 2017). The poisoning caused by plants having cyanide has been known for millennia. The first cyanide poisoning description of the bitter almonds was published by Wepfer in the year 1679. Cyanide poisoning caused due to the inhalation of cyanide-containing gas like that of hydrogen cyanide well as the dust contains liquid or solid cyanide. Some of the typical sources were industrial sources like those of acrylic manufacturing, silver and gold mining, electroplating, synthetic rubber, plastic combustion, and many more. Cyanogenic glycoside is a chemical compound that is present in foods that release hydrogen cyanide while chewing/digestion and the act of chewing as well as digestion leads to hydrolysis of those substances, causing the cyanide to be released (Reade et al., 2012). Cassava/*Manihot esculenta* is one of the major economic crops of Thailand produced for the animal food industry. It has been commonly known among Thai people that fresh cassava root is toxic (Ruangkanchanasetr et al., 1999) (Table 19.2, Figure 19.11).

The drainage of a cyanide solution into the surrounding water bodies leads to volatising at a high rate and then detoxifies and releases hydrogen cyanide into the air. At the time of the leaching process, hydrogen cyanide may volatilise from tailing solutions after extraction of gold and is reported to be destroyed by the process of photolysis or oxidation from that of the surrounding atmosphere of gold mines as well as sunlight (Brüger et al., 2018) (Figure 19.12).

The exposure of cyanide is done by breathing contaminated air, drinking water, soil contaminated with cyanide, contaminated food, etc. (Kwaansa-Ansah et al., 2017). The mild poisoning symptoms includes nausea, dizziness, headache, drowsiness, mucous membrane, anxiety, as well as hyperpnoea; later, hypotension, bradycardia, dyspnoea, periods of cyanosis, as well as in severe cases convulsions,

TABLE 19.2
Possible sources of cyanide poisoning (Bhattacharya and Flora, 2009)

1	Industrial exposure	Synthetic rubber production, metallurgy, plastics production, fumigation of pesticides, tanning in leather industry, textile, paper manufacture, and electroplating.
2	Fire smoke	Polyacrylonitriles, melamine resins, nylon, in accidents including industrial, car, residential, as well as in aircraft.
3	Dietary	Linseed, bamboo sprout, hydrangea, sorghum species, Rosaceae family (peach, plum, apple, pear as well as bitter almond), Linum specie.
4	Drugs	Laetrile, sodium nitroprusside, succinonitrile.
5	Others	Terrorist attack, homicide, nail polish remover ingestion, chemical warfare, suicide, cigarette smoking, chemical warfare.

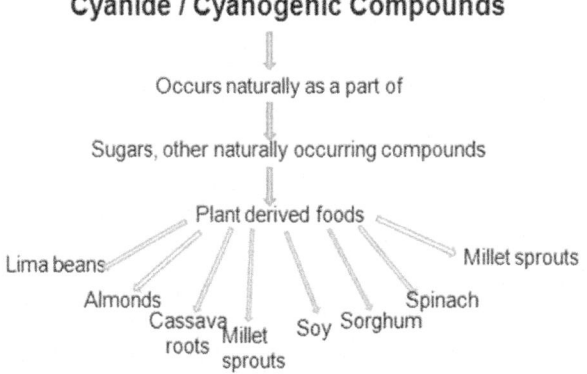

FIGURE 19.11 Different foods containing.

collapse of cardiovascular with shock, as well as pulmonary oedema and progressive coma also occurs with fatal outcomes (Beasley and Glass, 1998). It is difficult to detect cyanide poisoning and its most important clue for diagnosis is circumstantial evidence rather than symptoms or signs (Jethava et al., 2014). The cyanide toxicity results from two chemical species (CN- as well as HCN) that exist under biological conditions following exposure to gas (cyanide) as well as salts (Isom and Borowitz, 2015) (Figure 19.13).

The above figure defines the primary metabolic pathway of cyanide. According to ATSDR (2006), one of the important metabolic pathways for cyanide was the conversion to a low acutely toxic compound, thiocyanate, mainly by rhodanase, including some sort of conversions mediated by 3-mercaptopyruvate sulfur transferase. Conversion to thiocyanate may account for about 60–80% of cyanide doses. On the other side, the minor pathways consist of incorporation into a 1-carbon metabolic pool and its conversion into 2-aminothiazoline-4-carboxylic acid (Newhouse and Chiu, 2016) (Figure 19.14).

19.11 Methods of Cyanide Removal

Different techniques exist for cyanide remediation and there are several chemical, physical, adsorption, and complexation techniques which have been used in the past for separation or degradation procedures. These separation methods are used to recover concentrated cyanide from their sources and recycling, while destruction techniques are used to break triple bonds between carbon as well as nitrogen. This removes cyanide and converts toxic species to less-toxic/non-toxic compounds, as the carbon and nitrogen atoms change their state of oxidation and destruction procedures are mostly an oxidation reaction (Young and Jordan, 1995).

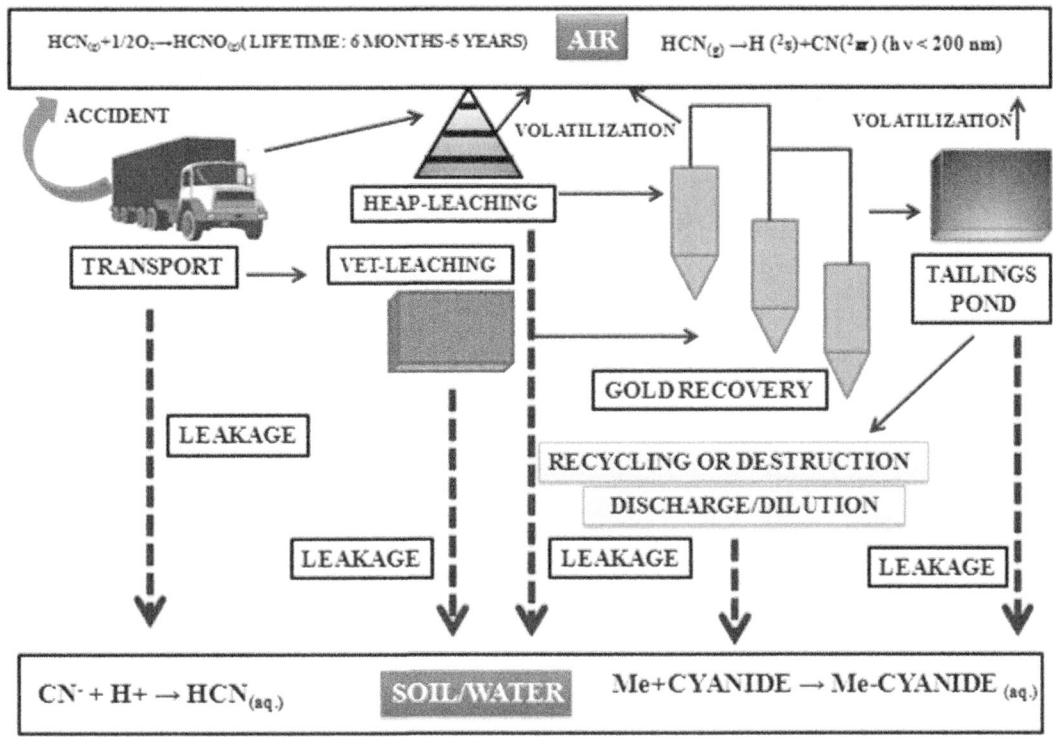

FIGURE 19.12 Cyanide discharge in soil, air, and water.

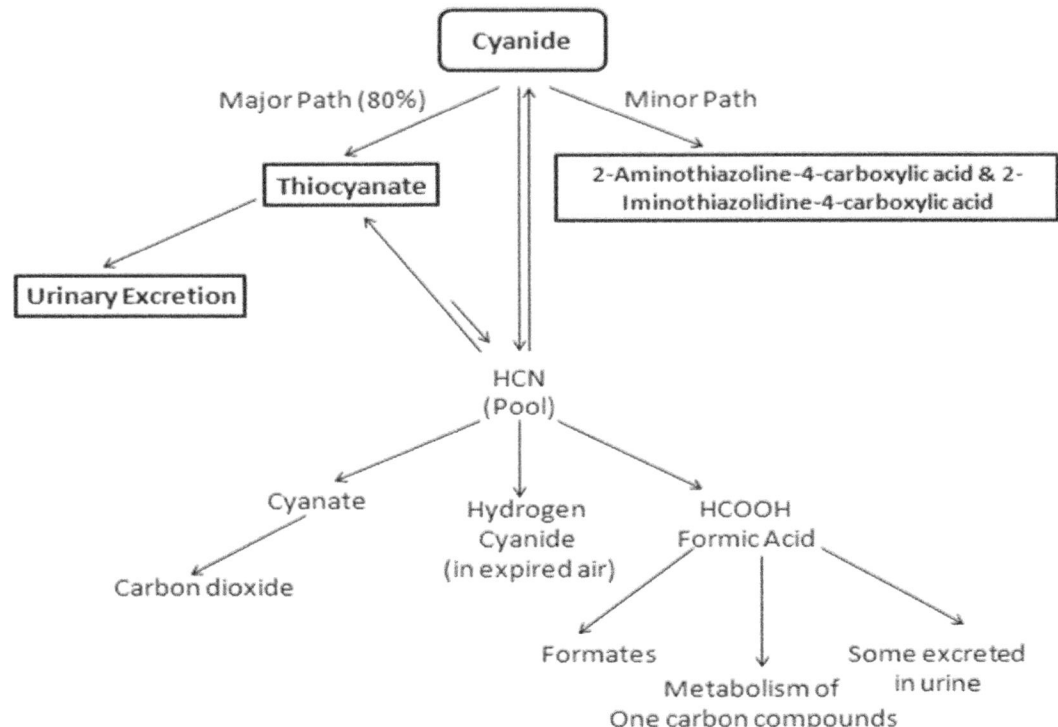

FIGURE 19.13 Primary metabolic pathway of cyanide.

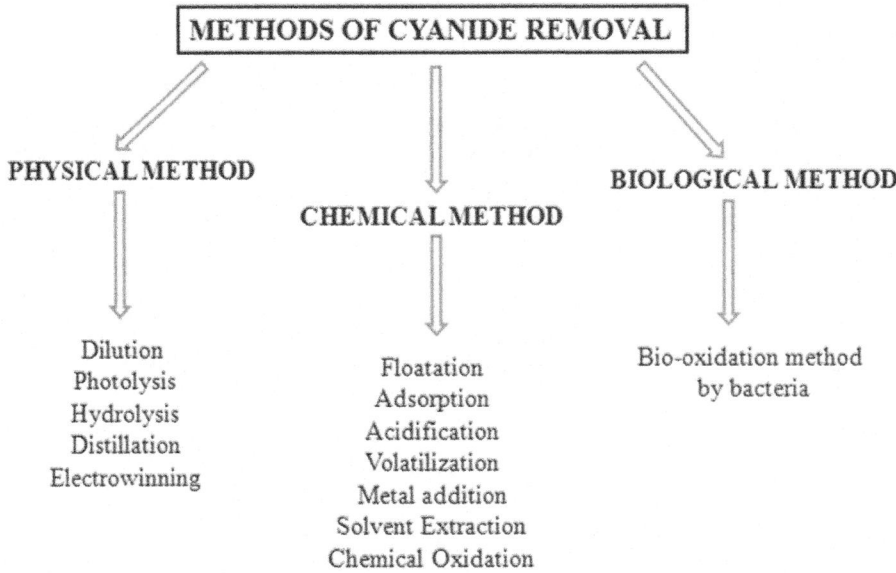

FIGURE 19.14 Methods used for the recovery of cyanide.

19.11.1 Remediation by Physical Methods

Physical methods that accomplished cyanide remediation are dilution, electrowinning, membranes, and hydrolysis followed by distillation (Mooiman and Miller, 1991).

19.11.1.1 Dilution Method

The procedure of dilution is not based upon separation or destruction of cyanide. It involves lowering the toxic cyanide concentration in waste and makes such effluent that yields a concentration within effluent discharge limits. Dilution is very simple and affordable. Dilution is generally not acceptable as the final amount of cyanide in discharge remains the same (Mooiman and Miller, 1991), precipitation and adsorption take place naturally, and results are concentrated in the presence of cyanide within surface and groundwater (Mooiman and Miller, 1991).

19.11.1.2 Photolysis

Photolysis provides electromagnetic radiation energy to catalyze redox reactions. Compounds absorbed radiation energy and transfer electrons from ground energy level to excited energy levels. Electrons are very sensitive to their surroundings and tend to precipitate chemical compounds during oxidation reduction reactions. When a chemical compound absorbs energy and tends to release electrons for another chemical species, it is known as photo-reduction. In contrast, photo-oxidation occurs when a chemical compound absorbs radiation and accepts electrons from another chemical species. Photolysis can be direct (without photosensitizers) or indirect (homogeneous or heterogeneous) photocatalysts. So, photolysis is a quite novel method and hence is mentioned as an advanced oxidation technique. In a study, pure hydrogen cyanide and mixture of hydrogen cyanide with water and ammonia at low-temperature 18°K with ice films of 0.1 μm thickness, are exposed to 0.8 MeV protons or ultraviolet rays of the photon range 110–250 nm for simulating the space environments. Photolysis yields products like ammonium ion, cyanide ion, isocyanic acid, and cyanate ion formamide, which are less toxic than hydrogen cyanide (Gerakines et al., 2004).

Cyanide: Sources, Health Issues 369

19.11.1.3 Membranes

Cyanide could be separated from water resources by utilising membranes in reverse osmosis or electrodialysis. Cyanide recovery takes place through membranes. No other secondary species of cyanide pollutants is generated in this operation. Energy requirements are low. Cyanide separation from water is done by applying membranes in association with either reverse osmosis or electrodialysis. The reverse osmosis procedure includes a pressure applied to a cyanide containing water and then it is forced to pass by a membrane that is impermeable for cyanide compounds. While an electrodialysis procedure involves two electrodes being apart by a membrane that is permeable for cyanide, then a potential is applied between the electrodes. Cyanide solution is filled in a cathode chamber with a negative charge. Cyanide also has a negative charge, so it diffuses via membranes and moves towards the anode/positive electrode (Bouhidel and Rumeau, 2004; Rosehart, 1973; Young and Jordan, 1995; Bodamer, 1977).

19.11.1.4 Electrowinning

For the metal recovery, like silver or gold, from the solution of gold cyanide or silver cyanide by applying a current across the solution. Electrons cause chemical reduction reaction and reduce the gold or silver ion, and produce a solid gold/silver compound at the cathode (Torre et al., 2006). Electrowinning is mostly benefitted from gold recovery and processing; though, in the case of cyanide remediation or regeneration, it is generally mentioned as the HAS or Celec process. Electrowinning worked efficiently with concentrated solutions; whereas hydrogen generation prevails while working with dilute concentrations. Advancement is ongoing to develop an electrowinning method economically feasible, so direct applications of dilute solutions might be possible for cyanide remediation.

19.11.1.5 Hydrolysis Followed by Distillation

Cyanide, which occurs in free form as cyanide ions, hydrolyzes inside water naturally and is converted into aqueous hydrogen cyanide:

$$CN^- + H^+ \rightarrow HCN(aq.)$$

The resulting aqueous form of hydrogen cyanide could go for distillation to obtain hydrocyanic gas as a volatile product (Roberts and Jackson, 1971).

$$HCN(aq.) \rightarrow HCN(gas)$$

HCN gas could be obtained and concentrated for distillation in conventional absorption-scrubbing towers. Obtained hydrocyanic gas can be concentrated for recycling purposes; also it is vented to an open environment as it had been found to occur naturally in tailings as well as ponds in warm environments. For such discharge of gas, it must be assured those environmental regulations are followed and fulfilled (Roberts and Jackson, 1971). Cyanide separation efficiency using distillation could be enhanced by increasing the air to solution ratio, surface area of interface, agitation rate, and raising the temperature (Roberts and Jackson, 1971).

19.11.2 Chemical Methods of Cyanide Removal

19.11.2.1 Acidification Along with Volatilisation

Low pH values (pH < 8) will trigger volatilization of cyanide, i.e., lowering the pH values tends to bring greater volatilization rates. Weak acid dissociation generates HCN gas if the pH value falls below 2. Similar results were obtained from strong acid dissociation and thiocyanate at a pH below 0. The pH range between 1.5 and 2 is generally accepted for conventional acidification/volatilisation processes. Volatilization rates can be improved by increasing the air to solution ratio, surface area of interface, agitation rate, and raising the temperature. Leach wastes are used to recover cyanide via acidification using SO_2/H_2SO_4; hydrogen cyanide undergoes volatilization with air stripping followed by absorption of HCN in NaOH or any other basic solution. The processes of acidification and volatilization are shown in the following chemical reactions (Gönen et al., 2004).

$$H^+ + CN^- \rightarrow HCN \text{ (by acidification)}$$

$$HCN \text{ (aqueous)} \rightarrow HCN \text{(gas)} \text{ (by volatilization)}$$

$$NaOH \text{ (aqueous)} + HCN \text{(gas)} \rightarrow CN^- \text{(aqueous)} + Na^+ \text{(aqueous)} \text{ (Adsorption)}$$

19.11.2.2 Solvent Extraction

In solvent extraction, an organic solvent is use to solubilize a minimum of one extractant of many organic compounds in a mixture by its chelating, solvating, and ion-exchange abilities. The solvent should have less density than water and must be immiscible. The extractant and aqueous phase is essential to remain in the solvent and have selectiveness towards the aqueous species being remediated. The solvent and extractant are jointly known as the organic phase. Cyanidation is a process to obtain precious metals; the solvent extraction is a process widely used to concentrate aurocyanide complexes from their dilute sources/solutions. The gold-cyanide extraction process usually involves quaternary ammonium salts or some other organic extractants. Quaternary ammonium salts dissolve in a diluent/supporting matrix, like octane or kerosene for the gold extraction (Alonso-González et al., 2010).

19.11.2.3 Adsorption

Adsorption is the most simple to operate and widely and efficiently used techniques in cyanide removal.

Activated carbon, minerals, and resins are useful adsorbents for cyanide adsorption. Packed-bed columns, agitation cells, loops, and elutriation columns are used as contact vessels for adsorption.

When the cyanide is adsorbed, adsorbent material is withdrawn from the solution by using several methods like flotation, gravity separation, or screening. Then cyanide-adsorbed beads are placed into a different column to desorb the concentrated cyanide solution (low volume).

Ultimately, the adsorbent is screened and again reactivated for its reuse (recycled). Adsorption does not depend on the target compound toxicity as well as does not require any hazardous chemicals. Adsorption is primarily used for concentration and followed by recovery of adsorbed compound. Plain/modified activated carbon is widely used in adsorption that shows less cyanide adsorption capability ranging from 0.4 to 29.6 mg/g (Moussavi and Khosravi, 2010).

19.11.3 Biological Method

Several microorganisms may produce cyanide (cyanogenesis) or degrade it. They degrade cyanide either to detoxify it or to use it as a source of nitrogen for growth. Significant amounts of cyanide were formed as a secondary metabolite by a wide range of fungi as well as bacteria by glycine decarboxylation. Biological treatment is most effective in the secondary treatment of mining effluents such as wastewaters, tailings, acidic mine drainage, etc. Some microbial species like *Pseudomonas* sp. are found effective in cyanide degradation and are converted into less toxic compounds. *Pseudomonas* uses cyanide compounds as sources of carbon and nitrogen, and produce carbonate along with ammonia under favorable conditions (Moradkhani et al., 2018).

19.11.4 Kinetics and Isotherms Used

To understand the adsorption phenomenon, there are two types of models, i.e., kinetics (on basis of contact time) as well as equilibrium model (on basis of initial cyanide concentration) (Maulana et al., 2018). Commonly, Langmuir isotherm was helpful in model adsorption phenomenon onto an entirely homogeneous surface with negligible interaction between the adsorbed molecules. Freundlich isotherm must be used for non-ideal as well as reversible adsorption; it is applicable for a heterogeneous surface as well as multilayer sorption (Maulana and Takahashi, 2018).

19.11.5 Cyanide Removal from Soil, Air, and Water

It has been reported that plant cells play a key role in the removal of free cyanide as plants give easily and large manageable biomasses, which are responsible for suitable candidates as bioreactors for removal of cyanide from those of gold mining wastewater (Larsen et al., 2004).

Tables 19.3, 19.4, and 19.5 show the removal of cyanide with several biosorbents from air, water, as well as soil. Hydrogen cyanide (HCN) is considered to be an acutely poisonous compound as it enters our body through breathing. To remove the HCN from air, the best-known adsorbents are the metal salt impregnated activated carbons that are usually obtained by wetting carbon (non-impregnated) with a solvent-carrying transition metal salts followed by drying (Oliver et al., 2005). The removal of cyanide from the wastewater takes place by passing the water through adsorbent porous solid substrates that have been treated with the water insoluble metal compounds, mainly copper sulfide (CuS) facilitates the cyanide adsorption (Yan, 1994). Cyanide has been used for more than 100 years to extract mined gold following the cyanide leaching recovery of gold process. In North America, gold mining consumes about 80 million kg of cyanide per year (Ebel et al., 2007).

19.12 Conclusion

As cyanide is a potent as well as fast-acting poison, and the sources of cyanide are also highly distributed, as well as there are many fatal diseases that have taken place due to its toxicity. The detection

TABLE 19.3

Removal of cyanide from soil by several adsorbents

Cyanide-Contaminated Soil	Biosorbent	Removal Efficiency	pH	References
Cyanide	*Fusarium solani*	–	9.2–10.7	(Dumestre et al., 1997)
Iron-cyano complexes	*Pseudomonas aeruginosa*	90%	4.0–10.0	(Aronstein et al., 1994)
Potassium cyanide	*Sambucus chinensis*	–	–	(Yu et al., 2004)
Potassium cyanide (KCN)	*Salix babylonica L.*	–	–	(Yu et al., 2005)
Cyanide	*Manihot esculenta*	–	–	(Siller and Winter, 1998)

TABLE 19.4

Removal of cyanide from air by several adsorbents

Cyanide-Contaminated Air	Biosorbent	Removal Efficiency	pH	References
Hydrogen cyanide from air	Synthetic activated carbons	–	–	(Oliver et al., 2005)
Cyanide from air	Sulfur dioxide	–	8–10	(Devuyst et al., 1989)
Hydrogen cyanide	Activated carbons impregnated with Cu and Zn	–	–	(Nickolov and Mehandjiev, 2004)

TABLE 19.5

Removal of cyanide from water by several adsorbents

Cyanide-Contaminated Water	Biosorbent	Removal Efficiency	pH	References
Cyanide	Commercial granular activated carbon	–	9.7 & 5	(Dash et al., 2009)
Synthetic cyanide wastes (dilute solutions of sodium as well as copper cyanide)	Illuminated TiO_2 catalyst	78%	11	(Barakat et al., 2004)
Cyanide from wastewater	Advanced oxidation methods	95%	10	(Kepa et al., 2008)
Cyanide from wastewater	*Eichhornia crassipes*	40%	–	(Ebel et al., 2007)
Cyanide from aqueous solution	Nonthermal plasma reactor	92%	11	(Hijosa-Valsero et al., 2013)
Cyanide from wastewater	Pistachio hull wastes	99%	10	(Moussavi and Khosravi, 2010)
Cyanide from wastewater	Activated carbon	100%	–	(Yan, 1994)
Cyanide from wastewater	Modified activated carbon	–	–	(Monser and Adhoum, 2002)

of cyanide is very critical as the clinical manifestation for this element is non-specific. There are several treatment techniques that have been used to detect these elements from air, soil, and water that are toxic for humans and animals. To remove this element from the atmosphere, biosorption is one of the most widely applicable methods due it is efficiency and cost effectiveness.

REFERENCES

Agarwal, S., Pramanick, S., Rahaman, S.A., Ghanta, K.C., & Dutta, S. (2019). A cost-effective approach for abatement of cyanide using iron-impregnated activated carbon: kinetic and equilibrium study. *Applied Water Science, 9*, 74.

Alonso-González, O., Nava-Alonso, F., Uribe-Salas, A., & Dreisinger, D. (2010). Use of quaternary ammonium salts to remove copper–cyanide complexes by solvent extraction. *Minerals Engineering, 23*, 765–770.

Anseeuw, K., Delvau, N., Burillo-Putze, G., De Iaco, F., Geldner, G., Holmström, P., Lambert, Y., & Sabbe, M. (2013). Cyanide poisoning by fire smoke inhalation: a European expert consensus. *European Journal of Emergency Medicine, 20*, 2–9.

Aronstein, B.N., Maka, A., & Srivastava, V.J. (1994). Chemical and biological removal of cyanides from aqueous and soil-containing systems. *Applied Microbiology and Biotechnology, 41*, 700–707.

Barakat, M.A., Chen, Y.T., & Huang, C.P. (2004). Removal of toxic cyanide and Cu (II) Ions from water by illuminated TiO_2 catalyst. *Applied Catalysis B: Environmental, 53*, 13–20.

Beasley, D.M.G., & Glass, W.I. (1998). Cyanide poisoning: pathophysiology and treatment recommendations. *Occupational Medicine, 48*, 427–431.

Bhattacharya, R., & Flora, S.J. (2009). Cyanide toxicity and its treatment. In *Handbook of Toxicology of Chemical Warfare Agents* (pp. 255–270). Academic Press.

Bodamer, G.W. (1977). Electrodialysis for Closed Loop Control of Cyanide Rinse Waters.

Borron, S.W., Baud, F.J., Mégarbane, B., & Bismuth, C. (2007). Hydroxocobalamin for severe acute cyanide poisoning by ingestion or inhalation. *The American Journal of Emergency Medicine, 25*, 551–558.

Bouhidel, K.E., & Rumeau, M. (2004). Ion-exchange membrane fouling by boric acid in the electrodialysis of nickel electroplating rinsing waters: generalization of our results. *Desalination, 167*, 301–310.

Brüger, A., Fafilek, G., & Rojas-Mendoza, L. (2018). On the volatilisation and decomposition of cyanide contaminations from gold mining. *Science of the Total Environment, 627*, 1167–1173.

Cummings, T.F. (2004). The treatment of cyanide poisoning. *Occupational Medicine, 54*, 82–85.

Dash, R.R., Balomajumder, C., & Kumar, A. (2009). Removal of cyanide from water and wastewater using granular activated carbon. *Chemical Engineering Journal, 146*, 408–413.

Dash, R.R., Gaur, A., & Balomajumder, C. (2009). Cyanide in industrial wastewaters and its removal: a review on biotreatment. *Journal of Hazardous Materials, 163*, 1–11.

Devuyst, E.A., Conard, B.R., Vergunst, R., & Tandi, B. (1989). A cyanide removal process using sulfur dioxide and air. *JOM, 41*, 43–45.

Dose, BMD Benchmark, and DIAMOND DIetAry Modelling Of Nutritional Data. "Survey of cyanogenic glycosides in plant-based foods in Australia and New Zealand."

Dumestre, A., Chone, T., Portal, J., Gerard, M., & Berthelin, J. (1997). Cyanide degradation under alkaline conditions by a strain of fusarium solani isolated from contaminated soils. *Journal of Applied Environment and Microbiology, 63*, 2729–2734.

Dwivedi, N., Balomajumder, C., & Mondal, P. (2016). Comparative investigation on the removal of cyanide from aqueous solution using two different bioadsorbents. *Water Resources and Industry, 15*, 28–40.

Ebel, M., Evangelou, M.W., & Schaeffer, A. (2007). Cyanide phytoremediation by water hyacinths (*Eichhornia crassipes*), *Chemosphere, 66*, 816–823.

Egekeze, J.O., & Oehme, F.W. (1980). Cyanides and their toxicity: a literature review. *Veterinary Quarterly, 2*, 104–114.

Eisnor, D.L., & Morgan, B.W. (2019). Ocular toxicology in military and civilian disaster environments. In *Ophthalmology in Military and Civilian Casualty Care* (pp. 171–208). Cham: Springer.

Gebresemati, M., Gabbiye, N., & Sahu, O. (2017). Sorption of cyanide from aqueous medium by coffee husk: response surface methodology. *Journal of Applied Research and Technology, 15*, 27–35.

Gerakines, P.A., Moore, M.H., & Hudson, R.L. (2004). Ultraviolet photolysis and proton irradiation of astrophysical ice analogs containing hydrogen cyanide. *Icarus, 170*, 202–213.

Gönen, N., Kabasakal, O.S., & Özdil, G. (2004). Recovery of cyanide in gold leach waste solution by volatilization and absorption. *Journal of Hazardous Materials, 113*, 231–236.

Graham, D.L., Laman, D., Theodore, J., & Robin, E.D. (1977). Acute cyanide poisoning complicated by lactic acidosis and pulmonary edema. *Archives of Internal Medicine, 137*, 1051–1055.

Hall, A., Kulig, K.W., & Rumack, B.H. (1989). Suspected cyanide poisoning in smoke inhalation: complications of sodium nitrite therapy. *Journal de Toxicologie Clinique et Experimentale, 9*, 3–9.

Hamel, J. (2011). A review of acute cyanide poisoning with a treatment update. *Critical Care Nurse, 31*, 72–82.

Hijosa-Valsero, M., Molina, R., Schikora, H., Müller, M., & Bayona, J.M. (2013). Removal of cyanide from water by means of plasma discharge technology. *Water Research, 47*, 1701–1707.

Holland, M.A., & Kozlowski, L.M. (1986). Clinical features and management of cyanide poisoning. *Clinical Pharmacy, 5*, 737–741.

Isom, G.E., & Borowitz, J.L. (2015). Biochemical mechanisms of cyanide toxicity. *Toxicology of Cyanides and Cyanogens: Experimental, Applied and Clinical Aspects*, 70–81.

Jaszczak, E., Polkowska, Z., Narkowicz, S., & Namieśnik, J. (2017). Cyanides in the environment—analysis—problems and challenges. *Environmental Science and Pollution Research, 24*, 15929–15948.

Jennings, A.A. (2012). Worldwide regulatory guidance values for surface soil exposure to carcinogenic or mutagenic polycyclic aromatic hydrocarbons. *Journal of Environmental Management, 110*, 82–102.

Jethava, D., Gupta, P., Kothari, S., Rijhwani, P., & Kumar, A. (2014). Acute cyanide Intoxication: a rare case of survival. *Indian Journal of Anaesthesia, 58*, 312.

Kepa, U., Stanczyk-Mazanek, E., & Stepniak, L. (2008). The use of the advanced oxidation process in the ozone+ hydrogen peroxide system for the removal of cyanide from water. *Desalination, 223*, 187–193.

Kjeldsen, P. (1999). Behaviour of cyanides in soil and groundwater: a review. *Water, Air, and Soil Pollution, 115*, 279–308.

Kwaansa-Ansah, E.E., Amenorfe, L.P., Armah, E.K., & Opoku, F. (2017). Human health risk assessment of cyanide levels in water and tuber crops from Kenyasi, a mining community in the Brong Ahafo Region of Ghana. *International Journal of Food Contamination, 4*, 16.

Larsen, M., Trapp, S., & Pirandello, A. (2004). Removal of cyanide by woody plants. *Chemosphere, 54*, 325–333.

Lawson-Smith, P., Jansen, E.C., & Hyldegaard, O. (2011). Cyanide intoxication as part of smoke inhalation-a review on diagnosis and treatment from the emergency perspective. *Scandinavian Journal of Trauma, Resuscitation and Emergency Medicine, 19*, 14.

Ma, J., & Dasgupta, P.K. (2010). Recent developments in cyanide detection: a review. *Analytica Chimica Acta, 673*, 117–125.

Martin, T., & Patel, J.A. (1969). Determination of sodium nitroprusside in aqueous solution. *American Journal of Health-System Pharmacy*, 26, 51–53.

Maulana, I., & Takahashi, F. (2018). Cyanide removal study by raw and iron-modified synthetic zeolites in batch adsorption experiments. *Journal of Water Process Engineering, 22*, 80–86.

Monosson, E. (2004). Chemical mixtures: considering the evolution of toxicology and chemical assessment. *Environmental Health Perspectives, 113*, 383–390.

Monser, L., & Adhoum, N. (2002). Modified activated carbon for the removal of copper, zinc, chromium and cyanide from wastewater. *Separation and Purification Technology, 26*, 137–146.

Mooiman, M.B., & Miller, J.D. (1991). The chemistry of gold solvent extraction from alkaline cyanide solution by solvating extractants. *Hydrometallurgy, 27*, 29–46.

Moradkhani, M., Yaghmaei, S., & Nejad, Z.G. (2018). Biodegradation of cyanide under alkaline conditions by a strain of Pseudomonas putida isolated from gold mine soil and optimization of process variables through response surface methodology (RSM). *Periodica Polytechnica Chemical Engineering, 62*, 265–273.

Mousavi, S.R., Balali-Mood, M., Riahi-Zanjani, B., & Sadeghi, M. (2013). Determination of cyanide and nitrate concentrations in drinking, irrigation, and wastewaters. *Journal of Research in Medical Sciences, 18*, 65.

Moussavi, G., & Khosravi, R. (2010). Removal of cyanide from wastewater by adsorption onto pistachio hull wastes: Parametric experiments, kinetics and equilibrium analysis. *Journal of Hazardous Materials, 183*, 724–730.

Newhouse, K., & Chiu, N. (2016). *Toxicological Review of Hydrogen Ccyanide and Cyanide Salts*. Washington, DC: EPA http://www. epa. gov/iris/toxreviews/0060tr. pdf.

Nickolov, R.N., & Mehandjiev, D.R. (2004). Comparative study on removal efficiency of impregnated carbons for hydrogen cyanide vapors in air depending on their phase composition and porous textures. *Journal of Colloid and Interface Science, 273*, 87–94.

Oliver, T.M., Jugoslav, K., Aleksandar, P., & Nikola, D. (2005). Synthetic activated carbons for the removal of hydrogen cyanide from air. *Chemical Engineering and Processing: Process Intensification, 44*, 1181–1187.

Ostrowski, S.R., Wilbur, S., Chou, C.H.S.J., Pohl, H.R., Stevens, Y.W., Allred, P.M., Roney, N., Fay, M., & Tylenda, C.A. (1999). Agency for Toxic Substances and Disease Registry's 1997 priority list of hazardous substances. Latent effects—carcinogenesis, neurotoxicology, and developmental deficits in humans and animals. *Toxicology and Industrial Health, 15*, 602–644.

Raza, S.K., & Jaiswal, D.K. (1994). Mechanism of cyanide toxicity and efficacy of its antidotes. *Defence Science Journal, 44*, 331–340.

Razanamahandry, L.C., Karoui, H., Andrianisa, H.A., & Yacouba, H. (2017). Bioremediation of soil and water polluted by cyanide: a review. *African Journal of Environmental Science and Technology, 11*, 272–291.

Reade, M.C., Davies, S.R., Morley, P.T., Dennett, J., Jacobs, I.C. , & Australian Resuscitation Council (2012). Management of cyanide poisoning. *Emergency Medicine Australasia, 24*, 225–238.

Roberts, R.F., & Jackson, B. (1971). The determination of small amounts of cyanide in the presence of ferrocyanide by distillation under reduced pressure. *Analyst, 96*, 209–212.

de Rosa, C. (2005). International Programme on Chemical Safety (IPCS), Concise International Chemical Assessment Document, 60 (CICAD): Chlorobenzenes other than Hexachlorobenzene: Environmental Aspects. *Indian Journal of Medical Research, 122*, 180.

Rosehart, R.G. (1973). Mine water purification by reverse osmosis. *The Canadian Journal of Chemical Engineering, 51*, 788–789.

Ruangkanchanasetr, S., Wananukul, V., & Suwanjutha, S. (1999). Cyanide poisoning, 2 cases report and treatment review. *Journal-Medical Association of Thailand, 82*, S162–S167.

Salkowski, A.A., & Penney, D.G. (1994). Cyanide poisoning in animals and humans: a review. *Veterinary and Human Toxicology, 36*, 455–466.

Sharma, S., & Bhattacharya, A. (2017). Drinking water contamination and treatment techniques. *Applied Water Science, 7,* 1043–1067.

Shen, Z., Han, B., & Wickramasinghe, S.R. (2006). Cyanide removal from industrial praziquantel wastewater using integrated coagulation–gas-filled membrane absorption. *Desalination, 195,* 40–50.

Siller, H., & Winter, J. (1998). Degradation of cyanide in agroindustrial or industrial wastewater in an acidification reactor or in a single-step methane reactor by bacteria enriched from soil and peels of cassava. *Applied Microbiology and Biotechnology, 50,* 384–389.

Thompson, J.P., & Marrs, T.C. (2012). Hydroxocobalamin in cyanide poisoning. *Clinical Toxicology, 50,* 875–885.

Torre, M., Bachiller, D., Rendueles, M., Menéndez, C.O., & Díaz, M. (2006). Cyanide recovery from gold extraction process waste effluents by ion exchange I. Equilibrium and kinetics. *Solvent Extraction and Ion Exchange, 24,* 99–117.

US Department of Health and Human Services (1984). Agency for Toxic Substances and Disease Registry. 1997. *Toxicological Profile for Benzene, 423.*

Yan, T.Y., ExxonMobil Oil Corp (1994). *Wastewater treatment by catalytic oxidation.* U.S. Patent 5,338,463.

Young, C.A., & Jordan, T.S. (1995), May. Cyanide remediation: current and past technologies. In *Proceedings of the 10th Annual Conference on Hazardous Waste Research* (pp. 104–129). Manhattan, KS: Kansas State University.

Young, C.A., & Jordan, T.S. (1995), May. Cyanide remediation: current and past technologies. In *Proceedings of the 10th Annual Conference on Hazardous Waste Research* (pp. 104–129). Manhattan, KS: Kansas State University.

Yu, X., Trapp, S., & Zhou, P. (2005). Phytotoxicity of cyanide to weeping willow trees. *Environmental Science and Pollution Research, 12,* 109–113.

Yu, X., Trapp, S., Zhou, P., Wang, C., & Zhou, X. (2004). Metabolism of cyanide by Chinese vegetation. *Chemosphere, 56,* 121–126.

20 Biosurfactant: An Alternative Towards Sustainability

Sanchayita Rajkhowa[1] and Jyotirmoy Sarma[2]
[1]*Department of Chemistry, Jorhat Institute of Science & Technology, Jorhat, Assam, India*
[2]*Department of Chemistry, Kaziranga University, Jorhat, Assam, India*

CONTENTS

- 20.1 Introduction 378
- 20.2 Properties of Biosurfactants 379
  - 20.2.1 Surface Activity 382
  - 20.2.2 Emulsification and De-emulsification 383
  - 20.2.3 Biodegradability 383
  - 20.2.4 Low Toxicity 383
  - 20.2.5 Tolerance to Temperature, Ionic Strength, and pH 384
  - 20.2.6 Antiadhesive Activity 384
- 20.3 Classification of Biosurfactants 384
  - 20.3.1 Glycolipids 386
    - 20.3.1.1 Rhamnolipids 386
    - 20.3.1.2 Trehalolipids 386
    - 20.3.1.3 Sophorolipids 386
  - 20.3.2 Lipopeptides and Lipoproteins 386
    - 20.3.2.1 Surfactin 387
    - 20.3.2.2 Lichenysin 387
  - 20.3.3 Fatty Acids, Phospholipids, and Neutral Lipids 387
  - 20.3.4 Polymeric Biosurfactants 387
  - 20.3.5 Particulate Biosurfactants 388
- 20.4 Sources of Biosurfactants 388
  - 20.4.1 Bacterial Biosurfactants 388
  - 20.4.2 Fungal Biosurfactants 388
  - 20.4.3 Non-pathogenic/Probiotic Biosurfactants 388
- 20.5 Raw Materials for Biosurfactant Production 389
- 20.6 Factors That Influence the Synthesis of Biosurfactants 389
  - 20.6.1 Environmental Factors 389
  - 20.6.2 Nutritional Factors 390
    - 20.6.2.1 Carbon Substrates 390
    - 20.6.2.2 Nitrogen Substrates 390
- 20.7 Applications of Biosurfactants 391
  - 20.7.1 Bioremediation of Hydrocarbon Pollutants Using Biosurfactants 391
  - 20.7.2 Application of Biosurfactants in Metal Bioremediation 392
  - 20.7.3 Application of Biosurfactants in CO_2 Reduction 393
  - 20.7.4 Application of Biosurfactants in the Ore and Mineral Industry 394
  - 20.7.5 Other Applications 395

DOI: 10.1201/9781003004684-20

20.8	Advantages of Biosurfactants	396
20.9	Disadvantages of Biosurfactants	396
20.10	Future Trends	396
20.11	Biosurfactants and Environmental Sustainability Prospective	397
20.12	Conclusion	398
References		398

20.1 Introduction

SURFace ACTive AgeNTS or simply 'SURFACTANTS' are among a few of the versatile compounds whose application domain is diverse from detergents, cosmetics, petroleum, paints and emulsions, paper products, food, pharmaceuticals, to water treatment (Güçlü-Üstündağ and Mazza, 2007; Gupta et al., 2013; Elazzazy et al., 2015; Lu et al., 2017; Mahamallik and Pal, 2017; Varjani and Upasani, 2017). Surfactants can also have advanced applications in enhanced oil recovery, improved pesticides and fertilizers in agriculture, treatment of lung diseases, and fire extinguishers for flammable liquids (Kovalchuk and Simmons, 2020). According to a report, household detergents still remain the most dominant sector for surfactants, accounting for approximately 46% of global consumption, which is equivalent to 16.8 MMT[1] of global demand, with an average annual growth rate of 2% within the next five years (Bland, 2019). The growth of the surfactant market is driven by the population and increasing urbanisation. Moreover, the growing awareness about products like handwash and hand sanitizer during the COVID-19 pandemic is another contributory factor for its demand. Commercially available synthetic surfactants are mostly derived from animal fats and petrochemicals. Therefore, the search for novel strategies that replace the conventional non-renewable fossil fuels like coal, petroleum, natural gas, etc. with biodegradable, renewable, as well as sustainable sources for the fabrication of surfactants has been continuing in recent years. Although petroleum is a source of numerous household and daily consumed products, the extensive exploitation of petroleum sources deteriorates the human and ecological health with significant contributions in pollution. On the other hand, production of petroleum and its products is expected to decline within the next few decades (Frumkin et al., 2009). Thus, a quest for environmentally and industrially sustainable alternatives of synthetic surfactants have been accomplished with "biosurfactants" produced from cheap renewable feedstocks, potentially fulfilling all the physicochemical properties of their synthetic counterparts. Interestingly, biosurfactants are stable at relatively high temperatures, tolerant to high salt concentrations and in robust environments, they are easily biodegradable when disposed of to the environment. At the same time, the production and cost of biosurfactants greatly depend on the availability of the raw materials, their storage, and transportation. Nevertheless, the inclination of manufacturers towards the development of commercially feasible eco-friendly products makes biosurfactants an attractive alternative, progressively substituting the conventional surfactants from the market (Marchant and Banat, 2012). The first report on biosurfactants was made in 1957 and wide investigation and application of such compounds has expanded since then (Hauser and Karnovsky, 1957). In general, natural surfactants or biosurfactants are derived from various plants, marine organisms, and microorganisms as secondary metabolites (Siñeriz et al., 2001). They are found in almost all components of plants viz., buds, flowers, leaves, fruits, seeds, and roots. Natural surfactants are also called "saponins", a Latin origin meaning foam-producing agents derived from plants, which are soluble in water and molecular weights in the range of 600–2000 Da[2] (Piispanen et al., 2004).

Biosurfactants are amphiphilic compounds composed of two distinctly different parts: an ionic hydrophilic head, comprising an acid, peptide cations, or anions, mono-, di-, or polysaccharides; and a polar hydrophobic tail that consists of unsaturated/saturated hydrocarbon chains or fatty acids, as shown in Figure 20.1 (Banat et al., 2010). Hyrodophilic and hydrophobic terms are lyophilic and lyophobic,

---

[1] MMT stands for Million Metric Tons.

[2] Dalton, symbol Da is a unit of molar mass, especially used in biochemistry, with the definition 1 Da = 1 g/mol.

Biosurfactant

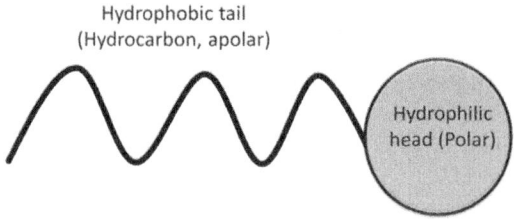

FIGURE 20.1 Structural representation of a surfactant moiety having a hydrophobic (or apolar) tail and a hydrophilic (or polar) head group.

respectively, if the solvent is other than water in which surfactants are dissolved. Any surfactant (either synthetic or natural origin) has an inherent tendency to get adsorbed at the surfaces and/or interfaces. An "interface" is nothing but the boundary that separates any two mutually immiscible phases, viz. liquid–liquid, solid–liquid, and gas–liquid. Surfactants can either interact with an interface, mostly a gas–liquid interface, or they can self assemble to form various aggregates in the solution phase (Mulligan, 2009). Adsorption of a surfactant to an interface is driven by a thermodynamic parameter, i.e., the free-energy change at the phase boundary that attains a minimum value upon addition of a surfactant. The amount of work required to stretch the interface is represented by free energy per unit area at the interface that is commonly known as surface tension. When a surfactant is added to a liquid (say water), the phase boundary will be occupied by the surfactant molecules and, subsequently, there is a significant reduction in surface tension. The higher the surfactant concentration at the interface, the larger is the decrease in surface tension value. Surfactants may get adsorbed at almost all types of interfaces mentioned previously. The stronger the tendency to get adsorbed at the interface, the better the surfactant is. A surfactant is considered to be good if it can reduce the surface tension of water from 72 up to 30 mN/m (De et al., 2015). It is worth remarking that the degree of interfacial adsorption of a surfactant primarily depends on two factors: (i) chemical structure of the surfactant molecule and (ii) the nature of the two immiscible phases. Practically, there is no surfactant that can universally be suitable for all purposes. The choice of surfactant invariably depends on its application. Biosurfactants are primarily produced, but not exclusively, by microorganisms: yeast, bacteria, and fungi from a number of substrates like alkanes, sugars, oils, sugars, and wastes, etc. Sophorolipids are the first commercially available microbial surfactants (De et al., 2015). Several biosurfactants are used as emulsifiers while some others are in use to reduce oil-water or similar system's interfacial tension. Moreover, the wettability and foaming capability make these molecules suitable for detergency application. Remediation of organic compounds (primarily hydrocarbons) produced by industrial activities, inorganic compounds containing heavy metals is being accomplished by washing treatment with biosurfactants, enhanced oil recovery by reducing the oil–water interfacial tension, increasing the membrane permeability for drug molecules during cell lysis; application of biosurfactants as effective and safe therapeutic antimicrobial agents as well as treatment in the treatment of autoimmune disease and transplantation are several noted fields where use of biosurfactants have recently been practiced (Banat et al., 2010).

The escalating interest of the synthesis, development, and application of biosurfactants due to their diversity, selectivity, large-scale production prospective, performance under adverse conditions, and most significantly their low toxicity and biodegradability will make them a wise alternative to synthetic surfactants in the coming years. In view of this, our chapter is an attempt to discuss biosurfactants and their features in relation to environmental sustainability.

20.2 Properties of Biosurfactants

The physicochemical properties of biosurfactants, such as reduction of surface as well as interfacial tension, low CMC, detergency, foaming, emulsifying, and stabilizing capacities, are of great significance in terms of performance and also the selection of microorganisms for the potential production of a particular biosurfactant. Although each biosurfactant has distinctive chemical compositions, properties, and source of origin, a few common characteristics are displayed by a majority of the biosurfactants.

Due to the unique structure of biosurfactant (or surfactant) molecules, they tend to reside at the interfacial region, orienting themselves in such a manner that the polar groups are in contact with the polar phase while aligning the non-polar tail away from it (Figure 20.2) and thus reducing the surface tension of the solvent/medium in which they are dissolved.

All the activities of biosurfactants are governed by theirconcentration until a particular value is reached. The minimum surfactant concentration that is required to form certain organised molecular self-assemblies like micelles, vesicles, etc., is termed critical micelle concentration (CMC). On plotting surface tension values against surfactant concentration, the point when a surfactant obtains the lowest surface tension corresponds to the CMC of that particular surfactant, as shown in Figure 20.3.

Above the CMC, monomer molecules aggregate to form a variety of structures like micelles, vesicles, and multilayers (Figure 20.4). These aggregates can easily be tailor-made by the simple addition of organic solvents, inorganic salts, and change in pH, temperature, or irradiation. This characteristic of aggregate formation facilitates the lowering of interfacial/surface tension and hence, simultaneously increasing the solubility of organic compounds (Whang et al., 2008). The CMC value determines the efficacy of the surfactant; the lower its value the more effective the surfactant is, which means less

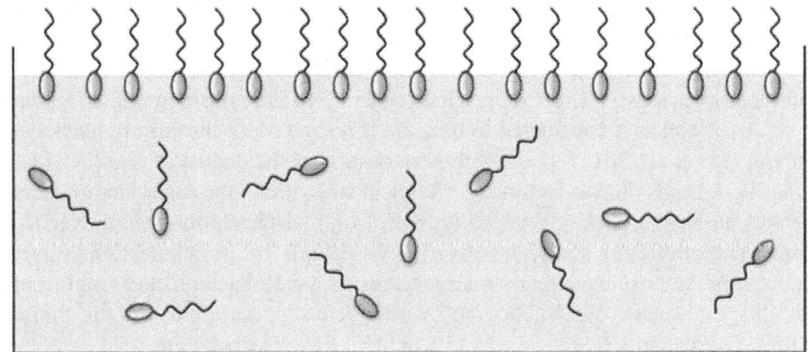

FIGURE 20.2 Schematic illustration of adsorption of biosurfactants at air-water interface.

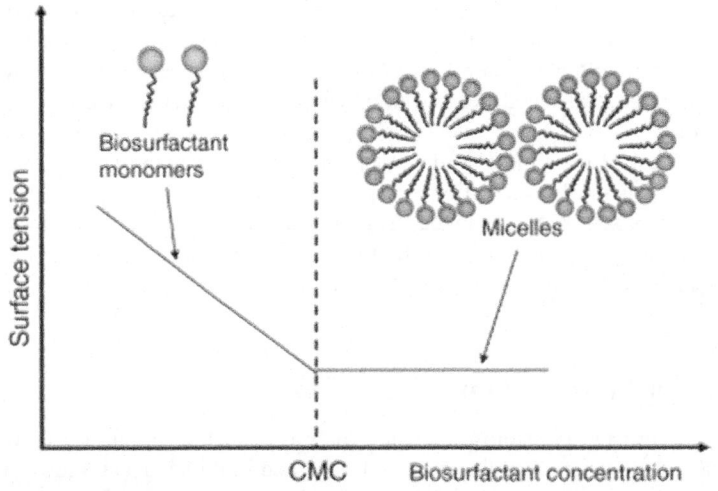

FIGURE 20.3 A typical graph showing CMC and formation of micelles beyond CMC (Reprinted with permission from Akbari et al., 2018).

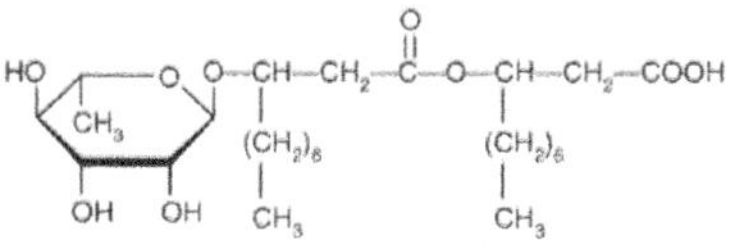

Monorhamnolipid

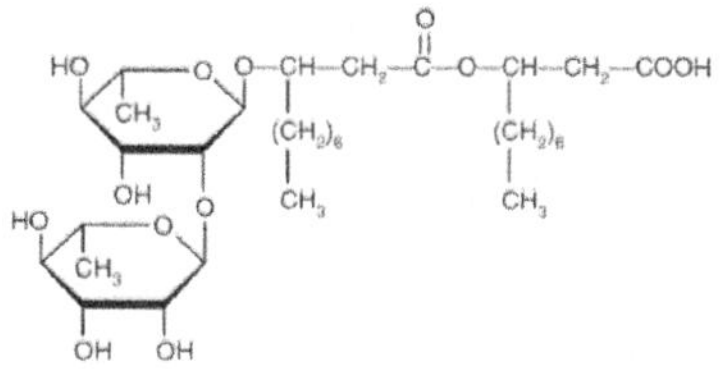

Dirhamnolipid

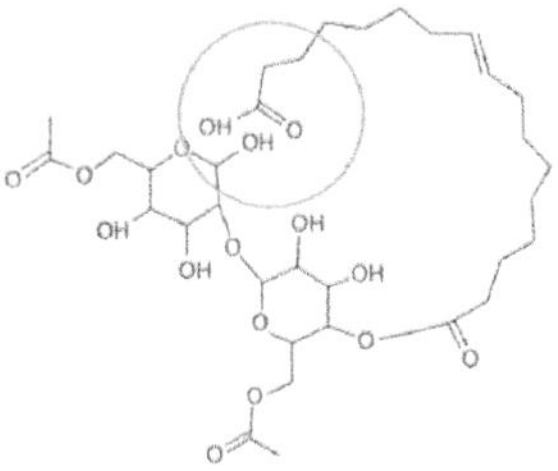

Acidic Sophorolipid

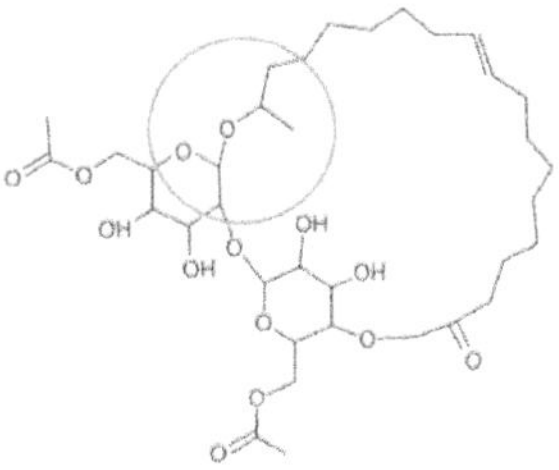

Lactonic Sophorolipid

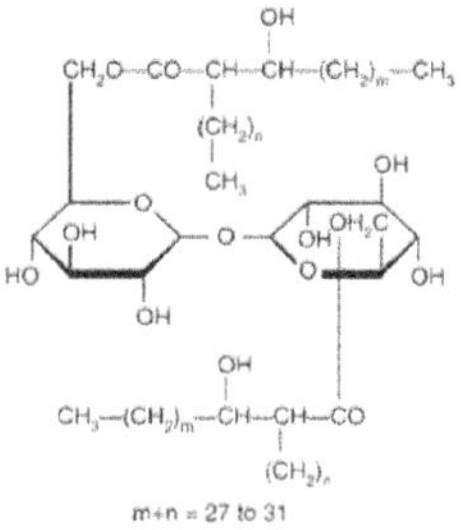

Trehalose dimycolates

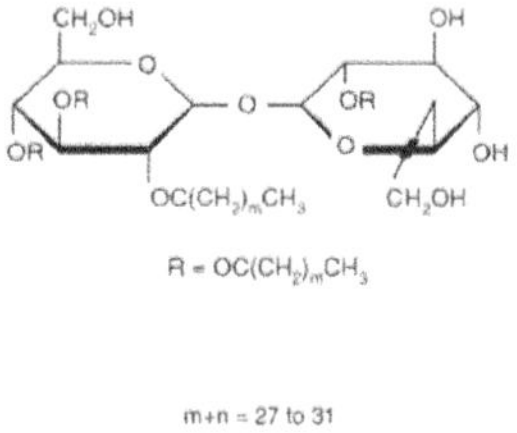

Trehalose monomycolates

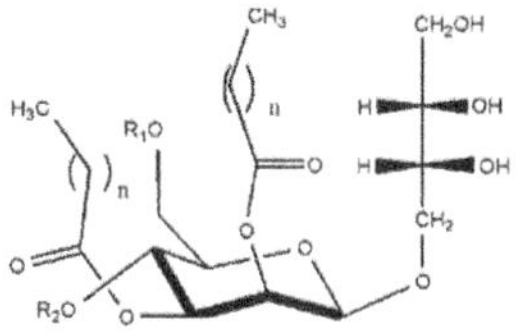

Mannosylerythritol lipids

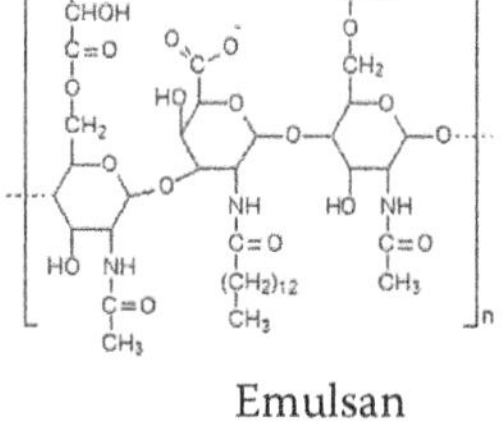

Emulsan

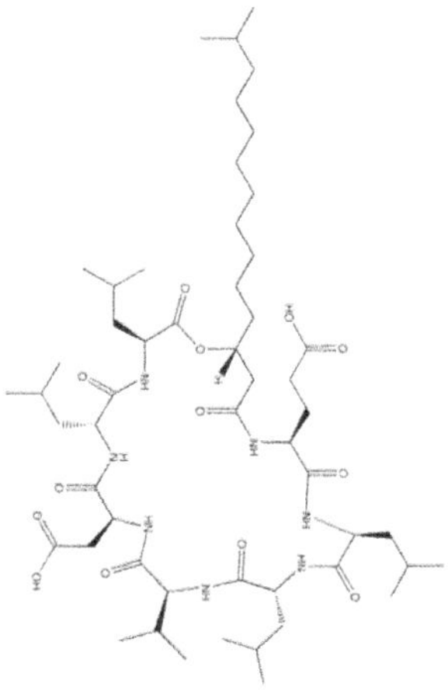

Surfactin

FIGURE 20.4 Structure of various common biosurfactants (Perfumo et al., 2010; Mulligan et al., 2014).

amounts of biosurfactants will be sufficient to lower the surface tension significantly. Another parameter, hydrophilic-lipophilic balance (HLB) value, quantifies the affinity of biosurfactants towards oil or water and can decide the type of emulsion formed by the biosurfactant, i.e., oil-in-water or water-in-oil. The lower empirical HLB value means greater affinity for oil (stabilize water-in-oil emulsion), whereas a higher value suggests more affinity to water (stabilize oil-in-water emulsion) (Gadhave, 2014). The surface activity of biosurfactants is analogous to that of their synthetic counterparts. For instance, biosurfactants can effectively lower the surface tension of water from 72 mN/m to 29.0 mN/m (at the CMC), whereas SDS, a synthetic surfactant, reduces the same up to 28.6 mN/m (Pornsunthorntawee et al., 2008). A diverse range of properties possessed by biosurfactants, e.g., foaming, cleansing, emulsification, wetting, phase separation, surface activity, and reduction in viscosity of crude oil, are amongst the most significant ones; this makes them a wise choice for researchers and scientists.

A few of the distinctive features are briefly discussed in the following sections.

20.2.1 Surface Activity

A good surfactant is characterised by its efficiency and effectiveness, the former being a measure of the CMC while the latter is related to interfacial tension. Biosurfactants are found to be promising candidates in this regard. The CMC of biosurfactants falls within the range from 1 to 2,000 mg/L. The oil–water interfacial as well as surface tension of biosurfactants are recorded at ~1 and 30 mN/m, respectively (Sobrinho et al., 2013). A compound is considered a good surfactant only if the surface tension of water is reduced from 72 to 35 mN/m and the interfacial tension of water-in-hexadecane system is reduced from 40 to 1 mN/m (Santos et al., 2016). In general, biosurfactants are found to be better with more efficiency and effectiveness, and CMC values of about 10–40 times lower compared to the chemically synthesized surfactants. Table 20.1 summarises a comparison of effectiveness and efficiency of biosurfactants and synthetic surfactants.

TABLE 20.1

Comparison of effectiveness and efficiency of a few common synthetic surfactants with biosurfactants

Type	Surfactant Name	Effectiveness in Lowering Surface Tension of Water (mN/m)	Efficiency (CMC)	References
Synthetic Surfactant	*Sodium dodecyl sulfate*	33	8.2 mM	(Rajkhowa et al., 2017; Satpute et al., 2017)
	Cetyltrimethylammonium bromide	34	0.92–1.0 mM	(Rajkhowa et al., 2017; Satpute et al., 2017)
	Triton X 100	30	0.28 mM	(Satpute et al., 2017)
Biosurfactant	*Surfactin*	27	25 mg/L	(Dhanarajan and Sen, 2014)
	Lichenysin A	28	12 mg/L	(Dhanarajan and Sen, 2014)
	Arthrofactin	24	1×10^{-5} M	(Dhanarajan and Sen, 2014)
	Rhamnolipid	26	24.26 mg/L	(Kłosowska-Chomiczewska et al., 2017)
	Sophorolipid	34	27.17 mg/L	(Dhanarajan and Sen, 2014)

20.2.2 Emulsification and De-emulsification

Biosurfactants have a special property of behaving as either emulsifiers or de-emulsifiers. The process of dispersion of one immiscible liquid (dispersed phase) into another (dispersed medium), where the dispersed phase forms droplets of microscopic size (of diameter > 1 μm), is known as emulsification. On the other hand, de-emulsification breaks the emulsion by rupturing the firm surface formed between the dispersed phase and dispersed medium. Emulsions are heterogeneous in nature with two distinct types: oil-in-water (o/w) and water-in-oil (w/o). The addition of biosurfactants to emulsions stabilizes them over several years; otherwise they have a short life span (Velikonja and Kosaric, 1993). Low-molecular-weight biosurfactants are effective surface as well as interfacial active agents, whereas the high-molecular-weight biosurfactants can be efficient in emulsification and de-emulsification. For example, sorphorolipids are known to minimise the surface and interfacial tension; however, they are not used as emulsifiers (Cavalero and Cooper, 2003). Liposan, a water-soluble emulsifier, does not reduce surface tension but has the ability to form stable edible oil/water emulsions (Cirigliano and Carman, 1985). Emulsifiers can also be characterised by HLB values and low HLB value-indicated o/w emulsification, while higher HLB values infer opposite kinds of emulsification (i.e., w/o emulsifiers).

20.2.3 Biodegradability

The growing ecological concern draws the attention toward biosurfactants, which are easily biodegradable compared to their synthetic counterparts and thus widely applied for bioremediation/biosorption. The biodegradability tests carried out for sophorolipids within the OECD Guidelines for Testing of Chemicals (301C Modified MITI Test) have displayed immediate biodegradability of natural surfactants after cultivating while the chemically synthesised ones remain dormant even after 8 days (Hirata et al., 2009). Biosurfactants produced from marine microorganisms are used for biosorption of polycyclic hydrocarbons present in aquatic surfaces. The removal efficiency of these organic solvents by sophorolipids is 90% within 30 minutes of treatment (Gharaei-Fathabad, 2011; Guerra-Santos et al., 1984). Biodegradability of rhamnolipids was found to be 92% when incubated separately in red soil and black loamy soil for 1 week (Xiaohong et al., 2009). A comparison of biosurfactants produced by five various bacterial sources with sodium dodecyl sulfate (SDS) revealed superior biodegradability of those biosurfactants over SDS (Lima et al., 2011). Another research also explained the aerobic and anaerobic biodegradability of rhamnolipid, while Triton X-100 (a synthetic surfactant) is moderately biodegradable only in aerobic circumstances but remains non-biodegradable under anaerobic conditions (Mohan et al., 2006).

20.2.4 Low Toxicity

Although vast literature has not been found that extensively investigates the adverse impacts and poisonous quality of biosurfactants, there is no denial for them to be non-harmful and low toxic, which is suitable for applications in food, pharmaceutics, and cosmetics. Studies have ascertained the high toxicity of the chemically synthesised surfactant that exhibited a LC50[3] against *Photobacterium phosphoreum,* which was ten times less than that of rhamnolipids (Poremba et al., 1991). Another report made a comparative study of the toxicity and mutagenicity profiles of *Pseudomonas aeruginosa*–or-originated biosurfactants and chemical surfactants and confirmed the non-existence of any such characteristics in biosurfactants (Flasz et al., 1998). Interestingly, the low-toxicity profile of sorphorolipids makes them suitable for use in food industries. The cytotoxicity of MEL derived from *C. Antarctica* was found to be less toxic than its synthetic counterparts without having any adverse effect on human eyes and skin; thus, it can be applicable in cosmetics and/or personal care products (Kim et al., 2002). A handful number of reports suggested low toxicity of biosurfactants after comparing this property with the chemical counterparts (Dehghan-Noudeh et al., 2005; Edwards et al., 2003; Hirata et al., 2009).

---

[3] LC50, Lethal concentration 50 is the dosage or inhaled concentration of a substance that will lead to the deaths of 50% of the dosed population.

20.2.5 Tolerance to Temperature, Ionic Strength, and pH

The efficiency of bio-based surfactants mainly depends on (i) surface activeness and (ii) stability under vigorous environmental conditions like temperature variation, high salt concentration, pH, and so forth. Biosurfactants produced from microorganisms and extremophiles have gained much attention in recent years for their use in commercial purposes. Surfactin displays reasonable surface activeness even at around 100°C, over a wide range of pH from 5 to 12, and is tolerant to 20% NaCl and 0.5% $CaCl_2$ salt concentrations (Gong et al., 2009). Similarly, lichenysin produced from *Bacillus licheniformis* was found to be active up to a temperature of 50°C, 4.5–9.0 pH, and $[Na^+]$ = 50 g/L and $[Ca^{2+}]$ = 25 g/L (McInerney et al., 1990). A lipopeptide biosurfactant from *B. licheniformis* 86 is resistant to a temperature range 25–120°C, pH 4–13, and 30% NaCl. Sophorolipids from *Candida bombicola* displayed effective surface activity after incubating under boiling water for 2 h, over a wide pH from 2 to 10, and NaCl concentration of up to 20% (Xia et al., 2011). *Arthrobacter protophormiae* produces a biosurfactant that is thermostable over 30–100°C as well as pH-stable from 2 to 12 (Singh and Cameotra, 2004). In a recent study, a biosurfactant originated from *P. aeruginosa B0406* was found unhindered within –20°C to 120°C for a range of pH of 2–12, and up to NaCl concentrations of 20% (Somoza-Coutiño et al., 2020). Biosurfactants are, in general, tolerant to 10% salt concentrations; however, 2% NaCl can effectively inactivate conventional synthetic surfactants (Sobrinho et al., 2013). Thus, stability of biosurfactants after being exposed to these factors at extreme conditions is necessary in order to perform under industrial processes.

20.2.6 Antiadhesive Activity

A biofilm is formed by colonisation of an assembly of microbes or any other organic matter on a surface. Bacterial adherence on any surface initiates bioflim formation that can be dependent on several other parameters like type of microorganism, biosurfactant's hydrophobicity, surface charges, and potential production of extracellular polymers by microorganisms that will help the cells to grapple to the surfaces as well as environmental conditions (Garrett et al., 2008). Alteration of hydrophobicity of the surfaces with the help of biosurfactants ultimately reflects on the adherence of microorganisms over those surfaces. Surfactants produced from *Streptococcus thermophilus* delay the colonization process of thermophilic strains of *Streptococcus* over a steel surface that imparts a foul smell to steel. Likewise, another biosurfactant originated from *Pseudomonas fluorescens* hindered the adherence of *Listeria monocytogenes* over such surfaces (Chakrabarti, 2012).

20.3 Classification of Biosurfactants

Chemically synthesised surfactants are commonly categorised according to ionic charge of the polar group as cationic, anionic, non-ionic, and zwitterionic. Biosurfactants can be classified on the basis of their microbial origin, chemical composition, molecular weight, mode of action, and physicochemical properties. Structurally, biosurfactants are amphiphilic moeities constituting a hydrophilic part (may be an alcohol, acid, phosphate, simple ester, peptide, cation, or anion; mono-, di-, or polysaccharides; and also proteins) and a hydrophobic part (comprising unsaturated or a saturated hydrocarbon chains or fatty acids, fatty alcohols having 8–18 carbon atoms) (Sharma, 2016). These two parts are connected by amide linkage (single and peptide), ester linkage with acids (organic or inorganic), or glycosidic linkages (sugar-hydroxy and sugar-sugar fatty acids). The nature of the hydrophilic part decides the rate of solubility, while the hydrophobic (or lipophilic) part is accountable for the capillary action of biosurfactants. Biosurfactants are broadly divided into two categories according to their molecular mass (Rosenberg and Ron, 1999).

i. **Low-molecular-weight (LMW) microbial surfactants**: These LWM biosurfactants are significantly effective in reducing the surface as well as interfacial tensions; for example, phospholipids, glycolipids, and lipopeptides.

ii. **High-molecular-weight (HMW) surfactants**: They are very effective and useful in stabilising o/w emulsion and primarily include proteins, lipoproteins, amphipathic polysaccharides, lipopolysaccharides, and polymeric and particulate surfactants.

Furthermore, biosurfactants are generally classified on the basis of their biochemical nature, structure, or their microbial origin, as shown in Table 20.2.

TABLE 20.2

Classification of biosurfactants and their microbial origin. Source: (De et al., 2015) with permission (Pacwa-Płociniczak et al., 2011; Shakeri et al., 2021)

Biosurfactant			Microorganisms	Applications in Environmental Biotechnology
Group	Sub-group	Class		
Low-molecular-weight surface active agents	Glycolipids	Rhamnolipids	*P. aeruginosa, Pseudomonas* sp.	Enhancement of the degradation and dispersion of different classes of hydrocarbons; emulsification of hydrocarbons and vegetable oils; removal of metals from soil
		Sophorolipids	*T. bombicola, T. apicola*	Recovery of hydrocarbons from dregs and muds; removal of heavy metals from sediments; enhancement of oil recovery
		Trehalolipids	*R. erythropolis, Mycobacterium* sp.	Enhancement of the bioavailability of hydrocarbons
	Lipopeptides and lipoproteins	Surfactin	*B. subtilis*	Enhancement of the biodegradation of hydrocarbons and chlorinated pesticides; removal of heavy metals from a contaminated soil, sediment, and water; increasing the effectiveness of phytoextraction
		Viscosin	*P. fluorescens*	Antimicrobial activity
		Peptide-lipid (Lichenysin)	*B. licheniformis*	Enhancement of oil recovery (EOR), antibacterial activity chelating properties that might explain the membrane-disrupting effect of lipopeptides
		Serrawettin	*Serratia marcescens*	Chemorepellent
		Subtilisn	*Bacillus subtilis*	Antimicrobial activity
		Gramicidin	*Brevibacterium brevis*	Antibiotic, disease control
	Fatty acids, neutral lipids and phospholipids	Fatty acids	*C. lepus*	Enhancement of bitumen recovery Increasing the tolerance of bacteria to heavy metal
		Neutral lipids	*N. erythropolis*	
		Phospholipids	*T. thiooxidans*	
High-molecular-weight surface active agents	Polymeric surfactants	Emulsan	*A. calcoaceticus*	Stabilisation of the hydrocarbon-in-water emulsions
		Alasan	*A. radioresistens*	
		Biodispersan	*A. calcoaceticus*	Dispersion of limestone in water
		Liposan	*C. lipolytica*	Stabilisation of hydrocarbon-in-water emulsions
		Mannoprotein	*Saccharomyces cerevisiae*	
	Particulate surfactants	Vesicles and fimbriae	*A. calcoaceticus*	Degradation and removal of hydrocarbons
		Whole cells	Variety of bacteria	

20.3.1 Glycolipids

This category of biosurfactants is more common due to their carbohydrate-based skeleton and wide range of applications; hence, they are the most extensively studied microbial-originated biosurfactants. The carbohydrate segments of glycolipids are connected to long-chain hydroxyl fatty acids through a linkage of ester or ether group. In general, glycolipids are constituted of mono- or oligosaccharides connected to a lipid molecule resulting in LMW biosurfactants. The sugar part is primarily made up of glucose, rhamnose, galactose, or xylose moelcues while the lipid part is composed of either hydroxyl fatty acids or saturated/unsaturated fatty acids. The most commonly investigated biosurfactants and their structures are given in Figure 20.4.

20.3.1.1 Rhamnolipids

Rhamnolipids are glycolipids containing rhamnose sugar moieties bonded with one or two molecules of fatty acids, i.e., β-hydroxydecanoic acid. This primary category of glycolipids is first produced by *Pseudomonas aeruginosa* (Jarvis and Johnson, 1949). Rhamnolipids are mostly prepared from α-L-rhamnose sugar moieties, linked through *O*-glycosidic bond to hydroxyl fatty acid parts where the hydroxyl fatty acids with number of carbon atoms from 8 to 16. It is also reported that species of *Burkholderia plantarii* are able to produce another type of extracellular rhamnolipids with a primary fatty acid, i.e., hydroxyl tetradecanoic acid, with distinctive structure and properties different from that of the *Pseudomonas aeruginosa* rhamnolipids (Hörmann et al., 2010).

20.3.1.2 Trehalolipids

This type of glycolipids is non-reducing sugar made up of two glucose units bonded via an α,α-1,1-glycosidic bond and are produced by various microorganisms such as *Mycobacterium, Rhodococcus, Nocardia,* and *Corynebacterium*. However, trehalose lipids derived from *Rhodococcus erythropolis* and *Arthrobacter* spp. are extensively studied due to their ability to lower the surface tension (ST) and IFT of the broth where they are cultured up to 25–40 and 1–5 mN/m, respectively. *Rhodococcus erythropolis* is known to produce a composite mixture of biosurfactants which are chiefly constituted of trehalose mycolates. Due to their potential lowering of surface tension and IFT, trehalolipids have attracted the attention of researchers and scientists for their potential applications in various fields (Franzetti et al., 2010).

20.3.1.3 Sophorolipids

These glycolipids are produced by non-pathogenic yeast and composed of dimeric carbohydrate sophorose covalently associated to a long-chain hydroxyl fatty acid via glycosidic linkage. They are usually a combination of at least 6–9 different hydrophobic sophorolipids, available in the lactonic or the acid form where the former is preferable for numerous applications. Sophorolipids were reported as microbial surface-active agents for the first time in 1961 (Gorin et al., 1961). Since then, they have been immensely explored for being less toxic, biodegradable surface-active agents in tune with environmental sustainability. Among a number of sophorolipid-producing yeasts, *Candida torulopsis* and *Candida bombicola* are predominantly efficient. Although sorphorolipids play an ambiguous role in yeast, their basic role is to provide carbon supply along with resisting the contending microorganisms. These biosurfactants are not best options to utilize as effective emulsifying agents. Nevertheless, they have excellent antifungal, antibacterial, and spermicidal activities in cosmetic formulations.

20.3.2 Lipopeptides and Lipoproteins

They are composed of a fatty acid linked to cyclic lipopeptides. Many of these biosurfactants exhibit antimicrobial activity against pathogens (bacteria, fungi, and virus) and algae along with excellent interfacial properties. Itutrins, cyclic lipopeptides mainly produced by *Bacillus* sp., are reported to be used in biomedicine and biocontrol due to their hemolytic and antifungal properties (Dang et al., 2019). This biosurfactant remain active within pH range of 5–11 and stable at –18°C for 6 months even after being autoclaved (Nitschke and Pastore, 2006). Another cyclic lipopeptide, surfactin, produced by *Bacillus subtilus,* reduces the ST of water up to 27.9 mN/m even if its concentration is very low. Lipopeptides of *B. subtilis* origin display a variety of antimicrobial activities including antifungal, antibacterial, antiviral, and antimycoplasmal. These biosurfactants are extracellularly produced by a number of microorganisms, viz. *Bacillus* sp., *Lactobacillus* sp., and some actinomycetes.

20.3.2.1 Surfactin

Surfactin is the first lipopeptide biosurfactant produced from *B. subtilis* and is considered among the most powerful ones. It is composed of seven amino acids in a ring structure coupled to the carboxyl and hydroxyl groups of a 14-carbon acid. Surfactin can reduce the surface tension of water up to 27 mN/m with concentrations as low as 0.005%, IFT can be less than 1 mN/m, and the CMC is 10 mg/L (Mulligan et al., 2014). It also has blood clotting properties as well as inactivation capabilities of herpes and retrovirus.

20.3.2.2 Lichenysin

The lipopeptide lichenysin produced by *Bacillus licheniformis* is quite stabile at a high temperature and pH and tolerant to high salt concentrations. It structurally resembles surfactin. However, a glutaminyl residue present at the first position of the peptide chain in lichenysin differentiates it from surfactin, imparting significant properties to it. Lichenysin is known to have higher surface activities that reflect in the ability to lower the ST and IFT of water up to 27 and 0.36 mN/m, respectively (McInerney et al., 1990).

20.3.3 Fatty Acids, Phospholipids, and Neutral Lipids

Several pathogens are responsible for the production of this particular class of molecules in abundance by consuming hydrophobic nutrients like alkanes. Phospholipids, being primary constituents of microbial cell membranes, are produced when organic substrates are degraded by microorganisms. Strains of *Acinetobacter* sp. and *Aspergillus* sp. fall into this category of microorganisms to produce such kinds of biosurfactants. For instance, *Acinetobacter* sp. produces vesicles that are mainly formed by 1-N, phosphatidyl ethanolamine and these vesicles are known to form a transparent micro-emulsion in water. Moreover, the microorganisms can secrete fatty acids with alkyl branches of OH functional groups; e.g., corynomucolic acids. The length of the hydrocarbon chain greatly influences the hydrophilic or lipophilic nature of the fatty acids. Saturated fatty acids with 12–14 carbon atoms can actively reduce the surface and interfacial tensions. On the other hand, *Thiobacillus thiooxidans* is responsible for production of phospholipids quantitatively which are used as wetting agents for elemental Sulphur. Phosphatidylethanolamine, a product of *Rhodococcus erythropolis* when cultured in the presence of n-alkane, reduces the IFT up to 1 mN/m and CMC of 30 mg/L of a water-hexadecane interface (Kretschmer et al., 1982). These biosurfactants found their applicability in medical practices. For example, deficiency of the phospholipid protein complex could be a prime reason for the respiratory organ failure in prematurely born children (Gautam and Tyagi, 2006).

20.3.4 Polymeric Biosurfactants

This category enlists a number of polymeric biosurfactants of microbial origin that includes liposan, lipomanan, alasan, emulsan, and polysaccharide–protein complexes that are the widely investigted. *Acinetobacter calcoaceticus* is a well-documented microorganism that produces polymeric biosurfactants with a heteropolysaccharide skeleton to which fatty acids are covalently associated (Rosenberg et al., 1988). Emulsan is effective in reducing the surface tension of water even at concentrations as low as 0.001–0.01% and also exhibit excellent emulsifying properties. Liposan, an extracellular biosurfactant, is used as water-soluble emulsifier (in food and cosmetic industries), is produced by *Candida lipolytica,* and is composed of a carbohydrate–protein mixture at 83% and 17%, respectively (Chakrabarti, 2012). Likewise, alasan and biodispersan are commonly employed as emulsifiers in food industries.

20.3.5 Particulate Biosurfactants

These biosurfactants are present in two distinct forms: (i) extracellular vesicles and (ii) whole microbial cell. The extracellular vesicles separate the hydrocarbons to produce microemulsions that facilitate the microbial cells to consume alkanes. Vesicles produced by the *Acinetobacter* sp. strain HO1-N are made up of lipopolysaccharides, proteins, and phospholipids. These vesicles have a size of 20–50 nm range diameter with buoyant density of 1.158 g/cm^3 (Chakrabarti, 2012). Several cases are witnessed where the entire bacterial cell itself works as a biosurfactant.

20.4 Sources of Biosurfactants

A large number of microorganisms have been identified as producers of biosurfactants with unique molecular structures by hydrocarbon degradation. These producers of biosurfactants can be of bacterial, fungal, or nonpathogenic/probiotic origin. Some of the common biosurfactants and the related microorganisms are listed in Table 20.2.

20.4.1 Bacterial Biosurfactants

Microorganisms depend on various organic substrates for carbon and energy sources, as those are essential for their development and growth. If the organic compound is insoluble in an aqueous medium (e.g., hydrocarbon), then microorganisms produce a typical substance, i.e., biosurfactant, which helps the organic compound to diffuse into their cell. Several bacteria and yeast are known to excrete ionic biosurfactants that help in emulsifying the hydrocarbon present in broth cultures. For example, rhamnolipids and sophorolipids are produced by various *Pseudomonas* spp. and *Torulopsis* spp., respectively. Interestingly, several other microorganisms have a unique ability to modulate their cell membrane for producing lipopolysaccharides or nonionic biosurfactants. *Rhodococcus erythropolis*, *Mycobacterium* spp., and *Arthrobacter* spp. belong to such a category of microorganisms that can form non-ionic trehalose corynomycolates whenever required. Emulsan, a lipopolysaccharide, produced by *Acinetobacter* spp. and surfactin and subtilisin (lipoproteins), originated from *B. subtilis* and is an example of a bacterial biosurfactant.

20.4.2 Fungal Biosurfactants

Unlike biosurfactants produced by bacterial microorganisms, there are limited reports on fungi-originated biosurfactants. Nevertheless, *Candida bombicola*, *Aspergillus ustus*, *Candida ishiwadae*, *Candida lipolytica*, *Trichosporon ashii*, and are a few fungi among those explored. Many of these can produce biosurfactants from cheaper raw materials. For instance, *Candida lipolytica* generates cell-wall-bound lipopolysaccharides on n-alkanes. Sophorolipids (glycolipids) are one of the major kinds of biosurfactants produced by these fungi.

20.4.3 Non-pathogenic/Probiotic Biosurfactants

Although, biosurfactants produced by microbes have been widely explored and applied in environmental and industrial treatments, their use in healthcare and therapeutic treatments is dubious and uncertain until today. Most of these biosurfactants are produced from soilborne microorganisms that may become opportunistic at times. On the contrary, biosurfactants obtained from probiotic microorganisms are known to possess potential antimicrobial, antiviral, antibiofilm, antimycoplasma, and antitumour properties. Moreover, these biosurfactants have additional benefits over the conventional biosurfactants of microbial origin (especially the soil bacteria) since the probiotics and/or non-pathogenic microorganisms produce harmless biosurfactants that are essential for human health. In the current scenario, most of the pathogenic bacteria can acquire resistance against antibiotics and remain unaffected for prolonged exposure to conventional antibiotics present at significantly high concentrations. Biosurfactants as metabolites derived from probiotic bacteria are able to eliminate pathogens with ease (Oelschlaeger, 2010). Although probiotic biosurfactants are gaining recognition in biomedical and therapeutic applications, only a little is known about their applications in food processing. Pathogen-originated biosurfactants have been raising concern over human health, toxicity, and environmental vitality during the last several years (Sharma and Malik, 2012; Sharma and Saharan, 2016). Biosurfactants produced by non-pathogenic microorganisms/probiotic bacteria/recombinant strains, as congener types and non-toxic to living cells, demand further research for novel biosurfactants to work towards the common goal of environmental sustainability. There are a number of agroindustrial wastes that act as substrates to microorganisms to produce biosurfactants in a cost-effective way, e.g., probiotic lactic acid bacteria can produce biosurfactants from several substrates like cheese whey, grape marc, and lignocellulosic hydrolysate, etc.

20.5 Raw Materials for Biosurfactant Production

Despite possessing a variety of properties and advantages over their chemical counterparts, the production of biosurfactants has been restricted due to their high production cost and low yields. Therefore, use of cheaper and various biodegradable waste products as initial raw materials can satisfactorily overcome a share of economic constraints associated with the biosurfactant production process (Mukherjee et al., 2006). However, waste products should be chosen in such a way that they assure an appropriate nutrient balance for the survival and growth of microbes with significant yield of biosurfactants that will be experimentally as well as economically viable. Any waste with high carbohydrates or lipid content is considered to be an ideal substrate. Agroindustrial waste is one such substrate for beneficial biosurfactant production at an industrial scale and, thus, investigations on various waste materials with further implantation are being optimised for feasible outcomes. A variety of agricultural and industrial wastes are found to be served as starting materials for biosurfactant production. These agroindustrial wastes, loaded with high carbohydrates or lipid components, are found to be cost effective and readily available substrates for adequate production of biosurfactants. In literature, various sources of such substrates viz., whey, fruit/vegetable peels (like orange, banana, potato, etc.), molasses, bagasse, vegetable oils, starchy effluents, vegetable fat, oily effluents, animal fat, vegetable cooking oil waste, dairy industry waste, cassava flour wastewater, corn steep liquor, oil distillery waste, glycerol, and oily sludge from refineries are considered for biosurfactant production (Sobrinho et al., 2013).

20.6 Factors That Influence the Synthesis of Biosurfactants

Several factors affecting the process of biosurfactant production can be broadly categorised as environmental and nutritional.

20.6.1 Environmental Factors

Among numerous environmental factors, temperature, salinity, pH, and growth media are the most common parameters that affect the production of biosurfactants. These environmental factors may vary according to the characteristics of the microorganism that produces the biosurfactant. For example, certain bacteria can only grow and maximize the biosurfactant production in n-hexadecane, while several others lose their activity in the hydrocarbon medium. Bacteria used in a microbially enhanced oil recovery (MEOR) in-situ process can grow even under adverse conditions, such as increased temperature and pressure, high salt concentration, and less oxygen content, that usually prevail in oil reservoirs. Additionally, it has been reported that MEOR 171 and MEOR 172 (*Pseudomonas* strains) could produce biosurfactants that remain unaffected by temperature, pH, and high concentrations of Ca^{2+} and Mg^{2+} ions. Nevertheless, environmental factors like temperature, pH, agitation, salinity, and oxygen availability greatly influence the growth rate of bacteria and, thereby, the production of biosurfactants (Desai and Banat, 1997). The pH of bacterial growth mediums plays a crucial role in sophorolipid production by *T. bombicola*. Rhamnolipid production by *Pseudomonas* spp. is found to be at a maximum within a pH ranging 6–6.5 and decreases sharply afterwards above pH 7. Alternatively, synthesis of di- and penta-saccharide lipids by *N. corynbacteroides* remains intact within the pH range of 6.5–8. Heat treatment to some of the biosurfactants does not alter their properties like the reduction of surface tension, IFT, and the emulsification efficiency significantly. Salt concentration interferes the cellular activities of microorganisms and can consequently affect biosurfactant production. A small number of biosurfactants, however, may appear to be unaffected up to 10% (wt/v) salt concentrations, though a small decrease in their CMC values is observed. Likewise, agitation speed could be another deciding factor in biosurfactant production. In general, the yield of bacterial biosurfactants reduces with the speed of stirring as in case of biosurfactant production by *A. calcoaceticus* RAG-1. But, the yield of biosurfactants produced from yeast increases if the stirring rate increases along with a sufficient supply of air. During stirring, a sufficient amount of oxygen is supplied to the microbial growth and the process of biosurfactant production is sped up, e.g., surfactin produced by *B. subtilis* (Desai and Banat, 1997).

20.6.2 Nutritional Factors

There are two principal nutritional factors involved in the process of biosurfactant production.

20.6.2.1 Carbon Substrates

A wide range of carbon substrates have been gaining attention due to their potential applicability in biosurfactant production. Indeed, the nature of the substrate greatly influences the quality, quantity, and type of biosurfactant being produced. Crude oil and diesel are considerable sources of carbon for the production of biosurfactants. There are several other water-soluble compounds such as glucose, sucrose, glycerol, ethanol, and mannitol that are used as sources of carbon substrates for production of rhamnolipid biosurfactants by *Pseudomonas* spp. (Desai and Banat, 1997). Carbon nutrient concentration, pH, and aging of the culture medium have defined the yields of rhamnolipid production. Soluble acetate and sparingly soluble hexadecane are also used as carbon sources for *Gordonia amarae* growth and eventually produce biosurfactants in large-scale wastewater treatment batch reactors (Pagilla et al., 2002). The biosurfactants, however, obtained from these sources, are inferior in terms of yield (12–36 mg/g substrate) to those acquired from hydrophobic/water-insoluble substrates (100–165 mg/g substrate) such as corn oil, olive oil, lard (saturated and unsaturated fats), n-alkanes. and long-chain alcohols (Rahman and Gakpe, 2008; Syldatk et al., 1985a). Notably, the biosurfactant composition produced by *Pseudomonas* spp. may depend on the nature of the carbon substrate, but substrates with different chain lengths do not exhibit any significant effect on the fatty acid chain length produced under glycolipids. Alternatively, the number of carbons present in an alkane chain influences the quality of biosurfactant produced in H13-A and H01-N (*Acinetobacter* spp.) strains. On a positive note, *Pseudomonas aeruginosa* can easily be obtained from a number of carbon-containing sources like olive oil, glycerol, glucose, mannitol, fructose, citrate, succinate, pyruvate, and C11 and C12 alkanes (Robert et al., 1989). It is, therefore, apparent that the carbon substrates play a crucial role in biosurfactant production.

20.6.2.2 Nitrogen Substrates

Nitrogen is the second most vital nutritional factor necessary for biosurfactant production. When nitrogen levels drop, metabolites are being produced by bacteria. On the other hand, the presence of excessive nitrogen will be consumed in the synthesis of bacterial cellular material. Among the variety of inorganic salts being tested, urea and ammonium salts were the most common nitrogen sources used by

Arthrobacter paraffineus, whereas NaNO$_3$ resulted in a higher surfactant yield than (NH$_4$)$_2$SO$_4$ by *P. aeruginosa* and *Rhodococcus* spp. for biosurfactant production (Duvnjak et al., 1983; Santa Anna et al., 2001). Reports have been found on the enhancement of biosurfactant yield by *A. paraffineus* on adding L-amino acids such as glutamic acid, aspartic acid, glycine, and asparagines into the broth culture (Duvnjak et al., 1983). In addition, the structure of surfactin is affected by the amount (concentration) of L-amino acids present in the culture that decides the production of either Val-7 or Leu-7 surfactins. Likewise, the addition of L-glutamic acid and L-asparagine to the medium escalates the formation of lichenysin-A biosurfactant in *B. licheniformis* BAS50 by two and four times, respectively (Yakimov et al., 1996). Nitrate is an excellent source of N for the production of biosurfactant by the *Pseudomonas* strain 44T1 cultured in olive oil and the *Rhodococcus* strain ST-5 in paraffin. There is a gradual rise in rhamnolipid formation by *P. aeruginosa*, and an increase in glutamine synthetase activity when the culture media reaches a threshold of nitrogen amount (known as nitrogen limitation). Nitrogen limitation can also ameliorate the biosurfactant production in *C. tropicalis* IIP-4, *P. aeruginosa*, and the *Nocardia* strain SFC-D (Desai and Banat, 1997). An overproduction of biosurfactants with different compositions is also achieved when the nitrogen limitation is reached (Syldatk et al., 1985b). Rhamnolipid production maximizes when the nitrogen limitation is carried out at a C:N ratio within the range of 16:1–18:1, but no surfactants are produced if that ratio is maintained below 11:1 without nitrogen limitation in the culture.

Finally, it has been observed that the absolute amount of nitrogen rather than its concentration is the decisive factor in optimum biosurfactant yield, while the concentration of the hydrophobic substrate (containing carbon) ascertains the yield of biosurfactants from the carbon content (Desai and Banat, 1997).

20.7 Applications of Biosurfactants

A number of studies and research on applications of biosurfactants have been going on to meet various aspects at the personal and commercial level. Petroleum-based chemical surfactants, though, have captured the commercial market and are fulfilling the daily needs; their non-biodegradability, bioaccumulation, and hazardous nature to the environment enforce our inclination towards a possible alternative, i.e., biosurfactants to protect the ecosystem and maintain environmental regulations. Biosurfactants are potentially performance-effective compounds used in numerous domains, gaining the status of "multifunctional materials" of the 21st century for industrial applications and novel uses. These molecules are enormously used in EOR, emulsion, antimicrobial activity, and degradation of hydrocarbons (Table 20.2). Industrial activities such as accidental or indented release of waste compounds (organic as well as inorganic) directly into the environment are the reasons behind environmental contamination that leads to remediation problems because those compounds tend to readily bind to soil particles with further percolation into water sources. Degradation of organic compounds, especially the hydrocarbons, and then enhancing their bioavailability via mobilisation and removal of contaminants through emulsification and pseudo-solubilisation and emulsification are among the priority areas of biosurfactants' applications. Several other applications of biosurfactants are mentioned in Table 20.2.

20.7.1 Bioremediation of Hydrocarbon Pollutants Using Biosurfactants

Bioremediation is a natural biodegradative processes that involves removal of hazardous contaminants from soil and water surface/subsurface with the help of microorganisms and plants, thereby improving the availability of nutrient materials, environmental conditions, and the existing microorganisms. Therefore, bioremediation in general is the use of nitrogen- and phosphorus-based fertilizers, adjustment of medium pH and also the amount of water, if required, providing oxygen with the regular addition of microorganisms. If any bacteria, fungi, yeast, or their enzymes are employed to convert the hazardous pollutants present in the environment into less toxic and/or non-toxic materials that can be integrated into a natural biogeochemical cycle, then it is defined as microbial bioremediation. On the other hand, phytoremediation is an in-situ process where plants and their associated microorganisms are utilized to detoxify pollutants in order to treat contaminated soil, water, and sediments. The addition of emulsifiers often stimulates the process if the bacterial growth is slow (generally cold weather, presence of high amount of contaminants) or not easily degradable compounds as contaminants like polycyclic aromatic hydrocarbons (PAHs). With advanced genetic technologies, the addition of overproducing bioemulsifier bacteria is a better option to increase the bioemulsifier concentration during the bioremediation process. For example, *A. radioresistens* produces an alasan bioemulsifier, which can be added to a mixture of oil-degrading bacteria to EOR via the bioremediation process (Vijayakumar and Saravanan, 2015).

Persistent organic pollutants (POPs) such as PAHs (phenanthrene, crysene) are major components of oil present in sediment and wastewater. These POPs are hydrophobic in nature and thus their solubility in water usually decreases with the increasing number of rings in their basic skeleton. This property ultimately leads to their low bioavailability, making it crucial for bioremediation of PAHs. Nevertheless, solubility of a few PAHs in water can be enhanced by adding biosurfactants by several folds. Moreover, most of these hydrocarbons are strongly adsorbed into soil and, hence, result in their removal efficiency limited in low mass transfer phases. Interestingly, the addition of biosurfactants as solubilising agents to the medium improves the bioavailability of less soluble and easily adsorbed hydrocarbons.

The bioremediation process can be accelerated with the addition of biosurfactants by means of emulsification (improved by high molar mass), mobilisation (promoted low-molar mass), and solubilisation. If the concentration of biosurfactants is below the CMC, then mobilisation takes place. Under these conditions, biosurfactants are able to reduce the ST as well as IFT in the air (or soil)/water interfaces. Due to these phenomena, interactions of biosurfactants with the soil/oil system increase the contact angle and hence reduce the holding capacity of soil for oil. On the other hand, above the CMC of the biosurfactant, the solubilisation process can occur when biosurfactant molecules assemble together to form micelles that dramatically enhance the solubility of oil and extraction of oil thus becomes possible (Figure 20.5). Hydrophobic organic compounds (HOCs) mainly include saturated hydrocarbons like alkanes, cyclic hydrocarbons (cycloalkanes and aromatics), and aromatic hydrocarbons with saturates (asphaltenes, resins, etc.). Biosurfactants are known to improve the bioavailability of these HOCs through a variety of mechanisms:

i. Transfer of the contaminants from a solid phase (soil) to water via the interaction of surfactants with hydrocarbons/contaminants and then mobilization of these contaminants in soil to water,
ii. Enhance the solubility of contaminants (HOCs), and
iii. Emulsify the liquid contaminants in the non-aqueous phase by lowering the IFT at aqueous/non-aqueous interfaces; hence, it increases the mass transport, the contact area, and transportation of non-aqueous-phase contaminants.

Biosurfactants indeed facilitate the adsorption of microorganisms onto soil particles contaminated with pollutants, thereby reducing the diffusion time between adsorption sites and bioadsorption sites by microorganisms.

The addition of biosurfactants results in enhancing the solubility of HOCs and the consequent degradation process depends on the presence of nature of microorganisms, water, composition of hydrocarbons, inorganic nutrients, pH, temperature, and oxygen content in soil.

20.7.2 Application of Biosurfactants in Metal Bioremediation

Heavy metals (HMs) are one of the major soil contaminants that have been taken up by plants and vegetation, along with required nutrients. Soil contaminated with potentially toxic metals like zinc,

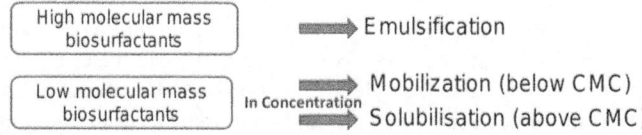

FIGURE 20.5 Hydrocarbon removal process by biosurfactants (adapted from Płociniczak et al., 2011).

cadmium, lead, chromium, etc. needs to undergo a remediation process prior to vegetation, exhumation, and/or transportation for landfilling. Surface and subsurface soils contaminated with HMs were under remediation using microorganisms in situ due to their low toxicity and cost effectiveness compared to conventional remediation processes. Similar to the bioremediation mechanism for hydrocarbons, use of biosurfactants in metal remediation is also to increase the solvency of the metal contaminants and then to facilitate their removal by biodegradation or flushing. However, metal contaminants differ from organic contaminants in several aspects. First, heavy metals are non-biodegradable in nature and thus they can only be converted from one chemical form to another, which results in alteration of their mobility rate and state of toxicity. Reduction or alkylation of these metals is a feasible alternative. Microorganisms can accumulate metals via metabolism-dependent uptake that ultimately influences the mobility of metals indirectly either by pH adjustment or stimulation of mobility-induced substances present in the soil. *Serratia marcescens* (bacteria), *Penicillium vermiculatum*, *Arthrobotrys conoides*, *Rhizopus stolonifer* (fungi), and *Cryptococcus terreus* (yeast) can readily reduce the nickel toxicity. Albeit these organisms can reduce the toxicity of nickel metal, frequent metal transformation might increase their toxicity (Babich and Stotzky, 1983; Miller, 1995). This phenomenon can be explained in the fact that there is an increase in adsorption of metal ions by microorganisms under high pH conditions.

Secondly, hydrocarbons are mostly neutral species, whereas metals are most often present in the form of cations. Since the extent of sorption of contaminants is determined by the chemical nature of the contaminant and soil molecules, therefore an appropriate surfactant must be chosen for complex formation with metal contaminants. There are two approaches by which desorption of HMs from soil could be promoted by the addition of a biosurfactant: (i) formation of complex with the free metal ions, (ii) under reduced interfacial tension, accumulation of biosurfactants at the solid-solution interface allows a direct contact between the sorbed metal and biosurfactant. Moreover, soils contain innumerable cations that may compete with contaminant metals for complexation with biosurfactants. Thus, the selectivity of a biosurfactant for contaminant HMs present in the soil and/or solution medium should always be pre-investigated. In addition, percolation of biosurfactant-HM complexes in soil is greatly influenced by the structure, size, and charge of the biosurfactant (Miller, 1995). Removal of heavy metals by biosurfactants involves three steps: (i) biosurfactant binding to soil molecules as well as to the metals at the same time, (ii) percolation of HMs from the solid (soil) to solution phase, and (iii) trapping HMs into micelles formed by biosurfactants. Encapsulation of HMs by micelle structures takes place due to the electrostatic interactions among two distinct entities (i.e., HM and the charge bearing micelles) and then those HMs can easily be recovered through physicochemical processes such as precipitation or membrane separation techniques (Figure 20.6).

The following list contains the bacterial genera that are involved in hydrocarbon remediation.

Gordonia	*Micrococcus*	*Mycobacterium*	*Ochrobactrum*	*Pseudomonas*
Rhodococcus	*Sphingomonas*	*Aeromonas*	*Alcaligenes*	*Acinetobacter*
Arthobacter xanthomonas	*Bacillus*	*Brevibacterium*	*Flavobacterium*	*Geobacillus*

20.7.3 Application of Biosurfactants in CO_2 Reduction

The greenhouse effect is a natural process that maintains the earth's ambient temperature; anthropogenic production of atmospheric gases like CO_2, water vapour, and methane in excess impacts the energy balance of the earth by trapping long-wave radiations (i.e., infra red) emitted by earth's surface, which ultimately causes global warming. Efforts have continuously been made to reduce greenhouse gas emissions and several measures have also been adopted by policy makers in various international climate change treaties to tackle health concerns and negative ecological impacts related to global warming. The 1997 UNFCCC Kyoto Protocol and the Conference of the Parties (COP21) are such instances where a number of countries agreed to control the production of greenhouse gases (especially focusing on CO_2) and putting a step forward to sustainable development by ensuring the rise in global mean temperature below 2°C (UNFCCC, 1998; IEA, 2015). Studies have revealed that biosurfactants play a key role in minimizing CO_2 emission from earth's atmosphere (Rahman and Gakpe, 2008). If the

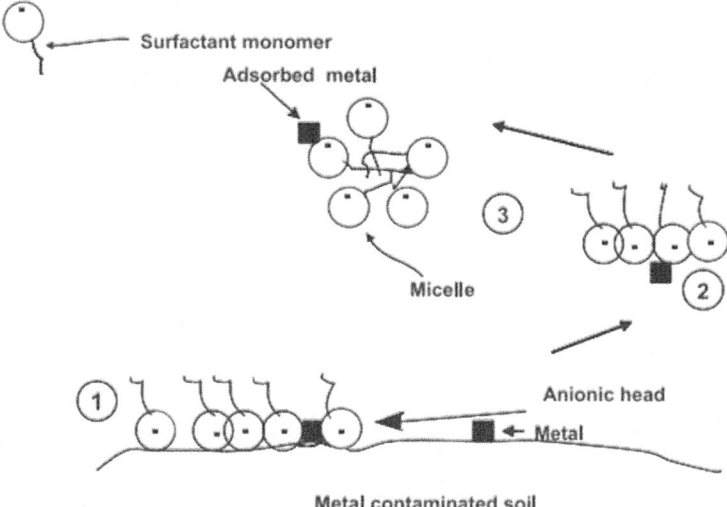

1. Accumulation of surfactant as hemimicelles or admicelles at soil interface
2. Removal of metal by lowering of interfacial tension and electrostatic attraction
3. Incorporation of metal into micelle

FIGURE 20.6 Steps of metal remediation by biosurfactants (reproduced from Mulligan, 2005 with permission).

total surfactant production hadn't changed until 2010 in the EU, and then the amount of oleochemical[4] surfactants would have increased from approximately 880 kt in 1998 to 1,100 kt in 2010 with an overall increment of 24%t. This arrangement of substitution of synthetic surfactants with natural ones also reduces the emission of CO_2 life cycle from surfactants by 8%; however, theoretically a 37% reduction could be achieved by this substitution. On a positive note, an estimated amount of 1.5 million tons of CO_2 emission could be prevented through the generation of oleochemical surfactants (Patel, 2004). Viewing a future sustainability to make a balance between socioeconomic development and environment by considering expected reduction in CO_2 emissions to exceed 8% from surfactant substitution will address a long-term impact.

20.7.4 Application of Biosurfactants in the Ore and Mineral Industry

In mineral processing industries, the role of surface-active agents is vital for mineral beneficiation. Mineral beneficiation works on the separation of active minerals from gangue minerals. This can be achieved by washing ores with techno-economically cheap and eco-friendly surfactants in industries. The surface-active agent gets adsorbed onto the mineral surface and the hydrophobic molecules present, if any, are washed out along with the surfactant. However, keeping in mind the environment sustainability use of biosurfactants is noteworthy. In this context, starch and sodium extracts of humic acid (i.e., sodium humate, Fig. 20.7) play a significant role. Humic acid is a natural organic matter (NOM) analogue containing primarily hydroxyl (–OH) and carboxyl (–COOH) functional groups. These types of functional groups are responsible for the interaction with ore and mineral surfaces.

As an example, application of sodium humate as a surface-active agent in the beneficiation of Indian iron ore is found effective. The surfactants, either simple or complex, get adsorbed onto the hematite

---

[4] Oleochemicals are derived from animal or vegetable oils or fats and include fatty acids, fatty alcohols, glycerine, and biodiesel.

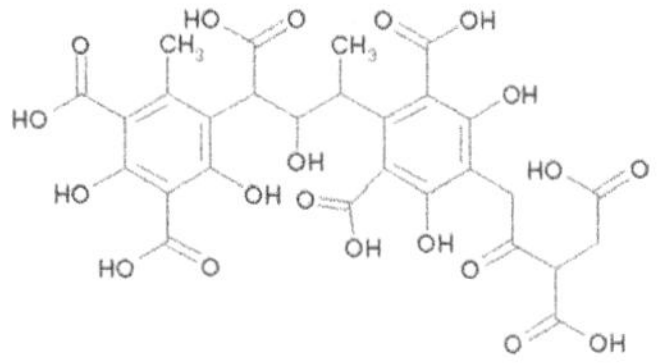

FIGURE 20.7 Structure of humic acid with –OH and –COOH functional groups (Source: Stevenson, 1994).

surfaces, making the hematite particles to flocculate of and alumino-silicate bearing gangue minerals will be dispersed. Various surfactants can result in a stable, dispersed phase containing gangue minerals that are liberated from feed iron ore fines and slime in aqueous slurry. Humic acid is composed of a number of aromatic acids viz., mono-, di-hydroxybenzoic, and benzenecarboxylic acids. These molecules are widely studied for their complex formation ability with minerals (Ali and Dzombak, 1996; Evanko and Dzombak, 1998). Decomposition of the vegetative matters results in the production of such aromatic acids, which are an integral part of the aquatic ecosystem (Blum, 1996; Sparling and Vaughan, 1981). Nevertheless, the phenolic constituents of humic acid are regarded as secondary metabolites found in plant-based foods and trees (Meulenberg, 2009). The phenolic compounds representing NOM exhibit surface-active properties due to their hydrophilic functional groups. The carboxylic and hydroxyl (i.e., –COOH and –OH) functionalities present in the phenolic compounds are mainly responsible for the complex formation between organic acids and mineral molecules (Guan et al., 2006).

20.7.5 Other Applications

Biosurfactants have a wide range of other applications in food, emulsion, agriculture, pesticide, laundry, cosmetic, and pharmaceutical industries. Fatty acid esters with glycerol, ethylene glycol or sorbitan, lectin and its derivatives, ethoxylated derivatives of monoglycerides, and oligopeptides are used as food additives (emulsifiers). These additives impart flavour, improve texture, and enhance the taste and quality of food items with reduced health effects. Bioemulsifiers isolated from the marine *Enterobacter cloaceae* strain are used as viscosity enhancers in food items that contain edible acids viz., ascorbic acid or citric acid (Iyer et al., 2006). In the agriculture industry, biosurfactants are effective against several pathogens and thus can be used as pesticides. For example, rhamnolipids are used against zoosporic plant pathogens viz., *Phythium aphanidermatum, Plasmopara lactucea-radicis,* and *Phytophthora capsici*. The rhamnolipid biosurfactants from *Pseudomonas* spp. possesses potential antimicrobial activity. Zonix^{TM} biogungicide, a commercial product of a mixture of two rhamnolipid biosurfactants, has been a successful biofungicide that controls and prevents pathogenic fungal effects on agricultural and horticultural crops. Moreover, no negative impact on humans or the environment makes rhamnolipid biosurfactants suitable for prolonged application. Besides, reports on 30–45% biodegradation of α- and β-endosulfan (chlorinated pesticides) by employing a biosurfactant produced from *B. subtilis* MTCC 1427 gives an extra edge to their application with environment protection (Awasthi et al., 1999). From a study, it has been observed that cyclic lipopeptide (CLP) synthesised by *B. subtilis* strains retain their surface-active property even at a basic pH range from 7.0 to 12.0 and also after heating at a temperature as high as 80°C for 60 minutes. Crude CLP biosurfactants display excellent emulsification with vegetable oils with superior compatibility and stability compared to synthetic laundry detergents, which favours their inclusion in laundry detergent formulations (Mukherjee, 2007). Biosurfactants are effective against various viruses, bacteria, fungi, and algae demonstrating their antimicrobial activity. Rhamnolipids from *P. aeruginosa*, lipopeptides from *B. subtilis*, and *B. licheniformis* and mannosylerythritol lipids from *C. antarctica* are known to possess antimicrobial activities. Iturin, originated from *B. subtilis,* is known to possess antifungal properties. Pumilacidin, a surfactin analog, is found to inhibit herpes simplex virus 1 (HSV-1), gastic ulcers, and H^+/ K^+- Atpase in vivo (Mukherjee and Das, 2010). Surfactin, on the other hand, displays potency against the human immunodeficiency virus 1 (HIV-1) and may serve as a reliable molecule for the synthesis of anti-HIV drugs. Interestingly, sophorolipid produced by *C. bombicola* also has spermicidal, cytotoxic, and anti-HIV activities, whereas the acidic (lactonic) form of sophorolipids can be used in cosmetology and dermatology.

20.8 Advantages of Biosurfactants

With the advancements in biosurfactant formulation in recent years, biosurfactants have become advantageous over chemically synthesized surfactants in several aspects. Some of their notable advantages are:

- Biodegradability
- Low toxicity
- Biocompatibility and digestibility
- Easy availability of raw material
- Emulsion forming/breaking
- Specificity
- Economically viable and environmentally sustainable

20.9 Disadvantages of Biosurfactants

In spite of notable advantages, biosurfactants are known to exhibit the following demerits:

- A high production cost is involved during large-scale manufacturing of biosurfactants required in petroleum and environmental applications. Nevertheless, this might be minimized by assimilating the production process to utilize waste or cheaper raw materials, thereby reducing the pollution caused due to these substrates.
- For obtaining a pure form of biosurfactants utilized in food, cosmetics, and pharmaceutical applications, the process of purification (downstream) must undergo a series of steps to achieve high concentrations and yields in bioreactors.
- Overproducing strains of bacteria are scarce with low productivity. Besides, a complex culture media is required for their production.

20.10 Future Trends

- Due to resistance development ability and adaptation of pathogens against conventional antimicrobial drugs, there's always a constant challenge to formulate an alternate by researchers. In this context, microbial surfactants with potential antimicrobial activity would be a better option for effective therapeutic applications.
- Chemically fabricated emulsifying agents are unable to meet the consumer demands in terms of health and well-being. Therefore, unconventional emulsifying agents like microbial surfactants could be a solution. In the near future, biosurfactants are expected to be used in stable emulsions, nano-emulsions, and thickening agents in food formulations.
- Another issue related to the food processing industry is the biofilm control, where inadequate and ineffective measures are practiced to control microbial pathogen growth in the present scenario. Thus, application of biosurfactants in cleaning and removal of biofilm would definitely be more greener and environmentally benign.
- Biosurfactants are found to replace harmful synthetic counterparts in many of the cosmetics, topical, as well as healthcare products; e.g., rhamnolipids-based toothpaste, personal care products, and other cleaning agents have already been commercialised (Tiso et al., 2017). The antimicrobial potential of biosurfactants makes them suitable for use as biopreservatives in food preservation.

- Biosurfactants have diverse effects on the immune response and, based on their therapeutic utility, the latest advances in immunomodulation potential of biosurfactants is explored, considering their better effectiveness, selectivity, environmental compatibility, and stability compared to many synthetic compounds.
- With the recent pandemic outbreak in severe acute respiratory syndrome – coronavirus-2 (SARS-CoV-2), biosurfactants have proven to be ideal materials as novel solutions in resolving the pandemic of this scale. The amphiphilic nature of biosurfactants facilitates their interaction with the hydrophobic viral membrane significantly enough to disrupt it, which ultimately results in the virus' structure breakdown and final deactivation. Due to this mechanism, biosurfactants will be used in the formulation of handwashes and cleaning agents to prevent the spread of the virus, in drug delivery systems, and also their use in other major fields such as antiviral face masks are noteworthy (Smith et al., 2020).
- Recently, the ability of biosurfactants to form molecular self-assembly is utilised in nanotechnology for synthesizing nanoparticles and microemulsions, e.g., rhamnolipids are natural green biosurfactants.
- It is anticipated that biosurfactants can be a solution for cleaning in place (CIP) in industrial cleaning and future sanitization processes. Biosurfactant-based formulations will also find their application in food industries for their ability to prevent biofouling.

20.11 Biosurfactants and Environmental Sustainability Prospective

Sustainable production of biosurfactants relies on their economic viability in the future that includes their potential applications with greater market share. In this context, selection, design, and method of production techniques must be cost effective with low capital investment. However, some of these factors may be vitiated during large production of biosurfactants. Cost, functionality, and production capacity are the deciding parameters for industrial applications of biosurfactants. Rufino et al. reiterated the indispensable need of cheaper raw substrates for successful production of biosurfactants (Rufino et al., 2014). With this an additional parameter, i.e., the recovery process must be addressed while discussing the biosurfactants' future with sustainable development. However, it is a challenge to maintain environmental sustainability along with economic viability of industrial applications of biosurfactants aiming to achieve the high product yield from low-cost substrates. The proposed or expected field of biosurfactant applications as well as the target market will be other deciding factors that determine the overall production cost. In terms of economic growth, use of renewable substrates will minimize the dependency on non-renewable resources and build a sustainable development with higher profits. In 2015, the European Commission has established a "circular economy" strategy focusing primarily on waste prevention or utilization of wastes if generated in bioeconomical ways. Particularly biorefineries are the efficient ones to treat and/or process wastes and offer sustainable methods following the concept of "zero waste" society (Morais and Bogel-Lukasik, 2013). With the latest document entitled "European Biorefinery Joint Strategic Research Roadmap for 2020" provides necessary information and tools to enable policy makers to constitute a framework for the European bioeconomy based on sustainable development, with a network of biorefineries playing a central role (Star-COLIBRI, 2011). Nonetheless, the oil sector is one of the most prominent areas for consuming biosurfactants during microbial EOR from oil sludge and oil wells, washing of oil storage tanks, and bioremediation processes (Banat et al., 2014). Therefore, it is substantive to invest in novel strategies that will promote the production of biosurfactants commercially. Novel and efficient methods of biosurfactant isolation during the production process will confirm economic benefits with the development of clean processing technologies and research in the near future.

20.12 Conclusion

Biosurfactants, a revolutionary development in terms of clean surface-active agents, are bringing an inclination towards cleaning up environmental pollutants produced across industries. Due to their versatile nature and numerous advantages, these microbial products have a major role in environmental sustainability and thus also ensure an economic sustainability. Ongoing research on harnessing the cheaper wastes as substrate materials emphasizes the vast potential related to the applications of biosurfactants. However, there is demand for further research to develop production processes that increase the yield synchronous with production cost. With the discovery of novel, low-temperature biosurfactants, a new area for research and industry has paved its way, aiming their potential applications in cold regions and also the development of low-energy manufacturing processes. Life cycle sustainability assessment of biosurfactants is another area identified for further research and investigation. Utilization of green chemistry and genetic engineering approaches is the need of the hour while designing new strategies and technologies to enhance the yield, quality of the biosurfactant product, and minimizing the production cost with cheaper and renewable substrates. At the same time, proper care must be taken to keep intact the nutritional values of such substrates. Wastes from animal fats, food processing, and dairy industrial sectors will open the doors for tremendous scope in biosurfactant production. The focus of all the research should be on production-consumption-disposal stages and their impacts, such as carbon footprint, toxicity, and resource depletion. Biosurfactants are, certainly, multifunctional materials with all the necessary characteristics to replace their synthetic counterparts, which will maintain environmental sustainability in the 21st century and beyond.

REFERENCES

Akbari, S., Abdurahman, N.H., Yunus, R.M., Fayaz, F., & Alara, O.R. (2018). Biosurfactants-a new frontier for social and environmental safety: a mini review. *Biotechnology Research and Innovation*, *2*(1), 81–90.

Ali, M.A., & Dzombak, D.A. (1996). Interactions of copper, organic acids, and sulfate in goethite suspensions. *Geochimica et Cosmochimica Acta*, *60*(24), 5045–5053.

Awasthi, N., Kumar, A., Makkar, R., & Cameotra, S.S. (1999). Biodegradation of soil-applied endosulfan in the presence of a biosurfactant. *Journal of Environmental Science and Health, Part B*, *34*(5), 793–803.

Babich, H., & Stotzky, G. (1983). Temperature, pH, salinity, hardness, and particulates mediate nickel toxicity to eubacteria, an actinomycete, and yeasts in lake, simulated estuarine, and sea waters. *Aquatic Toxicology*, *3*, 195–208.

Banat, I.M., Franzetti, A., Gandolfi, I., Bestetti, G., Martinotti, M.G., Fracchia, L., Smyth, T.J., & Marchant, R. (2010). Microbial biosurfactants production, applications and future potential. *Applied Microbiology and Biotechnology*, *87*, 427–444.

Banat, I.M., Satpute, S.K., Cameotra, S.S., Pati, R., & Nyayanit, N.V. (2014). Cost effective technologies and renewable substrates for biosurfactants production. *Frontiers in Microbiology*, *5*, 1–18.

Bland, A.S. (2019). Global Surfactants Industry. *IHS Markit*. Retrieved November 18, 2020 (https://ihsmarkit.com/research-analysis/global-surfactants-industry.html).

Blum, U. (1996). Allelopathic interactions involving phenolic acids. *Jornal of Nematology*, *28*(3), 259–267.

Cavalero, D.A., & Cooper, D.G. (2003). The effect of medium composition on the structure and physical state of sophorolipids produced by *Candida bombicola* ATCC 22214. *Journal of Biotechnology*, *103*, 31–41.

Cavalero, D.A.,& Cooper, D.G. (2003). The effect of medium composition on the structure and physical state of sophorolipids produced by *Candida bombicola* ATCC 22214." *Journal of Biotechnology*, *103*, 31–41.

Chakrabarti, S. (2012). *Bacterial biosurfactant: Characterization, antimicrobial and metal remediation properties."* M.Sc. Dissertation, Rourkela. Odisha: Department of Life Science, National Institute of Technology.

Cirigliano, M.C., & Carman, G.M. (1985). Purification and characterization of liposan, a bioemulsifier from *Candida lipolytica*. *Applied and Environmental Microbiology*, *50*, 846–850.

Dang, Y., Zhao, F., Liu, X., Fan, X., Huang, R., Gao, W., Wang, S., & Yang, C. (2019). Enhanced production of antifungal lipopeptide iturin A by *Bacillus amyloliquefaciens* LL3 through metabolic engineering and culture conditions optimization. *Microbial Cell Factories*, *18*, 68.

Dehghan-Noudeh, G., Housaindokht, M., & Bazzaz, B.S.F. (2005). Isolation, characterization, and investigation of surface and hemolytic activities of a lipopeptide biosurfactant produced by *Bacillus subtilis* ATCC 6633. *Journal of Microbiology*, *43*(3), 272–276.

De, S., Malik, S., Ghosh, A., Saha, R., & Saha, B. (2015). A review on natural surfactants. *RSC Advances*, *5*(81), 65757–65767.

Desai, J.D., & Banat, I.M. (1997). Microbial production of surfactants and their commercial potential. *Microbiology and Molecular Biology Reviews*, *61*(1), 47–64.

Dhanarajan, G., & Sen, R. (2014). Amphiphilic molecules of microbial origin. pp. 31–48 in *Biosurfactants: research trends and applications*, edited by C.N. Mulligan, S.K. Sharma, & A. Mudhoo. Boca Raton: CRC Press.

Duvnjak, Z., Cooper, D.G., & Kosaric, N. (1983). Effect of nitrogen source on surfactant production by Arthrobacter paraffines ATCC 19558. P. 66–72 in *Microbial enhanced oil recovery*, edited by J.E. Zajic, D.G. Cooper, T.R. Jack, & N. Kosaric. Tulsa, Okla: Pennwell Books.

Edwards, K.R., Lepo, J.E., & Lewis, M.A. (2003). Toxicity comparison of biosurfactants and synthetic surfactants used in oil spill remediation to two estuarine species. *Marine Pollution Bulletin*, *46*, 1309–1316.

Elazzazy, A.M., Abdelmoneim, T.S., & Almaghrabi, O.A. (2015). Isolation and characterization of biosurfactant production under extreme environmental conditions by alkali-halo-thermophilic bacteria from Saudi Arabia. *Saudi Journal of Biological Sciences*, *22*, 466–475.

Evanko, C.R., & Dzombak, D.A. (1998). Influence of structural features on sorption of NOM-analogue organic acids to goethite. *Environmental Science & Technology*, *32*(19), 2846–2855.

Flasz, A., Rocha, C.A., Mosquera, B., & Sajo, C. (1998). A comparative study of the toxicity of a synthetic surfactant and one produced by Pseudomonas aeruginosa ATCC 55925. *Medical Science Research*, *26*, 181–185.

Franzetti, A., Gandolfi, I., Bestetti, G., Smyth, T.J., & Banat, I.M. (2010). Production and applications of trehalose lipid biosurfactants. *European Journal of Lipid Science and Technology*, *112*(6), 617–627.

Frumkin, H., Hess, J., & Vindigni, S. (2009). Energy and public health: the challenge of peak petroleum. *Public Health Reports*, *124*(1), 5–19.

Gadhave, A. (2014). Determination of hydrophilic-lipophilic balance value. *International Journal of Science and Research*, *3*(4), 573–575.

Garrett, T.G., Bhakoo, M., & Zhang, Z. (2008). Bacterial adhesion and biofilms on surfaces. *Progress in Natural Science*, *18*, 1049–1056.

Gautam, K.K., & Tyagi, V.K. (2006). Microbial surfactants: a review. *Journal of Oleo Science*, *55*, 155–166.

Gharaei-Fathabad, E. (2011). Biosurfactants in pharmaceutical industry: a mini review. *American Journal of Drug Discovery and Development*, *1*(1), 58–69.

Gong, G., Zheng, Z., Chen, H., Yuan, C., Wang, P., Yao, L., & Yu, Z. (2009). Enhanced production of surfactin by *Bacillus subtilis* E8 mutant obtained by ion beam implantation. *Food Technology and Biotechnology*, *47*(1), 27–31.

Gorin, P.A.J., Spencer, J.F.T., & Tulloch, A.P. (1961). Hydroxy fatty acid glycosides of sophorose from *Torulopsis magnoliae*. *Canadian Journal of Chemistry*, *39*(4), 846–855.

Guan, X.H., Shang, C., & Chen, G.H. (2006). ATR-FTIR investigation of the role of phenolic groups in the interaction of some NOM model compounds with aluminum hydroxide. *Chemosphere*, *65*(11), 2074–2081.

Güçlü-Üstündağ, Ö., & Mazza, G. (2007). Saponins: properties, applications and processing. *Critical Reviews in Food Science and Nutrition*, *47*(3), 231–258.

Guerra-Santos, L.H., Kappeli, O., & Fiechter, A. (1984). *Pseudomonas aeruginosa* biosurfactant production in continuous culture with glucose as carbon source. *Applied and Environmental Microbiology*, *48*, 301–305.

Gupta, P., Khanday, W.A., Majid, S.A., Kushwa, V., Tomar, S., & Tomar, R. (2013). Study of sorption of metal oxoanions from waste water on surfactant modified analog of laumontite. *Journal of Environmental Chemical Engineering*, *1*(3), 510–515. Retrieved 2020.

Hauser, G., & Karnovsky, M.L. (1957). Rhamnose and rhamnolipide biosynthesis by pseudomonas aeruginosa. *Journal of Biological Chemistry*, *224*(1), 91–105.

Hirata, Y., Ryu, M., Oda, Y., Igarashi, K., Nagatsuka, A., Furuta, T., & Sugiura, M. (2009). Novel characteristics of sophorolipids, yeast glycolipid biosurfactants, as biodegradable low-foaming surfactants. *Journal of Bioscience and Bioengineering*, *108*(2), 142–146.

Hirata, Y., Ryu, M., Oda, Y., Igarashi, K., Nagatsuka, A., Furuta, T., & Sugiura, M. (2009). Novel characteristics of sophorolipids, yeast glycolipid biosurfactants, as biodegradable low-foaming surfactants. *Journal of Bioscience and Bioengineering*, *108*(2), 142–146.

Hörmann, B., Müller, M.M., Syldatk, C., & Hausmann, R. (2010). Rhamnolipid production by *Burkholderia plantarii* DSM 9509T. *European Journal of Lipid Science and Technology, 112*(6), 674–680.

IEA. (2015). Energy amd climate change. *International Energy Agency*. Retrieved November 21, 2020 (https://www.arpa.veneto.it/temi-ambientali/energia/file-e-allegati/WEO2015SpecialReportonEnergyandClimateChange.pdf).

Iyer, A., Mody, K., & Jha, B. (2006). Emulsifying properties of a marine bacterial exopolysaccharide. *Enzyme and Microbial Technology, 38*, 220–222.

Jarvis, F.G., & Johnson, M.J. (1949). A glyco-lipide produced by *Pseudomonas aeruginosa*. *Journal of the American Chemical Society, 71*(12), 4124–4126.

Kim, H.S., Jeon, J.W., Kim, S.B., Oh, H.M., Kwon, T.J., & Yoon, B.D. (2002). Surface and physico-chemical properties of a glycolipid biosurfactant, mannosylerythritol lipid, from *Candida antarctica*. *Biotechnology Letters, 24*, 1637–1641.

Kłosowska-Chomiczewska, I.E., Mędrzycka, K., Hallmann, E., Karpenko, E., Pokynbroda, T., Macierzanka, A., & Jungnickel, C. (2017). Rhamnolipid CMC prediction. *Journal of Colloid and Interface Science, 488*, 10–19.

Kovalchuk, N.M., & Simmons, M.J.H. (2020). Surfactant-mediated wetting and spreading: recent advances and applications. *Current Opinion in Colloid and Interface Science*, 10.1016/j.cocis.2020.07.004.

Kretschmer, A., Bock, H., & Wagner, F. (1982). Chemical and physical characterization of interfacial active lipids from *Rhodococcus erythropolis* grown on n-alkanes. *Applied and Environmental Microbiology, 44*(4), 864–870.

Lima, T.M.S., Procopio, L.C., Brandao, F.D., Carvalho, A.M.X., Totola, M.R., & Borges, A.C. (2011). Biodegradability of bacterial surfactants. *Biodegradation, 22*, 585–592.

Lu, B., Miao, Y., Vigneron, P., Chagnault, V., Grand, E., Wadouachi, A., Postel, D., Pezron, I., Egles, C., & Vayssade, M. (2017). Measurement of cytotoxicity and irritancy potential of sugar-based surfactants on skin-related 3D models. *Toxicology In Vitro, 40*, 305–312.

Mahamallik, P., & Pal, A. (2017). Degradation of textile wastewater by modified photo-Fenton process: application of Co (II) adsorbed surfactant-modified alumina as heterogeneous catalyst. *Journal of Environmental Chemical Engineering, 5*(3), 2886–2893.

Marchant, R., & Banat, I.M. (2012). Biosurfactants: a sustainable replacement for chemical surfactants? *Biotechnology Letters, 34*, 1597–1605.

McInerney, M.J., Javaheri, M., & Nagle, D.P. Jr. (1990). Properties of the biosurfactant produced by *Bacillus licheniformis* strain JF-2. *Journal of Industrial Microbiology and Biotechnology, 5*, 95–101.

Meulenberg, E.P. (2009). Phenolics: occurrence and immunochemical detection in environment and food. *Molecules, 14*(1), 439–473.

Miller, R.M. (1995). Biosurfactant-facilitated remediation of metal-contaminated soils. *Environmental Health Perspectives, 103*, 59–62.

Mohan, P.K., Nakhla, G., & Yanful, E.K. (2006). Biodegradability of surfactants under aerobic, anoxic, and anaerobic conditions. *Journal of Environmental Engineering, 132*(2), 279–283.

Morais, A.R.C., & Bogel-Lukasik, R. (2013). Green chemistry and the biorefinery concept. *Sustainable Chemical Processes, 1*(18), 56–69.

Mukherjee, A.K. (2007). Potential application of cyclic lipopeptide biosurfactants produced by *Bacillus subtilis* strains in laundry detergent formulations. *Letters in Applied Microbiology, 47*, 330–335.

Mukherjee, A.K., & Das, K. (2010). Microbial surfactants and their potential applications: an overview. pp. 54–64 in *Biosurfactants. Advances in Experimental Medicine and Biology*, edited by R. Sen. New York: Springer.

Mukherjee, S., Das, P., & Sen, R. (2006). Towards commercial production of microbial surfactants (Review). *Trends in Biotechnology, 24*(11), 509–515.

Mulligan, C.N. (2005). Environmental applications for biosurfactants. *Environmental Pollution, 133*, 183–198.

Mulligan, C.N. (2009). Recent advances in the environmental applications of biosurfactants. *Current Opinion in Colloid & Interface Science, 14*, 372–378.

Mulligan, C.N., Sharma, S.K., & Mudhoo, A. (2014). *Biosurfactants: Research trends and applications*. Boca Raton, Florida: CRC Press (Taylor & Francis Group).

Nitschke, M., & Pastore, G.M. (2006). Production and properties of a surfactant obtained from *Bacillus subtilis* grown on cassava wastewater. *Bioresource Technology, 97*(2), 336–341.

Oelschlaeger, T.A. (2010). Mechanisms of probiotic actions—a review. *International Journal of Medical Microbiology*, *300*(1), 57–62.

Pacwa-Płociniczak, M., Płaza, G.A., Piotrowska-Seget, Z., & Cameotra, S.S. (2011). Environmental applications of biosurfactants: recent advances. *International Journal of Molecular Sciences*, *12*, 633–654.

Pagilla, K.R., Sood, A., & Kim, H. (2002). Gordonia (Nocardia) amarae foaming due to biosurfactant production. *Water Science & Technology*, *46*(1–2), 519–524.

Patel, M. (2004). Surfactants based on renewable raw materials: carbon dioxide reduction potential and policies and measures for the European Union. *Journal of Industrial Ecology*, *7*(3–4), 47–62.

Perfumo, A., Smyth, T., Marchant, R., & Banat, I. (2010). Production and roles of biosurfactants and bioemulsifiers in accessing hydrophobic substrates. pp. 1502–1512. in *Handbook of Hydrocarbon and Lipid Microbiology*, edited by K.N. Timmis. Berlin, Heidelberg: Springer. 10.1007/978-3-540-775 87-4_103.

Piispanen, P.S., Persson, M., Claesson, P., & Norin, T. (2004). Surface properties of surfactants derived from natural products. part 1: syntheses and structure/property relationships—solubility and emulsification. *Journal of Surfactants and Detergents*, *7*(2), 147–159.

Płociniczak, M.P., Płaza, G.A., Piotrowska-Seget, Z., & Cameotra, S.S. (2011). Environmental applications of biosurfactants: recent advances . *International Journal of Molecular Sciences*, *12*, 633–654.

Poremba, K., Gunkel, W., Lang, S., & Wagner, F. (1991). Toxicity testing of synthetic and biogenic surfactants on marine microorganisms. *Environmental Toxicology and Water Quality*, *6*(2), 157–163.

Pornsunthorntawee, O., Wongpanit, P., Chavadej, S., Abe, M., & Rujiravanit, R. (2008). Structural and physicochemical characterization of crude biosurfactant produced by *Pseudomonas aeruginosa* SP4 isolated from petroleum-contaminated soil. *Bioresource Technology*, *99*(6), 1589–1595.

Rahman, P.K., & Gakpe, E. (2008). Production, characterisation and applications of biosurfactants-Review *Biotechnology*, *7*(2), 360–370.

Rajkhowa, S., Mahiuddin, S., & Ismail, K. (2017). An assessment of the aggregation and adsorption behavior of the sodium dodecylsulfate–cetyltrimethylammonium bromide mixed surfactant system in aqueous medium. *Journal of Solution Chemistry*, *46*, 11–24.

Robert, M., Mercade, M.E., Bosch, M.P., Parra, J.L., Espuny, M.J., Manresa, M.A., & Guinea, J. (1989). Effect of the carbon source on biosurfactant production by *Pseudomonas aeruginosa* 44T1. *Biotechnology Letters*, *11*(12), 871–874.

Rosenberg, E., & Ron, E.Z. (1999). High- and low-molecular-mass microbial surfactants. *Applied Microbiology and Biotechnology*, *52*(2), 154–162.

Rosenberg, E., Rubinovitz, C., Gottlieb, A., Rosenhak, S., & Ron, E.Z. (1988). Production of biodispersan by *Acinetobacter calcoaceticus*, A2. *Applied and Environmental Microbiology*, *54*(2), 317–322.

Rufino, R.D., De Luna, J.M., De Campos Takaki, G.M., & Sarubbo, L.A. (2014). Characterization and properties of the biosurfactant produced by *Candida lipolytica* UCP 0988. *Electronic Journal of Biotechnology*, *17*, 34–38.

Santa Anna, L.M., Sebastian, G.V., Pereira, N., Alves, T.L.M., Menezes, E.P., & Freire, D.M.G. (2001). Production of biosurfactant from a new and promising strain of *Pseudomonas aeruginosa*, PA1. *Applied Biochemistry and Biotechnology*, *91*(1–9), 459–467.

Santos, D.K.F., Rufino, R.D., Luna, J.M., Santos, V.A., & Sarubbo, L.A. (2016). Biosurfactants: multifunctional biomolecules of the 21st century. *International Journal of Molecular Sciences*, *17*(3), 401.

Satpute, S.K., Płaza, G.A., & Banpurkar, A.G. (2017). Biosurfactants' production from renewable natural resources: example of innovative and smart technology in circular bioeconomy. *Industrial Engineering and Management Systems*, *25*(1), 46–54.

Shakeri, F., Babavalian, H., Amoozegar, M.A., Ahmadzadeh, Z., Zuhuriyanizadi, S., & Afsharian, S.P. (2021). Production and application of biosurfactants in biotechnology. *Biointerface Research in Applied Chemistry*, *11*(3), 10446–10460. 10.33263/BRIAC113.1044610460.

Sharma, D. (2016). Classification and properties of biosurfactants. pp. 21–42 in *Biosurfactants in Food*, edited byR.W. Hartel. Switzerland: Springer International Publishing.

Sharma, D., & Malik, A. (2012). Incidence and prevalence of antimicrobial resistant Vibrio cholerae from dairy farms . *African Journal of Microbiology Research*, *6*(25), 5331–5334.

Sharma, D., & Saharan, B.S. (2016). Functional characterization of biomedical potential of biosurfactant produced by *Lactobacillus helveticus*. *Biotechnological Reports*, *11*, 27–35.

Siñeriz, F., Hommel, R.K., & Kleber, H.P. (2001). Production of biosurfactants. *Encyclopedia of Life Support Systems*Oxford: Eolls Publishers.

Singh, P. & Cameotra, S.S. (2004). Potential applications of microbial surfactants in biomedical sciences. *Trends in Biotechnology*, 22, 142–146.

Smith, M.L., Gandolfi, S., Coshall, P.M., & Rahman, P.K. (2020). Biosurfactants: a Covid-19 perspective. *Frontiers in Microbiology*, 11, 1341, 10.3389/fmicb.2020.01341.

Sobrinho, H.B., Luna, J.M., Rufino, R.D., Porto, A.L.F., & Sarubbo, L.A. (2013). Biosurfactants: classification, properties and environmental applications. *Record Developmental Biotechnology*, 11(14), 1–29.

Sobrinho, H.B., Luna, J.M., Rufino, R.D., Porto, A.L.F., & Sarubbo, L.A. (2013). Biosurfactants: classification, properties and environmental applications. *Record Developmental Biotechnology*, 11(14), 1–29.

Somoza-Coutiño, G., Wong-Villarreal, A., Blanco-González, C., Pérez-Sariñana, B., Mora-Herrera, M., Mora-Herrera, S.I., Rivas-Caceres, R.R., De la Portilla-López, N., Lugo, J., Vaca-Paulín, R., & Del Águila, P. (2020). A bacterial strain of *Pseudomonas aeruginosa* B0406 pathogen opportunistic, produce a biosurfactant with tolerance to changes of pH, salinity and temperature. *Microbial Pathogenesis*, 139, 103869.

Sparling, G.P., & Vaughan, D. (1981). Soil phenolic-acids and microbes in relation to plant-growth. *Journal of the Science of Food and Agriculture*, 32(6), 625–626.

Star-COLIBRI. (2011). European biorefinery joint strategic research roadmap. *Strategic Targets for 2020 – Collaboration Initiative on Biorefinerie*. Retrieved November 26, 2020 (http://beaconwales.org/uploads/resources/Vision_2020_-_European_Biorefinery_Joint_Strategic_Research_Roadmap.pdf).

Stevenson, F.J. (1994). *Humus Chemistry: Genesis, Composition, Reactions*. 2nd ed. John Wiley & Sons.

Syldatk, C., Lang, S., Matulovic, U., & Wagner, F.Z. (1985a). Production of four interfacial active rhamnolipids from n-alkanes or glycerol by resting cells of *Pseudomonas* sp. DSM 2874. *Zeitschrift für Naturforschung*, 40C(1–2), 61–67.

Syldatk, C., Lang, S., & Wagner, F. (1985b). Chemical and physical characerization of four interfacial-active rhamnolipids from *Pseudomonas* sp. DSM 2874 grown on n-alkanes. *Zeitschrift für Naturforschung*, 40C, 51–60.

Tiso, T., Thies, S., Müller, M., Tsvetanova, L., Carraresi, L., Bröring, S., Jaeger, K.-E., & Blank, L.M. (2017). Rhamnolipids: production, performance, and application. pp. 587–622 in *Consequences of Microbial Interactions with Hydrocarbons, Oils, and Lipids: Production of Fuels and Chemicals. Handbook of Hydrocarbon and Lipid Microbiology*, edited by S. Lee. Springer, Cham.

UNFCCC. (1998). Kyoto protocol to the united nations framework convention on climate change. *UNFCCC*. Retrieved November 21, 2020 (https://unfccc.int/resource/docs/convkp/kpeng.pdf).

Varjani, S.J. & Upasani, V.N. (2017). Critical review on biosurfactant analysis, purification and characterization using rhamnolipid as a model biosurfactant. *Bioresource Technology* 232, 389–397.

Velikonja, J., & Kosaric, N. (1993). Biosurfactants in food applications. pp. 419–448 in *Biosurfactants: Production, Properties, Applications*, edited byN. Kosaric & F.V. Sukan. New York, NY, USA: CRC Press.

Vijayakumar, S., & Saravanan, V. (2015). Biosurfactants-types, sources and applications. *Research Journal of Microbiology*, 10(5), 181–192.

Whang, L.M., Liu, P.W.G., Ma, C.C., & Cheng, S.S. (2008). Application of biosurfactants, rhamnolipid, and surfactin, for enhanced biodegradation of diesel-contaminated water and soil. *Journal of Hazardous Materials*, 151(1), 155–163.

Xia, W., Dong, H., Yu, L., & Yu, D. (2011). Comparative study of biosurfactant produced by microorganisms isolated from formation water of petroleum reservoir. *Colloids and Surfaces A: Physicochemical and Engineering Aspects*, 392, 124–130.

Xiaohong, P., Xinhua, Z., & Lixiang, Z. (2009). Effect of biosurfactant on the sorption of phenanthrene onto original and H_2O_2-treated soils. *Journal of Environmental Sciences*, 21(10), 1378–1385.

Yakimov, M.M., Fredrickson, H.L., & Timmis, K.N. (1996). Effect of heterogeneity of hydrophobic moieties on surface activity of lichenysin A, a lipopeptide biosurfactant from *Bacillus licheniformis* BAS50. *Biotechnology and Applied Biochemistry*, 23, 13–18.

Index

Note: *Italicized* page numbers refer to figures, **bold** page numbers refer to tables.

A

Acanthus ilicifolius, 256
Achillea millefolium, 165
Achromobacter, 173
acid mine drainage, 2
acid mine drainage (AMD), 2
acid-catalysed hydrolysis, 353
acidic sophorolipid, *381*
acidification, 369–370
Acinetobacter, 387, 388, 390
activated sludge process, 8
adsorption, 95, **223**, 370
aerobic composting, 15–17
    composition, 16
    moisture content, 16
    nutrients, 16
    oxygen and aeration, 15–16
    particle size, 16
    pH, 16
    porosity, 16
    temperature, 16–17
    time, 16–17, **17**
aerogels, 302–303, *304*
Ageratina altissima, **334**
agricultural wastes, 252–253
agriculture wastewater, 135
air pollution, **160**
    expedients for reducing, 159–161
    indoor, 168–173
    origin, 156–158, 158–159
    outdoor, 163–168
    overview, 156–158
    particulate matters, 158–159
    trace gases, 159
airborne particulate matter, 163–164
airflow rate, 13
algae-based water treatment, 135–137
    advantages and disadvantages of, 137
    algae with metal remediation properties, **137**
    algal growth requirement, 135–136
    bioethanol production, 141–142
    and biofuel, 136–137
alginates, 297, **313**, **350**. *See also* biopolymers
alkali-catalysed polyester hydrolysis, 353
Alkaline chlorination, 358
Alpinia calcarat, 257
Alpinianigra, **333**
Amaranthus hybridus, 6
Amaranthus spinosus, 6
anaerobic composting, 17–18, 18

    dry fermentation, 18
    wet fermentation, 18
anaerobic digestion process, 138–139
anthropogenic contaminants, 28–29
    personal care compounds, 29
    pesticides, 28
    pharmaceuticals, 28–29
    by-products of water treatment, 29
antiadhesive activity, 384
aquatic macrophytes, 215–216
aquatic plants biosorbents, 71–82
    for inorganic pollutants, 79
    macrophytes, 74–75, 79–80
    for organic pollutants, 79
    overview, 73
    phytoremediation, 75
    plant types
        duckweed, 78–79
        Eichhornia crassipes, 78
        Pistia stratiotes L., 76–77
        Salvinia auriculatais, 77–78
    role of, 76
    substantial metals in plants and animals, 72–73
aqueous biphasic systems (ABSs), 266
Arabidopsis halleri, 200, 241
Arthrobacter paraffineus, 391
Arthrobotrys conoides, 393
arthrofactin, **382**
Ascophyllum, 297
Ascophyllum nodosum, 58
Aspergillus, 387
Aspergillus ustus, 388
Aster gymnocephalus, 165
attenuation, 112
autotrophic microorganisms, 282
Azalea indica, 173
Azolla caroliniana, 79, 216, 244
Azolla filiculoides, 61, 256
Azolla pinnata, 61

B

Bacillus cereus, 172
Bacillus licheniformis, 387, 391, 395
Bacillus macerans, 298
Bacillus marisflavi YCISMN 5, **336**
Bacillus megaterium, 298
Bacillus mucilaginous, 289
Bacillus subtilis, 387, 395
bacterial biosurfactants, 388
Berteroa incana, 165

bioaugmentation, 40, 99–100, 113–114, 127
   for composting, 9–10
   for hydrocarbon remediation, 7–8
   process and limitations, **102**
   for treatment of industrial wastewater, 8, **9**
bioavailability, 102
bio-based/biodegradable polymer, 343–355
   advantages of, 345–346
   applications, 347–349
   biodegradation mechanisms, 353–354
      acid-catalysed hydrolysis, 353
      alkali-catalysed polyester hydrolysis, 353
      degradation of biodegradable polymers, 353
      enzymatic degradation, 353–354
      hydrolytic degradation, 353
   challenges and limitations, 354
   characteristics of, 347
   classification, 346
   future of, 354–355
   natural, 347, **350**
   overview, 344
   pollutant removal using, 351–353, **352**
   production capacities, *345*
   sources of, 346
   synthetic, 347, **348–349**
   versus synthetic conventional polymers, 344
   types of, 347
   for wastewater treatment, 346–347
      bio-nanocomposites, 349–350
      cellulose nanofibers, 350
      chitosan biopolymers, 349
      foam membranes, 351
      nano crystalline cellulose, 351
      polyhydroxyalkanoates, 350
biochar, 219–220
biochemical oxygen demand (BOD), 268
biodegradation mechanisms, 353–354
   acid-catalysed hydrolysis, 353
   alkali-catalysed polyester hydrolysis, 353
   degradation of biodegradable polymers, 353
   enzymatic degradation, 353–354
   hydrolytic degradation, 353
biodiesel, 139–140
   integrated biogas–biodiesel production approach, 147–148
   process efficiency, 146
bioelectricity, 144
bioethanol
   preparation process, 141
   process efficiency, 146
   production using wastewater, 140–143
      from food industry, 142
      from softdrink industry, 142–143
      using algae grown in wastewater, 141–142
biofiltration, 173–176
biofuel, 265
   and algal wastewater treatment, 136–137
   biodiesel, 139–140
   bioethanol, 140–143
   biogas, 137–139
   production using wastewater, 137–143
biogas, 137–139
   anaerobic digestion process, 138–139
   calorific value, 139
   integrated biogas–biodiesel production approach, 147–148
   process efficiency, 145–146
   production process, 138
biohydrogen, 143–144
   multi-stage bioreactors, 148
   process efficiency, 146–147
bioinventing, 115–116
biological testing, 30–31
biomineralisation, 37. *See also* bioremediation
bio-nanocomposites, 349–350. *See also* bio-based/biodegradable polymer
biopiles, 33, 114–115. *See also* ex-situ bioremediation
bioplastics, 145
biopolymers, 295–313. *See also* bio-based/biodegradable polymer
   in adsorption, 307–311
      gas adsorption, 310–311
      removal of dyes, 307–308
      removal of metals, 308–309
      removal of micropollutants, 309–310
   alginate, 297
   cellulose, 297
   characterisation of, 298–302
      Fourier transform infrared spectroscopy, 300, *301*
      nuclear magnetic resonance spectroscopy, 301, *302*
      optical microscopy, 303, *304*
      scanning electron microscopy, 299, *299*
      thermal gravimetric analysis, 301–302, *303*
      X-ray diffraction, 299–300, *300*
   chitin, 297
   chitosan, 297
   cyclodextrin, 298
   dextran, 298
   from food waste, 311–313, *312*
   hybrid systems of, 306
   overview, 295–296
   in photocatalysis, 311
   processing of, 302–306
      aerogels, 302–303, *304*
      hydrogels, 303–305, *305*
      nano biopolymers/nanofibrilation, 305–306, *306*
   production of, 144–145
      L-lactic acid, 144–145
      polyhydroxy alkanoates, 145
   starch, 298
   types of, *296*
bioreactors, 32–33. *See also* ex-situ bioremediation
   multi-stage, 148
   types of, 115
bioreduction, 327–329
bioremediation, 25–43, 98–103
   advances in biotechnology for, 36–40
      biomineralisation, 37
      biosorption, 36
      biostimulation, 38

cyanoremediation, 39–40
dendroremediation, 37
genoremediation, 39
mycoremediation, 38–39
rhizoremediation, 37–38
applications of, 40–41
bioavailability, 102
contaminants, 27–31
defined, 98
environmental factors, 102
ex-situ, 32–33, 101, **102**, 111–112, 125
   biopiles, 33, 114–115
   bioreactors, 32–33, 115, 148
   land farming, 33, 112–113
factors of, 35
   environmental, 35
   microbial population, 35
future of, 41–42
of hydrocarbon pollutants, 391–392
nanotechnology in, 128–129
overview, 24–26
process of, *100*
in-situ, 34–35, 99–101, 112, 125
   bioaugmentation, 99–100
   bioslurping, 34–35
   biosparging, 34, 101
   biostimulation, 7–8, 38, 101, **102**, 114–127
   bioventing, 10–13, 34, 40, 100–101
strategies, 31–35
technologies and combination of other technologies, 125–129
using biosurfactants, 391–392
bioslurping, 34–35, 116. *See also* in-situ bioremediation
biosorption, 36, 116. *See also* bioremediation
biosparging, 13–15, 34, 41, 101, 116
   advantages, 14
   applications, 14
   disadvantages, 14–15
   process and limitations, **102**
biostimulation, 38, 101, 114, 127. *See also* in-situ bioremediation
   for hydrocarbon remediation, 7–8
   process and limitations, **102**
biosurfactants, 127, 377–398
   adsorption of, *380*
   advantages of, 396
   applications of, 391–395
      bioremediation of hydrocarbon pollutants, 391–392
      carbon dioxide reduction, 393–394
      metal bioremediation, 392–393, *394*
      in ore and mineral industry, 394–395
   bacterial, 388
   classification of, 384–388, **385**
      fatty acids, 387
      glycolipids, 386
      lichenysin, 387
      lipopeptides, 386–387
      lipoproteins, 386–387
      neutral lipids, 387
      particulate biosurfactants, 388
      phospholipids, 387
      polymeric biosurfactants, 387–388
      surfactin, 387
   defined, 377
   disadvantages of, 396
   and environmental sustainability, 397
   fungal, 388
   future trends in, 396–397
   high-molecular-weight, 385
   low-molecular-weight, 384
   microbial origins, **385**
   non-pathogenic/probiotic, 388–389
   overview, 377–378
   properties of, 379–384
      antiadhesive activity, 384
      biodegradability, 383
      emulsification and de-emulsification, 383
      low toxicity, 383
      surface activity, 382
      tolerance to temperature, ionic strength, and pH, 384
   raw materials for production of, 389
   sources of, 388–389
   structural representation of moiety, *379*
   structures of, *381*
   synthesis of, 389–391
      carbon substrates, 390
      environmental factors, 389–390
      nitrogen substrates, 390–391
      nutritional factors, 390
   versus synthetic surfactants, **382**
bioventing, 10–13, 34, 40, 100–101. *See also* in-situ bioremediation
   airflow rate, 13
   designing bioventing system, 13
   nutrients, 13
   process and limitations, **102**
   soil moisture, 13
   typical system, *12*
bisphenol S (BPA), 311
Brassica juncea, 2–5, 243
Brassica rapa, 6
brilliant green (BG), 311
brownfield, 2–5
Bryophyta, 74
BTEX, 14–15
Burkholderia plantarii, 386
by-products of water treatment, 29

C

Cabomba piauhyensis, 60
caffeine, 29
calcium alginate, 307
calcium carbonate, 289
Canavalia ensiformes, 256
Candida acidithermophilium, 141
Candida antartica, 383, 395
Candida bombicola, 386, 388, 395
Candida ishiwadae, 388
Candida lipolytica, 388

Candida torulopsis, 386
carbon biosequestration, 281–292
    advantages of, **285**
    carbon dioxide-fixing microorganisms, **286**, 286–287
    designed, 284–285, **290**
    Escherichia coli in, 287–289
    future research, 291
    limitations, 289–291
    limitations of, 285, **285**
    natural, 282–284, **283**
    overview, 281–282
carbon dioxide reduction, 393–394
carbon dioxide-fixing microorganisms, 282, **286**, 286–287
carbon nanotubes (CNTs), 126
carbonic anhydrate, 289
Carex pendula, 61
Carpobrotus acinaciformis, **334**
Catharanthus roseus, 241, **333**
Caulerpa lentillifera, 57
cellulose, 297, **313**, **350**. *See also* biopolymers
cellulose nanofibers, 350. *See also* bio-based/biodegradable polymer
Ceratophyllum demersum, 60, 243
cetyltrimethylammonium bromide, **382**
Chamaedorea elegans, 170
Chara vulgaris, 79
chelators, 219
chemical analysis, 30–31
chemical oxygen demand (COD), 36, 57, 77, 138, 144–145, 266, 268
chemical precipitation, **223**
chitin, 297, **350**
chitosan, 297, **313**, 349. *See also* bio-based/biodegradable polymer; biopolymers
Chlorella, 144
Chlorella vulgaris, 55
chlorine, 29
chlorobenzene, 14
Chlorophyta, 74
Chlorophytum comosum, 170
Chlorophytum elatum, 171
chromium, 79
Chrysopogon zizanioides, 2–5
ciprofloxacin, 311, 312
Cladophora, 57
Cladosporium oxysponum, **335**
Clean Water Act, 8
Clostridium saccharoperbutylacetonicum, 142
composting, 15–18, 112. *See also* in-situ bioremediation
    aerobic, 15–17
    anaerobic, 17–18
    bioaugmentation for, 9–10
    conditions for, 15
    process and limitations, **102**
Conference of the Parties (COP21), 393
conjugation, 98
contaminants, 27–31
    heavy metals, 199, **199**
    impact on the ecosystem, 29
    sources of, 198
    sustainable management of, 253–254
    types of, 27–29
        anthropogenic, 28–29
        natural, 28
contaminated soil, ecological risk assessment of, 29–31
contamination, 251–252
    sources of, 252–254
        agricultural wastes, 252–253
        industrial wastes, 253
        natural sources, 252
copper oxide nanoparticles, **333**, **334**
critical micelle concentration (CMC), *380*, 380
Cryptococcus tepidarius, 141
Cryptococcus terreus, 393
culture-based approach, 113
cyanide, 357–372
    biological method of removal, 370
    chemical methods of removal, 369–370
        acidification with volatilisation, 369–370
        adsorption, 370, **371–372**
        solvent extraction, 370
    clinical manifestation of poisoning, 358–359, *359*
    contamination
        in air, 363, *364*
        in soil, *365*, 365–366
        in water, 363, *364*
    distribution in liver and its elimination, *362*
    distribution in several parts of the body, *362*
    exposure to, 359–360
    fates in the body, 361, *361*
    foods containing, 366
    forms of, *358*
    industrial utilisation of, 363
    kinetics and isotherms, 371
    overview, 357–358
    physiochemical analysis of, 360, *361*
    primary metabolic pathway of, *367*
    recovery methods, 368
    remediation by physical method, 368–369
        dilution method, 368
        electrowinning, 369
        hydrolysis followed by distillation, 369
        membranes, 369
        photolysis, 368
    removal from soil, air, and water, 371, **371–372**
    removal methods, 366–371
    sources of, 363
    sources of cyanide poisoning, **366**
    symptoms of exposure to, **360**
Cyanobacteria, 74
cyanoremediation, 39–40. *See also* bioremediation
cyclodextrin, 298, **313**. *See also* biopolymers
Cymbopogon citrates, 257
Cyphomandra betacea, **334**
cytochrome oxidase, 357

D

decision making, 30–31
de-emulsification, 383

Dendrobium phalaenopsis, 171
dendroremediation, 37. *See also* bioremediation
detergents, 378
dextran, 298, **313**. *See also* biopolymers
Dichomitus squalens, 271
diclofenac, 310
Dieffenbachia compacta, 170
diethylene triamine penta–acetic acid (DTPA), 219
diffusion, 95
dilution method, 368
dirhamnolipid, *381*
dissolved oxygen (DO), 14
distillation, 369
Dracaena deremensis, 170
Dracaena sanderiana, 170
dry fermentation, 18
duckweed, 78–79
Dunaliella, 55
dyes, removal of, 307–308
Dypsis lutescens, 170

E

Echinochloa crusgalli, 221
Echinodorus amazonicus, **80**
ecological contaminants, 80
ecological risk assessment (ERA), 29–31
    biological testing, 30–31
    chemical analysis, 30–31
    decision making, 30–31
    ecological survey, 30
    ecotoxicity testing, 30
    steps in, *30*
ecological survey, 30
ecotoxicity testing, 30
Egeria densa, 60
Eichhornia crassipes, 2, 60, 78, 79, 80, **80**, 215, 244
electro deionising (EDI), 75–76
electro kinetic-enhanced phytoremediation, 221–222
electro-coagulation (EC), **223**
electro-deposition (ED), **223**
electro-floatation (EF), **223**
electrospun cellulose fibers (ECFs), 303
electrowinning, 369
Elodea densa, 60
emulsan, *381*
emulsification, 383
Enterobacter cloaceae, 395
environmental remediation
    bioremediation, 98–103
    conventional versus biobased techniques, **91**
    life forms for, *91*, **91**, 91–92
    overview, 89–90
    phytoremediation, 93–98
    sources and impacts of contaminants, **90**
enzymatic degradation, 353–354
Equisetum ramosisti, 244
Erechtites hieracifolia, 168
Erwinia herbicola, **336**
Escherichia coli, 282, 287–289

Escherichia coli K12, **335**
ethanol
    bioethanol production from waatewater, 140–143
    Brazil's production of, 265
    octane rating, 141
    production process, *268*
ethylene diamine tetra–acetic acid (EDTA), 137, 219, 309
ethylene glycol tetra–acetic acid (AGTA), 219
Eucalyptus globulus, 168, 222
Eucalyptus grandis, 168
Euphorbia milii, 173
exogenous phytohormones, 221
ex-situ bioremediation, 32–33, 101, 111–112, 125. *See also* in-situ bioremediation
    biopiles, 33, 114–115
    bioreactors, 32–33, 115, 148
    land farming, 33, 112–113
    process and limitations, **102**
extracellular synthesis, *326*

F

fatty acids, 387. *See also* biosurfactants
$Fe_2O_3@SiO_2$ nanoparticles, **333**
Fe_3O_4 nanoparticles, **333**
fermentation, 18
ferrous oxide nanoparticles, **334**
Ficus benjamina, **333**
Flaveria trinervia, 165
foam membranes, 351. *See also* bio-based/biodegradable polymer
food waste, 311
Fourier transform infrared spectroscopy (FTIR), 300, *301*
Fucus spiralis, 58
fulvic acid, 29
fungal biosurfactants, 388

G

Ganoderma lucidum, 271
genomes, 126
genoremediation, 39. *See also* bioremediation
ghatti, 311
Gluconacetobacter xylinium, **333**
glycidyl methacrylate (GMA), 307
glycolipids, 386. *See also* biosurfactants
glycolysis, 288
gold nanoparticles, **333**, **335**
Gordonia amarae, 390
Gracilaria changii, 57
Gracilaria edulis, 57
Gracilaria lemaneiformis, 57
granular activated carbon (GAC), 10–13
graphene oxide, 309
green synthesis
    advantages of, *324*
    via plants and microorganisms, 325
    vs. chemical synthesis, 324
greenwall, 173–176, **175–176**

H

heavy metal transporting ATPases (HMAs), 200
heavy metals, 72
    bioremediation, 392–393, *394*
    cellular mechanism for detoxification and tolerance of, 77
    effects of, 199, **199**
    effects on human health, **216**
    factors affecting the mobility, absorption, and accumulation in plants, 216–217
    nanophytoremediation of, 258
    overview, 208–209
    phytoremediation of, 5–6, 215–216
        biochar, 219–220
        chelators, 219
        exogenous phytohormones, 221
        microbes, 221
        nanoparticles, 222
        transgenic plants, **220**, 220–221
    removal of, 75–76, 308–309
    removal using biopolymers, 351–353
    sources of, 199, **215**
    trophic transfer of, *73*
    uptake and translocation mechanisms, 200, *201*
Helianthus annuus, 243, 257
heterotrophic microorganisms, 282
high-molecular-weight biosurfactants, 385
household wastewater, 135
humic acid, applications of, in ore and mineral industry, 394–395
humid acid, 29
Hybanthus austrocaledonicus, 240
hybrid approaches, 117
hybrid membrane bioreactors, 118
Hydrilla verticillata, 60, 243
hydrocarbon pollutants, 391–392
hydrocarbon remediation, 7–8
hydrogels, 303–305, *305*
Hydrogen cyanide, 358
hydrolysis, 369
hydrolytic degradation, 353
hydrophilic-lipophilic balance (HLB) value, 382
hydrophobic organic compounds (HOCs), 392
hydroxyl radicals, 322
hyper-accumulator plants, 240–242

I

ibuprofen, 310
Impatiens balsamina, 258
indoor air pollution, 168–173
induced phytoremediation, 212
Industrial Emissions Directive (IED), 8
industrial wastes, 253
industrial wastewater, 8, 135
inorganic air contaminants, 164–168
in-situ bioremediation, 34–35, 99–101, 112, 125.
    See also ex-situ bioremediation

bioaugmentation, 99–100, **102**
bioslurping, 34–35, 116
biosparging, 34, 101, **102**
biostimulation, 7–8, 38, 101, **102**, 114–127
bioventing, 10–13, 34, 40, 100–101, **102**
composting, 15–18, 112
intracellular synthesis, *326*
intrinsic attenuation, 112
ion exchange, **223**
Ipomoea aquatica, 60
Ipomoea batatas, 171
Ipomonea aquatica, **80**
Isoetes taiwanensis, **80**

J

Jatropha curcas, **334**
Juncus effusus, 244
Juniperus chinensis, 170

K

Kappaphycus alvarezii, 58
ketoprofen, *310*
Klebsiella pneumoniae, 298
Kluveromyces, 141
Kummerowia striata, 221
Kyoto Protocol, 393

L

laccase, 269–270
    applications, 273–274
    indicators used in production, **271**
    production, 270–272
    purification, **272**, 272–273
    three-dimensional structure of, 269, *270*
lactonic sophorolipid, *381*
Laminaria, 297
Laminaria hyperborea, 58
land farming, 33, 41, 112–113. *See also* ex-situ bioremediation
Landoltia, 78
Leersia hexandra, 244
Lemna, 78
Lemna gibba, 60, 79
Lemna minor, 79, **80**, 215, 216, 243
Leuconostoc lactis, **335**
Leuconostoc mesenteroides, 298
lichenysin, **382**, 387. *See also* biosurfactants
lipopeptides, 386–387. *See also* biosurfactants
lipoproteins, 386–387. *See also* biosurfactants
Listeria monocytogenes, 384
L-lactic acid, 145
locust bean gum, 307
Lolium perenne, 258
Lolium perenne L., 5–6
low-molecular-weight biosurfactants, 384
lxeris denticulate, 221

M

macroalgae, 57–58
Macrocystis, 297
macrophytes, 74–75, 79–80
Magnolia kobus, 168
membrane bioreactors, 118
membrane filtration, **223–224**
membrane technologies, 118
membranes, 369
metabolomics, 126
Metal cyanide, 358
metal nanoparticles, **328**
metal transporter proteins (MTPs), 200
metatranscriptomics, 113
microalgae, 55–56
microbes, in phytoremediation of heavy metals, 221
microbial electrochemical technologies, 116–117
microbial fuel cell (MFC), 144
microbial remediation, 52–54
microbially enhanced oil recovery (MEOR), 389–390
microorganisms, 35
micropollutants, 309–310
monoliths, 117–118
monorhamnolipid, *381*
multi-stage bioreactors, 148
multi-walled carbon nanotubes (MWCNTs), 307
Musa balbisiana, **333**
mycoremediation, 38–39, 54–55. *See also* bioremediation
Myriopyllum spicatum, 60, 79

N

NADH, 288
NADH-dependent reductase, 329
Najas marina, 215
nano biopolymers/nanofibrilation, 305–306, *306*
nano crystalline cellulose, 351. *See also* bio-based/biodegradable polymer
nanobioremediation, 119
nanocellulose-metal oxides, **352**
nanofibrilation, 305–306, *306*
nanoparticle synthesis, 321–336
 applications of, 332, **333–336**
 bottom-up approach, *323*, 323–324
 green synthesis, 324–325
 intracellular and extracellular, *326*
 overview, 321–323
 top-down approach, *323*, 323–324
 using microorganisms, 325
  mechanisms of, 326–329
   mediated reduction by NADH-dependent reductase, 329
   mediated reduction by nitrate/nitrite reductase, 329
   metal nanoparticles synthesized, **328**
   oxidoreductase enzymes for bioreduction, 327–329
  via different plant parts, 329–332
   advantages of, *329*
   mechanisms of, 331–332, *332*
   metal nanoparticles synthesized, **330–331**
nanoparticles, 251–259
 applications of, *322*
 list of, **259**
nanophytoremediation, 251–259
 contaminants, 251–255
 defined, 255
 of organic contaminants, 257
 strategies, 255–256, **256**
 of toxic heavy metals, 258
nanoproxen, 310
nanoremediation, 255
nanotechnology, 126, 128–129, 254–255
natural attenuation, 112
natural contaminants, 28
naturally resistant associated macrophage proteins (NRAMPs), 200
nature-based solutions (NBSs) seems, 3
Neisseria gonorrheae, 289
neutral lipids, 387. *See also* biosurfactants
Ni@Fe$_3$O$_4$ nanoparticles, **334**
nicotinamide adenine dinucleotide (NAD$^+$), 311
nicotine, 29
nitrate reductase, 329
nitrilotriacetic acid (NTA), 219
nitrite reductase, 329
N-nitrosodimethylamine (NDMA), 29
Nocardia, 391
Noccaea caerulescens, 241
non-aqueous phase liquids (NAPLs), 13
non-pathogenic/probiotic biosurfactants, 388–389
nuclear magnetic resonance (NMR) spectroscopy, 301, *302*
nutrients, 13

O

Ocimum sanctum, 257
oil contamination, 6–7
omics-based approach, 113
optical microscopy, 303, *304*
organic pollutants, photocatalytic degradation of, 352
outdoor air pollution, 163–168
 airborne particulate matter, 163–164
 inorganic air contaminants, 164–168
 volatile organic compounds, 163–164
oxidoreductase, 327–329

P

Padina, 58
palladium nanoparticles, **335**
Panicum maximum, 257
particulate biosurfactants, 388. *See also* biosurfactants
particulate matters, 158–159
Penicillium vermiculatum, 393
permeable reactive barrier, 117
persistent organic pollutants (POPs), 392
personal care compounds, 29
pesticides, 28
Phaeocystis, 55
pharmaceuticals, 28–29
Phlebia fascicularia, 271

phosphatidylethanolamine, 387
phospholipids, 387. *See also* biosurfactants
phosphoribulokinase, 289
Photobacterium phosphoreum, 383
photocatalysis, **224**, 311, 322
photocatalysts, **352**
photolysis, 368
Phragmites australis, 60, 74
Phragmites communis, 79
phycoremediation, 55–58
    algal candidates, **56–57**
    macroalgae, 57–58
    mechanism of, 58
    microalgae, 55–56
Phythium aphanidermatum, 395
phytoaccumulation, 93–94
phytodegradation, 94, 202, 214–215, 242, **256**
phytodesalination, 201–202
phytodetoxification
    avoidance mechanism, 239
    purpose of, 239
    tolerance mechanism, 239–240
phytoextraction, 93–94, 200–201, 212–213, 240–242
    characteristics of plants used in, 213
    classification, *212*
    defined, 1–2
    mechanism, **256**
    plant species for, **4**
phytofiltration, 213, 242–243
phytohydraulic containment, 202–203
phytoimmobilization *see* phytostabilisation
Phytophthora capsici, 395
phytoplanktons, 282–284
phytoremediation, 1–7, 58–61, 93–98, 114, 161–168, 197–205
    acid mine drainage, 2
    advantages and limitations of, 81, 98, **99**, 244–245
    advantages of, 210
    in aerial and underground plant parts, *96*
    aquatic plants in, 75
    brownfield redevelopment, 2–5
    case studies, 243–244
    challenges in, 217–218
    chemical-assisted, 219
    contaminant uptake and breakdown mechanism, 95–98
        adsorption, 95
        breakdown, 96–98
        conjugation, 98
        diffusion, 95
        sequestration, 98
    defined, 1, 93, 209–210
    electro kinetic-enhanced, 221–222
    enhancing plant capability for, 203
    as environmental sustainability tool, 222
    future trends in, 222–224
    of heavy metals, 5–6, 215–216
        biochar, 219–220
        chelators, 219
        exogenous phytohormones, 221
        microbes, 221
        nanoparticles, 221
        transgenic plants, **220**, 220–221
    hyper-accumulator plants, 240–242, **241**
    of indoor air pollution, 168–173
        active botanical biofiltration with green, 173–176
        airborne particulates, 170
        potted plant/static chamber experiments, 173–174
        volatile organic compounds, 170–171
    induced, 212
    list of plants, **59**
    mechanism of, *60*, 60–61
    microbial-mediated, 172–173
    for oil contamination, 6–7
    of outdoor air pollution, 163–168
        airborne particulate matter, 163–164
        inorganic air contaminants, 164–168
        volatile organic compounds, 163–164
    plant area part, 171–172
    and plant roots, 218
    role of, 76
    strategies/techniques, 93–95, 210–215, *211*, 240–243
        phytodegradation, 94, **97**, 161–162, 202, 214–215, 242
        phytodesalination, 201–202
        phytoextraction, 93–94, **97**, 161–162, 200–201, 212–213, 242
        phytofiltration, 242–243
        phytohydraulic containment, 202–203
        phytostabilisation, 94, **97**, 162, 202, 213–214, 242–243
        phytostimulation, 202–203
        phytovolatilisation, 94, **97**, 161–162, 201, 214
        rhizodegradation, 163
        rhizofiltration, 94, **97**, 163, 202–203, 213
    system of, 75
    wetlands for, 211–212
phytostabilisation, 94, 202, 213–214, 242–243, **256**
    defined, 1–2
    mechanism, **256**
    plant species for, **4**
phytostimulation, 202–203
phytotransformation, 214–215
phytovolatilisation, 94, 201, 214, **256**
Pichia stipitis, 141
Pistia stratiotes, 2, 60, 61, 76–77, **80**
plant roots, 218
Plantago major, 257
Plasmopara lactucea-radicis, 395
Pleurotus, 267
poly(ethylene terephthalate) (PET), **348–349**
polyacrylamide, 309
polybutylene succinate(PBS), **348**
polycaprolactone (PCL), **348–349**
polycyclic aromatic hydrocarbons (PAHs), 392
polydopamine, 309
polyethylenefuranoate (PEF), **348**
polyglycolide (PGA), **348**
Polygonum aviculare, 165
polyhydroxyalkanoates (PHA), 144–145, **348**, 350.
    See also bio-based/biodegradable polymer

Index

polylactic acid (PLA), **348**
polymeric biosurfactants, 387–388. *See also* biosurfactants
polymers
    factors affecting, 344
    properties of triangles of creativity, 344
    synthetic, 344
polyvinyl alcohols, 308–309
polyvinylpyrrolidone, 309
Populus, 61
Populus nigra, 168
Potamogeton pectinatus, 74
Potamogeton pusillus, 60
potted plant/static chamber experiments, 173–174
probiotic biosurfactants, 388–389
process intensification, 110–111
Propioibacterium pentosaceum, 282
protein polymers, **350**
proteomics, 126
Pseudomonas, 389–390, 395
Pseudomonas aeruginosa, 383, 386, 390, 395
Pseudomonas fluorescens, 384
Pseudomonas putida, 173
Psidium guajava, **333**
Pteridophyta, 74
Pteris cretica, 61
Pteris vittata, 61, 243

R

reduced graphene oxide, 309
rhamnolipids, *381*, **382**, 386, 395
rhizofiltration, 94, 202–203, 213
Rhizopus stolonifer, 393
rhizoremediation, 37–38. *See also* bioremediation
Rhodococcus, **335**, 391
Rhodococcus erythropolis, 386, 387
Rhodophyta, 74
Robinia pseudo-acacia, 168
Ruellia tuberosa, **334**
ryegrass (*Lolium perenne* L.), **4**

S

Saccharina japonica, 58
Saccharomyces cerevisiae, 140, 142, 282, **335**
Sagittaria montevidensis, 60
Salix, 61
Salvinia auriculata, 60
Salvinia auriculatais, 77–78
Salvinia molesta, 61, 257
Sardina pilchardus, 72
Sargassum, 58
Sargassum horneri, 57, 58
Sargassum vulgare, 58
scanning electron microscopy (SEM), 299, *299*
Scenedesmus, 144
Schefflera actinophylla, 170
Schefflera arboricola, 170
Scindapsus aureus, 171

Secale montanum, 258
secondary plant metabolites (SPMEs), 92
Sedirea japonica, 171
semi-volatile organic compounds (SVOCs), 12
Senecio coronatus, 241
separate hydrolysis and fermentation (SHF), 141
sequestration, 98
Serratia marcescens, 393
Sesuvium portulacastrum, 256
silver nanoparticles, **333**, **334**, **335**
simultaneous saccharification and fermentation (SSF), 141
sodium dodecyl sulfate, **382**
soil moisture, 13
soil vapor extraction (SVE), 10–11, *11*, *12*
solvent extraction, 370
Sophora japonica, 168
sophorolipids, *381*, **382**, 386
sparfloxacin, 312
Spathiphyllum wallisii, 170
Spermatophyta, 74
Spirodela polyrhiza, 2, 79
Spirogyra, 57
stannic oxide nanoparticles, **334**, **336**
starch, 298, **313**, **350**
Stigeoclonium, 142
Stilizobium aterrimum, 256
Streptanthus polygaloides, 241
Streptococcus thermophilus, 384
sugarcane residue, 267–269
surfactants, 377
surfactin, *381*, 387. *See also* biosurfactants
Syngonium aureus, 171
Syngonium podophyllum, 171

T

Tamarix smyrnensis, 212
Thalassiosira pseudonana, 57
Thalassiosira weissflogii, 57
thermal gravimetric analysis (TGA), 301–302, *303*
Thermoanaerobacter, 298
Thiobacillus thiooxidans, 387
Thlaspi caerulescens, 241
titanium dioxide nanoparticles, **334**
trace gases, 159
Trameter multicolor, 271
Trameter pubescens, 271
Trametes versicolor, 271
transcriptomics, 126
transgenic plants, **220**, 220–221
trehalolipids, 386
trehalose dimycolates, *381*
trehalose monomycolates, *381*
Trichosporon ashii, 388
triclosan, 29
trinitrotoluene (TNT), 257
Triton X, 100, **382**
Typha latifolia, 61, 74

U

ultrafine particle (UFP), 170
Ulva lactuca, 57, **335**
Upflow Anaerobic Sludge Blanket (UASB) reactor, 138

V

Vaccinium macrocarpon, **333**
vermicomposting, 41
volatile fatty acids (VFA), 145
volatile organic compounds, 163–164, 174
volatile organic compounds (VOCs), 10–11, 159
volatilisation, 369–370

W

wastewater, 133–149
   algae-based treatment, 135–137
   bioelectricity production using, 144
   biofuel production, 137–143
   biohydrogen production using, 143–144, 148
   biopolymer production using, 144–145
   conversion process efficiency, 145–147
   integrated biogas–biodiesel production approach, 147–148
   large-scale technologies, 147–148
   reclamation, 51–61
     microbial remediation, 52–54
     mycoremediation, 54–55
     phycoremediation, 55–58
     phytoremediation, 58–61
   sources of, 135–136
wastewater treatment
   bio-based/biodegradable polymer for, 346–347
     bio-nanocomposites, 349–350
     cellulose nanofibers, 350
     chitosan biopolymers, 349
     foam membranes, 351
     nano crystalline cellulose, 351
     polyhydroxyalkanoates, 350
   conventional techniques for, **223–224**
   electrochemical, **223**
   physico-chemical, **223–224**
   by-products of, 29
water hyacinth, 78
water pollution, sources of, 351
weather of rocks, 252
wet fermentation, 18
wetlands, 211–212
windrows, 115
Wolffia, 78
Wolffia globosa, 215
Wolffiella, 78

X

Xanthophyta, 74
xenobiotics, 41
X-ray diffraction (XRD), 299–300, *300*

Z

Zamioculas zamiifolia, 173
ZRT1RT- like proteins (ZIP), 200
ZSM-5 zeolite, 307
Zymomonas, 140